Prestressed Concrete

Prentice Hall International Series in Civil Engineering
and Engineering Mechanics
William J. Hall, Editor

Prestressed Concrete
A Fundamental Approach

Dr. Edward G. Nawy, P.E.

Distinguished Professor
Department of Civil and Environmental Engineering
Rutgers University—The State University of New Jersey

PRENTICE HALL
Englewood Cliffs, New Jersey 07632

Library of Congress Cataloging-in-Publication Data

Nawy, Edward G.
    Prestressed concrete.

    (Prentice-Hall international series in civil
engineering and engineering mechanics)
    Bibliography: p.
    Includes index.
    1. Prestressed concrete construction.   I. Title.
II. Series.
TA683.9.N39 1989   624.1'83412   88–25397
    ISBN 0–13–698375–8

Production supervision: The Book Company
Cover design: Diane Saxe
Manufacturing buyer: Mary Noonan

COVER PHOTO CREDITS:

Prestressed Concrete T-Beam Cracking in Flexure: Prof. Edward G. Nawy

Marina Island Office Building, San Mateo, California: Robert Englekirk, Los Angeles, California

2 ½-million gallon tendon prestressed concrete tank: Ib Falk Jorgensen, Denver, Colorado

Sunshine Skyway Bridge, Tampa Bay, Florida: Figg and Muller Inc., Tallahassee, Florida

 ©1989 by Prentice-Hall, Inc.
A Division of Simon & Schuster
Englewood Cliffs, New Jersey 07632

The author and publisher of this book have used their best efforts in preparing
this book. These efforts include the development, research, and testing of the
theories and programs to determine their effectiveness. The author and
publisher make no warranty of any kind, expressed or implied, with regard to
these programs or the documentation contained in this book. The author and
publisher shall not be liable in any event for incidental or consequential
damages in connection with, or arising out of, the furnishing, performance, or
use of these programs.

Printed in the United States of America
10 9 8 7 6 5 4 3 2 1

ISBN 0-13-698375-8

Prentice-Hall International (UK) Limited, *London*
Prentice-Hall of Australia Pty. Limited, *Sydney*
Prentice-Hall Canada Inc., *Toronto*
Prentice-Hall Hispanoamericana, S.A., *Mexico*
Prentice-Hall of India Private Limited, *New Delhi*
Prentice-Hall of Japan, Inc., *Tokyo*
Simon & Schuster Asia Pte. Ltd., *Singapore*
Editora Prentice-Hall do Brasil, Ltda., *Rio de Janeiro*

To Rachel E. Nawy

*For her high limit state of stress endurance over
the years which made the writing of this book a reality.*

Contents

Contents

Contents   **xi**

Contents

Contents

Preface

Prestressed concrete is a widely used material in construction. Hence, graduates of every civil engineering program must have, as a minimum requirement, a basic understanding of the fundamentals of linear and circular prestressed concrete. The high-technology advancements in the science of materials have made it possible to construct and assemble large-span systems such as cable-stayed bridges, nuclear reactor vessels, and offshore oil drilling platforms, work hitherto impossible to undertake.

Reinforced concrete's tensile strength is limited, while its compressive strength is extensive. Consequently, prestressing becomes essential in many applications in order to fully utilize that compressive strength and, through proper design, to eliminate or control cracking and deflection due to tensile stresses. Additionally, design of the members of a total structure is achieved only by trial and adjustment: assuming a section and then analyzing it. Hence, design and analysis are combined in this work in order to make it simpler for the student first introduced to the subject of prestressed concrete design.

The text is the outgrowth of the author's lecture notes developed in teaching the subject at Rutgers University over the past 28 years and the experience accumulated over the years in teaching and research in the areas of reinforced and prestressed concrete up to the Ph.D. level. The material is presented in such a manner that the student can become familiarized with the properties of plain concrete and its components prior to embarking on the study of structural behavior. The book is uniquely different from other textbooks on the subject in that the major topics of material behavior, prestress loss, flexure, and shear are self-contained and can be covered in one semester at the senior year and the graduate level. The in-depth discussions of these topics permit the advanced undergraduate and graduate student, as well as the design engineer, to develop a profound understanding of fundamentals of prestressed concrete structural behavior and performance.

The concise discussion presented in Chapters 1 through 3 on basic principles, the historical development of prestressed concrete, the properties of constituent materials, the long-term basic behavior of such materials, and the evaluation of prestress losses should give an adequate introduction to the subject of prestressed concrete. It should also aid in developing fundamental knowledge regarding the reliability of performance of prestressed structures, a concept to which every engineering student should be exposed today.

Chapters 4 and 5 on flexure, shear, and torsion, with the step-by-step logic of trial and adjustment as well as the flowcharts shown, give the student and the engineer a basic understanding of both the service load and the limit state of load at failure, producing a basic understanding of the reserve strength and the safety factors inherent in the design expressions. The discussion on torsional behavior covers two approaches: the modified skew bending-space truss analogy theory and the compression strut theory. Design examples are given, and a comparison between the two is made.

Chapter 6 on indeterminate prestressed concrete structures covers in detail continuous prestressed beams as well as portal frames. Numerous detailed examples illustrate the use of the basic concepts method, the C-line method, and the load-balancing method presented in Chapter 1. Chapter 7 discusses in detail the design for camber, deflection, and crack control considering both short- and long-term effects using three different approaches: the PCI multipliers method, the incremental time steps method, and the approximate time steps method, including the evaluation of deflections of composite beams. A state-of-the-art discussion is presented, based on the author's work, of the evaluation and control of flexural cracking in partially prestressed beams. Several design examples are included in the discussion. Chapter 8 covers the proportioning of prestressed compression and tension members, including the buckling behavior and design of prestressed columns and piles.

Chapter 9 presents a thorough analysis of the service load behavior and yield-line behavior of two-way action slabs and plates. The service load behavior utilizes, with extensive examples, the equivalent frame method of flexural design (analysis) and deflection evaluation. Extensive coverage is presented of the yield-line failure mechanisms of all the usual combinations of loads on floor slabs and boundary conditions, including the design expressions for these various conditions. Chapter 10 on connections for prestressed concrete elements covers the design of connections for dapped-end beams, ledge beams, and bearings, in addition to the design of the beams and corbels presented in Chapter 5 on shear and torsion.

This book is also unique in that Chapter 11 gives a detailed account of the analysis and design of prestressed concrete tanks and their shell roofs. Presented are the basics of the membrane and bending theories of cylindrical shells for use in the design of prestressed tanks for the various wall boundary conditions of fixed, semifixed, hinged, and sliding wall bases, as well as the incorporation of vertical prestressing. Chapter 11 also discusses the theory of axisymmetrical shells and domes that are used in the design of domed roofs for circular tanks.

It is important to emphasize that in this field the use of computers prevails today. Access to affordable transportable personal computers and handheld computers has made it possible for almost every student and engineer to be equipped with such a tool.

Accordingly, Appendix A-1 presents one typical computer program in BASIC for IBM microcomputers for the evaluation of time-dependent losses in prestress. Other programs as described in the appendix can be procured from SOFTWARE, Box 161, East Brunswick, New Jersey 08816. The inclusion of extensive flowcharts throughout the book and the discussion of the logic involved in them makes it possible for the reader to develop or use such programs without difficulty with any microcomputer.

Selected photographs involving various areas of the structural behavior of concrete elements at failure are included in all the chapters. They are taken from research work published by the author with many of his M.S. and Ph.D. students at Rutgers University over the past two decades. Additionally, photographs of some major prestressed concrete "landmark" structures, mainly in the United States, are included throughout the book to illustrate the versatility of design in pretensioned and post-tensioned prestressed concrete. Appendices have also been included, with nomograms and tables on standard properties, sections and charts of flexural and shear evaluation of sections. Conversion to SI metric units are included in the examples throughout the book.

The topics of the book have been presented in as concise a manner as possible without sacrificing the need for instructional details. The major portions of the text can be used without difficulty in an advanced senior-level course as well as at the graduate level for any student who has had a prior course in reinforced concrete. The contents should also serve as a valuable guideline for the practicing engineer who has to keep abreast of the state of the art in prestressed concrete, as well as the designer who seeks a concise treatment of the fundamentals of linear and circular prestressing.

Edward G. Nawy
Rutgers University
The State University of New Jersey
New Brunswick, New Jersey

Acknowledgements

Grateful acknowledgement is due to Dr. Edward J. Bloustein, President of Rutgers University, for his support and encouragement over the years; to Dean Ellis H. Dill for his interest and encouragement, and to the American Concrete Institute, the Prestressed Concrete Institute and the Post-Tensioning Institute for their gracious support in permitting generous quotations from the ACI 318 and other relevant Codes and Reports and the numerous illustrations and tables from so many PCI and PTI publications. Special mention has to be made of the author's original mentor, the late Professor A.L.L. Baker of London University's Imperial College of Science and Technology who inspired him with the affection that he has developed for systems constructed of reinforced and prestressed concrete. Grateful acknowledgement is also made to the author's many students, both undergraduate and graduate, who had much to do with generating the writing of this book; to the many who assisted in his research activities over the past 30 years; and to his colleague at Rutgers University, Dr. P.N. Balaguru, who reviewed portions of this manuscript.

Special thanks are due to the distinguished panel of authoritative reviewers: Professor Carl E. Ekberg of Iowa State University; Clifford L. Freyermuth, Executive Director of the Post-Tensioning Institute: Professor Thomas T.C. Hsu of the University of Houston; Daniel P. Jenny, Vice President of the Prestressed Concrete Institute; and Professor A. Fattah Shaikh of the University of Wisconsin—with deep gratitude for agreeing to critically review the entire manuscript and for contributing extensive suggestions and advice which considerably improved the content. Thanks are also due for their valuable suggestions to Ib Falk Jorgensen, President, Jorgensen, Hendrickson and Close, Inc., Denver, Colorado, and to Professor Maher K. Tadros of the University of Nebraska for their detailed review of Chapter 11 on prestressed tanks and to Edward Cohen, chairman of the board, Amman and Whitney, for his continuous professional advice and support.

Thanks to Ph.D. candidate Ahmed Ezeldin for his diligence, input to the manuscript and its computer programs, and technical assistance in processing the contents; to Engineer Regina Silveira Rocha Souza of Brazil, a former outstanding M.S. graduate, for reviewing several portions of the work; to M.S. candidate Hala Elnajar for her diligent work on the book contents, and example and problem solutions; to M.S. candidate Manik Charmarthy for his extensive work on the solutions manual, and to Robert M. Nawy, Rutgers Engineering Class of 1983, for overall review. Thanks are also due to Doug Humphrey, Senior Engineering Editor who ably guided the development of this manuscript; Dr. Brian Baker, the superb Production Editor; Colleen Brosnan, Supervisory Editor; and Alice Dworkin, Supplements Editor, all of Prentice Hall; and George Calmenson of The Book Company for their input and contributions.

CHAPTER 1

Basic Concepts

1.1 INTRODUCTION

Concrete is strong in compression, but weak in tension: its tensile strength varies from 8 to 14 percent of its compressive strength. Due to such a low tensile capacity, flexural cracks develop at early stages of loading. In order to reduce or prevent such cracks from developing, a concentric or eccentric force is imposed in the longitudinal direction of the structural element. This force prevents the cracks from developing by eliminating or considerably reducing the tensile stresses at the critical midspan and support sections at service load, thereby raising the bending, shear, and torsional capacities of the sections. The sections are then able to behave elastically, and almost the full capacity of the concrete in compression can be efficiently utilized across the entire depth of the concrete sections when all loads act on the structure.

Such an imposed longitudinal force is called a *prestressing force,* i.e., a compressive force that prestresses the sections along the span of the structural element prior to the application of the transverse gravity dead and live loads or transient horizontal live loads. The type of prestressing force involved, together with its magnitude, are determined mainly on the basis of the type of system to be constructed and the span length and slenderness desired. Since the prestressing force is applied longitudinally along or parallel to the axis of the member, the prestressing principle involved is commonly known as *linear* prestressing.

Circular prestressing, used in liquid containment tanks, pipes, and pressure reactor vessels, essentially follows the same basic principles as does linear pre-

The Diamond Baseball Stadium, Richmond, Virginia. Situ cast and precast post-tensioned prestressed structure. (*Courtesy,* Prestressed Concrete Institute.)

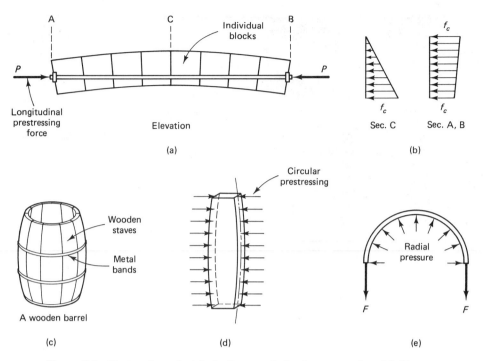

Figure 1.1 Prestressing principle in linear and circular prestressing. (a) Linear prestressing of a series of blocks to form a beam. (b) Compressive stress on midspan section C and end section A or B. (c) Circular prestressing of a wooden barrel by tensioning the metal bands. (d) Circular hoop prestress on one wooden stave. (e) Tensile force F on half of metal band due to internal pressure, to be balanced by circular hoop prestress.

stressing. The circumferential hoop, or "hugging" stress on the cylindrical or spherical structure, neutralizes the tensile stresses at the outer fibers of the curvilinear surface caused by the internal contained pressure.

Figure 1.1 illustrates, in a basic fashion, the prestressing action in both types of structural systems and the resulting stress response. In (a), the individual concrete blocks act together as a beam due to the large compressive prestressing force P. Although it might appear that the blocks will slip and vertically simulate shear slip failure, in fact they will not because of the longitudinal force P. Similarly, the wooden staves in (c) might appear to be capable of separating as a result of the high internal radial pressure exerted on them. But again, because of the compressive prestress imposed by the metal bands as a form of circular prestressing, they will remain in place.

1.1.1 Comparison with Reinforced Concrete

From the preceding discussion, it is plain that permanent stresses in the prestressed structural member are created before the full dead and live loads are applied in order to eliminate or considerably reduce the net tensile stresses caused by these loads. With reinforced concrete, it is assumed that the tensile strength of the concrete is negligible

Bay Area Rapid Transit (BART), San Francisco and Oakland, California. Guideways consist of prestressed precast simple-span box girders 70 ft long and 11 ft wide. (*Courtesy,* Bay Area Rapid Transit District, Oakland, California.)

Chaco-Corrientes Bridge, Argentina. The longest precast prestressed concrete cable-stayed box girder bridge in South America. (*Courtesy,* Ammann & Whitney.)

and disregarded. This is because the tensile forces resulting from the bending moments are resisted by the bond created in the reinforcement process. Cracking and deflection are therefore essentially irrecoverable in reinforced concrete once the member has reached its limit state at service load.

The reinforcement in the reinforced concrete member does not exert any force of its own on the member, contrary to the action of prestressing steel. The steel

Park Towers, Tulsa, Oklahoma. (*Courtesy*, Prestressed Concrete Institute.)

required to produce the prestressing force in the prestressed member actively preloads the member, permitting a relatively high controlled recovery of cracking and deflection. Once the flexural tensile strength of the concrete is exceeded, the prestressed member starts to act like a reinforced concrete element.

By controlling the amount of prestress, a structural system can be made either flexible or rigid without influencing its strength. In reinforced concrete, such a flexibility in behavior is considerably more difficult to achieve if considerations of economy are to be observed in the design. Flexible structures such as fender piles in wharves have to be highly energy absorbent, and prestressed concrete can provide the required resiliency. Structures designed to withstand heavy vibrations, such as machine foundations, can easily be made rigid through the contribution of the prestressing force to the reduction of their otherwise flexible deformation behavior.

1.1.2 Economics of Prestressed Concrete

Prestressed members are shallower in depth than their reinforced concrete counterparts for the same span and loading conditions. In general, the depth of a prestressed concrete member is usually about 65 to 80 percent of the depth of the equivalent reinforced concrete member. Hence, the prestressed member requires less concrete, and about 20 to 35 percent of the amount of reinforcement. Unfortunately, this saving in material weight is balanced by the higher cost of the higher quality materials needed in prestressing. Also, regardless of the system used, prestressing operations themselves result in an added cost: formwork is more complex, since the geometry of prestressed sections is usually composed of flanged sections with thin webs.

In spite of these additional costs, if a large enough number of precast units are

Basic Concepts   Chap. 1

manufactured, the difference between at least the initial costs of prestressed and reinforced concrete systems is usually not very large. And the indirect long-term savings are quite substantial, because less maintenance is needed, a longer working life is possible due to better quality control of the concrete, and lighter foundations are achieved due to the smaller cumulative weight of the superstructure.

Once the beam span of reinforced concrete exceeds 70 to 90 feet, the dead weight of the beam becomes excessive, resulting in heavier members and, consequently, greater long-term deflection and cracking. Thus, for larger spans, prestressed concrete becomes mandatory since arches are expensive to construct and do not perform as well due to the severe long-term shrinkage and creep they undergo. Very large spans such as segmental bridges or cable-stayed bridges can *only* be constructed through the use of prestressing.

1.2 HISTORICAL DEVELOPMENT OF PRESTRESSING

Prestressed concrete is not a new concept, dating back to 1872, when P. H. Jackson, an engineer from California, patented a prestressing system that used a tie rod to construct beams or arches from individual blocks. (See Figure 1.1(a).) In 1888, C. W. Doehring of Germany obtained a patent for prestressing slabs with metal wires. But these early attempts at prestressing were not really successful because of the loss of the prestress with time. J. Lund of Norway and G. R. Steiner of the United States tried early in the twentieth century to solve this problem, but to no avail.

After a long lapse of time during which little progress was made because of the

Wiscasset Bridge, Maine. (*Courtesy,* Post-Tensioning Institute.)

Executive Center, Honolulu, Hawaii. (*Courtesy,* Post-Tensioning Institute.)

Stratfjord "B" Condeep offshore oil drilling platform, Norway. (*Courtesy,* Ben C. Gerwick.)

unavailability of high-strength steel to overcome prestress losses, R. E. Dill of Alexandria, Nebraska, recognized the effect of the shrinkage and creep (transverse material flow) of concrete on the loss of prestress. He subsequently developed the idea that successive post-tensioning of *unbonded* rods would compensate for the time-dependent loss of stress in the rods due to the decrease in the length of the member because of creep and shrinkage. In the early 1920s, W. H. Hewett of Minneapolis developed the principles of circular prestressing. He hoop-stressed horizontal reinforcement around walls of concrete tanks through the use of turnbuckles to prevent cracking due to internal liquid pressure, thereby achieving watertightness. Thereafter, prestressing of tanks and pipes developed at an accelerated pace in the United States, with thousands of tanks for water, liquid, and gas storage built and much mileage of prestressed pressure pipe laid in the two to three decades that followed.

Linear prestressing continued to develop in Europe and in France, in particular through the ingenuity of Eugene Freyssinet, who proposed in 1926–28 methods to overcome prestress losses through the use of high-strength and high-ductility steels. In 1940, he introduced the now well-known and well-accepted Freyssinet system comprising the conical wedge anchor for 12-wire tendons.

During World War II and thereafter, it became necessary to reconstruct in a prompt manner many of the main bridges that were destroyed by war activities. G. Magnel of Ghent, Belgium, and Y. Guyon of Paris extensively developed and used the concept of prestressing for the design and construction of numerous bridges in western and central Europe. The Magnel system also used wedges to anchor the prestressing wires. They differed from the original Freyssinet wedges in that they were flat in shape, accommodating the prestressing of two wires at a time.

P. W. Abeles of England introduced and developed the concept of partial prestressing between the 1930s and 1960s. F. Leonhardt of Germany, V. Mikhailov of Russia, and T. Y. Lin of the United States also contributed a great deal to the art and science of the design of prestressed concrete. Lin's load-balancing method deserves particular mention in this regard, as it considerably simplified the design process, particularly in continuous structures. These twentieth-century developments have led to the extensive use of prestressing throughout the world, and in the United States in particular.

Today, prestressed concrete is used in buildings, underground structures, TV towers, floating storage and offshore structures, power stations, nuclear reactor vessels, and numerous types of bridge systems including segmental and cable-stayed bridges. Note the variety of prestressed structures in the photos throughout the book; they demonstrate the versatility of the prestressing concept and its all-encompassing

Sunshine Skyway Bridge, Tampa Bay, Florida. Designed by Figg and Muller Engineers, Inc., the bridge has a 1,200 ft cable-stayed main span with a single pylon, 175 ft vertical clearance, and total length of 21,878 ft. It has twin 40 ft roadways and has 135-ft spans in precast segmental trestle for trestle and high approaches to elevation +130 ft. (*Courtesy,* Figg and Muller Engineers, Inc.)

applications. The success in the development and construction of all these landmark structures has been due in no small measure to the advances in the technology of materials, particularly prestressing steel, and the accumulated knowledge in estimating the short- and long-term losses in the prestressing forces.

1.3 BASIC CONCEPTS OF PRESTRESSING

1.3.1 Introduction

The prestressing force P that satisfies the particular conditions of geometry and loading of a given element (see Figure 1.2) is determined from the principles of mechanics and of stress-strain relationships. Sometimes simplification is necessary, as when a prestressed beam is assumed to be homogeneous and elastic.

Consider, then, a simply supported rectangular beam subjected to a *concentric* prestressing force P as shown in Figure 1.2(a). The compressive stress on the beam cross section is uniform and has an intensity

$$f = -\frac{P}{A_c} \tag{1.1}$$

where $A_c = bh$ is the cross-sectional area of a beam section of width b and total depth h. A *minus* sign is used for compression and a *plus* sign for tension throughout the text. Also, bending moments are drawn on the tensile side of the member.

If external transverse loads are applied to the beam, causing a maximum moment M at midspan, the resulting stress becomes

$$f^t = -\frac{P}{A} - \frac{Mc}{I_g} \tag{1.2 a}$$

and

$$f_b = -\frac{P}{A} + \frac{Mc}{I_g} \tag{1.2 b}$$

where f^t = stress at the top fibers
f_b = stress at the bottom fibers
$c = \frac{1}{2}h$ for the rectangular section
I_g = gross moment of inertia of the section ($bh^3/12$ in this case)

Equation 1.2b indicates that the presence of prestressing-compressive stress $-P/A$ is reducing the tensile flexural stress Mc/I to the extent intended in the design, either eliminating tension totally (even inducing compression), or permitting a level of tensile stress within allowable code limits. The section is then considered uncracked and behaves elastically: the concrete's inability to withstand tensile stresses is effectively compensated for by the compressive force of the prestressing tendon.

The compressive stresses in Equation 1.2a at the top fibers of the beam due to prestressing are compounded by the application of the loading stress $-Mc/I$, as seen in Figure 1.2(b). Hence, the compressive stress capacity of the beam to take a substantial external load is reduced by the *concentric* prestressing force. In order to avoid this limitation, the prestressing tendon is placed *eccentrically* below the neutral

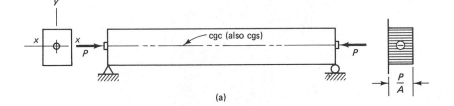

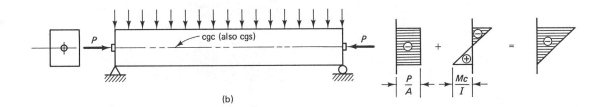

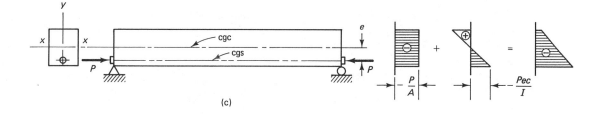

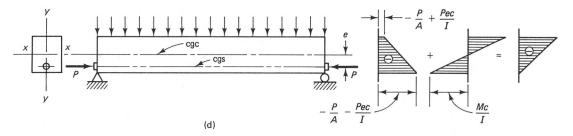

Figure 1.2 Concrete fiber stress distribution in a rectangular beam with straight tendon. (a) Concentric tendon, prestress only. (b) Concentric tendon, self-weight added. (c) Eccentric tendon, prestress only. (d) Eccentric tendon, self-weight added.

axis at midspan, to induce tensile stresses at the top fibers due to prestressing. (See Figure 1.2(c), (d).) If the tendon is placed at eccentricity e from the center of gravity of the concrete, termed the *cgc line,* it creates a moment Pe, and the ensuing stresses at midspan become

$$f^t = -\frac{P}{A_c} + \frac{Pec}{I_g} - \frac{Mc}{I_g} \qquad (1.3\text{ a})$$

$$f_b = -\frac{P}{A_c} - \frac{Pec}{I_g} + \frac{Mc}{I_g} \qquad (1.3\text{ b})$$

Douglas Bridge Crossing, Gastineau Channel, Juneau and Douglas, Alaska. (*Courtesy,* Prestressed Concrete Institute.)

Since the support section of a simply supported beam carries no moment from the external transverse load, high tensile fiber stresses at the top fibers are caused by the eccentric prestressing force. To limit such stresses, the eccentricity of the prestressing tendon profile, the *cgs line,* is made less at the support section than at the midspan section, or eliminated altogether, or else a negative eccentricity above the cgc line is used.

1.3.2 Basic Concept Method

In the basic concept method of designing prestressed concrete elements, the concrete fiber stresses are *directly* computed from the external forces applied to the concrete by longitudinal prestressing and the external transverse load. Equations 1.3a and b can be modified and simplified for use in calculating stresses at the initial prestressing stage and at service load levels. If P_i is the initial prestressing force before stress losses, and P_e is the effective prestressing force after losses, then

$$\gamma = \frac{P_e}{P_i}$$

can be defined as the residual prestress factor. Substituting r^2 for I_g/A_c in Equations 1.3, where r is the radius of gyration of the gross section, the expressions for stress can be rewritten as follows:

(a) Prestressing Force Only

$$f^t = -\frac{P_i}{A_c}\left(1 - \frac{ec_t}{r^2}\right) \tag{1.4 a}$$

$$f_b = -\frac{P_i}{A_c}\left(1 + \frac{ec_b}{r^2}\right) \tag{1.4 b}$$

where c_t and c_b are the distances from the center of gravity of the section (the cgc line) to the extreme top and bottom fibers, respectively.

Tianjin Yong-He cable-stayed prestressed concrete bridge, Tianjin, China, the largest span bridge in Asia, with a total length of 1,673 ft and a suspended length of 1,535 ft, was completed in 1988. (*Credits* Owner: Tianjin Municipal Engineering Bureau. General contractor: Major Bridge Engineering Bureau of Ministry of Railways of China. Engineer for project design and construction control guidance: Tianjin Municipal Engineering Survey and Design Institute, Chief Bridge Engineer, Bang-yan Yu.)

(b) Prestressing Plus Self-weight

If the beam self-weight causes a moment M_D at the section under consideration, Equations 1.4a and b respectively become

$$f^t = -\frac{P_i}{A_c}\left(1 - \frac{ec_t}{r^2}\right) - \frac{M_D}{S^t}$$ (1.5 a)

and

$$f_b = -\frac{P_i}{A_c}\left(1 + \frac{ec_b}{r^2}\right) + \frac{M_D}{S_b}$$ (1.5 b)

where S^t and S_b are the moduli of the sections for the top and bottom fibers, respectively.

The change in eccentricity from the midspan to the support section is obtained by raising the prestressing tendon either abruptly from the midspan to the support, a process called harping, or gradually in a parabolic form, a process called draping.

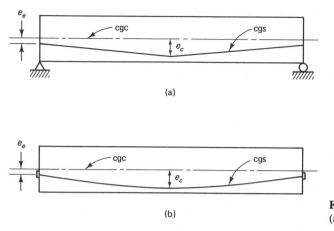

(a)

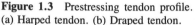

(b)

Figure 1.3 Prestressing tendon profile. (a) Harped tendon. (b) Draped tendon.

Figure 1.3(a) shows a harped profile usually used for pretensioned beams and for concentrated transverse loads. Figure 1.3(b) shows a draped tendon usually used in post-tensioning.

Subsequent to erection and installation of the floor or deck, live loads act on the structure, causing a superimposed moment M_s. The full intensity of such loads normally occurs after the building is completed and some time-dependent losses in prestress have already taken place. Hence, the prestressing force used in the stress equations would have to be the effective prestressing force P_e. If the total moment due to gravity loads is M_T, then

$$M_T = M_D + M_{SD} + M_L \qquad (1.6)$$

where M_D = moment due to self-weight

M_{SD} = moment due to superimposed dead load, such as flooring

M_L = moment due to live load, including impact and siesmic loads if any

Equations 1.5 then become

$$f^t = -\frac{P_e}{A_c}\left(1 - \frac{ec_t}{r^2}\right) - \frac{M_T}{S^t} \qquad (1.7\text{ a})$$

$$f_b = -\frac{P_e}{A_c}\left(1 + \frac{ec_b}{r^2}\right) + \frac{M_T}{S_b} \qquad (1.7\text{ b})$$

Some typical elastic concrete stress distributions at the critical section of a prestressed flanged section are shown in Figure 1.4. The tensile stress in the concrete in part (c) permitted at the extreme fibers of the section cannot exceed the maximum permissible in the code, e.g., $f_t = 6\sqrt{f_c'}$ in the ACI code. If it is exceeded, bonded non-prestressed reinforcement proportioned to resist the total tensile force has to be provided to control cracking at service loads.

1.3.3 C-Line Method

In this line-of-pressure or thrust concept, the beam is analyzed as if it were a plain concrete elastic beam using the basic principles of statics. The prestressing force is considered an external compressive force, with a constant tensile force T in the tendon

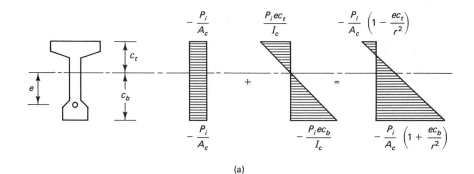

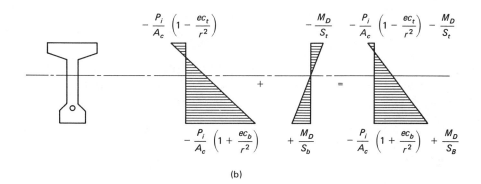

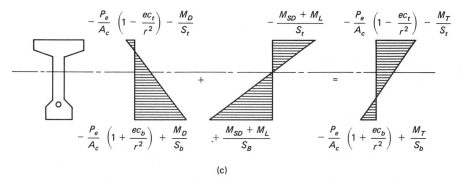

Figure 1.4 Elastic fiber stresses due to the various loads in a prestressed beam. (a) Initial prestress before losses. (b) Addition of self-weight. (c) Service load at effective prestress.

throughout the span. In this manner, the effects of external gravity loads are disregarded. Equilibrium equations $\Sigma H = 0$ and $\Sigma M = 0$ are applied to maintain equilibrium in the section.

Figure 1.5 shows the relative line of action of the compressive force C and the tensile force T in a reinforced concrete beam as compared to that in a prestressed concrete beam. It is plain that in a reinforced concrete beam, T can have a finite value only when transverse and other external loads act. The moment arm a remains basically constant throughout the elastic loading history of the reinforced concrete

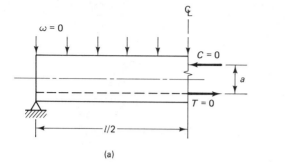

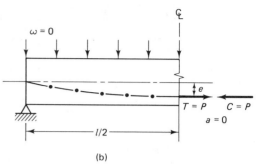

(a)

(b)

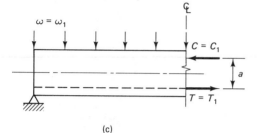

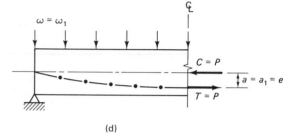

(c)

(d)

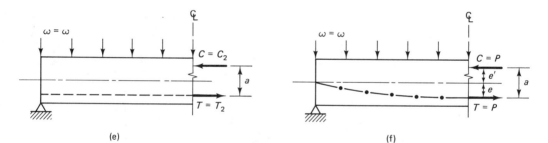

(e)

(f)

Figure 1.5 Comparative free-body diagrams of a reinforced concrete (R.C.) beam and a prestressed concrete (P.C.) beam. (a) R.C. beam with no load. (b) P.C. beam with no load. (c) R.C. beam with load w_1. (d) P.C. beam with load w_1. (e) R.C. beam with typical load w. (f) P.C. beam with typical load w.

beam while it changes from a value $a = 0$ at prestressing to a maximum at full superimposed load.

Taking a free-body diagram of a segment of a beam as in Figure 1.6, it is evident that the C-line, or center-of-pressure line, is at a varying distance a from the T-line. The moment is given by

$$M = Ca = Ta \tag{1.8}$$

and the eccentricity e is known or predetermined, so that in Figure 1.6,

$$e' - u = e \tag{1.9 a}$$

Since $C = T$, $a = M/T$, giving

East Huntington Bridge over Ohio River. A segmentally assembled precast prestressed concrete cable-stayed bridge spanning 200–900–608 ft. (*Courtesy,* Arvid Grant and Associates, Inc.)

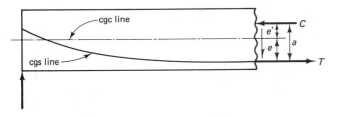

Figure 1.6 Free-body diagram for the C-line (center of pressure).

$$e' = \frac{M}{T} - e \qquad (1.9\text{ b})$$

From the figure,

$$f^t = -\frac{C}{A_c} - \frac{Ce'c_t}{I_c} \qquad (1.10\text{ a})$$

$$f_b = -\frac{C}{A_c} + \frac{Ce'c_b}{I_c} \qquad (1.10\text{ b})$$

But in the tendon the force T equals the prestressing force P_e; so

$$f^t = -\frac{P_e}{A_c} - \frac{P_e e'c_t}{I_c} \qquad (1.11\text{ a})$$

$$f_b = -\frac{P_e}{A_c} + \frac{P_e e'c_b}{I_c} \qquad (1.11\text{ b})$$

Since $I_c = A_c r^2$, Equations 1.11a and b can be rewritten as

$$f^t = -\frac{P_e}{A_c}\left(1 + \frac{e'c_t}{r^2}\right) \qquad (1.12\text{ a})$$

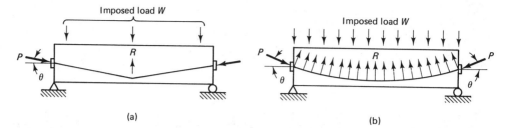

Figure 1.7 Load-balancing forces. (a) Harped tendon. (b) Draped tendon.

$$f_b = -\frac{P_e}{A_c}\left(1 - \frac{e'c_b}{r^2}\right) \tag{1.12 b}$$

Equations 1.12a and b and Equations 1.7a and b should yield identical values for the fiber stresses.

1.3.4 Load-Balancing Method

A third useful approach in the design (analysis) of continuous prestressed beams is the load-balancing method developed by Lin and mentioned earlier. This technique is based on utilizing the vertical force of the draped or harped prestressing tendon to counteract or balance the imposed gravity loading to which a beam is subjected. Hence, it is applicable to nonstraight prestressing tendons.

Figure 1.7 demonstrates the balancing forces for both harped- and draped-tendon prestressed beams. The load balancing reaction R is equal to the vertical component of the prestressing force P. The horizontal component of P, as an approximation in long-span beams, is taken to be equal to the full force P in computing the concrete fiber stresses *at midspan* of the simply supported beam. At other sections, the actual horizontal component of P is used.

1.3.4.1. Load-balancing distributed loads and parabolic tendon profile
Consider a parabolic tendon as shown in Figure 1.8. Let the parabolic function

$$Ax^2 + Bx + C = y \tag{1.13}$$

represent the tendon drape; the force T denotes the pull to which the tendon is subjected. Then for $x = 0$, we have

$$y = 0 \qquad C = 0$$

$$\frac{dy}{dx} = 0 \qquad B = 0$$

and for $x = \ell/2$,

$$y = a \qquad A = \frac{4a}{\ell^2}$$

But from calculus, the load intensity is

$$q = T\frac{\partial^2 y}{\partial x^2} \tag{1.14}$$

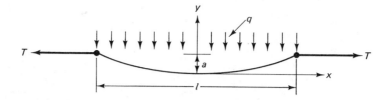

Figure 1.8 Sketched tendon subjected to transverse load intensity q.

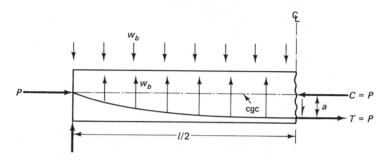

Figure 1.9 Load-balancing force on free-body diagram.

Finding $\partial^2 y/\partial x^2$ in Equation 1.13 and substituting into Equation 1.14 yields

$$q = T\frac{4a}{\ell^2} \times 2 = \frac{8Ta}{\ell^2} \tag{1.15 a}$$

or

$$T = \frac{q\ell^2}{8a} \tag{1.15 b}$$

$$Ta = \frac{q\ell^2}{8} \tag{1.15 c}$$

Hence, if the tendon has a parabolic profile in the prestressed beam and the prestressing force is denoted by P, the balanced-load intensity, from Equation 1.15a, is

$$w_b = \frac{8Pa}{\ell^2} \tag{1.16}$$

Figure 1.9 gives a free-body diagram of the forces acting on a prestressed beam with a parabolic tendon profile. Clearly, the two sets of equal and opposite transverse loads w_b *cancel* each other, and no bending stress is produced. This is reasonable to expect in the load-balancing method, since it is always the case that $T = C$, and C has to cancel T to satisfy the equilibrium requirement that $\Sigma H = 0$. As there is no bending, the beam remains straight, without having a convex shape, or camber, at the top face.

The concrete fiber stress across the depth of the section at midspan becomes

$$f_b^t = -\frac{P'}{A} = -\frac{C}{A} \tag{1.17}$$

This stress, which is constant, is due to the force $P' = P \cos \theta$. Figure 1.10 shows

Walt Disney World Monorail, Orlando, Florida. A series of hollow precast pre-stressed concrete 100-foot box girders individually post-tensioned to provide six-span continuous structure. Designed by ABAM Engineers, Tacoma, Washington. (*Courtesy*, Walt Disney World Corporation.)

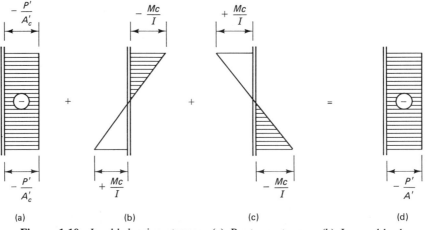

Figure 1.10 Load-balancing stresses. (a) Prestress stresses. (b) Imposed-load stresses. (c) Balanced-load stresses. (d) Net stress.

the superposition of stresses to yield the net stress. Note that the prestressing force in the load-balancing method has to act at the center of gravity (cgc) of the support section in simply supported beams and at the cgc of the free end in the case of a cantilever beam. This condition is necessary in order to prevent any eccentric unbalanced moment.

When the imposed load exceeds the balancing load w_b such that an additional *unbalanced* load w_{ub} is applied, a moment $M_{ub} = w_{ub}\ell^2/8$ results at midspan. The corresponding fiber stresses at midspan become

$$f_b^t = -\frac{P'}{A_c} \mp \frac{M_{ub}c}{I_c} \qquad (1.18)$$

Equation 1.18 can be rewritten as the two equations

$$f^t = -\frac{P'}{A_c} - \frac{M_{ub}}{S^t} \qquad (1.19\ a)$$

and

$$f_b = -\frac{P'}{A_c} + \frac{M_{ub}}{S_b} \qquad (1.19\ b)$$

Equations 1.19 will yield the same values of fiber stresses as Equations 1.7 and 1.12. Keep in mind that P' is taken to be equal to P at the midspan section because the prestressing force is horizontal at this section, i.e., $\theta = 0$.

1.4 COMPUTATION OF FIBER STRESSES IN A PRESTRESSED BEAM BY THE BASIC METHOD

Example 1.1

A pretensioned simply supported T-beam has a span of 64 ft (19.51 m) and the geometry shown in Figure 1.11. It is subjected to a uniform intensity of superimposed gravity dead-load intensity W_{SD} and live-load intensity W_L summing to 420 plf (6.13 kN/m). The initial prestressing stress before losses is $f_{pi} \cong 0.70 f_{pu} = 189,000$ psi (1,303 MPa), and the effective prestress after losses is $f_{pe} = 150,000$ psi (1,034 MPa). Compute the extreme fiber stresses at the midspan section due to

(a) the initial full prestress and no external gravity load.

(b) the final service load conditions when prestress losses have taken place.

Allowable stress data are as follows:

$f_c' = 6,000$ psi, normal weight (41.3 MPa)

$f_{pu} = 270,000$ psi, stress relieved (1,862 MPa) = specified tensile strength of the tendons

$f_{py} = 220,000$ psi (1,517 MPa) = specified yield strength of the tendons

$f_{pe} = 150,000$ psi (1,034 MPa)

$f_t = 12\sqrt{f_c'} = 930$ psi (6.4 MPa) = maximum allowable tensile stress in concrete

$f_{ci}' = 4,800$ psi (33.1 MPa) = concrete compressive strength at time of initial prestress

$f_{ci} = 0.6 f_{ci}' = 2,880$ psi (19.9 MPa) = maximum allowable stress in concrete at initial prestress

$f_c = 0.45 f_c' =$ maximum allowable compressive stress in concrete at service

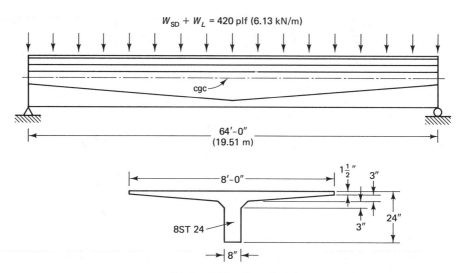

$W_{SD} + W_L = 420\ \text{plf (6.13 kN/m)}$

cgc

64'-0"
(19.51 m)

8'-0"

$1\tfrac{1}{2}''$ 3"

24"

3"

8ST 24

8"

Figure 1.11 Example 1.1

Assume that twelve $\tfrac{1}{2}$-in.-dia seven-wire-strand (twelve 12.7-mm-dia strand) tendons are used to prestress the beam, and take

$A_c = 474\ \text{in}^2\ (3{,}058\ \text{cm}^2)$

$I_c = 21{,}540\ \text{in}^4\ (896{,}674\ \text{cm}^4)$

$r^2 = I_c/A_c = 45.44\ \text{in}^2$

$c_b = 18.06\ \text{in.}\ (459\ \text{mm})$

$c_t = 5.94\ \text{in.}\ (151\ \text{mm})$

$e_c = 14.6\ \text{in.}\ (371\ \text{mm})$

$e_e = 0$

$S_b = 1{,}193\ \text{in}^3\ (19{,}550\ \text{cm}^3)$

$S^t = 3{,}626\ \text{in}^3\ (59{,}419\ \text{cm}^3)$

$W_D = 494\ \text{plf}\ (6.13\ \text{kN/m})$

Solution

(i) Initial Conditions at Prestressing

$A_{ps} = 12 \times 0.153 = 1.836\ \text{in}^2$

$P_i = A_{ps}f_{pi} = 1.836 \times 189{,}000 = 347{,}004\ \text{lb}\ (1{,}544\ \text{kN})$

$P_e = 1.836 \times 150{,}000 = 275{,}400\ \text{lb}\ (1{,}225\ \text{kN})$

The midspan self-weight dead-load moment is

$$M_D = \frac{W_D\ell^2}{8} = \frac{494(64)^2}{8} \times 12 = 3{,}035{,}136\ \text{in.-lb}\ (343\ \text{kNm})$$

From Equations 1.7,

$$f^t = -\frac{P_i}{A_c}\left(1 - \frac{ec_t}{r^2}\right) - \frac{M_D}{S^t}$$

Prestressed lightweight concrete mid-body for Arctic offshore drilling platform, Global Marine Development. (*Courtesy*, Ben C. Gerwick.)

$$= -\frac{347{,}004}{474}\left(1 - \frac{14.6 \times 5.94}{45.44}\right) - \frac{3{,}035{,}136}{3{,}626}$$

$$= +665.8 - 837.0 = -171.2 \text{ psi } (C)$$

$$f_b = -\frac{P_i}{A_c}\left(1 + \frac{ec_b}{r^2}\right) + \frac{M_D}{S_b}$$

$$= -\frac{347{,}004}{474}\left(1 + \frac{14.6 \times 18.06}{45.44}\right) + \frac{3{,}035{,}136}{1{,}193}$$

$$= -4{,}980.1 + 2{,}544.1 = -2{,}436.0 \text{ psi } (C)$$

$$< f_{ci} = -2{,}880 \text{ psi allowed, O.K.}$$

(ii) Final Condition at Service Load   Midspan moment due to superimposed dead and live load is

$$M_{SD} + M_L = \frac{420(64)^2}{8} \times 12 = 2{,}580{,}480 \text{ in.-lb}$$

Total moment $M_T = 3{,}035{,}136 + 2{,}580{,}480$

$$= 5{,}615{,}616 \text{ in.-lb (635 kNm)}$$

$$f^t = -\frac{P_e}{A_c}\left(1 - \frac{ec_t}{r^2}\right) - \frac{M_T}{S^t}$$

$$= -\frac{275{,}400}{474}\left(1 - \frac{14.6 \times 5.94}{45.44}\right) - \frac{5{,}615{,}616}{3{,}626}$$

$$= +527.9 - 1{,}548.7 = -1{,}020.8 \text{ psi } (C) \text{ (7 MPa)}$$

$$< f_c = 0.45 \times 6{,}000 = 2{,}700 \text{ psi, O.K.}$$

$$f_b = -\frac{P_e}{A_c}\left(1 + \frac{ec_b}{r^2}\right) + \frac{M_T}{S_b}$$

$$= -\frac{275,400}{474}\left(1 + \frac{14.6 \times 18.06}{45.44}\right) + \frac{5,615,616}{1,193}$$

$$= -3,952.5 + 4,707.1 = +755 \text{ psi } (T) \text{ (5.2 MPa)}$$

$$<f_t = 12\sqrt{f_c'} = 930 \text{ psi, O.K.}$$

1.5 C-LINE COMPUTATION OF FIBER STRESSES

Example 1.2

Solve Example 1.1 for the final service-load condition by the line-of-thrust, C-line method.

Solution

$$P_e = 275,400 \text{ lb}$$

$$M_T = 5,615,616 \text{ in.-lb}$$

$$a = \frac{M_T}{P_e} = \frac{5,615,616}{275,400} = 20.39 \text{ in.}$$

$$e' = a - e = 20.39 - 14.6 = 5.79 \text{ in.}$$

From Equations 1.12,

$$f^t = -\frac{P_e}{A_c}\left(1 + \frac{e'c_t}{r^2}\right)$$

Dauphin Island Bridge, Mobile County, Alabama. (*Courtesy*, Post-Tensioning Institute.)

$$= -\frac{275,400}{474}\left(1 + \frac{5.79 \times 5.94}{45.44}\right) = -1,020.8 \text{ psi } (C)$$

$$f_b = -\frac{P_e}{A_c}\left(1 - \frac{e'c_b}{r^2}\right)$$

$$= -\frac{275,400}{474}\left(1 - \frac{5.79 \times 18.06}{45.44}\right) = +755 \text{ psi } (T)$$

Notice how the C-line method is shorter than the basic method used in Example 1.1.

1.6 LOAD-BALANCING COMPUTATION OF FIBER STRESSES

Example 1.3

Solve Example 1.1 for the final service-load condition after losses using the load-balancing method.
Solution

$$P' = P_e = 275,400 \text{ lb at midspan}$$

$$a = 14.6 \text{ in.} = e = 1.217 \text{ ft}$$

For the balancing load, we have

$$W_b = 8\frac{Pa'}{\ell^2} = \frac{8 \times 275,400 \times 1.217}{(64)^2}$$

$$= 654.43 \text{ plf } (9.55 \text{ kN/m})$$

Thus, if the total gravity load would have been 654.43 plf, only the axial load P'/A would act if the beam had a parabolically draped tendon with no eccentricity at the supports. This is because the gravity load is balanced by the tendon at the midspan. Hence,

$$\text{Total load to which the beam is subjected} = W_D + W_L$$

$$= 494 + 420 = 914 \text{ plf}$$

$$\text{Unbalanced load } W_{ub} = 914 - 654.43 = 259.57 \text{ plf}$$

$$\text{Unbalanced moment } M_{ub} = \frac{W_{ub}(1^2)}{8} = \frac{259.57(64)^2}{8} \times 12$$

$$= 1,594,798 \text{ in.-lb}$$

From Equations 1.19,

$$f^t = -\frac{P'}{A_c} - \frac{M_{ub}}{S^t} = -\frac{275,400}{474} - \frac{1,594,798}{3,626}$$

$$= -581.0 - 439.8 = -1,020.8 \text{ psi } (C)$$

$$f_b = -\frac{P'}{A_c} + \frac{M_{ub}}{S_b} = -\frac{275,400}{474} + \frac{1,594,798}{1,193}$$

$$= -581.0 + 1,336.8 \cong 755 \text{ psi } (T)$$

$$< f_t = 930 \text{ psi allowed, O.K.}$$

REFERENCES

1.1 Freyssinet, E. *The Birth of Prestressing.* London: Public Translation, Cement and Concrete Association, 1954.

1.2 Guyon, Y. *Limit State Design of Prestressed Concrete,* vol. 1. New York: Halsted-Wiley, 1972.

1.3 Gerwick, B. C., Jr. *Construction of Prestressed Concrete Structures.* New York: Wiley-Interscience, 1971.

1.4 Nilson, A. H. *Design of Prestressed Concrete.* New York: Wiley, 1987.

1.5 Lin, T. Y., and Burns, N. H. *Design of Prestressed Concrete Structures.* 3d ed. New York: Wiley, 1981.

1.6 Nawy, E. G. *Reinforced Concrete—A Fundamental Approach.* Englewood Cliffs, N. J.: Prentice-Hall, 1985.

1.7 Dobell, C. "Patents and Code Relating to Prestressed Concrete." *Journal of the American Concrete Institute* 46 (1950): 713–724.

1.8 Naaman, A. E. *Prestressed Concrete Analysis and Design.* New York: McGraw Hill, 1982.

1.9 Dill, R. E. "Some Experience with Prestressed Steel in Small Concrete Units." *Journal of the American Concrete Institute* 38 (1942): 165–168.

1.10 Institution of Structural Engineers. "First Report on Prestressed Concrete." *Journal of the Institution of Structural Engineers,* September 1951.

1.11 Magnel, G. *Prestressed Concrete.* London: Cement and Concrete Association, 1948.

1.12 Abeles, P. W., and Bardhan-Roy, B. K. *Prestressed Concrete Designer's Handbook.* 3d ed. London: Viewpoint Publications, 1981.

1.13 Libby, James R. *Modern Prestressed Concrete.* 3d ed. New York: Van Nostrand Reinhold, 1984.

PROBLEMS

1.1 An AASHTO prestressed simply supported I-beam has a span of 34 ft (10.4 m) and is 36 in. (91.4 cm) deep. Its cross section is shown in Figure P1.1. It is subjected to a live-load

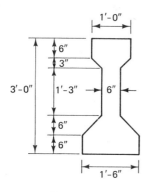

Figure P1.1.

intensity $W_L = 3,600$ plf (52.6 kN/m). Determine the required $\frac{1}{2}$-in.-dia stress-relieved seven-wire strands to resist the applied gravity load and the self-weight of the beam, assuming that the tendon eccentricity at midspan is $e_c = 13.12$ in. (333 mm). Maximum permissible stresses are as follows:

$$f'_c = 6,000 \text{ psi (41.4 MPa)}$$
$$f_c = 0.45 f'_c$$
$$= 2,700 \text{ psi (26.7 MPa)}$$
$$f_t = 12\sqrt{f'_c} = 930 \text{ psi (6.4 MPa)}$$
$$f_{pu} = 270,000 \text{ psi (1,862 MPa)}$$
$$f_{pi} = 189,000 \text{ psi (1,303 MPa)}$$
$$f_{pe} = 145,000 \text{ psi (1,000 MPa)}$$

The section properties, given these stresses, are:

$$A_c = 369 \text{ in}^2$$
$$I_g = 50,979 \text{ in}^4$$
$$r^2 = I_c/A_c = 138 \text{ in}^2$$
$$c_b = 15.83 \text{ in.}$$
$$S_b = 3,220 \text{ in}^3$$
$$S^t = 2,527 \text{ in}^3$$
$$W_D = 384 \text{ plf}$$
$$W_L = 3,600 \text{ plf}$$

Solve the problem by each of the following methods:

(a) Basic concept

(b) C-line

(c) Load balancing

1.2 Solve problem 1.1 for a 45 ft (13.7 m) span and a superimposed live load W_L = 2,000 plf (29.2 kN/m).

1.3 A simply supported pretensioned T-beam for a floor has a span of 82 ft (25 m) and the geometrical dimensions shown in Figure P1.3. It is subjected to a gravity live-load

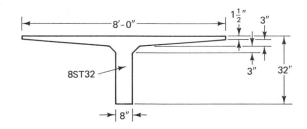

Figure P1.3.

intensity W_L = 480 plf (7 kN/m), and the prestressing tendon has an eccentricity at midspan of e_c = 19.44 in. (494 mm). Compute the concrete extreme fiber stresses in this beam at transfer and at service load, and verify whether they are within the permissible limits. Assume that all permissible stresses and materials used are the same as in Example 1.1. The section properties are:

A_c = 538 in^2

I_g = 49,329 in^4

$r^2 = I_c/A_c$ = 91.7 in^2

c_b = 23.44 in.

c_t = 8.56 in.

S_b = 2,105 in^3

S^t = 5,763 in^3

W_D = 560 plf

e_c = 19.44 in.

Design the prestressing steel needed using $\frac{1}{2}$-in.-dia stress-relieved seven-wire strands. Use the three methods of analysis discussed in this chapter in your solution.

1.4 A T-shaped simply supported beam has the cross section shown in Figure P1.4. It has a

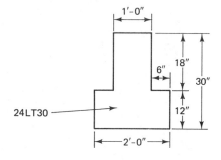

Figure P1.4.

span of 36 ft (11 m), is loaded with a gravity live-load unit intensity W_L = 2,500 plf (36.5 kN/m), and is prestressed with twelve $\frac{1}{2}$-in-dia (twelve 12.7-mm-dia) seven-wire stress-relieved strands. Compute the concrete fiber stresses at service load by each of the following methods:

(a) Basic concept

(b) C-line

(c) Load balancing

Assume that the tendon eccentricity at midspan is e_c = 9.6 in. (244 mm). Then given that

$$f'_c = 5,000 \text{ psi } (34.5 \text{ MPa})$$
$$f_t = 12\sqrt{f'_c} = 849 \text{ psi } (5.9 \text{ MPa})$$
$$f_{pe} = 165,000 \text{ psi } (1,138 \text{ MPa})$$

the section properties are as follows:

A_c = 504 in^2

I_c = 37,059 in^4

$r^2 = I_c/A_c$ = 73.5 in^2

c_b = 12.43 in.

S_b = 2,981 in^3

S^t = 2,109 in^3

W_D = 525 plf

e_c = 9.6 in.

A_{ps} = twelve $\frac{1}{2}$-in.-dia, seven-wire stress-relieved strands

1.5 Solve problem 1.4 if f'_c = 7,000 psi (48.3 MPa) and f_{pe} = 160,000 psi (1,103 MPa).

CHAPTER 2

Materials
and
Systems
for
Prestressing

2.1 CONCRETE

2.1.1 Introduction

Concrete, particularly high-strength concrete, is a major constituent of all prestressed concrete elements. Hence, its strength and long-term endurance have to be achieved through proper quality control and quality assurance at the production stage. Numerous texts are available on concrete production, quality control, and code requirements. The following discussion is intended to highlight the topics directly related to concrete in prestressed elements and systems; it is assumed that the reader is already familiar with the fundamentals of concrete and reinforced concrete.

2.1.2 Parameters Affecting the Quality of Concrete

Strength and endurance are two major qualities that are particularly important in prestressed concrete structures. Long-term detrimental effects can rapidly reduce the prestressing forces and could result in unexpected failure. Hence, measures have to be taken to ensure strict quality control and quality assurance at the various stages of production and construction as well as maintenance. Figure 2.1 shows the various factors that result in good-quality concrete.

Pasco-Kennewick Intercity Bridge. Segmentally assembled prestressed concrete cable-stayed bridge, spans 407–981–407 ft. (*Courtesy*, Arvid Grant and Associates, Inc.)

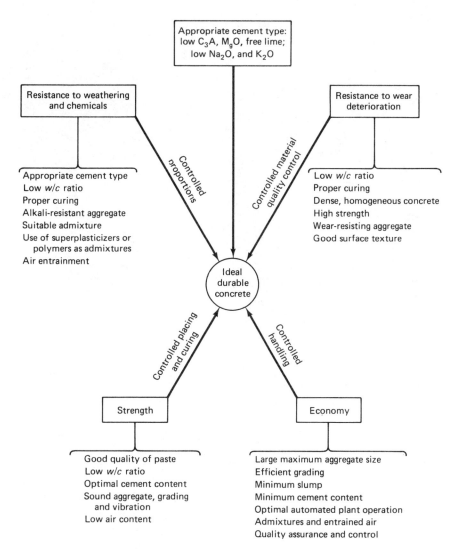

Figure 2.1 Principal properties of good concrete.

2.1.3 Properties of Hardened Concrete

The mechanical properties of hardened concrete can be classified into two categories: short-term or instantaneous properties, and long-term properties. The short-term properties are strength in compression, tension, and shear; and stiffness, as measured by the modulus of elasticity. The long-term properties can be classified in terms of creep and shrinkage. The following subsections present some details on these properties.

2.1.3.1 Compressive strength. Depending on the type of mix, the properties of aggregate, and the time and quality of the curing, compressive strengths of concrete can be obtained up to 18,000 psi or more. Commercial production of concrete

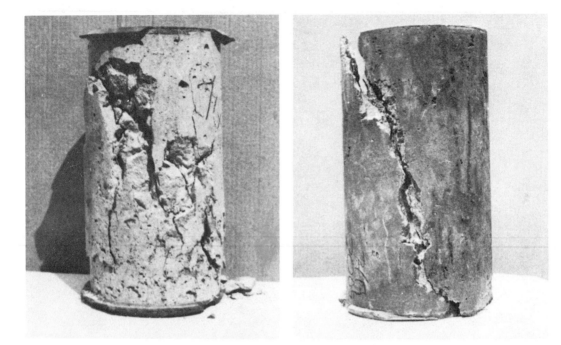

Concrete cylinders tested to failure in compression. Specimen A, low-epoxy-cement content; specimen B, high-epoxy-cement content. (Tests by Nawy, Sun, and Sauer.)

with ordinary aggregate is usually in the range 300 to 10,000 psi, with the most common concrete strengths being in the range 3,000 to 6,000 psi.

The compressive strength f'_c is based on standard 6 in. by 12 in. cylinders cured under standard laboratory conditions and tested at a specified rate of loading at 28 days of age. The standard specifications used in the United States are usually taken from ASTM C-39. The strength of concrete in the actual structure may not be the same as that of the cylinder because of the difference in compaction and curing conditions.

For a strength test, the ACI code specifies using the average of two cylinders from the same sample tested at the same age, which is usually 28 days. As for the frequency of testing, the code specifies that the strength of an individual class of concrete can be considered as satisfactory if (1) the average of all sets of three consecutive strength tests equals or exceeds the required f'_c, and (2) no individual strength test (average of two cylinders) falls below the required f'_c by more than 500 psi. The average concrete strength for which a concrete mix must be designed should exceed f'_c by an amount that depends on the uniformity of plant production.

Note that the design f'_c should not be the average cylinder strength, but rather the minimum conceivable cylinder strength.

2.1.3.2 Tensile strength.　　The tensile strength of concrete is relatively low. A good approximation for the tensile strength f_{ct} is $0.10f'_c < f_{ct} < 0.20f'_c$. It is more difficult to measure tensile strength than compressive strength because of the gripping problems with testing machines. A number of methods are available for tension

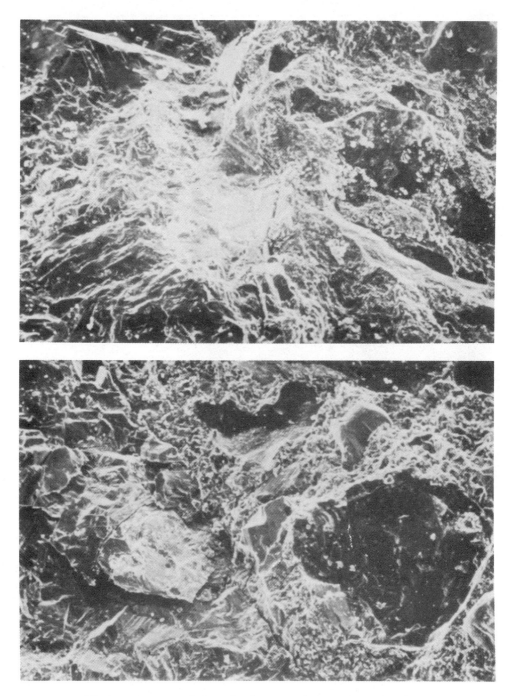

Electron microscope photographs of concrete from specimens A and B in the preceding photograph. (Tests by Nawy et al.)

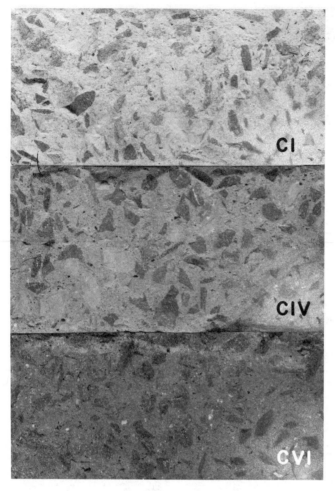

Fracture surfaces in tensile splitting tests of concretes with different w/c contents. Specimens CI and CIV have higher w/c content, hence more bond failures than specimen CVI. (Tests by Nawy et al.)

testing, the most commonly used method being the cylinder splitting, or Brazilian, test.

For members subjected to bending, the value of the modulus of rupture f_r rather than the tensile splitting strength f_t' is used in design. The modulus of rupture is measured by testing to failure plain concrete beams 6 in. square in cross section, having a span of 18 in., and loaded at their third points (ASTM C-78). The modulus of rupture has a higher value than the tensile splitting strength. The ACI specifies a value of $7.5\sqrt{f_c'}$ for the modulus of rupture of normal-weight concrete.

In most cases, lightweight concrete has a lower tensile strength than does normal-weight concrete. The following are the code stipulations for lightweight concrete:

1. If the splitting tensile strength f_{ct} is specified,

$$f_r = 109f_{ct} \leq 7.5\sqrt{f_c'} \tag{2.1}$$

2. If f_{ct} is not specified, use a factor of 0.75 for all-lightweight concrete and 0.85 for sand-lightweight concrete. Linear interpolation may be used for mixtures of natural sand and lightweight fine aggregate.

2.1.3.3 Shear strength. Shear strength is more difficult to determine experimentally than the tests discussed previously because of the difficulty in isolating shear from other stresses. This is one of the reasons for the large variation in shear-strength values reported in the literature, varying from 20 percent of the compressive strength in normal loading to a considerably higher percentage of up to 85 percent of the compressive strength in cases where direct shear exists in combination with compression. Control of a structural design by shear strength is significant only in rare cases, since shear stresses must ordinarily be limited to continually lower values in order to protect the concrete from failure in diagonal tension.

2.2 STRESS-STRAIN CURVE OF CONCRETE

Knowledge of the stress-strain relationship of concrete is essential for developing all the analysis and design terms and procedures in concrete structures. Figure 2.2 shows a typical stress-strain curve obtained from tests using cylindrical concrete specimens loaded in uniaxial compression over several minutes. The first portion of the curve, to about 40 percent of the ultimate strength f_c', can essentially be considered linear for all practical purposes. After approximately 70 percent of the failure stress, the material loses a large portion of its stiffness, thereby increasing the curvilinearity of the diagram. At ultimate load, cracks parallel to the direction of loading become distinctly visible, and most concrete cylinders (except those with very low strengths) fail suddenly shortly thereafter. Figure 2.3 shows the stress-strain curves of concrete of various strengths reported by the Portland Cement Association. It can be observed that (1) the lower the strength of concrete, the higher the failure strain; (2) the length of the initial relatively linear portion increases with the increase in the compressive strength of concrete; and (3) there is an apparent reduction in ductility with increased strength.

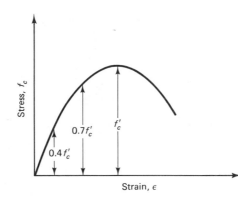

Figure 2.2 Typical stress-strain curve of concrete.

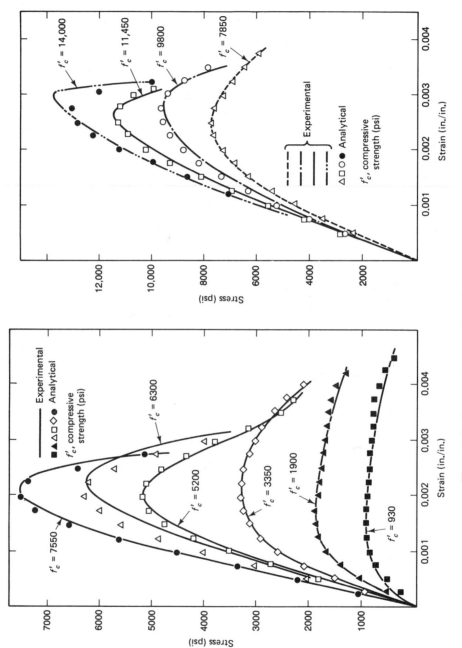

Figure 2.3 Stress-strain curves for various concrete strengths.

2.3 MODULUS OF ELASTICITY AND CHANGE IN COMPRESSIVE STRENGTH WITH TIME

Since the stress-strain curve shown in Figure 2.4 is curvilinear at a very early stage of its loading history, Young's modulus of elasticity can be applied only to the tangent of the curve at the origin. The initial slope of the tangent to the curve is defined as the initial tangent modulus, and it is also possible to construct a tangent modulus at any point of the curve. The slope of the straight line that connects the origin to a given stress (about $0.4f_c'$) determines the secant modulus of elasticity of concrete. This value, termed in design calculation the *modulus of elasticity*, satisfies the practical assumption that strains occurring during loading can be considered basically elastic (completely recoverable on unloading), and that any subsequent strain due to the load is regarded as creep.

The ACI building code gives the following expressions for calculating the secant modulus of elasticity of concrete, E_c.

$$E_c = 33w_c^{1.5}\sqrt{f_c'} \qquad \text{for } 90 < w_c < 155 \text{ lb/ft}^3 \qquad (2.2)$$

where w_c is the density of concrete in pounds per cubic foot (1 lb/ft³ = 16.02 kg/m³) and f_c' is the compressive cylinder strength in psi. For normal-weight concrete,

$$E_c = 57,000\sqrt{f_c'} \text{ psi}$$

or

$$E_c = 4,730\sqrt{f_c'} \text{ N/mm}^2 \qquad (2.3)$$

Since prestressing is performed in most cases prior to concrete's achieving its 28 days' strength, it is important to determine the concrete compressive strength f_{ci}' at the prestressing stage as well as the concrete modulus E_c at the various stages in the loading history of the element. The general expression for the compressive strength as a function of time (Ref 2.18) is

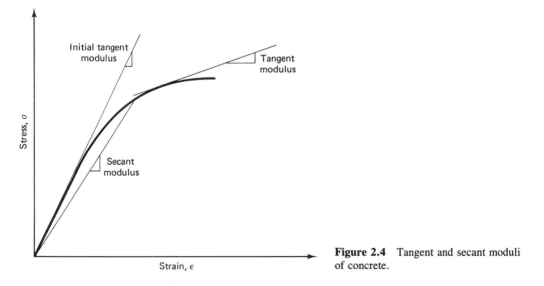

Figure 2.4 Tangent and secant moduli of concrete.

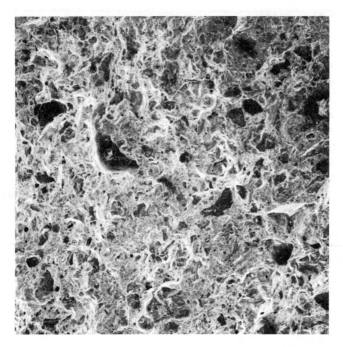

Scanning electron microscope photograph of concrete fracture surface. (Tests by Nawy, Sun, and Sauer.)

$$f'_{ci} = \frac{t}{\alpha + \beta t} f'_c \qquad (2.4\ a)$$

where f'_c = 28 days' compressive strength

t = time in days

α = factor depending on type of cement and curing conditions

= 4.00 for moist-cured type I cement and 2.30 for moist-cured type III cement

= 1.00 for steam-cured type I cement and 0.70 for steam-cured type III cement

β = factor depending on the same parameters for α giving corresponding values of 0.85, 0.92, 0.95, and 0.98, respectively

Hence, for a typical moist-cured type I cement concrete,

$$f'_{ci} = \frac{t}{4.00 + 0.85t} f'_c \qquad (2.4\ b)$$

The effective modulus of concrete, E'_c, is

$$E'_c = \frac{\text{stress}}{\text{elastic strain} + \text{creep strain}} \qquad (2.5)$$

and the ultimate effective modulus is given by

$$E_{cn} = \frac{E_c}{1 + \gamma_t} \qquad (2.6\ a)$$

where γ_t is the creep ratio defined as

$$\gamma_t = \frac{\text{ultimate creep strain}}{\text{elastic strain}}$$

The creep ratio γ_t has upper and lower limits as follows for prestresssd quality concrete:

$$\text{Upper:} \quad \gamma_t = 1.75 + 2.25\left(\frac{100 - H}{65}\right) \qquad (2.6\ b)$$

$$\text{Lower:} \quad \gamma_t = 0.75 + 0.75\left(\frac{100 - H}{50}\right) \qquad (2.6\ c)$$

where H is the mean humidity in percent.

It has to be pointed out that these expressions are valid only in general terms, since the value of the modulus of elasticity is affected by factors other than loads, such as moisture in the concrete specimen, the water/cement ratio, the age of the concrete, and temperature. Therefore, for special structures such as arches, tunnels, and tanks, the modulus of elasticity needs to be determined from test results.

Limited work exists on the determination of the modulus of elasticity in tension because the low tensile strength of concrete is normally disregarded in calculations. It is, however, valid to assume within those limitations that the value of the modulus in tension is equal to that in compression.

2.4 CREEP

Creep, or lateral material flow, is the increase in strain with time due to a sustained load. The initial deformation due to load is the *elastic strain*, while the additional strain due to the same sustained load is the *creep strain*. This practical assumption is quite acceptable, since the initial recorded deformation includes few time-dependent effects.

Figure 2.5 illustrates the increase in creep strain with time, and as in the case of shrinkage, it can be seen that creep rate decreases with time. Creep cannot be observed directly and can be determined only by deducting elastic strain and shrinkage strain from the total deformation. Although shrinkage and creep are not independent phenomena, it can be assumed that superposition of strains is valid; hence,

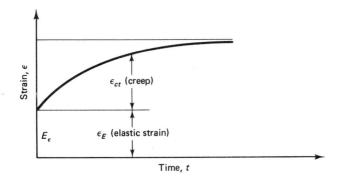

Figure 2.5   Strain-time curve.

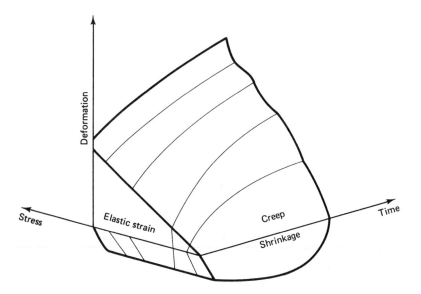

Figure 2.6 Three-dimensional model of time-dependent structural behavior.

$$\text{Total strain } (\epsilon_t) = \text{elastic strain } (\epsilon_e) + \text{creep } (\epsilon_c) + \text{shrinkage } (\epsilon_{sh})$$

An example of the relative numerical values of strain due to the foregoing three factors for a normal concrete specimen subjected to 900 psi in compression is as follows:

Immediate elastic strain, $\epsilon_e =$	250×10^{-6} in./in.
Shrinkage strain after 1 year, $\epsilon_{sh} =$	500×10^{-6} in./in.
Creep strain after 1 year, $\epsilon_c =$	750×10^{-6} in./in.
$\epsilon_t =$	$1,500 \times 10^{-6}$ in./in.

These relative values illustrate that stress-strain relationships for short-term loading lose their significance and long-term loadings become dominant in their effect on the behavior of a structure.

Figure 2.6 qualitatively shows, in a three-dimensional model, the three types of strain discussed that result from sustained compressive stress and shrinkage. Since creep is time dependent, this model has to be such that its orthogonal axes are deformation, stress, and time.

Numerous tests have indicated that creep deformation is proportional to applied stress, but the proportionality is valid only for low stress levels. The upper limit of the relationship cannot be determined accurately, but can vary between 0.2 and 0.5 of the ultimate strength f'_c. This range in the limit of the proportionality is due to the large extent of microcracks at about 40 percent of the ultimate load.

Figure 2.7(a) shows a section of the three-dimensional model in Fig. 2.6 parallel to the plane containing the stress and deformation axes at time t_1. It indicates that both elastic and creep strains are linearly proportional to the applied stress. In a similar manner, Fig. 2.7(b) illustrates a section parallel to the plane containing the time and strain axes at a stress f_1; hence, it shows the familiar relationships of creep with time and shrinkage with time.

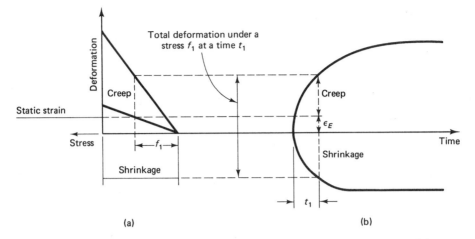

Figure 2.7 (a) Section parallel to the stress-deformation plane. (b) Section parallel to the deformation-time plane.

As in the case of shrinkage, creep is not completely reversible. If a specimen is unloaded after a period under a sustained load, an immediate elastic recovery is obtained which is less than the strain precipitated on loading. The instantaneous recovery is followed by a gradual decrease in strain, called *creep recovery*. The extent of the recovery depends on the age of the concrete when loaded, with older concretes presenting higher creep recoveries, while residual strains or deformations become frozen in the structural element (see Figure 2.8).

Creep is closely related to shrinkage, and as a general rule, a concrete that is resistant to shrinkage also presents a low creep tendency, as both phenomena are related to the hydrated cement paste. Hence, creep is influenced by the composition of the concrete, the environmental conditions, and the size of the specimen, but principally creep depends on loading as a function of time.

The composition of a concrete specimen can be essentially defined by the water/cement ratio, aggregate and cement types, and aggregate and cement contents. Therefore, like shrinkage, an increase in the water/cement ratio and in the cement content increases creep. Also, as in shrinkage, the aggregate induces a restraining effect such that an increase in aggregate content reduces creep.

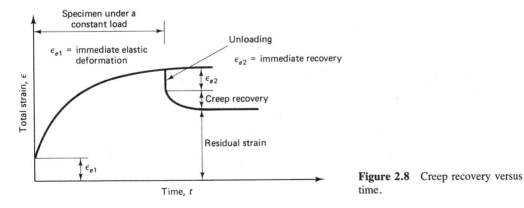

Figure 2.8 Creep recovery versus time.

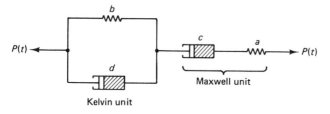

Figure 2.9 Burgers model.

2.4.1 Effects of Creep

As in shrinkage, creep increases the deflection of beams and slabs, and causes loss of prestress. In addition, the initial eccentricity of a reinforced concrete column increases with time due to creep, resulting in the transfer of the compressive load from the concrete to the steel in the section.

Once the steel yields, additional load has to be carried by the concrete. Consequently, the resisting capacity of the column is reduced and the curvature of the column increases further, resulting in overstress in the concrete, leading to failure.

2.4.2 Rheological Models

Rheological models are mechanical devices that portray the general deformation behavior and flow of materials under stress. A model is basically composed of elastic springs and ideal dashpots denoting stress, elastic strain, delayed elastic strain, irrecoverable strain, and time. The springs represent the proportionality between stress and strain, and the dashpots represent the proportionality of stress to the rate of strain. A spring and a dashpot in parallel form a Kelvin unit, and in series they form a Maxwell unit.

Two rheological models will be discussed: the Burgers model and the Ross model. The Burgers model in Figure 2.9 is shown since it can approximately simulate the stress-strain-time behavior of concrete at the limit of proportionality with some limitations. This model simulates the instantaneous recoverable strain, a; the delayed recoverable elastic strain in the spring, b; and the irrecoverable time-dependent strains in the dashpots, c and d. The weakness in the model is that it continues to deform at a uniform rate as long as the load is sustained by the Maxwell dashpot—a behavior not similar to concrete, where creep reaches a limiting value with time, as shown in Figure 2.5.

A modification in the form of the Ross rheological model in Figure 2.10 can eliminate this deficiency. A in this model represents the Hookian direct proportionality of stress-to-strain element, D is the dashpot, and B and C are the elastic springs that can transmit the applied load $P(t)$ to the enclosing cylinder walls by direct friction. Since each coil has a defined frictional resistance, only those coils whose resistance equals the applied load $P(t)$ are displaced; the others remain unstressed, symbolizing the irrecoverable deformation in concrete. As the load continues to increase, it overcomes the spring resistance of unit B, pulling out the spring from the dashpot and signifying failure in a concrete element. More rigorous models, such as Roll's, have been used to assist in predicting the creep strains. Mathematical expressions for such predictions can be very rigorous. One convenient expression due to Ross defines the creep C under load after a time interval t as

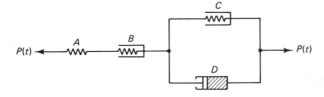

Figure 2.10 Ross model.

$$C = \frac{t}{a + bt} \tag{2.7}$$

where a and b are constants determinable from tests.

Work by Branson (Refs. 2.18 and 2.19) has simplified creep evaluation. The additional strain ϵ_{cu} due to creep can be defined as

$$\epsilon_{cu} = \rho_u f_{ci} \tag{2.8}$$

where ρ_u = unit creep coefficient, generally called *specific creep*
f_{ci} = stress intensity in the structural member corresponding to unit strain ϵ_{ci}

The ultimate creep coefficient, C_u, is given by

$$C_u = \rho_u E_c \tag{2.9}$$

Or average $C_u \simeq 2.35$.

Branson's model, verified by extensive tests, relates the creep coefficient C_t at any time to the ultimate creep coefficient (for standard conditions) as

$$C_t = \frac{t^{0.6}}{10 + t^{0.6}} C_u \tag{2.10}$$

or, alternatively,

$$\rho_t = \frac{t^{0.6}}{10 + t^{0.6}} \tag{2.11}$$

where t is the time in days and ρ_t is the time multiplier. Standard conditions as defined by Branson pertain to concretes of slump 4 in. (10 cm) or less and a relative humidity of 40 percent.

When conditions are not standard, creep correction factors have to be applied to Equations 2.10 or 2.11 as follows:

(a) For moist-cured concrete loaded at an age of 7 days or more,

$$k_a = 1.25t^{-0.118} \tag{2.12}$$

(b) For steam-cured concrete loaded at an age of 1–3 days or more,

$$k_a = 1.13t^{-0.095} \tag{2.13}$$

For greater than 40 percent relative humidity, a further multiplier correction factor of

$$k_{c_1} = 1.27 - 0.0067H \tag{2.14}$$

has to be applied in addition to those of Equations 2.12 and 2.13, where H = relative humidity value in percent.

Energy Center, New Orleans, Louisiana.
(*Courtesy,* Post-Tensioning Institute.)

2.5 SHRINKAGE

Basically, there are two types of shrinkage: plastic shrinkage and drying shrinkage. *Plastic shrinkage* occurs during the first few hours after placing fresh concrete in the forms. Exposed surfaces such as floor slabs are more easily affected by exposure to dry air because of their large contact surface. In such cases, moisture evaporates faster from the concrete surface than it is replaced by the bleed water from the lower layers of the concrete elements. *Drying shrinkage*, on the other hand, occurs after the concrete has already attained its final set and a good portion of the chemical hydration process in the cement gel has been accomplished.

Drying shrinkage is the decrease in the volume of a concrete element when it loses moisture by evaporation. The opposite phenomenon, that is, volume increase through water absorption, is termed *swelling*. In other words, shrinkage and swelling represent water movement out of or into the gel structure of a concrete specimen due to the difference in humidity or saturation levels between the specimen and the surroundings irrespective of the external load.

Shrinkage is not a completely reversible process. If a concrete unit is saturated with water after having fully shrunk, it will not expand to its original volume. Figure 2.11 relates the increase in shrinkage strain ϵ_{sh} with time. The rate decreases with time since older concretes are more resistant to stress and consequently undergo less shrinkage, such that the shrinkage strain becomes almost asymptotic with time.

Several factors affect the magnitude of drying shrinkage:

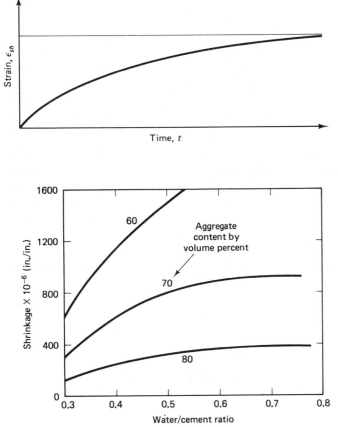

Figure 2.11 Shrinkage-time curve.

Figure 2.12 w/c ratio and aggregate content effect on shrinkage.

1. *Aggregate.* The aggregate acts to restrain the shrinkage of the cement paste; hence, concretes with high aggregate content are less vulnerable to shrinkage. In addition, the degree of restraint of a given concrete is determined by the properties of aggregates: those with a high modulus of elasticity or with rough surfaces are more resistant to the shrinkage process.

2. *Water/cement ratio.* The higher the water/cement ratio, the higher the shrinkage effects. Figure 2.12 is a typical plot relating aggregate content to water/cement ratio.

3. *Size of the concrete element.* Both the rate and the total magnitude of shrinkage decrease with an increase in the volume of the concrete element. However, the duration of shrinkage is longer for larger members since more time is needed for drying to reach the internal regions. It is possible that 1 year may be needed for the drying process to begin at a depth of 10 in. from the exposed surface, and 10 years to begin at 24 in. below the external surface.

4. *Medium ambient conditions.* The relative humidity of the medium greatly affects the magnitude of shrinkage; the rate of shrinkage is lower at high states of relative humidity. The environment temperature is another factor, in that shrinkage becomes stabilized at low temperatures.

5. *Amount of reinforcement.* Reinforced concrete shrinks less than plain concrete; the relative difference is a function of the reinforcement percentage.

6. *Admixtures.* This effect varies depending on the type of admixture. An accelerator such as calcium chloride, used to accelerate the hardening and setting of the concrete, increases the shrinkage. Pozzolans can also increase the drying shrinkage, whereas air-entraining agents have little effect.

7. *Type of cement.* Rapid-hardening cement shrinks somewhat more than other types, while shrinkage-compensating cement minimizes or eliminates shrinkage cracking if used with restraining reinforcement.

8. *Carbonation.* Carbonation shrinkage is caused by the reaction between the carbon dioxide (CO_2) present in the atmosphere and that present in the cement paste. The amount of the combined shrinkage varies according to the sequence of occurrence of carbonation and drying processes. If both phenomena take place simultaneously, less shrinkage develops. The process of carbonation, however, is dramatically reduced at relative humidities below 50 percent.

Branson (Ref 2.18) recommends the following relationships for the shrinkage strain as a function of time for standard conditions of humidity ($H \cong 40$ percent):

(a) For moist-cured concrete any time t after 7 days,

$$\epsilon_{SH,t} = \frac{t}{35 + t}(\epsilon_{SH,u})$$ (2.15)

where $\epsilon_{SH,u} = 800 \times 10^{-6}$ in/in if local data are not available.

(b) For steam-cured concrete after the age of 1–3 days,

$$\epsilon_{SH,t} = \frac{t}{55 + t}\epsilon_{SH,u}$$ (2.16)

For other than standard humidity, a correction factor has to be applied to Equations 1.4 and 1.5 as follows:

(a) For $40 < H \le 80$ percent,

$$k_{SH} = 1.40 - 0.010H$$ (2.17 a)

(b) For $80 < H \le 100$ percent,

$$k_{SH} = 3.00 - 0.30H$$ (2.17 b)

2.6 NONPRESTRESSING REINFORCEMENT

Steel reinforcement for concrete consists of bars, wires, and welded wire fabric, all of which are manufactured in accordance with ASTM standards. The most important properties of reinforcing steel are:

1. Young's modulus, E_s
2. Yield strength, f_y
3. Ultimate strength, f_u
4. Steel grade designation
5. Size or diameter of the bar or wire

To increase the bond between concrete and steel, projections called *deformations* are rolled onto the bar surface as shown in Fig. 2.13, in accordance with ASTM specifications. The deformations shown must satisfy ASTM Specification A616-76 for the bars to be accepted as deformed. Deformed wire has indentations pressed into the wire or bar to serve as deformations. Except for wire used in spiral reinforcement in columns, only deformed bars, deformed wires, or wire fabric made from smooth or deformed wire may be used in reinforced concrete under approved practice.

Figure 2.14 shows typical stress-strain curves for grades 40, 60, and 80 steels. These have corresponding yield strengths of 40,000, 60,000, and 80,000 psi (276, 345, and 517 N/mm², respectively) and generally have well-defined yield points. For steels that lack a well-defined yield point, the yield-strength value is taken as the strength corresponding to a unit strain of 0.005 for grades 40 and 60 steels, and 0.0035 for grade 80 steel. The ultimate tensile strengths corresponding to the 40, 60, and 80 grade steels are 70,000, 90,000, and 100,000 psi (483, 621, and 690 N/mm²), respectively, and some steel types are given in Table 2.1. The percent elongation at fracture, which varies with the grade, bar diameter, and manufacturing source, ranges from 4.5 to 12 percent over an 8-in. (203.2-mm) gage length.

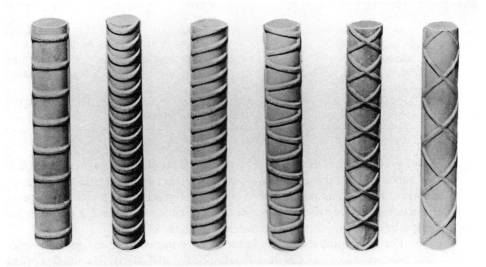

Figure 2.13 Various forms of ASTM-approved deformed bars.

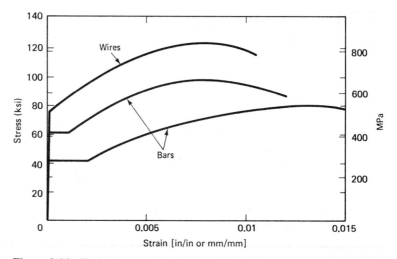

Figure 2.14 Typical stress-strain diagrams for various nonprestressing steels.

TABLE 2.1 REINFORCEMENT GRADES AND STRENGTHS

1982 Standard type	Minimum yield point or yield strength, f_y (psi)	Ultimate strength, f_u (psi)
Billet steel (A615)		
Grade 40	40,000	70,000
Grade 60	60,000	90,000
Axle steel (A617)		
Grade 40	40,000	70,000
Grade 60	60,000	90,000
Low-alloy steel (A706): Grade 60	60,000	80,000
Deformed wire		
Reinforced	75,000	85,000
Fabric	70,000	80,000
Smooth wire		
Reinforced	70,000	80,000
Fabric	65,000, 56,000	75,000, 70,000

Welded wire fabric is increasingly used for slabs because of the ease of placing the fabric sheets, the control over reinforcement spacing, and the better bond. The fabric reinforcement is made of smooth or deformed wires which run in perpendicular directions and are welded together at intersections. Table 2.2 presents geometrical properties of some standard wire reinforcement.

For most mild steels, the behavior is assumed to be elastoplastic and Young's modulus is taken as 29×10^6 psi $(200 \times 10^6$ MPa). Table 2.1 presents the reinforcement-grade strengths, and Table 2.3 presents geometrical properties of the various sizes of bars.

TABLE 2.2 STANDARD WIRE REINFORCEMENT

| W&D size | | U.S. customary | | | Area (in²/ft of width for various spacings) | | | | | | |
| Smooth | Deformed | Nominal diameter (in.) | Nominal area (in²) | Nominal weight (lb/ft) | Center-to-center spacing (in.) | | | | | | |
					2	3	4	6	8	10	12
W31	D31	0.628	0.310	1.054	1.86	1.24	0.93	0.62	0.465	0.372	0.31
W30	D30	0.618	0.300	1.020	1.80	1.20	0.90	0.60	0.45	0.366	0.30
W28	D28	0.597	0.280	0.952	1.68	1.12	0.84	0.56	0.42	0.336	0.28
W26	D26	0.575	0.260	0.934	1.56	1.04	0.78	0.52	0.39	0.312	0.26
W24	D24	0.553	0.240	0.816	1.44	0.96	0.72	0.48	0.36	0.288	0.24
W22	D22	0.529	0.220	0.748	1.32	0.88	0.66	0.44	0.33	0.264	0.22
W20	D20	0.504	0.200	0.680	1.20	0.80	0.60	0.40	0.30	0.24	0.20
W18	D18	0.478	0.180	0.612	1.08	0.72	0.54	0.36	0.27	0.216	0.18
W16	D16	0.451	0.160	0.544	0.96	0.64	0.48	0.32	0.24	0.192	0.16
W14	D14	0.422	0.140	0.476	0.84	0.56	0.42	0.28	0.21	0.168	0.14
W12	D12	0.390	0.120	0.408	0.72	0.48	0.36	0.24	0.18	0.144	0.12
W11	D11	0.374	0.110	0.374	0.66	0.44	0.33	0.22	0.165	0.132	0.11
W10.5		0.366	0.105	0.357	0.63	0.42	0.315	0.21	0.157	0.126	0.105
W10	D10	0.356	0.100	0.340	0.60	0.40	0.30	0.20	0.15	0.12	0.10
W9.5		0.348	0.095	0.323	0.57	0.38	0.285	0.19	0.142	0.114	0.095
W9	D9	0.338	0.090	0.306	0.54	0.36	0.27	0.18	0.135	0.108	0.09
W8.5		0.329	0.085	0.289	0.51	0.34	0.255	0.17	0.127	0.102	0.085
W8	D8	0.319	0.080	0.272	0.48	0.32	0.24	0.16	0.12	0.096	0.08
W7.5		0.309	0.075	0.255	0.45	0.30	0.225	0.15	0.112	0.09	0.075
W7	D7	0.298	0.070	0.238	0.42	0.28	0.21	0.14	0.105	0.084	0.07
W6.5		0.288	0.065	0.221	0.39	0.26	0.195	0.13	0.097	0.078	0.065
W6	D6	0.276	0.060	0.204	0.36	0.24	0.18	0.12	0.09	0.072	0.06
W5.5		0.264	0.055	0.187	0.33	0.22	0.165	0.11	0.082	0.066	0.055
W5	D5	0.252	0.050	0.170	0.30	0.20	0.15	0.10	0.075	0.06	0.05
W4.5		0.240	0.045	0.153	0.27	0.18	0.135	0.09	0.067	0.054	0.045
W4	D4	0.225	0.040	0.136	0.24	0.16	0.12	0.08	0.06	0.048	0.04
W3.5		0.211	0.035	0.119	0.21	0.14	0.105	0.07	0.052	0.042	0.035
W3		0.195	0.030	0.102	0.18	0.12	0.09	0.06	0.045	0.036	0.03
W2.9		0.192	0.029	0.098	0.174	0.116	0.087	0.058	0.043	0.035	0.029
W2.5		0.178	0.025	0.085	0.15	0.10	0.075	0.05	0.037	0.03	0.025
W2		0.159	0.020	0.068	0.12	0.08	0.06	0.04	0.03	0.024	0.02
W1.4		0.135	0.014	0.049	0.084	0.056	0.042	0.028	0.021	0.017	0.014

TABLE 2.3 WEIGHT, AREA, AND PERIMETER OF INDIVIDUAL BARS

Bar designation number	Weight per foot (lb)	1982 Standard nominal dimensions		
		Diameter, d_b[in. (mm)]	Cross-sectional area, A_b (in.^{2})	Perimeter (in.)
3	0.376	0.375 (9)	0.11	1.178
4	0.668	0.500 (13)	0.20	1.571
5	1.043	0.625 (16)	0.31	1.963
6	1.502	0.750 (19)	0.44	2.356
7	2.044	0.875 (22)	0.60	2.749
8	2.670	1.000 (25)	0.79	3.142
9	3.400	1.128 (28)	1.00	3.544
10	4.303	1.270 (31)	1.27	3.990
11	5.313	1.410 (33)	1.56	4.430
14	7.65	1.693 (43)	2.25	5.32
18	13.60	2.257 (56)	4.00	7.09

2.7 PRESTRESSING REINFORCEMENT

2.7.1 Types of Reinforcement

Because of the high creep and shrinkage losses in concrete, effective prestressing can be achieved by using very high-strength steels in the range of 270,000 psi or more (1,862 MPa or higher). Such high-stressed steels are able to counterbalance these losses in the surrounding concrete and have adequate leftover stress levels to sustain the required prestressing force. The magnitude of normal prestress losses can be expected to be in the range of 35,000 to 60,000 psi (241 to 414 MPa). The initial prestress would thus have to be very high, on the order of 180,000 to 220,000 psi (1,241 to 1,517 MPa). From the aforementioned magnitude of prestress losses, it can be inferred that normal steels with yield strengths $f_y = 60,000$ psi (414 MPa) would have little prestressing stress left after losses, obviating the need for using very high-strength steels for prestressing them.

Prestressing reinforcement can be in the form of single wires, strands composed of several wires twisted to form a single element, and high-strength bars. Three types commonly used in the United States are:

- Uncoated stress-relieved wires.
- Uncoated stress-relieved strands and low-relaxation strands.
- Uncoated high-strength steel bars.

Wires or strands that are not stress-relieved, such as the straightened wires or oil-tempered wires often used in other countries, exhibit higher relaxation losses than stress-relieved wires or strands. Consequently, it is important to account for the appropriate magnitude of losses once a determination is made on the type of prestressing steel required.

Prestressed concrete Valdez floating dock. Designed by ABAM Engineers, built in two pieces in Tacoma, Washington, then towed to Alaska by deployment. (*Courtesy, ABAM Engineers, Tacoma, Washington.*)

2.7.2 Stress-Relieved Wires and Strands

Stress-relieved wires are cold-drawn single wires conforming to ASTM standard A 421; stress-relieved strands conform to ASTM standard A 416. The strands are made from seven wires by twisting six of them on a pitch of 12 to 16 wire diameter around a slightly larger, straight control wire. Stress-relieving is done after the wires are woven into the strand. The geometrical properties of the wires and strands as required by ASTM are given in Tables 2.4 and 2.5, respectively.

To maximize the steel area of the seven-wire strand for any nominal diameter, the standard wire can be drawn through a die to form a *compacted* strand as shown in Figure 2.15(b); this is opposed to the standard seven-wire strand in Figure 2.15(a). ASTM standard A 779 requires the minimum strengths and geometrical properties given in Table 2.6.

Figure 2.16 shows a typical stress-strain diagram for wire and strand prestressing steels.

TABLE 2.4 STRESS-RELIEVED WIRE FOR PRESTRESSED CONCRETE (ASTM A 421)

Nominal diameter (in.)	Min. tensile strength (psi)		Min. stress at 1% extension (psi)	
	Type BA	Type WA	Type BA	Type WA
0.192		250,000		212,500
0.196	240,000	250,000	204,000	212,500
0.250	240,000	240,000	204,000	204,000
0.276	235,000	235,000	199,750	199,750

Source: Post-Tensioning Institute

TABLE 2.5 STRESS-RELIEVED SEVEN-WIRE STANDARD STRAND FOR PRESTRESSED CONCRETE (ASTM A 416)

Nominal diameter of strand (in.)	Breaking strength of strand (min. lb)	Nominal steel area of strand (sq in.)	Nominal weight of strands (lb per 1000 ft)*	Minimum load at 1% extension (lb)
GRADE 250				
$\frac{1}{4}$(0.250)	9 000	0.036	122	7 650
$\frac{5}{16}$(0.313)	14 500	0.058	197	12 300
$\frac{3}{8}$(0.375)	20 000	0.080	272	17 000
$\frac{7}{16}$(0.438)	27 000	0.108	367	23 000
$\frac{1}{2}$(0.500)	36 000	0.144	490	30 600
$\frac{3}{5}$(0.600)	54 000	0.216	737	45 900
GRADE 270				
$\frac{3}{8}$(0.375)	23 000	0.085	290	19 550
$\frac{7}{16}$(0.438)	31 000	0.115	390	26 350
$\frac{1}{2}$(0.500)	41 300	0.153	520	35 100
$\frac{3}{5}$(0.600)	58 600	0.217	740	49 800

*100,000 psi = 689.5 MPa
0.1 in. = 2.54 mm; 1 in² = 645 mm²
weight: mult. by 1.49 to obtain weight in kg per 1,000 m.
1,000 lb = 4,448 Newton
Source: Post-Tensioning Institute

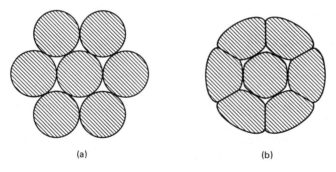

(a)　　　　　(b)

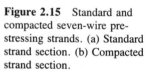

Figure 2.15 Standard and compacted seven-wire prestressing strands. (a) Standard strand section. (b) Compacted strand section.

TABLE 2.6 STRESS-RELIEVED SEVEN-WIRE COMPACTED STRAND FOR PRESTRESSED CONCRETE (ASTM A 779)

Nominal diameter (in.)	Breaking strength of strand (min. lb)*	Nominal steel area (in²)	Nominal weight of strand (per 1,000 ft-lb)
$\frac{1}{2}$	47,000	0.174	600
0.6	67,440	0.256	873
0.7	85,430	0.346	1176

* 1000 lb = 4,448 Newton
Grade 270: f_{pu} = 270,000 psi ult. strength (1,862 MPa)
1 in. = 25.4 mm   1 in² = 645 mm²

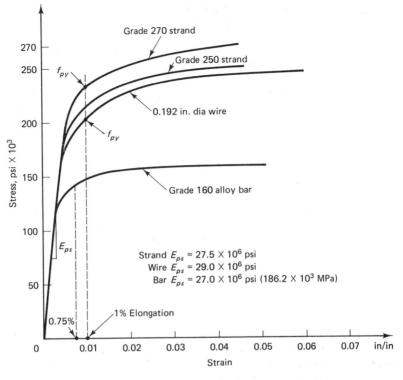

Figure 2.16 Stress-strain diagram for prestressing steel.

2.7.3 High-Tensile-Strength Prestressing Bars

High-tensile-strength alloy steel bars for prestressing are either smooth or deformed, and are available in nominal diameters from $\frac{3}{4}$ in. (19 mm) to $1\frac{3}{8}$ in. (35 mm). They must conform to ASTM standard A 722. Cold drawn in order to raise their yield strength, these bars are stress relieved as well to increase their ductility. Stress relieving is achieved by heating the bar to an appropriate temperature, generally below 500°C. Though essentially the same stress-relieving process is employed for bars as for strands, the tensile strength of prestressing bars has to be a minimum of 150,000 psi (1,034 MPa), with a minimum yield strength of 85 percent of the ultimate strength for smooth bars and 80 percent for deformed bars.

Table 2.7 lists the geometrical properties of the prestressing bars as required by ASTM standard A 722, and Figure 2.16 shows a typical stress-strain diagram for such bars.

2.7.4 Steel Relaxation

Stress relaxation in prestressing steel is the loss of prestress when the wires or strands are subjected to essentially constant strain. It is identical to creep in concrete, except that creep is a *change* in strain whereas steel relaxation is a *loss in steel stress*. Where t = time, in hours, after prestressing, the loss of stress due to relaxation in stress-relieved wires and strands can be evaluated from the expression

TABLE 2.7 STRESS-RELIEVED STEEL BARS FOR
PRESTRESSED CONCRETE

Bar type*	Nominal diameter (in.)	Nominal steel area (in²)
Smooth Alloy	0.750	0.442
Steel Grade	0.875	0.601
145 or 160	1.000	0.785
(ASTM A722)	1.125	0.994
	1.250	1.227
	1.375	1.485
Deformed	0.625	0.280
Bars	1.000	0.852
	1.250	1.295

* Grade 145: $f_{pu} = 145,000$ psi (1,000 MPa)
Grade 160: $f_{pu} = 160,000$ psi (1,103 MPa)
1 in. = 25.4 mm   1 in² = 645 mm²

$$\Delta f_R = f_{pi} \frac{\log t}{10}\left(\frac{f_{pi}}{f_{py}} - 0.55\right) \tag{2.18}$$

provided that $f_{pi}/f_{py} \geq 0.55$ and $f_{py} \cong 0.85 f_{pu}$ for stress-relieved strands and 0.90 for low-relaxation strands. Also, $f_{pi} = 0.82 f_{py}$ immediately after transfer but $f_{pi} \leq 0.74 f_{pu}$ for pretensioned, and $0.70 f_{pu}$ for post-tensioned, concrete. In general, $f_{pi} \cong 0.70 f_{pu}$.

It is possible to decrease stress relaxation loss by subjecting strands that are initially stressed to 70 percent of their ultimate strength f_{pu} to temperatures of 20°C to 100°C for an extended time in order to produce a permanent elongation—a process called *stabilization*. The prestressing steel thus produced is termed *low-relaxation steel* and has a relaxation stress loss that is 25 percent of that of normal stress-relieved steel.

The expression for stress relaxation in low-relaxation prestressing steels is

$$\Delta f_R = f_{pi} \frac{\log t}{45}\left(\frac{f_{pi}}{f_{py}} - 0.55\right) \tag{2.19}$$

Figure 2.17 shows the relative relaxation loss for stress-relieved and low-relaxation steels for seven-wire strands held at constant length at 29.5°C.

2.7.5 Corrosion and Deterioration of Strands

Protection against corrosion of prestressing steel is more critical than in the case of nonprestressed steel. Such precaution is necessary since the strength of the prestressed concrete element is a function of the prestressing force, which in turn is a function of the prestressing tendon area. Reduction of the prestressing steel area due to corrosion can drastically reduce the nominal moment strength of the prestressed section, which can lead to premature failure of the structural system. In pretensioned members, protection against corrosion is provided by the concrete surrounding the tendon, provided that adequate concrete cover is available. In post-tensioned members, pro-

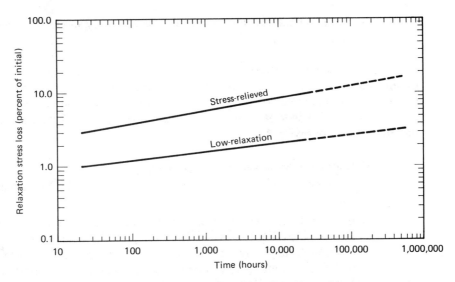

Figure 2.17 Relaxation loss vs. time for stress-relieved low-relaxation prestressing steels at 70% of the ultimate. (*Courtesy*, Post-Tensioning Institute.)

tection can be obtained by full grouting of the ducts after prestressing is completed or by greasing.

Another form of wire or strand deterioration is *stress corrosion*, which is characterized by the formation of microscopic cracks in the steel which lead to brittleness and failure. This type of reduction in strength can occur only under very high stress and, though infrequent, is difficult to prevent.

2.8 ACI MAXIMUM PREMISSIBLE STRESSES IN CONCRETE AND REINFORCEMENT

Following are definitions of some important mathematical terms used in this section:

f_{py} = specified yield strength of prestressing tendons, in psi

f_y = specified yield strength of nonprestressed reinforcement, in psi

f_{pu} = specified tensile strength of prestressing tendons, in psi

f_c' = specified compressive strength of concrete, in psi

f_{ci}' = compressive strength of concrete at time of initial prestress

2.8.1 Concrete Stresses in Flexure

Stresses in concrete immediately after prestress transfer (before time-dependent prestress losses) shall not exceed the following:

(a) Extreme fiber stress in compression .... $0.60 f_{ci}'$

(b) Extreme fiber stress in tension except as permitted in (c) .... $3\sqrt{f_{ci}'}$

(c) Extreme fiber stress in tension at ends of simply supported members .... $6\sqrt{f'_{ci}}$

Where computed tensile stresses exceed these values, bonded auxiliary reinforcement (nonprestressed or prestressed) shall be provided in the tensile zone to resist the total tensile force in concrete computed under the assumption of an uncracked section.

Stresses in concrete at service loads (after allowance for all prestress losses) shall not exceed the following:

(a) Extreme fiber stress in compression .... $0.45f'_c$

(b) Extreme fiber stress in tension in precompressed tensile zone .... $6\sqrt{f'_c}$

(c) Extreme fiber stress in tension in precompressed tensile zone of members (except two-way slab systems), where analysis based on transformed cracked sections and on bilinear moment-deflection relationships shows that immediate and long-time deflections comply with the ACI definition requirements and minimum concrete cover requirements .... $12\sqrt{f'_c}$

2.8.2 Prestressing Steel Stresses

Tensile stress in prestressing tendons shall not exceed the following:

(a) Due to tendon jacking force .... $0.94f_{py}$
but not greater than the lesser of $0.80f_{pu}$ and the maximum value recommended by the manufacturer of prestressing tendons or anchorages.

(b) Immediately after prestress transfer .... $0.82f_{py}$
but not greater than $0.74f_{pu}$.

(c) Post-tensioning tendons, at anchorages and couplers, immediately after tendon anchorage .... $0.70f_{pu}$

2.9 AASHTO MAXIMUM PERMISSIBLE STRESSES IN CONCRETE AND REINFORCEMENT

2.9.1 Concrete Stresses Before Creep and Shrinkage Losses

Compression
  Pretensioned members .... $0.60f'_{ci}$
  Post-tensioned members .... $0.55f'_{ci}$
Tension
  Precompressed tensile zone .... No temporary allowable stresses are specified.

Other Areas

In tension areas with
no bonded reinforcement .... 200 psi or $3\sqrt{f'_{ci}}$

Where the calculated tensile stress exceeds this value, bonded reinforcement shall be provided to resist the total tension force in the
concrete computed on the assumption of an uncracked section. The
maximum tensile stress shall not exceed .... $7.5\sqrt{f'_{ci}}$

2.9.2 Concrete Stresses at Service Load after Losses

Compression .... $0.40 f'_c$

Tension in the precompressed tensile zone
(a) For members with bonded reinforcement .... $6\sqrt{f'_c}$
For severe corrosive exposure conditions, such as
coastal areas .... $3\sqrt{f'_c}$
(b) For members without bonded reinforcement 0

Tension in other areas is limited by the allowable temporary stresses
specified in Section 2.8.1.

2.9.2.1 Cracking stresses

Modulus of rupture from tests or if not available.
For normal-weight concrete .... $7.5\sqrt{f'_c}$
For sand-lightweight concrete .... $6.3\sqrt{f'_c}$
For all other lightweight concrete .... $5.5\sqrt{f'_c}$

2.9.2.2 Anchorage-bearing stresses

Post-tensioned anchorage at service load .... 3,000 psi
(but not to exceed $0.9 f'_{ci}$)

2.9.3 Prestressing Steel Stresses

Temporary stress before loss due to creep and shrinkage ..... $0.70 f'_s$
Stress at service load after losses .... $0.80 f_y$

(Overstressing to $0.80 f'_s$ for short periods of time may be permitted provided the stress, after transfer to concrete in pretensioning or seating of
anchorage in post-tensioning, does not exceed $0.70 f'_s$.)

2.9.4 Relative Humidity Values

Figure 2.18 gives the mean annual relative humidity values for all regions in the
United States in percent, to be used for evaluating shrinkage losses in concrete.

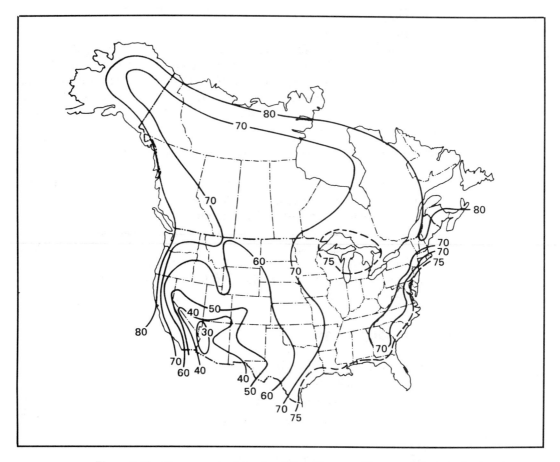

Figure 2.18 Mean annual relative humidity. (*Courtesy*, Prestressed Concrete Institute.)

2.10 PRESTRESSING SYSTEMS AND ANCHORAGES

2.10.1 Pretensioning

Prestressing steel is pretensioned against independent anchorages *prior to* the placement of concrete around it. Such anchorages are supported by large and stable bulkheads to support the exceedingly high concentrated forces applied to the individual tendons. The term "pretensioning" means pretensioning of the prestressing steel, not the beam it serves. Consequently, a *pretensioned beam* is a prestressed beam in which the prestressing tendon is tensioned prior to casting the section, while a *posttensioned beam* is one in which the prestressing tendon is tensioned after the beam has been cast and has achieved the major portion of its concrete strength. Pretensioning is normally performed at precasting plants, where a precasting stressing bed of a long reinforced concrete slab is cast on the ground with vertical anchor bulkheads or walls at its ends. The steel strands are stretched and anchored to the vertical walls, which are designed to resist the large eccentric prestressing forces. Prestressing can be

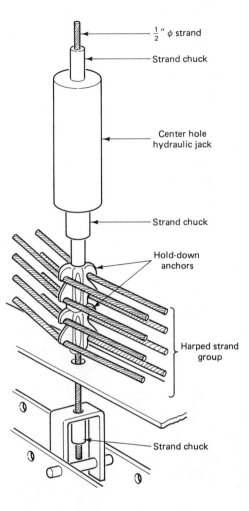

$\frac{1}{2}$″ ϕ strand

Strand chuck

Center hole hydraulic jack

Strand chuck

Hold-down anchors

Harped strand group

Strand chuck

Figure 2.19 Hold-down anchor for harping pretensioning tendons. (*Courtesy*, Post-Tensioning Institute.)

accomplished by prestressing *individual* strands, or *all* the strands at one jacking operation.

For harped tendon profiles, the prestressing bed is provided with hold-down devices as shown in Figure 2.19. Since the bed can be several hundred feet long, several precast prestressed elements can be produced in one operation, and the exposed prestressing strands between them can be cut after the concrete hardens. Pretensioning several elements in a prestressing bed is represented schematically in Figure 2.20, while harping of tendons in a prestressing bed system is shown in Figure 2.21.

In pretensioning, strands and single wires are anchored by several patented systems. One of these, a chuck system by Supreme Products, is used for anchoring tendons in post-tensioning. The gripping mechanism of this system is illustrated in Figure 2.22. A prestressing bed for moderately sized pretensioned beams up to 24 ft (7.32 m) long was developed and used by the author in Ref. 2.31 for his continuing work on the behavior of pretensioned and post-tensioned structural systems. Supreme Products anchorage chucks have been used together with the Freyssinet jack, where

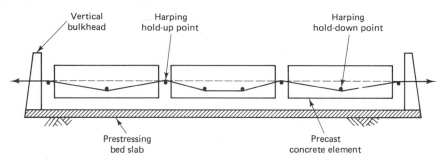

Figure 2.20 Schematic of pretensioning bed.

Figure 2.21 Harping of tendons in a prestressing bed system.

applicable. Figures 2.23 and 2.24 give details of the prestressing bed system also used for post-tensioning developed by Nawy and Potyondy at Rutgers University, while Figure 2.25 shows the dimensional details of the system.

2.10.2 Post-Tensioning

In post-tensioning, the strands, wires, or bars are tensioned after hardening of the concrete. The strands are placed in the longitudinal ducts within the precast concrete element. The prestressing force is transferred through end anchorages such as the Supreme Products chucks shown in Figure 2.22, or other systems shown in the Appendix. The tendons of strands should *not* be bonded or grouted prior to full prestressing.

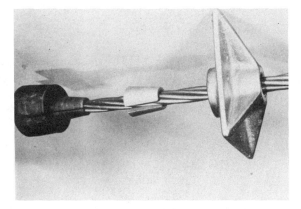

Strand anchor.

Monostrand anchor.

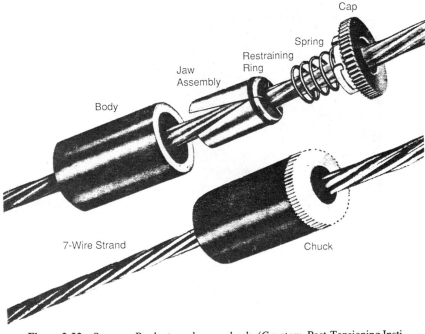

Figure 2.22 Supreme Products anchorage chuck. (*Courtesy*, Post-Tensioning Institute.)

2.10.3 Jacking Systems

One of the fundamental components of a prestressing operation is the jacking system applied, i.e., the manner in which the prestressing force is transferred to the steel tendons. Such a force is applied through the use of hydraulic jacks of capacity 10 to 20 tons and a stroke from 6 to 48 in., depending on whether pretensioning or post-tensioning is used and whether individual tendons are being prestressed or all the tendons are being stressed simultaneously. In the latter case large-capacity jacks are

Figure 2.23 Prestress tensioning arrangement (Nawy et al.).

Figure 2.24 Intermediate connections between frames of the prestressing system for continuous beams (Nawy et al.).

needed, with a stroke of at least 30 in. (762 mm). Of course, the cost will be higher than sequential tensioning. Figure 2.26 shows a 500-ton multistrand jack for simultaneous jacking through a center hole.

2.10.4 Grouting of Post-Tensioned Tendons

In order to provide permanent protection for the post-tensioned steel and to develop a bond between the prestressing steel and the surrounding concrete, the prestressing ducts have to be filled under pressure with the appropriate cement grout in an injection process.

2.10.4.1 Grouting materials

1. **Portland Cement.** Portland cement should conform to one of the following specifications: ASTM C150, Type I, II, or III.

    Cement used for grouting should be fresh and should not contain any lumps or other indications of hydration or "pack set."

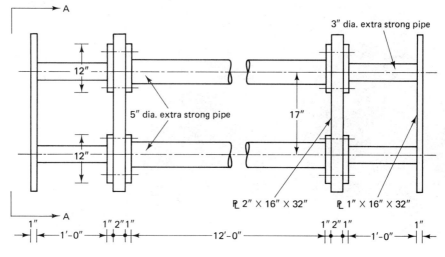

3″ dia. extra strong pipe

12″

5″ dia. extra strong pipe

17″

12″

℞ 2″ × 16″ × 32″   ℞ 1″ × 16″ × 32″

1″   1″ 2″1″   1″ 2″ 1″   1″
1′-0″   12′-0″   1′-0″

(a)

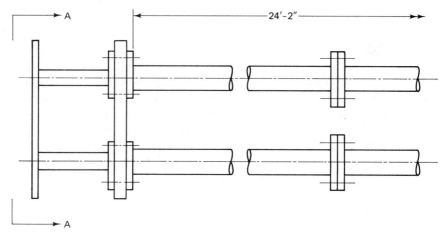

24′-2″

(b)

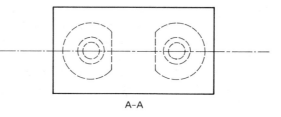

A-A

Figure 2.25 Dimensioning details of the pretensioning or post-tensioning laboratory system used for research at Rutgers (Nawy et al.).

Figure 2.26 Stresstek Multistrand 500-ton jack. (*Courtesy,* Post-Tensioning Institute.)

2. **Water.** The water used in the grout should be potable, clean, and free of injurious quantities of substances known to be harmful to portland cement or prestressing steel.

3. **Admixtures.** Admixtures, if used, should impart the properties of low water content, good flow, minimum bleed, and expansion if desired. Their formulation should contain no chemicals in quantities that may have a harmful effect on the prestressing steel or cement. Admixtures containing chlorides (as Cl in excess of 0.5 percent by weight of admixture, assuming 1 pound of admixture per sack of cement), fluorides, sulphites, or nitrates should not be used. Aluminum powder of the proper fineness and quantity, or any other approved gas-evolving material which is well dispersed through the other admixture, may be used to obtain 5 to 10 percent unrestrained expansion of the grout.

2.10.4.2 Ducts

1. **Forming.**
   (a) **Formed Ducts.** Ducts formed by sheath left in place should be of a type that does not permit the entrance of cement paste. They should transfer bond stresses as required and should retain their shape under the weight of the concrete. Metallic sheaths should be of a ferrous metal, and they may be galvanized.
   (b) **Cored Ducts.** Cored ducts should be formed with no constrictions which would tend to block the passage of grout. All coring material should be removed.

2. **Grout Openings or Vents.** All ducts should have grout openings at both ends. For draped cables, all high points should have a grout vent except where the cable curvature is small, such as in continuous slabs. Grout vents or drain holes should be provided at low points if the tendon is to be placed, stressed, and grouted in a freezing climate. All grout openings or vents should include provisions for preventing grout leakage.

Prestress conduit for a bridge deck.

3. **Duct Size.** For tendons made up of a plurality of wires, bars, or strands, the duct area should be at least twice the net area of the prestressing steel. For tendons made up of a single wire, bar, or strand, the duct diameter should be at least $\frac{1}{4}$ inch larger than the nominal diameter of the wire, bar, or strand.

4. **Placement of Ducts.** After the placement of ducts, reinforcement, and forming are complete, an inspection should be made to locate possible duct damage. Ducts should be securely fastened at close enough intervals to avoid displacement during concreting. All holes or openings in the duct must be repaired prior to placement of concrete. Grout openings and vents must be securely anchored to the duct and to either the forms or the reinforcing steel, to prevent displacement during concrete-placing operations.

2.10.4.3 Grouting process

1. Ducts with concrete walls (cored ducts) should be flushed to ensure that the concrete is thoroughly wetted.

2. All grout and high-point vent openings should be open when grouting starts. Grout should be allowed to flow from the first vent after the inlet pipe until any residual flushing water or entrapped air has been removed, at which time the vent should be capped or otherwise closed. Remaining vents should be closed in sequence in the same manner. The pumping pressure at the tendon inlet should not exceed 250 psig.

3. Grout should be pumped through the duct and continuously wasted at the outlet pipe until no visible slugs of water or air are ejected. The efflux time of the ejected grout should not be less than the injected grout. To ensure that the tendon remains filled with grout, the outlet and/or inlet should be closed. Plugs, caps, or valves thus required should not be removed or opened until the grout has set.

4. When one-way flow of grout cannot be maintained, the grout should be immediately flushed out of the duct with water.

5. In temperatures below 32°F, ducts should be kept free of water to avoid damage due to freezing.

6. The temperature of the concrete should be 35°F or higher from the time of grouting until job-cured 2-in. cubes of grout reach a minimum compressive strength of 800 psi.

7. Grout should not be above 90°F during mixing or pumping. If necessary, the mixing water should be cooled.

Additional details and specifications on grouting are given by the Post-Tensioning Institute in Ref 2.29.

2.11 CIRCULAR PRESTRESSING

Circular prestressing involves the development of hoop or hugging compressive stresses on circular or cylindrical containment vessels, including prestressed water tanks and pipes. It is usually accomplished by a wire-wound technique, in which the concrete pipe or tank is wrapped with continuous high-tensile wire tensioned to prescribed design levels. Such tension results in uniform radial compression that prestresses the concrete cylinder or core and prevents tensile stresses from developing in the concrete wall section under internal fluid pressure. Figure 2.27 shows a preload circular tank being prestressed by the wire-wrapping process along its height.

Figure 2.27 Prestressing of preload circular tank. (*Courtesy*, N.A. Legates, Pre-load Technology, Inc., New York.)

2.12 TEN PRINCIPLES

The following ten principles are taken from Abeles (Ref 2.32) and applicable not only to prestressing concrete, but to any endeavor that the engineer is called upon to undertake:

1. You cannot have everything. (Each solution has advantages and disadvantages that have to be tallied and traded off against each other.)

2. You cannot have something for nothing. (One has to pay in one way or another for something which is offered as a "free gift" into the bargain, notwithstanding a solution's being optimal for the problem.)

3. It is never too late (e.g., to alter a design, to strengthen a structure before it collapses, or to adjust or even change principles previously employed in the light of increased knowledge and experience).

4. There is no progress without considered risk. (While it is important to ensure sufficient safety, overconservatism can never lead to an understanding of novel structures.)

5. The proof of the pudding is in the eating. (This is in direct connection with the previous priniciple indicating the necessity of tests.)

6. Simplicity is always an advantage, but beware of oversimplification. (The latter may lead to theoretical calculations which are not always correct in practice, or to a failure to cover all conditions.)

7. Do not generalize, but rather qualify the specific circumstances. (Serious misunderstandings may be caused by unreserved generalizations.)

8. The important question is how good, not how cheap an item is. (A cheap price given by an inexperienced contractor usually results in bad work; similarly, cheap, unproved appliances may have to be replaced.)

9. We live and learn. (It is always possible to increase one's knowledge and experience.)

10. There is nothing completely new. (Nothing is achieved instantaneously, but only by step-by-step development.)

REFERENCES

2.1 American Society for Testing and Materials. *Annual Book of ASTM Standards: Part 14, Concrete and Mineral Aggregates*. Philadelphia: ASTM, 1983.

2.2 Popovices, S. *Concrete-Making Materials*. New York: McGraw-Hill, 1979.

2.3 ACI Committee 221. "Selection and Use of Aggregate for Concrete." *Journal of the American Concrete Institute*, Proc. Vol. 58, No. 5 (1961): 513–542.

2.4 American Concrete Institute. *ACI Manual of Concrete Practice 1983:* Part I: *Materials*. Detroit: American Concrete Institute, 1983.

2.5 Portland Cement Association. *Design and Control of Concrete Mixtures,* 12th ed. Skokie, Ill.: PCA, 1979.

2.6 ACI Committee 212. "Admixtures for Concrete," in *ACI Manual of Concrete Practice 1983*. Detroit: ACI, 1983, ACI 212.1 R-81.

2.7 Nawy, E. G., Ukadike, M. M., and Sauer, J. A. "High Strength Field Modified Concretes," *Journal of the Structural Division, ASCE,* 103, No. ST12 (December 1977): 2307–2322.

2.8 American Concrete Institute. *Super-plasticizers in Concrete,* ACI Special Publication SP-62. Detroit: ACI, 1979.

2.9 ACI Committee 211. "Standard Practice for Selecting Proportions for Normal, Heavyweight, and Mass Concrete," ACI 211.1–81. Detroit: American Concrete Institute.

2.10 ACI Committee 211. "Standard Practice for Selecting Proportions for Structural Lightweight Concrete," ACI 211.1-81. Detroit: American Concrete Institute.

2.11 Nawy, E. G. *Reinforced Concrete—A Fundamental Approach.* Englewood Cliffs, N.J.: Prentice-Hall, Inc., 1985.

2.12 ACI Committee 318. "Building Code Requirements for Reinforced Concrete 318–89." Detroit: American Concrete Institute, 1989, and the "Commentary on Building Code Requirements for Reinforced Concrete," 318R–89.

2.13 American Society for Testing and Materials. *Significance of Tests and Properties of Concrete and Concrete Making Materials,* Special Technical Publication 169B. Philadelphia: ASTM, 1978.

2.14 Ross, A. D. "The Elasticity, Creep and Shrinkage of Concrete," in *Proceedings of the Conference on Non-metallic Brittle Materials.* London: Interscience Publishers, 1958.

2.15 Neville, A. M. *Properties of Concrete,* 3rd ed. London: Pitman Books, 1981.

2.16 Freudenthal, A. M., and Roll, F. "Creep and Creep Recovery of Concrete under High Compressive Stress," *Journal of the American Concrete Institute,* Proc. 54 (June 1958): 1111–1142.

2.17 Ross, A. D., "Creep Concrete Data," *Proceedings, Institution of Structural Engineers,* London, 15 (1937): 314–326.

2.18 Branson, D. E. *Deformation of Concrete Structures.* New York: McGraw-Hill, 1977.

2.19 Branson, D. E. "Compression Steel Effects on Long Term Deflections," *Journal of the American Concrete Institute,* Proc. 68 (August 1971): 555–559.

2.20 Mindess, S., and Young, J. F. *Concrete.* Englewood Cliffs, N.J.: Prentice-Hall, Inc., 1981.

2.21 Nawy, E. G., and Balaguru, P. N. "High Strength Concrete," Chapter 5 in *Handbook of Structural Concrete.* London: Pitman Books, New York: McGraw-Hill, 1983.

2.22 Mehta, P. Kumar. *Concrete-Structure, Properties, and Materials.* Englewood Cliffs, N.J.: Prentice-Hall, Inc. 1986.

2.23 American Society for Testing and Materials. "Standard Specification for Deformed and Plain Billet-Steel Bars for Concrete Reinforcement, A6 15-79." Philadelphia: ASTM, 1980, 588–599.

2.24 American Society for Testing and Materials. "Standard Specification for Rail-Steel Deformed and Plain Bars for Concrete Reinforcement, A6 16-79." Philadelphia: ASTM, 1980, 600–605.

2.25 American Society for Testing and Materials. "Standard Specification for Axle Steel Deformed and Plain Bars for Concrete Reinforcement, A6 17-79." Philadelphia: ASTM, 1980, 606–611.

2.26 American Society for Testing and Materials. "Standard Specification for Cold-Drawn Steel Wire for Concrete Reinforcement, A8 2-79." Philadelphia: ASTM, 1980, 154–157.

2.27 American Society for Testing and Materials. "Standard Specification for Low-Alloy Steel Deformed Bars for Concrete Reinforcement, A706-79." Philadelphia: ASTM, 1980, 755–760.

2.28 ACI-ASCE Committee 423, "Recommendation for Concrete Members Prestressed with Unbonded Tendons" (ACI 423.3R-83). *Concrete International* 5 (1983): 61–76.

2.29 Post-Tensioning Institute. "Guide Specifications for Post-Tensioning Materials." In *Post-Tensioning Manual.* 4th ed. Phoenix, Ariz.: Post-Tensioning Institute, 1985.

2.30 AASHTO. *Standard Specifications For Highway Bridges.* 13th ed. Washington, D.C.: American Association of State Highway and Transportation Officials, 1983.

2.31 Nawy, E. G., and Potyondy, J. G. "Moment Rotation, Cracking, and Deflection of Spirally Bound Pretensioned Prestressed Concrete Beams." *Engineering Research Bulletin No. 51.* New Brunswick, N.J.: Bureau of Engineering Research, Rutgers University, 1970, pp. 1–97.

2.32 Abeles, P. W., and Bardhan-Roy, B. K. *Prestressed Concrete Designer's Handbook.* 3d ed. London: Viewpoint Publications, 1981.

2.33 Nawy, E. G. *Simplified Reinforced Concrete.* Englewood Cliffs, N.J.: Prentice Hall, 1986.

2.34 Nawy, E. G., "Concrete." In *Corrosion and Chemical Resistant Masonry Materials Handbook.* Park Ridge, N.J.: Noyes, 1986, pp. 57–73.

CHAPTER 3

Partial Loss of Prestress

3.1 INTRODUCTION

It is a well-established fact that the initial prestressing force applied to the concrete element undergoes a progressive process of reduction over a period of approximately five years. Consequently, it is important to determine the level of the prestressing force at each loading stage, from the stage of transfer of the prestressing force to the concrete, to the various stages of prestressing available at service load, up to the ultimate. Essentially, the reduction in the prestressing force can be grouped into two categories:

- Immediate elastic loss during the fabrication or construction process, including elastic shortening of the concrete, anchorage losses, and frictional losses.
- Time-dependent losses such as creep, shrinkage, and those due to temperature effects and steel relaxation, all of which are determinable at the service-load limit state of stress in the prestressed concrete element.

An exact determination of the magnitude of these losses—particularly the time-dependent ones—is not feasible, since they depend on a multiplicity of interrelated factors. Empirical methods of estimating losses differ with the different codes of practice or recommendations, such as those of the Prestressed Concrete Insitutute, the ACI-ASCE joint committee approach, the AASHTO lump-sum approach, the Comité Eurointernationale du Béton (CEB), and the FIP (Federation Internationale de la Précontrainte). The degree of rigor of these methods depends on the approach chosen and the accepted practice of record.

A very high degree of refinement of loss estimation is neither desirable nor

Executive Center, Honolulu, Hawaii. (*Courtesy,* Post-Tensioning Institute.)

warranted, because of the multiplicity of factors affecting the estimate. Consequently, lump-sum estimates of losses are more realistic, particularly in routine designs and under average conditions. Such lump-sum losses can be summarized in Table 3.1 of AASHTO and Table 3.2 of PTI. They include elastic shortening, relaxation in the prestressing steel, creep, and shrinkage, and are applicable only to routine, standard conditions of loading; normal concrete, quality control, construction procedures, and environmental conditions; and the importance and magnitude of the system. Detailed analysis has to be performed if these standard conditions are not fulfilled.

A summary of the sources of the separate prestressing losses and the stages of their occurrence is given in Table 3.3, in which the subscript i denotes "initial" and

TABLE 3.1 AASHTO LUMP-SUM LOSSES

Type of prestressing steel	Total loss	
	$f'_c = 4{,}000$ psi (27.6 N/mm^2)	$f'_c = 5{,}000$ psi (34.5 N/mm^2)
Pretensioning strand		45,000 psi (310 N/mm^2)
Post-tensioning[a] wire or strand	32,000 psi (221 N/mm^2)	33,000 psi (228 N/mm^2)
Bars	22,000 psi (152 N/mm^2)	23,000 psi (159 N/mm^2)

[a] Losses due to friction are excluded. Such losses should be computed according to Section 6.5 of the AASHTO specifications.

TABLE 3.2 APPROXIMATE PRESTRESS LOSS VALUES FOR POST-TENSIONING

Post-tensioning tendon material	Prestress loss, psi	
	Slabs	Beams and joists
Stress-relieved 270 strand and stress-relieved 240 wire	30,000 (207 N/mm^2)	35,000 (241 N/mm^2)
Bar	20,000 (138 N/mm^2)	25,000 (172 N/mm^2)
Low-relaxation 270K strand	15,000 (103 N/mm^2)	20,000 (138 N/mm^2)

Note: This table of approximate prestress losses was developed to provide a common post-tensioning industry basis for determining tendon requirements on projects in which the magnitude of prestress losses is not specified by the designer. These loss values are based on use of normal-weight concrete and on average values of concrete strength, prestress level, and exposure conditions. Actual values of losses may vary significantly above or below the table values where the concrete is stressed at low strengths, where the concrete is highly prestressed, or in very dry or very wet exposure conditions. The table values do not include losses due to friction.

Source: Post-Tensioning Institute.

TABLE 3.3 TYPES OF PRESTRESS LOSS

Type of prestress loss	Stage of occurrence		Tendon stress loss	
	Pretensioned members	Post-tensioned members	During time interval (t_i, t_j)	Total or during life
Elastic shortening of concrete (ES)	At transfer	At sequential jacking	....	Δf_{pES}
Relaxation of tendons (R)	Before and after transfer	After transfer	$\Delta f_{pR}(t_i, t_j)$	Δf_{pR}
Creep of concrete (CR)	After transfer	After transfer	$\Delta f_{pC}(t_i, t_j)$	Δf_{pCR}
Shrinkage of concrete (SH)	After transfer	After transfer	$\Delta f_{pS}(t_i, t_j)$	Δf_{pSH}
Friction (F)	....	At jacking	....	Δf_{pF}
Anchorage seating loss (A)	....	At transfer	....	Δf_{pA}
Total	Life	Life	$\Delta f_{pT}(t_i, t_j)$	Δf_{pT}

the subscript j denotes the loading stage after jacking. From this table, the total loss in prestress can be calculated for pretensioned and post-tensioned members as follows:

(i) Pretensioned Members

$$\Delta f_{pT} = \Delta f_{pES} + \Delta f_{pR} + \Delta f_{pCR} + \Delta f_{pSH} \tag{3.1 a}$$

where $\Delta f_{pR} = \Delta f_{pR}(t_0, t_{tr}) + \Delta f_{pR}(t_{tr}, t_s)$
t_0 = time at jacking
t_{tr} = time at transfer
t_s = time at stablilized loss

Hence, computations for steel relaxation loss have to be performed for the time interval t_1 through t_2 of the respective loading stages.

As an example, the transfer stage, say, at 18 hours would result in $t_{tr} = t_2 = 18$ hours and $t_0 = t_1 = 0$. If the next loading stage is between transfer and 5 years (17,520 hours), when losses are considered stabilized, then $t_2 = t_s = 17,520$ hours and $t_1 = 18$ hours. Then, If f_{pi} is the initial prestressing stress that the concrete element is subjected to and f_{pJ} is the jacking stress in the tendon, then

$$f_{pi} = f_{pJ} - \Delta f_{pR}(t_0, t_{tr}) - \Delta f_{pES} \tag{3.1 b}$$

(ii) Post-tensioned members

$$\Delta f_{pT} = \Delta f_{pA} + \Delta f_{pF} + \Delta f_{pES} + \Delta f_{pR} + \Delta f_{pCR} + \Delta f_{pSH} \tag{3.1 c}$$

where Δf_{pES} is applicable only when tendons are jacked sequentially, and not simultaneously.

In the post-tensioned case, computation of relaxation loss starts between the transfer time $t_1 = t_{tr}$ and the end of the time interval t_2 under consideration. Hence

$$f_{pi} = f_{pJ} - \Delta f_{pES} - \Delta f_{pF} \tag{3.1 d}$$

3.2 ELASTIC SHORTENING OF CONCRETE (ES)

Concrete shortens when a prestressing force is applied. As the tendons that are bonded to the adjacent concrete simultaneously shorten, they lose part of the prestressing force that they carry.

3.2.1 Pretensioned Elements

For pretensioned (precast) elements, the compressive force imposed on the beam by the tendon results in the longitudinal shortening of the beam, as shown in Figure 3.1. The unit shortening in concrete is $\epsilon_{ES} = \Delta_{ES}/L$, so

$$\epsilon_{ES} = \frac{f_c}{E_c} = \frac{P_i}{A_c E_c} \tag{3.2 a}$$

Since the prestressing tendon suffers the same magnitude of shortening,

$$\Delta f_{pES} = E_s \epsilon_{ES} = \frac{E_s P_i}{A_c E_c} = \frac{n P_i}{A_c} = n f_{cs} \tag{3.2 b}$$

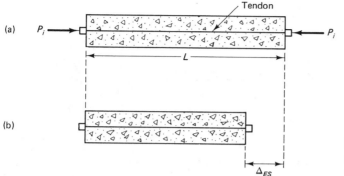

(a) P_i → ← P_i

← L →

(b)

Δ_{ES}

Figure 3.1 Elastic shortening. (a) Unstressed beam. (b) Longitudinally shortened beam.

The stress in the concrete at the centroid of the steel due to the initial prestressing is

$$f_{cs} = -\frac{P_i}{A_c} \tag{3.3}$$

If the tendon in Figure 3.1 has an eccentricity e at the beam midspan and the self-weight moment M_D is taken into account, the stress the concrete undergoes at the midspan section at the level of the prestressing steel becomes

$$f_{cs} = -\frac{P_i}{A_c}\left(1 + \frac{e^2}{r^2}\right) + \frac{M_D e}{I_c} \tag{3.4}$$

where P_i has a lower value after transfer of prestress. The *small* reduction in the value of P_J to P_i occurs because the force in the prestressing steel immediately after transfer is less than the initial jacking prestress force P_J. However, since it is difficult to accurately determine the reduced value of P_i, and since observations indicate that the reduction is only a few percentage points, it is possible to use the initial value of P_i before transfer in Equations 3.2–3.4, or reduce it by about 10 percent for refinement if desired.

3.2.1.1 Elastic shortening loss in pretensioned beams

Example 3.1

A pretensioned prestressed beam has a span of 50 ft (15.2 m), as shown in Figure 3.2. For this beam,

$f'_c = 6,000$ psi (41.4 MPa)

$f_{pu} = 270,000$ psi (1,862 MPa)

$f_{ci} = 4,500$ psi (31 MPa)

$A_{ps} = $ ⓪ $10\frac{1}{2}''$ dia. seven-wire-strand tendon

    $= $ ⑩ $\times 0.153 = 1.53$ in^2

$E_{ps} = 27 \times 10^6$ psi (1,862 MPa)

Calculate the concrete fiber stresses at transfer at the centroid of the tendon for the midspan section of the beam, and the magnitude of loss in prestress due to the effect of elastic shortening of the concrete. Assume that prior to transfer, the jacking force on the tendon was 75% f_{pu}.

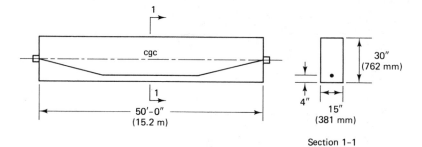

Figure 3.2 Beam in Example 3.1.

Solution

$$A_c = 15 \times 30 = 450 \text{ in}^2$$

$$I_c = \frac{15(30)^3}{12} = 33,750 \text{ in}^3$$

$$r^2 = \frac{I_c}{A_c} = 75 \text{ in}^2$$

$$A_{ps} = 10 \times 0.153 = 1.53 \text{ in}^2$$

$$e = \frac{30}{2} - 4 = 11 \text{ in}$$

$$P_i = 0.75 f_{pu} A_{ps} = 0.75 \times 270,000 \times 1.53 = 309,825 \text{ lb}$$

$$M_D = \frac{w\ell^2}{8} = \frac{15 \times 30}{144} \times 150 \frac{(50)^2}{8} \times 12 = 1,757,813 \text{ in.-lb}$$

From Equation 3.4, the concrete fiber stress at the steel centroid of the beam at the moment of transfer, assuming that $P_i \cong P_J$, is

$$f_{cs} = -\frac{P_i}{A_c}\left(1 + \frac{e^2}{r^2}\right) + \frac{M_D e}{I_c}$$

$$= -\frac{309,825}{450}\left(1 + \frac{11^2}{75}\right) + \frac{1,757,813 \times 11}{33,750}$$

$$= -1,799.3 + 572.9 = -1,226.4 \text{ psi } (8.50 \text{ MPa})$$

We also have

$$\text{Initial } E_{ci} = 57,000\sqrt{f'_{ci}} = 57,000\sqrt{4,500} = 3.824 \times 10^6 \text{ psi}$$

$$\text{Modular ratio } n = \frac{E_s}{E_{ci}} = \frac{27 \times 10^6}{3.824 \times 10^6} = 7.06$$

$$28 \text{ days' strength } E_c = 57,000\sqrt{6,000} = 4.415 \times 10^6 \text{ psi}$$

$$28 \text{ days' modular ratio } n = \frac{27 \times 10^6}{4.415 \times 10^6} = 6.12$$

From Equation 3.2b, the loss of prestrees due to elastic shortening is

$$\Delta f_{pES} = n f_{cs} = 7.06 \times 1,226.4 = 8,659.2 \text{ psi } (59.7 \text{ MPa})$$

If a reduced P_i is used with assumed 10 percent reduction,

$$\Delta f_{pES} = 0.90 \times 8,659.2 = 7,793.3 \text{ psi (53.7 MPa)}.$$

The difference of 865.9 psi in steel stress is insignificant compared to the total loss in prestress due to all factors of about 45,000 to 55,000 psi.

3.2.2 Post-tensioned Elements

In post-tensioned beams, the elastic shortening varies from zero if all tendons are jacked simultaneously to half the value calculated in the pretensioned case if several sequential jacking steps are used, such as jacking two tendons at a time. If n is the number of tendons or pairs of tendons sequentially tensioned, then

$$\Delta f_{pES} = \frac{1}{n} \sum_{j=1}^{n} (\Delta f_{pES})_j \tag{3.5}$$

where j denotes the number of jacking operations. Note that the tendon that was tensioned last does not suffer any losses due to elastic shortening, while the tendon that was tensioned first suffers the maximum amount of loss.

3.2.2.1 Elastic shortening loss in post-tensioned beam

Example 3.2

Solve Example 3.1 if the beam is post-tensioned and the prestressing operation is such that

   (a) Two tendons are jacked at a time.
   (b) One tendon is jacked at a time.
   (c) All tendons are simultaneously tensioned.

Solution

   (a) From Example 3.1, $\Delta f_{pE} = 8,659.2$ psi. Clearly, the last tendon suffers no loss of prestress due to elastic shortening. So only the first four pairs have losses, with the first pair suffering the maximum loss of 8,659.2 psi. From Equation 3.5, the loss due to elastic shortening in the post-tensioned beam is

$$\Delta f_{pES} = \frac{4/4 + 3/4 + 2/4 + 1/4}{5}(8,659.2)$$

$$= \frac{10}{20} \times (8,659.2) = 4,330 \text{ psi (29.9 MPa)}$$

   (b) $$\Delta f_{pES} = \frac{9/9 + 8/9 + \cdots + 1/9}{10}(8,659.2)$$

$$= \frac{45}{90} \times (8,659.2) = 4,330 \text{ psi (29.9 MPa)}$$

In both cases the loss in prestressing in the post-tensioned beam is half that of the pretensioned beam.

   (c) $\Delta f_{pES} = 0$

3.3 STEEL STRESS RELAXATION (R)

Stress-relieved tendons suffer loss in the prestressing force due to constant elongation with time, as discussed in Chapter 2. The magnitude of the decrease in the prestress depends not only on the duration of the sustained prestressing force, but also on the ratio f_{pi}/f_{py} of the initial prestress to the yield strength of the reinforcement. Such a loss in stress is termed *stress relaxation*. The ACI 318-89 Code limits the tensile stress in the prestressing tendons to the following:

(a) For stresses due to the tendon jacking force, $f_{pJ} = 0.94 f_{py}$, but not greater than the lesser of $0.80 f_{pu}$ and the maximum value recommended by the manufacturer of the tendons and anchorages.

(b) Immediately after prestress transfer, $f_{pi} = 0.82 f_{py}$, but not greater than $0.74 f_{pu}$.

(c) In post-tensioned tendons, at the anchorages and couplers immediately after tendon anchorage the stress f_{pJ} at jacking should not exceed $0.80 f_{pu}$.

The range of values of f_{py} is given by the following:

Prestressing bars: $f_{py} = 0.80 f_{pu}$
Stress-relieved tendons: $f_{py} = 0.85 f_{pu}$
Low-relaxation tendons: $f_{py} = 0.90 f_{pu}$

If f_{pR} is the remaining prestressing stress in the steel after relaxation, the following expression defines f_{pR} for stress relieved steel:

$$\frac{f_{pR}}{f_{pi}} = 1 - \left(\frac{\log t_2 - \log t_1}{10}\right)\left(\frac{f_{pi}}{f_{py}} - 0.55\right) \tag{3.6}$$

In this expression, $\log t$ in hours is to the base 10, f_{pi}/f_{py} exceeds 0.55, and $t = t_2 - t_1$. Also, for low-relaxation steel, the denominator of the log term in the equation is divided by 45 instead of 10. A plot of Equation 3.6 is given in Figure 3.3. An approximation of the term ($\log t_2 - \log t_1$) can be made in Equation 3.6 so

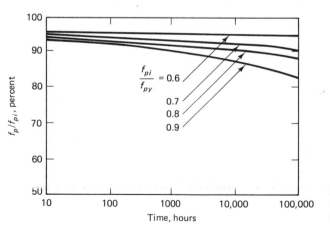

Figure 3.3 Stress-relaxation relationship in stress-relieved strands. (*Courtesy*, Post-Tensioning Institute.)

Partial Loss of Prestress   Chap. 3

that $t = \log(t_2 - t_1)$ without significant loss in accuracy. In that case, the stress-relaxation loss becomes

$$\Delta f_{pR} = f'_{pi}\frac{\log t}{10}\left(\frac{f'_{pi}}{f_{py}} - 0.55\right) \tag{3.7}$$

where f'_{pi} is the initial stress in steel to which the concrete element is subjected.

If a step-by-step loss analysis is necessary, the loss increment at any particular stage can be defined as

$$\Delta f_{pR} = f'_{pi}\left(\frac{\log t_2 - \log t_1}{10}\right)\left(\frac{f'_{pi}}{f_{py}} - 0.55\right) \tag{3.8}$$

where t_1 is the time at the beginning of the interval and t_2 is the time at the end of the interval from jacking to the time when the loss is being considered.

3.3.1 Relaxation Loss Computation

Example 3.3

Find the relaxation loss in prestress at the end of five years in Example 3.1, assuming that relaxation loss from jacking to transfer, from elastic shortening, and from long-term loss due to creep and shrinkage over this period is 20 percent of the initial prestress. Assume also that the yield strength $f_{py} = 230,000$ psi (1,571 MPa).

Solution From Equation 3.1b,

$$f_{pi} = f_{pJ} - \Delta f_{pR}(t_0, t_{tr}) - \Delta f_{pES}$$
$$= 0.75 \times 270,000 = 202,500 \text{ psi } (1,396 \text{ MPa})$$

The reduced stress for calculating relaxation loss is

$$f'_{pi} = (1 - 0.20) \times 202,500 = 162,000 \text{ psi } (1,170 \text{ MPa})$$

The duration of the stress-relaxation process is

$$5 \times 365 \times 24 \cong 44,000 \text{ hours}$$

From Equation 3.7,

$$\Delta f_{pR} = f'_{pi}\frac{\log t}{10}\left(\frac{f'_{pi}}{f_{py}} - 0.55\right)$$

$$= 162,000\frac{\log 44,000}{10}\left(\frac{162,000}{230,000} - 0.55\right)$$

$$= 162,000 \times 0.4643 \times 0.1543 = 11,606 \text{ psi } (80.0 \text{ MPa})$$

3.3.2 ACI-ASCE Method of Accounting for Relaxation Loss

The ACI-ASCE method uses the separate contributions of elastic shortening, creep, and shrinkage in the evaluation of the steel stress-relaxation loss by means of the equation

$$\Delta f_{pR} = [K_{re} - J(f_{pES} + f_{pCR} + f_{pSH})] \times C$$

The values of K_{re}, J, and C are given in Tables 3.4 and 3.5.

TABLE 3.4 VALUES OF C

$f_{pi}f_{pu}$	Stress-relieved strand or wire	Stress-relieved bar or low-relaxation strand or wire
0.80		1.28
0.79		1.22
0.78		1.16
0.77		1.11
0.76		1.05
0.75	1.45	1.00
0.74	1.36	0.95
0.73	1.27	0.90
0.72	1.18	0.85
0.71	1.09	0.80
0.70	1.00	0.75
0.69	0.94	0.70
0.68	0.89	0.66
0.67	0.83	0.61
0.66	0.78	0.57
0.65	0.73	0.53
0.64	0.68	0.49
0.63	0.63	0.45
0.62	0.58	0.41
0.61	0.53	0.37
0.60	0.49	0.33

Source: Post-Tensioning Institute

TABLE 3.5 VALUES OF K_{RE} AND J

Type of tendon[a]	K_{re}	J
270 Grade stress-relieved strand or wire	20,000	0.15
250 Grade stress-relieved strand or wire	18,500	0.14
240 or 235 Grade stress-relieved wire	17,600	0.13
270 Grade low-relaxation strand	5,000	0.040
250 Grade low-relaxation wire	4,630	0.037
240 or 235 Grade low-relaxation wire	4,400	0.035
145 or 160 Grade stress-relieved bar	6,000	0.05

[a] In accordance with ASTM A416-74, ASTM A421-76, or ASTM A722-75.

Source: Prestressed Concrete Institute

3.4 CREEP LOSS (CR)

Experimental work over the past half century indicates that flow in materials occurs with time when load or stress exists. This lateral flow or deformation due to the longitudinal stress is termed *creep*. A more detailed discussion is given in Ref. 3.9.

The deformation or strain resulting from this time-dependent behavior is a function of the magnitude of the applied load, its duration, the properties of the concrete including its mix proportions, curing conditions, the age of the element at first loading, and environmental conditions. Since the stress-strain relationship due to creep is essentially linear, it is feasible to relate the creep strain ϵ_{CR} to the elastic strain ϵ_{EL} such that a creep coefficient C_u can be defined as

$$C_u = \frac{\epsilon_{CR}}{\epsilon_{EL}} \qquad (3.9\ a)$$

Then the creep coefficient at any time t in days can be defined as

$$C_t = \frac{t^{0.60}}{1 + t^{0.60}} C_u \qquad (3.9\ b)$$

As discussed in Chapter 2, the value of C_u ranges between 2 and 4, with an average of 2.35 for ultimate creep. The loss in prestressed members due to creep can be defined for bonded members as

$$\Delta f_{pCR} = C_u \frac{E_{ps}}{E_c} f_{cs} \qquad (3.10)$$

where f_{cs} is the stress in the concrete at the level of the centroid of the prestressing tendon. In general, this loss is a function of the stress in the concrete at the section being analyzed. In post-tensioned, nonbonded members, the loss can be considered essentially uniform along the whole span. Hence, an average value of the concrete stress $\bar{f}_{cs}$ between the anchorage points can be used for calculating the creep in post-tensioned members.

The ACI-ASCE Committee expression for evaluating creep loss has essentially the same format as Equation 3.10, viz.,

$$\Delta f_{pCR} = K_{CR} \frac{E_{ps}}{E_c} (\bar{f}_{cs} - \bar{f}_{csd}) \qquad (3.11\ a)$$

or

$$\Delta f_{pCR} = n K_{CR} (\bar{f}_{cs} - \bar{f}_{csd}) \qquad (3.11\ b)$$

where K_{CR} = 2.0 for pretensioned members
$\qquad$ = 1.60 for post-tensioned members (both for normal concrete)
$\quad \bar{f}_{cs}$ = stress in concrete at level of steel cgs immediately after transfer
$\quad \bar{f}_{csd}$ = stress in concrete at level of steel cgs due to all superimposed dead loads applied after prestressing is accomplished
$\qquad n$ = modular ratio

Note that K_{CR} should be reduced by 20 percent for lightweight concrete.

3.4.1 Computation of Creep Loss

Example 3.4

Compute the loss in prestress due to creep in Example 3.1 given that the total super-imposed load, including the beam's own weight after transfer, is 375 plf (5.5 kN/m).

Solution At full concrete strength,

$$E_c = 57,000\sqrt{6,000} = 4.415 \times 10^6 \text{ psi } (30.4 \times 10^3 \text{ MPa})$$

$$n = \frac{E_s}{E_c} = \frac{27.0 \times 10^6}{4.415 \times 10^6} = 6.12$$

$$M_{SD} = \frac{375(50)^2}{8} \times 12 = 1,406,250 \text{ in.-lb } (158.9 \text{ kN-m})$$

$$\bar{f}_{csd} = \frac{M_{SD}e}{I_c} = \frac{1,406,250 \times 11}{33,750} = 458.3 \text{ psi } (3.2 \text{ MPa})$$

From Example 3.1,

$$\bar{f}_{cs} = 1,226.4 \text{ psi } (8.5 \text{ MPa})$$

Also, for normal concrete use, $K_{CR} = 2.0$ (pretensioned beam); so from Equation 3.11a,

$$\Delta f_{pCR} = nK_{CR}(\bar{f}_{cs} - \bar{f}_{csd})$$

$$= 6.12 \times 2.0(1,226.4 - 458.3)$$

$$= 9,401.5 \text{ psi } (64.8 \text{ MPa})$$

3.5 SHRINKAGE LOSS (SH)

As in concrete creep, the magnitude of the shrinkage of concrete is affected by several factors that are time dependent. They include mix proportions, type of aggregate, type of cement, curing time, time between the end of external curing and the application of prestressing, and environmental conditions. Approximately 80 percent of shrinkage takes place in the first year of life of the structure. The long-term shrinkage strain ϵ_{SH} can vary between 400×10^{-6} in/in and 800×10^{-6} in/in, with an average of 550×10^{-6} in/in. This average value is modified by the relative humidity RH and the volume-to-surface ratio such that the shrinkage unit strain can be defined as

$$\epsilon_{SH} = 550 \times 10^{-6}\left(1 - 0.06\frac{V}{S}\right)(1.5 - 0.015RH)$$

Simplifying, this expression becomes

$$\epsilon_{SH} = 8.2 \times 10^{-6}\left(1 - 0.06\frac{V}{S}\right)(100 - RH) \tag{3.12}$$

For pretensioned members,

$$\Delta f_{pSH} = \epsilon_{SH} \times E_r \tag{3.13}$$

For post-tensioned members, the loss in prestressing due to shrinkage is somewhat less

101/280/680 interchange connectors, South San Jose, California.

since some shrinkage has already taken place before post-tensioning. Hence, a general equation for both types of prestressing is

$$\Delta f_{pSH} = 8.2 \times 10^{-6} K_{SH} E_{ps} \left(1 - 0.06 \frac{V}{S} \right) (100 - RH) \tag{3.14}$$

where $K_{SH} = 1.0$ for pretensioned members. Table 3.6 gives the values of K_{SH} for post-tensioned members.

Detailed computations of shrinkage losses as a function of time t after seven days for moist curing and three days for steam curing can be obtained from the following expressions (see Chapter 2):

(a) Moist curing for seven days:

$$\epsilon_{SH,t} = \frac{t}{t + 35} \epsilon_{SH} \tag{3.15 a}$$

where $\epsilon_{SH} = 800 \times 10^{-6}$ in/in.

TABLE 3.6 VALUES OF K_{SH} FOR POST-TENSIONED MEMBERS

Time from end of moist curing to application of prestress, days	1	3	5	7	10	20	30	60
K_{sh}	0.92	0.85	0.80	0.77	0.73	0.64	0.58	0.45

Source: Prestressed Concrete Institute

Sec. 3.5   Shrinkage Loss (*SH*)

(b) Steam curing for one to three days:

$$\epsilon_{SH,t} = \frac{t}{t + 55} \epsilon_{SH} \qquad (3.15\ b)$$

where $\epsilon_{SH} = 730 \times 10^{-6}$ in/in.

3.5.1 Computation of Shrinkage Loss

Example 3.5

Compute the loss in prestress due to shrinkage in Examples 3.1 and 3.2 at seven days after moist curing using both the ultimate K_{SH} method of Equation 3.14 and the time-dependent method of Equation 3.15. Assume that the relative humidity RH is 70 percent and the volume-to-surface ratio is 2.0.

Solution A

   K_{SH} **method**

(a) Pretensioned beam, $K_{SH} = 1.0$:
   From Equation 3.14,

$$\Delta f_{pSH} = 8.2 \times 10^{-6} \times 1.0 \times 27 \times 10^{6}(1 - 0.06 \times 2.0)(100 - 70)$$

$$= 5{,}845.0 \text{ psi (40.3 MPa)}$$

(b) Post-tensioned beam, from Table 3.6, $K_{SH} = 0.77$:

$$\Delta f_{pSH} = 0.77 \times 5{,}845 = 4{,}500.7 \text{ psi (31.0 MPa)}$$

Solution B
   Time-dependent method
From Equation 3.15a,

$$\epsilon_{SH,t} = \frac{t}{35 + t} \epsilon_{SH} = \frac{7}{35 + 7} \times 800 \times 10^{-6} = 133.3 \times 10^{-6} \text{ in/in}$$

$$\Delta f_{pSH} = \epsilon_{SH,t} E_s = 133.3 \times 10^{-6} \times 27 \times 10^{6} = 3{,}599.1 \text{ psi (24.8 MPa)}$$

3.6 LOSSES DUE TO FRICTION (F)

Loss of prestressing occurs in post-tensioning members due to friction between the tendons and the surrounding concrete ducts. The magnitude of this loss is a function of the tendon form or alignment, called the *curvature effect,* and the local deviations in the alignment, called the *wobble effect.* The values of the loss coefficients are often refined while preparations are made for shop drawings by varying the types of tendons and the duct alignment. Whereas the curvature effect is predetermined, the wobble effect is the result of accidental or unavoidable misalignment, since ducts or sheaths cannot be perfectly placed.

3.6.1 Curvature Effect

As the tendon is pulled with a force F_1 at the jacking end, it will encounter friction with the surrounding duct or sheath such that the stress in the tendon will vary from the

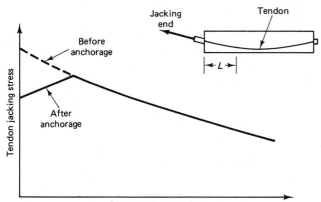

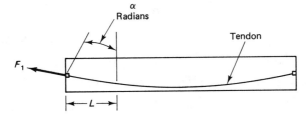

Figure 3.4 Frictional force stress distribution in tendon.

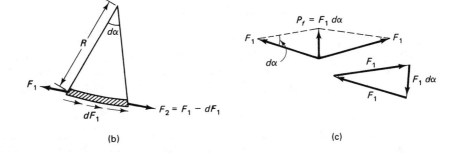

Figure 3.5 Curvature friction loss. (a) Tendon alignment. (b) Forces on infinitesimal length where F_1 is at the jacking end. (c) Polygon of forces assuming $F_1 = F_2$ over the infinitesimal length in (b).

jacking plane to a distance L along the span as shown in Figure 3.4. If an infinitesimal length of the tendon is isolated in a free-body diagram as shown in Figure 3.5, then, assuming that μ denotes the coefficient of friction between the tendon and the duct due to the curvature effect, we have

$$dF_1 = -\mu F_1 d\alpha$$

or

$$\frac{dF_1}{F_1} = -\mu d\alpha \qquad (3.16\ a)$$

Integrating both sides of this equation yields

$$\log_e F_1 = -\mu\alpha \tag{3.16 b}$$

If $\alpha = L/R$, then

$$F_2 = F_1 e^{-\mu\alpha} = F_1 e^{-\mu(L/R)} \tag{3.17}$$

3.6.2 Wobble Effect

Suppose that K is the coefficient of friction between the tendon and the surrounding concrete due to wobble effect or length effect. Friction loss is caused by imperfection in alignment along the length of the tendon, regardless of whether it has a straight or draped alignment. Then by the same principles described in developing Equation 3.16,

$$\log_e F_1 = -KL \tag{3.18}$$

or

$$F_2 = F_1 e^{-KL} \tag{3.19}$$

Superimposing the wobble effect on the curvature effect gives

$$F_2 = F_1 e^{-\mu\alpha - KL}$$

or, in terms of stresses,

$$f_2 = f_1 e^{-\mu\alpha - KL} \tag{3.20}$$

The frictional loss of stress Δf_{pF} is then given by

$$\Delta f_{pF} = f_1 - f_2 = f_1(1 - e^{-\mu\alpha - KL}) \tag{3.21}$$

Assuming that the prestress force between the start of the curved portion and its end is small ($\cong 15$ percent), it is sufficiently accurate to use the initial tension for the entire curve in Equation 3.21. Equation 3.21 can thus be simplified to yield

$$\Delta f_{pF} = -f_1(\mu\alpha + KL) \tag{3.22}$$

Since the ratio of the depth of beam to its span is small, it is sufficiently accurate to use the projected length of the tendon for calculating α. Assuming the curvature of the tendon to be based on that of a circular arc, the central angle α along the curved segment in Figure 3.6 is twice the slope at either end of the segment. Hence,

$$\tan\frac{\alpha}{2} = \frac{m}{x/2} = \frac{2m}{x}$$

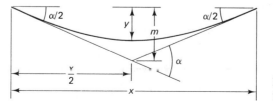

Figure 3.6 Approximate evaluation of the tendon's central angle.

TABLE 3.7 WOBBLE AND CURVATURE FRICTION COEFFICIENTS

Type of tendon	wobble coefficient, K per foot	curvature coefficient, μ
Tendons in flexible metal sheathing		
Wire tendons	0.0010–0.0015	0.15–0.25
7-wire strand	0.0005–0.0020	0.15–0.25
High-strength bars	0.0001–0.0006	0.08–0.30
Tendons in rigid metal duct		
7-wire strand	0.0002	0.15–0.25
Mastic-coated tendons		
Wire tendons and 7-wire strand	0.0010–0.0020	0.05–0.15
Pregreased tendons		
Wire tendons and 7-wire strand	0.0003–0.0020	0.05–0.15

Source: Prestressed Concrete Institute

If

$$y \cong \tfrac{1}{2}m \quad \text{and} \quad \alpha/2 = 4y/x$$

then

$$\alpha = 8y/x \text{ radian} \tag{3.23}$$

Table 3.7 gives the design values of the curvature friction coefficient μ and the wobble or length friction coefficient K adopted from the ACI 318-83 Commentary.

3.6.3 Computation of Friction Loss

Example 3.6

Assume that the alignment characteristics of the tendons in the post-tensioned beam of Example 3.2 are as shown in Figure 3.7. If the tendon is made of seven-wire uncoated

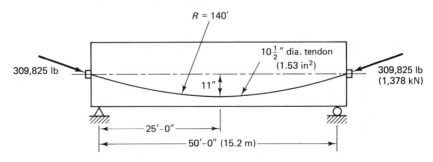

Figure 3.7 Prestressing tendon alignment.

strands in flexible metal sheathing, compute the frictional loss of stress in the prestressing wires due to the curvature and wobble effects.
Solution

$$P_i = 309,825 \text{ lb}$$

$$f_1 = \frac{309,825}{1.53} = 202,500 \text{ psi}$$

From Equation 3.23,

$$\alpha = \frac{8y}{x} = \frac{8 \times 11/12''}{50} = 0.1467 \text{ radian}$$

From Table 3.5, use $K = 0.0020$ and $\mu = 0.20$. From Equation 3.22, the prestress loss due to friction is

$$\Delta f_{pF} = f_{pi} (\mu\alpha + KL)$$

$$= 202,500 (0.20 \times 0.1467 + 0.0020 \times 50)$$

$$= 202,500 \times 0.1293 = 26,191 \text{ psi} (180.6 \text{ MPa})$$

This loss due to friction is 12.93 percent of the initial prestress.

3.7 ANCHORAGE-SEATING LOSSES (A)

Anchorage-seating losses occur in post-tensioned members due to the seating of wedges in the anchors when the jacking force is transferred to the anchorage. They can also occur in the prestressing casting beds of pretensioned members due to the adjustment expected when the prestressing force is transferred to these beds. A remedy for this loss can be easily effected during the stressing operations by overstressing. Generally, the magnitude of anchorage-seating loss ranges between $\frac{1}{4}$ in. and $\frac{3}{8}$ in. (6.35 mm and 9.53 mm) for two-piece wedges. The magnitude of the overstressing that is necessary depends on the anchorage system used since each system has its particular adjustment needs, and the manufacturer is expected to supply the data on the slip expected due to anchorage adjustment. If Δ_A is the magnitude of the slip, L is the tendon length, and E_{ps} is the modulus of the prestressing wires, then the prestress loss due to anchorage slip becomes

$$\Delta f_{pA} = \frac{\Delta_A}{L} E_{ps} \tag{3.24}$$

3.7.1 Computation of Anchorage-Seating Loss

Example 3.7

Compute the anchorage-seating loss in the post-tensioned beam of Example 3.2 if the estimated slip is $\frac{1}{4}$ in. (6.35 m).
Solution

$$E_s = 27 \times 10^6 \text{ psi}$$

$$\Delta_A = 0.25 \text{ in.}$$

Terracentre, Denver, Colorado. (*Courtesy*, Post-Tensioning Institute.)

$$\Delta f_{pA} = \frac{\Delta A}{L} E_{ps} = \frac{0.25}{50 \times 12} \times 27 \times 10^6 = 11,250 \text{ psi } (77.6 \text{ MPa})$$

Note that the percentage of loss due to anchorage slip becomes very high in short beam elements and thus becomes of major significance in short-span beams. In such cases it becomes difficult to post-tension such beams with high accuracy.

3.8 CHANGE OF PRESTRESS DUE TO BENDING OF A MEMBER (Δf_{pB})

As the beam bends due to prestress or external load, it becomes convex or concave depending on the nature of the load, as shown in Figure 3.8. If the unit compressive strain in the concrete along the level of the tendon is ϵ_c, then the corresponding change in prestress in the steel is

$$\Delta f_{pB} = \epsilon_c E_s$$

where E_s is the modulus of the steel. Note that any loss due to bending need not be taken into consideration if the prestressing stress level is measured after the beam has already bent, as is usually the case.

Figure 3.9 presents a flowchart for step-by-step evaluation of time-dependent prestress losses without deflection.

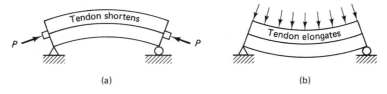

Figure 3.8 Change in beam longitudinal shape. (a) Due to prestressing. (b) Due to external load.

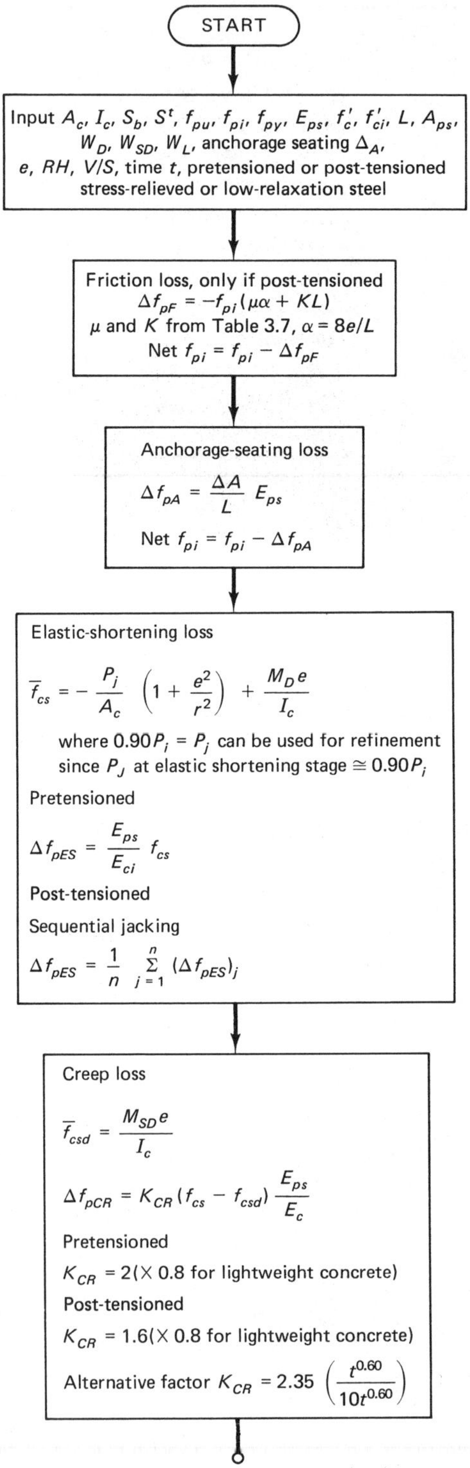

START

Input A_c, I_c, S_b, S^t, f_{pu}, f_{pi}, f_{py}, E_{ps}, f'_c, f'_{ci}, L, A_{ps}, W_D, W_{SD}, W_L, anchorage seating Δ_A, e, RH, V/S, time t, pretensioned or post-tensioned stress-relieved or low-relaxation steel

Friction loss, only if post-tensioned
$$\Delta f_{pF} = -f_{pi}(\mu\alpha + KL)$$
μ and K from Table 3.7, $\alpha = 8e/L$
Net $f_{pi} = f_{pi} - \Delta f_{pF}$

Anchorage-seating loss
$$\Delta f_{pA} = \frac{\Delta A}{L} E_{ps}$$
Net $f_{pi} = f_{pi} - \Delta f_{pA}$

Elastic-shortening loss
$$\overline{f}_{cs} = -\frac{P_i}{A_c}\left(1 + \frac{e^2}{r^2}\right) + \frac{M_D e}{I_c}$$
where $0.90 P_i = P_i$ can be used for refinement since P_J at elastic shortening stage $\cong 0.90 P_i$
Pretensioned
$$\Delta f_{pES} = \frac{E_{ps}}{E_{ci}} f_{cs}$$
Post-tensioned
Sequential jacking
$$\Delta f_{pES} = \frac{1}{n}\sum_{j=1}^{n}(\Delta f_{pES})_j$$

Creep loss
$$\overline{f}_{csd} = \frac{M_{SD} e}{I_c}$$
$$\Delta f_{pCR} = K_{CR}(f_{cs} - f_{csd})\frac{E_{ps}}{E_c}$$
Pretensioned
$K_{CR} = 2(\times 0.8$ for lightweight concrete)
Post-tensioned
$K_{CR} = 1.6(\times 0.8$ for lightweight concrete)
Alternative factor $K_{CR} = 2.35\left(\dfrac{t^{0.60}}{10t^{0.60}}\right)$

Figure 3.9 Flowchart for step-by-step evaluation of prestress losses.

Shrinkage loss

$$\Delta f_{pSH} = 8.2 \times 10^{-6} K_{SH} E_{ps}(1 - 0.06 V/S)(100 - RH)$$

$K_{SH} = 1$ for pretensioned

K_{SH} from Table 3.6 for post-tensioned

Alternatively

$$\Delta f_{pSH} = 8.0 \times 10^{-6} \left(\frac{t}{t + 35}\right) E_{ps} - \text{moist curing}$$

$$\Delta f_{pSH} = 7.3 \times 10^{-6} \left(\frac{t}{t + 55}\right) E_{ps} - \text{steam curing}$$

Relaxation of steel loss

(i) Stress-relieved strands

Pretensioned

$$f_{pi} = f_{pJ} - \Delta f_{pR}(t_0, t_{tr}) - \Delta f_{pES}$$
$$f_{pJ} - \Delta f_{pR}(t_0, t_{tr}) \cong 0.90 f_{pJ}$$
$$f_{pR} = f_{pi} \frac{(\log t_2 - \log t_1)}{10} \left(\frac{f_{pi}}{f_{py}} - 0.55\right)$$

where t_2 and t_1 are in hours

Post-tensioned

$$f_{pi} = f_{pJ} - \Delta f_{pF} - \Delta f_{pES}$$

where f_{pES} is for case of sequential jacking

$$\Delta f_{pR} = f_{pi} \frac{\log t}{10} \left(\frac{f_{pi}}{f_{py}} - 0.55\right)$$

where $\log t = \log(t_2 - t_1)$

(ii) Low-relaxation strands

Replace the denominator (10) in the $(\log t_2 - \log t_1)$ term for pretensioned and the $(\log t)$ term for post-tensioned by a denominator value of 45.

Add all losses Δf_{pT}

(i) Pretensioned

$$\Delta f_{pT} = \Delta f_{pES} + \Delta f_{pR} + \Delta f_{pCR} + \Delta f_{pSH}$$

(ii) Post-tensioned

$$\Delta f_{pT} = \Delta f_{pA} + \Delta f_{pF} + \Delta f_{pES} + \Delta f_{pR} + \Delta f_{pCR} + \Delta f_{pSH}$$

where Δf_{pES} is applicable only when tendons are jacked sequentially and not simultaneously

Calculate % of each type of loss

Add % of all losses

Figure 3.9 (*continued*)

END

3.9 STEP-BY-STEP COMPUTATON OF ALL TIME-DEPENDENT LOSSES IN A PRETENSIONED BEAM

Example 3.8

A simply supported pretensioned 70-ft-span lightweight steam-cured double T-beam as shown in Figure 3.10 is prestressed by twelve $\frac{1}{2}''$ diameter (twelve 12.7 mm dia) 270-K grade stress-relieved strands. The tendons are harped, and the eccentricity at midspan is 18.73 in. (476 mm) and at the end 12.98 in. (330 mm). Compute the prestress loss at the critical section in the beam of 0.40 span at

(a) stage I at transfer

(b) stage II after concrete topping is placed

(c) two years after concrete topping is placed

Suppose the topping is 2 in. (51 mm) normal-weight concrete cast at 30 days. Suppose also that prestress transfer occurred 18 hours after tensioning the strands. Given

$$f_c' = 5,000 \text{ psi, lightweight (34.5 MPa)}$$

$$f_{ci}' = 3,500 \text{ psi (24.1 MPa)}$$

and the following noncomposite section properties.

$$A_c = 615 \text{ in}^2 \ (3,968 \text{ cm}^2)$$

$$I_c = 59,720 \text{ in}^4 \ (2.49 \times 10^6 \text{ cm}^4)$$

$$c_b = 21.98 \text{ in. (55.8 cm)}$$

$$c_t = 10.02 \text{ in. (25.5 cm)}$$

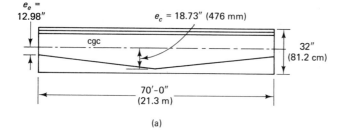

(a)

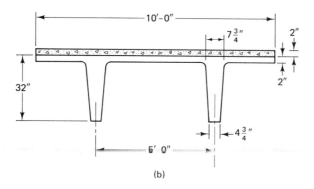

(b)

Figure 3.10 Double T pretensioned beam. (a) Elevation. (b) Pretensioned section.

$S_b = 2,717 \text{ in}^3$

$S^t = 5,960 \text{ in}^3$

$W_D \text{ (no topping)} = 491 \text{ plf } (7.2 \text{ kN/m})$

$W_{SD} \text{ (2'' topping)} = 250 \text{ plf } (3.65 \text{ kN/m})$

$W_L = 40 \text{ psf } (1,915 \text{ Pa})$

$f_{pu} = 270,000 \text{ psi } (1,862 \text{ MPa})$

$f_{py} = 0.85 f_{pu}$

$f_{pi} = 0.70 f_{pu} = 0.82 f_{py} = 0.85 \times 0.82 f_{pu} \cong 0.70 f_{pu}$
$\qquad = 189,000 \text{ psi } (1,303 \text{ MPa})$

$f_{py} = 230,000 \text{ psi } (1,589 \text{ MPa})$

$E_s = 28 \times 10^6 \text{ psi } (193.1 \times 10^3 \text{ MPa})$

Solution

$$f'_{ci} = 3,500 \text{ psi}$$

$$E_{ci} = 115^{1.5}(33\sqrt{3,500}) = 2.41 \times 10^6 \text{ psi}$$

$$E_c = 115^{1.5}(33\sqrt{5,000}) = 2.88 \times 10^6 \text{ psi}$$

Stage I: Stress Transfer

(a) *Elastic shortening.* Given critical section distance from support $= 0.40 \times 70 = 28$ ft, e at critical section $= 12.98 + 0.8(18.73 - 12.98) = 17.58$ in. Dead-load moment M_D at 0.40 of the span is

$$M_D = W_D \frac{X}{2}(L - X) = 491\left(\frac{28}{2}\right)(70 - 28)$$

$$= 288,708 \text{ ft-lb} = 3,464,496 \text{ in.-lb}$$

$$f_{pi} = 0.70 f_{pu} = 0.70 \times 270,000 = 189,000 \text{ psi}$$

Assume elastic shortening loss and steel relaxation loss $\cong 18,000$ psi; then net steel stress $f_{pi} = 189,000 - 18,000 = 171,000$ psi, and

$$P_i = A_{ps} f_{pi} = 12 \times 0.153 \times 171,000 = 313,956 \text{ lb}$$

$$r^2 = \frac{I_c}{A_c} = \frac{59,720}{615} = 97.11 \text{ in}^2$$

$$f_{cs} = -\frac{P_i}{A_c}\left(1 + \frac{e^2}{r^2}\right) + \frac{M_D}{I_c}$$

$$= -\frac{313,956}{615}\left(1 + \frac{(17.58)^2}{97.11}\right) + \frac{3,464,496 \times 17.58}{59,720}$$

$$= -2,135.2 + 1,019.9 = -1,115.3 \text{ psi}$$

$$n = \frac{E_s}{E_{ci}} = \frac{28 \times 10^6}{2.41 \times 10^6} = 11.62$$

$$\Delta f_{pES} = nf_{cs} = 11.62 \times 1,115.3 = 12,958 \text{ psi } (85.4 \text{ MPa})$$

If $f_{pi} = 189,000$ psi is used, then the net $f_{pi} = 189,000 - 12,958 = 176,042$ psi, and we have

$$f_{cs} = -2,135.20 \times \frac{189,000}{176,042} + 1,019.90 = -1,272.50 \text{ psi}$$

$$\Delta f_{pES} = nf_{cs} = 11.62 \times 1,272.57 = 14,786 \text{ psi } (102.0 \text{ MPa})$$

vs. 12,985 psi in the refined solution, a small difference of 11.6 percent. Thus, an assumption of 10 percent loss at the beginning in estimating $P_i \cong 0.9P_J$ would have been adequate.

(b) *Steel Stress Relaxation.* Calculate the steel relaxation at transfer.

$$f_{py} = 230,000 \text{ psi}$$

$$f_{pi} = 189,000 \text{ psi (or net } f_{pi} = 171,000 \text{ psi could be used)}$$

$$t = 18 \text{ hours}$$

$$\Delta f_{pR} = f'_{pi}\left(\frac{\log t_2 - \log t_1}{10}\right)\left(\frac{f'_{pi}}{f_{py}} - 0.55\right)$$

$$= 189,000\left(\frac{\log 18}{10}\right)\left(\frac{189,000}{230,000} - 0.55\right)$$

$$= 6,446.9 \text{ psi} \cong 6,447 \text{ psi}$$

$$\Delta f_{pES} + \Delta f_{pR} = 12,958 + 6,447 = 19,405 \cong 18,000 \text{ psi, assumed OK.}$$

(c) *Creep Loss*

$$\Delta f_{pCR} = 0$$

(d) *Shrinkage Loss*

$$\Delta f_{pSH} = 0$$

The stage I total losses are

$$\Delta f_{pT} = \Delta f_{pES} + \Delta f_{pR} + \Delta f_{pCR} + \Delta f_{pSH}$$

$$= 12,958 + 6,447 + 0 + 0 = 19,405 \text{ psi } (133.8 \text{ MPa})$$

The strand stress f_{pi} at the end of stage I $= 189,000 - 19,405 = 169,595$ psi (1,169 MPa).

Stage II: Transfer to Placement of Topping after 30 Days

(a) *Creep Loss*

$$E_c = 2.88 \times 10^6 \text{ psi}$$

$$E_s = 28 \times 10^6 \text{ psi}$$

$$n = \frac{E_c}{E_s} = \frac{28 \times 10^6}{2.88 \times 10^6} = 9.72$$

$$\bar{f}_{cs} = 1,115.3 \text{ psi}$$

Intensity of 2-in. normal-weight concrete topping:

$$W_{SD} = \frac{2}{12} \times 10 \times 150 = 250 \text{ plf}$$

The moment due to the 2-in. topping is

$$M_{SD} = W_{SD}\left(\frac{x}{2}\right)(L - x) = 250\left(\frac{28}{2}\right)(70 - 28) \times 12 = 1,764,000 \text{ in.-lb}$$

$$\bar{f}_{csd} = \frac{M_{SD}e}{I_c} = \frac{1,764,000 \times 17.58}{59,720} = 519.3 \text{ psi}$$

Although 30 days' duration is short for long-term effects, sufficient approximation can be justified in stage II using the creep factor K_{CR} of Equation 3.12 to account for stage III as well (see stage III creep calculations).

For lightweight concrete, use $K_{CR} = 2.0 \times 80\% = 1.6$. Then, from Equation 3.10, the prestress loss due to long-term creep is

$$\Delta f_{pCR} = nK_{CR}(f_{CS}f_{CSD}) = 9.72 \times 1.6(1,115.3 - 519.3) = 9,269 \text{ psi (63.3 MPa)}$$

(b) *Shrinkage Loss.* Assume relative humidity $RH = 70\%$. Then, from Equation 3.15, the prestress loss due to long-term shrinkage is

$$\Delta f_{pSH} = 8.2 \times 10^{-6}K_{SH}E_s\left(1 - 0.06\frac{V}{S}\right)(100 - RH)$$

$$\frac{V}{S} = \frac{615}{364} = 1.69$$

$K_{SH} = 1.0$ for pretensioned members; hence,

$$\Delta f_{pSH} = 8.2 \times 10^{-6} \times 1.0 \times 28 \times 10^6(1 - 0.06 \times 1.69)(100 - 70)$$

$$= 6,190 \text{ psi (42.7 MPa)}$$

(c) *Steel Relaxation Loss at 30 Days*

$t_1 = 18 \text{ hours}$

$t_2 = 30 \text{ days} = 30 \times 24 = 720 \text{ hours}$

$f_{ps} = 169,595 \text{ psi from stage I}$

$$\Delta f_{pR} = 169,595\left(\frac{\log 720 - \log 18}{10}\right)\left(\frac{169,595}{230,000} - 0.55\right) = 5,091 \text{ psi (35.1 MPa)}$$

Stage II total loss is

$$\Delta f_{pT} = \Delta f_{pCR} + \Delta f_{pSH} + \Delta f_{pR} = 9,269 + 6,190 + 5,091 = 20,550 \text{ psi (141.7 MPa)}$$

The increase in stress in the strands due to the addition of topping is

$$f_{SD} = n\bar{f}_{csd} = 9.72 \times 519.3 = 5,048 \text{ psi (34.8 MPa)}$$

Hence, the strand stress at the end of stage II is

$$f_{pe} = f_{ps} - \Delta f_{pT} + f_{SD} = 169,595 - 20,550 + 5,048 = 154,093 \text{ psi (1,062 MPa)}$$

Stage III: At end of two years. The values for long-term creep and long-term shrinkage evaluated for stage II are assumed not to have increased significantly, since the long-term values of K_{CR} for creep and K_{SH} for shrinkage were used in stage II computations. Accordingly,

$$f_{pe} = 154,093 \text{ psi (1,066 MPa)}$$

$$t_1 = 30 \text{ days} = 720 \text{ hours}$$

$$t_2 = 2 \text{ years} \times 365 \times 24 = 17{,}520 \text{ hours}$$

The steel relaxation stress loss is

$$\Delta f_{pR} = 154{,}093 \left(\frac{\log 17{,}520 - \log 720}{10} \right) \left(\frac{154{,}093}{230{,}000} - 0.55 \right) = 2{,}563 \text{ psi} \ (17.7 \text{ MPa})$$

So the strand stress f_{pe} at the end of stage III $\cong 154{,}093 - 2{,}563 = 151{,}530$ psi (1,033 MPa).

Summary of stresses

Stress level at various stages	Steel stress, psi	Percent
After tensioning ($0.70 f_{pu}$)	189,000	100.0
Elastic shortening loss	−12,958	−6.9
Creep loss	−9,269	−4.9
Shrinkage loss	−6,190	−3.3
Relaxation loss (6,447 + 5,091 + 2,563)	−14,101	−7.5
Increase due to topping	5,048	2.7
Final net stress f_{pe}	151,530 psi (1,045 MPa)	80.1

Percentages of total losses = $100 - 80.1 = 19.9\%$, say, 20% for this pretensioned beam.

3.10 STEP-BY-STEP COMPUTATION OF ALL TIME-DEPENDENT LOSSES IN A POST-TENSIONED BEAM

Example 3.9

Solve Example 3.8 assuming that the beam is post-tensioned. Assume also that the anchorage seating loss is $\frac{1}{4}$ in. and that all strands are simultaneously tensioned in a flexible duct.

Solution

(a) Anchorage seating loss

$$\Delta_A = \frac{1''}{4} = 0.25'' \qquad L = 70 \text{ ft}$$

From Equation 3.24, the anchorage slip stress loss is

$$\Delta f_{pA} = \frac{\Delta_A}{L} E_{Ps} = \frac{0.25}{70 \times 12} \times 28 \times 10^6 \cong 8{,}333 \text{ psi} \ (40.2 \text{ MPa})$$

(b) Elastic shortening. Since all jacks are simultaneously tensioned, the elastic shortening will precipitate during jacking. As a result, no elastic shortening stress loss takes place in the strands. Hence, $\Delta f_{pE} = 0$.

(c) Frictional loss. Assume that the parabolic tendon approximates the shape of an arc of a circle. Then, from Equation 3.23,

Linn Cove Viaduct, Grandfather Mountain, North Carolina. A 90° cantilever and a 10 percent superelevation in one direction to a full 10 percent in the opposite direction within 180 ft. Designed by Figg and Muller Engineers, Inc., Tallahassee, Florida. (*Courtesy*, Figg and Muller Engineers, Inc.)

$$\alpha = \frac{8y}{x} = \frac{8(18.73 - 12.98)}{70 \times 12} = 0.0548 \text{ radian}$$

From Table 3.7, use $K = 0.001$ and $\mu = 0.25$. Then, from Example 3.8,

$$f_{pi} = 189,000 \text{ psi (1,303 MPa)}$$

The stress in the strands after anchorage seating loss is

$$189,000 - 8,333 = 180,667 \text{ psi (1,241 MPa)}$$

From Equation 3.22, the stress loss in prestress due to friction is

$$\Delta f_{pF} = f_{pi}(\mu\alpha - KL)$$

$$= 180,667(0.25 \times 0.0548 + 0.001 \times 70)$$

$$= 15,122 \text{ psi (104.3 MPa)}$$

The stress remaining in the prestressing steel after all initial instantaneous losses is

$$f_{pi} = 189,000 - 8,333 - 0 - 15,122 = 165,545 \text{ psi (1,141 MPa)}$$

Hence, the net prestressing force is

$$P_i = 165,545 \times 12 \times 0.153 = 303,944 \text{ lb}$$

compared to $P_i = 313,956$ lb in the pretensioned case of Example 3.8.

Stage I: Stress at Transfer
(a) Anchorage Loss

$$\text{Anchorage loss} = 8,333 \text{ psi}$$

$$\text{Net stress} = 165{,}545 \text{ psi}$$

(b) Relaxation Loss

$$\Delta f_{pR} = 165{,}545 \left(\frac{\log 18}{10} \right) \left(\frac{165{,}545}{230{,}000} - 0.55 \right)$$

$$\cong 3{,}528 \text{ psi} \ (24.3 \text{ MPa})$$

(c) Creep Loss

$$\Delta f_{pCR} = 0$$

(d) Shrinkage Loss

$$\Delta f_{pSH} = 0$$

So the tendon stress f_{pi} at the end of stage I is

$$165{,}545 - 3{,}528 = 162{,}017 \text{ psi} \ (1{,}117 \text{ MPa})$$

Stage II: Transfer to Placement of Topping after 30 Days

(a) Creep Loss

$$P_i = 162{,}017 \times 12 \times 0.153 = 297{,}463 \text{ lb}$$

$$\bar{f}_{cs} = -\frac{P_i}{A_c} \left(1 + \frac{e^2}{r^2} \right) + \frac{M_D e}{I_c}$$

$$= -\frac{297{,}463}{615} \left(1 + \frac{(17.58)^2}{97.11} \right) + \frac{3{,}464{,}496 \times 17.58}{59{,}720}$$

$$= -2{,}023.00 + 1{,}020.00 = 1{,}003.0 \text{ psi} \ (6.99 \text{ MPa})$$

Hence, the creep loss is

$$\Delta f_{pCR} = n K_{CR} (\bar{f}_{cs} - \bar{f}_{csd})$$

$$= 9.72 \times 1.6 (1{,}003.0 - 519.3) \cong 7{,}523 \text{ psi} \ (51.9 \text{ MPa})$$

(b) Shrinkage Loss. From Example 3.8,

$$\Delta f_{pSH} = 6{,}190 \text{ psi} \ (42.7 \text{ MPa})$$

(c) Steel Relaxation Loss at 30 days

$$f_{ps} = 162{,}017 \text{ psi}$$

The relaxation loss in stress becomes

$$\Delta f_{pR} = 162{,}017 \left(\frac{\log 720 - \log 18}{10} \right) \left(\frac{162{,}017}{230{,}000} - 0.55 \right)$$

$$\cong 4{,}008 \text{ psi} \ (27.6 \text{ MPa})$$

Stage II: Total Losses

$$\Delta f_{pT} = \Delta f_p CR + \Delta f_{pSH} + \Delta f_{pR}$$

$$= 7{,}523 + 6{,}190 + 4{,}008 = 17{,}721 \text{ psi} \ (122.2 \text{ MPa})$$

From Example 3.8, the increase in stress in the strands due to the addition of topping, is $f_{SD} = 5{,}048$ psi (34.8 MPa); hence, the strand stress at the end of stage II is

$$f_{pe} = f_{ps} - \Delta f_{pT} + \Delta f_{SD} = 162,017 - 17,721 + 5,048 = 149,344 \text{ psi } (1,030 \text{ MPa})$$

Stage III: At End of Two Years

$$f_{pe} = 149,344 \text{ psi}$$

$$t_1 = 720 \text{ hours}$$

$$t_2 = 17,520 \text{ hours}$$

The steel relaxation stress loss is

$$\Delta f_{pR} = 149,344 \left(\frac{\log 17,520 - \log 720}{10}\right)\left(\frac{149,344}{230,000} - 0.55\right)$$

$$\cong 2,056 \text{ psi } (14.2 \text{ MPa})$$

The strand stress f_{pe} at the end of stage III is approximately

$$149,344 - 2,056 = 147,288 \text{ psi } (1,016 \text{ MPa})$$

Summary of stresses

Stress level at various stages	Steel stress psi	Percent
After tensioning $(0.70 f_{pu})$	189,000	100.0
Elastic shortening loss	0	0.0
Anchorage loss	−8,333	−4.4
Frictional loss	−15,122	−8.0
Creep loss	−7,523	−4.0
Shrinkage loss	−6,190	−3.3
Relaxation loss (3,528 + 4,008 + 2,056)	−9,592	−5.1
Increase due to topping	5,048	2.7
Final net stress f_{pe}	147,288	77.9

Percentage of total losses = 100 − 77.9 = 22.1%, say, 22% for this post-tensioned beam.

3.11 LUMP-SUM COMPUTATION OF TIME-DEPENDENT LOSSES IN PRESTRESS

Example 3.10

Solve Examples 3.8 and 3.9 by the approximate lump-sum method, and compare the results.

Solution for Example 3.8. From Table 3.1, the total loss $\Delta_{pT} = 33,000$ psi (227.5 MPa). So the net final strand stress by this method is

$$f_{pe} = 189,000 - 33,000 = 156,000 \text{ psi } (1,076 \text{ MPa})$$

Step-by-step f_{pe} value = 151,530

$$\text{Percent difference} = \frac{156,000 - 151,530}{189,000} = 2.4\%$$

Solution for Example 3.9. From Table 3.2, the total loss $\Delta_{pT} = 35,000$ psi (241.3 MPa). So the net final strand stress by the lump-sum method is

$$f_{pe} = 189,000 - 35,000 = 154,000 \text{ psi } (1,062 \text{ MPa})$$

Step-by-step f_{pe} value $= 147,282$ psi (1,016 MPa)

$$\text{Percent difference} = \frac{154,000 - 147,282}{189,000} = 3.5\%$$

In both cases the difference between the step-by-step "exact" method and the approximate lump-sum method is quite small, indicating that in normal, standard cases both methods are equally reliable.

REFERENCES

3.1 PCI Committee on Prestress Loss. "Recommendations for Estimating Prestress Loss." *Journal of the Prestressed Concrete Institute* 20 (1975): 43–75.

3.2 ACI-ASCE Joint Committee 423. "Tentative Recommendations for Prestressed Concrete." *Journal of the American Concrete Institute* 54, (1957): 548–578.

3.3 AASHTO Subcommittee on Bridges and Structures. *Interim Specifications Bridges*. Washington, D.C.: American Association of State Highway and Transportation Officials, 1975.

3.4 Prestressed Concrete Institute. *PCI Design Handbook*. Chicago: PCI, 1985.

3.5 Post-Tensioning Institute. *Post-Tensioning Manual*. 4th ed. Phoenix, Ariz.: Post-Tensioning Institute, 1985.

3.6 Lin, T. Y. "Cable Friction in Post-Tensioning." *Journal of Structure Division*. New York: ASCE, November 1956, pp. 1107–1 to 1107–13.

3.7 Tadros, M. K., Ghali, A., and Dilger, W. H. "Time Dependent Loss and Deflection in Prestressed Concrete Members." *Journal of the Prestressed Concrete Institute* 20 (1975): 86–98.

3.8 Branson, D. E. "The Deformation of Non-composite and Composite Prestressed Concrete Members." In *Deflection of Structures*. Detroit: American Concrete Institute, 1974, pp. 83–128.

3.9 Nawy, E. G. *Reinforced Concrete—A Fundamental Approach*. Englewood Cliffs, N.J.: Prentice-Hall, 1985.

3.10 Cohn, M. Z. *Partial Prestressing, From Theory to Practice*. NATO-ASI Applied Science Series, vols 1 and 2. Dordrecht, The Netherlands: Martinus Nijhoff, in Cooperation with NATO Scientific Affairs Division, 1986.

3.11 ACI Committee 318, *Building Code Requirements for Reinforced Concrete, ACI Standard 318–89*. Detroit: American Concrete Institute, 1989.

PROBLEMS

3.1 A simply supported pretensioned beam has a span of 75 ft (22.9 m) and the cross section shown in Figure P3.1. It is subjected to a uniform gravitational live-load intensity $W_L = 1,200$ plf (17.5 kN/m) in addition to its self-weight and is prestressed with 20 stress-relieved $\frac{1}{2}''$ dia (12.7 mm dia) seven-wire strands. Compute the total prestress losses by the step-by-step method, and compare them with the values obtained by the lump-sum method. Take the following values as given:

$f'_c = 6,000$ psi (41.4 MPa), normal-weight concrete
$f'_{ci} = 4,500$ psi (31 MPa)
$f_{pu} = 270,000$ psi (1,862 MPa)
$f_{pi} = 0.70 f_{pu}$
Relaxation time $t = 15$ years
$e_c = 19$ in. (483 mm)
Relative humidity $RH = 75\%$
$V/S = 3.0$ in. (7.62 cm)

3.2 Compute, by the detailed step-by-step method, the total losses in prestress of the

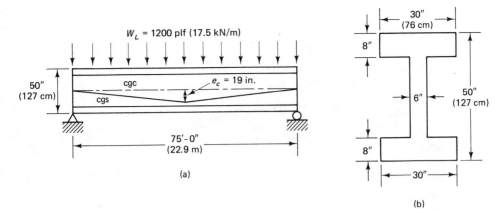

W_L = 1200 plf (17.5 kN/m)

50"
(127 cm)

cgc

cgs

e_c = 19 in.

75'-0"
(22.9 m)

(a)

30"
(76 cm)

8"

6"

8"

50"
(127 cm)

30"

50"
(127 cm)

(b)

Figure P3.1 (a) Elevation. (b) Section.

8-ft (2.44-m)-wide flange single T-beam in Example 1.1 which has a span of 64 ft (19.5 m) for a steel relaxation period of 7 years. Use RH = 70% and V/S = 3.5 in. (8.9 cm), and solve for both pretensioned and post-tensioned prestressing conditions.

3.3 Compute, by the detailed step-by-step method, the total losses of prestress in the AASHTO 36-in. (91.4 cm)-deep beam used in Problem 1.1 and which has a span of 34 ft (10.4 m) for both the pretensioned and the post-tensioned case. Use all the data of Problem 1.1 in your solution, and assume that the relative humidity RH = 70% and the volume-to-surface ratio V/S = 3.2. Determine the steel relaxation losses at the end

of the first year after erection and at the end of 4 years.

3.4 Compute, by the detailed step-by-step method, the total prestress losses of the simply supported double T-beam of Example 3.9 if it was post-tensioned using flexible ducts for the tendon. Assume that the tendon profile is essentially parabolic and that its radius R = 110 ft (33.5 m). Assume also that all strands are tensioned simultaneously and that the anchorage slip $\Delta_A = \frac{3}{8}$ in. (9.5 mm). All the data are identical to those of Example 3.8; the critical section is determined to be at a distance 0.4 times the span from the face of the support.

CHAPTER 4

Flexural Design of Prestressed Concrete Elements

4.1 INTRODUCTION

Flexural stresses are the result of external, or imposed, bending moments. In most cases, they control the selection of the geometrical dimensions of the prestressed concrete section regardless of whether it is pretensioned or post-tensioned. The design process starts with the choice of a preliminary geometry, and by trial and adjustment it converges to a final section with geometrical details of the concrete cross section and the sizes and alignments of the prestressing strands. The section satisfies the flexural (bending) requirements of concrete stress and steel stress limitations. Thereafter, other factors such as shear and torsion capacity, deflection, and cracking are analyzed and satisfied.

While the input data for the analysis of sections differ from the data needed for design, every design is essentially an analysis. One assumes the geometrical properties of the section to be prestressed and then proceeds to determine whether the section can safely carry the prestressing forces and the required external loads. Hence, a good understanding of the fundamental principles of analysis and the alternatives presented thereby significantly simplifies the task of designing the section. As seen from the discussion in Chapter 1, the basic mechanics of materials, principles of equilibrium of internal couples, and elastic principles of superposition have to be adhered to in all stages of loading.

It suffices in the flexural design of reinforced concrete members to apply only the limit states of stress at failure for the choice of the section, provided that other requirements such as serviceability, shear capacity, and bond are met. In the design of prestressed members, however, additional checks are needed at the load transfer

Maryland Concert Center parking garage, Baltimore. (*Courtesy,* Prestressed Concrete Institute.)

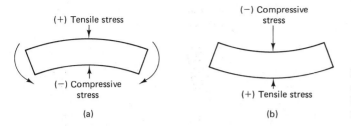

Figure 4.1 Sign convention for flexure stress and bending moment. (a) Negative bending moment. (b) Positive bending moment.

and limit state at service load, as well as the limit state at failure, with the failure load indicating the reserve strength for overload conditions. All these checks are necessary to ensure that at service load cracking is negligible and the long-term effects on deflection or camber are well controlled.

In view of the preceding, this chapter covers the major aspects of both the service-load flexural design and the ultimate-load flexural design check. The principles and methods presented in Chapter 1 for service load computations are extended into step-by-step procedures for the design of prestressed concrete linear elements, taking into consideration the impact of the magnitude of prestress losses discussed in Chapter 3. Note that a logical sequence in the design process entails *first* the service-load design of the section required in flexure, and then the analysis of the available moment strength M_n of the section for the limit state at failure. Throughout the book, a negative sign $(-)$ is used to denote compressive stress and a positive sign $(+)$ is used to denote tensile stress in the concrete section. A convex or hogging shape indicates negative bending moment, a concave or sagging shape positive bending moment, as shown in Figure 4.1.

Unlike the case of reinforced concrete members, the external dead load and partial live load are applied to the prestressed concrete member at varying concrete strengths at various loading stages. These loading stages can be summarized as follows:

- Initial prestress force P_i is applied; then, at transfer, the force is transmitted from the prestressing strands to the concrete.
- The full self-weight W_D acts on the member together with the initial prestressing force, provided that the member is simply supported, i.e., there is no intermediate support.
- The full superimposed dead load W_{SD} including topping for composite action, is applied to the member.
- Most short-term losses in the prestressing force occur, leading to a reduced prestressing force P_{eo}.
- The member is subjected to the full service load, with long-term losses due to creep, shrinkage, and steel strand relaxation taking place and leading to a net prestressing force P_e.
- Overloading of the member occurs under certain conditions up to the limit state at failure.

A typical loading history and corresponding stress distribution across the depth of the critical section are shown in Figure 4.2, while a schematic plot of load versus

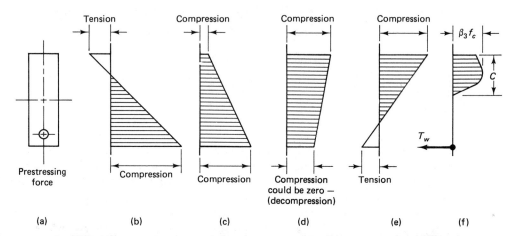

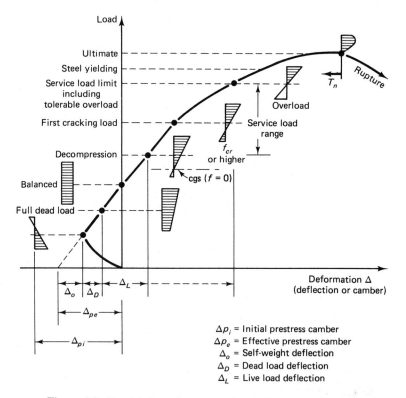

Figure 4.2 Flexural stress distribution throughout loading history. (a) Beam section. (b) Initial prestressing stage. (c) Self-weight and effective prestress. (d) Full dead load plus effective prestress. (e) Full service load plus effective prestress. (f) Limit state of stress at ultimate load for underreinforced beam.

Figure 4.3 Load-deformation curve of typical prestressed beam.

deformation (camber or deflection) is shown in Figure 4.3 for the various loading stages from the self-weight effect up to rupture.

4.2 SELECTION OF GEOMETRICAL PROPERTIES OF SECTION COMPONENTS

4.2.1 General Guidelines

Under service-load conditions, the beam is assumed to be homogeneous and elastic. Since it is also assumed (because expected) that the prestress compressive force transmitted to the concrete closes the crack that might develop at the tensile fibers of the beam, beam sections are considered uncracked. Stress analysis of prestressed beams under these conditions is no different from stress analysis of a steel beam, or, more accurately, a beam column. The axial force due to prestressing is always present regardless of whether bending moments do or do not exist due to other external or self-loads.

As seen from Chapter 1, it is advantageous to have the alignment of the prestressing tendons eccentric at the critical sections, such as the midspan section in a simple beam and the support section in a continuous beam. As compared to a rectangular solid section, a nonsymmetrical flanged section has the advantage of efficiently using the concrete material and of concentrating the concrete in the compressive zone of the section where it is most needed.

Equations 4.1, 4.2, and 4.3 to be subsequently presented are stress equations that are convenient in the analysis of stresses in the section once the section is chosen. For design, it is necessary to transpose the three equations into geometrical equations so that the student and the designer can readily choose the concrete section. A logical transposition is to define the minimum section modulus that can withstand all the loads after losses.

4.2.2 Minimum Section Modulus

To design or choose the section, a determination of the required minimum section modulus S_b^t has to be made first. If

f_{ci} = maximum allowable compressive stress in concrete immediately after transfer and prior to losses

= $0.60f_{ci}'$

f_{ti} = maximum allowable tensile stress in concrete immediately after transfer and prior to losses

= $3\sqrt{f_{ci}'}$ (the value can be increased to $6\sqrt{f_{ci}'}$ at the supports for simply supported members)

f_c = maximum allowable compressive stress in concrete after losses at service-load level

= $0.45f_c'$

Ninian Central oil drilling platform, Cheiron, England, and C. G. Doris, Scotland. (*Courtesy*, Ben C. Gerwick.)

f_t = maximum allowable tensile stress in concrete after losses at service load level

$= 6\sqrt{f_c'}$ (the value can be increased in one-way systems to $12\sqrt{f_c'}$ if long-term deflection requirements are met)

then the *actual* extreme fiber stresses in the concrete cannot exceed the values listed.

Using the uncracked unsymmetrical section, a summary of the equations of stress from Section 1.3 for the various loading stages is as follows.

Stress at Transfer

$$f^t = -\frac{P_i}{A_c}\left(1 - \frac{ec_t}{r^2}\right) - \frac{M_D}{S^t} \leq f_{ti} \tag{4.1 a}$$

$$f_b = -\frac{P_i}{A_c}\left(1 + \frac{ec_b}{r^2}\right) + \frac{M_D}{S_b} \leq f_{ci} \tag{4.1 b}$$

where P_i is the initial prestressing force. While a more accurate value to use would be the horizontal component of P_i, it is reasonable for all practical purposes to disregard such refinement.

Effective Stresses after Losses

$$f^t = -\frac{P_e}{A_c}\left(1 - \frac{ec_t}{r^2}\right) - \frac{M_D}{S^t} \leq f_t \tag{4.2 a}$$

$$f_b = -\frac{P_e}{A_c}\left(1 + \frac{ec_b}{r^2}\right) + \frac{M_D}{S_b} \leq f_c \tag{4.2 b}$$

Service-load Final Stresses

$$f^t = -\frac{P_e}{A_c}\left(1 - \frac{ec_t}{r^2}\right) - \frac{M_T}{S^t} \leq f_c \tag{4.3 a}$$

$$f_b = -\frac{P_e}{A_c}\left(1 + \frac{ec_b}{r^2}\right) + \frac{M_T}{S_b} \leq f_t \tag{4.3 b}$$

where $M_T = M_D + M_{SD} + M_L$

P_i = initial prestress

P_e = effective prestress after losses

t denotes the top, and b denotes the bottom fibers

e = eccentricity of tendons from the concrete section center of gravity, cgc

r^2 = square of radius of gyration

S^t/S_b = top/bottom section modulus value of concrete section

The *decompression stage* denotes the increase in steel strain due to the increase in load from the stage when the effective prestress P_e acts *alone* to the stage when the additional load causes the compressive stress in the concrete at the cgs level to reduce to zero (see Figure 4.3). At this stage, the *change* in concrete stress due to decompression is

$$f_{decomp} = \frac{P_e}{A_c}\left(1 + \frac{e^2}{r^2}\right) \qquad (4.3\ c)$$

This relationship is based on the assumption that the strain between the concrete and the prestressing steel bonded to the surrounding concrete is such that the gain in the steel stress is the same as the decrease in the concrete stress.

4.2.2.1 Beams with variable tendon eccentricity. Beams are prestressed with either draped or harped tendons. The maximum eccentricity is usually at the midspan controlling section for the simply supported case. Assuming that the effective prestressing force is

$$P_e = \gamma P_i$$

where γ is the residual prestress ratio, the loss of prestress is

$$P_i - P_e = (1 - \gamma)P_i \qquad (a)$$

If the actual concrete extreme fiber stress is equivalent to the maximum allowable stress, the change in this stress after losses, from Equations 4.1a and b, is given by

$$\Delta f^t = (1 - \gamma)\left(f_{ti} + \frac{M_D}{S^t}\right) \qquad (b)$$

$$\Delta f_b = (1 - \gamma)\left(-f_{ci} + \frac{M_D}{S_b}\right) \qquad (c)$$

From Figure 4.4, as the superimposed dead-load moment M_{SD} and live-load moment M_L act on the beam, the net stress at top fibers is

$$f_n^t = f_{ti} - \Delta f^t - f_c$$

or

$$f_n^t = \gamma f_{ti} - (1 - \gamma)\frac{M_D}{S^t} - f_c \qquad (d)$$

The net stress at the bottom fibers is

$$f_{bn} = f_t - f_{ci} - \Delta f_b$$

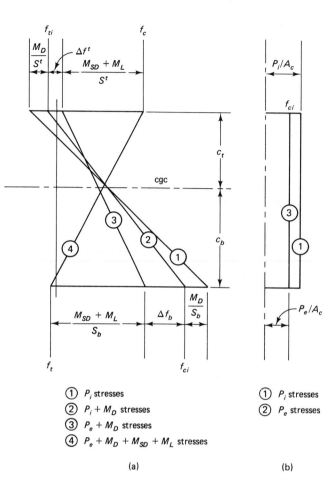

① P_i stresses
② $P_i + M_D$ stresses
③ $P_e + M_D$ stresses
④ $P_e + M_D + M_{SD} + M_L$ stresses

(a)

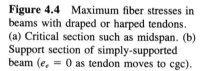

① P_i stresses
② P_e stresses

(b)

Figure 4.4 Maximum fiber stresses in beams with draped or harped tendons. (a) Critical section such as midspan. (b) Support section of simply-supported beam ($e_e = 0$ as tendon moves to cgc).

or

$$f_{bn} = f_t - \gamma f_{ci} - (1 - \gamma) \frac{M_D}{S_b} \qquad \text{(e)}$$

From Equations (d) and (e), the chosen section should have section moduli values

$$S^t \geq \frac{(1 - \gamma)M_D + M_{SD} + M_L}{\gamma f_{ti} - f_c} \qquad \text{(4.4 a)}$$

and

$$S_b \geq \frac{(1 - \gamma)M_D + M_{SD} + M_L}{f_t - \gamma f_{ci}} \qquad \text{(4.4 b)}$$

The required eccentricity of the prestressing tendon at the critical section, such as the midspan section, is

$$e_c = (f_{ti} - \bar{f}_{ci}) \frac{S^t}{P_i} + \frac{M_D}{P_i} \qquad \text{(4.4 c)}$$

where $\bar{f}_{ci}$ is the concrete stress at transfer at the level of the centroid cgc of the concrete section and

$$P_i = \bar{f}_{ci} A_c$$

Thus,

$$\bar{f}_{ci} = f_{ti} - \frac{c^t}{h}(f_{ti} - f_{ci}) \qquad (4.4\ d)$$

4.2.2.2 Beams with constant tendon eccentricity. Beams with constant tendon eccentricity are beams with straight tendons, as is normally the case in precast moderate-span simply supported beams. Because the tendon has a large eccentricity at the support, creating large tensile stresses at the top fibers without any reduction due to superimposed $M_D + M_{SD} + M_L$, in such beams smaller eccentricity of the tendon at midspan has to be used as compared to a similar beam with a draped tendon. In other words, the controlling section is the support section, for which the stress distribution at the support is shown in Figure 4.5. Hence,

$$\Delta f^t = (1 - \gamma)(f_{ti}) \qquad (a')$$

and

$$\Delta f_b = (1 - \gamma)(-f_{ci}) \qquad (b')$$

The net stress at the service-load condition after losses at the top fibers is

$$f_n^t = f_{ti} - \Delta f^t - f_c$$

or

$$f_n^t = \gamma f_{ti} - f_{cs} \qquad (c')$$

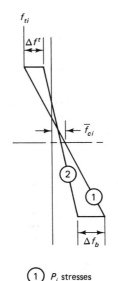

① P_i stresses

② P_e stresses

Figure 4.5 Maximum fiber stresses at support section of beams with straight tendons (stress distribution at midspan section similar to that of Figure 4.4).

where f_{cs} is the actual service-load stress in concrete. The net stress at service load after losses at the bottom fibers is

$$f_{bn} = f_t - f_{ci} - \Delta f_b$$

or

$$f_{bn} = f_t - \gamma f_{ci} \tag{d'}$$

From Equations (c) and (d), the chosen section should have section moduli values

$$S^t \geq \frac{M_D + M_{SD} + M_L}{\gamma f_{ti} - f_c} \tag{4.5 a}$$

and

$$S_b \geq \frac{M_D + M_{SD} + M_L}{f_t - \gamma f_{ci}} \tag{4.5 b}$$

The required eccentricity value at the critical section, such as the support for an ideal beam section having properties close to those required by Equations 4.5a and b, is

$$e_e = (f_{ti} - \bar{f}_{ci})\frac{S^t}{P_i} \tag{4.5 c}$$

A graphical representation of section moduli of nominal sections is shown in Figure 4.6. It may be used as a speedy tool for the choice of initial trial sections in the design process.

Table 4.1 gives the section moduli of standard PCI rectangular sections. Tables 4.2 and 4.3 give the geometrical outer dimensions of standard PCI T-sections and

TABLE 4.1 SECTION PROPERTIES AND MODULI OF STANDARD PCI RECTANGULAR SECTIONS

Designation	12RB16	12RB20	12RB24	12RB28	12RB32	12RB36	16RB32	16RB36	16RB40
Section modulus S, (in³)	512	800	1,152	1,568	2,048	2,592	2,731	3,456	4,267
Width, b (in.)	12	12	12	12	12	12	16	16	16
Depth, h (in.)	16	20	24	28	32	36	32	36	40

TABLE 4.2 GEOMETRICAL OUTER DIMENSIONS AND SECTION MODULI OF STANDARD PCI T-SECTIONS

Designation	Top-/bottom-section modulus, in³	Flange width b_f, in.	Flange depth t_f, in.	Total depth h, in.	Web width b_w, in.
8DT12	1,001/315	96	2	12	9.5
8DT14	1,307/429	96	2	14	9.5
8DT16	1,630/556	96	2	16	9.5
8DT20	2,320/860	96	2	20	9.5
8DT24	3,063/1,224	96	2	24	9.5
8DT32	5,140/2,615	96	2	32	9.5
10DT32	5,960/2,717	120	2	32	12.5
83T36	6,899/3,630	96	3	36	8
10ST48	13,194/4,803	120	3	48	8

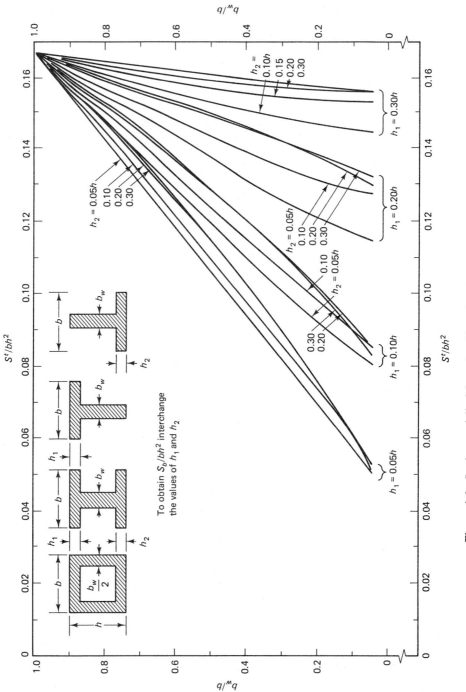

Figure 4.6 Section moduli of flanged and boxed sections (adapted from Ref. 4.15)

	AASHTO sections					
Designation	Type 1	Type 2	Type 3	Type 4	Type 5	Type 6
Top-/bottom-section modulus, in^3	1,476 1,807	2,527 3,320	5,070 6,186	8,908 10,544	16,790 16,307	20,587 20,157
Top flange width, b_f (in.)	12	12	16	20	42	42
Top flange average thickness, t_f (in.)	6	8	9	11	7	7
Bottom flange width, b_2 (in.)	16	18	22	26	28	28
Bottom flange average thickness, t_2 (in.)	7	9	11	12	13	13
Total depth, h (in.)	28	36	45	54	63	72
Web width, b_w (in.)	6	6	7	8	8	8
c_t/c_b (in.)	15.41 12.59	20.17 15.83	24.73 20.27	29.27 24.73	31.04 31.96	35.62 36.38

AASHTO I-sections, respectively, as well as the top-section moduli of those sections
needed in the preliminary choice of the section in the service-load analysis. Table 4.4
gives dimensional details of the *actual* "as-built" geometry of the standard PCI and
AASHTO sections. Additional details are presented in Refs. 4.9 and 4.11.

4.3 SERVICE-LOAD DESIGN EXAMPLES

4.3.1 Variable Tendon Eccentricity

Example 4.1

Design a simply supported pretensioned beam with harped tendon and with a span of
65 ft (19.8 m) using the ACI 318 Building Code allowable stresses. The beam has to
carry a superimposed service load of 1,100 plf (16.1 kN/m) and superimposed dead load
of 100 plf (1.5 kN/m), and has no concrete topping. Assume the beam is made of
normal-weight concrete with $f'_c = 5,000$ psi (34.5 MPa), and that the concrete strength
f'_{ci} at transfer is 75 percent of the cylinder strength. Assume also that the time-dependent
losses of the initial prestress are 18 percent of the initial prestress, and that $f_{pu} =
270,000$ psi (1,862 MPa) for stress-relieved tendons.

Solution

$$\gamma = 100 - 18 = 82\%$$

$$f'_{ci} = 0.75 \times 5,000 = -3,750 \text{ psi (25.9 MPa)}$$

TABLE 4.4 GEOMETRICAL DETAILS OF AS-BUILT PCI AND AASHTO SECTIONS

Designation	b_f (in.)	h_f (in.)	b_{w1} (in.)	b_{w2} (in.)	h (in.)	b (in.)
8DT12	96	2	5.75	3.75	12	48
8DT14	96	2	5.75	3.75	14	48
8DT16	96	2	5.75	3.75	16	48
8DT18	96	2	5.75	3.75	18	48
8DT20	96	2	5.75	3.75	20	48
8DT24	96	2	5.75	3.75	24	48
8DT32	96	2	7.75	4.75	32	48
10DT32	120	2	7.75	4.75	32	60

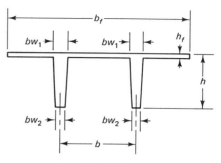

Actual double-T sections

Designation	b_f (in.)	x_1 (in.)	x_2 (in.)	b_w (in.)	h (in.)
8ST36	96	1.5	3	8	36
10ST48	120	1.5	4	8	48

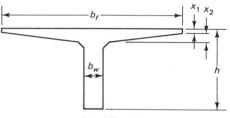

Actual T sections

Designation	b_f (in.)	x_1 (in.)	x_2 (in.)	b_2 (in.)	x_3 (in.)	x_4 (in.)	b_w (in.)	h (in.)
AASHTO 1	12	4	3	16	5	5	6	28
AASHTO 2	12	6	3	18	6	6	6	36
AASHTO 3	16	7	4.5	22	7.5	7	7	45
AASHTO 4	20	8	6	26	9	8	8	54
AASHTO 5	42	5	7	28	10	8	8	63
AASHTO 6	42	5	7	28	10	8	8	72

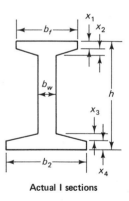

Actual I sections

$$f_{ci} = -0.60 \times 3{,}750 = -2{,}250 \text{ psi (15.5 MPa)}$$

$$f_{ti} = 3\sqrt{3{,}750} = 184 \text{ psi (midspan)}$$

$$= 6\sqrt{3{,}760} = 368 \text{ psi (support)}$$

$$f_c = -0.45 \times 5{,}000 = -2{,}250 \text{ psi (15.5 MPa)}$$

Use $f_t = 6\sqrt{5,000} = 425$ psi (2.95 MPa) as the maximum stress in tension, and assume a self-weight of approximately 850 plf (12.4 kN/m). Then the self-weight moment is given by

$$M_D = \frac{wl^2}{8} = \frac{850(65)^2}{8} \times 12 = 5,386,875 \text{ in.-lb (608.7 kN-m)}$$

and the superimposed load moment is

$$M_{SD} + M_L = \frac{(1,100 + 100)(65)^2}{8} \times 12 = 7,605,000 \text{ in.-lb (859.4 kN-m)}$$

Since the tendon is harped, the critical section is close to the midspan, where dead-load and superimposed dead-load moments reach their maximum. The critical section is in many cases taken at $0.40L$ from the support, where L is the beam span. From Equations 4.4a and b,

$$S^t \geq \frac{(1 - \gamma)M_D + M_{SD} + M_L}{\gamma f_{ti} - f_c}$$

$$\geq \frac{(1 - 0.82)5,386,875 + 7,605,000}{0.82 \times 184 + 2,250} = 3,572 \text{ in}^3 \text{ (58,535 cm}^3\text{)}$$

$$S_b \geq \frac{(1 - \gamma)M_D + M_{SD} + M_L}{f_t - \gamma f_{ci}}$$

$$\geq \frac{(1 - 0.82)5,386,875 + 7,605,000}{425 + (0.82 \times 2,250)} = 3,777 \text{ in}^3 \text{ (61.892 cm}^3\text{)}$$

From the PCI design Handbook, select a nontopped single-T 10ST48 with strand pattern 128-D1, since it has the bottom-section modulus value S_b closest to the required value. The section properties of the concrete are as follows:

$A_c = 782 \text{ in}^2$ $\qquad$ $I_c = 169,020 \text{ in}^4$ $\qquad$ $r^2 = \dfrac{I_c}{A_c} = 216.14 \text{ in}^2$

$c_b = 35.19 \text{ in.}$ $\qquad$ $c_t = 12.81 \text{ in.}$ $\qquad$ $S_b = 4,803 \text{ in}^3$

$S^t = 13,194 \text{ in}^3$ $\qquad$ $W_D = 815 \text{ plf}$ $\qquad$ $\dfrac{V}{S} = 2.33 \text{ in.}$

$e_c = 33.14 \text{ in.}$ $\qquad$ $e_e = 23.78 \text{ in.}$

$A_{ps} = $ twelve $\frac{1}{2}$-in.-dia (12.7-mm-dia) seven-wire 270 K stress-relieved strands

Design of Strands and Check of Stresses. From Figure 4.7, the assumed

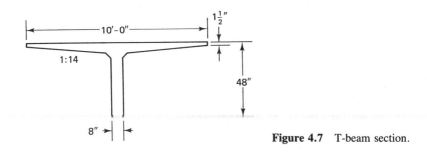

Figure 4.7 T-beam section.

self-weight is close to the actual self-weight. Hence, use $M_D = 5,386,875$ in.-lb

$$f_{pi} = 0.70 \times 270,000 = 189,000 \text{ psi}$$

$$f_{pe} = 0.82 f_{pi} = 0.82 \times 189,000 = 154,980 \text{ psi}$$

(a) Analysis of Stresses at Transfer. From Equation 4.1a,

$$f^t = -\frac{P_i}{A_c}\left(1 - \frac{ec_t}{r^2}\right) - \frac{M_D}{S^t} \leq f_{ti} = 184 \text{ psi}$$

Then

$$184 = -\frac{P_i}{782}\left(1 - \frac{33.14 \times 12.81}{216.14}\right) - \frac{5,386,875}{13,194}$$

or

$$184 = \frac{0.964 P_i}{782} - 408.3$$

$$P_i = 480,476 \text{ lb } (2137 \text{ kN})$$

Required number of strands $= \dfrac{480,476}{189,000 \times 0.153} = 16.62 \frac{1}{2}''$ strands. Try twelve $\frac{1}{2}''$
strands that the standard section has:

$$A_{ps} = 12 \times 0.153 = 1.836 \text{ in}^2 \text{ (11.9 cm}^2)$$

$$P_i = 1.836 \times 189,000 = 347,000 \text{ lb } (1,544 \text{ kN})$$

$$P_e = 1.836 \times 154,980 = 284,543 \text{ lb } (1,266 \text{ kN})$$

(b) Analysis of Stresses at Service Load

$$P_e = 284,543 \text{ lb}$$

$$M_{SD} = \frac{100(65)^2 12}{8} = 633,750 \text{ in.-lb } (72 \text{ kN-m})$$

$$M_L = \frac{1,100(65)^2 12}{8} = 6,971,250 \text{ in.-lb } (788 \text{ kN-m})$$

Total moment $M_T = M_D + M_{SD} + M_L = 5,386,875 + 7,605,000$

$$= 12,991,875 \text{ in.-lb } (1,468 \text{ kN-m})$$

From Equation 4.3a,

$$f^t = -\frac{P_e}{A_c}\left(1 - \frac{ec_t}{r^2}\right) - \frac{M_T}{S^t}$$

$$= -\frac{284,543}{782}\left(1 - \frac{33.14 \times 12.81}{216.14}\right) - \frac{12,991,875}{13,194}$$

$$= +350.8 - 987.4 = -636.6 \text{ psi} < f_c = -2,250 \text{ psi, O.K.}$$

From Equation 4.3b,

$$f_b = -\frac{P_e}{A_c}\left(1 + \frac{ec_b}{r^2}\right) + \frac{M_T}{S_b}$$

$$= -\frac{284{,}543}{782}\left(1 + \frac{33.14 \times 35.19}{216.14}\right) + \frac{12{,}991{,}875}{4{,}803}$$

$$= -2{,}327 + 2{,}705 = +378 \text{ psi } (T) < f_t = +425 \text{ psi, O.K.}$$

(c) Support Section Stresses

$$e_e = 23.75 \text{ in. } (604 \text{ m})$$

$$f_{ti} = 6\sqrt{f'_{ci}} = 6\sqrt{3{,}750} \cong 367 \text{ psi}$$

$$f_t = 6\sqrt{f'_c} = 6\sqrt{5{,}000} = 425 \text{ psi}$$

(i) At Transfer

$$f^t = -\frac{374{,}004}{782}\left(1 - \frac{23.78 \times 12.81}{216.14}\right) - 0 = +196 \text{ psi } (T)$$

$$< f_{ti} = 367.4 \text{ psi, O.K.}$$

$$f_b = -\frac{374{,}004}{782}\left(1 + \frac{23.78 \times 35.19}{216.14}\right) + 0 = -2{,}330 \text{ psi } (C)$$

$$> f_{ci} = 2{,}250 \text{ psi, no good}$$

Reduce the support eccentricity e_e to 20 in (508 mm):

$$f_b = -\frac{374{,}004}{782}\left(1 + \frac{20.0 \times 35.19}{216.14}\right) + 0 = -2{,}035.2 \text{ psi } (C) < 2{,}250 \text{ psi, O.K.}$$

$$f^t = -\frac{374{,}004}{782}\left(1 - \frac{20.0 \times 12.81}{216.14}\right) - 0 = +89 \text{ psi} < 367 \text{ psi, O.K.}$$

(ii) At Service Load

$$f^t = -\frac{284{,}543}{782}\left(1 - \frac{20.0 \times 12.81}{216.14}\right) - 0 = +78.7 \text{ psi } (T) < f_t = 425 \text{ psi, O.K.}$$

$$f_b = -\frac{284{,}543}{782}\left(1 + \frac{20.0 \times 35.19}{216.14}\right) + 0 = -1{,}669.2 \text{ psi } (C)$$

$$< f_c = -2{,}250 \text{ psi, O.K.}$$

Hence, accept the section for service-load conditions using twelve $\frac{1}{2}''$ (1.7 mm) strands with midspan eccentricity $e_c = 33.14$ in. (842 mm) and end eccentricity $e_e = 20.0$ in. (508 mm).

Observe that if $12\sqrt{f'_c}$ were allowed at service load in tension in the concrete in this example instead of $6\sqrt{f'_c}$, 10 strands instead of 12 would have been adequate.

4.3.2 Variable Tendon Eccentricity with No Height Limitation

Example 4.2

Assume there are no restrictions on the type of section that is chosen for the pretensioned beam in Example 4.1. Select the ideal section for minimum section moduli required in that example. Assume that the same losses occurred in this beam.

Crack development is prestressed T-beam (Nawy et al.).

Solution

$$\text{Required } S^t = 3{,}572 \text{ in}^3 \ (58{,}535 \text{ cm}^3)$$

$$\text{Required } S_b = 3{,}187 \text{ in}^3 \ (52{,}226 \text{ cm}^3)$$

Since the section moduli at the top and bottom fibers are almost equal, a symmetrical section is adequate. Next, analyze the section in Figure 4.8 chosen by trial and adjustment.

Analysis of Stresses at Transfer. From Equation 4.4d,

$$\bar{f}_{ci} = f_{ti} - \frac{c_t}{h}(f_{ti} - f_{ci})$$

$$= +184 - \frac{21.16}{40}(+184 + 2{,}250) \cong -1{,}104 \text{ psi } (C) \ (7.6 \text{ MPa})$$

$$P_i = A_c \bar{f}_{ci} = 377 \times 1{,}104 = 416{,}208 \text{ lb } (1{,}851 \text{ kN})$$

$$M_D = \frac{393(65)^2}{8} \times 12 = 2{,}490{,}638 \text{ in.-lb } (281 \text{ kN-m})$$

From Equation 4.4c, the eccentricity required at the section of maximum moment at midspan is

$$e_c = (f_{ti} - \bar{f}_{ci})\frac{S^t}{P_i} + \frac{M_D}{P_i}$$

$$= (184 + 1{,}104)\frac{3{,}340}{416{,}208} + \frac{2{,}490{,}638}{416{,}208}$$

$$= 10.34 + 5.98 = 16.32 \text{ in. } (452 \text{ mm})$$

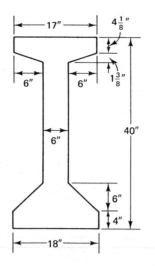

Figure 4.8 I-beam section in Example 4.2.

Since $c_b = 18.84$ in., and assuming a cover of 3.75 in., try $e_c = 18.84 - 3.75 \cong$ 15.0 in. (381 mm).

$$\text{Required area of tendons } A_p = \frac{P_i}{f_{pi}} = \frac{416{,}208}{189{,}000} = 2.2 \text{ in}^2 \ (14.2 \text{ cm}^2)$$

$$\text{Number of strands} = \frac{2.2}{0.153} = 14.38$$

Try thirteen $\frac{1}{2}''$ strands, $A_{ps} = 1.99$ in^2 (12.8 cm^2), and an actual $P_i = 189{,}000 \times 1.99 = 376{,}110$ lb (1,673 kN), and check the concrete extreme fiber stresses. From Equation 4.1a,

$$f^t = -\frac{P_i}{A_c}\left(1 - \frac{ec_t}{r^2}\right) - \frac{M_D}{S^t}$$

$$= -\frac{376{,}110}{377}\left(1 - \frac{15.0 \times 21.16}{187.5}\right) - \frac{2{,}490{,}638}{3{,}340}$$

$$= +691.2 - 745.7 = -54.50 \text{ psi } (C), \text{ no tension at transfer, O.K.}$$

From Equation 4.1.b,

$$f_b = -\frac{P_i}{A_c}\left(1 + \frac{ec_b}{r^2}\right) + \frac{M_D}{S_b}$$

$$= -\frac{376{,}110}{377}\left(1 + \frac{15 \times 18.84}{187.5}\right) + \frac{2{,}490{,}638}{3{,}750}$$

$$= -2{,}501.3 + 664.2 = -1{,}837.1 \text{ psi } (C) < f_{ci} = 2{,}250 \text{ psi, O.K.}$$

Analysis of Stresses at Service Load   From Equation 4.3a,

$$f^t = -\frac{P_e}{A_c}\left(1 - \frac{ec_t}{r^2}\right) - \frac{M_T}{S^t}$$

$$P_e - 13 \times 0.153 \times 154{,}980 - 308{,}255 \text{ lb } (1{,}371 \text{ kN})$$

Total moment $M_T = M_D + M_{SD} + M_L = 2{,}490{,}638 + 7{,}605{,}000$

$$= 10{,}095{,}638 \text{ in.-lb } (1{,}141 \text{ kN-m})$$

$$f^t = -\frac{308{,}225}{377}\left(1 - \frac{15.0 \times 21.16}{187.5}\right) - \frac{10{,}095{,}638}{3{,}340}$$

$$= +566.5 - 3{,}022.6 = -2{,}456.1 \text{ psi } (C) > f_c = -2{,}250 \text{ psi}$$

Hence, either enlarge the depth of the section or use higher strength concrete. Using $f'_c = 6{,}000$ psi,

$$f_c = 0.45 \times 6{,}000 = -2{,}700 \text{ psi, O.K.}$$

$$f_b = -\frac{P_e}{A_c}\left(1 + \frac{ec_b}{r^2}\right) + \frac{M_T}{S_b} = -\frac{308{,}255}{377}\left(1 + \frac{15.0 \times 18.84}{187.5}\right) + \frac{10{,}095{,}638}{3{,}750}$$

$$= -2{,}050 + 2{,}692.2 = 642 \text{ psi } (T), \text{ O.K.}$$

$6\sqrt{f'_c} = 12\sqrt{6000}$

$= 929$

Check Support Section Stresses

$$f'_{ci} = 0.75 \times 6{,}000 = 4{,}500 \text{ psi}$$

$$f_{ci} = 0.60 \times 4{,}500 = 2{,}700 \text{ spi}$$

$$f_{ti} = 3\sqrt{f'_{ci}} = 201 \text{ psi for support}$$

$$f_{ti} = 6\sqrt{f'_{ci}} = 402 \text{ psi for span}$$

$$f_c = 0.45\sqrt{f'_c} = 2{,}700 \text{ psi}$$

$$f_{t1} = 6\sqrt{f'_c} = 465 \text{ psi}$$
$$f_{t2} = 12\sqrt{f'_c} = 930 \text{ psi}$$

(a) At Transfer. Support section compressive fiber stress,

$$f_b = -\frac{P_i}{A_c}\left(1 + \frac{ec_b}{r^2}\right) + 0$$

$$P_i = 376{,}110 \text{ lb}$$

or

$$-2{,}700 = -\frac{376{,}110}{377}\left(1 + \frac{e \times 18.84}{187.5}\right)$$

so that

$$e_e = 16.98 \text{ in.}$$

Accordingly, try $e_e = 12.49$ in.:

$$f^t = -\frac{376{,}110}{377}\left(1 - \frac{12.49 \times 21.16}{187.5}\right) - 0$$

$$= 408.6 \text{ psi } (T) > f_{ti} = 402 \text{ psi}$$

Thus, use mild steel at the top fibers at the support section to take all tensile stresses in the concrete, or use a higher strength concrete for the section, or reduce the eccentricity.

(b) At Service Load

$$f^t = -\frac{308{,}255}{377}\left(1 - \frac{12.49 \times 21.16}{187.5}\right) - 0 = 334.9 \text{ psi } (T) < 465 \text{ psi, O.K.}$$

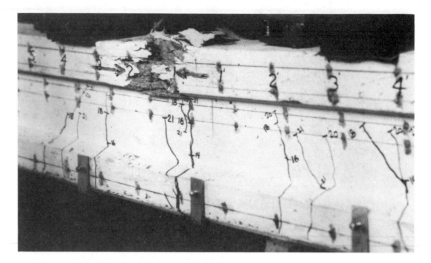

Prestressed beam at failure. Note crushing of concrete on top fibers (Nawy, Potyondy, et al.).

$$f_b = -\frac{308,255}{377}\left(1 + \frac{12.49 \times 18.84}{187.5}\right) + 0 = -1,843.8 \text{ psi } (C) < -2,700 \text{ psi, O.K.}$$

Hence, adopt the 40-in. (102-cm)-deep I-section prestressed beam of f_c' equal to 6,000 psi (41.4 MPa) normal-weight concrete with thirteen $\frac{1}{2}''$ strands having midspan eccentricity $e_c = 15.0$ in. (381 mm) and end section eccentricity $e_e = 12.5$ in. (318 m).

An alternative to this solution is to continue using $f_c' = 5,000$ psi, but change the number of strands and eccentricities.

4.3.3 Constant Tendon Eccentricity

Example 4.3

Solve Example 4.2 assuming that the prestressing tendon has constant eccentricity. Use $f_c' = = 5,000$ psi (34.5 MPa) normal-weight concrete, permitting a maximum concrete tensile stress $f_t = 12\sqrt{f_c'} = 849$ psi.

Solution Since the tendon has constant eccentricity, the dead-load and superimposed dead- and live-load moments at the support section of the simply supported beam are zero. Hence, the support section controls the design. The required section modulus at the support, from Equation 4.5a, is

$$S^t \geq \frac{M_D + M_{SD} + M_L}{\gamma f_{ti} - f_c}$$

$$S_b \geq \frac{M_D + M_{SD} + M_L}{f_t - \gamma f_{ci}}$$

Assume $W_D = 425$ plf. Then

$$M_D = \frac{425 \times (65)^2}{8} \times 12 = 2,693,438 \text{ in.-lb (304 kN-m)}$$

$$M_{SD} + M_L = 7,605,000 \text{ in.-lb (859 kN-m)}$$

Thus, the total moment $M_T = 10,298,438$ in.-lb (1,164 kN-m), and we also have

$$f_{ci} = -2,250 \text{ psi} \quad = 0.6 \cdot f'_{ci}$$

$$f'_{ci} = -3,750 \text{ psi}$$

$$f_{ti} = 6\sqrt{f'_{ci}} \text{ for support section} = 367 \text{ psi}$$

$$f_c = -2,250 \text{ psi (15.5 MPa)} \quad = 0.45 \cdot f'_c$$

$$f_t = +849 \text{ psi} \quad = 12\sqrt{f'_c} \quad \text{for support section}$$

$$\gamma = 0.82$$

$$\text{Required } S' = \frac{10,298,438}{0.82 \times 367 + 2,250} = 4,035.8 \text{ in}^3 \ (61,947 \text{ cm}^3)$$

$$\text{Required } S_b = \frac{M_D + M_{SD} + M_L}{f_t - \gamma f_{ci}} = \frac{10,298,438}{849 + 0.82 \times 2,250}$$

$$= 3,823.0 \text{ in}^3 \ (62,713 \text{ cm}^3)$$

First Trial. Since the required $S' = 4,035.8$, which is greater than the available S' in Example 4.2, choose the next larger I-section with $h = 44$ in. as shown in Figure 4.9. The section properties are:

$$I_c = 92,700 \text{ in}^4$$

$$r^2 = 228.9 \text{ in}^2$$

$$A_c = 405 \text{ in}^2$$

$$c_t = 23.03 \text{ in.}$$

$$S' = 4,030 \text{ in}^3$$

$$c_b = 20.97 \text{ in.}$$

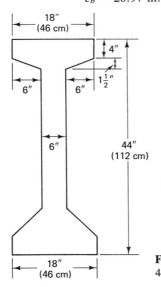

Figure 4.9 I-beam section in Example 4.3.

$$S_b = 4,420 \text{ in}^3$$

$$W_D = 422 \text{ plf}$$

From Equation 4.5c, the required eccentricity at the critical section at the support is

$$e_e = (f_{ti} - \bar{f}_{ci})\frac{S^t}{P_i}$$

where

$$\bar{f}_{ci} = f_{ti} - \frac{c_t}{h}(f_{ti} - f_{ci})$$

$$= 367 - \frac{23.03}{44}(367 + 2,250) = -1,002 \text{ psi (6.9 MPa)}$$

and

$$P_i = A_c \bar{f}_{ci} = 405 \times 1,002 = 405,810 \text{ lb (1,805 kN)}$$

Hence,

$$e = (367 + 1,002)\frac{4,030}{405,810} = 13.60 \text{ in. (346 mm)}$$

The required prestressed steel area is

$$A_p = \frac{P_i}{f_{pi}} = \frac{405,810}{189,000} = 2.15 \text{ in}^2 \text{ (14.4 cm}^2)$$

So we try $\frac{1}{2}''$ strands. The required number of strands is $2.15/0.153 = 14.05$. Accordingly, use fourteen $\frac{1}{2}''$ (12.7 mm) strands. As a result,

$$P_i = 14 \times 0.153 \times 189,000 = 404,838 \text{ lb (1,801 kN)}$$

(a) Analysis of Stresses at Transfer at End Section. From Equation 4.1a,

$$f^t = -\frac{P_i}{A_c}\left(1 - \frac{ec_t}{r^2}\right) - \frac{M_D}{S^t} = -\frac{404,838}{405}\left(1 - \frac{13.60 \times 23.03}{228.9}\right) - 0$$

$$= +368.2 \text{ psi } (T) \cong f_{ti} = 367, \text{ O.K.}$$

From Equation 4.2b,

$$f_b = -\frac{P_i}{A_c}\left(1 + \frac{ec_b}{r^2}\right) + \frac{M_D}{S_b} = -\frac{404,838}{405}\left(1 + \frac{13.6 \times 20.97}{228.9}\right) + 0$$

$$= -2,245.0 \text{ psi } (C) \cong f_{ci} = -2,250, \text{ O.K.}$$

(b) Analysis of Final Service-Load Stresses at Support

$$P_e = 14 \times 0.153 \times 154,980 = 331,967 \text{ lb (1,477 kN)}$$

Total moment $M_T = M_D + M_{SD} + M_L = 0$

From Equation 4.3a,

$$f^t = -\frac{P_e}{A_c}\left(1 - \frac{ec_t}{r^2}\right) - \frac{M_T}{S^t}$$

$$= -\frac{331{,}967}{405}\left(1 - \frac{13.60 \times 23.03}{228.9}\right) + 0 = 301.9 \text{ psi } (T) < f_t = 849 \text{ psi, O.K.}$$

From Equation 4.3b

$$f_b = -\frac{P_e}{A_c}\left(1 + \frac{ec_b}{r^2}\right) + \frac{M_T}{S_b}$$

$$= -\frac{331{,}967}{405}\left(1 + \frac{13.60 \times 20.97}{228.9}\right) + 0$$

$$= -1{,}840.9 \text{ psi } (12.2 \text{ MPa}) \ (C) < f_c = -2{,}250 \text{ psi, O.K.}$$

(c) Analysis of Final Service-Load Stresses at Midspan. From before, the total moment $M_T = M_D + M_{SD} + M_L = 10{,}298{,}438$ in.-lb. So the extreme concrete fiber stress due to M_T is

$$f_1^t = \frac{M_T}{S^t} = -\frac{10{,}298{,}438}{4{,}030} = -2{,}555.4 \text{ psi } (C) \ (17.6 \text{ MPa})$$

$$f_{1b} = \frac{M_T}{S_b} = \frac{10{,}298{,}438}{4{,}420} = +2{,}330.0 \text{ psi } (T) \ (16.1 \text{ MPa})$$

Hence, the final midspan fiber stresses are

$$f^t = +301.9 - 2{,}555.4 = -2{,}254 \text{ psi } (C) \cong f_c = -2{,}250 \text{ psi, accept}$$

$$f_b = -1{,}840.9 + 2{,}330 = +489.1 \text{ psi } (T) < f_t = 849 \text{ psi, O.K.}$$

Consequently, accept the trial section with a constant eccentricity $e = 13.60$ in. (345 mm) for the fourteen $\frac{1}{2}''$ (12.7 mm dia.) strands of tendon.

4.4 PROPER SELECTION OF BEAM SECTIONS AND PROPERTIES

4.4.1 General Guidelines

Unlike steel rolled sections, prestressed sections are not yet fully standardized. In most cases the design engineer has to select the type of section to be used in the particular project. In the majority of simply supported beam designs, the distance between the cgc and cgs lines, viz., the eccentricity e, is proportional to the required prestressing force. Since the midspan moment usually controls the design, the larger the eccentricity at midspan the smaller is the needed prestressing force, and consequently the more economical is the design. For a large eccentricity, a large concrete area at the top fibers is needed. Hence, a T-section or a wide-flange I-section becomes suitable. The end section is usually solid in order to avoid large eccentricities at planes of zero moment, and also in order to increase the shear capacity of the support section.

Another popular section in wide use is the double-T section. This section adds the advantages of the single-T section to its own ease of handling and erection inherent in its stability. Figure 4.10 shows typical sections in general usage. Other sections such as hollow-core slabs and nonsymmetrical sections are also commonly used. Note

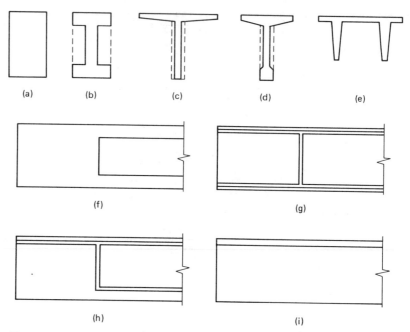

Figure 4.10 Typical prestressed concrete sections. (a) Rectangular beam section. (b) I-beam section. (c) T-beam section. (d) T-section with heavy bottom flange. (e) Double-T section. (f) End part of beam in (b). (g) End part of beam in (c). (h) End part of beam in (d). (i) End part of beam in (e).

that flanged sections can replace rectangular solid sections of the same depth without any loss of flexural strength. Rectangular sections, however, are used as short-span supporting girders or ledger beams.

I-sections are used as typical floor beams with composite slab topping action in long-span parking structures. T-sections with heavy bottom flanges, such as that in Figure 4.10(d), are generally used in bridge structures. Double-T sections are widely used in floor systems in buildings and also in parking structures, particularly because of the composite action advantage of the top wide flange, which is 10 feet wide in many cases.

Hollow-core cast and extruded sections are shallow one-way beam strips that serve as easily erectable floor slabs. Large, hollow box girders are used as bridge girders for very large spans in what are known as *segmental bridge deck systems*. These segmental girders have large torsional resistance, and their flexural strength-to-weight ratio is relatively higher than in other types of prestressing systems.

4.4.2 Gross Area, the Transformed Section, and the Presence of Ducts

In general, the gross cross-sectional area of the concrete section is adequate for use in the service-load design of prestressed sections. While some designers prefer refining their designs through the use of the transformed section in their solutions, the accuracy gained in accounting for the contribution of the area of the reinforcement to the stiffness of the concrete section is normally not warranted. In post-tensioned beams, where ducts are grouted, the gross cross section is still adequate for all practical design considerations. It is only in cases of large-span bridge and industrial prestressed beams, where the area of the prestressing reinforcement is large, that the transformed section or the net concrete area excluding the duct openings has to be used.

4.4.3 Envelopes for Tendon Placement

The tensile stress in the extreme concrete fiber under service-load conditions cannot exceed the maximum allowable by codes such as the ACI, PCI, AASHTO, or CEB-FIP. It is therefore important to establish the limiting zone in the concrete section, i.e., an envelope within which the prestressing force can be applied without causing tension in the extreme concrete fibers. From Equation 4.1a, we have

$$f_t = 0 = -\frac{P_i}{A_c}\left(1 - \frac{ec_t}{r^2}\right)$$

for the prestressing force part only, giving $e = r^2/c_t$. Hence, the *lower kern point*

$$k_b = \frac{r^2}{c_t} \qquad (4.6\ a)$$

Similarly, from Equation 4.1b, if $f_b = 0$, $-e = r^2/c_b$, where the negative sign represents measurements upwards from the neutral axis, since positive eccentricity is positive downwards. Hence, the *upper kern point*

$$k_t = \frac{r^2}{c_b} \qquad (4.6\ b)$$

From the determination of the upper and lower kern points, it is clear that

Super CIDS offshore platform under tow to Arctic, Global Marine Development. (*Courtesy*, Ben C. Gerwick.)

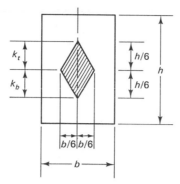

Figure 4.11 Central kern area for a rectangular section.

(a) If the prestressing force acts below the lower kern point, tensile stresses result at the extreme upper concrete fibers of the section.

(b) If the prestressing force acts above the upper kern point, tensile stresses result at the extreme lower concrete fibers of the section.

In a similar manner, kern points can be established for the right and left of the vertical line of symmetry of a section so that a central kern or core area for load application can be established, as Figure 4.11 shows for a rectangular section.

4.4.4 Advantages of Curved or Harped Tendons

Although straight tendons are widely used in precast beams of moderate span, the use of curved tendons is more common in in-situ-cast post-tensioned elements. Nonstraight tendons are of two types:

(a) Draped: gradually curved alignment such as parabolic forms, used in beams subjected primarily to uniformly distributed external loading.

(b) Harped: inclined tendons with a discontinuity in alignment at planes of concentrated load applications, used in beams subjected primarily to concentrated transverse loading.

Figures 4.12, 4.13, and 4.14 describe the alignment bending moment and stress distribution for beams that are prestressed with straight, draped, and harped tendons, respectively. These diagrams are intended to illustrate the economic advantages of the draped and harped tendons over the straight tendons. In Figure 4.12, at section 1-1 undesirable tensile stress in the concrete is shown at the top fibers. Section 1-1 in Figures 4.13 and 4.14 show the uniform compression if the tendon acts at the cgc of the section at the support. Another advantage of draped and harped tendons is that they allow the prestressed beams to carry heavy loads because of the balancing effect of the vertical component of the prestressing nonstraight tendon. In other words, the required prestressing forces P_p for the parabolic tendon in Figure 4.13 and P_h for the harped tendon in Figure 4.14 are smaller at the midspan than the force required in the straight tendon of Figure 4.12. Hence, for the same stress level, a smaller number of strands are needed in the case of draped or harped tendons, and sometimes smaller

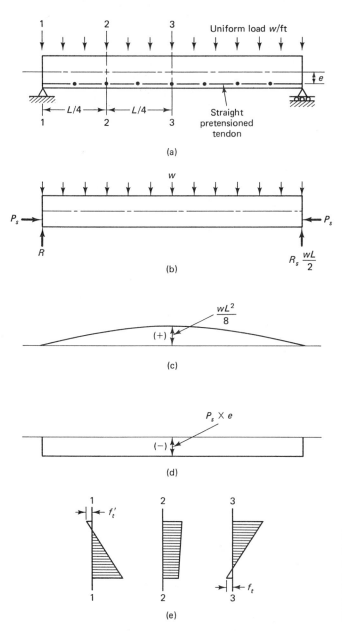

Figure 4.12 Beam with straight tendon. (a) Beam elevation. (b) Free-body diagram. (c) External load balancing moment diagram. (d) Prestressing force bending moment diagram. (e) Typical stress distribution in sections 1, 2, and 3 (Equation 4.3).

concrete sections can be used with the resulting efficiency in the design. (Compare Examples 4.2 and 4.3 again.)

4.4.5 Limiting-Eccentricity Envelopes

It is desirable that the designed eccentricities of the tendon along the span be such that limited or no tension develops at the extreme fibers of the beam controlling sections. If it is desired to have no tension along the span of the beam in Figure 4.6 with the

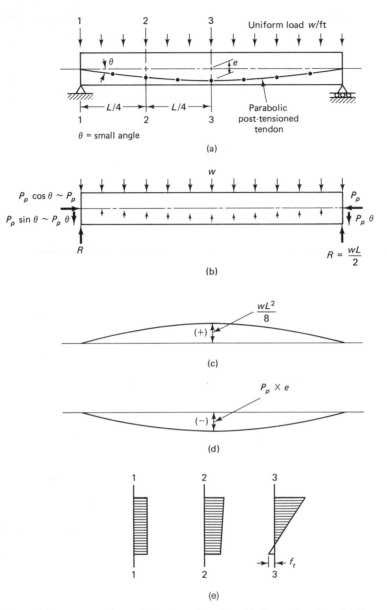

Figure 4.13 Beam with parabolic draped tendon. (a) Beam elevation. (b) Free-body diagram. (c) External load bending moment diagram. (d) Prestressing force bending moment diagram. (e) Typical stress distribution on sections 1, 2, and 3 (Equation 4.3).

draped tendon, the controlling eccentricities have to be determined at the sections that follow along the span. If M_D is the self-weight dead-load moment and M_T is the total moment due to all transverse loads, then the arms of the couple composed of the center-of-pressure line (C-line) and the center of the prestressing tendon line (cgs line) due to M_D and M_L are a_{max} and a_{min}, respectively, as shown in Figure 4.15.

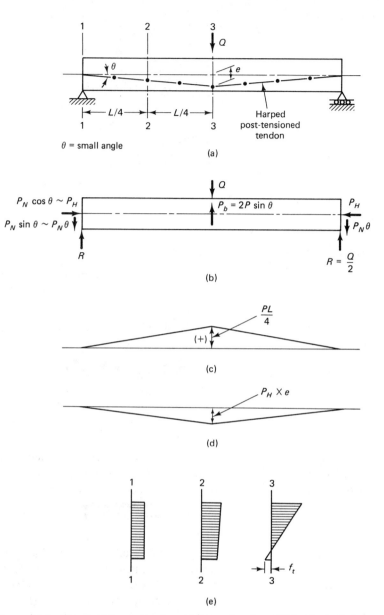

Figure 4.14 Beam with harped tendon. (a) Beam elevation. (b) Free-body diagram. (c) External load bending moment diagram. (d) Prestressing force bending moment diagram. (e) Typical stress distribution on sections 1, 2, and 3 (Equation 4.3).

Lower cgs Envelope. The minimum arm of the tendon couple is

$$a_{min} = \frac{M_D}{P_i} \tag{4.7 a}$$

This defines the maximum distance below the *bottom* kern where the cgs line is to be

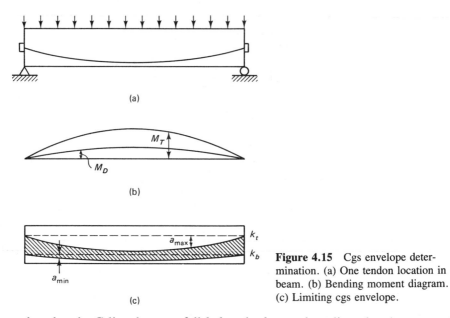

Figure 4.15 Cgs envelope determination. (a) One tendon location in beam. (b) Bending moment diagram. (c) Limiting cgs envelope.

located so that the C-line does not fall *below* the *bottom* kern line, thereby preventing tensile stresses at the *top* extreme fibers. Hence, the limiting bottom eccentricity is

$$e_b = (a_{\min} + k_b) \tag{4.7 b}$$

Upper cgs Envelope. The maximum arm of the tendon couple is

$$a_{\max} = \frac{M_T}{P_e} \tag{4.7 c}$$

This defines the minimum distance below the *top* kern where the cgs line is to be located so that the C-line does not fall *above* the *top* kern, thereby preventing tensile stresses at the *bottom* extreme fibers. Hence, the limiting top eccentricity is

$$e_t = (a_{\max} - k_t) \tag{4.7 d}$$

Limited tensile stress is allowed in some codes both at transfer and at service-load levels. In such cases, it is possible to allow the cgs line to fall slightly outside the two limiting cgs envelopes described in Equations 4.7a and c.

If an additional eccentricity e_b', e_t' is superimposed on the cgs-line envelope that results in limited tensile stress at both the top and bottom extreme concrete fibers, the additional top stress $f^{(t)}$ and bottom stress $f_{(b)}$ would be

$$f^{(t)} = \frac{P_i e_b' c_t}{I_c} \tag{4.8 a}$$

and

$$f_{(b)} = \frac{P_e e_t' c_b}{I_c} \tag{4.8 b}$$

where t and b denote the top and bottom fibers, respectively. From Equation 4.6, the additional eccentricities to be added to Equations 4.7b and d would be

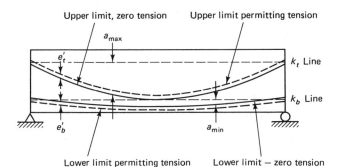

Upper limit, zero tension   Upper limit permitting tension

a_{max}

e'_t

k_t Line

k_b Line

e'_b a_{min}

Lower limit permitting tension   Lower limit – zero tension

Figure 4.16 Envelope permitting tension in concrete extreme fibers.

$$e'_b = \frac{f^{(t)} A_c k_b}{P_i} \qquad (4.9\ a)$$

and

$$e'_t = \frac{f_{(b)} A_c k_t}{P_e} \qquad (4.9\ b)$$

The envelope allowing limited tension is shown in Figure 4.16.

4.4.6 Prestressing Tendon Envelopes

Example 4.4

Suppose that the beam in Example 4.2 is a post-tensioned bonded beam and that the prestressing tendon is draped in a parabolic shape. Determine the limiting envelope for tendon location such that the limiting concrete fiber stresses are at no time exceeded. Assume that the magnitude of prestress losses is the same as in Example 4.2.

Solution   The design stresses of the I-beam in Example 4.2 can be summarized together with the section properties needed here:

$$P_i = 376,110 \text{ lb } (1,673 \text{ kN})$$

$$P_e = 308,255 \text{ lb } (1,371 \text{ kN})$$

$$M_D = 2,490,638 \text{ in.-lb } (281 \text{ kN-m})$$

$$M_{SD} + M_L = 7,605,000 \text{ in.-lb } (859 \text{ kN-m})$$

$$M_T = M_D + M_{SD} + M_L = 10,095,638 \text{ in.-lb } (1,141 \text{ kN-m})$$

$$A_c = 377 \text{ in}^2 \ (2,536 \text{ cm}^2)$$

$$f'_c = 6,000 \text{ psi}$$

$$r^2 = 187.5 \text{ in}^2 \ (1,210 \text{ cm}^2)$$

$$c_t = 21.16 \text{ in. } (537 \text{ mm})$$

$$c_b = 18.84 \text{ in. } (479 \text{ mm})$$

Since bending moments in this example are due to a uniformly distributed load, the shape of the bending moment diagram is parabolic, with the moment value being zero at the simply supported ends. Hence, quarter-span moments are

$$M_D = 0.75 \times 2,490,635 = 1,867,979 \text{ in.-lb } (211 \text{ kN-m})$$

$$M_{SD} + M_L = 0.75 \times 10,095,638 = 7,571,729 \text{ in.-lb } (856 \text{ kN-m})$$

From Equations 4.6a and b, the kern point limits are

$$k_t = \frac{r^2}{c_b} = \frac{187.5}{18.84} = 9.95 \text{ in. } (253 \text{ mm})$$

$$k_b = \frac{r^2}{c_t} = \frac{187.5}{21.16} = 8.86 \text{ in. } (225 \text{ mm})$$

From Equation 4.7a, the minimum distance that the cgs line is to be placed below the *bottom* kern to prevent tensile stress at the top fibers is determined as follows:

(i) *Midspan*

$$a_{min} = \frac{M_D}{P_i} = \frac{2,490,638}{376,110} = 6.62 \text{ in. } (168 \text{ mm})$$

giving

$$e_1 = k_b + a_{min} = 8.86 + 6.62 = 15.48 \text{ in. } (393 \text{ mm})$$

(ii) *Quarter span*

$$a_{min} = \frac{1,867,979}{376,110} = 4.97 \text{ in. } (126 \text{ mm})$$

giving

$$e_2 = 8.86 + 4.97 = 13.83 \text{ in. } (351 \text{ mm})$$

(iii) *Support*

$$a_{min} = 0$$

giving

$$e_3 = 8.86 + 0 = 8.86 \text{ in. } (225 \text{ mm})$$

From Equation 4.7b, the minimum distance that the cgs line is to be placed below the *top* kern to prevent tensile stress at the bottom extreme fibers is determined as follows:

(i) *Midspan*

$$a_{max} = \frac{M_T}{P_E} = \frac{10,095,638}{308,255} = 32.75 \text{ in. } (832 \text{ mm})$$

$$e_1 = a_{max} - k_t = 32.75 - 9.95 = 22.80 \text{ in. } (579 \text{ mm})$$

But $c_b = 18.84$ in. Hence, e_1 cannot exceed $c_b - 3.0 = 18.84 - 3 = 15.84$ in. (482 mm). Accordingly, use $e_1 = 15.84$ in.

(ii) *Quarter span*

$$a_{max} = \frac{7,571,729}{308,255} = 24.56 \text{ in. } (624 \text{ mm})$$

$$e_2 = 24.56 - 9.95 = 14.61 \text{ in. } (371 \text{ mm})$$

Flexural Design of Prestressed Concrete Elements   Chap. 4

(iii) *Support*

$$a_{\max} = 0$$

$$e_3 = 0 - 9.95 = -9.95 \text{ in. } (-253 \text{ mm}) \qquad (9.95 \text{ in. above cgs line})$$

Now, assume for practical purposes that the maximum fiber tensile stresses under working-load conditions for the purpose of constructing the cgs envelopes does not exceed $f_t = 6\sqrt{f'_c} = 465$ psi for both top and bottom fibers, since $f'_c = 6{,}000$ psi from Example 4.2. From Equation 4.9a, this additional eccentricity to add to the *lower* cgs envelope in order to allow limited tension at the *top* fibers is

$$e'_b = \frac{f^{(t)} A_c k_c}{P_i} = \frac{465 \times 377 \times 8.86}{376{,}110} = 4.13 \text{ in. } (105 \text{ mm})$$

Similarly, from Equation 4.9b, the additional eccentricity to add to the *upper* cgs envelope in order to allow limited tension at the *bottom* fibers is

$$e'_t = \frac{f_{(b)} A_c k_t}{P_e} = \frac{465 \times 377 \times 9.95}{308{,}255} = 5.66 \text{ in. } (144 \text{ mm})$$

We thus have the following summary of cgs envelope eccentricities:

	Zero tension, in.	Allowable tension, in.
Midspan		
Lower envelope	15.48 + 4.13	19.61
Upper envelope	15.84 − 5.66	10.18
Quarter span		
Lower envelope	13.83 + 4.13	17.96
Upper envelope	14.61 − 5.66	8.95
Support		
Lower envelope	8.86 + 4.13	12.99
Upper envelope	−9.95 − 5.66	−15.61

Actual midspan eccentricity $e_c = 15$ in. < 19.61 in.
Hence, tendon is inside envelope at midspan.
Actual support eccentricity $e_e = 11.94$ in. < 12.99 in.
Hence, tendon is also inside envelope at support.

Figure 4.17 illustrates the band of the cgs envelopes for both zero and limited tension in the concrete.

4.4.7 Reduction of Prestress Force Near Supports

As seen from Example 4.3 and Sections 4.4.3 and 4.4.5, straight tendons in pretensioned members can cause high tensile stresses in the concrete extreme fibers at the support sections because of the absence of bending moment stresses due to self-weight and superimposed loads and the dominance of the moment due to the prestressing force alone. Two common and practical methods of reducing the stresses at the support section due to the prestressing force are:

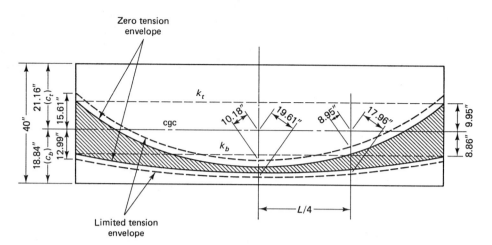

Figure 4.17 Cgs-line envelopes for the prestressing tendon (1 in. = 25.4 mm).

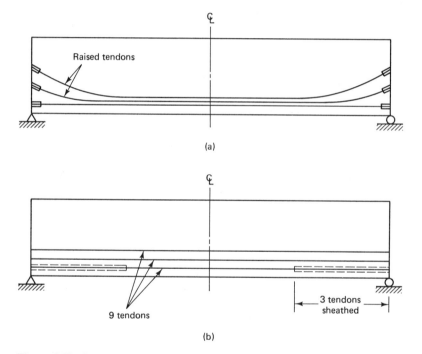

Figure 4.18 Reduction of prestressing force near supports. (a) Raising part of the tendons. (b) Sheathing part of the tendons.

1. Changing the eccentricity of some of the cables by raising them towards the support zone, as shown in Figure 4.18(a). This reduces the moment values.

2. Sheathing some of the cables by plastic tubing towards the support zone, as shown in Figure 4.18(b). This eliminates the prestress transfer of part of the cables at some distance from the support section of the simply supported prestressed beam.

Note that raised cables are also used in long-span post-tensioned prestressed beams, theoretically discontinuing part of the tendons where they are no longer needed by raising them upwards. Additional frictional losses due to these additional curvatures have to be accounted for in the design (analysis) of the section.

4.5 END BLOCKS AT SUPPORT ANCHORAGE ZONES

4.5.1 Stress Distribution

A large concentration of compressive stress in the longitudinal direction occurs at the support section on a small segment of the face of the beam end, both in pretensioned and post-tensioned beams, due to the large tendon prestressing forces. In the *pretensioned beams* the concentrated load transfer of the prestressing force to the surrounding concrete gradually occurs over a length ℓ_t from the face of the support section until it becomes essentially uniform.

In *post-tensioned beams,* this manner of gradual load distribution and transfer is not possible since the force acts directly on the face of the end of the beam through bearing plates and anchors. Also, some or all of the tendons in the post-tensioned beams are raised or draped towards the top fibers through the web part of the concrete section.

As the nongradual transition of the longitudinal compressive stress from concentrated to linearly distributed produces high transverse *tensile* stresses in the vertical (transverse) direction, longitudinal bursting cracks develop at the anchorage zone. When the stresses exceed the modulus of rupture of the concrete, the end block has to split (crack) longitudinally unless appropriate vertical reinforcement is provided. The location of the concrete bursting stresses and the resulting bursting cracks as well as the surface spalling cracks would thus have to be dependent on the location and distribution of the horizontal concentrated forces applied by the prestressing tendons to the end bearing plates.

It can sometimes become necessary to increase the area of the section towards the support by a gradual transition of the web to a width at the support equal to the

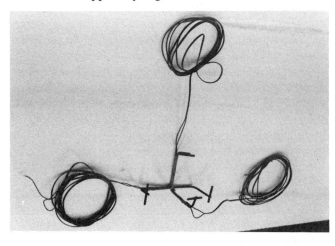

Three-dimensional instrumentation for end-block stress determination (Nawy et al.).

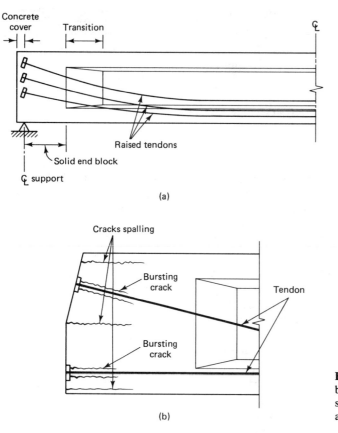

(a)

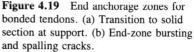

(b)

Figure 4.19 End anchorage zones for bonded tendons. (a) Transition to solid section at support. (b) End-zone bursting and spalling cracks.

flange width, in order to accommodate the raised tendons (see Figure 4.19(a)). Such an increase in the cross-sectional area does not contribute, however, to preventing bursting or spalling cracks, and has no effect on reducing the transverse tension in the concrete. In fact, both test results and the theoretical analysis of this three-dimensional stress problem demonstrate that the tensile stresses are often increased if a solid end-block section is used.

Consequently, it is reasonable to maintain the size of the web to the end support wherever possible in post-tensioned as well as pretensioned beams, while providing the necessary anchorage reinforcement in the load transfer zone in the form of *closed* ties or stirrups enclosing all the main prestressing and mild nonprestressed longitudinal reinforcement. However, it is at the same time advisable to insert reinforcing vertical mats close to the end face behind the bearing plates in the case of post-tensioned beams. If the design has to follow AASHTO requirements for bridges, end blocks are still required as in Figure 4.19(a). Typical stress contours of equal vertical stress based on three-dimensional analysis and test results from Ref. 4.6 are shown in Figure 4.20, and an idealization of the tensile and compressive stress paths is shown in Figure 4.21.

4.5.2 Development and Transfer Length in Pretensioned Members

As the jacking force is released in pretensioned members, the prestressing force is dynamically transferred through the bond interface to the surrounding concrete. The

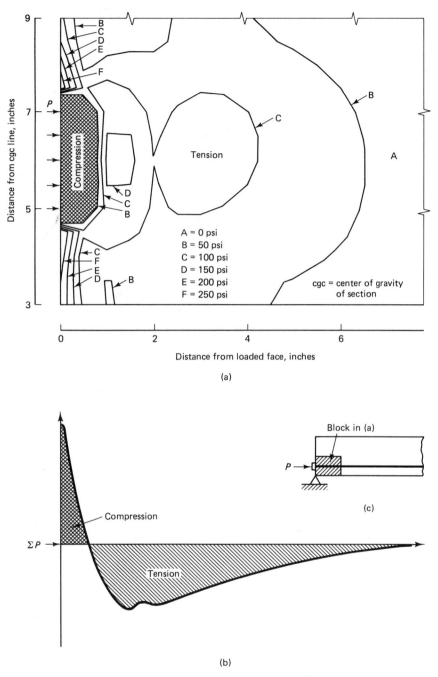

Figure 4.20 Principal tensile stress contours of equal vertical stress at anchorage zone (eccentricity $e_e = 6$ in.). (a) Contours of stress. (b) Stress distribution at 4.5 in. above base. (c) Segment of beam elevation.

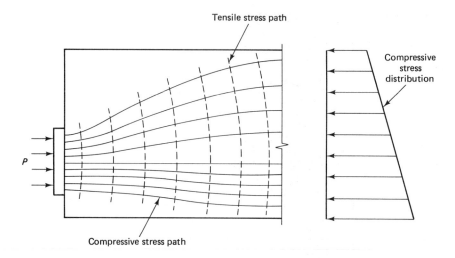

Figure 4.21 Idealized tensile and compressive stress paths at end blocks.

interlock or adhesion between the prestressing tendon circumference and the concrete over a finite length of the tendon gradually transfers the concentrated prestressing force to the entire concrete section at planes away from the end block and towards the midspan. The length of embedment determines the magnitude of prestress that can be developed along the span: the larger the embedment length, the higher is the prestress developed.

As an example for a half-inch seven-wire strand, an embedment of 40 in. (102 cm) develops a stress of 180,000 psi (1,241 MPa), whereas an embedment of 70

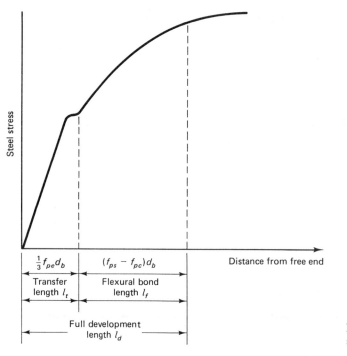

Figure 4.22 Development length for prestressing strand.

End block of post-tensioned I-beam at ultimate load (Nawy et al.).

Anchorage block instrumentation (Nawy et al.).

in. (178 cm) develops a stress of 206,000 psi (1,420 MPa). From Figure 4.22, it is plain that the embedment length ℓ_d that gives the full development of stress is a combination of the transfer length ℓ_t and the flexural bond length ℓ_f. These are given respectively by

$$\ell_t = \frac{1}{1000}\left(\frac{f_{pe}}{3}\right)d_b \qquad (4.10\ \text{a})$$

or

$$\ell_t = \frac{f_{pe}}{3000}d_b \qquad (4.10\ \text{b})$$

and

$$\ell_f = \frac{1}{1,000}(f_{ps} - f_{pe})d_b \qquad (4.10\ \text{c})$$

where f_{ps} = stress in prestressed reinforcement at nominal strength (psi)
f_{pe} = effective prestress after losses (psi)
d_b = nominal diameter of prestressing tendon (in.).

Combining Equations 4.10a and 4.10c gives

$$\text{Min } \ell_d = \frac{1}{1,000}\left(f_{ps} - \frac{2}{3}f_{pe}\right)d_b \qquad (4.10\text{ d})$$

Equation 4.10d gives the minimum required development length for prestressing strands. If part of the tendon is sheathed towards the beam end to reduce the concentration of bond stesses near the end, the stress transfer in that zone is eliminated and an increased adjusted development length ℓ_d is needed.

4.5.3 Design of Reinforcement

4.5.3.1 Post-tensioned beams. Figure 4.23 shows the end-block forces and the fiber stresses due to the prestressing force P_i, as well as the bending moment value for each possible crack height y above the beam bottom CD. The maximum moment value M_{max} determines the potential position of the horizontal bursting crack. This moment is resisted by the couple provided by the tensile force T of the vertical anchorage zone reinforcement and the compressive force C provided by the end-block concrete, while the horizontal shear force V at the crack split surface is resisted by the aggregate interlock forces. From practical observations, the vertical anchorage zone stirrups that provide the force T should be distributed over a zone width $h/2$ from the end face of the beam such that X in Figure 4.23 can vary between $h/4$ and $h/5$.
From equilibrium of moments,

$$T = \frac{M_{max}}{h - x} \qquad (4.11)$$

and the total required area of vertical steel reinforcement becomes

$$A_t = \frac{T}{f_s} \qquad (4.12)$$

where the steel stress f_s used in the calculation should not exceed 20,000 psi (138.5 MPa) for crack width control purposes. Furthermore, the permissible bearing stress f_b due to the high stress concentration under the anchorage bearing plates should not exceed the limiting values allowed both at the initial prestress and after losses as follows:

Initial Prestress

$$f_b = 0.8f'_{ci}\sqrt{A_2/A_1 - 0.2} \le 1.25f'_{ci} \qquad (4.13)$$

After Losses

$$f_b = 0.6f'_c\sqrt{A_2/A_1} \le f'_c \qquad (4.14)$$

where A_1 = bearing area of anchor plate for post-tensioned tendons
A_2 = maximum area of the portion of the anchorage surface that is geometrically similar to and *concentric* with the area of the anchor plate of the post-tensioning tendons

The width of area A_2 would normally be the width of the beam web.

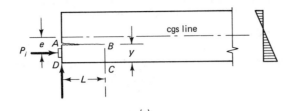

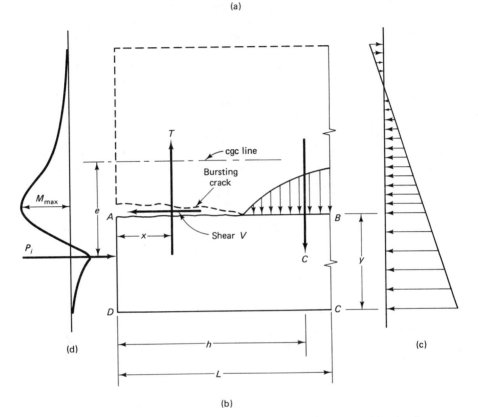

Figure 4.23 Post-tensioned beam end-block forces. (a) Beam elevation, showing transfer length L. (b) Free-body diagram ABCD. (c) Fiber stress distribution across beam block depth. (d) Moment values on crack surface AB for all possible locations γ across beam block depth.

4.5.3.2 Pretensioned beams. Based on laboratory tests, empirical expressions developed by Mattock et al. give the total stirrup force F as

$$F = 0.0106 \frac{P_i h}{\ell_t} \tag{4.15}$$

where h is the pretensioned beam depth and ℓ_t is the transfer length. If the average stress in a stirrup is taken as *half* the maximum permissible steel stress f_s, then $F = \frac{1}{2}A_t f_s$. Substituting this for F in Equation 4.15 gives

$$A_t = 0.021 \frac{P_i h}{f_s \ell_t} \tag{4.16}$$

where A_t is the total area of the stirrups and $f_s \leq 20,000$ psi (138 MPa) for crack-control purposes.

4.5.4 Design of End Anchorage for Post-tensioned Beams

Example 4.5

Design an end anchorage reinforcement for the post-tensioned beam in Example 4.4, giving the size, type, and distribution of reinforcement. Use $f'_c = 5,000$ psi (34.5 MPa) normal-weight concrete, and maintain the web section throughout the span.

Solution   Divide the beam depth into 4-in. increments of height as shown in Figure 4.24, and assume that the concrete stress at the center of each increment is uniform across the depth of the increment. Then calculate the incremental moments due to these internal stresses and due to the external prestressing force P_i about each horizontal plane in order to determine the *net* moment on the section. The net maximum moment will determine the position of the potential horizontal bursting crack and the reinforcement that has to be provided to prevent the crack from developing. Using a plus (+) sign for clockwise moment, the initial prestressing force before losses, from Example 4.4, is $P_i = 376,110$ lb (1,673 kN). From Figure 4.24, the concrete internal moment at the plane 4 in. from the bottom fibers is

$$M_{c4} = 2,117 \times 4 \times 18 \times (2 \text{ in.}) = 304,848 \text{ in.-lb}$$

$$= 0.3 \times 10^6 \text{ in.-lb (34.4 kN-m)}$$

and that at the plane 8 in. from the bottom fibers is

$$M_{c8} = 2,117 \times 4 \times 18 \times (6 \text{ in.}) + 1,851 \times 4 \times \frac{18 + 10}{2} \times (2 \text{ in.})$$

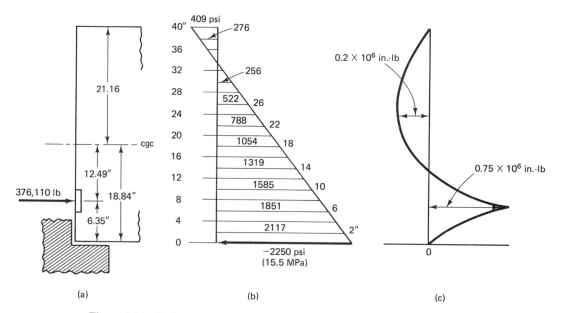

Figure 4.24   Anchorage zone stresses and moments in Example 4.5. (a) Post-tensioned end-block zone. (b) Concrete stress distribution across depth. (c) Net anchorage zone moments on potential horizontal cracking planes along the beam depth.

$$= 1,121,856 \text{ in.-lb} = 1.12 \times 10^6 \text{ in.-lb (126.8 kN-m)}$$

The prestressing force moment at the plane 8 in. from the bottom fibers is

$$M_{p8} = 376,110 \times (8 - 6.35) = -620,582 \text{ in.-lb}$$

$$= -0.62 \times 10^6 \text{ in.-lb (70.1 kN-m)}$$

The net moment is then $1.12 \times 10^6 - 0.62 \times 10^6 = 0.50 \times 10^6$ in.-lb (56.6 kN-m).

In a similar manner, we can find the net moment for all the other incremental planes at 4-in. increments to get the values tabulated in Table 4.5. From the table, the maximum net moments are $+M_{max} = 0.75 \times 10^6$ in.-lb (84.6 kN-m) at the horizontal plane 6.35 in. above the beam bottom fibers (bursting potential crack effect) and $-M_{max} = -0.20 \times 10^6$ in.-lb at the horizontal plane 24 in. (61 cm) above the beam bottom fibers (spalling potential crack effect).

From Equation 4.11, assuming that the center of the tensile vertical force T is at a distance $x = h/4 = 40/4 = 10$ in., we obtain

$$T = \frac{M_{max}}{h - x} = \frac{+0.75 \times 10^6}{40 - 10} = 25,000 \text{ lb (111 kN)}$$

Allowing a maximum steel stress, $f_s = 20,000$ psi.

The bursting zone reinforcement is

$$A_s = \frac{T_b}{f_s} = \frac{25,000}{20,000} = 1.25 \text{ in.}^2 \text{ (139.5 cm}^2)$$

TABLE 4.5 ANCHORAGE ZONE MOMENTS FOR EXAMPLE 4.5

Moment plane dist. d from bottom	Section width	Stress at plane *$(\bar{d} - 2.0)$	Concrete resistance force at *$(\bar{d} - 2.0)$	Moment M_p for a Pi about horiz. plane in col. (1)	Moment M_c of concrete in col. (4) about horiz. plane in col. (1)	Net moment $(M_c - M_p)$ col. (6) − col. (5)
in.	in.	psi	lb	in.-lb$\times 10^6$	in.-lb$\times 10^6$	in.-lb$\times 10^6$
(1)	(2)	(3)	(4)	(5)	(6)	(7)
0	8	−2,250	162,000	0	0	0
4	8	−2,117	152,424	0	+0.30	+0.30
6.35	13.3	−1,851	103,656	0	+0.75	+0.75
8	10	−1,585	63,400	−0.62	+1.12	+0.50
12	6	−1,319	31,656	−2.13	+2.25	+0.12
16	6	−1,054	25,296	−3.63	+3.54	−0.09
20	6	−788	18,912	−5.13	+4.94	−0.19
24	6	−522	12,528	−6.64	+6.44	−0.20
28	6	−256	6,144	−8.14	+7.99	−0.15
32	6	~0	~0	−9.65	+9.61	−0.04
36	17	+276	−18,768	−11.15	+11.13	−0.02
40	17	+409	−27,812	−12.66	+12.65	~0

*$\bar{d}$ = distance from plane about which moment is taken less half the depth of one slice (in this example slice depth = 4 in.).

Typical bursting crack of the anchorage zone (Nawy et al.).

So

Try #3 closed ties ($A_s = 2 \times 0.11 = 0.22$ in^2).

Required no. of stirrups $= \dfrac{1.25}{0.22} = 5.68$

Use six #3 ties in addition to what is required for shear.

The spalling zone force

$$T_s = \frac{-0.2 \times 10^6}{40 - 10} = 6{,}671 \text{ lb}$$

So

$$A_s = \frac{T_s}{f_s} = \frac{6{,}667}{20{,}000} = 0.33 \text{ in}^2 \ (2.2 \text{ cm}^2)$$

Hence, we have

$$\text{Req no. of \#3 stirrups} = \frac{0.33}{0.22} = 1.5$$

So use two #3 additional stirrups. Then

$$\text{Total number of stirrups} = 6 + 2 = 8$$

$$\text{Stirrup } zone = h/2 = 40/2 = 20 \text{ in. } (50.8 \text{ cm})$$

Space the #3 closed ties at 3 in. center to center with the first stirrup starting 3 in. from the beam end, as shown in Figure 4.25. Also, provide four #3 0–10 in. long bars at 3 in. center to center each way 2 in. from the end face at the anchor location, since cracking can occur vertically and horizontally. Add spiral reinforcement under the anchors if the manufacturer specifies that such a reinforcement is useful.

Next, check the bearing plate stresses.

At Transfer. From Equation 4.13,

$$f_b = 0.8 f'_{ci} \sqrt{A_2/A_1 - 0.2} \le 1.25 f'_{ci}$$

Try $A_1 = 5 \times 5 = 25$ in^2 (161 cm^2). Then the width of the web is 6 in. So try $A_2 = 6 \times 6 = 36$ in^2. Then

Flexural Design of Prestressed Concrete Elements   Chap. 4

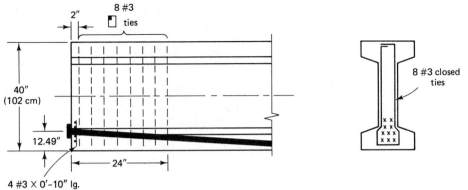

Figure 4.25 End anchorage reinforcement in Example 4.5. (a) Anchorage zone. (b) Beam cross section.

$$f'_{ci} = 3{,}750 \text{ psi}$$

$$1.25 f'_{ci} = 4{,}688 \text{ psi (32.3 MPa)}$$

$$f_b = 0.8 \times 3{,}750 \sqrt{36/25} - 0.2 = 3{,}340 \text{ psi} < 4{,}688 \text{ psi, O.K.}$$

At Service Load. From Equation 4.14,

$$f_b = 0.6 f'_c \sqrt{A_2/A_1} \le f'_c$$

$$= 0.6 \times 5{,}000 \sqrt{36/25} = 3{,}600 \text{ psi (24.8 MPa)} < 4{,}688 \text{ psi, O.K.}$$

Consequently, adopt bearing plate size 5" × 5" (12.7 cm × 12.7 cm).

4.5.5 Design of End Anchorage for Pretensioned Beams

Example 4.6

Assume the beam in Example 4.5 is pretensioned. Design the anchorage reinforcement needed to prevent bursting or spalling cracks from developing.
Solution

$$P_i = 376{,}110 \text{ lb (1,673 kN)}$$

From Equation 4.16,

$$A_t = 0.021 \frac{P_i h}{f_s \ell_t}$$

From Equation 4.10b, the transfer length is $\ell_t = (f_{pe}/3{,}000)d_b$. So since $f_{pe} = 154{,}980$ psi and $d_b = \frac{1}{2}$ in., we have

$$\ell_t = \frac{154{,}980}{3{,}000} \times 0.5 = 25.83 \text{ in. (66 cm)}$$

Now,

$$At = 0.021 \frac{P_t h}{f_s \ell_t}$$

So since $f_s \le 20{,}000$ psi, we get

$$A_t = 0.021 \frac{376,110 \times 40}{2,000 \times 25.83}$$

$$= 0.61 \text{ in}^2 \ (3.9 \text{ cm}^2)$$

Then we have:

Trying #3 closed ties,

$$2 \times 0.11 = 0.22 \text{ in}^2 \ (9.5 \text{ mm dia}) \text{ ties}$$

$$\text{Min no. of stirrups} = \frac{0.61}{0.22} = 2.78$$

Use three #3 ties to provide the envelope for all the main longitudinal reinforcement.

4.6 FLEXURAL DESIGN OF COMPOSITE BEAMS

Composite sections are normally precast, prestressed supporting elements over which sit situ-cast top slabs that act integrally with them. Composites have the advantage of the precast part becoming in many cases the falsework for supporting the situ-cast top slab and topping in bridges and industrial buildings, as shown in Figure 4.26. Sometimes the precast, prestressed element is shored during the placement and curing of the situ-cast top slab. In such a case, the slab weight acts only on the composite section, which has a substantially larger section modulus than the precast section. Hence, the concrete stress calculations have to take this situation into account in the design. The concrete stress distribution due to composite action can be seen in Figure 4.27, in part (e) of which the load taken by the cured composite section is the sum of $W_{SD} + W_L$.

4.6.1 Unshored Precast Beam Case

From Equations 4.2.a and b, the extreme concrete fiber stress equations before casting the top slab are

$$f^t = -\frac{P_e}{A_c}\left(1 - \frac{ec_t}{r^2}\right) - \frac{M_D + M_{SD}}{S^t} \tag{4.18 a}$$

and

$$f_b = -\frac{P_e}{A_c}\left(1 + \frac{ec_b}{r^2}\right) + \frac{M_D + M_{SD}}{S_b} \tag{4.18 b}$$

where S^t and S_b are the section moduli of the precast section only, and M_{SD} is any additional superimposed moment.

After the situ-cast slab hardens and composite action takes place, new higher moduli S_c^t and S_{cb} are available, with the cgc line moving upwards towards the top fibers. The concrete fiber stress counterparts to Equations 4.18 a and b for the extreme top and bottom fibers of the *precast* part of the composite section (level AA in Figure 4.27(e)) are

$$f^t = -\frac{P_e}{A_c}\left(1 - \frac{ec_t}{r^2}\right) - \frac{M_D + M_{SD}}{S^t} - \frac{M_{CSD} + M_L}{S_c^t} \tag{4.19 a}$$

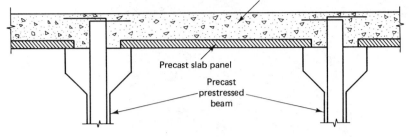

Figure 4.26 Composite prestressed concrete construction.

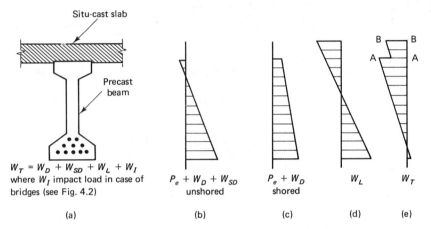

Figure 4.27 Flexural stress distribution in composite beams. (a) Composite beam. (b) Concrete stress distribution. (c) Concrete stress distribution with precast beam shored. (d) Live-load stress for shored case, or live load plus superimposed dead load for unshored case. (e) Final service-load stress due to all loads.

and

$$f_b = -\frac{P_e}{A_c}\left(1 + \frac{ec_b}{r^2}\right) + \frac{M_D + M_{SD}}{S_b} + \frac{M_{CSD} + M_L}{S_{cb}} \qquad (4.19 \text{ b})$$

where M_{CSD} is the additional composite superimposed dead load after erection, i.e., at service, and S_c^t and S_{cb} are the section moduli of the composite section at the level of the top and bottom fibers, respectively, of the precast section.

The fiber stresses at the level of the top and bottom fibers of the situ-cast slab (levels BB and AA of Figure 4.27(e)) are

$$f^{ts} = -\frac{M_{CSD} + M_L}{S_{cb}^t} \qquad (4.20 \text{ a})$$

and

$$f_{bs} = -\frac{M_{CSD} + M_L}{S_{bcb}} \qquad (4.20 \text{ b})$$

where $M_{CSD} + M_L$ are the incremental moments added after composite action has

Jacking tendons of post-tensioned beam
(Nawy et al.).

developed, and S_{cb}^t and S_{bcb} are the section moduli of the composite section for the top and bottom fibers AA and BB, respectively, of the slab in Figure 4.27(e).

4.6.2 Fully Shored Precast Beam Case

In cases where the span of a precast section is fully shored until composite action develops, the concrete fiber stresses before shoring and top slab casting become, from Equations 4.18a and b,

$$f^t = -\frac{P_e}{A_c}\left(1 - \frac{ec_t}{r^2}\right) - \frac{M_D}{S^t} \tag{4.21 a}$$

and

$$f_b = -\frac{P_e}{A_c}\left(1 + \frac{ec_b}{r^2}\right) + \frac{M_D}{S_b} \tag{4.21 b}$$

After the top slab is situ cast and full composite action is developed when the concrete hardens, Equations 4.19a and b become, for the beam shored after erection,

$$f^t = -\frac{P_e}{A_c}\left(1 - \frac{ec_t}{r^2}\right) - \frac{M_D + M_{SD} + M_{CSD} + M_L}{S_c^t} \tag{4.22 a}$$

and

$$f_b = -\frac{P_e}{A_c}\left(1 + \frac{ec_b}{r^2}\right) + \frac{M_D + M_{SD} + M_{CSD} + M_L}{S_{cb}} \tag{4.22 b}$$

Note that adequate check has to be made for the horizontal interface shear stresses between the situ-cast and the precast beams, as will be discussed in Chapter 5.

4.6.3 Effective Flange Width

In order to determine the theoretical composite action that resists the flexural stresses, a determination has to be made of the slab width that can effectively contribute to the stiffness increase resulting from composite action.

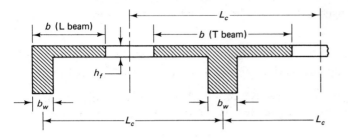

Figure 4.28 Effective flange width of composite section.

TABLE 4.6 VALUES OF EFFECTIVE FLANGE WIDTH

Width b as the least of the tabulated values
(Modify to $b_M = n_c b$, where $n_c = E_{ct}/E_c$ when
flange concrete is of different strength from that of the precast web

	End beam	Intermediate beam
ACI	$b_w + 6h_f$ $\frac{1}{2}(b_w + L_c)$ $b_w + \frac{L}{12}$	$b_w + 16h_f$ L_c $\frac{L}{4}$
AASHTO	$b_w + 6h_f$ $\frac{1}{2}(b_w + L_c)$ $b_w + \frac{L}{12}$	$b_w + 12h_f$ L_c $\frac{L}{4}$

L = span of end or intermediate beam

Figure 4.28 and Table 4.6 give the ACI and AASHTO requirements for determining the effective top flange width of the composite section. If the topping concrete is of different strength than that of the precast section, the width b has to be modified to account for the difference in the moduli of the two concretes in order to ensure that the strains in both materials at the interface are compatible. The modified width of the composite topping for calculating the composite I_{cc} is

$$b_m = \frac{E_{ct}}{E_c}(b) = n_c b \qquad (4.23)$$

where E_{ct} = modulus of the topping concrete
E_c = modulus of the precast concrete

Once the modified width b_m is defined, the entire composite section is considered to be of the higher strength concrete.

4.7 SUMMARY OF STEP-BY-STEP TRIAL-AND-ADJUSTMENT PROCEDURE FOR THE SERVICE-LOAD DESIGN OF PRESTRESSED MEMBERS

1. Given the superimposed dead-load intensity W_{SD}, the live-load intensity W_L, the span and the height limitation, the material strengths f_{pu}, f'_c, the type of concrete, and whether the prestress type is pretensioning or post-tensioning,

2. Assume the intensity of self-weight W_D, and calculate moments M_d, M_{SD}, and M_L.

3. Calculate $f_{pi}, f'_{ci}, f_{ti}, f_t$ and f_c, where $f_{pi} = 0.70 f_{pu}, f_{ci} = 0.60 f'_{ci}, f_{ti} = 3\sqrt{f'_{ci}}$, and $f_{ti} = 6\sqrt{f'_{ci}}$, for the support section, $f_c = 0.45 f'_c$ and $f_t = 6\sqrt{f'_c}$ to $12\sqrt{f'_c}$.

4. Calculate the prestress losses $\Delta f_{pT} = \Delta f_{pES} + \Delta f_{pR} + \Delta f_{pSH} + \Delta f_{pCR} + \Delta f_{pE} + \Delta f_{pA} + \Delta f_{pB}$ for the type of prestressing used. Determine net stresses $f_{pe} = f_{ps} - \Delta f_{pT}$.

5. Find the minimum required section modulus of the minimum efficient section for evaluating the concrete fiber stresses at the top and bottom fibers.

(a) For harped or draped tendons, use the midspan controlling section:

$$S^t \geq \frac{(1 - \gamma)M_D + M_{SD} + M_L}{\gamma f_{ti} - f_c}$$

$$S_b \geq \frac{(1 - \gamma)M_D + M_{SD} + M_L}{f_t - \gamma f_{ci}}$$

where

$$\gamma = \frac{P_e}{P_i}$$

(b) For straight tendons, use the end-support controlling section:

$$S^t \geq \frac{M_D + M_{SD} + M_L}{\gamma f_{ti} - f_c}$$

$$S_b \geq \frac{M_D + M_{SD} + M_L}{f_t - \gamma f_{ci}}$$

6. Select a trial section with section modulus properties close to those required in step 5 to be checked later for composite section fiber stress requirements.

7. For (a) the controlling section in the span (usually the midspan or at 0.4 of span), (b) the controlling section at the support, and (c) any other section along the span if both straight and draped tendons are used, analyze the concrete fiber stresses expected at stress transfer immediately before such transfer:

$$f^t = -\frac{P_i}{A_c}\left(1 - \frac{ec_t}{r^2}\right) - \frac{M_D}{S^t}$$

$$f_b = -\frac{P_i}{A_c}\left(1 + \frac{ec_b}{r^2}\right) + \frac{M_D}{S_b}$$

If the stresses exceed the allowable values, enlarge the section, or change the eccentricity e_c or e_e, or both.

8. Analyze the concrete fiber stresses for the service-load conditions, as in step 7.

$$f^t = -\frac{P_e}{A_c}\left(1 - \frac{ec_t}{r^2}\right) - \frac{M_T}{S^t}$$

$$f_b = -\frac{P_e}{A_c}\left(1 + \frac{ec_b}{r^2}\right) + \frac{M_T}{S_b}$$

where $M_T = M_D + M_{SD} + M_L$. If the stresses exceed the allowable values, enlarge the section or change the eccentricity e_c or e_e, or both.

9. For cases where many strands have to be used, establish the envelopes of limiting eccentricities for zero tension $e_b = (k_b + a_{min})$ and $e_t = (a_{max} - k_t)$, where $a_{min} = M_D/P_i$ and $a_{max} = M_T/P_e$. If the tension in the concrete is used in the design, add

$$e_b' = \frac{f_{(t)}A_c k_b}{P_i}$$

and

$$e_t' = \frac{f_{(b)}A_c k_t}{P_e}$$

to the bottom and top envelopes, respectively, where $f_{(t)}$ and $f_{(b)}$ are the extreme fiber stresses calculated to be in the concrete.

10. Investigate the end-block anchorage zone stresses, and design the necessary reinforcement to prevent bursting or spalling cracks. Determine the minimum development length

$$\ell_d = \frac{1}{1,000}\left(f_{ps} - \frac{2}{3}f_{pe}\right)d_b$$

of which the transfer length $\ell_t = \dfrac{f_{pe}}{3,000}d_b$, where f_{pe} is in psi units.

Post-tensioned Anchorage Zone

$$T = \frac{M_{max}}{h - x}$$

and the total vertical stirrups area $A_t = T/f_s$, where $f_s \le 20,000$ psi. The area of the end bearing plate A_1 is determined by bearing stress limitation:

$$f_b \le 0.8f_{ci}'\sqrt{A_2A_1} - 0.2 \le 1.25f_{ci}'$$

and

$$f_b \le 0.6f_c'\sqrt{A_2/A_1} \le f_c'$$

Pretensioned Prestress Transfer Zone

$$A_t = 0.021\frac{P_i h}{f_s \ell_t}$$

11. Determine the composite action stresses, and revise the section if these stresses exceed the maximum allowable concrete fiber stresses both in the precast section and the situ-cast top slab. Use the modified effective width $b_m = (E_{ct}/E_c)b$ for the top composite flange when calculating the section modulus of the composite section.

Unshored Precast Beam Case.   Before the top slab is situ cast:

$$f^t = -\frac{P_e}{A_c}\left(1 - \frac{ec_t}{r^2}\right) - \frac{M_D + M_{SD}}{S^t}$$

$$f_b = -\frac{P_e}{A_c}\left(1 + \frac{ec_b}{r^2}\right) + \frac{M_D + M_{SD}}{S_b}$$

After the top slab is cast and cured to develop full composite action, the stresses at top and bottom fibers of the *precast* part of the composite section will be

$$f^t = -\frac{P_e}{A_c}\left(1 - \frac{ec_t}{r^2}\right) - \frac{M_D + M_{SD}}{S^t} - \frac{M_{CSD} + M_L}{S_c^t}$$

$$f_b = -\frac{P_e}{A_c}\left(1 + \frac{ec_b}{r^2}\right) + \frac{M_D + M_{SD}}{S_b} + \frac{M_{CSD} + M_L}{S_{cb}}$$

where S_c^t and S_{cb} are the section moduli of the composite section at the level of the top and bottom fibers of the precast section. Also, M_L includes M_I if impact stresses exist.

The fibers stresses at the level of the top and bottom fibers of the situ-cast hardened slab are

$$f^{ts} = -\frac{M_{CSD} + M_L}{S_{cb}^t}$$

$$f_{bs} = -\frac{M_{CSD} + M_L}{S_{bcb}}$$

Photo layout of bridge deck prestressing tendon.

where S_{cb}^t and S_{bcb} are the section moduli of the composite section at the level of the top and bottom of the situ-cast slab.

Fully Shored Precast Beam Stem. Before shoring, and before the topping is situ-cast,

$$f^t = -\frac{P_e}{A_c}\left(1 - \frac{ec_t}{r^2}\right) - \frac{M_D}{S^t}$$

$$f_b = -\frac{P_e}{A_c}\left(1 + \frac{ec_b}{r^2}\right) + \frac{M_D}{S_b}$$

After the situ-cast slab is cured and *full composite action develops,*

$$f^t = -\frac{P_e}{A_c}\left(1 - \frac{ec_t}{r^2}\right) - \frac{M_D + M_{SD} + M_{CSD} + M_L}{S_c^t}$$

$$f_b = -\frac{P_e}{A_c}\left(1 + \frac{ec_b}{r^2}\right) + \frac{M_D + M_{SD} + M_{CSD} + M_L}{S_{cb}}$$

The effective width of the top flange of the composite section is determined in accordance with the applicable ACI or AASHTO specifications.

12. Proceed to determine the strength of the section for the limit state at failure and for shear and torsional strength.

Figure 4.29 shows a flowchart for the service-load flexural design of prestressed beams.

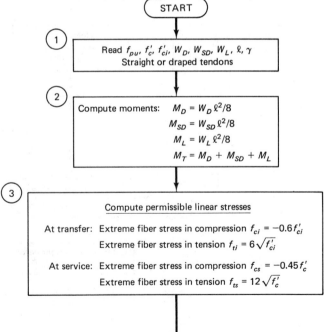

Figure 4.29 Flowchart for service-load flexural design of prestressed beams.

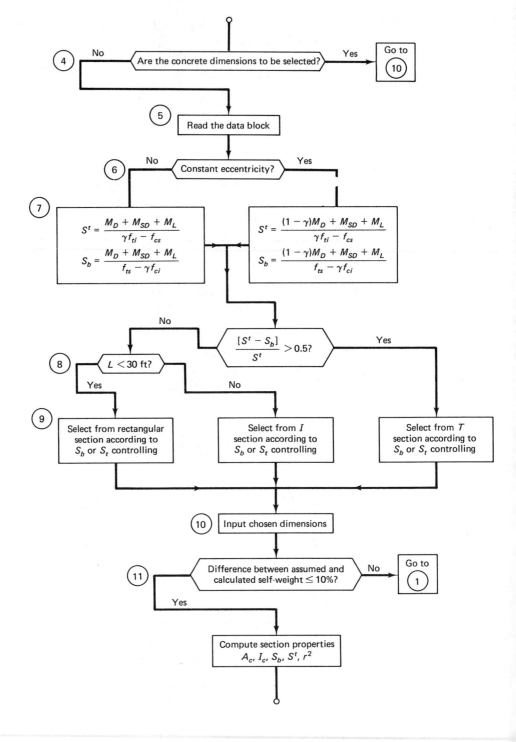

Figure 4.29 (*continued*)

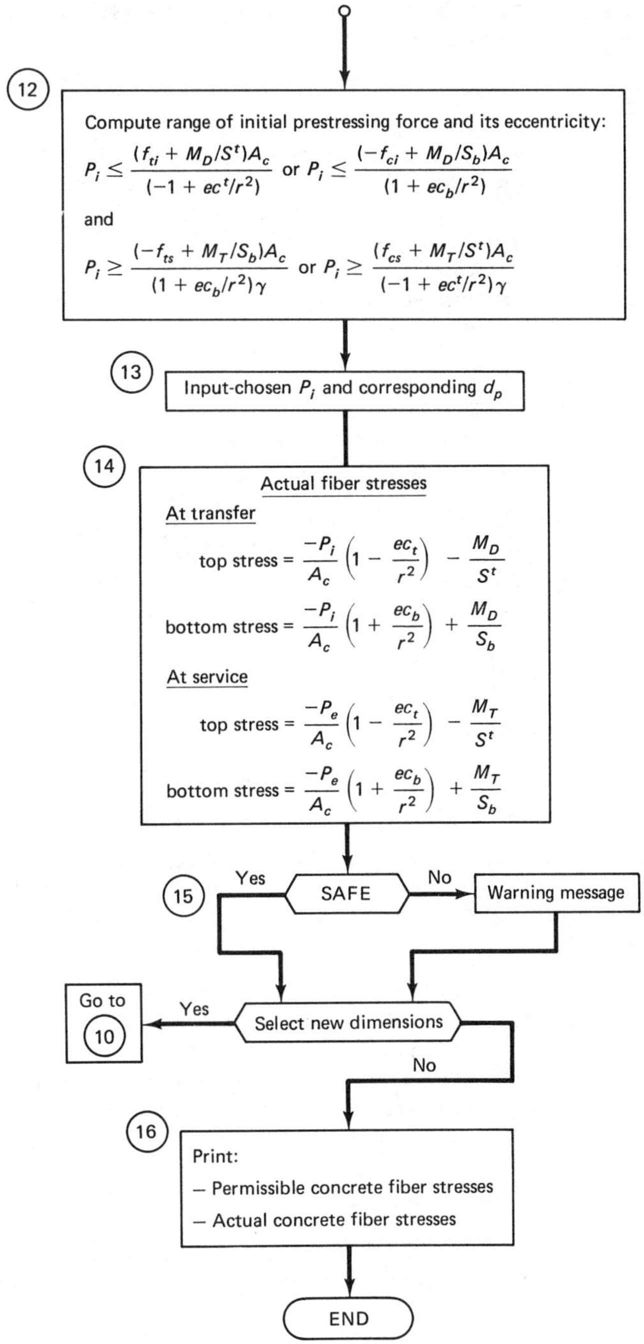

12

Compute range of initial prestressing force and its eccentricity:

$$P_i \leq \frac{(f_{ti} + M_D/S^t)A_c}{(-1 + ec^t/r^2)} \quad \text{or} \quad P_i \leq \frac{(-f_{ci} + M_D/S_b)A_c}{(1 + ec_b/r^2)}$$

and

$$P_i \geq \frac{(-f_{ts} + M_T/S_b)A_c}{(1 + ec_b/r^2)\gamma} \quad \text{or} \quad P_i \geq \frac{(f_{cs} + M_T/S^t)A_c}{(-1 + ec^t/r^2)\gamma}$$

13 — Input-chosen P_i and corresponding d_p

14

Actual fiber stresses

At transfer

$$\text{top stress} = \frac{-P_i}{A_c}\left(1 - \frac{ec_t}{r^2}\right) - \frac{M_D}{S^t}$$

$$\text{bottom stress} = \frac{-P_i}{A_c}\left(1 + \frac{ec_b}{r^2}\right) + \frac{M_D}{S_b}$$

At service

$$\text{top stress} = \frac{-P_e}{A_c}\left(1 - \frac{ec_t}{r^2}\right) - \frac{M_T}{S^t}$$

$$\text{bottom stress} = \frac{-P_e}{A_c}\left(1 + \frac{ec_b}{r^2}\right) + \frac{M_T}{S_b}$$

15 — Yes ← SAFE → No — Warning message

Go to **10** ← Yes — Select new dimensions — No

16

Print:
- Permissible concrete fiber stresses
- Actual concrete fiber stresses

END

Figure 4.29 (*continued*)

4.8 DESIGN OF COMPOSITE POST-TENSIONED PRESTRESSED SIMPLY SUPPORTED SECTION

Example 4.7

Supported Section. A two-lane simply supported bridge has a 64 ft (19.5 m) span, center to center, of bearings. The width of the bridge is such that the exterior beams are 28 ft (8.54 m) center to center. The spacing of the interior beams is at 7 ft center to center. Design the supporting interior post-tensioned beams, unshored during construction to carry AASHTO HS20-44 loading; establish the tendon eccentricities and tendon envelopes; and design the anchorage block and reinforcement, given the following information:

Concrete

Precast beam f'_c = 5,000 psi normal-weight concrete

$5\frac{3}{4}$ in. deck f'_c = 3,000 psi normal-weight concrete

f'_{ci} = 4,000 psi (27.6 MPa)

f_{ci} = 0.55f'_{ci} = −2,200 psi (15.2 MPa)

f_c = 0.40f'_c = −2,000 psi (13.8 MPa)
f_{ti} = 212 psi = 3$\sqrt{f'_c}$
f_t = 6$\sqrt{f'_c}$ = 424 psi

Prestressing Steel

f_{pu} = 270,000 psi (1,862 MPa)

f_{py} = 0.90f_{pu} = 243,000 psi (1,675 MPa)

f_{pi} = 0.70f_{pu} = 189,000 psi (1,303 MPa)

f_{pe} after losses = 0.80f_{pi} = 151,200 psi (1,043 MPa)

Locate and draw the distribution of strands in both the midspan and end sections. Use a live-load moment value including impact due to AASHTO HS20-44 loading for one interior beam of the 64 ft span bridge = 9,300,000 in.-lb (1,051 kN-m).
Solution

Bending Moments and Net Allowable Stresses (Steps 1–4). Since the beam spacing and span length are known, the moments due to situ-cast slab and diaphragms can be initially determined. The clear distance between the webs of the beams is 7 ft, 0 in. − 6 in. = 6 ft, 6 in. Assume a deck slab $5\frac{3}{4}$ in. (14.6 cm) thick made of $1\frac{3}{4}$ in. precast panels and 4 in. situ-cast topping and a diaphragm 8 in. (20 cm) thick at midspan and 48 in. (122 cm) deep, cast integrally with the deck slab. We have:

$$\text{Diaphragm weight} = \frac{8}{12} \times \frac{48}{12} \times 6.5 \times 150 = 2,600 \text{ lb (11.6 kN)}$$

$$1\tfrac{3}{4} \text{ in. precast formwork panel weight} = \frac{1.75}{12} \times 7 \times 150 = 153 \text{ plf (2.2 kN/m)}$$

$$4 \text{ in. situ-cast topping weight} = \frac{4}{12} \times 7 \times 150 = 350 \text{ plf (5.1 kN/m)}$$

$$M_{gD} = \frac{153(64)^2}{8} \times 12 = 940,032 \text{ in.-lb}$$

$$M_{CSD} = \frac{PL}{4} + \frac{W_{SD}L^2}{8}$$

$$= \frac{2,600 \times 64 \times 12}{4} + \frac{350(64)^2 12}{8}$$

$$= 499,200 + 2,150,400 = 2,649,600 \text{ in.-lb}$$

$$M_{SD} + M_{CSD} = 940,032 + 2,649,600 = 3,589,632 \text{ in.-lb (406 kN-m)}$$

Minimum Section Moduli and Choice of Trial Section (Steps 5–6)

$$\gamma = \frac{151,200}{189,000} = 0.80$$

$$S^t \geq \frac{(1 - \gamma)M_D + M_{SD} + M_L}{\gamma f_{ti} - f_c}$$

$$S_b \geq \frac{(1 - \gamma)M_D + M_{SD} + M_L}{f_t - \gamma f_{ci}}$$

Assume that the self-weight of the precast beam element is approximately 583 plf (8.5 kN/m). Then

$$M_D = \frac{583(64)^2 \times 12}{8} = 3,581,952 \text{ in.-lb (405 kN-m)}$$

$$\text{Min } S^t = \frac{(1 - 0.80)3,581,952 + 3,589,632 + 9,300,000}{0.80(212) - (-2,000)} = 6,271 \text{ in}^3$$

$$(103,222 \text{ cm}^3)$$

$$\text{Min } S_b = \frac{(1 - 0.80)3,581,952 + 3,589,632 + 9,300,000}{424 - 0.80(-2,200)} = 6,229 \text{ in}^3$$

$$(102,075 \text{ cm}^3)$$

The expected actual section modulus for the top fibers is usually considerably larger than the section modulus for the bottom fibers of the composite section. So choose the precast element based on $S_b = 6,299 \text{ in}^3$ (100,813 cm³).

AASHTO type III is chosen as the closest trial section for $S_b = 6,299 \text{ in}^3$ since AASHTO type IV has a much larger section modulus. We have as in Fig. 4.30:

$$S^t = 5,070 \text{ in}^3 \ (83,082 \text{ cm}^3)$$

$$S_b = 6,186 \text{ in}^3 \ (101,370 \text{ cm}^3)$$

$$I_c = 125,390 \text{ in}^4 \ (5.2 \times 10^6 \text{ cm}^4)$$

$$A_c = 560 \text{ in}^2 \ (3,613 \text{ cm}^2)$$

$$r^2 = 223.9 \text{ in}^2 \ (1,445 \text{ cm}^2)$$

$$c_t = 24.73 \text{ in. } (62.8 \text{ cm})$$

$$c_b = 20.27 \text{ in. } (51.5 \text{ cm})$$

$$W_D = 560 \text{ plf } (8.2 \text{ kN/m})$$

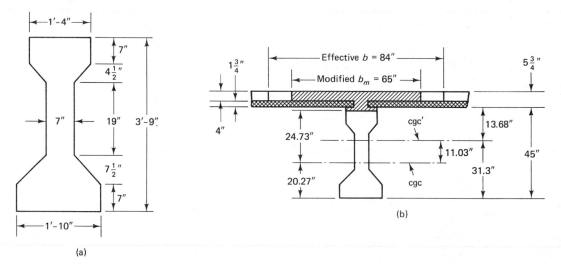

Figure 4.30 Example 4.7. (a) Section (AASHTO-III). (b) Composite section properties.

Try A_{ps} = twenty-two $\frac{1}{2}$ in. dia (12.7 mm) seven-wire stress-relieved strands.

Stresses at Transfer

$$M_T = 3{,}581{,}952 + 3{,}589{,}632 + 9{,}300{,}000 = 16{,}471{,}584 \text{ in.-lb (1,861 kN-m)}$$

$$A_{ps} = 22 \times 0.153 = 3.366 \text{ in}^2 \ (21.7 \text{ cm}^2)$$

$$P_i = A_{ps}f_{pi} = 3.366 \times 0.70 \times 270{,}000 = 636{,}174 \text{ lb (2,830 kN)}$$

$$P_e = 0.80P_i = 0.80 \times 636{,}174 = 508{,}940 \text{ lb (2,264 kN)}$$

$$k_t = \frac{r^2}{c_b} = \frac{223.91}{20.27} = 11.05 \text{ in. (28.1 cm)}$$

$$k_b = \frac{r^2}{c_t} = \frac{223.91}{24.73} = 9.05 \text{ in. (23.0 cm)}$$

$$a_{min} = \frac{M_D}{P_i} = \frac{3{,}581{,}952}{636{,}174} = 5.63 \text{ in.}$$

$$a_{max} = \frac{M_T}{P_e} = \frac{16{,}471{,}584}{508{,}940} = 32.36 \text{ in.}$$

$$e_b = a_{min} + k_b = 5.63 + 9.05 = 14.68 \text{ in.}$$

$$e_t = a_{max} - k_t = 32.36 - 11.05 = 21.31 \text{ in.}$$

Hence, use tendon eccentricities

$$e_c = 16.27 \text{ in. (413 mm)} \qquad e_e = 10 \text{ in. (254 mm)}$$

(a) Midspan Section

$$f^t = -\frac{P_i}{A_c}\left(1 - \frac{e_c c_t}{r^2}\right) - \frac{M_D}{S^t}$$

$$= -\frac{636{,}174}{560}\left(1 - \frac{16.27 \times 24.73}{223.91}\right) - \frac{3{,}581{,}952}{5{,}070}$$

$$= 905.4 - 706.5 = 198.9 \text{ psi } (T) < 200 \text{ psi, O.K.}$$

$$f_b = -\frac{P_i}{A_c}\left(1 + \frac{e_c c_b}{r^2}\right) + \frac{M_D}{S_b}$$

$$= -\frac{636{,}174}{560}\left(1 + \frac{16.27 \times 20.27}{223.91}\right) + \frac{3{,}581{,}952}{6{,}186}$$

$$= -2{,}809.3 + 579.0 = -2{,}230.3 \ (C) \ (15.4 \text{ MPa}) \qquad \cong f_{ci} = -2{,}200 \text{ psi, O.K.}$$

(b) Support Section

$$f^t = \frac{P_i}{A_c}\left(1 - \frac{e_e c_t}{r^2}\right) = -\frac{636{,}174}{560}\left(1 - \frac{10 \times 24.73}{223.91}\right)$$

$$= +118.7 \text{ psi } (T) < 212 \text{ psi, O.K.}$$

$$f_b = -\frac{P_i}{A_c}\left(1 + \frac{e_e c_b}{r^2}\right) = -\frac{636{,}174}{560}\left(1 + \frac{10 \times 20.27}{223.9}\right)$$

$$= -2{,}164.4 \text{ psi } (C) < 2{,}200 \text{ psi, O.K.}$$

Composite Section Properties

$$\frac{E_c(\text{topping})}{E_c(\text{precast})} = \frac{57{,}000\sqrt{3{,}000}}{57{,}000\sqrt{5{,}000}} = 0.77$$

Effective flange width = 7 ft = 84 in. (213 cm)

Modified effective flange width = $0.77 \times 84 = 65$ in. (165 cm)

$$c_b' = \frac{(5.75 \times 65)(47.875) + (560 \times 20.27)}{(5.75)(65) + 560} = 31.32 \text{ in.}$$

$$I_c' = 125{,}390 + 560(31.32 - 20.27)^2 + \frac{65(5.75)^3}{12} + 65 \times 5.75(16.56)^2$$

$$= 297{,}044 \text{ in}^4$$

$$r^2 = 318.12 \text{ in}^2$$

$$S_{cb} = 9{,}490 \text{ in}^3$$

$$S_c^t = 21{,}714 \text{ in}^3$$

Precast $c^t = 45 - 31.32 = 13.68$ in.

$$S_c^t = \frac{297{,}044}{13.68} = 21{,}714 \text{ in}^3$$

slab top $c^t = 13.68 + 5.75 = 19.43$ in.

$$S_{cs}^t = \frac{297{,}044}{19.43} = 15{,}288 \text{ in}^3$$

$$S_{bcs} = \frac{297{,}044}{(19.43 - 4)} = 19{,}251 \text{ in}^3$$

Stresses after $1\frac{3}{4}$ In. Precast Panel Is Erected As Formwork, W_{SD} (Step 11a)

(a) Midspan Section

$$f^t = -\frac{P_e}{A_c}\left(1 - \frac{e_c c_t}{r^2}\right) - \frac{M_D + M_{SD}}{S^t}$$

$$M_D + M_{SD} = 3,581.952 + 940,032 = 4,521,984 \text{ in.-lb}$$

$$f^t = -\frac{508,940}{560}\left(1 - \frac{16.27 \times 24.73}{223.9}\right) - \frac{4,521,984}{5,070}$$

$$= +724.3 - 891.9 = -167.6 \text{ psi } (C), \text{ no tension, O.K.}$$

$$f_b = -\frac{P_e}{A_c}\left(1 + \frac{e_c c_b}{r^2}\right) + \frac{M_D + M_{SD}}{S_b}$$

$$= -\frac{508,940}{560}\left(1 + \frac{16.27 \times 20.27}{223.9}\right) + \frac{4,521,984}{6,186}$$

$$= -2,247.4 + 731.0 = -1,516.4 \text{ psi } (C) < 2,200 \text{ psi, O.K.}$$

(b) Support Section

$$f^t = -\frac{508,940}{560}\left(1 - \frac{10 \times 24.73}{223.9}\right) = 94.9 \text{ psi } (T) < f_t = 424 \text{ psi, O.K.}$$

$$f_b = -\frac{508,940}{560}\left(1 + \frac{10 \times 20.27}{223.9}\right) = -1,731.6 \text{ psi } (11.9 \text{ MPa})(C)$$

$$< -2,000 \text{ psi, O.K.}$$

Stresses at Service Load (Step 11)

(a) Midspan Section

$$f^t = -\frac{P_e}{A_c}\left(1 - \frac{e_c c_t}{r^2}\right) - \frac{M_D + M_{SD}}{S^t} - \frac{M_{CSD} + M_L}{S_c^t}$$

$$M_D + M_{SD} = 4,521,984 \text{ in.-lb}$$

$$M_{CSD} + M_L = 2,649,600 + 9,300,000 = 11,949,600 \text{ in.-lb}$$

$$f^t = -\frac{508,940}{560}\left(1 - \frac{16.27 \times 24.73}{223.9}\right) - \frac{4,521,984}{5,070} - \frac{11,949,600}{21,714}$$

$$= 727.3 - 891.9 - 550.3 = -717.9 \text{ psi } (C) < -2,000 \text{ psi, O.K.}$$

From the previous step, $f_b = -1,516.4$ psi (C). So

$$f_b = -1,516.4 + \frac{M_{CSD} + M_L}{S_{cb}}$$

$$= -1,516.4 + \frac{11,949,600}{9,492} = -1,516.4 + 1,259.2 = -257.5 \text{ psi } (1.8 \text{ MPa}) (C)$$

No tension at bottom fibers, O.K.

(b) Support Section. Same kind of calculations as in previous step. Result is

$$f^t = +94.9 \text{ psi } (T)$$

$$f_b = -1,731.6 \text{ psi } (C)$$

Check Stresses Immediately After Casting Concrete Slab Topping: Midspan Section

$$f^t = -\frac{P_e}{A_c}\left(1 - \frac{e_c c_t}{r^2}\right) - \frac{M_D + M_{SD} + M_{CSD}}{S^t}$$

$$M_D + M_{SD} + M_{CSD} = 4,521,984 + 2,649,600 = 7,171,584 \text{ in.-lb}$$

$$f^t = -\frac{508,940}{560}\left(1 - \frac{16.27 \times 24.73}{223.9}\right) - \frac{7,171,584}{5,070}$$

$$= 724.3 - 1,414.5 = -690.2 \text{ psi } (C) < -2,000 \text{ psi, O.K.}$$

$$f_b = -\frac{P_e}{A_c}\left(1 + \frac{e_c c_t}{r^2}\right) + \frac{M_D + M_{SD} + M_{CSD}}{S_b}$$

$$= -\frac{508,940}{560}\left(1 + \frac{16.27 \times 20.27}{223.9}\right) + \frac{7,171,584}{6,186}$$

$$= -2,247.4 + 1,159.3 = -1,088.1 \text{ psi } (C)$$

$$< -2,000 \text{ psi } (7.5 \text{ MPa} < 13.8 \text{ MPa}), \text{ O.K.}$$

The support section stresses are the same as in the previous step.

Tendon Envelope

(a) Midspan Section

$$a_{max} = 32.36 \text{ in.}$$

$$a_{min} = 5.63 \text{ in.}$$

(b) Quarter Section

$$M_D = 2,686,464$$

$$a_{min} = \frac{M_D}{P_i} = 4.22 \text{ in.}$$

$$M_T = 12,355,248$$

$$a_{max} = \frac{M_T}{P_e} = 24.28 \text{ in.}$$

See Figure 4.31 for the tendon envelope and Figure 4.32 for the stress distribution. Figure 4.33 gives the anchorage zone stresses and the net moments along the depth of the beam.

Design of End-Block Anchorage

$$P_i = 636,174 \text{ lb}$$

$$e_e = 10 \text{ in.}$$

$$f^t = +119 \text{ psi } (T)$$

$$f_b = -2,164 \text{ psi } (C)$$

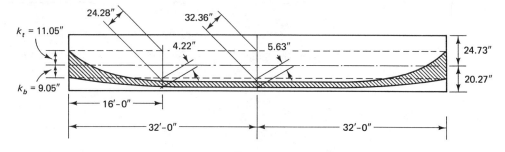

Figure 4.31 Prestressing tendon envelope.

All stresses are psi (1000 psi = 6.895 MPa)

$\ominus$ = compression; + = tension

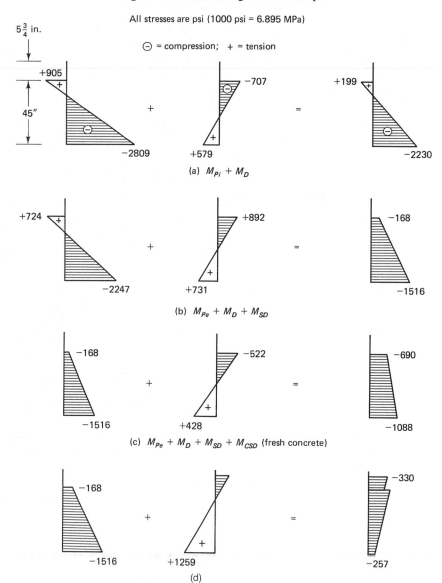

Figure 4.32 Midspan concrete fiber stresses in Example 4.7. (a) Transfer. (b) Erection. (c) Topping cast. (d) Service-load stage.

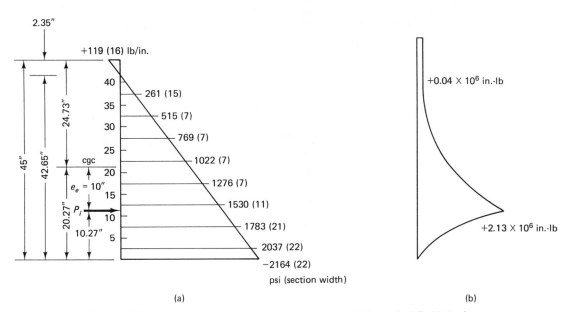

Figure 4.33 Anchorage zone stresses and net moments in Example 4.7. (a) Anchorage zone stresses. Bracketed values are section widths (in.) along which anchorage stresses are acting. (b) Net bursting or splitting moments ($M_p - M_c$) at 5 in. depth intervals.

(a) Bursting Crack Reinforcement

$$h = 45 \text{ in.}$$

Use

$$x = h/4 = 11.25 \text{ in.}$$

$$T_b = \frac{M_{max}}{h - x} = \frac{2.15 \times 10^6}{45 - 11.25} = 63{,}704 \text{ lb } (280.7 \text{ kN})$$

$$A_s = \frac{T_b}{F_s} = \frac{63{,}704}{20{,}000} = 3.18 \text{ in}^2 \ (20.6 \text{ cm}^2)$$

(b) Spalling Crack Reinforcement

$$T_{sp} = \frac{M_{min}}{h - x} = \frac{+0.04 \times 10^6}{45 - 11.25} = 1.185 \text{ lb}$$

$$A_s = \frac{T_{sp}}{20{,}000} = \frac{1{,}185}{20{,}000} = 0.06 \text{ in}^2$$

Total reinforcement $= 3.18 + 0.06 = 3.24 \text{ in}^2 \ (21.7 \text{ cm}^2)$

Try #4 vertical reinforcement (12.7 mm dia.):

$$\text{Number required} = \frac{3.24}{0.20 \times 2} = 8.10$$

Use ten #4 closed ties enclosing all the prestressing strands equally spaced over a zone extending from the beam ends to 45 in. (114 cm) into the span.

Arrangement of strands. Use a 2 in. × 2 in. (25 mm × 25 mm) grid for arranging the distribution of strands. Eccentricities are:

$$e_c = 16.27 \text{ in. (41.3 cm)}$$

$$e_e = 10.0 \text{ in. (25.4 cm)}$$

The arrangement of the strands to develop the required tendon eccentricities are shown in Fig 4.34.

The anchorage zone moments at the various planes at 5-in. intervals along the height of the section are given in Table 4.7.

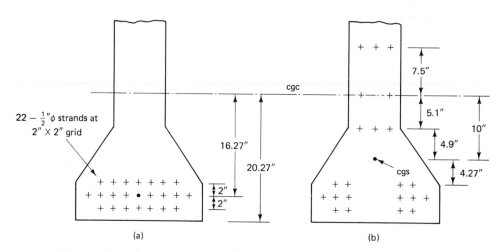

Figure 4.34 Strand arrangement in Example 4.7. (a) Midspan ($e_c = 16.27$ in.). (b) End section ($e_e = 10.0$ in.).

TABLE 4.7 ANCHORAGE ZONE MOMENTS FOR EXAMPLE 4.7.

Moment plane dist. d from bottom (in.)	Section width (in.)	Stress at plane $(d - 2.5)$ (psi)	Concrete resist. force at $(d - 2.5)$ (lb)	Moment M_p of force P_i about horiz. plane in col. (1) (in.-lb × 10^6)	Moment M_C of Concrete in col. (4) about horiz. plane in col. (1) (in.-lb × 10^6)	Net Moment $(M_c - M_p)$ col. (6) − col. (5) (in.-lb × 10^6)
(1)	(2)	(3)	(4)	(5)	(6)	(7)
0	22	−2,164	0	0	0	0
5	22	−2,037	237,040	0	+0.56	+0.56
10.27	15	−1,783	133,725	0	+2.15	+2.15
15	7	−1,530	53,550	−3.01	+4.42	+1.41
20	7	−1,276	44,660	−6.19	+7.00	+0.81
25	7	−1,022	35,770	−9.37	+9.80	+0.43
30	7	−769	26,915	−12.55	+12.76	+0.19
35	12	−515	25,750	−15.73	+15.80	+0.07
40	16	−216	20,880	−18.91	+18.95	+0.04
45	16	+119	+9,520	22.09	+22.15	+0.06

4.9 ULTIMATE-STRENGTH FLEXURAL DESIGN

4.9.1 Cracking-Load Moment

As mentioned in Chapter 1, one of the fundamental differences between prestressed and reinforced concrete is the continuous shift in prestress beams of the compressive C-line away from the tensile cgs line as the load increases. In other words, the moment arm of the internal couple continues to increase with the load without any appreciable change in the stress f_{pe} in the prestressing steel. As the flexural moment continues to increase when the full superimposed dead load and live load act, a loading stage is reached where the concrete compressive stress at the bottom-fibers reinforcement level of a simply supported beam becomes zero. This stage of stress is called the limit state of *decompression:* any additional external load or overload results in cracking at the bottom face, where the modulus of rupture of concrete f_r is reached due to the cracking moment M_{cr} caused by the first cracking load. At this stage, a sudden increase in the steel stress takes place and the tension is dynamically transferred from the concrete to the steel.

Figure 4.35 relates load to steel stress at the various loading stages. It shows not only the load-deformation curve, including its abrupt change of slope at the first cracking load, but also the dynamic dislocation in the load-stress diagram at the first cracking load after decompression in a bonded prestressed beam. Beyond that dislocation point the beam can no longer be considered to behave elastically, and the rise

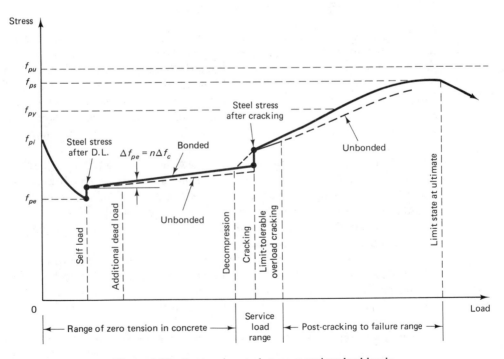

Figure 4.35 Prestressing steel stress at various load levels.

in the compressive C-line stabilizes and stops so that the section starts to behave like a reinformced concrete section with *constant* moment resistance arm.

It is important to evaluate the first cracking load, since the section stiffness is reduced and hence an increase in deflection has to be considered. Also, the crack width has to be controlled in order to prevent reinforcement corrosion or leakage in liquid containers.

The concrete fiber stress at the tension face is

$$f_b = -\frac{P_e}{A_c}\left(1 + \frac{ec_b}{r^2}\right) + \frac{M_{cr}}{S_b} = f_r' \tag{4.24}$$

where the modulus of rupture $f_r = 7.5\sqrt{f'_c}$ and the cracking moment M_{cr} is the moment due to all loads at that load level $(M_D + M_{SD} + M_L)$. From Eq. 4.24,

$$M_{cr} = f_r'S_b + P_e\left(e + \frac{r^2}{c_b}\right) \tag{4.25}$$

Note that the term r^2/c_b is the upper kern value k_b, so that $P_e r^2/c_b$ denotes the elastic moment required to raise the C-line from the prestressing steel level to the upper kern point giving zero tension at the bottom fibers. Consequently, the term $f_r S_b$ is that additional moment required to cause the development of the first crack at the extreme tension fibers due to overload, such as at the bottom fibers at midspan of a simply supported beam.

4.9.2 Partial Prestressing

"Partial prestressing" is a controversial term, since it is *not* intended to denote that a beam is prestressed partially, as might seem to be the case. Rather, partial prestressing describes prestressed beams wherein limited cracking is permitted through the use of additional mild nonprestressed reinforcement to control the extent and width of the cracks and to assume part of the ultimate flexural moment strength. Two major advantages of partial prestressing are the efficient use of all constituent materials and the control of excessive camber due to the long-term creep of concrete under compression.

Reinforced concrete beams always have to be designed as underreinforced to ensure ductile failure by yielding of the reinforcement. Prestressed beams can be either *underreinforced* using a relatively small percentage of mild nonprestressed steel, leading to rupture of the tensile steel at failure, or essentially *overreinforced,* using a large perentage of steel, resulting in crushing of the concrete at the compression top fibers in a somewhat less ductile failure.

Another type of, say, premature failure occurs at the first cracking load level, where M_{cr} approximates the nominal moment strength M_n of the section. This type of failure can occur in members that are prestressed and reinforced with very small amounts of steel, or in members that are concentrically prestressed with small amounts of steel, or in hollow members.

It is generally advisable to evaluate the magnitude of the cracking moment M_{cr} in order to determine the reserve strength and overload limits that the designed section has.

4.9.3 Cracking Moment Evaluation

Example 4.8

Calculate the cracking moment M_{cr} in the I-beam of Example 4.2, and evaluate the magnitude of the overlaod moment that the beam can tolerate at the modulus of rupture of concrete. Also, determine what safety factor the beam has against cracking due to overload. Given is $f_r = 7.5\sqrt{f'_c} = 7.5\sqrt{5,000} = 530$ psi (3.7 MPa).

Solution From Example 4.2,

$$P_e = 308,255 \text{ lb } (1,371 \text{ kN})$$

$$r^2/c_b = 187.5/18.84 = 9.95 \text{ in. } (25.3 \text{ cm})$$

$$S_b = 3,750 \text{ in}^3 \ (61,451 \text{ cm}^3)$$

$$e_c = 14 \text{ in. } (35.6 \text{ cm})$$

$$M_D + M_{SD} = 2,490,638 \text{ in.-lb } (281 \text{ kN-m})$$

$$M_L = 7,605,000 \text{ in.-lb } (859 \text{ kN-m})$$

From Equation 4.25,

$$M_{cr} = f_r S_b + P_e\left(e + \frac{r^2}{c_b}\right)$$

$$= 530 \times 3750 + 308,255(14 + 9.95)$$

$$= 9,370,207 \text{ in.-lb } (1,509 \text{ kN-m})$$

$$M_T = M_D + M_{SD} + M_L = 2,490,638 + 7,605,000$$

$$= 10,095,638 \text{ in.-lb } (1,141 \text{ kN-m})$$

$$\text{Overload moment} = M_T - M_{cr} = 10,095,638 - 9,370,207$$

$$= 725,431 \text{ in.-lb } (83 \text{ kN-m})$$

Since $M_T > M_{cr}$, the beam had tensile cracks at service load, as the design in Example 4.2 presupposed. The safety factor against cracking is given by

$$\frac{M_{cr}}{M_T} = \frac{9,370,207}{10,095,638} = 0.93$$

If the service load M_T is less than M_{cr}, the safety factor against cracking will be greater than 1, as in nonpartially prestressed members.

Note that where nonprestressed reinforcment is used to develop a partially prestressed section, the total factored moment $\phi M_n \geq 1.2 M_{cr}$, as required by the ACI code.

4.10 LOAD AND STRENGTH FACTORS

4.10.1 Reliability and Structural Safety of Concrete Components

Three developments in recent decades have had a major influence on present and future design procedures: the vast increase in the experimental and analytical evaluation of concrete elements, the probabilistic approach to the interpretation of behavior, and the digital computational tools available for rapid analysis of safety and reliability of systems. Until recently, most safety factors in design have had an empirical background based on local experience over an extended period of time. As additional experience is accumulated and more knowledge is gained from failures as well as familiarity with the properties of concrete, factors of safety are adjusted and in most cases lowered by the codifying bodies.

In 1956, Baker (Ref. 4.22) proposed a simplified method of safety factor determination, as shown in Table 4.8, based on probabilistic evaluation. This method expects the design engineer to make critical choices regarding the magnitudes of safety margins in a design. The method takes into consideration that different weights should be assigned to the various factors affecting a design. The weighted failure effects W_t for the various factors of workmanship, loading conditions, results of failure, and resistance capacity are tabulated in the table.

The safety factor against failure is

$$\text{S.F.} = 1.0 + \frac{\sum W_t}{10} \tag{4.26}$$

where the maximum total weighted value of all parameters affecting performance equals 10.0. In other words, for the worst combination of conditions affecting structural performance, the safety factor S.F. = 2.0.

This method assumes adequate prior performance data similar to a design in progress. Such data in many instances are not readily available for determining safe weighted values W_t in Eq. 4.26. Additionally, if the weighted factors are numerous, a probabilistic determination of them is more difficult to codify. Hence, an undue value-judgment burden is probably placed on the design engineer if the full economic benefit of the approach is to be achieved.

Another method with a smaller number of probabilistic parameters deals primarily with loads and resistances. Its approaches for steel and concrete structures are

TABLE 4.8 BAKER'S WEIGHTED SAFETY FACTOR

Weighted failure effect		Maximum W_t
1. Results of failure: 1.0 to 4.0		
Serious, either human or economic		4.0
Less serious, only the exposure of nondamageable material	1.0	
2. Workmanship: 0.5 to 2.0		
Cast in place		2.0
Precast "factory manufactured"	0.5	
3. Load conditions: 1.0 to 2.0		2.0
(high for simple spans and overload possibilities; low for load combinations such as live loads and wind)		
4. Importance of member in structure		0.5
(beams may use lower value than columns)		
5. Warning of failure		1.0
6. Depreciation of strength		0.5
	Total $= \sum W_t =$	10.0

$$\text{S.F.} = 1.0 + \frac{\sum W_t}{10}$$

Post-tensioning bridge deck conduits.

generally similar: both the load-and-resistance-factor-design method (LRFD) and first-order second-moment method (FOSM) propose general reliability procedures for evaluating probability-based factored load design criteria (see Refs. 4.23 and 4.24). They are intended for use in proportioning structural members on the basis of load types such that the resisting strength levels are *greater* than the factored load or moment distributions. As these approaches are basically load oriented, they reduce the number of individual variables that have to be considered, such as those listed in Table 4.8.

Assume that ϕ_i represents the resistance factors of a concrete element and that γ_i represents the load factors for the various types of load. If R_n is the nominal resistance of the concrete element and W_i represents the load effect for various types of superimposed load,

$$\phi_i R_n \geq \gamma_i W_i \qquad (4.27)$$

where i represents the load in question, such as dead, live, wind, earthquake, or time-dependent effects.

Figures 4.36(a) and (b) show a plot of the separate frequency distributions of the actual load W and the resistance R with mean values $\bar{R}$ and $\bar{W}$. Figure 4.36(c) gives the two distributions superimposed and intersecting at point C.

It is recognized that safety and reliable integrity of the structure can be expected to exist if the load effect W falls at a point to the left of intersection C on the W curve, and to the right of intersection C on the resistance curve R. Failure, on the other hand, would be expected to occur if the load effect or the resistance fall within the shaded area in Fig. 4.36(c). If β is a safety index, then

$$\beta = \frac{\bar{R} - \bar{W}}{\sqrt{\sigma_R^2 + \sigma_W^2}} \qquad (4.28)$$

where σ_R and σ_W are the standard deviations of the resistance and the load, respectively.

A plot of the safety index β for a hypothetical structural system against the probability of failure of the system is shown in Fig. 4.37. One can observe that such

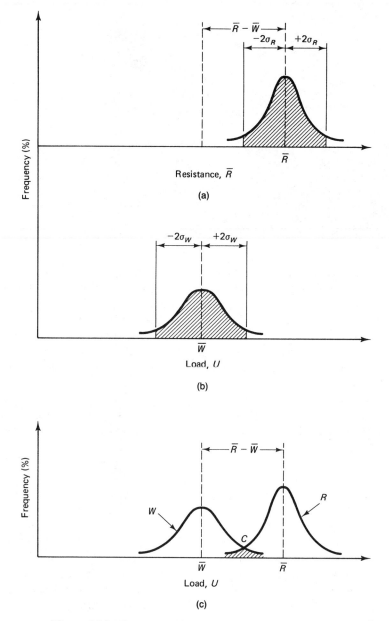

Figure 4.36 Frequency distribution of loads vs. resistance.

a probability is reduced as the difference between the mean resistance $\bar{R}$ and load effect $\bar{W}$ is increased, or the variability of resistance as measured by their standard deviations σ_R and σ_W is decreased, thereby reducing the shaded area under intersection C in Fig. 4.36(c).

The extent of increasing the difference $(\bar{R} - \bar{W})$ or decreasing the degree of scatter of σ_R or σ_W is naturally dictated by economic considerations. It is economically unreasonable to design a structure for zero failure, particularly since types of risk other

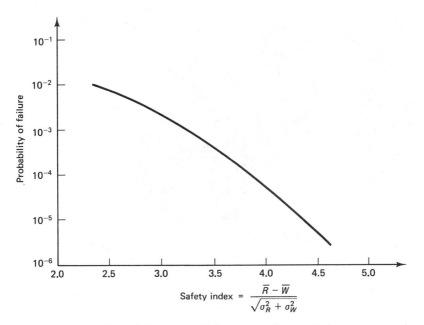

Figure 4.37 Probability of failure vs. safety index.

than load are an accepted matter, such as the risks of severe earthquake, hurricane, volcanic eruption, and fire. Safety factors and corresponding load factors would thus have to disregard those types or levels of load, stress, and overstress whose probability of occurrence is very low. In spite of this, it is still possible to achieve reliable safety conditions by choosing such a safety index value β through a proper choice of R_n and W_i using the appropriate resistance factors ϕ_i and load factors γ_i in Equation 4.27. A safety index β having the value 1.75 to 3.2 for concrete structures is suggested where the lower value accounts for load contributions from wind and earthquake.

If the factored external load is expressed as U_i, then $\Sigma\ \gamma_i W = U_i$ for the different loading combinations. The following are values of U_i recommended in Ref. 4.23 to choose a maximum U for use in Eq. 4.27, that is, $\phi_i R_n \geq \gamma_i W_i \geq U_{i(\text{max})}$:

$$\gamma_i U_w = \text{max.} \begin{cases} 1.4D_n \\ 1.2D_n + 1.6L_n \\ 1.2D_n + 1.6S_n + (0.5L_n \text{ or } 0.8W_n) \\ 1.2D_n + 1.3W_n + 0.5L_n \\ 1.2D_n + 1.5E_n + (0.5L_n \text{ or } 0.2S_n) \\ 0.9D_n - (1.3W_n \text{ or } 1.5E_n) \end{cases} \tag{4.29}$$

where the subscript n stands for the nominal value of the variable working load, i.e.,

D_n = dead load L_n = live load S_n = snow load

W_n = wind load E_n = earthquake load

ϕ_i and γ_i are considered to have optimal values; a resistance factor ϕ of 0.7 to

Tendon stressing with Freyssinet jack.

0.85 is recommended. In this probabilistic approach, the present ACI code would require that

$$\phi_i R_n = \text{max.} \begin{cases} 1.4D_n + 1.7L_n \\ 0.9D_n - 1.3W_n \end{cases} \tag{4.30}$$

As more substantive records of performance are compiled, the details of the foregoing approach to reliability, safety, and reserve strength evaluation of structural components will be more universally accepted and extended beyond the treatment of the component elements to the treatment of the total structural system.

4.10.2 ACI Load Factors and Safety Margins

The general concepts of safety and reliability of performance presented in the preceding section are inherent in a more simplified but less accurate fashion in the ACI code. The load factors γ and the strength reduction factors ϕ give an overall safety factor based on load types such that

$$S.F. = \frac{\gamma_1 D + \gamma_2 L}{D + L} \times \frac{1}{\phi} \tag{4.31}$$

where ϕ is the strength reduction factor and γ_1 and γ_2 are the respective load factors for the dead load D and the live load L. Basically, a single common factor is used for dead load and another for live load. Variation in resistance capacity is accounted for in ϕ. Hence, the method is a simplified empirical approach to safety and the reliability of structural performance that is not economically efficient for every case and not fully adequate in other instances, such as combinations of dead and wind loads.

The ACI factors are termed *load factors*, as they restrict the estimation of reserve strength to the loads only as compared to the other parameters listed in Table 4.8. The estimated service or working loads are magnified by the coefficients, such as a coefficient of 1.4 for dead loads and 1.7 for live load. The types of normally occurring loads can be identified as dead load, D; live load, L; wind load, W; loads due to lateral pressure such as from soil in a retaining wall, H; lateral fluid pressure loads, F; loads due to earthquake, E; and loads due to time-dependent effects, such as creep or shrinkage.

The basic combination of vertical gravity loads is dead load plus live load. The *dead load*, which constitutes the weight of the structure and other relatively permanent features, can be estimated more accurately than the live load. The *live load* is estimated using the weight of nonpermanent loads, such as people and furniture. The transient nature of live loads makes them difficult to estimate more accurately. Therefore, a higher load factor is normally used for live loads than for dead loads. If the combination of loads consists only of live and dead loads, the ultimate load can be taken as

$$U = 1.4D + 1.7L \qquad \text{(4.32 a)}$$

Structures are seldom subjected to dead and live loads alone; wind load is often present. For structures in which wind load should be considered, the recommended combination is

$$U = 0.75(1.4D + 1.7L + 1.7W) \qquad \text{(4.32 b)}$$

Maximum dead, live, and wind loads rarely, if ever, occur simultaneously. Hence, the total factored load has to be reduced using a reduction factor of 0.75 Since wind load is applied laterally, it is possible that the absence of vertical live load while wind load is present can produce maximum stress. The following load combination should also be used to arrive at the maximum value of the factored load U:

$$U = 0.9D + 1.3W \qquad \text{(4.32 c)}$$

Structures that have to resist lateral pressure due to earth fill or fluid pressure should be designed for the worst of the following combination of factored loads:

$$U = 1.4D + 1.7L + 1.7H \qquad \text{(4.33 a)}$$

$$U = 0.9D + 1.7H \qquad \text{(4.33 b)}$$

$$U = 1.4D + 1.7L \qquad \text{(4.33 c)}$$

$$U = 1.4D + 1.7L + 1.4F \qquad \text{(4.33 d)}$$

$$U = 0.9D + 1.4F \qquad \text{(4.33 e)}$$

$$U = 1.4D + 1.7L \qquad \text{(4.33 f)}$$

The largest of the following combinations must be considered for earthquake loading:

$$U = 0.75(1.4D + 1.7L + 1.87E) \qquad \text{(4.34 a)}$$

$$U = 0.9D + 1.43E \qquad \text{(4.34 b)}$$

$$U \geq 1.4D + 1.7L \qquad \text{(4.34 c)}$$

The philosophy used for combining the various load components for earthquake loading is essentially the same as that used for wind loading.

4.10.3 Design Strength vs. Nominal Strength: Strength Reduction Factor ϕ

The strength of a particular structural unit calculated using the current established procedures is termed *nominal strength*. For example, in the case of a beam, the

TABLE 4.9 RESISTANCE OR STRENGTH REDUCTION FACTOR ϕ

Structural element	Factor ϕ
Beam or slab: bending or flexure	0.9
Columns with ties	0.7
Columns with spirals	0.75
Columns carrying very small axial loads (refer to Chapter 9 for more details)	0.7–0.9, or 0.75–0.9
Beam: shear and torsion	0.85

resisting moment capacity of the section calculated using the equations of equilibrium and the properties of concrete and steel is called the *nominal strength moment M_n* of the section. This nominal strength is reduced using a strength reduction factor ϕ to account for inaccuracies in construction, such as in the dimensions or position of reinforcement or variations in properties. The reduced strength of the member is defined as the design strength of the member.

For a beam, the design moment strength ϕM_n should be at least equal to, or better, slightly greater than, the external factored moment M_u for the worst condition of factored load U. The factor ϕ varies for the different types of behavior and for the different types of structural elements. For beams in flexure, for instance, ϕ is 0.9.

For tied columns that carry dominant compressive loads, the factor ϕ equals 0.7. The smaller strength reduction factor used for columns is due to the structural importance of the columns in supporting the total structure compared to other members, and to guard against progressive collapse and brittle failure with no advance warning of collapse. Beams, on the other hand, are designed to undergo excessive deflections before failure. Hence, the inherent capability of the beam for advanced warning of failure permits the use of a higher strength reduction factor or resistance factor.

Table 4.9 summarizes the resistance factors ϕ for various structural elements as given in the ACI code. A comparison of these values to those given in Ref. 4.24 indicates that the ϕ values in this table, as well as the load factors of Equation 4.33, are in some cases more conservative than they should be. In cases of earthquakes, wind, and shear forces, the probability of load magnitude and reliability of performance are subject to higher randomness, and hence a higher coefficient of variation, than the other types of loading.

4.10.4 AASHTO Strength Reduction Factors

Flexure

For factory-produced precast prestressed concrete members, $\phi = 1.0$

For post-tensioned cast-in-place concrete members, $\phi = 0.95$

Shear and Torsion

Reduction factor for prestressed members, $\phi = 0.90$

4.10.5 ANSI Alternate Load and Strength Reduction Factors

The 1982 A58.1 Standard, "Minimum Design Loads for Buildings and Other Structures," presents a set of load factors and load combinations for use in the strength design of concrete, steel, or masonry buildings. These factors can be summarized as follows:

$$U = 1.4D + 1.0L \qquad (4.35\text{ a})$$

$$U = 1.2D + 1.6L \qquad (4.35\text{ b})$$

$$U = 1.2D + 1.3W + 0.5L \qquad (4.35\text{ c})$$

$$U = 1.2D + 1.5E + 0.5L \qquad (4.35\text{ d})$$

$$U = 0.9D + (1.3W \quad \text{or} \quad 1.5E) \qquad (4.35\text{ e})$$

The load factors for live load L in Equation 4.35c and d are to be taken as 1.0 for areas occupied in places of public assembly and all areas where specified live loads exceed 100 psf.

Where deformation-induced forces T are significant in design, structural members and connections are to be designed for the following load combinations, which are modified versions of Equations 4.35a and b:

$$U = 1.4D + 1.0L + 1.2T \qquad (4.36\text{ a})$$

$$U = 1.2(D + T) + 1.6L \qquad (4.36\text{ b})$$

Estimates of differential settlement, creep, shrinkage, and temperature changes are to be based on realistic assessment of such effects occurring in service.

The corresponding strength recduction factors to be used in conjunction with Equations 4.35 and 4.36 are as follows:

Flexure and or axial tension: 0.85

Axial compression members spirally reinforced: 0.70

Other axial compression members: 0.65

Shear and torsion: 0.75

Bearing on concrete: 0.60

There is little significant difference between using the ACI Code factors and the alternative ANSI Code appendix factors in the design of members. As an example comparing the flexural strength by both methods, we have

$$\text{ACI Code: } N = \frac{U}{\phi} = \frac{1.4D + 1.7L}{0.90} = 1.56D + 1.89L$$

$$\text{ANSI Appendix: } N = \frac{U}{\phi} = \frac{1.2D + 1.6L}{0.85} = 1.41D + 1.88L$$

The comparison shows that the reserve strength for live load in both methods is almost the same (a factor of 1.89 in the ACI versus 1.88 in the ANSI). For the dead load, which normally does not exceed 20–30 percent of the live load, the difference is less than 10 percent, with the cumulative effect on the total for all types of load being less than 2 to 3 percent. Hence, design by either approach would be almost identical to that by the other for all practical purposes. The ACI Code factors are the mandatory ones to use.

4.11 LIMIT STATE IN FLEXURE AT ULTIMATE LOAD IN BONDED MEMBERS: DECOMPRESSION TO ULTIMATE LOAD

4.11.1 Introduction

As discussed in Section 4.9.1, the prestressed concrete beam starts to behave like a reinforced concrete beam when the value of the flexural moment is well beyond the cracking moment M_{cr} and the total service load moment M_T. The ultimate theory in flexure and the principles and concepts underlying it are thus equally applicable to prestressed concrete. A detailed fundamental treatment of this subject is given in chapter 5 of Ref. 4.3. The same fundamental format of equations will be given here, modified to reflect the characteristics of the different reinforcing materials and the geometry peculiar to prestressed concrete.

Cracking develops when the tensile stress in the concrete at the extreme fibers of the critical section exceeds the maximum stress level $f_r \cong 7.5\sqrt{f_c'}$. Prior to attaining this level, the overload causes the compressive stress in the concrete *at the level of the prestressing steel* to continually decrease until it becomes zero at a load level termed the *decompression load*. The stress level in the tendons is correspondingly termed the *decompression stress* (see Figures 4.3 and 4.38).

Some investigators define the decompression load as the load at which the first crack appears at the extreme fibers of the critical section, such as the bottom midspan fibers of a simply supported beam. A minor difference results in the analysis using this definition of the decompression load.

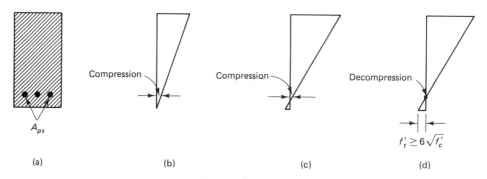

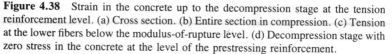

Figure 4.38 Strain in the concrete up to the decompression stage at the tension reinforcement level. (a) Cross section. (b) Entire section in compression. (c) Tension at the lower fibers below the modulus-of-rupture level. (d) Decompression stage with zero stress in the concrete at the level of the prestressing reinforcement.

To follow the loading step-by-step stages, suppose that the effective prestress f_{pe} at service load due to all loads results in a strain ϵ_1 such that

$$\epsilon_1 = \epsilon_{pe} = \frac{f_{pe}}{E_{ps}} \tag{4.37 a}$$

At decompression, i.e., when the compressive stress in the surrounding concrete at the level of the prestressing tendon is neutralized by the tensile stress due to overload, a decompression strain $\epsilon_{decomp} = \epsilon_2$ results such that

$$\epsilon_2 = \epsilon_{decomp} = \frac{P_e}{A_c E_c}\left(1 + \frac{e^2}{r^2}\right) \tag{4.37 b}$$

Figures 4.38 and 4.39 illustrate the stress distribution at and after the decompression stage where the behavior of the prestressed beam starts to resemble that of a reinforced concrete beam.

As the load approaches the limit state at ultimate, the additional strain ϵ_3 in the steel reinforcement follows the linear triangular distribution shown in Fig. 4.39b, where the maximum compressive strain at the extreme compression fibers is $\epsilon_c = 0.003$ in./in. In such a case, the steel strain increment due to overload above the decompression load is

$$\epsilon_3 = \epsilon_c\left(\frac{d - c}{c}\right) \tag{4.37 c}$$

where c is the depth of the neutral axis. Consequently, the total strain in the prestressing steel at this stage becomes

$$\epsilon_s = \epsilon_1 + \epsilon_2 + \epsilon_3 \tag{4.37 d}$$

The corresponding stress f_{ps} at nominal strength can be easily obtained from the stress-strain diagram of the steel supplied by the producer.

4.11.2 The Equivalent Rectangular Block and Nominal Moment Strength

It is important to be able to evaluate the reserve strength in the prestressed beam up to failure, as discussed in Chapter 1. Hence, the total design would have to incorporate the moment strength of the prestressed section in addition to the service-load level checks described in detail in Sections 4.1 through 4.7. The following assumptions are made in defining the behavior of the section at ultimate load:

1. The strain distribution is assumed to be linear. This assumption is based on Bernoulli's hypothesis that plane sections remain plane before bending and perpendicular to the neutral axis after bending.

2. The strain in the steel and the surrounding concrete is the same prior to cracking of the concrete or yielding of the steel as after such cracking or yielding.

3. Concrete is weak in tension. It cracks at an early stage of loading at about 10 percent of its compressive strength limit. Consequently, concrete in the tension zone of the section is neglected in the flexural analysis and design computations, and the tension reinforcement is assumed to take the total tensile force.

To satisfy the equilibrium of the horizontal forces, the compressive force C in the concrete and the tensile force T in the steel should balance each other—that is,

$$C = T \qquad (4.38)$$

The terms in Figure 4.39 are defined as follows:

b = width of the beam at the compression side

d = depth of the beam measured from the extreme compression fiber to the centroid of steel area

h = total depth of the beam

4.11.2.1 Rectangular sections.

The actual distribution of the compressive stress in a section at failure has the form of a rising parabola, as shown in Figure 4.39(c). It is time consuming to evaluate the volume of the compressive stress block if it has a parabolic shape. An equivalent rectangular stress block due to Whitney can be used with ease and without loss of accuracy to calculate the compressive force and hence the flexural moment strength of the section. This equivalent stress block has a depth a and an average compressive strength $0.85f_c'$. As seen from Figure 4.39(d), the value of $a = \beta_1 c$ is determined by using a coefficient β_1 such that the area of the equivalent rectangular block is approximately the same as that of the parabolic compressive block, resulting in a compressive force C of essentially the same value in both cases.

The value $0.85f_c'$ for the average stress of the equivalent compressive block is based on the core test results of concrete in the structure at a minimum age of 28 days. Based on exhaustive experimental tests, a maximum allowable strain of 0.003 in./in. was adopted by the ACI as a safe limiting value. Even though several forms of stress blocks, including the trapezoidal, have been proposed to date, the simplified equivalent rectangular block is accepted as the standard in the analysis and design of reinforced concrete. The behavior of the steel is assumed to be elastoplastic.

Using all the preceding assumptions, the stress distribution diagram shown in Figure 4.39(c) can be redrawn as shown in Figure 4.39(d). One can easily deduce that the compression force C can be written $0.85f_c'ba$—that is, the *volume* of the compressive block at or near the ultimate when the tension steel has yielded ($\epsilon_s > \epsilon_y$). The tensile force T can be written $A_{ps}f_{ps}$; thus, the equilibrium Equation 4.38 can be rewritten as

$$A_{ps}f_{ps} = 0.85f_c'ba \qquad (4.39)$$

A little algebra yields

$$a = \beta_1 c = \frac{A_{ps}f_{ps}}{0.85f_c'b}$$

The nominal moment strength is obtained by multiplying C or T by the moment arm $(d_p - a/2)$, yielding

$$M_n = A_{ps}f_{ps}\left(d_p - \frac{a}{2}\right) \qquad (4.40\text{ a})$$

where d_p is the distance from the compression fibers to the center of the prestressed

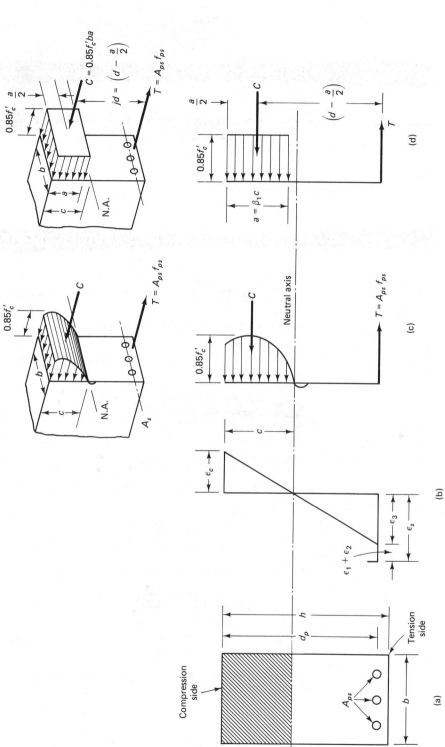

Figure 4.39 Stress and strain distribution across beam depth. (a) Beam cross section. (b) Strains. (c) Actual stress block. (d) Assumed equivalent stress block.

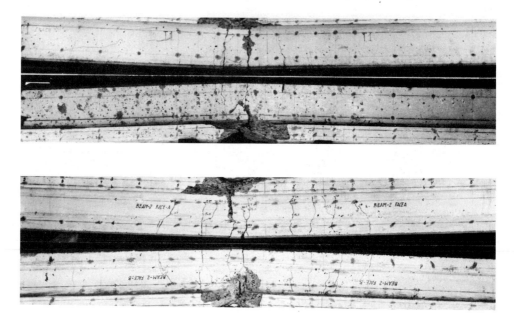

Flexural cracks at failure of prestressed T-beams (Nawy, Potyondy).

reinforcement. The steel percentage $\rho_p = A_{ps}/bd_p$ gives nominal strength of the prestressing steel only as follows

$$M_n = \rho_p f_{ps} bd_p^2 \left(1 - 0.59\rho_p \frac{f_{ps}}{f_c'}\right) \qquad \text{(4.40 b)}$$

If ω_p is the reinforcement index $= \rho_p(f_{ps}/f_c')$, Equation 4.40b becomes

$$M_n = \rho_p f_{ps} bd_p^2 (1 - 0.59\omega_p) \qquad \text{(4.40 c)}$$

The contribution of the mild steel tension reinforcement should be similarly treated, so that the depth a of the compressive block is

$$a = \frac{A_{ps}f_{ps} + A_s f_y}{0.85 f_c' b} \qquad \text{(4.41 a)}$$

If $c = a/\beta_1$, the strain at the level of the mild steel is (Fig. 4.39)

$$\epsilon_s = \epsilon_c \left(\frac{d - c}{c}\right) \qquad \text{(4.41 b)}$$

Equation 4.40b, for rectangular sections but with mild tension steel and no compression steel accounted for, becomes

$$M_n = \rho_p f_{ps} bd_p^2 \left(1 - 0.59\rho_p \frac{f_{ps}}{f_c'}\right) + \rho f_y bd^2 \left(1 - 0.59\frac{f_y}{f_c'}\right) \qquad \text{(4.42 a)}$$

or can be rewritten as either

$$M_n = A_{ps}f_{ps}\left\{1 - 0.59\left(\omega_p + \frac{d}{d_p}\omega\right)\right\} + A_s f_y\left\{1 - 0.59\left(\frac{d_p}{d}\omega_p + \omega\right)\right\} \qquad \text{(4.42 b)}$$

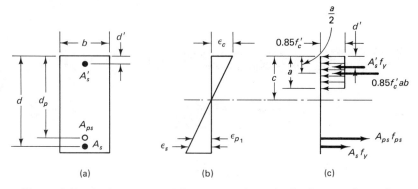

Figure 4.40 Strain, stress, and forces across beam depth of rectangular section. (a) Beam section. (b) Strain. (c) Stresses and forces.

where $\omega = \rho(f_y/f_c')$, or

$$M_n = A_{ps}f_{ps}\left(d_p - \frac{a}{2}\right) + A_s f_y\left(d - \frac{a}{2}\right) \qquad (4.42\text{ c})$$

The contribution from compression reinforcement can be taken into account provided it has been found to have yielded

$$a = \frac{A_{ps}f_{ps} + A_s f_y - A_s' f_y}{0.85 f_c' b} \qquad (4.43)$$

where b is the section width at the compression face of the beam.

Taking moments about the center of gravity of the compressive block in Fig. 4.40, the nominal moment strength in Equation 4.42b becomes

$$M_n = A_{ps}f_{ps}\left(d_p - \frac{a}{2}\right) + A_s f_y\left(d - \frac{a}{2}\right) + A_s' f_y\left(\frac{a}{2} - d'\right) \qquad (4.44)$$

4.11.2.2 Nominal moment strength of flanged sections. When the compression flange thickness h_f is less than the neutral axis depth c and equivalent rectangular block depth a, the section can be treated as a flanged section as in Fig. 4.41. From the figure,

$$T_p + T_s = T_{pw} + T_{pf} \qquad (4.45)$$

where T_p = total prestressing force = $A_{ps}f_{ps}$
T_s = ultimate force in the nonprestressed steel = $A_s f_y$
T_{pw} = part of the total force in the tension reinforcement required to develop the web = $A_{pw}f_{ps}$
A_{pw} = total reinforcement area corresponding to the force T_{pw}
T_{pf} = part of the total force in the tension reinforcement required to develop the flange = $C_f = 0.85 f_c'(b - b_w)h_f$
$C_w = 0.85 f_c' b_w a$

Substituting in Equation 4.45, we obtain

$$T_{pw} = A_{ps}f_{ps} + A_s f_y - 0.85 f_c'(b - b_w)h_f \qquad (4.46)$$

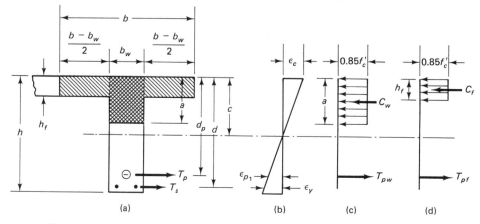

Figure 4.41 Strain, stress, and forces in flanged sections. (a) Beam section. (b) Strain. (c) Web stress and forces. (d) Flange stress and force.

Summing up all forces in Figures 4.41(c) and (d), we have

$$T_{pw} + T_{pf} = C_w + C_f$$

giving

$$a = \frac{A_{pw}f_{ps}}{0.85f'_c b_w} \tag{4.47 a}$$

or

$$a = \frac{A_{ps}f_{ps} + A_s f_y - 0.85f'_c(b - b_w)h_f}{0.85f'_c b_w} \tag{4.47 b}$$

Eq. 4.44 for a beam with compression reinforcement can be rewritten to give the nominal moment strength for a flanged section where the neutral axis falls outside the flange and $a > h_f$ as follows, taking moments about the center of the prestressing steel:

$$M_n = A_{pw}f_{ps}\left(d_p - \frac{a}{2}\right) + A_s f_y(d - d_p) + 0.85f'_c(b - b_w)h_f\left(d_p - \frac{h_f}{2}\right) \tag{4.48}$$

The design moment in all cases would be

$$M_u = \phi M_n \tag{4.49}$$

where $\phi = 0.90$ for flexure.

In order to determine whether the neutral axis falls outside the flange, requiring a flanged section analysis, one has to determine, as discussed in Ref. 4.3, whether the total compressive force C_n is larger or smaller than the total tensile force T_n. If $T_p + T_s$ in Figure 4.41 is larger than C_f, the neutral axis falls outside the flange and the section has to be treated as a flanged section. Otherwise it should be treated as a rectangular section of the width b of the compression flange.

Another method of determining whether the section can be considered flanged is to calculate the value of the equivalent rectangular block depth a from Eq. 4.47b, thereby determining the neutral axis depth $c = a/\beta_1$.

4.11.2.3 Determination of prestressing steel nominal failure stress f_{ps}.

The value of the stress f_{ps} of the prestressing steel at failure is not readily available. However, it can be determined by *strain compatibility* through the various loading stages up to the limit state at failure, as discussed in Section 4.8.6 and defined in Equations 4.40a, b, and c. Such a procedure is required if

$$f_{pe} = \frac{P_e}{A_{ps}} < 0.50 f_{pu} \qquad (4.50\ a)$$

Approximate determination is allowed by the ACI 318 building code provided that

$$f_{pe} = \frac{P_e}{A_{ps}} \geq 0.50 f_{pu} \qquad (4.50\ b)$$

with separate equations for f_{ps} given for bonded and nonbonded members.

Bonded Tendons. The empirical expression for bonded members is

$$f_{ps} = f_{pu}\left(1 - \frac{\gamma_p}{\beta_1}\left[\rho_p\frac{f_{pu}}{f'_c} + \frac{d}{d_p}(\omega - \omega')\right]\right) \qquad (4.51)$$

where the reinforcement index for the compression nonprestressed reinforcement is $\omega' = \rho'(f_y/f'_c)$. If the compression reinforcement is taken into account when calculating f_{ps} by Eq. 4.51, the term $[\rho_p(f_{pu}/f'_c) + (d/d_p)(\omega - \omega')]$ should not be less than 0.17 and d' should not be greater than $0.15d_p$. Also,

$$\gamma_p = 0.55 \text{ for } f_{py}/f_{pu} \text{ not less than } 0.80$$
$$= 0.40 \text{ for } f_{py}/f_{pu} \text{ not less than } 0.85$$
$$= 0.28 \text{ for } f_{py}/f_{pu} \text{ not less than } 0.90$$

The value of the factor γ_p is based on the criterion that $f_{py} = 0.80f_{pu}$ for high-strength prestressing bars, 0.85 for stress-relieved strands, and 0.90 for low-relaxation strands.

Unbonded Tendons. For a span-to-depth ratio of 35 or less,

$$f_{ps} = f_{pe} + 10,000 + \frac{f'_c}{100\rho_p} \qquad (4.52\ a)$$

where f_{ps} shall not be greater than f_{py} or $(f_{pe} + 60,000)$.
For a span-to-depth ratio greater than 35,

$$f_{ps} = f_{pe} + 10,000 + \frac{f'_c}{300\rho_p} \qquad (4.52\ b)$$

where f_{ps} shall not be greater than f_{py} or $(f_{pe} + 30,000)$. Figure 4.42, from Ref. 4.9, shows seating losses for typical unbonded tendons.

4.11.2.4 Limiting values of the reinforcement index.

The reinforcement index ω_p, a measure of the percentage of reinforcement in the section, is given by

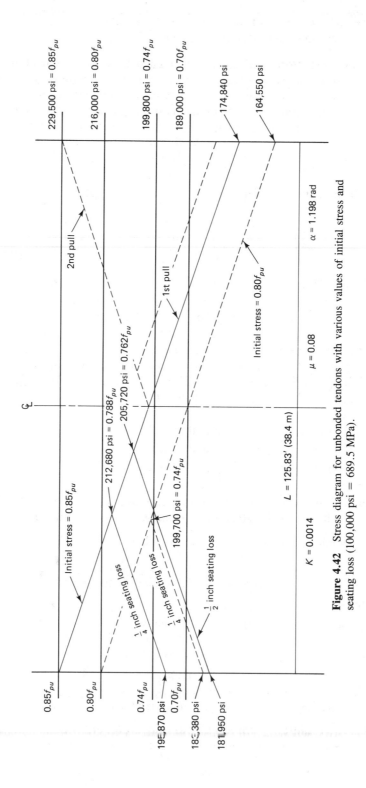

Figure 4.42 Stress diagram for unbonded tendons with various values of initial stress and seating loss (100,000 psi = 689.5 MPa).

$$\omega_p = \frac{A_{ps}f_{ps}}{bdf'_c} = \rho_p \frac{f_{ps}}{f'_c} \qquad (4.53)$$

Minimum Reinforcement. If the percentage of reinforcement is too small, the concrete section will be too weak to resist the tensile stress level after cracking and the section will behave almost as a plain section, with premature abrupt failure through rupture of the reinforcement. Hence, a minimum percentage ρ_{min} with a minimum $\omega_{p_{min}}$ has to be observed in the design in order to prevent such a failure. The total amount of prestressed and nonprestressed reinforcement required by the ACI should not be less than that required to develop a factored moment $M_u = \phi M_n$ such that

$$M_u \geq 1.2M_{cr} \qquad (4.54 \text{ a})$$

where M_{cr} is based on a modulus of rupture $f_r = 7.5\sqrt{f'_c}$. An exception can be made where the flexural member has shear and flexural strengths at least twice those of the factored loads in Equations 4.33. Also, the minimum area of bonded nonprestressed reinforcement in beams in accordance with the ACI code has to be computed as

$$\text{Min } A_s = 0.004A \qquad (4.54 \text{ b})$$

where A is that part of the cross section between the flexural tension face and the center of gravity cgc of the gross section (in^2). This reinforcement has to be uniformly distributed over the precompressed tensile zone as close as possible to the extreme tension fibers.

In two-way flat plates, where the tension stress in the concrete at service load exceeds $2\sqrt{f'_c}$, bonded nonprestressed steel is required such that

$$A_s = \frac{N_c}{0.5f_y} \qquad (4.55 \text{ a})$$

where $f_y \leq 60,000$ psi
N_c = tensile force in the concrete due to unfactored dead plus live loads $(D + L)$.

In the negative-moment areas of slabs at column support, the minimum area of nonprestressed steel in each direction should be

$$A_s = 0.00075h\ell \qquad (4.55 \text{ b})$$

where h = total slab thickness
ℓ = span length in the direction parallel to that of the reinforcement being determined.

The reinforcement A_s has to be distributed with a slab width between lines that are $1.5h$ outside the faces of the column support, with at least four bars or wires to be provided in each direction and a spacing not to exceed 12 in.

Maximum Reinforcement. If the percentage of reinforcement is too large, the concrete section behaves as if it were overreinforced. As a result, a nonductile failure would occur by initial crushing of the concrete at the compression fibers since the reinforcement at the tension side cannot yield first. Reinforced concrete beams always have to be designed as underreinforced, with a maximum reinforcement percentage ρ not exceeding 75 percent of the balanced percentage ρ_b denoting the conditions where the tensile steel yields simultaneously with crushing of the concrete at the compression fibers.

In prestressed beams, however, it is not always possible to impose an under-reinforced condition. The prestressing forces P_i and P_e at transfer and service load also control the value of the area of the tensile reinforcement needed, including the area of the nonprestressed reinforcement. Additionally, the yield strength, and in turn the yield strain value of the prestressing steel is not well defined. Consequently, the prestressed beam designed to satisfy all the service-load requirements could behave as either underreinforced or overreinforced at the limit state of ultimate-load design, particularly if it is a partially prestressed beam.

In order to ensure ductility of behavior, the ACI code limits the percentage of reinforcement in such manner that the reinforcement index does not exceed $0.36\beta_1$ as a measure of underreinforced behavior, where the concrete strength effect

$$\beta_1 = 0.85 - \frac{0.05(f'_c - 4,000)}{1,000} \geq 0.65 \tag{4.56}$$

For a rectangular section with prestressing steel only,

$$\omega_p = \rho_p \frac{f_{ps}}{f'_c} \leq 0.36\beta_1 \tag{4.57 a}$$

For rectangular sections with compression steel,

$$\left[\omega_p + \frac{d}{d_p}(\omega - \omega')\right] \leq 0.36\beta_1 \tag{4.57 b}$$

where

$$\omega = \frac{A_s f_y}{bd f'_c}$$

and

$$\omega' = \frac{A'_s f_y}{bd f'_c}$$

Finally, for flanged sections,

$$\left[\omega_{pw} + \frac{d}{d_p}(\omega_w - \omega'_w)\right] \leq 0.36\beta_1 \tag{4.57 c}$$

where ω_{pw}, ω_w, and ω'_w are computed in the same manner as in Equations 4.57a, b, except that the web width b_w is used in the denominators of these equations. Note that the terms ω_p, $(\omega_p + (d/d_p)(\omega - w'))$, and $(\omega_{pw} + (d/d_p)(\omega_w - \omega'_w))$ are each equal to $0.85a/d_p$, where a is the depth of the equivalent rectangular concrete compressive block as follows:

(a) In rectangular sections and in flanged sections in which $a \leq h_f$,

$$\left[\omega_p + \frac{d}{d_p}(\omega - \omega')\right] = \left[\frac{A_{ps}}{bd_p} \cdot \frac{f_{ps}}{f'_c} + \frac{d}{d_p}\left(\frac{A_s}{bd} \cdot \frac{f_y}{f'_c} - \frac{A'_x}{bd} \cdot \frac{f_y}{f'_c}\right)\right]$$

$$= \frac{A_{ps}f_{ps} + A_s f_y - A'_s f_y}{bd_p \cdot f'_c} = \frac{0.85 f'_c ab}{bd_p f'_c} = \frac{0.85a}{d_p}$$

(b) In flanged sections in which $a > h_f$, let C_F be the resultant concrete compression force in outstanding flanges. Then,

$$\left[\omega_{pw} + \frac{d}{d_p}(\omega_w - \omega'_w) \right] = \left[\frac{(A_{ps}f_{ps} - C_F)}{b_w d_p f'_c} + \frac{d}{d_p}\left(\frac{A_s}{b_w d} \cdot \frac{f_y}{f'_c} - \frac{A'_s}{b_w d} \cdot \frac{f_y}{f'_c} \right) \right]$$

$$= \frac{A_{ps}f_{ps} + A_s f_y - A'_s f_y - C_F}{b_w d_p f'_c}$$

$$= \frac{\text{compression force in web}}{b_w d_p f'_c}$$

$$= \frac{0.85 f'_c b_w a}{b_w d_p f'_c} = \frac{0.85a}{d_p}$$

An exception can be made to the ACI requirements in Equations 4.57a, b, and c, provided that the design moment strength does not exceed the moment strength based on the compression portion of the moment couple. In other words, unless a strain compatibility analysis is performed, the overreinforced prestressed beam moment strength should be determined from the empirical expression

$$M_n = 0.25 f'_c b d^2 \tag{4.58 a}$$

for rectangular sections, and

$$M_n = 0.25 f'_c b_w d^2 + 0.85 f'_c (b - b_w) h_f (d - 0.5 h_f) \tag{4.58 b}$$

for flanged sections. These equations can be modified as follows:

(a) For the overreinforced rectangular section,

$$M_n = f'_c b d_p^2 (0.36\beta_1 - 0.08\beta_1^2) \tag{4.59 a}$$

(b) For the overreinforced flanged section,

$$M_n = f'_c b_w d_p^2 (0.36\beta_1 - 0.08\beta_1^2) + 0.85 f'_c (b - b_w) h_f (d_p - 0.5 h_f) \tag{4.59 b}$$

4.11.2.5 Limit state in flexure at ultimate load in nonbonded tendons.
The discussion presented in Sections 4.11.1–4.11.8 defines the design and analysis process for pretensioned beams, where the concrete is cast around the prestressed tendons, thereby achieving full bond, as well as for post-tensioned beams, where the tendons are fully grouted under pressure after the tendons are prestressed.

Post-tensioned tendons that are not grouted or that are asphalt coated (many in the United States) are nonbonded tendons. Consequently, as the superimposed load acts on the beam, slip results between the tendons and the surrounding concrete, permitting a uniform deformation along the entire length of the prestressing tendon. As cracks develop at the critical high-moment zones, the increase in the steel tensile stress is not concentrated at the cracks, but is uniformly distributed along the freely sliding tendon. As a result, the net increase in strain and stress is less in the nonbonded case than in the case of bonded tendons as the load continues to increase to the ultimate. Hence, a lesser number of cracks, but of larger width, develops in non-bonded prestressing (Refs. 4.4, 4.5). The final stress in the prestressed tendons at ultimate load would be only slightly higher than the effective prestress f_{pe}.

In order to ensure a structure with good serviceability performance, a reasonable

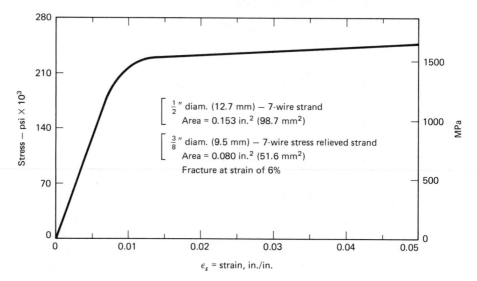

Figure 4.43 Typical stress-strain relationship of seven-wire 270 K prestressing strand.

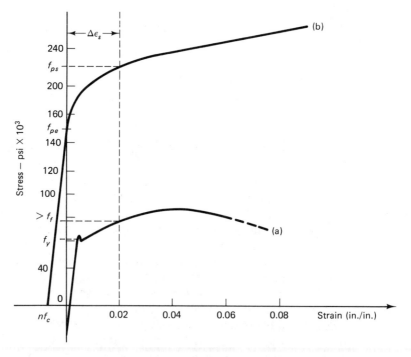

Figure 4.44 Stress-strain diagrams for reinforcement. (a) Nonprestressed steel. (b) Prestressed steel (100,000 psi = 689.5 MPa).

percentage of nonprestressed steel has to be used, within the limitations mentioned in Section 4.11.2.4. The nonprestressed reinforcement controls the flexural crack development and width, and contributes to substantially increasing the moment strength capacity M_n of the section. It undergoes a strain larger than its yield strain, since its deformation at the postelastic range has to be compatible with the deformation of the adjacent prestressing strands. Hence, the stress level in the nonprestressed steel will always be higher than its yield strength at ultimate load. Figure 4.43 shows a typical stress-strain diagram for a 270K seven-wire $\frac{1}{2}$ in. prestressing strand, while Fig. 4.44 schematically illustrates the relative stresses of the prestressed and the nonprestressed steel and seating losses.

From this discussion, it can be concluded that the expressions presented for the M_n calculations of the nominal moment strength for bonded beams can be equally used for nonbonded elements. Note that while it is always advisable to grout the post-tensioned tendons, it is sometimes not easy to do so, as, for example, in two-way slab systems or shallow-box elements, where the concrete thickness is small. Also, consideration has to be given to the cost of pressure grouting in cases where there is a congestion of tendons.

4.12 PRELIMINARY ULTIMATE-LOAD DESIGN

If the preliminary design starts at the ultimate-load level, the required design moment $M_u = \phi M_n$ has to be at least equal to the factored moment M_u. The first trial depth has to be based on a reasonable span-to-depth ratio, with the top flange width determined by whether the beam is for residential floors or parking garages, where a single-T or double-T section or a hollow-box shallow section is preferable, or whether the beam is intended to support a bridge deck with spacing decided by load and the number of lanes, where an I-section might be preferable.

As a rule of thumb, the average depth of a prestressed beam is about 75 percent of the depth of a comparably loaded reinforced concrete beam. Another guideline for an initial trial is to use 0.6 in. of depth per foot of span. Once a first-trial depth is chosen, a determination is made of the other geometrical properties of the section.

Assume that the center of gravity of the prestressing steel is approximately $0.85h$ from the middepth of the flange. Then the lever arm of the moment couple

Bridge deck prestressing reinforcement set up prior to concreting.

Full-scale bridge test (Nawy, Goodkind).

$jd \cong 0.80h$. Assume also that the nominal strength of the prestressed steel is $f_{ps} \cong 0.90 f_{pu}$. Then the area A_{ps} of the prestressing tendons is

$$A_{ps} = \frac{M_u/\phi}{0.9 f_{pu}(0.85h)} \qquad (4.60\ a)$$

or

$$A_{ps} = \frac{M_n}{0.76 f_{pu} h} \qquad (4.60\ b)$$

If the compressive block depth a equals the flange thickness h_f, the volume of the compressive block of Figure 4.39(d) in terms of the area $ba = A_c$ is

$$C = 0.85 f_c' A_c$$

$$T = 0.9 f_{pu} A_{ps} = \frac{M_n}{0.8h}$$

From the equilibrium of forces, $C = T$. Hence, the area of the compression flange is

$$A_c' = \frac{M_n}{0.85 f_c'(0.8h)} = \frac{M_n}{0.68 f_c' h} \qquad (4.61)$$

Once the width of the flange is chosen for the first trial and the beam depth is known, the web thickness can be chosen based on the shear requirements to be discussed in Chapter 5. Thereafter, by trial and adjustment, one can select the ideal section for the particular design requirement conditions and proceed to analyze the stresses for the service-load conditions.

4.13 SUMMARY STEP-BY-STEP PROCEDURE FOR LIMIT-STATE-AT-FAILURE DESIGN OF THE PRESTRESSED MEMBERS

1. Determine whether or not partial prestressing is to be chosen, using an effective percentage of nonprestressed steel. Choose a trial depth h based on either 0.6 in. per ft of span or 75 percent of the depth needed for reinforced concrete sections after calculating the required nominal strength $M_n = M_u/\phi$.

2. Select a trial flange thickness such that the total concrete area of the flange $A'_c \cong M_n/0.68f'_c h$, based on choosing a flange width dictated by planning requirements and spacing of beams. Choose a preliminary area of prestressing steel $A_{ps} = M_n/0.76 f_{pu} h$.

3. Use a reasonable value for the steel stress f_{ps} at failure for a first trial. If $f_{pe} < 0.5 f_{pu}$, strain compatibility analysis would thereafter be needed. Determine whether the tendons are bonded or nonbonded. Use the value of the effective prestress f_{pe} from the service-load analysis if that design was already made. If $f_{pe} > 0.5 f_{pu}$, use the approximate values from the following applicable cases:

(a) *Bonded tendons*

$$f_{ps} = f_{pu}\left(1 - \frac{\gamma_p}{\beta_1}\left\{\rho_p \frac{f_{pu}}{f'_c} + \frac{d}{d_p}(\omega - \omega')\right\}\right)$$

(b) *Nonbonded tendons, span/depth ratio ≤ 35*

$$f_{ps} = f_{pe} + 10{,}000 + \frac{f'_c}{100\,\rho_p}$$

(c) *Nonbonded tendons, span/depth ratio > 35*

$$f_{ps} = f_{pe} + 10{,}000 + \frac{f'_c}{300\,\rho_p}$$

4. Determine whether the trial section chosen should be considered rectangular or flanged by determining the position of the neutral axis, $c = a/\beta_1$. If rectangular,

$$a = \frac{A_{ps}f_{ps} + A_s f_y - A'_s f_y}{0.85 f'_c b}$$

If flanged,

$$a = \frac{A_{pw}f_{ps}}{0.85 f'_c b_w}$$

where $A_{pw} = A_{ps}f_{ps} + A_s f_y - 0.85 f'_c (b - b_w)h_f$.

5. If h_f is larger than c and a, analyze the element as a rectangular section singly or doubly reinforced.

6. Find the reinforcement indices ω_p, ω, and ω' for the case $a < h_f$ (neutral axis within the flange; hence, use for a rectangular section).

(a) Rectangular sections with prestressing steel only:

$$\omega_T = \omega_p = \rho_p \frac{f_{ps}}{f'_c} = \frac{A_{ps}}{bd}\frac{f_{ps}}{f'_c}$$

(b) Rectangular sections with compression steel in addition to nonprestressed tension steel:

$$\omega_T = \omega_p + \frac{d}{d_p}(\omega - \omega')$$

If the total index in (a) or (b) is less than or equal to $0.36\beta_1$, then the moment strength is

$$M_n = A_{ps}f_{ps}\left(d_p - \frac{a}{2}\right) + A_s f_y\left(d - \frac{a}{2}\right) + A'_s f_y\left(\frac{a}{2} - d'\right)$$

7. Find the reinforcement indices ω_{pw}, ω_w and ω'_w for the case $a > h_f$ (neutral axis outside the flange), with the total index

$$\omega_T = \omega_{pw} + \frac{d}{d_p}(\omega_w - \omega'_w)$$

The indices are calculated on the basis of the web width b_w. If the total index $\omega_T < 0.36\beta_1$, then

$$M_n = A_{pw}f_{ps}\left(d_p - \frac{a}{2}\right) + A_s f_y(d - d_p) + 0.85f'_c(b - b_w)h_f\left(d_p - \frac{h_f}{2}\right)$$

where

$$a = \frac{A_{pw}f_{ps}}{0.85f'_c b_w}$$

and

$$A_{pw}f_{ps} = A_{ps}f_{ps} + A_s f_y - 0.85f'_c(b - b_w)h_f$$

If the total index $\omega_T > 0.36\beta_1$, the section is overreinforced and the nominal strength is

$$M_n = f'_c b_w d_p^2(0.36\beta_1 - 0.08\beta_1^2) + 0.85f'_c(b - b_w)h_f(d_p - 0.5h_f)$$

8. Check for the minimum required reinforcement $A_s > 0.004A$. Also, check whether $M_u \geq 1.2M_{cr}$ to ensure the use of adequate nonprestressed tension steel, particularly in nonbonded tendons.

9. Select the size and spacing of the nonprestressed tension reinforcement, and compression reinforcement where applicable.

10. Verify that the design moment $M_u = \phi M_n$ is equal to or larger than the factored moment M_u. If not, adjust the design.

A flowchart for programming the step-by-step trial-and-adjustment procedure in analyzing the nominal flexural strength of rectangular and flanged prestressed sections taking d_p as the single-layer cgs depth of tendon is shown in Fig. 4.45. Similarly, a flowchart for programming the nominal flexural strength of prestressed beams using strain-compatibility analysis of *multilayered* strand depths d_{p_1} to d_{p_n} is given in Fig. 4.46. Both charts are applicable to fully prestressed beams that use no mild steel and that allow no tension in the concrete, as well as to "partially prestressed" beams where limited tensile stress is permitted in the concrete through the use of nonprestressed reinforcement. A computer program based on the flowcharts in the two figures can be equally used for a single effective depth d_p of the cgs tendon profile.

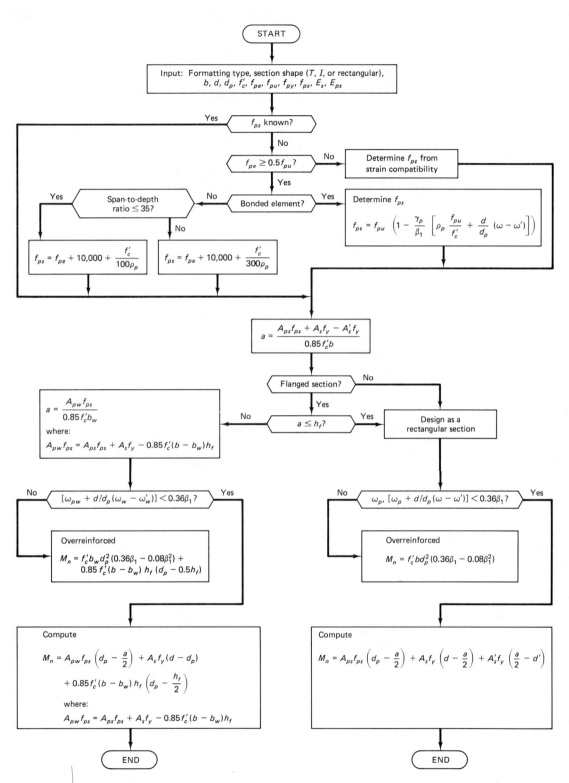

Figure 4.45 Flowchart for flexural analysis of rectangular and flanged prestressed sections based on cgs profile depth.

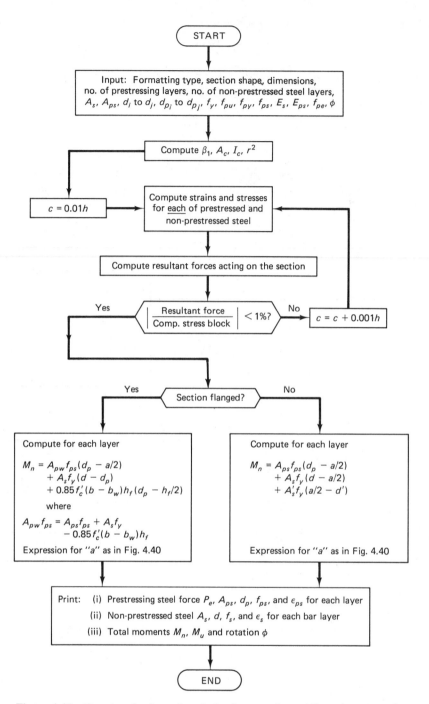

Figure 4.46 Flowchart for flexural analysis of rectangular and flanged prestressed sections using compatibility analysis for individual layers of strands and bars.

4.14 ULTIMATE-STRENGTH DESIGN OF PRESTRESSED SIMPLY SUPPORTED BEAM BY STRAIN COMPATIBILITY

Example 4.9

Design the bonded beam in Example 4.2 by the ultimate-load theory using non-prestressed reinforcement to *partially* carry part of the factored loads. Use strain compatibility to evaluate f_{ps}, given

$f_{pu} = 270,000$ psi (1,862 MPa)

$f_{py} = 0.85 f_{pu}$ for stress-relieved strands

$f_y = 60,000$ psi (414 Mpa)

$f'_c = 5,000$ psi normal-weight concrete (34.5 MPa)

Use seven-wire $\frac{1}{2}$ in. dia strands. The nonprestressed partial mild steel is to be placed with a $1\frac{1}{2}$ in. clean cover, and no compression steel is to be accounted for. No wind or earthquake is taken into consideration.

Solution   From Example 4.2,

$$\text{Service } W_L = 1,100 \text{ plf (16.1 kN/m)}$$

$$\text{Service } W_{SD} = 100 \text{ plf (1.46 kN/m)}$$

$$\text{Assumed } W_D = 393 \text{ plf (5.74 kN/m)}$$

$$\text{Beam span} = 65 \text{ ft (19.8 m)}$$

1. Factored moment (step 1)

$$W_u = 1.4(W_D + W_{SD}) + 1.7 W_L$$

$$= 1.4(100 + 393) + 1.7(1,100) = 2,560 \text{ plf (37.4 kN/m)}$$

The factored moment is given by

$$M_u = \frac{w_u \ell^2}{8} = \frac{2,560(65)^2 12}{8} = 16,224,000 \text{ in.-lb (1,833 kN-m)}$$

and the required nominal moment strength is

$$M_n = \frac{M_u}{\phi} = \frac{16,224,000}{0.9} = 18,026,667 \text{ in.-lb (2,037 kN-m)}$$

2. Choice of preliminary section (step 2)

Assuming a depth of 0.6 inch per foot of span, we can have a trial section depth $h = 0.6 \times 65 \cong 40$ in. (102 cm). Then assume a mild partial steel $4\#6 = 4 \times 0.44 = 1.76 \text{ in}^2$ (11.4 cm^2). From Equation 4.61,

$$A'_c = \frac{M_n}{0.68 f'_c h} = \frac{18,026,667}{0.68 \times 5,000 \times 40} = 132.5 \text{ in}^2 \text{ (855 cm}^2\text{)}$$

Assume a flange width of 18 in. Then the average flange thickness = $132.5/18 \cong 7.5$ in. (191 mm). So suppose the web $b_w = 6$ in. (152 mm), to be subsequently verified for shear requirements. Then from Equation 4.60b,

$$A_{ps} = \frac{M_n}{0.76 f_{pu} h} = \frac{18,026,667}{0.76 \times 270,000 \times 40} = 2.19 \text{ in}^2 \ (14 \text{ cm}^2)$$

and the number of $\frac{1}{2}$ in. stress-relieved wire strands $= 2.19/0.153 = 14.31$. So try thirteen $\frac{1}{2}$ in. strands:

$$A_{ps} = 13 \times 0.153 = 1.99 \text{ in}^2 \ (12.8 \text{ cm}^2)$$

3. **Calculate the stress f_{ps} in the prestressing tendon at nominal strength using the strain-compatibility approach (step 3)**

The geometrical properties of the trial section are very close to the assumed dimensions for the depth h and the top flange width b. Hence, use the following data for the purpose of the example:

$A_c = 377 \text{ in}^2$

$c_t = 21.16 \text{ in.}$

$d_p = 15 + e_t = 15 + 21.16 = 36.16 \text{ in.}$

$r^2 = 187.5 \text{ in}^2$

$e = 15 \text{ in. at midspan}$

$e^2 = 225 \text{ in}^2$

$e^2/r^2 = 225/187.5 = 1.20$

$E_c = 57,000\sqrt{5,000} = 4.03 \times 10^6 \text{ psi} \ (27.8 \times 10^3 \text{ MPa})$

$E_p = 28 \times 10^6 \text{ psi} \ (193 \times 10^3 \text{ MPa})$

The maximum allowable compressive strain ϵ_c at failure $= 0.003$ in./in. Assume that the effective prestress at service load is $f_{pe} \cong 155,000 \text{ psi} \ (1,069 \text{ MPa})$.

(a)
$$\epsilon_1 = \epsilon_{pe} = \frac{f_{pe}}{E_p} = \frac{155,000}{28 \times 10^6} = 0.0055 \text{ in./in.}$$

$$P_e = 13 \times 0.153 \times 155,000 = 308,295 \text{ lb}$$

The increase in prestressing steel strain as the concrete is decompressed by the increased external load (see Figure 4.3 and Equation 4.3c) is given as

$$\epsilon_2 = \epsilon_{decomp} = \frac{P_e}{A_c E_c}\left(1 + \frac{e^2}{r^2}\right)$$

$$= \frac{308,295}{377 \times 4.03 \times 10^6}(1 + 1.20) = 0.0004 \text{ in./in.}$$

(b) Assume that the stress $f_{ps} \cong 205,000 \text{ psi}$ as a first trial. Suppose the neutral axis inside the flange is verified. Then, from Equation 4.30, (4.41a)

$$a = \frac{A_{ps} f_{ps} + A_s f_y}{0.85 f_c' b} = \frac{1.99 \times 205,000 + 1.76 \times 60,000}{0.85 \times 5,000 \times 18}$$

$$= 6.71 \text{ in. } (17 \text{ cm}) < h_f = 7.5 \text{ in.}$$

Hence, the equivalent compressive block is inside the flange and the section has to be treated as rectangular.

Accordingly, for 5,000 psi concrete,

$$\beta_1 = 0.85 - 0.05 = 0.8$$

$$c = \frac{a}{\beta_1} = \frac{6.71}{0.80} = 8.39 \text{ in. (22.7 cm)}$$

$$d = 40 - (1.5 + \tfrac{1}{2} \text{ in. for stirrups} + \tfrac{5}{16} \text{ in. for bar}) \cong 37.6 \text{ in.}$$

The increment of strain due to overload to the ultimate, from Equation 4.31b, is

$$\epsilon_3 = \epsilon_c\left(\frac{d - c}{c}\right) = 0.003\left(\frac{37.6 - 8.39}{8.39}\right) = 0.0104 \text{ in./in.}$$

and the total strain is

$$\epsilon_{ps} = \epsilon_1 + \epsilon_2 + \epsilon_3$$

$$= 0.0055 + 0.0004 + 0.0104 = 0.0163 \text{ in./in.}$$

From the stress-strain diagram in Figure 4.42, the f_{ps} corresponding to $\epsilon_{ps} = 0.0163$ is 230,000 psi.

Second trial for f_{ps} value

Assume $f_{ps} = 229,000 \text{ psi}$

$$a = \frac{1.99 \times 229,000 + 1.76 \times 60,000}{0.85 \times 5,000 \times 18} = 7.34 \text{ in.}$$

$$c = \frac{7.34}{0.80} = 9.17 \text{ in.}$$

$$\epsilon_3 = 0.003\left(\frac{37.6 - 9.17}{9.17}\right) = 0.0093$$

Then the total strain is $\epsilon_{ps} = 0.0055 + 0.0004 + 0.0093 = 0.0152 \text{ in./in.}$

From Figure 4.42, $f_{ps} = 229,000 \text{ psi (1,579 MPa); use}$

$$A_s = 4 \, \# \, 6 = 1.76 \text{ in.}$$

4. **Available moment strength (steps 6 through 10)**
   From Equation 4.42c,

$$M_n = 1.99 \times 229,000\left(36.16 - \frac{7.34}{2}\right) + 1.76 \times 60,000\left(37.6 - \frac{7.34}{2}\right)$$

$$= 14,806,017 + 3,583,008 = 18,389,025 \text{ in.-lb (2,078 kN-m)}$$

$$> \text{required } M_n = 18,026,667 \text{ in.-lb, O.K.}$$

The percentage of moment resisted by the nonprestressed steel is

$$\frac{3,583,008}{18,026,667} \cong 20\%$$

5. **Check for minimum and maximum reinforcement (steps 6 and 9)**
   (a) Min $A_s = 0.004A$

   where A is the area of the part of the section between the tension face and the cgc.
   From the cross section of Figure 4.8,

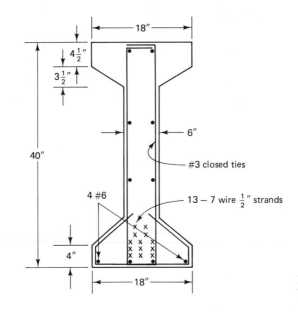

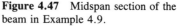

Figure 4.47 Midspan section of the beam in Example 4.9.

$$A = 377 - 17\left(4.125 + \frac{1.375}{2}\right) - 6(21.16 - 5.5) \cong 201 \text{ in}^2$$

Min $A_s = 0.004 \times 201 = 0.80 \text{ in}^2 < 1.76$ used, O.K.

(b) The maximum steel index, from Equation 4.57b, is

$$\omega_p + \frac{d}{d_p}(\omega - \omega') \le 0.36\beta_1 \le 0.29 \text{ for } \beta_1 = 0.80$$

and the actual total reinforcement index is

$$\omega_T = \frac{1.99 \times 229,000}{18 \times 36.16 \times 5,000} + \frac{37.6}{36.16}\left(\frac{1.76 \times 60,000}{18 \times 37.6 \times 5,000}\right)$$

$$= 0.14 + 0.03 = 0.17 < 0.29, \text{ O.K.}$$

6. Choice of section for ultimate load (step 11)

From steps 1–5 of the design, the section in Example 4.2 with the modifications shown in Fig. 4.47 has the normal moment strength M_n that can carry the factored load, provided that four #6 nonprestressed bars are used at the tension side as a partially prestressed section.

So one can adopt the section for flexure, as it also satisfies the service-load flexural stress requirements both at midspan and at the support. Note that the section could only develop the required nominal strength $M_n = 18,026,667$ in.-lb by the addition of the nonprestressed bars at the tension face to resist 20 percent of the total *required* moment strength. Note also that this section is adequate with a concrete $f'_c = 5,000$ psi, while the section in Example 4.2 has to have $f'_c = 6,000$ psi strength in order not to exceed the allowable service-load concrete stresses. Hence, ultimate-load computations are necessary in prestressed concrete design to ensure that the constructed elements can carry all the factored load and are thus an integral part of the total design.

4.15 STRENGTH DESIGN OF BONDED PRESTRESSED BEAM USING APPROXIMATE PROCEDURES

Example 4.10

Design the beam in Example 4.9 as a partially prestressed beam using the ACI approximate procedures if permissible. Use the exact standard section used in Example 4.2 with (a) bonded prestressing steel, and (b) nonbonded prestressing steel. Neglect the contribution of the compressive nonprestressed steel.

Solution

1. **Section properties (steps 1 and 2)**

   The width of the top flange in Example 4.2 is $b = 17$ in., and its average thickness is

   $$h_f = 4\tfrac{1}{8} + \frac{1\tfrac{3}{8}}{2} = 4.81 \text{ in. (12.2 cm)}$$

   Try (four) #6 (four 12.7 mm dia) nonprestressed tension steel bars in this cycle in addition to the prestressed steel beams.

2. **Stress f_{ps} in prestressing steel at nominal strength (step 3)**

   From the solution in Example 4.8, the neutral axis in this example falls outside the flange; hence, design as a flanged section with

   $$f_{pe} \cong 155,000 \text{ psi}$$

   $$0.5 f_{pu} = 0.50 \times 270,000 = 135,000 \text{ psi}$$

   $$f_{pe} > 0.5 f_{pu}$$

   Hence, one can use the ACI approximate procedure for determining f_{ps}.

   Bonded case

   From Equation 4.51,

   $$f_{ps} = f_{pu}\left(1 - \frac{\gamma_p}{\beta_1}\left[\rho_p \frac{f_{pu}}{f_c'} + \frac{d}{d_p}(\omega - \omega')\right]\right)$$

   $$\frac{f_{ps}}{f_{pu}} = \frac{229,000}{270,000} = 0.85, \quad \text{use } \gamma_p = 0.40$$

   $$A_{ps} = 13 \times 0.153 = 1.99 \text{ in}^2$$

   $$A_s = 4 \times 0.44 = 1.76 \text{ in}^2$$

   $$\rho_p = \frac{A_{ps}}{bd_p} = \frac{1.99}{17 \times 37.16} = 0.0032$$

   $$\omega = \frac{A_s}{bd} \times \frac{f_y}{f_c'} = \frac{1.76}{17 \times 37.6} \times \frac{60,000}{5,000} = 0.033$$

   For $\omega' = 0$,

   $$f_{ps} = 270,000\left(1 - \frac{0.04}{0.80}\left[0.0032 \times \frac{270,000}{5,000} + \frac{37.6}{36.16}(0.033)\right]\right)$$

   $$= 270,000 \times 0.896 = 241,920 \text{ psi (1,668 MPa)}$$

4.47a

From Equation 4.36a, the depth of the compressive block for a T-section is

$$a = \frac{A_{pw}f_{ps}}{0.85f'_c b_w}$$

where $A_{pw}f_{ps} = A_{ps}f_{ps} + A_s f_y - 0.85f'_c(b - b_w)h_f$

$$= 1.99 \times 241{,}920 + 1.76 \times 60{,}000 - 0.85$$

$$\times 5{,}000(17 - 6)4.81$$

$$= 481{,}421 + 105{,}600 - 224{,}868 = 362{,}153 \text{ lb (604 kN)}$$

$$a = \frac{362{,}153}{0.85 \times 5{,}000 \times 6} = 14.2 \text{ in. (36 cm)} > h_f$$

Hence, the neutral axis is outside the flange, and analysis has to be based on a T-section. We thus have

$$\rho_p = \frac{A_{ps}}{b_w d_p} = \frac{1.99}{6 \times 36.16} = 0.0092$$

If the neutral axis were inside the flange, the width b would be used instead of b_w for calculating ρ_p. Consequently,

$$\omega_w = \frac{A_s}{b_w d} \times \frac{f_y}{f'_c} = \frac{1.76}{6 \times 37.6} \times \frac{60{,}000}{5{,}000} = 0.0936$$

$$f_{ps} = 270{,}000\left(1 - \frac{0.40}{0.80}\left[0.0092 \times \frac{270{,}000}{5{,}000} + \frac{37.6}{36.16}(0.0936 - 0)\right]\right)$$

$$= 189{,}793 \text{ psi (1,309 MPa)}$$

$$A_{pw}f_{ps} = A_{ps}f_{ps} + A_s f_y - 0.85f'_c(b - b_w)h_f$$

$$= 1.99 \times 189{,}793 + 1.76 \times 60{,}000 - 0.85 \times 5{,}000(17 - 6)$$

$$\times 4.81$$

$$= 377{,}688 + 105{,}600 - 224{,}868 = 258{,}420 \text{ lb}$$

$$a = \frac{258{,}420}{0.85 \times 5{,}000 \times 6} = 10.13 \text{ in. (25.7 cm)}$$

3. **Available nominal moment strength (steps 4–8)**

$$M_n = A_{pw}f_{ps}\left(d_p - \frac{a}{2}\right) + A_s f_y(d - d_p) + 0.85f'_c(b - b_w)h_f\left(d_p - \frac{h_f}{2}\right)$$

$$= 258{,}420\left(36.16 - \frac{10.13}{2}\right) + 1.76(60{,}000)(37.6 - 36.16)$$

$$+ 0.85(5{,}000)(17 - 6) \times 4.81\left(36.16 - \frac{4.81}{2}\right) = 15{,}781{,}745 \text{ in.-lb}$$

$(1{,}783 \text{ kN-m}) <$ required $M_n = 18{,}026{,}667$ in.-lb (2,037 kN-m),

section no good

Proceed to another trial and adjustment cycle using more nonprestressed reinforcement. Try five #8s (five 25.4 mm dia), $A_s = 3.95$ in^2 (25.5 cm^2). We have

$$\omega_w = \frac{3.95}{6 \times 37.6} \times \frac{60,000}{5,000} = 0.21$$

giving $f_{ps} = 173,438$ psi and $A_{pw}f_{ps} = 357,303$ lb (1,589 kN). So

$$a = \frac{357,303}{0.85 \times 5,000 \times 6} = 14 \text{ in. (35.6 cm)}$$

$$M_n = 357,274 \left(36.16 - \frac{14}{2}\right) + 3.95 \,(60,000)(37.6 - 36.16)$$

$$+ 0.85 \,(5,000)(17 - 6) \times 4.81 \left(36.16 - \frac{4.81}{2}\right)$$

$$= 18,350,637 \text{ in.-lb (2,074 kN-m)}$$

$$> \text{Required } M_n = 18,026,667 \text{ in.-lb, O.K.}$$

Hence, use five #8 nonprestressed bars at the bottom fibers, and adopt the design for the bonded case.

Nonbonded case

$$\text{Span-to-depth ratio} = \frac{65 \times 12}{40} = 19.5 < 35$$

Hence, from Equation 4.52a,

$$f_{ps} = f_{pe} + 10,000 + \frac{f'_c}{100\,\rho_p} = 155,000 + 10,000 + \frac{5,000}{100 \times 1.99/(6 \times 36.16)}$$

$$= 170,451 \text{ psi (1,175 MPa)}$$

Notice that $b_w = 6$ in. is used here for ρ_p, since it is now known that the section behaves like a T-beam, as the neutral axis is below the flange. Thus,

$$f_{ps} = 170,451 \text{ psi (1,175 MPa)}$$

4. **Selection of nonprestressed steel**

Try five #8 nonprestressed tension reinforcements to resist part of the factored moment:

$$A_s = 5 \times 0.79 = 3.95 \text{ in}^2 \text{ (25.4 cm}^2\text{)}$$

$$A_{pw}f_{ps} = 1.99 \times 170,421 + 3.95 \times 60,000 - 0.85 \times 5,000(17 - 6)4.81$$

$$= 351,331 \text{ psi}$$

$$a = \frac{A_{pw}f_{ps}}{0.85 f'_c b_w} = \frac{351,331}{0.85 \times 5,000 \times 6} = 13.8 \text{ in. (35.1 cm)}$$

5. **Available moment strength** (steps 4–8)

From Equation 4.48,

$$\text{Available } M_n = 351,331 \left(36.16 - \frac{13.8}{2}\right) + 3.95 \times 60,000(37.6 - 36.16)$$

$$+ 0.85 \times 5,000(17 - 6) \times 4.81 \times \left(36.16 - \frac{4.81}{2}\right)$$

Diaphragm anchorage.

$$= 18{,}221{,}612 \text{ in.-lb } (2{,}059 \text{ kN-m})$$

$$> \text{Req. } M_n = 18{,}026{,}667 \text{ in.-lb, O.K.}$$

6. **Check for minimum reinforcement**

From Equation 4.25, the cracking moment M_{cr} is given by

$$M_{cr} = f_r S_b + P_e \left(e + \frac{r^2}{c_b} \right)$$

From Example 4.2, $f_r = 7.5\sqrt{5{,}000} = 530.3$ psi (3.7 MPa). So since $S_b = 3{,}750 \text{ in}^3$, $e = 15$ in., $r^2/c_b = 187.5/18.84 = 9.95$ in., and $P_e = 308{,}255$ lb (1,371 kN), we get

$$M_{cr} = 530.3 \times 3{,}750 + 308{,}295(15 + 9.95)$$

$$= 9{,}680{,}585 \text{ in.-lb } (1{,}090 \text{ kN-m})$$

$$1.2M_{cr} = 1.2 \times 9{,}680{,}585 = 11{,}616{,}702 \text{ in.-lb } (1{,}313 \text{ kN-m})$$

$$M_u = \phi M_n = 0.90 \times 18{,}026{,}667$$

$$= 16{,}224{,}000 \text{ in.-lb } (1{,}833 \text{ kN-m})$$

Finally, from Equation 4.54a,

$$M_u > 1.2M_{cr}$$

Hence, the requirement for minimum reinforcement is satisfied for both the nonbonded and the bonded case.

Accordingly, adopt the design that uses the concrete section in Example 4.2 and include five #8 nonprestressed steel bars at the tension side. Note that the moment strength capacity of the nonbonded section for the same area of nonprestressed steel is less than the moment strength capacity of the bonded section, which is expected (18,221,612 in.-lb vs. 18,349,792 in.-lb).

If $f'_c = 6,000$ psi would have been used in the strength design in this example, as it was in the service-load design of this section in Example 4.2, less mild steel reinforcement would have been needed.

4.16 USE OF THE ANSI LOAD AND STRENGTH REDUCTION FACTORS IN EXAMPLE 4.10

Using ANSI load factors for an alternate solution,

$$U = 1.2D + 1.6L$$

$$\phi = 0.85 \text{ for flexure}$$

$$w_u = 1.2(100 + 397) + 1.6(1,100) = 2,352 \text{ plf}$$

$$M_u = \frac{2,352(65)^2 \times 12}{8} = 14,905,800 \text{ in.-lb}$$

$$M_n = \frac{M_u}{\phi} = \frac{14,905,800}{0.85} = 17,536,236 \text{ in.-lb}$$

M_n by the ACI Code load factors $= 18,026,667$ in.-lb

$$\text{Percentage difference} = \frac{18,026,667 - 17,536,236}{18,026,667} = 2.7\%$$

Such a small percentage difference has no significant effect on the design.

REFERENCES

4.1 ACI Committee 318, *Building Code Requirements for Reinforced Concrete, ACI Standard 318–89.* Detroit: American Concrete Institute, 1989.

4.2 ACI Committee 318, *Commentary on Building Code Requirements for Reinforced Concrete, ACI 318R–89.* Detroit: American Concrete Institute, 1989.

4.3 Nawy, E. G., *Reinforced Concrete—A Fundamental Approach.* Englewood Cliffs, N.J.: Prentice Hall, 1985.

4.4 Nawy, E. G., and Chiang, J. Y., "Serviceability Behavior of Post-Tensioned beams." *Journal of the Prestressed Concrete Institute* 25, Jan.-Feb. 1980: 74–85.

4.5 Nawy, E. G., and Potyondy, G. J., "Moment Rotation, Cracking, and Deflection of Spirally Bound Pretensioned Prestressed Concrete

Beams." *Engineering Research Bulletin No 51.* New Brunswick, N.J.: Bureau of Engineering Research, Rutgers University, 1970, pp. 1–97.

4.6 Nawy, E. G., Yong, Y. K., and C.P. Gadegbekn. "Anchorage Zone Stresses of Post-Tensioned Prestressed Beams Subjected to Shear Forces." *Journal of the Structural Division,* ASCE 113, No. 8 (1987): 1789–1805.

4.7 Nawy, E. G., and Goodkind, H. "Longitudinal Crack in Prestressed Box Beam." *Journal of the Prestressed Concrete Institute* 14 (1969): 38–42.

4.8 Nawy, E. G., and Huang, P. T. "Crack and Deflection Control of Pretensioned Prestressed Beams." *Journal of the Prestressed Concrete Institute* 22 (1977): 30–47.

4.9 Prestressed Concrete Institute. *PCI Design Handbook, Precast and Prestressed Concrete.* Chicago: Prestressed Concrete Institute, 1985.

4.10 Nawy, E. G. "Flexural Cracking Behavior of Pretensioned and Post-Tensioned Beams—The State of the Art." *Journal of the American Concrete Institute,* December 1985, pp. 890–900.

4.11 American Association of State Highway and Transportation Officials. *AASHTO Standard Specifications for Highway Bridges.* Washington, D.C.: AASHTO, 1984.

4.12 Nilson, A. H. "Flexural Design Equations for Prestressed Concrete Members." *Journal of the Prestressed Concrete Institute* 14 (1969): 62–71.

4.13 Lin, T. Y., and Burns, N. H. *Design of Prestressed Concrete Structures.* New York: Wiley, 1981.

4.14 Nilson, A. H. *Design of Prestressed Concrete.* New York: Wiley, 1987.

4.15 Naaman, A. E. *Prestressed Concrete Analysis and Design.* New York: McGraw Hill, 1982.

4.16 Libby, J. R. *Modern Prestressed Concrete.* New York: Van Nostrand Reinhold, 1984.

4.17 Abeles, P. W., and Bardhan-Roy, B. K. *Prestressed Concrete Designer's Handbook.* 3d ed. London: Viewpoint Publications, 1981.

4.18 Guyon, Y. *Limit State Design of Prestressed Concrete; Vol. 1, Design of Section.* New York: Wiley, 1972.

4.19 Marshall, W. T., and Mattock, A. H. "Control of Horizontal Cracking in the Ends of Pretensioned Prestressed Concrete Girders." *Journal of the Prestressed Concrete Institute* 7 (1962): 56–74.

4.20 Ezeldin, A., and Balaguru, P. N. "Analysis of Partially Prestressed Beams for Strength and Ser- viceability Using Microcomputers." Paper presented at the First Canadian Conference on Computer Applications in Civil Engineering, MacMaster University, Hamilton, Ontario, Canada, May 1986.

4.21 Gergely, P., and Sozen, M. A., "Design of Anchorage Zone Reinforcement in Prestressed Concrete Beams." *Journal of the Prestressed Concrete Institute* 12 (1967): 63–75.

4.22 Baker, A. L. L. *The Ultimate Load Theory Applied to the Design of Reinforcement and Prestressed Concrete Frames.* London: Concrete Publications, 1956.

4.23 Ellingwood, B., McGregor, J. G., Galambos, T. V., and Cornell, C. A. "Probability Based Load Criteria: Load Factors and Load Combinations." *Journal of the Structural Division, ASCE* 108 (1982): 978–997.

4.24 American National Standards Institute. *Minimum Design Loads for Buildings and Other Structures, ANSI A58.* New York: ANSI, 1982, p. 100.

4.25 Stone, W. C., and Breen, J. E. "Design of Post-Tensioned Girder Anchorage Zones." *Journal of the Prestressed Concrete Institute* 29 (1984): 64–109.

4.26 Nawy, E. G., Yong, Y. K., and Gadebeku, B.K. "Anchorage Zone Stresses of Post-Tensioned Prestressed Beams." *Proceedings of the International Conference on Structural Mechanics of Reinforced and Prestressed Concrete,* Chinese Academy of Sciences, Nanjing Institute of Technology, 1986.

PROBLEMS

4.1 Design, for service-load and ultimate-load conditions, a pretensioned symmetrical I-section beam to carry a superimposed dead load of 750 plf (10.95 kN/m) and a service live load of 1,500 plf (21.90 kN/m) on a 50 ft (15.2 m) simply supported span. Assume that the sectional properties are $b = 0.5h$, $h_f = 0.2h$, and $b_w = 0.40b$, and suppose as given the following data:

$$f_{pu} = 270,000 \text{ psi } (1,862 \text{ MPa})$$
$$E_{ps} = 28.5 \times 10^6 \text{ psi}$$
$$(196 \times 10^3 \text{ MPa})$$
$$f'_c = 5,000 \text{ psi } (34.5 \text{ MPa}) \text{ normal-weight concrete}$$

$$f'_{ci} = 3,500 \text{ psi } (24.1 \text{ MPa})$$
$$f_t = 12\sqrt{f'_c} \text{ assuming deflection is not critical}$$

Sketch the design details, including the anchorage zone reinforcement, and arrangement of strands for (a) straight-tendon case, and (b) a harped tendon at the third span points with end eccentricity zero. Assume total prestress losses of 22 percent.

4.2 Solve Problem 4.1 if the beam is post-tensioned bonded and the tendon is draped. Use strain compatibility to determine the value of the tendon stress f_{ps} at nominal strength.

4.3 A double-T pretensioned roof beam is shown in Figure P4.3. It has a simple span of 74 ft (22.6 m) and carries superimposed service live and dead loads of 60 psf (2,873 Pa; W_{SD} part of load = 25 psf). It also carries a 2 in. (5.1 cm) concrete topping. Design the prestressing reinforcement and the appropriate eccentricities using 270 grade prestressing strands (f_{pu} = 1,862 MPa) with a total prestress loss of 20 percent. Use the appropriate percentage of nonprestressed mild steel for partial prestressing behavior at the limit state at failure. Assume the strands to be harped at midspan, and sketch the reinforcing details including the anchorage zone strands. Also, draw the distribution of stresses for the various loading stages in your solution. The following data are given:

f_{pu} = 270,000 psi, stress-
   relieved strands
   (1,862 MPa)

E_{ps} = 28 × 10⁶ psi
   (193 × 10³ MPa)

f'_c = 5,000 psi (34.5 MPa) normal-
   weight concrete

f'_c for topping = 3,000 psi
   (20.7 MPa)
   normal-weight
   concrete

f'_{ci} = 4,000 psi

V/S = 1.79 in.

e_c = 17.71 in.

	Untopped	Topped
A_c	567 in² (3,658 cm²)	—
I_c	55,464 in⁴	71,886 in⁴
c_b	21.21 in.	23.66 in.
c_t	10.79 in.	10.34 in.
S_b	2,615 in³	3,038 in³
S^t	5,140 in³	6,952 in³
W_D	591 plf	791 plf

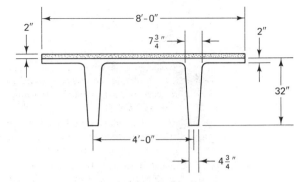

Figure P4.3 Double-T cross section.

4.4 A bridge girder has a simple span of 55 ft (16.8 m). It is subjected to a total superimposed service load of 4,600 plf (67.2 kN/m). Design the section as a post-tensioned bonded beam using an AASHTO standard section with parabolically draped tendons. Assume a 7 in. situ-cast slab over the precast section and the following data:

f_{pu} = 270,000 psi stress relieved
   (1,862 MPa)

E_{ps} = 28 × 10⁶ psi
   (193 × 10³ MPa)

f'_c = 5,000 psi (34.5 MPa)
   normal-weight concrete

slab f'_c = 3,000 psi (20.7 MPa)
   normal-weight concrete

f'_{ci} = 4,000 psi (27.6 MPa)

Design the bridge section for service-load and ultimate-load conditions, and detail all the reinforcement including the anchorage zone steel and the nonprestressed tensile steel. Assume the section to be constant throughout the span, and, for the limit state at failure analysis, find the stress f_{ps} at nominal strength by (a) the ACI approximate procedure, and (b) strain compatibility.

4.5 Solve problem 4.4 if the draped tendons are nonbonded, and compare the two solutions.

CHAPTER 5

Shear and Torsional Strength Design

5.1 INTRODUCTION

This chapter presents procedures for the design of prestressed concrete sections to resist shear and torsional forces resulting from externally applied loads. Since the strength of concrete in tension is considerably lower than its strength in compression, design for shear and torsion becomes of major importance in all types of concrete structures.

The behavior of prestressed concrete beams at failure in shear or combined shear and torsion is distinctly different from their behavior in flexure: they fail abruptly without sufficient advance warning, and the diagonal cracks that develop are considerably wider than the flexural cracks. Both shear and torsional forces result in shear stress. Such a stress can result in principal tensile stresses at the critical section which can exceed the tensile strength of the concrete.

Photos in this chapter show typical beam shear failure and torsion failure. Notice the curvelinear plane of twist depicting torsional failure caused by the imposed torsional moments. As will be discussed in subsequent sections, the shearing stresses in regular beams are caused, not by direct shear or pure torsion, but by a combination of external loads and moments. This leads to *diagonal tension,* or flexural shear stresses, in the member. Only in special applications in certain structural systems are direct shear or pure torsion applied. Examples of such cases are corbels or brackets involving direct shear, or a cantilever balcony involving essentially a direct twist on the supporting beam.

Empire State Performing Arts Center, Albany, New York, Ammann & Whitney design, prestressed concrete shell ring. (*Courtesy,* New York Office of General Services.)

5.2 BEHAVIOR OF HOMOGENEOUS BEAMS IN SHEAR

Consider the two infinitesimal elements A_1 and A_2 of a rectangular beam in Figure 5.1(a) made of homogeneous, isotropic, and linearly elastic material. Figure 5.1(b) shows the bending stress and shear stress distributions across the depth of the section. The tensile normal stress f_t and the shear stress v are the values in element A_1 across plane a_1–a_1 at a distance $\bar{y}$ from the neutral axis. From the principles of classical mechanics, the normal stress f and the shear stress v for element A_1 can be written as

$$f = \frac{My}{I} \tag{5.1}$$

or

$$v = \frac{VA\bar{y}}{Ib} = \frac{VQ}{Ib} \tag{5.2}$$

where M and V = bending moment and shear force at section a_1–a_1

$\quad\quad\quad A$ = cross-sectional area of the section at the plane passing through the centroid of element A_1

$\quad\quad\quad y$ = distance from the element to the neutral axis

$\quad\quad\quad \bar{y}$ = distance from the centroid of A to the neutral axis

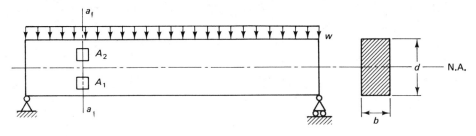

(a)

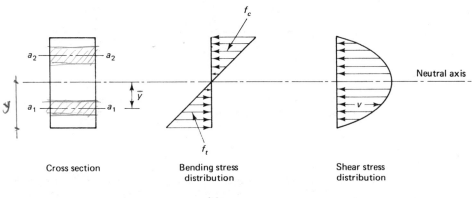

(b)

Figure 5.1 Stress distribution for a typical homogeneous rectangular beam.

Typical diagonal tension (flexure shear) failure at rupture load level. (Test by Nawy et al.)

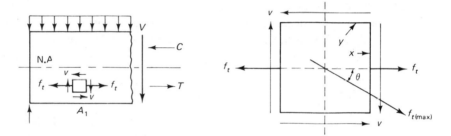

(a)

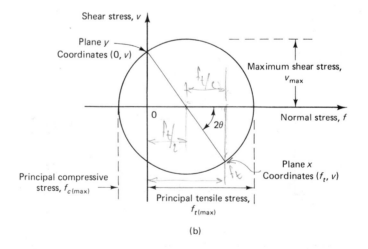

(b)

Figure 5.2 Stress state in elements A_1 and A_2. (a) Stress state in element A_1. (b) Mohr's circle representation, element A_1. (c) Stress state in element A_2. (d) Mohr's circle representation, element A_2.

Shear and Torsional Strength Design   Chap. 5

I = moment of inertia of the cross section

Q = statical moment of the cross-sectional area above or below that level about the neutral axis

b = width of the beam

Figure 5.2 shows the internal stresses acting on the infinitesimal elements A_1 and A_2. Using Mohr's circle in Figure 5.2(b), the principal stresses for element A_1 in the tensile zone below the neutral axis become

$$f_{t(max)} = \frac{f_t}{2} + \sqrt{\left(\frac{f_t}{2}\right)^2 + v^2} \qquad \text{principal tension} \qquad \text{(5.3 a)}$$

$$f_{c(max)} = \frac{f_t}{2} - \sqrt{\left(\frac{f_t}{2}\right)^2 + v^2} \qquad \text{principal compression} \qquad \text{(5.3 b)}$$

and

$$\tan 2\theta_{max} = \frac{v}{f_t/2} \qquad \text{(5.3 c)}$$

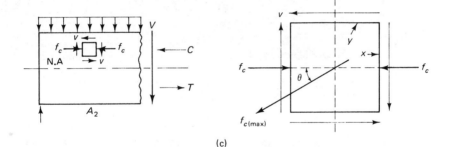

(c)

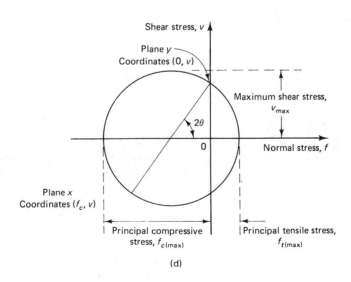

(d)

Figure 5.2 (*continued*)

Simply supported beam prior to developing diagonal tension crack in flexure shear (load stage 11). (Test by Nawy et al.)

Principal diagonal tension crack at failure of beam in the preceding photograph (load stage 12).

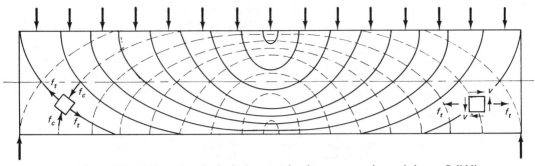

Figure 5.3 Trajectories of principal stresses in a homogeneous isotropic beam. Solid lines are tensile trajectories, dashed lines are compressive trajectories.

5.3 BEHAVIOR OF CONCRETE BEAMS AS NONHOMOGENEOUS SECTIONS

The behavior of reinforced and prestressed concrete beams differs from that of steel beams in that the tensile strength of concrete is about one-tenth of its strength in compression. The compression stress f_c in element A_2 of Figure 5.2(b) above the neutral axis prevents cracking, as the maximum principal stress in the element is in compression. For element A_1 below the neutral axis, the maximum principal stress is in tension; hence cracking ensues. As one moves toward the support, the bending moment and hence f_t decreases, accompanied by a corresponding increase in the shear stress. The principal stress $f_{t(max)}$ in tension acts at an approximately 45° plane to the normal at sections close to the support, as seen in Figure 5.3. Because of the low tensile strength of concrete, diagonal cracking develops along planes perpendicular to the planes of principal tensile stress—hence the term *diagonal tension cracks*. To prevent such cracks from opening, special diagonal tension reinforcement has to be provided.

If f_t close to the support in Figure 5.3 is assumed equal to zero, the element becomes nearly in a state of pure shear, and the principal tensile stress, using Equation 5.3 b, would be equal to the shear stress v on a 45° plane. It is this diagonal tension stress that causes the inclined cracks.

Definitive understanding of the correct shear mechanism in reinforced concrete is still incomplete. However, the approach of ACI-ASCE Joint Committee 426 gives a systematic empirical correlation of the basic concepts developed from extensive test results.

5.4 CONCRETE BEAMS WITHOUT DIAGONAL TENSION REINFORCEMENT

In regions of large bending moments, cracks develop almost perpendicular to the axis of the beam. These cracks are called *flexural cracks.* In regions of high shear due to the diagonal tension, the inclined cracks develop as an extension of the flexural crack and are termed *flexure shear cracks.* Figure 5.4 portrays the types of cracks expected

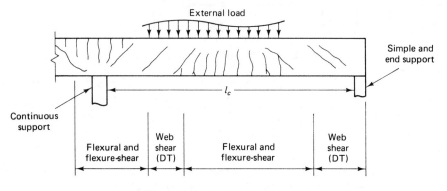

Figure 5.4 Crack categories.

in a reinforced concrete beam with or without adequate diagonal tension reinforcement.

In prestressed beams, the section is mostly in compression at service load. From Figures 5.2 c and d, the principal stresses for element A_2 would be

$$f_{t(\text{max})} = -\frac{f_c}{2} + \sqrt{(f_c/2)^2 + v^2} \qquad \text{principal tension} \qquad (5.4\text{ a})$$

$$f_{c(\text{max})} = -\frac{f_c}{2} - \sqrt{(f_c/2)^2 + v^2} \qquad \text{principal compression} \qquad (5.4\text{ b})$$

and

$$\tan 2\theta_{\text{max}} = \frac{v}{f_c/2} \qquad (5.4\text{ c})$$

5.4.1 Modes of Failure of Beams without Diagonal Tension Reinforcement

The slenderness of the beam, that is, its shear span-to-depth ratio, determines the failure mode of the beam. Figure 5.5 demonstrates schematically the failure patterns for the different slenderness ratio limits. The shear span a for concentrated load is the distance between the point of application of the load and the face of support. For distributed loads, the shear span l_c is the clear beam span. Fundamentally, three modes of failure or their combinations occur: flexural failure, diagonal tension failure, and shear compression failure (web shear). The more slender the beam, the stronger the tendency toward flexural behavior, as seen from the following discussion.

5.4.2 Flexural Failure (F)

In the region of flexural failure, cracks are mainly vertical in the middle third of the beam span and perpendicular to the lines of principal stress. These cracks result from a very small shear stress v and a dominant flexural stress f which results in an almost horizontal principal stress $f_{t(\text{max})}$. In such a failure mode, a few very fine vertical cracks

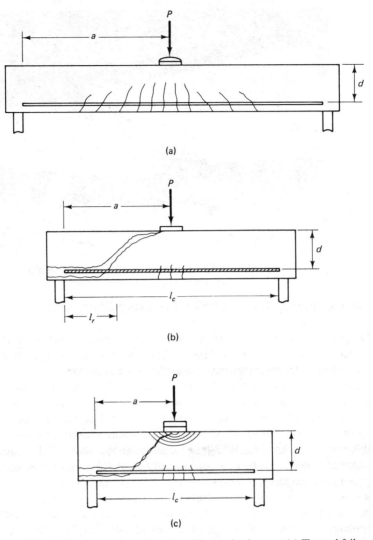

Figure 5.5 Failure patterns as a function of beam slenderness. (a) Flexural failure. (b) Diagonal tension failure (flexure shear). (c) Shear compression failure (web shear).

start to develop in the midspan area at about 50 percent of the failure load in flexure. As the external load increases, additional cracks develop in the central region of the span and the initial cracks widen and extend deeper toward the neutral axis and beyond, with a marked increase in the deflection of the beam. If the beam is under-reinforced, failure occurs in a ductile manner by initial yielding of the main longitudinal flexural reinforcement. This type of behavior gives ample warning of the imminence of collapse of the beam. The shear span-to-depth ratio for this behavior exceeds a value of 5.5 in the case of concentrated loading, and in excess of 16 for distributed loading.

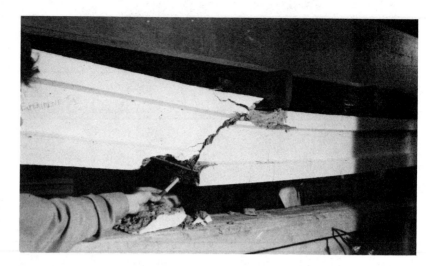

Shear failure in prestressed I-beam (Nawy et al.).

5.4.3 Diagonal Tension Failure (Flexure Shear, FS)

Diagonal tension failure precipitates if the strength of the beam in diagonal tension is lower than its strength in flexure. The shear span-to-depth ratio is of *intermediate* magnitude, varying between 2.5 and 5.5 for the case of concentrated loading. Such beams can be considered of intermediate slenderness. Cracking starts with the development of a few fine vertical flexural cracks at midspan, followed by the destruction of the bond between the reinforcing steel and the surrounding concrete at the support. Thereafter, without ample warning of impending failure, two or three diagonal cracks develop at about $1\frac{1}{2}d$ to $2d$ distance from the face of the support in the case of reinforced concrete beams, and usually at about a quarter of the span in the case of prestressed concrete beams. As they stabilize, one of the diagonal cracks widens into a principal diagonal tension crack and extends to the top compression fibers of the beam, as seen in Figures 5.5(b) and 5.7(c). Notice that the flexural cracks do not propagate to the neutral axis in this essentially brittle failure mode, which has relatively small deflection at failure.

Although the maximum external shear is at the support, the critical location of the maximum principal stress in tension is not. It is considerably reduced at that section because of the high compression force of the prestressing tendon, in addition to the vertical compression force of the beam reaction at the supports. This is the reason why the stabilized diagonal crack is located further into the span, depending on the magnitude of the prestressing force and the variation in its eccentricity, with an average value of about one-quarter of the span in flanged prestressed beams. In sum, the diagonal tension failure is the result of the combination of the flexural and shear stresses, taking into account the balancing contribution of the vertical component of the prestressing force, and noted by a combination of flexural and diagonal cracks. It is best termed *flexure shear* in the case of prestressed beams, and is more common to account for than *web shear*, discussed next.

5.4.4 Shear Compression Failure (Web Shear, WS)

Beams that are most subject to shear compression failure have a small span-to-depth ratio of magnitude 2.5 for the case of concentrated loading and less than 5.0 for distributed loading. As in the diagonal tension case, a few fine flexural cracks start to develop at midspan and stop propagating as destruction of the bond occurs between the longitudinal bars and the surrounding concrete at the support region. Thereafter, an inclined crack steeper than in the diagonal tension case suddenly develops and proceeds to propagate toward the neutral axis. The rate of its progress is reduced with the crushing of the concrete in the top compression fibers and a redistribution of stresses within the top region. Sudden failure takes place as the principal inclined crack dynamically joins the crushed concrete zone, as illustrated in Figure 5.5(c). This type of failure can be considered relatively less brittle than the diagonal tension failure due to the stress redistribution. Yet it is, in fact, a brittle type of failure with limited warning, and such a design should be avoided completely.

A concrete beam or element is not homogeneous, and the strength of the concrete throughout the span is subject to a normally distributed variation. Hence, one cannot expect that a stabilized failure diagonal crack occurs at both ends of the beam. Also, because of these properties, overlapping combinations of flexure–diagonal tension failure and diagonal tension–shear compression failure can occur at overlapping shear span-to-depth ratios. If the appropriate amount of shear reinforcement is provided, brittle failure of horizontal members can be eliminated with little additional cost.

Installation of precast prestressed double-T floor beams in a multifloor office structure.

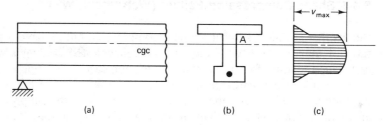

Figure 5.6 Maximum horizontal shear stress distribution across depth. (a) Beam elevation. (b) Beam cross section. (c) Shear stress.

It should be emphasized that most failures tend to occur by diagonal tension, which is a combination of flexure and shear effects. The shear-compression type of failure, with the resulting crushing of the top compressive area of the concrete and failure to resist the flexural forces, leads to separation of the tension flange from the web in the flanged section as the inclined crack extends towards the support. Crushing of the web of the section causes the beam to resemble a tied arch. This type of failure in prestressed beams can be better described as web-shear failure. It is important to evaluate *both* the flexure-shear capacity and the web-shear capacity of each critical section in order to determine which type predominates in determining the shear strength of the concrete section.

The distribution of the maximum horizontal shearing stress in an uncracked flanged section is shown in Figure 5.6. Because of the abrupt change of section width at the corner A, a check of the capacity of the section at critical locations along the span becomes necessary, particularly for web-shear failure.

5.5 SHEAR AND PRINCIPAL STRESSES IN PRESTRESSED BEAMS

As mentioned in Section 5.4, flexure shear in prestressed concrete beams includes the effect of the externally applied compressive prestressing force that the reinforced concrete beam does *not* have. The vertical component of the prestressing tendon force reduces the vertical shear caused by the external transverse load, and the *net* transverse load to which a beam is subjected is markedly less in prestressed than in reinforced concrete beams.

Additionally, the compressive force of the prestressing tendon, even in cases of straight tendons, considerably reduces the effect of the tensile flexural stresses, so that the extent and magnitude of flexural cracking in prestressed members are reduced. As a result, the shear forces and the resulting principal stresses in a prestressed beam are considerably lower than those same forces and stresses in reinforced concrete beams, all else being equal. Consequently, the basic equations developed for prestressed concrete in shear are, mutatis mutandis, identical to those developed for reinforced concrete and described in detail in Refs. 5.3 and 5.4. Figure 5.7 illustrates the contributions of the vertical component of the tendon force in counterbalancing part or most of the vertical shear V caused by the external transverse load. The net shearing force V_c carried by the concrete is

$$V_c = V - V_p \tag{5.5}$$

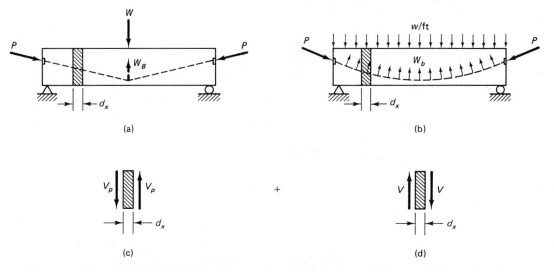

Figure 5.7 Balancing load to counteract vertical shear. (a) Beam with harped tendon. (b) Beam with draped tendon. (c) Internal shear vector V_p due to prestressing force p on infinitesimal element dx. (d) Internal shear vector V due to external load W on infinitesimal element dx.

From equation 5.2, the net unit shearing stress v at any depth of the cross section is

$$v_c = \frac{V_c Q}{Ib} \tag{5.6}$$

The compressive fiber-stress distribution f_c due to the external bending moment is

$$f_c = \frac{P_e}{A_c} \pm \frac{P_e ec}{I_c} \pm \frac{M_T c}{I_c} \tag{5.7}$$

and the principal tensile stress, from Equation 5.4a, is

$$f_t' = \sqrt{(f_c/2)^2 + v_c^2} - \frac{f_c}{2} \tag{5.8}$$

5.5.1 Flexure-Shear Strength (V_{ci})

To design for shear, it is necessary to determine whether flexure shear or web shear controls the choice of concrete shear strength V_c. The inclined stabilized crack at a distance $d/2$ from a flexural crack that develops at the first cracking load in flexure shear is shown in Figure 5.8. If the effective depth is d_P, the depth from the compression fibers to the centroid of the longitudinal reinforcement, the change in moment between sections 2 and 3 is

$$M - M_{cr} \cong \frac{V d_p}{2} \tag{5.9 a}$$

or

$$V = \frac{M_{cr}}{M/V - d_p/2} \tag{5.9 b}$$

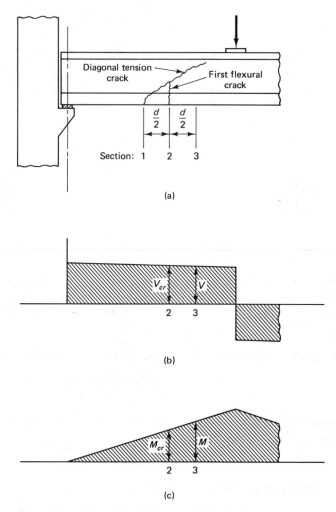

Figure 5.8 Flexure-shear crack development. (a) Crack pattern and types. (b) Shear diagram due to external load with frictional shear force V_{cr} ordinate at section 2. (c) Moment diagram with first cracking moment M_{cr} ordinate at section 2.

where V is the shear at the section under consideration. Extensive experimental tests indicate that an additional vertical shear force of magnitude $0.6 b_w d_p \sqrt{f'_c}$ is needed to fully develop the inclined crack in Figure 5.8 (Ref. 5.4). Hence, the total vertical shear acting at plane 3 of Figure 5.8 is

$$V_{ci} = \frac{M_{cr}}{M/V - d_p/2} + 0.6 b_w d_p \sqrt{f'_c} + V_D \tag{5.10}$$

where V_D is the vertical shear due to self-weight. The vertical component V_p of the prestressing force is disregarded in Equation 5.10, since it is small along the span sections where the prestressing tendon is not too steep.

The value of V in Equation 5.10 is the factored shear force V_i at the section under consideration due to externally applied loads occurring simultaneously with the maximum moment M_{max} occurring at that section, i.e.,

$$V_{ci} = 0.6\lambda\sqrt{f'_c}\,b_w d_p + V_d + \frac{V_i}{M_{max}}(M_{cr}) \geq 1.7\lambda\sqrt{f'_c}\,b_w d_p \qquad (5.11)$$

where λ = 1.0 for normal-weight concrete
$\quad\quad$ = 0.85 for sand-lightweight concrete
$\quad\quad$ = 0.75 for all-lightweight concrete
$\quad V_d$ = shear force at section due to unfactored dead load due to self-weight only
$\quad V_{ci}$ = nominal shear strength provided by the concrete when diagonal tension cracking results from combined vertical shear and moment
$\quad V_i$ = factored shear force at section due to externally applied load occurring simultaneously with M_{max}.

For lightweight concrete, $\lambda = f_{ct}/6.7\sqrt{f'_c}$ if the value of the tensile splitting strength f_{ct} is known. Note that the value $\sqrt{f'_c}$ should not exceed 100.

The equation for M_{cr}, the moment causing flexural cracking due to external load, is given by

$$M_{cr} = \frac{I_c}{y_t}(6\sqrt{f'_c} + f_{ce} - f_d) \qquad (5.12)$$

where f_{ce} = concrete compressive stress due to effective prestress after losses at extreme fibers of section where tensile stress is caused by external load, psi
$\quad f_d$ = stress due to unfactored dead load at extreme fiber of section resulting from self-weight only where tensile stress is caused by externally applied load, psi
$\quad y_t$ = distance from centroidal axis to extreme fibers in tension.

For simplicity, S_b may be substituted for I_c/y_c.

A plot of Equation 5.10 is given in Figure 5.9 with experimental data from Ref 5.6. Compare this plot with an analogous one in Figure 6.6 of Ref 5.3 for reinforced concrete where an asymptotic horizontal value of shear is achieved along the span.

Note that in shear design of composite sections, the same design stipulations used for precast sections apply. This is because design for shear is based on the limit state at failure due to factored loads. Although the entire composite section resists the factored shear as a monolithic section, calculation of the shear strength V_c should be based on the properties of the *precast* section since most of the shear strength is provided by the web of the precast section. Consequently, f_{ce} and f_d in Equation 5.12 are calculated using the precast section geometry.

5.5.2 Web-Shear Strength (V_{cw})

The web-shear crack in the prestressed beam is caused by an indeterminate stress that can best be evaluated by calculating the principal tensile stress at the critical plane from Equation 5.8. The shear stress v_c can be defined as the web shear v_{cw} and is maximal near the centroid cgc of the section where the actual diagonal crack develops, as extensive tests to failure have indicated. If v_{cw} is substituted for v_c and $\bar{f}_c$, which denotes the concrete stress f_c due to effective prestress *at the cgc level,* is substituted

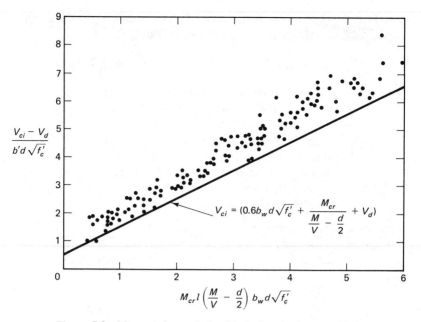

Figure 5.9 Moment-shear relationship in flexure-shear cracking.

for f_c in the equation, the expression equating the principal tensile stress in the concrete to the direct tensile strength becomes

$$f'_t = \sqrt{(\bar{f}_c/2)^2 + v_{cw}^2} - \frac{\bar{f}_c}{2} \qquad (5.13)$$

where $v_{cw} = V_{cw}/(b_w d_p)$ is the nominal shear stress in the concrete due to all loads causing a nominal strength vertical shear force V_{cw} in the web. Solving for v_{cw} in Equation 5.13 gives

$$v_{cw} = f'_t \sqrt{1 + \bar{f}_c/f'_t} \qquad (5.14 \text{ a})$$

Using $f'_t = 3.5\sqrt{f'_c}$ as a reasonable value of the tensile stress on the basis of extensive tests, Equation 5.14 a becomes

$$v_{cw} = 3.5\sqrt{f'_c}(\sqrt{1 + \bar{f}_c/3.5\sqrt{f'_c}}) \qquad (5.14 \text{ b})$$

which can be further simplified to

$$v_{cw} = 3.5\sqrt{f'_c} + 0.3\bar{f}_c \qquad (5.14 \text{ c})$$

In the ACI code, $\bar{f}_c$ is termed f_{pc}. The notation used herein is intended to emphasize that this is the stress in the concrete, and not the prestressing steel. The nominal shear strength V_{cw} provided by the concrete when diagonal cracking results from *excessive principal tensile stress* in the web becomes

$$V_{cw} = (3.5\lambda\sqrt{f'_c} + 0.3\bar{f}_c)b_w d_p + V_p \qquad (5.15)$$

where V_p = the vertical component of the effective prestress at the particular
        section contributing to added nominal strength
    λ = 1.0 for normal-weight concrete, and less for lightweight concrete

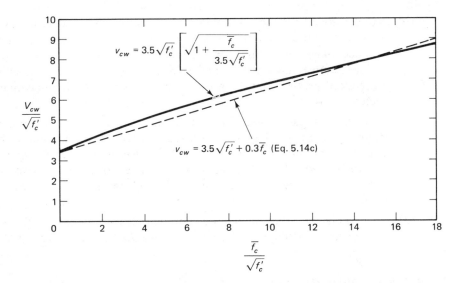

Figure 5.10 Centroidal compressive stress vs. nominal shear stress in web-shear cracking.

d_p = distance from the extreme compression fiber to the centroid of prestressed steel, or $0.8h$, whichever is greater.

The ACI code stipulates the value of $\bar{f_c}$ to be the resultant concrete compressive stress at either the centroid of the section or the junction of the web and the flange when the centroid lies within the flange. In case of composite sections, $\bar{f_c}$ is calculated on the basis of stresses caused by prestress and moments resisted by the precast member acting *alone*. A plot relating the nominal web shear stress v_{cw} to the centroidal compressive stress in the concrete is given in Figure 5.10. Note the similarity between the plots of Equations 5.14b and c, showing that the approximation used in the latter linearized equation is justified.

5.5.3 Controlling Values of V_{ci} and V_{cw} for the Determination of Web Concrete Strength V_c

The ACI code has the following additional stipulations for calculating V_{ci} and V_{cw} in order to choose the required value of V_c in the design:

(a) In pretensioned members where the section at a distance $h/2$ from the face of the support is closer to the end of the member than the transfer length of the prestressing tendon, a reduced prestressed value has to be considered when computing V_{cw}. This value of V_{cw} has to be taken as the maximum limit of V_c in the expression

$$V_{cw} = \left(0.6\lambda \sqrt{f_c'} + 700 \frac{V_u d_p}{M_u} \right) b_w d_p \geq 2\lambda \sqrt{f_c'}\, b_w d_p$$

$$\leq 5\lambda \sqrt{f_c'}\, b_w d_p \tag{5.16}$$

The value $V_u d_p / M_u$ cannot exceed 1.0.

(b) In pretensioned members where bonding of some tendons does not extend to the end of the member, a reduced prestress has to be considered when computing V_c in accordance with Equation 5.16 or with the *lesser* of the two values of V_c obtained from Equations 5.11 and 5.15. Also, the value of V_{cw} calculated using the reduced prestress must consequently be taken to be the maximum limit of Equation 5.16.

(c) Equation 5.16 can be used in determining V_c for members where the effective prestress force is not less than 40 percent of the tensile strength of the flexural reinforcement, *unless* a more detailed analysis is performed using Equations 5.11 for V_{ci} and 5.15 for V_{cw} and choosing the lesser of these two as the limiting V_c value to be used as the capacity of the web in designing the web reinforcement.

(d) The first plane for the total required nominal shear strength $V_n = V_u/\phi$ to be used for web steel calculation is also at a distance $h/2$ from the face of the support.

5.6 WEB-SHEAR REINFORCEMENT

5.6.1 Web Steel Planar Truss Analogy

In order to prevent diagonal cracks from developing in prestressed members, *whether due to flexure-shear or web-shear action,* steel reinforcement has to be provided, ideally in the form of the solid lines depicting tensile stress trajectories in Figure 5.3. However, practical considerations preclude such a solution, and other forms of reinforcement are improvised to neutralize the tensile stresses at the critical shear failure planes. The mode of failure in shear reduces the beam to a simulated arched section in compression at the top and tied at the bottom by the longitudinal beam tension bars, as seen in Figure 5.11(a). If one isolates the main concrete compression element shown in Figure 5.11(b), it can be considered as the compression member of a triangular truss, as shown in Figure 5.11(c), with the polygon of forces C_c, T_b, and T_s representing the forces acting on the truss members—hence the expression *truss analogy.* Force C_c is the compression in the simulated concrete strut, force T_b is the tensile force increment of the main longitudinal tension bar, and T_s is the force in the bent bar. Figure 5.12(a) shows the analogy truss for the case of using vertical stirrups instead of inclined bars, with the forces polygon having a vertical tensile force T_s instead of the inclined one in Figure 5.11(c).

As can be seen from the previous discussion, the shear reinforcement basically performs four main functions:

1. It carries a portion of the external factored shear force V_u.

2. It restricts the growth of the diagonal cracks.

3. It holds the longitudinal main reinforcing bars in place so that they can provide the dowel capacity needed to carry the flexural load.

4. It provides some confinement to the concrete in the compression zone if the stirrups are in the form of closed ties.

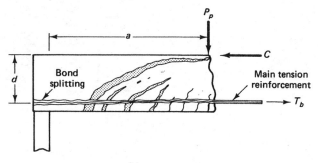

(a)

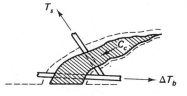

(b)

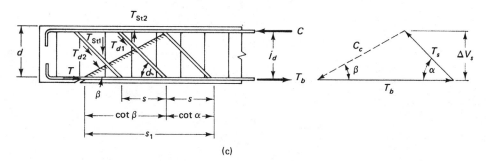

(c)

Figure 5.11 Diagonal tension failure mechanism. (a) Failure pattern. (b) Concrete simulated strut. (c) Planar truss analogy.

5.6.2 Web Steel Resistance

If V_c, the nominal shear resistance of the plain web concrete, is less than the nominal total vertical shearing force $V_u/\phi = V_n$, web reinforcement has to be provided to carry the difference in the two values; hence,

$$V_s = V_n - V_c \qquad (5.17)$$

Here, V_c is the *lesser* of V_{ci} and V_{cw}. V_c can be calculated from Equation 5.11 or 5.15, and V_s can be determined from equilibrium analysis of the bar forces in the analogous triangular truss cell. From Fig. 5.11(c),

$$V_s = T_s \sin \alpha = C_c \sin \beta \qquad (5.18 \text{ a})$$

where T_s is the force resultant of all web stirrups across the diagonal crack plane and

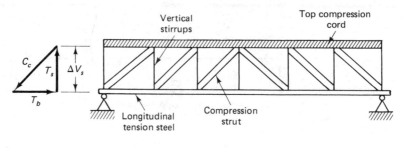

(a)

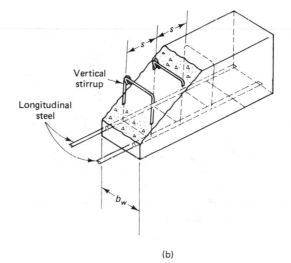

(b)

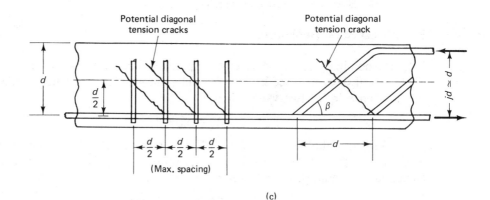

(c)

Figure 5.12 Web steel arrangement. (a) Truss analogy for vertical stirrups. (b) Three-dimensional view of vertical stirrups. (c) Spacing of web steel.

n is the number of spacings s. If $s_1 = ns$ in the bottom tension chord of the analogous truss cell, then

$$s_1 = jd(\cot \alpha + \cot \beta) \qquad (5.18\ b)$$

Assuming that moment arm $jd \simeq d$, the stirrup force per unit length from Equations 5.18 a and b, where $s_1 = ns$, becomes

$$\frac{T_s}{s_1} = \frac{T_s}{ns} = \frac{V_s}{\sin \alpha}\frac{1}{d(\cot \beta + \cot \alpha)} \qquad (5.18\ c)$$

If there are n inclined stirrups within the length s_1 of the analogous truss chord, and if A_v is the area of one inclined stirrup, then

$$T_s = nA_v f_y \qquad (5.19\ a)$$

Hence,

$$nA_v = \frac{V_s ns}{d \sin \alpha(\cot \beta + \cot \alpha)f_y} \qquad (5.19\ b)$$

But it can be assumed that in the case of diagonal tension failure the compression diagonal makes an angle $\beta = 45°$ with the horizontal; so Equation 5.19 b becomes

$$V_s = \frac{A_v f_y d}{s}[\sin \alpha(1 + \cot \alpha)]$$

or

$$V_s = \frac{A_v f_y d}{s}(\sin \alpha + \cos \alpha) \qquad (5.20\ a)$$

or, solving for s and using the fact that $V_s = V_n - V_c$,

$$s = \frac{A_v f_y d}{V_n - V_c}(\sin \alpha + \cos \alpha) \qquad (5.20\ b)$$

If the inclined web steel consists of a single bar or a single group of bars all bent at the same distance from the face of the support, then

$$V_s = A_v f_y \sin \alpha \le 3.0\sqrt{f_c'}\ b_w d$$

If vertical stirrups are used, angle α becomes $90°$, giving

$$V_s = \frac{A_v f_y d}{s} \qquad (5.21\ a)$$

or

$$s = \frac{A_v f_y d}{(V_u/\phi) - V_c} = \frac{A_v \phi f_y d}{V_u - \phi V_c} \qquad (5.21\ b)$$

In Equations 5.21 a and b, d_p is the distance from the extreme compression fibers to the centroid of the prestressing reinforcement, and d is the corresponding distance to the centroid of the nonprestressed reinforcement. The value of d_p need not be less than $0.80h$.

5.6.3 Limitation on Size and Spacing of Stirrups

Equations 5.20 and 5.21 give an inverse relationship between the spacing of the stirrups and the shear force or shear stress they resist, with the spacing s decreasing with the increase in $(V_n - V_c)$. In order for every *potential* diagonal crack to be resisted by a vertical stirrup, as shown in Figure 5.11(c), maximum spacing limitations are to be applied for the vertical stirrups as follows:

(a) $s_{max} \leq \frac{3}{4}h \leq 24$ in., where h is the total depth of the section.

(b) If $V_s > 4\lambda\sqrt{f'_c}\,b_w d_p$, the maximum spacing in (a) shall be reduced by half.

(c) If $V_s > 8\lambda\sqrt{f'_c}\,b_w d_p$, enlarge the section.

(d) If $V_u = \phi V_n > \frac{1}{2}\phi V_c$, a minimum area of shear reinforcement has to be provided. This area may be computed by the equation

$$A_v = \frac{50 b_w s}{f_y} \qquad (5.22\ a)$$

If the effective prestress force P_e is equal to or greater than 40 percent of the tensile strength of the flexural reinforcement, the equation

$$A_v = \frac{A_{ps} f_{pu} s}{80 f_y d}\sqrt{\frac{d_p}{b_w}} \qquad (5.22\ b)$$

which gives a lesser required minimum A_v, may be used instead.

(e) The web reinforcement must develop the full required development length in order to be effective. This means that the stirrups or mesh should extend to the

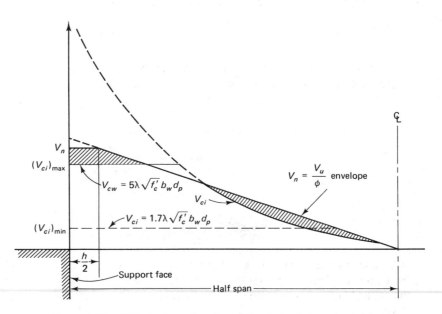

Figure 5.13 Web steel envelope for uniformly loaded prestressed beam.

Shear and Torsional Strength Design   Chap. 5

compression and tension surfaces of the section, less the clear concrete cover requirement and a 90° or 135° hook used at the compression side.

A typical qualitative diagram showing the zone along the span of a uniformly loaded prestressed beam for which web reinforcement has to be provided is given in Figure 5.13. The shaded area is the envelope of excess shear V_s requiring web steel.

5.7 HORIZONTAL SHEAR STRENGTH IN COMPOSITE CONSTRUCTION

Full transfer of horizontal shear forces has to be assumed at the contact surfaces of the interconnected elements.

5.7.1 Service-Load Level

The maximum horizontal shear stress v_h can be evaluated from the basic principles of mechanics and the equation

$$v_h = \frac{VQ}{I_c b_v} \qquad (5.23)$$

where V = unfactored design vertical shear acting on the composite section
Q = moment of area about cgc of the segment above or below cgc
I_c = moment of inertia of entire composite section
b_v = contact width of precast section web, or width of section at which horizontal shear is being calculated.

Equation 5.23 can be simplified to

$$v_h = \frac{V}{b_v d_{pc}} \qquad (5.24)$$

where d_{pc} is the effective depth from the extreme compression fibers of the composite section to the centroid cgs of the prestressing reinforcement.

5.7.2 Ultimate-Load Level

Direct Method. In the limit state at failure, Equation 5.24 can be modified such that the factored load V_u can be substituted for V to give

$$v_{uh} = \frac{V_u}{b_v d_{pc}} \qquad (5.25\ a)$$

or, in terms of the nominal vertical shear strength V_n,

$$v_{nh} = \frac{V_u/\phi}{b_v d_{pc}} = \frac{V_n}{b_v d_{pc}} \qquad (5.25\ b)$$

where $\phi = 0.85$. If V_{nh} is the nominal horizontal shear strength, then $V_u \leq V_{nh}$ and the total nominal shear strength is

$$V_{nh} = v_{nh} b_v d_{pc} \qquad (5.25\ c)$$

The ACI code limits v_{nh} to 80 psi if no dowels or vertical ties are provided and the contact surface is roughened, or if minimum vertical ties are provided but there is no roughening of the surface of contact. v_{nh} can go up to 350 psi; otherwise the friction theory, with the following assumptions, has to be used:

(a) When no vertical ties are provided, but the contact surface of the precast element is intentionally roughened, use

$$V_{nh} \leq 80A_c \leq 80b_v d_{pc} \qquad (5.26\ a)$$

where A_c is the area of concrete resisting shear $= b_v d_{pc}$.

(b) When minimum vertical ties are provided, where $A_v = 50(b_w s)/f_y$, but the contact surface of precast elements is *not* roughened, use

$$V_{nh} \leq 80b_v d_{pc}$$

(c) If the contact surface of the precast element is roughened to a full amplitude of $\frac{1}{4}$ in., and minimum vertical steel in (b) is provided, use

$$V_{nh} \leq 350b_v d_{pc} \qquad (5.26\ b)$$

(d) If the factored shear $V_u > \phi(350b_v d_p)$, the shear friction theory can be used to design the dowel reinforcement. In this case, all horizontal shear has to be taken by ties in the perpendicular plane such that

$$V_{nh} = \mu A_{vf} f_y \qquad (5.27)$$

where A_{vf} = area of shear-friction reinforcement, in^2
$\qquad f_y$ = design yield strength, not to exceed 60,000 psi
$\qquad \mu$ = coefficient of friction
$\qquad\quad$ = 1.0λ for concrete placed against intentionally roughened concrete surface
$\qquad\quad$ = 0.60λ for concrete placed against unroughened concrete surface
$\qquad \lambda$ = factor for type of concrete.

In all cases, the nominal shear strength $V_n \leq 0.20f'_c A_c \leq 800A_c$, where A_c is the concrete contact area resisting shear transfer. Note that in most cases the shear stress v_{nh} resulting from the factored shear force does not exceed 350 psi. Hence, the shear friction theory is not normally necessary in designing the dowel reinforcement for composite action.

The maximum allowable spacing of the dowels or ties for horizontal shear is the smaller of four times the supported dimension and 24 inches.

Basic Method. The ACI code allows an alternative method wherein horizontal shear is investigated by computing the actual change in compressive or tensile force in any plane and transferring that force as *horizontal* shear to the supporting elements. The area-of-contact surface A_{cc} is substituted for $b_v d_{ps}$ in Equations 5.25 b and c to give

$$V_{nh} = v_{nh} A_{cc} \qquad (5.28)$$

where $V_{nh} \geq F_h$, the horizontal shear force, and is at least equal to the compressive force C or tensile force T in Figure 5.14. (See Equation 5.30 for the value of F_h.)

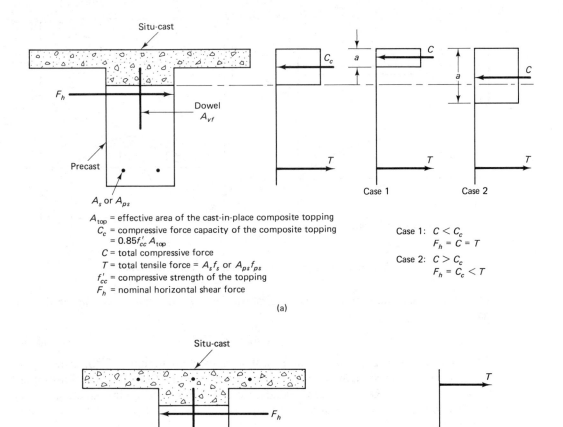

A_{top} = effective area of the cast-in-place composite topping
C_c = compressive force capacity of the composite topping
$= 0.85 f'_{cc} A_{\text{top}}$
C = total compressive force
T = total tensile force = $A_s f_s$ or $A_{ps} f_{ps}$
f'_{cc} = compressive strength of the topping
F_h = nominal horizontal shear force

Case 1: $C < C_c$
$\qquad F_h = C = T$

Case 2: $C > C_c$
$\qquad F_h = C_c < T$

(a)

(b)

Figure 5.14 Composite action forces (F_h acts longitudinally along the beam span). (a) Positive-moment section. (b) Negative-moment section.

The value of the contact area A_{cc} can be defined as

$$A_{cc} = b_v \ell_{vh} \qquad (5.29)$$

where ℓ_{vh} is the horizontal shear length defined in Figures 5.15(a) and (b) for simple span and continuous span members, respectively.

5.7.3 Design of Composite-Action Dowel Reinforcement

Ties for horizontal shear may consist of single bars or wires, multiple leg stirrups, or vertical legs of welded wire fabric. The spacing cannot exceed four times the least dimension of the support element or 24 in., whichever is less. If μ is the coefficient

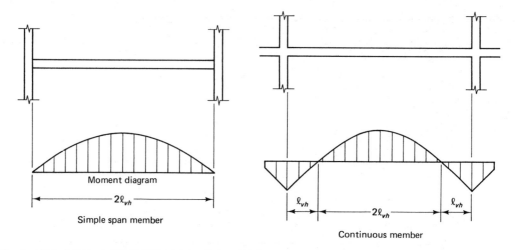

Figure 5.15 Horizontal shear length in composite action.

of friction, then the nominal horizontal shear force F_h in Figure 5.14 can be defined as

$$F_h = \mu A_{vf} f_y \le V_{nh} \tag{5.30}$$

The ACI values of μ are based on a limit shear-friction strength of 800 psi, a quite conservative value as demonstrated by extensive testing (Ref 5.7). The Prestressed Concrete Institute (Ref 5.8) recommends, for concrete placed against an intentionally roughened concrete surface, a maximum $\mu_e = 2.9$ instead of $\mu = 1.0\lambda$, and a maximum design shear force

$$V_u \le 0.25\lambda^2 f'_c A_c \le 1,000\ \lambda^2 A_{cc} \tag{5.31 a}$$

with a required area of shear-friction steel of

$$A_{vf} = \frac{V_{uh}}{\phi f_y \mu_e} \tag{5.31 b}$$

or

$$A_{vh} = \frac{V_{nh}}{\mu_e f_y} = \frac{F_h}{\mu_e f_y} \tag{5.31 c}$$

Using the PCI less conservative values, Equation 5.31 c becomes

$$F_h \le \mu_e A_{vf} f_y \le V_{nh} \tag{5.32}$$

with

$$\mu_e = \frac{1,000\ \lambda^2 b_v \ell_{vh}}{F_h} \le 2.9$$

where $b_v \ell_{vh} = A_{cc}$. The minimum reinforcement is

$$A_v = \frac{50 b_v s}{f_y} = \frac{50 b_v \ell_{vd}}{f_y} \tag{5.33}$$

5.8 WEB REINFORCEMENT DESIGN PROCEDURE FOR SHEAR

The following is a summary of a recommended sequence of design steps:

1. Determine the required nominal shear strength value $V_n = V_u/\phi$ at a distance $h/2$ from the face of the support.

2. Calculate the nominal shear strength V_c that the web has by one of the following two methods.

(a) *ACI conservative method if $f_{pe} > 0.40 f_{pu}$*

$$V_c = \left(0.60\lambda\sqrt{f_c'} + \frac{700 V_u d_p}{M_u}\right) b_w d_p$$

where $2\lambda\sqrt{f_c'}\, b_w d_p \leq V_c \leq 5\lambda\sqrt{f_c'}\, b_w d_p$ and where $V_u d_p/M_u \leq 1.0$ and V_u is calculated at the same section for which M_u is calculated.

If the average tensile splitting strength f_{ct} is specified for lightweight concrete, then $\lambda = f_{ct}/6.7\sqrt{f_c'}$ with $\sqrt{f_c'}$ not to exceed a value of 100.

(b) *Detailed analysis where V_c is the lesser of V_{ci} and V_{cw}*

$$V_{ci} = 0.60\lambda\sqrt{f_c'}\, b_w d_p + V_d + \frac{V_i}{M_{max}}(M_{cr}) \geq 1.7\lambda\sqrt{f_c'}\, b_w d_p$$

$$V_{cw} = (3.5\lambda\sqrt{f_c'} + 0.3\,\bar{f_c})b_w d_p + V_p$$

using d_p or $0.8h$, whichever is larger, and

where $M_{cr} = (I_c/y_t)(6\lambda\sqrt{f_c'} + f_{ce} - f_d)$
or $M_{cr} = S_b(6\lambda\sqrt{f_c'} + f_{ce} - f_d)$

$\quad V_i$ = factored shear force at section due to externally applied loads occurring simultaneously with M_{max}

$\quad f_{ce}$ = compressive stress in concrete after occurrence of all losses at extreme fibers of section where external load causes tension

3. If $V_u/\phi \leq \frac{1}{2}V_c$, no web steel is needed. If $V_u/\phi > \frac{1}{2}V_c \leq V_c$, provide minimum reinforcement. If $V_u/\phi > V_c$ and $V_s = V_u/\phi - V_c \leq 8\lambda\sqrt{f_c'}\, b_w d_p$, design the web steel. If $V_s = V_u/\phi - V_c > 8\lambda\sqrt{f_c'}\, b_w d_p$, or if $V_u > \phi(V_c + 8\lambda\sqrt{f_c'}\, b_w d_p)$, enlarge the section.

4. Calculate the required minimum web reinforcement. The spacing is $s \leq 0.75h$ or 24 in., whichever is smaller.

$$\text{Min } A_v = \frac{50 b_w s}{f_y} \qquad \text{(conservative)}$$

If $f_{pe} \geq 0.40 f_{pu}$, a less conservative Min A_v is the smaller of

$$\frac{A_{ps} f_{pu} s}{80 f_y d_p} \sqrt{\frac{d_p}{b_w}}$$

where $d_p \geq 0.80h$, and

$$A_v = 50 b_w s/f_y$$

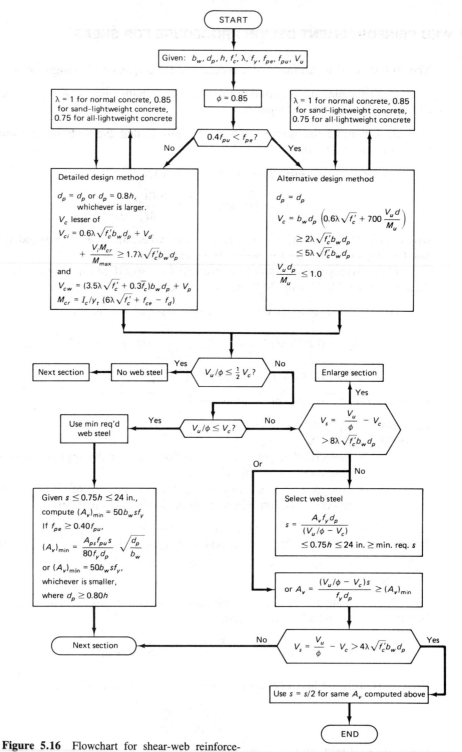

Figure 5.16 Flowchart for shear-web reinforcement.

5. Calculate the required web reinforcement size and spacing. If $V_s = (V_u/\phi - V_c) \le 4\lambda\sqrt{f_c'}\, b_w d_p$, then the stirrup spacing s is as required by the design expressions in step 6, to follow. If $V_s = (V_u/\phi - V_c) > 4\lambda\sqrt{f_c'}\, b_w d_p$, then the stirrup spacing s is half the spacing required by the design expressions in step 6.

6.

$$s = \frac{A_v f_y d_p}{(V_u/\phi) - V_c} = \frac{A_v \phi f_y d_p}{V_u - \phi V_c} \le 0.75h \le 24 \text{ in.} \ge \text{maximum } s \text{ from step 4}$$

7. Draw the shear envelope over the beam span, and mark the band requiring web steel.

8. Sketch the size and distribution of web stirrups along the span using #3 or #4 size stirrups as preferable, but no larger size than #6 stirrups.

9. Design the vertical dowel reinforcement in cases of composite sections.

(a) $V_{nh} \le 80 b_v d_{pc}$ for both roughened contact and no vertical ties or dowels, and nonroughened but with minimum vertical ties, use

$$A_v = \frac{50 b_w s}{f_y} = \frac{50 b_v \ell_{vh}}{f_y}$$

(b) $V_{nh} \le 350 b_v d_{pc}$ for a roughened contact surface with full amplitude $1/4$ in.

(c) For cases where $V_{nh} > 350 b_v d_{pc}$, design vertical ties for $V_{nh} = A_{vf} f_y \mu$,

where A_{vf} = area of frictional steel dowels
μ = coefficient of friction = 1.0λ for intentionally roughened surface, where $\lambda = 1.0$ for normal-weight concrete. In all cases, $V_n \le V_{nh} \le 0.2 f_c' A_c \le 800 A_{cc}$, where $A_{cc} = b_v \ell_{vh}$.

An alternative method of determining the dowel reinforcement area A_{vf} is by computing the horizontal force F_h at the concrete contact surface such that

$$F_h \le \mu_e A_{vf} f_y \le V_{nh}$$

where

$$\mu_e = \frac{1{,}000\,\lambda^2 b_v \ell_{vh}}{F_h} \le 2.9$$

Figure 5.16 outlines the foregoing steps in flowchart form.

5.9 PRINCIPAL TENSILE STRESSES IN FLANGED SECTIONS AND DESIGN OF DOWEL-ACTION VERTICAL STEEL IN COMPOSITE SECTIONS

Example 5.1

A prestressed concrete T-beam section has the distribution of compressive service load shown in Figure 5.17. The unfactored design external vertical shear $V = 120,000$ lb (554 kN), and the factored vertical shear $V_u = 190,000$ lb (845 kN).

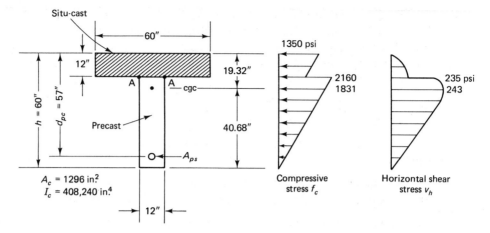

Figure 5.17 Beam cross section in Example 5.1.

(a) Compute the principal tensile stress at the centroidal cgc axis and at the reentrant corner A of the web-flange junction, and calculate the maximum horizontal shearing stresses at service load for these locations.

(b) Compute the required nominal horizontal shear strength at the interaction surface A–A between the precast web and the situ-cast flange, and design the necessary vertical ties or dowels to prevent fracture slip at A–A, thereby ensuring complete composite action. Use the ACI direct method, and assume that the contact surface is intentionally roughened. Given data are as follows:

f'_c for web = 6,000 psi normal-weight concrete

f'_c for flange = 3,000 psi normal-weight concrete

Effective width of flange b_m = 60 in. (152.4 cm)

where b_m is the modified width to account for the difference in moduli of the concrete of the precast and topping parts.

Solution

Service-Load Horizontal Shear Stresses. The maximum horizontal shear stress is

$$v_h = \frac{VQ}{Ib_v}$$

Now,

$$Q_A = 60 \times 12(19.32 - 6) = 9,590 \text{ in}^3 \text{ (157,152 cm}^3\text{)}$$

$$Q_{cgc} = 60 \times 12(19.32 - 6) + \frac{12 \, (7.32)^2}{2} = 9,912 \text{ in}^3 \text{ (162,429 cm}^3\text{)}$$

So the horizontal shear stresses at service load are

$$v_h \text{ at A} = \frac{120,000 \times 9590}{408,240 \times 12} = 235 \text{ psi (1.6 MPa)}$$

and

$$v_h \text{ at cgc} = \frac{120,000 \times 9,912}{408,240 \times 12} = 243 \text{ psi (1.7 MPa)}$$

From Equation 5.13, the corresponding principal tensile stresses are, at A,

$$f'_t = \sqrt{\left(\frac{f_{CA}}{2}\right)^2 + v_h^2} - \frac{f_{cA}}{2}$$

$$= \sqrt{\left(\frac{2,160}{2}\right)^2 + (235)^2} - \frac{2,160}{2} = 25 \text{ psi (111 Pa)}$$

and at cgc,

$$f'_t = \sqrt{\left(\frac{1,831}{2}\right)^2 + (243)^2} - \frac{1,831}{2} = 32 \text{ psi (221 Pa)}$$

Thus, the principal tensile stresses are low and should not cause any cracking at service load.

The horizontal shear stress $v_h = 235$ psi at contact surface A–A has to be checked to verify whether it is within acceptable limits. In accordance with AASHTO, the maximum allowed is 160 psi < 235 psi, and hence special provision for added vertical ties or dowels has to be made if AASHTO requirements are applicable.

Dowel Reinforcement Design

$$V_u = 190,000 \text{ lb}$$

$$\text{Req. } V_{nh} = \frac{V_u}{\phi} = \frac{190,000}{0.85} = 223,530 \text{ lb (994 kN)}$$

$$b_v = 12 \text{ in. (30.5 cm)} \qquad d_{pc} = 57 \text{ in. (145 cm)}$$

From Equation 5.26 b,

Available $V_{nh} = 350 b_v d_{pc} = 350 \times 12 \times 57 = 239,400$ lb (1,065 kN) > 223,530 lb

Hence, specify roughening the contact surface of the precast web fully to $\frac{1}{4}$ in. amplitude. Now,

$$\text{min unit} \frac{A_v}{\ell_{vh}} = \frac{50 b_v}{f_y} = \frac{50 \times 12}{60,000} = 0.01 \text{ in}^2/\text{in. along span}$$

So using #3 vertical stirrups, $A_v = 2 \times 0.11 = 0.22$ in^2, and $s = 0.22/0.01 = 22$ in. (56 cm) center to center < 24 in., O.K. Vertical-web steel reinforcement for shear in the web would most probably require smaller spacing. Hence, extend all web stirrups into the situ-cast top slab.

5.10 DOWEL STEEL DESIGN FOR COMPOSITE ACTION

Example 5.2

Using (a) the ACI friction coefficient and (b) the PCI friction coefficient design the dowel reinforcement of Example 5.1 for full composite action by the alternative method, assuming a simply supported beam of effective 65 ft (19.8 m) span.

Solution From Figures 5.14 and 5.17,

$$A_{top} = 60 \times 12 = 720 \text{ in}^2 \ (4{,}645 \text{ cm}^2)$$

$$C_c = 0.85 f'_{cc} A_{top} = 0.85 \times 3{,}000 \times 720 = 1{,}836{,}000 \text{ lb } (8{,}167 \text{ kN})$$

Assume $A_{ps} f_{ps} > C_c$, since the prestressing force is not given. Then

$$F_h = 1{,}836{,}000 \text{ lb}$$

$$\ell_{vh} = \frac{65 \times 12}{2} = 390 \text{ in.}$$

$$b_v = 12 \text{ in.}$$

$$80 b_v \ell_{vh} = 80 \times 12 \times 390 = 374{,}400 \text{ lb } (1{,}665 \text{ kN}) < 1{,}836{,}000 \text{ lb}$$

Hence, vertical ties are needed.

For roughened surface to full $\frac{1}{4}$ in. amplitude and minimum reinforcement,

$$V_{nh} = 350 b_v d = 350 \times 12 \times 57 = 239{,}400 \text{ lb}$$

$$\text{Req.} \frac{V_u}{\phi} = \frac{190{,}000}{0.85} = 223{,}529 \text{ lb} < V_{nh} = 239{,}400 < \text{available } F_h = 1{,}836{,}000 \text{ lb}$$

Use $F_h = 223{,}529$ lb for determining the required composite action reinforcement.

Using ACI μ Value. From Equation 5.27 and $\mu = 1.0$, with $\ell_{vh} = 390$ in.,

$$\text{Total } A_{vf} = \frac{223{,}529}{1.0 \times 60{,}000} = 3.73 \text{ in}^2 \ (24.1 \text{ cm}^2)$$

$$\text{Min } A_{vf} = \frac{50 b_v \ell_{vh}}{f_y} = \frac{50 \times 12 \times 390}{60{,}000} = 3.90 \text{ in}^2 \ (25.1 \text{ cm}^2)$$

From Example 5.1, min $A_{vf} = 0.01 \text{ in}^2/\text{in.} = 0.12 \text{ in}^2/12 \text{ in.}$, controls. So, trying #3 stirrups, we obtain $A_v = 2 \times 0.11 = 0.22 \text{ in}^2$, and

$$s = \frac{\ell_{vh} A_v}{A_{vf}} = \frac{390 \times 0.22}{3.90} = 22 \text{ in. center to center}$$

$$< 24 \text{ in.} < \text{max. allow. } 4 \times 12 = 48 \text{ in.}$$

So use #3 U ties @ 22 in. center to center.

Using PCI μ_e Value

$$\lambda = 1.0$$

$$\mu_e = \frac{1{,}000 \lambda^2 b_v \ell_{vh}}{F_h} \leq 2.9$$

$$= \frac{1{,}000 \times 1 \times 12 \times 390}{1{,}836{,}000} = 2.55 < 2.9$$

So use $\mu_e = 2.55$; then, from Equation 5.32,

$$\text{Req. } A_{vf} = \frac{233{,}529}{2.55 \times 60{,}000} = 1.46 \text{ in}^2 < \text{min. } A_{vf} = 3.90 \text{ in}^2, \text{ controls}$$

So use #3 U ties at 22 in. center to center (9.5 mm dia @ 56 cm).

5.11 DOWEL REINFORCEMENT DESIGN FOR COMPOSITE ACTION IN AN INVERTED T-BEAM

Example 5.3

A simply supported inverted T-beam has an effective 24 ft (7.23 m) span length. The beam, shown in cross section in Figure 5.18, has a 2 in. (5.1 cm) situ-cast topping on a nonroughened surface. Design the required dowel action stirrups to develop full composite behavior, assuming that the factored shear V_u to which the beam is subjected at the critical section is 160,000 lb (712 kN). Given data are as follows:

f'_c (precast) = 6,000 psi (41.4 MPa), normal-weight concrete

f'_{cc} (topping) = 3,000 psi (20.7 MPa), normal-weight concrete

Prestressing steel:

Twelve $\frac{1}{2}$ in. dia 270 k strands

f_{pu} = 270,00 psi (1,862 MPa)

f_{ps} = 242,000 psi (1,669 MPa)

Tie steel f_y = 60,000 psi (414 MPa)

Use both the ACI direct method and the alternative method with effective μ_e to carry out your design.

Solution

$$d_p = 2 + 2 + 10 + 12 - 3 = 23 \text{ in.}$$

$$A_{ps} = 12 \times 0.153 = 1.836 \text{ in}^2$$

$$T_n = A_{ps}f_{ps} = 1.836 \times 242,000 = 444,312 \text{ lb } (1,976 \text{ kN})$$

$$b_v = 12 \text{ in.}$$

$$\ell_{vh} = \frac{24 \times 12}{2} = 144 \text{ in.}$$

$$A_{top} = 2 \times 48 + 2 \times 12 = 120 \text{ in}^2$$

$$C_c = 0.85 f'_{cc} A_{top} = 0.85 \times 3,000 \times 120 = 306,000 \text{ lb } (1,361 \text{ kN})$$

$$< T_n = 444,312 \text{ lb}$$

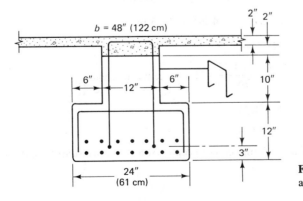

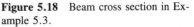

Figure 5.18 Beam cross section in Example 5.3.

Accordingly, use $F_h = 306,000$ lb (1,361 kN). Then, for a nonroughened surface,

$$\text{Available } V_{nh} = 80b_v \ell_{vh} = 80 \times 12 \times 144 = 138,240 \text{ lb (615 kN)}$$

$$< C_c = 306,000 \text{ lb}$$

Hence, ties are required for developing full composite action using $\lambda = 1.0$.

ACI Direct Method

$$\text{Req. } V_{nh} = \frac{V_u}{\phi} = \frac{160,000}{0.85} = 188,235 \text{ lb (837 kN)}$$

Use $\mu = 1.0$. Then, from Equation 5.26 a, with dowel reinforcement, we have

$$\text{Available } V_{nh} = 80b_v d_{pc} = 80 \times 12 \times 23 = 22,000 \text{ lb} \ll \text{required } V_{nh}$$

From Equation 5.27, for an unroughened surface $\mu = 0.6\lambda = 0.60$. Then

$$\text{Req. total } A_{vf} = \frac{V_n}{\mu f_y} = \frac{188,235}{0.60 \times 60,000} = 5.23 \text{ in}^2$$

$$\text{Req. min } A_{vf} = \frac{50b_v \ell_{vh}}{f_y} = \frac{50 \times 12 \times 144}{60,000} = 1.44 \text{ in}^2 < 5.23 \text{ in}^2$$

So use $A_{vf} = 5.23$ in^2 (33.7 cm^2), and try #3 inverted U ties. Then $A_{vf} = 2 \times 0.11 = 0.22$ in^2 (1.4 cm^2) and the spacing is

$$s = \frac{\ell_{vh} A_v}{A_{vf}} = \frac{144 \times 0.22}{5.23} = 6.05 \text{ in. (15.4 cm)}$$

The maximum allowable spacing is $s = 4(2 + 2) = 16$ in. < 24 in. Thus, use #3 inverted U ties 6 in. (15 cm) center to center over the entire simply supported span.

Alternative Method Using μ_e

$$F_h = 306,000 \text{ lb}$$

$$\mu_e = \frac{1,000\lambda^2 b_v \ell_{vh}}{F_h} = \frac{1,000 \times 1.0 \times 12 \times 144}{306,000} = 5.65 > 2.9$$

So use $\mu_e = 2.9$; then, from Equation 5.31 c, we get

$$\text{Req. } A_{vf} = \frac{F_h}{\mu_e f_y} = \frac{306,000}{2.9 \times 60,000} = 1.76 \text{ in}^2$$

$$\text{Req. min. } A_{vf} \text{ from (a)} = 1.44 \text{ in}^2 < 1.76 \text{ in}^2$$

So use $A_{vf} = 1.76$ in^2. Then the spacing is

$$s = \frac{\ell_{vh} A_v}{A_{vf}} = \frac{144 \times 0.22}{1.76} = 18 \text{ in. center to center}$$

and the maximum allowable spacing is

$$s = 4(2 + 2) = 16 \text{ in.} < 24 \text{ in.}$$

Hence, use #3 inverted U ties at 16 in. center to center over the entire simply supported span.

5.12 SHEAR STRENGTH AND WEB-SHEAR STEEL DESIGN IN A PRESTRESSED BEAM

Example 5.4

Design the bonded beam of Example 4.8 to be safe against shear failure, and proportion the required web reinforcement.

Solution

Data and Nominal Shear Strength Determination

f_{pu} = 270,000 psi (1,862 MPa)

f_y = 60,000 psi (414 MPa)

f_{pe} = 155,000 psi (1,069 MPa)

f'_c = 5,000 psi, normal-weight concrete

A_{ps} = 13 seven-wire $\frac{1}{2}$ in. strands = 1.99 in^2 (12.8 cm^2)

A_s = 4 #6 bars = 1.76 in^2 (11.4 cm^2)

Span = 65 ft (19.8 m)

Service W_L = 1,100 plf (16.1 kN/m)

Service W_{SD} = 100 plf (1.46 kN/m)

Service W_D = 393 plf (5.7 kN/m)

h = 40 in. (101.6 cm)

d_p = 36.16 in. (91.8 cm)

d = 37.6 in. (95.5 cm)

b_w = 6 in. (15 cm)

e_c = 15 in. (31 cm)

e_e = 12.5 in. (32 cm)

I_c = 70,700 in^4 (18.09 × 10^6 cm^4)

A_c = 377 in^2 (2,432 cm^2)

r^2 = 187.5 in^2 (1,210 cm^2)

c_b = 18.84 in. (48 cm)

c_t = 21.16 in. (54 cm)

P_e = 308,255 lb (1,371 kN)

Factored load W_U = 1.4D + 1.7L

$$= 1.4(100 + 393) + 1.7 \times 1,100 = 2,560 \text{ plf}$$

Factored shear force at face of support = $V_u = W_U L/2$

$$= (2,560 \times 65)/2 = 83,200 \text{ lb}$$

Req. $V_n = V_u/\phi$ = 83,200/0.85 = 97,882 lb at support

Plane at $\frac{1}{2} d_p$ from Face of Support

1. *Nominal shear strength V_c of web (steps 2, 3)*

$$\frac{1}{2} d_p = \frac{36.16}{2 \times 12} \cong 1.5 \text{ ft}$$

$$V_n = 97,882 \times \frac{[(65/2) - 1.5]}{65/2} = 93,364 \text{ lb}$$

$$V_u \text{ at } \tfrac{1}{2}d_p = 0.85 \times 93,364 = 79,359 \text{ lb}$$

$$f_{pe} = 155,000 \text{ psi}$$

$$0.40f_{pu} = 0.40 \times 270,000 = 108,000 \text{ psi (745 MPa)}$$

$$< f_{pe} = 155,000 \text{ psi (1,069 MPa)}$$

Use ACI alternate method

Since $d_p > 0.8h$, use $d_p = 36.16$ in., assuming that part of the prestressing strands continue straight to the support. From Equation 5.16,

$$V_c = \left(0.60\lambda \sqrt{f_c'} + 700\frac{V_u d_p}{M_u} \right) b_w d_p \geq 2\lambda \sqrt{f_c'} \, b_w d_p$$

$$\leq 5\lambda \sqrt{f_c'} \, b_w d_p$$

$$\lambda = 1.0 \text{ for normal-weight concrete}$$

$$M_u \text{ at } d/2 \text{ from face} = \text{reaction} \times 1.5 - \frac{W_u(1.5)^2}{2}$$

$$= 83,200 \times 1.5 - \frac{2,560(1.5)^2}{2} = 121,920 \text{ ft-lb} = 1,463,040 \text{ in.-lb}$$

$$\frac{V_u d_p}{M_u} = \frac{79,359 \times 36.16}{1,463,040} = 1.96 > 1.0$$

So use $V_u d_p / M_u = 1.0$. Then

$$\text{Min. } V_c = 2\lambda \sqrt{f_c'} \, b_w d_p = 2 \times 1.0\sqrt{5,000} \times 6 \times 36.16 = 30,683 \text{ lb}$$

$$\text{Max. } V_c = 5\lambda \sqrt{f_c'} \, b_w d_p = 76,707 \text{ lb}$$

$$V_c = (0.60 \times 1.0\sqrt{5,000} + 700 \times 1.0)6 \times 36.16$$

$$= 161,077 \text{ lb} > \text{max } V_c = 76,707 \text{ lb}$$

Then $V_c = 76,707$ lb and controls (341 kN). Also, $V_u/\phi > \tfrac{1}{2}V_c$; hence, web steel is needed. Accordingly,

$$V_s = \frac{V_u}{\phi} - V_c = 97,882 - 76,707 = 21,175 \text{ lb}$$

$$8\lambda \sqrt{f_c'} \, b_w d_p = 8 \times 1.0\sqrt{5,000} \times 6 \times 36.16 = 122,713 \text{ lb (546 kN)}$$

$$> V_s = 21,175 \text{ lb}$$

So the section depth is adequate.

2. *Minimum web steel (step 4)*

From Equation 5.22 b,

$$\text{Min. } \frac{A_v}{s} = \frac{A_{ps}f_{pu}}{80f_y d_p} \sqrt{\frac{d_p}{b_w}}$$

$$= \frac{1.99 \times 270,000}{90 \times 60,000 \times 36.16} \sqrt{\frac{36.16}{6}} = 0.0076 \text{ in}^2/\text{in.}$$

3. *Required web steel (steps 5, 6)*
From Equation 5.21 b,

$$s = \frac{A_v f_y d_p}{V_u/\phi - V_c} \le 0.75h \le 24 \text{ in.}$$

or

$$\frac{A_v}{s} = \frac{V_s}{f_y d_p} = \frac{21,175}{60,000 \times 36.16} = 0.0098 \text{ in}^2/\text{in.}$$

Then the minimum required web-shear steel $A_v/s = 0.0098$ in^2/in. So, trying #3 U stirrups, $A_v = 2 \times 0.11 = 0.22$ in^2, and we get $0.0098 = 0.22/s$, so that the maximum spacing is

$$s = \frac{0.22}{0.0098} = 22.4 \text{ in. (5.7 cm)}$$

and

$$4\lambda\sqrt{f'_c}\, b_w d_p = 4 \times 1.0\sqrt{5,000} \times 6 \times 36.16 = 61,366 \text{ lb} > V_s$$

Hence, we do not need to use $\frac{1}{2}s$. Now,

$$0.75h = 0.75 \times 40.0 = 30.0 \text{ in.}$$

Thus, use #3 U web-shear reinforcement at 22 in. center to center (9.5 mm dia at 56 cm center to center).

Plane at Which No Web Steel Is Needed. Assume such a plane is at distance x from support. By similar triangles,

$$V_c = 76,707 = 97,882 \times \frac{65/2 - x}{65/2}$$

or

$$\frac{65}{2} - x = \frac{76,707}{97,882} \times \frac{65}{2}$$

giving

$$x = 7.03 \text{ ft (2.14 m)} \approx 84 \text{ in.}$$

Therefore, adopt the design in question, using #3 U at 22 in. center to center over a stretch length of approximately 84 in., with the first stirrup to start at 18 in. from the face of support. Extend the stirrups to the midspan if composite action doweling is needed.

5.13 WEB-SHEAR STEEL DESIGN BY DETAILED PROCEDURES

Example 5.5

Solve Example 5.4 by detailed procedures, determining the value of V_c as the smaller of the flexure shear V_{ci} and the web shear V_{cw}. Assume that the tendons are harped at midspan and not draped.

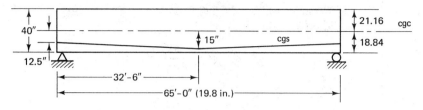

Figure 5.19 Tendon profile in Example 5.5.

Solution The profile of the prestressing strands is shown in Figure 5.19.

Plane at d/2 from Face of Support. From Example 5.4, $V_n = 93,364$ lb.

1. *Flexure-shear cracking, V_{ci} (step 2)*
From Equation 5.11,

$$V_{ci} = 0.60\lambda \sqrt{f_c'} \, b_w d_p + V_d + \frac{V_i}{M_{max}}(M_{cr}) \geq 1.7\lambda \sqrt{f_c'} \, b_w d_p$$

From Equation 5.12, the cracking moment is

$$M_{cr} = \frac{I_c}{y_t}(6\lambda \sqrt{f_c'} + f_{ce} - f_d)$$

where $I_c/y_t = S_b$ since y_t is the distance from the centroid to the extreme tension fibers. Now,

$$I_c = 70,700 \text{ in}^4$$

$$c_b = 18.84 \text{ in.}$$

$$P_e = 308,255 \text{ lb}$$

$$S_b = 3,753 \text{ in}^3$$

$$r^2 = 187.5 \text{ in}^2$$

So from Equation 4.3 b, the concrete stress at the extreme bottom fibers *due to prestress only* is

$$f_{ce} = -\frac{P_e}{A_c}\left(1 + \frac{ec_b}{r^2}\right)$$

and the tendon eccentricity at $d/2 \cong 1.5$ ft from the face of the support is

$$e = 12.5 + (15 - 12.5)\frac{1.5}{65/2} = 12.62 \text{ in.}$$

Thus,

$$f_{ce} = -\frac{308,255}{377}\left(1 + \frac{12.62 \times 18.84}{187.5}\right) \cong -1,855 \text{ psi (12.8 MPa)}$$

From Example 4.2, the unfactored dead load due to self-weight $W_D = 393$ plf (5.7 kN/m) is

$$M_{d/2} = \frac{W_D x(\ell - x)}{2} = \frac{393 \times 1.5(65 - 1.5) \times 12}{2} = 224,600 \text{ in.-lb (25.4 kN-m)}$$

and the stress due to the unfactored dead load at the extreme concrete fibers where tension is created by the external load is

$$f_d = \frac{M_{d/2}c_b}{I_c} = \frac{224,600 \times 18.84}{70,700} = 60 \text{ psi}$$

Also,

$$M_{cr} = 3,753(6 \times 1.0 \times \sqrt{6,000} + 1,855 - 60)$$

$$= 8,480,872 \text{ in.-lb } (958 \text{ kN-m})$$

$$V_d = W_D\left(\frac{\ell}{2} - x\right) = 393\left(\frac{65}{2} - 1.5\right) = 12,183 \text{ lb } (54.2 \text{ kN})$$

$$W_{SD} = 100 \text{ plf}$$

$$W_L = 1,000 \text{ plf}$$

$$W_U = 1.4 \times 100 + 1.7 \times 1100 = 2,010 \text{ plf}$$

The factored shear force at the section due to externally applied loads occurring simultaneously with M_{max} is

$$V_i = W_U\left(\frac{\ell}{2} - x\right) = 2,010\left(\frac{65}{2} - 1.5\right) = 62,310 \text{ lb } (277 \text{ kN})$$

and

$$M_{max} = \frac{W_U x(\ell - x)}{2} = \frac{2,010 \times 1.5(65 - 1.5)}{2} \times 12$$

$$= 1,148,715 \text{ in.-lb } (130 \text{ kN-m})$$

Hence,

$$V_{ci} = 0.6 \times 1.0\sqrt{6,000} \times 6 \times 36.16 + 12,183 + \frac{62,310}{1,148,715}(8,480,872)$$

$$= 482,296 \text{ lb } (54.5 \text{ kN-m})$$

$$1.7\lambda \sqrt{f'_c}\, b_w d_p = 1.7 \times 1.0\sqrt{6,000} \times 6 \times 36.16 = 28,569 \text{ lb } (127 \text{ kN})$$

$$< V_{ci} = 482,296 \text{ lb}$$

Hence, $V_{ci} = 482,296$ lb (214.5 kN).

2. *Web-shear cracking, V_{cw} (step 2)*
From Equation 5.15,

$$V_{cw} = (3.5\sqrt{f'_c} + 0.3\bar{f}_c)b_w d_p + V_p$$

$$\bar{f}_c = \text{compressive stress in concrete at the cgc}$$

$$= \frac{P_e}{A_c} = \frac{308,255}{377} \cong 818 \text{ psi } (5.6 \text{ MPa})$$

$$V_p = \text{vertical component of effective prestress at section}$$

$$= P_e \tan \theta$$

where θ is the angle between the inclined tendon and the horizontal. So

$$V_p = 308,255 \frac{(15 - 12.5)}{65/2 \times 12} = 1,976 \text{ lb } (8.8 \text{ kN})$$

Hence, $V_{cw} = (3.5\sqrt{6,000} + 0.3 \times 818) \times 6 \times 36.16 + 1,976 = 114,038 \text{ lb}$ (507 kN). In this case, web-shear cracking controls (i.e., $V_c = V_{cw} = 114,038 \text{ lb}$ (507 kN) is used for the design of web reinforcement). Compare this value with $V_c = 76,707 \text{ lb } (341 \text{ kN})$ obtained in Example 5.4 by the more conservative alternative method.

Now, from Example 5.4,

$$V_s = \frac{V_u}{\phi} - V_c = (97,882 - 114,038) \text{ lb}$$

So no web steel is needed unless $V_u/\phi > \frac{1}{2}V_c$. Accordingly, we evaluate the latter:

$$\frac{1}{2}V_c = \frac{114,038}{2} = 57,019 \text{ lb } (254 \text{ kN}) < 97,882 \text{ lb } (435 \text{ kN})$$

Since $V_u/\phi > \frac{1}{2}V_c$ but $< V_c$, use minimum web steel in this case.

3. *Minimum web steel (step 4)*
From Example 5.4,

$$\text{Req.} \frac{A_v}{s} = 0.0076 \text{ in}^2/\text{in.}$$

So, trying #3 U stirrups, we get $A_v = 2 \times 0.11 = 0.22 \text{ in}^2$, and it follows that

$$s = \frac{A_v}{\text{Req. } A_v/s} = \frac{0.22}{0.0076} = 28.94 \text{ in. } (73 \text{ cm})$$

We then check for the minimum A_v as the lesser of the two values given by

$$A_v = 50b_w s/f_y$$

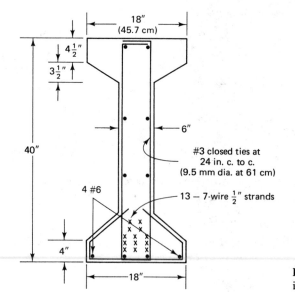

Figure 5.20 Web reinforcement details in Example 5.5.

or

$$A_v = \frac{A_{ps} \, f_{pu} S}{80 \, f_y d} \sqrt{\frac{d}{d_w}}$$

So the maximum allowable spacing $\leq 0.75h \leq 24$ in. Then use #3 U stirrups at 24 in. center to center over a stretch length of 84 in. from the face of the support, as in Example 5.4.

Details of the section reinforcement (step 8) are shown in Figure 5.20.

5.14 DESIGN OF WEB REINFORCEMENT FOR A PCI STANDARD SINGLE COMPOSITE T-BEAM

Example 5.6

A simply supported PCI standard single 8LST36 T-beam has a span of 70 ft (21.3 m). It is subjected to a total uniform service dead load of 655 plf (9.6 kN/m), including its self-weight $W_D = 455$ plf and a service live load $W_L = 720$ plf. Design the web reinforcement needed to prevent shear cracking at the quarter-span section 17 ft, 6 in. (5.3 m) from the support, Calculating the nominal web-shear strength V_c by the detailed design method. Also, design any dowel reinforcement if necessary, assuming that the top surface of the precast T-beam is intentionally unroughened. The section properties are shown in Figure 5.21 and are as follows:

Section property	Untopped	Topped
A_c	570 in^2	762 in^2
I_c	68,917 in^4	88,260 in^4
r^2	120.9 in^2	115.8 in^2
c_b	26.01 in.	29.09 in.
c_t	9.99 in.	8.91 in.
S_b	2,650 in^3	3,034 in^4
S^t	6,899 in^3	9,906 in^3
W_D	455 plf	655 plf

Other data are:

f'_c (precast) = 5,000 psi (34.5 MPa), normal-weight concrete

f'_{cc} (topping) = 3,000 psi (20.7 MPa), normal-weight concrete

f'_{ci} = 4,000 psi

f_{pu} = 270,000 psi (1,862 MPa), low-relaxation steel

f_{ps} = 240,000 psi (1,655 MPa)

f_{pe} = 148,000 psi (1,020 MPa)

e_e = 11.15 in. (28.3 cm)

e_c = 22.51 in. (57.2 cm)

A_{ps} = 14 $\frac{1}{2}$-in. (12.7 mm) dia strands

f_y for stirrups = 60,000 psi (414 MPa)

Use the same value for the effective depth d_p for the midspan as well as other sections.

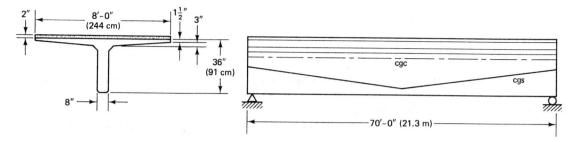

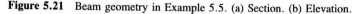

Figure 5.21 Beam geometry in Example 5.5. (a) Section. (b) Elevation.

Solution

$$W_u = 1.4 \times 655 + 1.7 \times 720 \cong 2{,}141 \text{ plf } (31.3 \text{ kN/m})$$

$$V_u \text{ at face of support} = \frac{2141 \times 70}{2} = 74{,}935 \text{ lb}$$

$$V_n \text{ at 17 ft, 6 in. from the face of support} = \frac{1}{0.85}\left(74{,}935 \times \frac{(35 - 17.5)}{35}\right)$$

$$= 44{,}079 \text{ lb } (196 \text{ kN})$$

1. *Flexure-shear cracking, V_{ci} (step 2)*

$$d_p = 36 - 26.01 + 22.51 = 32.5 \text{ in. } (82.6 \text{ cm})$$

$$P_e = 14 \times 0.153 \times 148{,}000 = 317{,}016 \text{ lb } (1{,}410 \text{ kN})$$

$$e \text{ at 17 ft, 6 in. from support} = 11.5 + (22.51 - 11.15)\frac{17.5}{35} = 16.83 \text{ in.}$$

Use the precast section properties for computing f_{ce} and f_d as discussed in Section 5.5:

$$f_{ce} = -\frac{P_e}{A_c}\left(1 + \frac{ec_b}{r^2}\right) = -\frac{317{,}016}{570}\left(1 + \frac{16.83 \times 26.01}{120.9}\right)$$

$$= 2{,}570 \text{ psi } (17.7 \text{ MPa})$$

Note that although $f_{ce} > 0.45 f'_c$, this should not affect the shear strength since f_{ce} is due to prestress only, and the inclusion of self-weight reduces it to less than $0.45 f'_c$. We thus have

Self-weight $W_D = 455$ plf

$$M_{17.5} = \frac{W_D \times (\ell - x)}{2} = \frac{455 \times 17.5(70 - 17.5)}{2} \times 12$$

$$= 2{,}508{,}188 \text{ in.-lb } (283 \text{ kN-m})$$

$$f_d = \frac{Mc_b}{I_c} = \frac{2{,}508{,}188 \times 26.01}{68{,}917} = 947 \text{ psi } (6.5 \text{ MPa})$$

$$M_{cr} = S_b(6\lambda \sqrt{f'_c} + f_{ce} - f_d)$$

$$= 2{,}650(6 \times 1.0\sqrt{5{,}000} + 2{,}570 - 947)$$

$$= 5{,}425{,}250 \text{ lb } (613 \text{ kN-m})$$

$$V_d = W_D\left(\frac{\ell}{2} - x\right) = 455\left(\frac{70}{2} - 17.5\right) = 7,963 \text{ lb}$$

$$W_{SD} = 655 - 455 = 200 \text{ plf}$$

$$W_L = 720 \text{ plf}$$

$$W_U = 1.4 \times 200 + 1.7 \times 720 = 1,504 \text{ plf (22.0 kN/m)}$$

$$V_i = W_u\left(\frac{\ell}{2} - x\right) = 1,504\left(\frac{70}{2} - 17.5\right) = 26,320 \text{ lb (117 kN)}$$

$$M_{max} = W_u x \frac{(\ell - x)}{2} = \frac{1,504 \times 17.5(70 - 17.5)}{2} \times 12$$

$$= 8,290,800 \text{ in.-lb (937 kN-m)}$$

$$V_{ci} = 0.6\lambda\sqrt{f'_c}\,b_w d_p + V_d + \frac{V_i}{M_{max}}(M_{cr}) \geq 1.7\lambda\sqrt{f'_c}\,b_w d_p$$

$$= 0.6 \times 1.0 \times \sqrt{5,000} \times 8 \times 32.5 + 7,963$$

$$+ \frac{26,320}{8,290,800}(5,425,250)$$

$$= 36,217 \text{ lb (161 kN)}$$

$$1.7\lambda\sqrt{f'_c}\,b_w d_p = 1.7 \times 1.0\sqrt{5,000} \times 8 \times 32.5 = 31,254 \text{ lb} < 36,217 \text{ lb}$$

Hence, $V_{ci} = 36,217$ lb controls.

2. *Web-shear cracking, V_{cw} (step 2)*

$$\bar{f}_c = \frac{P_e}{A_c} = \frac{317,016}{570} = 556 \text{ psi (3.8 MPa)}$$

For the vertical component of the prestress force,

$$V_p = P_e \tan\theta = 317,016\frac{(22.51 - 11.15)}{70/2 \times 12} = 8,575 \text{ lb (38.1 kN)}$$

$$V_{cw} = (3.5\sqrt{f'_c} + 0.3\bar{f}_c)b_w d_{cp} + V_p$$

$$= (3.5\sqrt{5,000} + 0.3 \times 556) \times 8 \times 32.5 + 8,575$$

$$= 116,290 \text{ lb vs. } V_{ci} = 36,217 \text{ lb}$$

Now, V_c is the smaller of V_{ci} and V_{cw}; hence,

$$V_c = V_{ci} = 36,217 \text{ lb (161 kN)}$$

3. *Design of web reinforcement (steps 3–8)*
From above,

$$V_c = 36,217 \text{ lb}$$

So

$$\frac{1}{2}V_c = 19,109 \text{ lb}$$

Now, V_u/ϕ at a section 17.5 ft from support $= 44,120$ lb $> V_c > \frac{1}{2}V_c$; hence, design of stirrups is necessary. If $V_u/\phi < V_c > \frac{1}{2}V_c$, only minimum-web steel is needed. So, using $A_v/s = 50b_w/f_y$, we obtain

$$\text{Req.}\ \frac{A_v}{s} = \frac{V_n - V_c}{f_y d_p} = \frac{(V_u/\phi) - V_c}{f_y d_p} = \frac{44{,}120 - 36{,}217}{60{,}000 \times 32.5}$$

$$= 0.0041\ \text{in}^2/\text{in. spacing}$$

Consequently, use $d \simeq d_p = 32.5$ in. and $b_w = 8$ in.:

$$\text{Min.}\left(\frac{A_v}{s}\right) = \frac{A_{ps}\ f_{pu}}{80\ f_y d} \sqrt{\frac{d}{d_w}} = \frac{14 \times 0.153}{80} \times \frac{270{,}000}{60{,}000 \times 32.5} \sqrt{\frac{32.5}{8}} = 0.0075$$

Or

$$\text{Min.}\left(\frac{A_v}{s}\right) = \frac{50b_w}{f_y} = \frac{50 \times 8}{60{,}000} = 0.0067\ \text{in}^2/\text{in.}$$

The *lesser* of the two values applies; hence, use

$$\frac{A_v}{s} = 0.0067\ \text{in}^2/\text{in.}$$

Then, try #3 U stirrups:

$$A_v = 2 \times 0.11 = 0.22\ \text{in}^2$$

$$s = \frac{A_v}{\text{Req}(A_v/s)} = \frac{0.22}{0.008} = 27.5\ \text{in. center to center}$$

The maximum allowable spacing is, then, $0.75h \le 24$ in. So we have

$$0.75h = 0.75 \times 36 = 27\ \text{in.}$$

Accordingly, use #3 U stirrups at 24 in. center to center at the quarter-span section (9.5 mm dia at 61 cm center to center).

Note, in comparing the solution for V_{ci} and V_{cw} in Example 5.6, that V_{ci} has its highest value close to the support and *rapidly* decreases towards the midspan, while V_{cw} has a lesser variation in its value, as can be seen from Figure 5.13. It is important to calculate the flexure shears V_{ci} and V_{cw} at several sections along the span in order to determine the most efficient distribution of the web steel. A computer program facilitates finding these values at constant intervals of, say, one-tenth of the span, and a plot can be made similar to the one in Figure 5.13 showing the variation of the shear strengths of the web along the span.

4. *Design of dowel steel for full composite section (step 9)*
   Section at $\frac{1}{2}d_p$ from face of support

$$V_u\ \text{at support} = 75{,}005\ \text{lb (334 kN)}$$

$$\frac{1}{2}d_p = \frac{32.5}{2 \times 12} = 1.35\ \text{ft (41 cm)}$$

$$V_u = 75{,}005 \times \frac{(35 - 1.35)}{35} = 72{,}112\ \text{lb (321 kN)}$$

$$\text{Req}\ V_{nh} = \frac{V_u}{\phi} = \frac{72{,}112}{0.85} = 84{,}838\ \text{lb (377 kN)}$$

$$b_v = 8\ \text{in.}$$

From Figure 5.14,

$$C_c = 0.85 f'_{cc} A_{\text{top}} = 0.85 \times 3000 \times 8 \times 12 \times 2 = 489{,}600 \text{ lb } (2{,}178 \text{ kN})$$

$$T_s = A_{ps} f_{ps} = 14 \times 0.153 \times 240{,}000 = 514{,}080 \text{ lb } (2{,}287 \text{ kN})$$

$$> C_c = 489{,}600 \text{ lb}$$

Then

$$F_h = 489{,}600 \text{ lb } (2{,}178 \text{ kN})$$

$$\ell_{vh} = \frac{70 \times 12}{2} = 420 \text{ in. } (1{,}067 \text{ cm})$$

$$b_v = 96 \text{ in. } (244 \text{ cm})$$

$$80 b_v \ell_{vh} = 80 \times 96 \times 420 = 3{,}225{,}600 \text{ lb } (14{,}347 \text{ kN}) \gg 489{,}620 \text{ lb}$$

No dowel reinforcement is needed to extend to the topping for full composite action to be developed.

5.15 BRACKETS AND CORBELS

Brackets and corbels are short-haunched cantilevers that project from the inner face of columns to support heavy concentrated loads or beam reactions. They are very important structural elements for supporting precast beams, gantry girders, and any other forms of precast structural systems. Precast and prestressed concrete is becoming increasingly dominant, and larger spans are being built, resulting in heavier shear loads at supports. Hence, the design of brackets and corbels has become increasingly important. The safety of the total structure could depend on the sound design and construction of the supporting element, in this case the corbel, necessitating a detailed discussion of this subject.

In brackets and corbels, the ratio of the shear arm or span to the corbel depth is often less than 1.0. Such a small ratio changes the state of stress of a member into a two-dimensional one. Shear deformations would hence affect the nonlinear stress behavior of the bracket or corbel in the elastic state and beyond, and the shear strength becomes a major factor. Corbels also differ from deep beams in the existence of potentially large horizontal forces transmitted from the supported beam to the corbel or bracket. These horizontal forces result from long-term shrinkage and creep deformation of the supported beam, which in most cases is anchored to the bracket.

The cracks are usually mostly vertical or steeply inclined pure shear cracks. They often start from the point of application of the concentrated load and propagate toward the bottom reentrant corner junction of the bracket to the column face, as in Figure 5.22(a). Or they start at the upper reentrant corner of the bracket or corbel and proceed almost vertically through the corbel toward its lower fibers, as shown in Figure 5.22(b). Other failure patterns in such elements are shown in Figure 5.22(c) and (d). They can also develop through a combination of the ones illustrated. Bearing failure can also occur by crushing of the concrete under the concentrated load-bearing plate, if the bearing area is not adequately proportioned.

As will be noticed in the subsequent discussion, detailing of the corbel or bracket

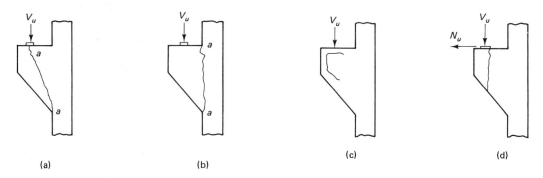

Figure 5.22 Failure patterns. (a) Diagonal shear. (b) Shear friction. (c) Anchorage splitting. (d) Vertical splitting.

reinforcement is of major importance. Failure of the element can be attributed in many cases to incorrect detailing that does not realize full anchorage development of the reinforcing bars.

5.15.1 Shear Friction Hypothesis for Shear Transfer in Corbels

Corbels cast at different times than the main supporting columns can have a potential shear crack at the interface between the two concretes through which shear transfer has to develop. The smaller the ratio a/d, the larger the tendency for pure shear to occur through essentially vertical planes. This behavior is accentuated in the case of corbels with a potential interface crack between two dissimilar concretes.

The shear friction approach in this case is recommended by the ACI, as shown in Figure 5.22(b). An assumption is made of an already cracked vertical plane (a–a in Figure 5.23) along which the corbel is considered to slide as it reaches its limit state of failure. A coefficient of friction μ is used to transform the horizontal resisting forces of the well-anchored closed ties into a vertical nominal resisting force larger than the external factored shear load. Hence, the nominal vertical resisting shear force

$$V_n = A_{vf} f_u \mu \tag{5.34 a}$$

to give

$$A_{vf} = \frac{V_n}{f_y \mu} \tag{5.34 b}$$

where A_{vf} is the total area of the horizontal, anchored closed shear ties.

The external factored vertical shear has to be $V_u \leq \phi V_n$, where for normal concrete,

$$V_n \leq 0.2 f'_c b_w d \tag{5.35 a}$$

or

$$V_n \leq 800 b_w d \tag{5.35 b}$$

whichever is smaller. The required effective depth d of the corbel can be determined from Equation 5.35 a, or b, whichever gives a larger value.

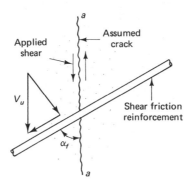

Figure 5.23 Shear-friction reinforcement at crack.

For all-lightweight or sand-lightweight concretes, the shear strength V_n should not be taken greater than $(0.2 - 0.07a/d)f'_c b_w d$, or $(800 - 280a/d)b_w d$ in pounds.

If the shear friction reinforcement is inclined to the shear plane such that the shear force produces some tension in the shear friction steel,

$$V_n = A_{vf}f_y(\mu \sin \alpha_f + \cos \alpha_f) \qquad (5.35 \text{ c})$$

where α_f is the angle between the shear friction reinforcement and the shear plane. The reinforcement area becomes

$$A_{vf} = \frac{V_n}{f_y(\mu \sin \alpha_f + \cos \alpha_f)} \qquad (5.35 \text{ d})$$

The assumption is made that all the shear resistance is due to the resistance at the crack interface between the corbel and the column. The ACI coefficient of friction μ has the following values:

Concrete case monolithically 1.4λ

Concrete placed against hardened roughened concrete 1.0λ

High-strength concrete corbel at failure (Nawy et al.).

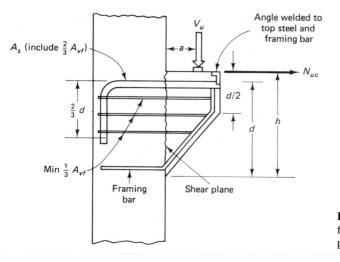

A_s (include $\frac{2}{3} A_{vf}$)

V_u

Angle welded to top steel and framing bar

N_{uc}

a

$d/2$

$\frac{2}{3} d$

d

h

Min $\frac{1}{3} A_{vf}$

Framing bar

Shear plane

Figure 5.24 Reinforcement schematic for corbel design by shear friction hypothesis.

Concrete placed against unroughened hardened concrete	0.6λ
Concrete anchored to structural steel	0.7λ

$\lambda = 1.0$ for normal-weight concrete, 0.85 for sand-lightweight concrete, and 0.75 for all-lightweight concrete. The PCI values are less conservative than the ACI values based on comprehensive tests.

If considerably higher strength concretes, such as polymer-modified concretes, are used in the corbels to interface with the normal concrete of the supporting columns, higher values of μ could logically be used for such cases as those listed above. Work in the field (Ref. 5.7) substantiates the use of higher values.

Part of the horizontal steel A_{vf} is incorporated in the top tension tie, and the remainder of A_{vf} is distributed along the depth of the corbel as in Figure 5.24. Evaluation of the top horizontal primary reinforcement layer A_s will be discussed in the next section.

5.15.2 Horizontal External Force Effect

When the corbel or bracket is cast monolithically with the supporting column or wall and is subjected to a large horizontal tensile force N_{uc} produced by the beam supported by the corbel, a modified approach is used, often termed the *strut theory approach*. In all cases, the horizontal factored force N_{uc} cannot exceed the vertical factored shear V_u. As shown in Figure 5.25, reinforcing steel A_n has to be provided to resist the force N_{uc}. We have

$$A_n = \frac{N_{uc}}{\phi f_y} \tag{5.36}$$

and

$$A_f = \frac{V_u a + N_{uc}(h - d)}{\phi f_y jd} \tag{5.37}$$

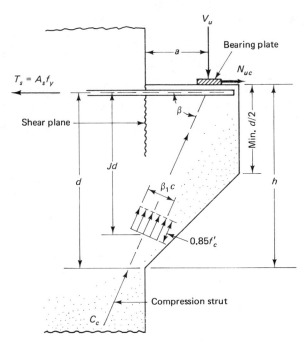

Figure 5.25 Compression strut in corbel.

Reinforcement A_f also has to be provided to resist the bending moments caused by V_u and N_{uc}.

The value of N_{uc} considered in the design should not be less than $0.20V_u$. The flexural steel area A_f can be obtained approximately by the usual expression for the limit state at failure of beams, that is,

$$A_f = \frac{M_u}{\phi f_y jd} \tag{5.38}$$

where $M_u = V_u a + N_{uc}(h - d)$. The axis of such an assumed section lies along a compression strut inclined at an angle β to the tension tie A_s, as shown in the figure. The volume of the compressive block is

$$C_c = 0.85 f_c' \beta_1 cb = \frac{T_s}{\cos \beta} = \frac{A_s f_y}{\cos \beta} = \frac{V_u}{\sin \beta} \tag{5.39 a}$$

for which the depth $\beta_1 c$ of the block is obtained perpendicular to the direction of the compressive strut, i.e.,

$$\beta_1 c = \frac{A_s f_y}{0.85 f_c' b \cos \beta} \tag{5.39 b}$$

The effective depth d minus $\beta_1 c/2 \cos \beta$ in the vertical direction gives the lever arm jd between the force T_s and the horizontal component of C_c in Figure 5.25. Therefore,

$$jd = d - \frac{\beta_1 c}{2 \cos \beta} \tag{5.39 c}$$

If the right-hand side is substituted for jd in Equation 5.38, then

$$A_f = \frac{M_u}{\phi f_y (d - \beta_1 c / 2 \cos \beta)} \qquad (5.40)$$

To eliminate several trials and adjustments, the lever arm jd from Equation (5.39 c) can be approximated for all practical purposes in most cases as

$$jd \cong 0.85d \qquad (5.41\ a)$$

so that

$$A_f = \frac{M_u}{0.85 \phi f_y d} \qquad (5.41\ b)$$

The area A_s of the primary tension reinforcement (tension tie) can now be calculated and placed as shown in Figure 5.26:

$$A_s \geq \tfrac{2}{3} A_{vf} + A_n \qquad (5.42)$$

or

$$A_s \geq A_f + A_n \qquad (5.43)$$

whichever is larger. Then

$$\rho = \frac{A_s}{bd} \geq 0.04 \frac{f_c'}{f_y}$$

If A_h is assumed to be the total area of the closed stirrups or ties parallel to A_s, then

$$A_h \geq 0.5(A_s - A_n) \qquad (5.44)$$

The bearing area under the external load V_u on the bracket should not project beyond the straight portion of the primary tension bars A_s, nor should it project beyond the interior face of the transverse welded anchor bar shown in Figure 5.26.

5.15.3 Sequence of Corbel Design Steps

As discussed in the preceding section, a horizontal factored force N_{uc}, a vertical factored force V_u, and a bending moment $[V_u a + N_{uc}(h - d)]$ basically act on the corbel. To prevent failure, the corbel has to be designed to resist these three parameters simultaneously by one of the following two methods, depending on the type of corbel construction sequence, that is, whether the corbel is cast monolithically with the column or not:

(a) For a monolithically cast corbel with the supporting column, by evaluating the steel area A_h of the closed stirrups which are placed below the primary steel ties A_s. Part of A_h is due to the steel area A_n from Equation 5.36 resisting the horizontal force N_{uc}.

(b) By calculating the steel area A_{vf} by the shear friction hypothesis if the corbel and the column are *not* cast simultaneously, using part of A_{vf} along the depth of the

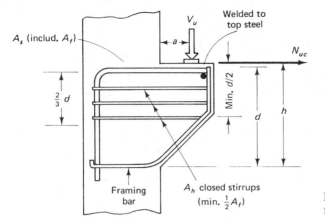

Framing
bar

A_h closed stirrups
(min. $\frac{1}{2}A_f$)

Figure 5.26 Reinforcement schematic for corbel design by strut theory.

corbel stem and incorporating the balance in the area A_s of the primary top steel reinforcing layer.

The primary tension steel area A_s is the major component of both methods. Calculations of A_s depend on whether Equation 5.42 or 5.43 governs. If Equation 5.42 controls, $A_s = \frac{2}{3}A_{vf} + A_n$ is used and the remaining $\frac{1}{3}A_{vf}$ is distributed over a depth $\frac{2}{3}d$ adjacent to A_s. If Equation 5.43 controls, $A_s = A_f + A_n$, with the addition of $\frac{1}{2}A_f$ provided as closed stirrups parallel to A_s and distributed within $\frac{2}{3}d$ vertical distance adjacent to A_s.

In both cases, the primary tension reinforcement plus the closed stirrups automatically yield the total amount of reinforcement needed for either type of corbel. Since the mechanism of failure is highly indeterminate and randomness can be expected in the propagation action of the shear crack, it is sometimes advisable to choose the larger calculated value of the primary top steel area A_s in the corbel regardless of whether the corbel element is cast simultaneously with the supporting column.

As seen from the foregoing discussions, the horizontal closed stirrups are also a major element in reinforcing the corbel. Occasionally, additional inclined closed stirrups are also used.

The following sequence of steps is proposed for the design of the corbel:

1. Calculate the factored vertical force V_u and the nominal resisting force V_n of the section such that $V_n \geq V_u/\phi$, where $\phi = 0.85$ for all calculations. V_u/ϕ should be $\leq 0.20f'_c b_w d$, or $\leq 800b_w d$ for normal-weight concrete. If not, the concrete section at the support should be enlarged.

2. Calculate $A_{vf} = V_n/f_y\mu$ for resisting the shear friction force, and use in the subsequent calcuation of the primary tension top steel A_s.

3. Calculate the flexural steel area A_f and the direct tension steel area A_n, where

$$A_f = \frac{v_u a + N_{uc}(h - d)}{\phi f_y j_d}.$$

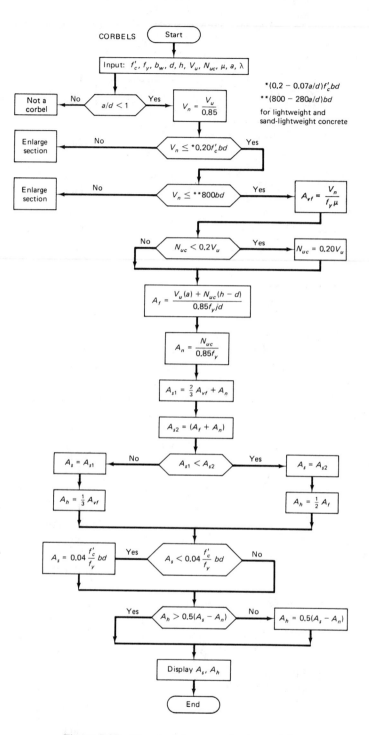

Figure 5.27 Flowchart for proportioning corbels.

and

$$A_n = \frac{N_{uc}}{\phi f_y}$$

4. Calculate the primary steel area from (a) $A_s = \frac{2}{3}A_{vf} + A_n$ and (b) $A_s = A_f + A_n$, whichever is larger. If case (a) controls, the remaining $\frac{1}{3}A_{vf}$ has to be provided as closed stirrups parallel to A_s and distributed with a $\frac{2}{3}d$ distance adjacent to A_s, as in Figure 5.24.

   If case (b) controls, use in addition $\frac{1}{2}A_f$ as closed stirrups distributed within a distance $\frac{2}{3}d$ adjacent to A_s, as in Figure 5.26. Then

$$A_h \geq 0.5(A_s - A_n)$$

and

$$\rho = \frac{A_s}{bd} \geq 0.04 \frac{f_c'}{f_y}$$

or

$$\text{Min } A_s = 0.04 \frac{f_c'}{f_y} bd$$

5. Select the size and spacing of the corbel reinforcement with special attention to the detailing arrangements, as many corbel failures are due to incorrect detailing.

Figure 5.27 shows a flowchart for proportioning corbels.

5.15.4 Design of a Bracket or Corbel

Example 5.7

Design a corbel to support a factored vertical load $V_u = 90,000$ lb (180 kN) acting at a distance $a = 5$ in. (127 mm) from the face of the column. The corbel has a width $b = 10$ in. (254 mm), a total thickness $h = 18$ in. (457.2 mm), and an effective depth $d = 14$ in. (355.6 mm). The following data are given:

$$f_c' = 5,000 \text{ psi (34.5 MPa), normal-weight concrete}$$

$$f_y = 60,000 \text{ psi (414 MPa)}$$

Assume the corbel to be either cast after the supporting column was constructed, or cast simultaneously with the column. Neglect the weight of the corbel.

Solution

   Step 1

$$V_n \geq \frac{V_u}{\phi} = \frac{90,000}{0.85} = 105,882 \text{ lb}$$

$$0.2f_c' b_w d = 0.2 \times 5,000 \times 10 \times 14 = 140,000 \text{ lb} > V_n$$

$$800b_w d = 800 \times 10 \times 14 = 112,000 \text{ lb} > V_n, \text{ O.K.}$$

Step 2

(a) Monolithic construction, normal-weight concrete $\mu = 1.4\lambda$:

$$A_{vf} = \frac{V_u}{\phi f_y \mu} = \frac{105,882}{60,000 \times 1.4} = 1.261 \text{ in}^2 \ (813.3 \text{ mm}^2)$$

(b) Nonmonolithic construction, $\mu = 1.0\lambda$:

$$A_{vf} = \frac{105,882}{60,000 \times 1.0} = 1,765 \text{ in}^2 \ (1138.2 \text{ mm}^2)$$

Choose the larger $A_{vf} = 1.765 \text{ in}^2$ as controlling.

Step 3. Since no value of the horizontal external force N_{uc} transmitted from the superimposed beam is given, use

$$\text{Min } N_{uc} = 0.20V_u = 0.2 \times 90,000 = 18,000 \text{ lb}$$

$$A_f = \frac{M_u}{\phi f_y jd} = \frac{V_u a + N_{uc}(h - d)}{\phi f_y jd} \quad (\text{where } jd \cong 0.85d)$$

$$= \frac{90,000 \times 5 + 18,000(18 - 14)}{0.90 \times 60,000(0.85 \times 14)} = 0.812 \text{ in}^2 \ (523.9 \text{ mm}^2)$$

$$A_n = \frac{N_{uc}}{\phi f_y} = \frac{18,000}{0.85 \times 60,000} = 0.353 \text{ in}^2 \ (277.6 \text{ mm}^2)$$

Step 4. Check the controlling area of primary steel A_s:

(a) $A_s = (\frac{2}{3}A_{vf} + A_n) = \frac{2}{3} \times 1.765 + 0.353 = 1.529 \text{ in}^2$

(b) $A_s = A_f + A_n = 0.812 + 0.353 = 1.165 \text{ in}^2$

$$\text{Min } A_s = 0.04\frac{f_c'}{f_y} bd = 0.04 \times \frac{5,000}{60,000} \times 10 \times 14 = 0.47 \text{ in}^2$$

$$< 1.529, \text{ O.K.}$$

Provide $A_s = 1.529 \text{ in}^2 \ (986.2 \text{ mm}^2)$.
Horizontal closed stirrups:
Since case (a) controls,

$$\tfrac{1}{3}A_{vf} = \tfrac{1}{3} \times 1.765 = 0.588 \text{ in}^2$$

$$A_h = 0.5(A_s - A_n) = 0.5(1.529 - 0.353) = 0.588 \text{ in}^2$$

Use the larger of the two values $\frac{1}{3}A_{vf}$ and A_h.

Step 5. Select bar sizes:

(a) Required $A_s = 1.529 \text{ in.}^2$; use three No. 7 bars $= 1.80 \text{ in}^2$ (three bars of diameter 22.2 mm gives $1,161 \text{ mm}^2$) A_s.

(b) Required $A_h = 0.588 \text{ in}^2$; use three No. 3 closed stirrups $= 2 \times 3 \times 0.11 = 0.66 \text{ in}^2$ spread over $\frac{2}{3}d = 9.33$ in. vertical distance. Hence, use three No. 3 closed stirrups at 3 in. center to center. Also use three framing size No. 3 bars and one welded No. 7 anchor to bar.

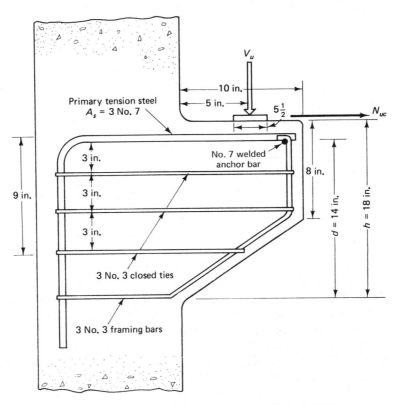

Figure 5.28 Corbel reinforcement details (Example 5.7).

Details of the bracket reinforcement are shown in Fig. 5.28. The bearing area under the load has to be checked, and the bearing pad designed such that the bearing stress at the factored load V_u should not exceed $\phi(0.85 f_c' A_1)$, where A_1 is the pad area. We have

$$V_u = 90,000 \text{ lb} = 0.70(0.85 \times 5,000)$$

$$A_1 = \frac{90,000}{0.70 \times 0.85 \times 5000} = 30.25 \text{ in}^2 \ (19,516 \text{ mm}^2)$$

Use a plate $5\frac{1}{2}$ in. $\times$ $5\frac{1}{2}$ in. Its thickness has to be designed based on the manner in which V_u is applied.

5.16 TORSIONAL BEHAVIOR AND STRENGTH

5.16.1 Introduction

Torsion occurs in monolithic concrete construction primarily where the load acts at a distance from the longitudinal axis of the structural member. An end beam in a floor panel, a spandrel beam receiving load from one side, a canopy or a bus-stand roof

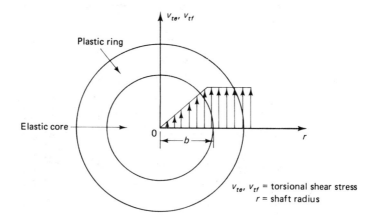

Figure 5.29 Torsional stress distribution through circular section.

projecting from a monolithic beam on columns, and peripheral beams surrounding a floor opening are all examples of structural elements subjected to twisting moments. These moments occasionally cause excessive shearing stresses. As a result, severe cracking can develop well beyond the allowable serviceability limits unless special torsional reinforcement is provided. Photos in this section illustrate the extent of cracking at failure of a beam in torsion. They show the curvilinear plane of twist caused by the imposed torsional moments. In actual spandrel beams of a structural system, the extent of damage due to torsion is usually not as severe. This is due to the redistribution of stresses in the structure. However, loss of integrity due to torsional distress should always be avoided by proper design of the necessary torsional reinforcement.

An introduction to the subject of torsional stress distribution has to start with the basic elastic behavior of simple sections, such as circular or rectangular sections. Most concrete beams subjected to twist are components of rectangles, for example, flanged sections such as T-beams and L-beams. Although circular sections are rarely a consideration in normal concrete construction, a brief discussion of torsion in circular sections serves as a good introduction to the torsional behavior of other types of sections.

Shear stress is equal to shear strain times the shear modulus at the elastic level in circular sections. As in the case of flexure, the stress is proportional to its distance from the neutral axis (i.e., the axis through the center of the circular section) and is maximum at the extreme fibers. If r is the radius of the element, $J = \pi r^4/2$ its polar moment of inertia, and v_{te} the elastic shearing stress due to an elastic twisting moment T_e, then

$$v_{te} = \frac{T_e r}{J} \tag{a}$$

When deformation takes place in the circular shaft, the axis of the circular cylinder is assumed to remain straight. All radii in a cross section also remain straight (i.e., there is no warping) and rotate through the same angle about the axis. As the circular element starts to behave plastically, the stress in the plastic outer ring becomes constant while the stress in the inner core remains elastic, as shown in Figure 5.29.

254 Shear and Torsional Strength Design   Chap. 5

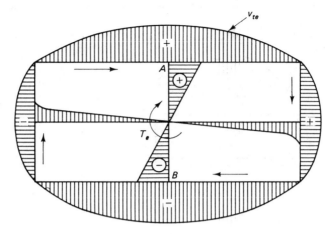

Figure 5.30 Pure torsional stress distribution in a rectangular section.

As the whole cross section becomes plastic, $b = 0$ and the shear stress

$$v_{tf} = \frac{3}{4} \frac{T_p r}{J} \tag{b}$$

where v_{tf} is the nonlinear shear stress due to an ultimate twisting moment T_p. (The subscript f denotes failure.)

In rectangular sections, the torsional problem is considerably more complicated. The originally plane cross sections undergo warping due to the applied torsional moment. This moment produces axial as well as circumferential shear stresses with zero values at the corners of the section and the centroid of the rectangle, and maximum values on the periphery at the middle of the sides, as shown in Figure 5.30. The maximum torsional shearing stress would occur at midpoints A and B of the larger dimension of the cross section. These complications plus the fact that the reinforced and prestressed concrete sections are neither homogeneous nor isotropic make it difficult to develop exact mathematical formulations based on physical models such as Equations (a) and (b) for circular sections.

For over 60 years, the torsional analysis of concrete members has been based on either (1) the classical theory of elasticity developed through mathematical formulations coupled with membrane analogy verifications (St.-Venant's), or (2) the theory of plasticity represented by the sand-heap analogy (Nadai's). Both theories were applied essentially to the state of pure torsion. But it was found experimentally that the elastic theory is not entirely satisfactory for the accurate prediction of the state of stress in concrete in pure torsion. The behavior of concrete was found to be better represented by the plastic approach. Consequently almost all developments in torsion as applied to prestressed concrete and to reinforced concrete have been in the latter direction.

5.16.2 Pure Torsion in Plain Concrete Elements

5.16.2.1 Torsion in elastic materials.

In 1853, St.-Venant presented his solution to the elastic torsional problem with warping due to pure torsion which develops in noncircular sections. In 1903, Prandtl demonstrated the physical

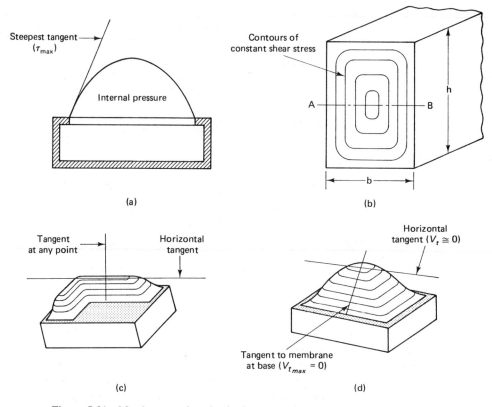

Figure 5.31 Membrane analogy in elastic pure torsion. (a) Membrane under pressure. (b) Contours in a real beam or in a membrane. (c) L-section. (d) Rectangular section.

significance of the mathematical formulations by his membrane analogy model. The model establishes particular relationships between the deflected surface of the loaded membrane and the distribution of torsional stresses in a bar subjected to twisting moments. Figure 5.31 shows the membrane analogy behavior for rectangular as well as L-shaped forms.

For small deformations, it can be proved that the differential equation of the deflected membrane surface has the same form as the equation that determines the stress distribution over the cross section of the bar subjected to twising moments. Similarly, it can be demonstrated that (1) the tangent to a contour line at any point of a deflected membrane gives the direction of the shearing stress at the corresponding cross section of the actual membrane subjected to twist; (2) the maximum slope of the membrane at any point is proportional to the magnitude of shear stress τ at the corresponding point in the actual member; and (3) the twisting moment to which the actual member is subjected is proportional to *twice* the volume under the deflected membrane.

It can be seen from Figures 5.30 and 5.31(b) that the torsional shearing stress is inversely proportional to the distance between the contour lines. The closer the lines, the higher the stress, leading to the previously stated conclusion that the maximum torsional shearing stress occurs at the middle of the longer side of the rectangle.

From the membrane analogy, this maximum stress has to be proportional to the steepest slope of the tangents at points A and B.

If δ is the maximum displacement of the membrane from the tangent at point A, then from basic principles of mechanics and St.-Venant's theory,

$$\delta = b^2 G\theta \tag{5.45 a}$$

where G is the shear modulus and θ is the angle of twist. But $v_{t(max)}$ is proportional to the slope of the tangent; hence,

$$v_{t(max)} = k_1 bG\theta \tag{5.45 b}$$

where k_1 is a constant. The corresponding torsional moment T_e is proportional to *twice* the volume under the membrane, or

$$T_e \propto 2(\tfrac{2}{3}\delta bh) = k_2 \delta bh$$

where, again, k_2 is a constant. From Or yet again,

$$T_e = k_3 b^3 hG\theta \tag{5.45 c}$$

with k_3 constant. From Equations 5.45 a and b,

$$v_{t(max)} = \frac{T_e b}{kb^3 h} \simeq \frac{T_e b}{J_1} \tag{5.45 d}$$

The denominator kb^3h in 5.45 d represents the polar moment of inertia J_1 of the section. Comparing this equation to Equation (a) for the circular section shows the similarity of the two expressions, except that the factor k in the equation for the rectangular section takes into account the shear strains due to warping. Equation 5.45 d can be further simplified to give

$$v_{t(max)} = \frac{T_e}{kb^2 h} \tag{5.46}$$

It can also be written to give the stress at planes inside the section, such as an inner concentric rectangle of dimensions x and y, where x is the shorter side, so that

$$v_{t(max)} = \frac{T_e}{kx^2 y} \tag{5.47}$$

It is important to note in using the membrane analogy approach that the torsional shear stress changes from one point to another along the same axis as AB in Figure 5.31, because of the changing slope of the analogous membrane, rendering the torsional shear stress calculations lengthy.

5.16.2.2 Torsion in plastic materials. As indicated earlier, the plastic sand-heap analogy provides a better representation of the behavior of brittle elements such as concrete beams subjected to pure torsion than does the elastic analogy. The torsional moment is also proportional to *twice* the volume under the heap, and the maximum torsional shearing stress is proportional to the slope of the sand heap. Figure 5.32 is a two- and three-dimensional illustration of the sand heap. The torsional moment T_p in part (d) of the figure is proportional to twice the volume of the rectangular heap shown in parts (b) and (c). It can also be recognized that the slope of the

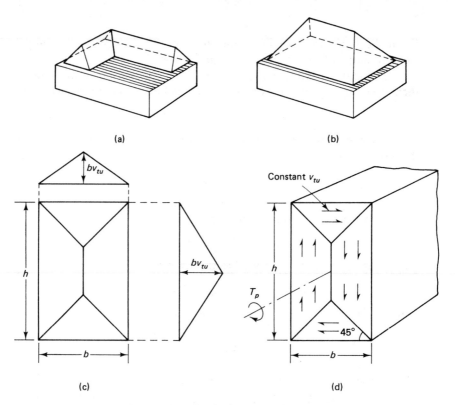

Figure 5.32 Sand-heap analogy in plastic pure torsion. (a) Sand-heap L-section. (b) Sand-heap rectangular section. (c) Plan of rectangular section. (d) Torsional shear stress.

sand-heap sides as a measure of the torsional shearing stress is *constant* in the sand-heap analogy approach, whereas it is continuously variable in the membrane analogy approach. This characteristic of the sand heap considerably simplifies the solutions.

5.16.2.3 Sand-heap analogy applied to L-beams. Most concrete elements subjected to torsion are flanged sections, most commonly L-beams comprising the external wall beams of a structural floor. The L-beam in Figure 5.33 is chosen in applying the plastic sand-heap approach to evaluate its torsional moment capacity and shear stresses to which it is subjected.

The sand heap is broken into three volumes:

V_1 = pyramid representing a square cross-sectional shape = $y_1 b_w^2/3$

V_2 = tent portion of the web, representing a rectangular cross-sectional shape = $y_1 b_w (h - b_w)/2$

V_3 = tent representing the flange of the beam, transferring part *PDI* to *NQM* = $y_2 h_f (b - b_w)/2$

The torsional moment is proportional to twice the volume of the sand heaps; hence,

$$T_p \cong \left[\frac{y_1 b_w^2}{3} + \frac{y_1 b_w (h - b_w)}{2} + \frac{y_2 h_f (b - b_w)}{2} \right] \tag{5.48}$$

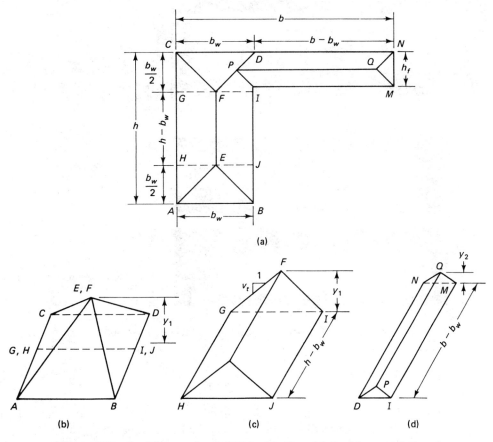

Figure 5.33 Sand-heap analogy of flanged section (a) Sand heap on L-shaped cross section. (b) Composite pyramid from web (V_1). (c) Tent segment from web (V_2). (d) Transformed tent of beam flange (V_3).

Also, the torsional shear stress is proportional to the slope of the sand heaps; hence,

$$y_1 = \frac{v_t b_w}{2} \tag{5.49}$$

$$y_2 = \frac{v_t h_f}{2} \tag{5.50}$$

Substituting y_1 and y_2 from Equations 5.49 and 5.50 into Equation 5.48 gives us

$$v_{t(\text{max})} = \frac{T_p}{(b_w^2/6)(3h - b_w) + (h_f^2/2)(b - b_w)} \tag{5.51}$$

If both the numerator and denominator of Equation 5.51 are divided by $(b_w h)^2$ and the terms rearranged, we have

$$v_{t(\text{max})} = \frac{T_p h/(b_w h)^2}{[\frac{1}{6}(3 - b_w/h)] + \frac{1}{2}(h_f/b_w)2(b/h - b_w/h)} \tag{5.52 a}$$

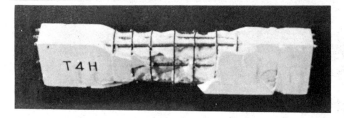

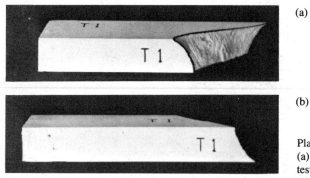

Reinforced plaster beam at failure in pure torsion. (Rutgers tests: Law, Nawy et al.)

(a)

(b)

Plain mortar beam in pure torsion. (a) Top view. (b) Bottom view. (Rutgers tests: Law, Nawy et al.)

If one assumes that C_t is the denominator in this equation and that $J_E = C_t/(b_w h)^2$, the equation becomes

$$v_{t(\max)} = \frac{T_p h}{J_E} \qquad (5.52\ b)$$

where J_E is the equivalent polar moment of inertia, a function of the shape of the beam cross section. Note that Equation 5.52 b is similar in format to Equation 5.45 d from the membrane analogy, except for the different values of the denominators J and J_E. Equation 5.52 a can be readily applied to rectangular sections by setting $h_f = 0$.

It must also be recognized that concrete is not a perfectly plastic material; hence, the actual torsional strength of the plain concrete section has a value lying between the membrane analogy and the sand-heap analogy values.

Equation 5.52 b can be rewritten designating $T_p = T_c$ as the nominal torsional resistance of the plain concrete and $v_{t(\max)} = v_{tc}$ using ACI terminology, so that

$$T_c = k_2 b^2 h v_{tc} \qquad (5.53\ a)$$

or

$$T_c = k_2 x^2 y v_{tc} \qquad (5.53\ b)$$

where x is the smaller dimension of the rectangular section.

Extensive work on reinforced concrete beams by Hsu and confirmed by others has established that k_2 can be taken as $\frac{1}{3}$. This value originated from research in the skew-bending theory of plain concrete. It was also established that $6\sqrt{f_c'}$ can be considered as a limiting value of the pure torsional strength of a member without torsional reinforcement. Using a reduction factor of 2.5 for the first cracking torsional

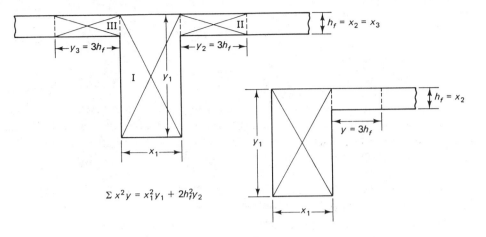

$$\Sigma\, x^2 y = x_1^2 y_1 + 2h_f^2 y_2$$

Figure 5.34 Component rectangles for calculation of T_c.

load $v_{tc} = 2.4\sqrt{f'_c}$, and using $k_2 = \frac{1}{3}$ in Equation 5.53, results in

$$T_c = 0.8\sqrt{f'_c}\, x^2 y \qquad\qquad (5.54\text{ a})$$

where x is the shorter side of the rectangular section. The high reduction factor of 2.5 is used to offset any effect of bending moments that might be present.

If the cross section is a T- or L-section, the area can be broken into component rectangles as in Figure 5.34, such that

$$T_c = 0.8\sqrt{f'_c}\, \Sigma\, x^2 y \qquad\qquad (5.54\text{ b})$$

5.17 TORSION IN REINFORCED AND PRESTRESSED CONCRETE ELEMENTS

Torsion rarely occurs in concrete structures without being accompanied by bending and shear. The foregoing should give a sufficient background on the contribution of the plain concrete in the section toward resisting *part* of the combined stresses resulting from torsional, axial, shear, or flexural forces. The capacity of the plain concrete to resist torsion when in combination with other loads could, in many cases, be lower than when it resists the same factored external twisting moments alone. Consequently, torsional reinforcement has to be provided to resist the excess torque.

Inclusion of longitudinal and transverse reinforcement to resist part of the torsional moments introduces a new element in the set of forces and moments in the section. If

T_n = required total nominal torsional resistance of the section, including the reinforcement

T_c = nominal torsional resistance of the plain concrete

and

T_s = torsional resistance of the reinforcement

then

$$T_n = T_c + T_s \qquad (5.55 \text{ a})$$

or

$$T_s = T_n - T_c \qquad (5.55 \text{ b})$$

In order to study the contribution of the longitudinal steel bars and the closed transverse bars so that T_s can be evaluated, one has to analyze the system of forces acting on the warped cross sections of the structural element at the limit state of failure. Basically, three approaches are presently acceptable:

1. The skew-bending theory, which is based on the plane deformation approach to plane sections subjected to bending and torsion.

2. The space truss analogy theory, which is a modification of the planar truss analogy for shear. According to this theory, the space truss composed of longitudinal bars and diagonal concrete struts is subjected to twist in which the stirrups and longitudinal bars are considered the tension members and the diagonal concrete struts at an angle θ between the cracks are considered the compression members, generally idealized to 45 degrees.

3. Compression field theory, which is a powerful modification of the truss analogy theory and is based on a more realistic evaluation of the inclination angle θ of the compression struts between the inclined cracks.

5.17.1 Skew-Bending Theory

Skew-bending theory considers in detail the internal deformational behavior of the series of transverse warped surfaces along the beam. Initially proposed by Lessig, it had subsequent contributions from Collins, Hsu, Zia, Gesund, Mattock, and Elfgren among the several researchers in this field. Hsu made a major contribution experimentally to the development of the skew-bending theory as it presently stands. In his book (Ref. 5.9), Hsu details the development of the theory of torsion as applied to concrete structures and how the skew-bending theory formed the basis of the ACI code provisions on torsion. The complexity of the torsional problem permits here only the brief discussion that follows.

The failure surface of the normal beam cross section subjected to bending moment M_u remains plane after bending, as shown in Figure 5.35(a). If a twisting moment T_u is also applied exceeding the capacity of the section, cracks develop on three sides of the beam cross section, and compressive stresses appear on portions of the fourth side along the beam. As torsional loading proceeds to the limit state at failure, a skewed failure surface results due to the combined torsional moment T_u and bending moment M_u. The neutral axis of the skewed surface and the shaded area in Figure 5.35(b) denoting the compression zone would no longer be straight, but subtend a varying angle θ with the original plane cross sections.

Prior to cracking, neither the longitudinal bars nor the closed stirrups have any appreciable contribution to the torsional stiffness of the section. At the postcracking stage of loading, the stiffness of the section is reduced, but its torsional resistance is considerably increased, depending on the amount and distribution of *both* the longitu-

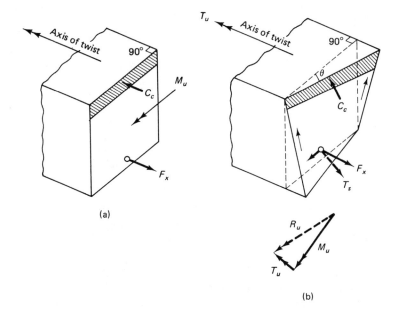

Figure 5.35 Skew bending due to torsion. (a) Bending before twist. (b) Bending and torsion.

dinal bars and the transverse *closed* ties. It has to be emphasized that little additional torsional strength can be achieved beyond the capacity of the plain concrete in the beam unless both longitudinal torsion bars and transverse ties are used.

The skew-bending theory idealizes the compression zone by considering it to be of uniform depth. It assumes the cracks on the remaining three faces of the cross section to be uniformly spread, with the steel ties (stirrups) at those faces carrying the tensile forces at the cracks and the longitudinal bars resisting shear through dowel action with the concrete. Figure 5.36(a) shows the forces acting on the skewly bent plane. The polygon in Figure 5.36(b) gives the shear resistance F_c of the concrete, the force T_l of the active longitudinal steel bars in the compression zone, and the normal compressive block force C_c.

The torsional moment T_c of the resisting shearing force F_c generated by the shaded compressive block area in Figure 5.36(a) is thus

$$T_c = \frac{F_c}{\cos 45°} \times \text{its arm about forces } F_v \text{ in the figure}$$

or

$$T_c = \sqrt{2}\, F_c(0.8x) \qquad (5.56\ a)$$

where x is the shorter side of the beam. Extensive tests (Refs. 5.9 and 5.10) to evaluate F_c in terms of the internal stress in concrete $k_1\sqrt{f_c'}$ and the geometrical torsional constants of the section $k_2 x^2 y$ led to the expression

$$T_c = \frac{2.4}{\sqrt{x}}\, x^2 y \sqrt{f_c'} \qquad (5.56\ b)$$

The dowel forces F_x and F_y are assumed to be proportional to the cross-sectional

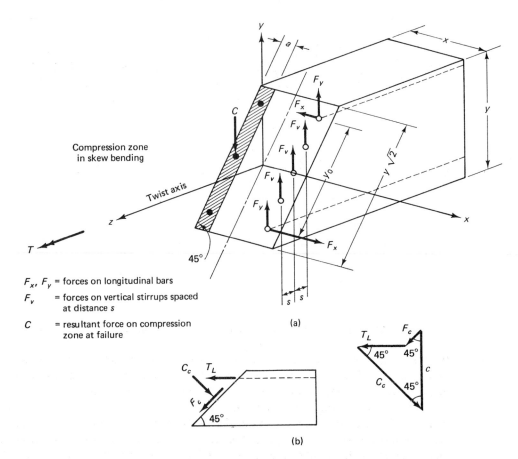

F_x, F_y = forces on longitudinal bars

F_v = forces on vertical stirrups spaced at distance s

C = resultant force on compression zone at failure

(a)

(b)

Figure 5.36 Forces on the skew-bent planes. (a) All forces acting on skew plane at failure. (b) Vector forces on compression zone.

areas of these bars. If a ratio is established between the proportion of torsional resistance given by the dowel forces and the torsional resistance of the hoop forces F_v, torsional moments are the summations

$$\Sigma\, F_v(\tfrac{1}{2}x_1),\ \Sigma F_x(\tfrac{1}{2}y_0),\ \Sigma\, F_y(\tfrac{1}{2}x_0),\ \Sigma\, T_1(\tfrac{1}{2}x_0)$$

The dimensions x_1 and y_1 are, respectively, the shorter and the longer center-to-center dimensions of the closed rectangular stirrups, and the dimensions x_0 and y_0 are the corresponding center-to-center dimensions of the longitudinal bars at the corners of the stirrups. The resulting expression for the torsional strength T_s provided by the hoops and the longitudinal steel in the rectangular section is

$$T_s = \alpha_1 \frac{x_1 y_1 A_t f_y}{s} \tag{5.57}$$

where $\alpha_1 = 0.66 + 0.33 y_1/x_1$, so that the total nominal torsional moment of resistance is $T_n = T_c + T_s$, or

$$T_n = \frac{2.4}{\sqrt{x}} x^2 y \sqrt{f'_c} + \left(0.66 + 0.33\frac{y_1}{x_1}\right) \frac{x_1 y_1 A_t f_y}{s} \tag{5.58}$$

5.17.2 Space Truss Analogy Theory

Space truss analogy theory was originally developed by Rausch and later extended by Lampert and Collins, with additional work by Hsu, Thurliman, Elfregren, and others. The space truss analogy is an extension of the model used in the design of the shear-resisting stirrups, in which the diagonal tension cracks, once they start to develop, are resisted by the stirrups. Because of the nonplanar shape of the cross sections due to the twisting moment, a space truss composed of the stirrups is used as the diagonal tension members, and the idealized concrete strips at an angle θ generally idealized to 45 degrees between the cracks are used as the compression members, as shown in Figure 5.37.

It is assumed in this theory that the concrete beam behaves in torsion similarly to a thin-walled box with a constant shear flow in the wall cross section, producing a constant torsional moment. The use of hollow-walled sections rather than solid sections proved to give essentially the same ultimate torsional moment, provided that the walls were not too thin. Such a conclusion is borne out by tests which have shown that the torsional strength of the solid sections is composed of the resistance of the closed stirrup cage, consisting of the longitudinal bars and transverse stirrups, and the idealized concrete inclined compression struts in the plane of the cage wall. The compression struts are the inclined concrete strips between the cracks in Figure 5.37.

The CEB-FIP code is based on the space truss model. In this code, the effective wall thickness of the hollow beam is taken as $\frac{1}{6}D_0$, where D_0 is the diameter of the

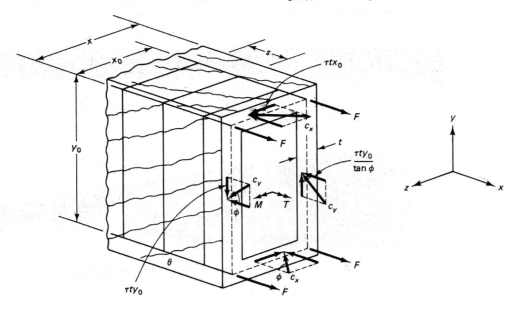

F = tensile force in each longitudinal bar

c_x = inclined compressive force on horizontal side

c_y = inclined compressive force on vertical side

τt = shear flow force per unit length of wall

Figure 5.37 Forces on hollow-box concrete surface by truss analogy.

Reinforced concrete beams in torsion, testing setup. (*Courtesy, Thomas T. C. Hsu.*)

Closeup of torsional cracking of beams in the preceding photograph. (*Courtesy, Thomas T.C. Hsu.*)

circle inscribed in the rectangle connecting the corner longitudinal bars, namely, $D_0 = x_0$ in Figure 5.37. In summary, the absence of the core does not affect the strength of such members in torsion—hence, the acceptability of the space truss analogy approach based on hollow sections.

If the shear flow in the walls of the box section is τt, where τ is the shear stress, and F is the tensile force in each longitudinal bar at the corner, the force equilibrium equation is

$$4F = 2\frac{\tau t x_0}{\tan \phi} + 2\frac{\tau t y_0}{\tan \phi} \tag{5.59}$$

and the moments due to the shear flow forces are

$$T_n = \tau t y_0 x_0 + \tau t x_0 y_0 \tag{5.60}$$

If A_t is the area of the stirrup cross section, and f_y is the yield strength of the stirrup spaced at a distance s, then

$$A_t f_y = \tau t s \tan \phi \tag{5.61 a}$$

Also, if A_l is the total area of the four longitudinal bars at the corners, then

$$F = \tfrac{1}{4} A_t f_y \tag{5.61 b}$$

Solving Equations 5.59, 5.60, and 5.61 a leads to

$$T_n = 2x_0 y_0 \sqrt{\frac{A_l f_y A_t f_y}{2s(x_0 + y_0)}} \tag{5.62}$$

For the case of equal volumes of longitudinal steel and transverse stirrups (i.e., $\phi = 45°$), the torsional moment of resistance T_n at failure is

$$T_n = 2\frac{A_t f_y}{s} x_0 y_0 \tag{5.63}$$

Note the similarity in format between Equation 5.57, developed by the skew-bending theory, and Equation 5.63, developed by the space truss analogy theory.

5.17.3 Compression Field Theory

Compression field theory can be considered a special case of the general truss model theory. Elfgren proposed the compression fields to describe components of the plasticity truss model currently used in the European (CEB) Code, and Collins and Mitchell modified the approach, proposing the terms to be subsequently discussed. The angle of inclination θ in Fig. 5.37 of the diagonal cracks or the compression struts between the diagonal cracks is not idealized to 45 degrees, but rather uses limits based on the areas of the longitudinal tension steel and the transverse torsional web steel (inclined or vertical closed stirrups or ties). Figure 5.38 points up the fact that the torsional force is resisted by the tangential components of the diagonal compression struts, which produce a shear flow q around the perimeter (Ref. 5.11).

Assuming that concrete carries no tension after cracking, and that torsional shear is carried by the field of diagonal compression struts, the angle of inclination θ of these struts can be defined as

$$\tan \theta^2 = \frac{\epsilon_\ell + \epsilon_d}{\epsilon_t + \epsilon_d} \tag{5.64}$$

where ϵ_ℓ = longitudinal tensile strain in the main bars A
ϵ_t = transverse tensile strain in bars B
ϵ_d = diagonal compression strain

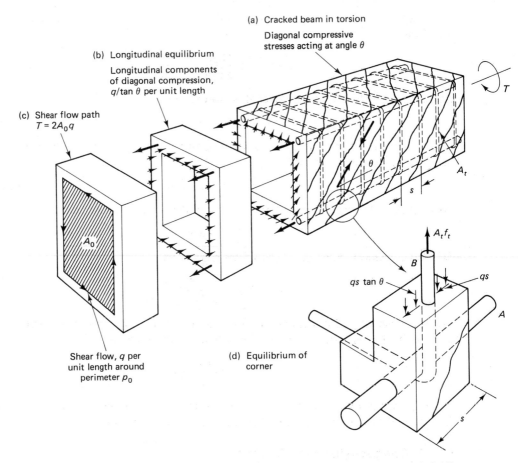

(a) Cracked beam in torsion

Diagonal compressive stresses acting at angle θ

(b) Longitudinal equilibrium

Longitudinal components of diagonal compression, $q/\tan \theta$ per unit length

(c) Shear flow path $T = 2A_0 q$

A_o

A_t

θ

s

$A_t f_t$

B

$qs \tan \theta$

qs

A

Shear flow, q per unit length around perimeter p_0

(d) Equilibrium of corner

s

Figure 5.38 Compression field truss model by Collins and Mitchell (Ref. 5.11).

The area A_o in the figure enclosed by the shear flow q can be obtained as

$$A_o = A_{oh} - \frac{a_0}{2} p_h \qquad (5.65)$$

where A_{oh} = area enclosed by the centerline of the hoop
p_h = hoop centerline perimeter
a_o = compression block depth (identical to the depth a of the equivalent rectangular block in flexure)

The equivalent wall thickness t_d in the analysis of the twisted beam is shown in Figure 5.39, and the depth a_0 of the compressive block is defined in Equation 5.68.

The diagonal torsional cracks, as well as the exposed transverse ties after spalling of the concrete cover at torsion failure, are demonstrated in Figure 5.40.

The transverse and longitudinal strains in the steel at the nominal torsional moment T_n can be respectively defined as

$$\epsilon_t = \left(\frac{0.85 \beta_1 f'_c A_o}{\tau_n A_{oh} \tan \theta} - 1 \right) 0.003 \qquad (5.66\ a)$$

Shear and Torsional Strength Design   Chap. 5

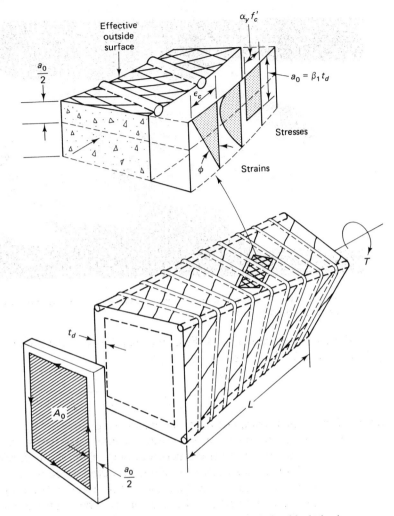

Figure 5.39 Effective thickness t_d and compression block depth a_o.

$$\epsilon_\ell = \left(\frac{0.85\beta_1 f'_c A_o}{\tau_n A_{oh}} \tan\theta - 1\right)0.003 \tag{5.66 b}$$

where the nominal torsional shear stress is

$$\tau_n = \frac{T_n p_h}{A_{oh}^2} \tag{5.67}$$

The area A_0 enclosed by the shear flow can be obtained from Equation 5.60 and the following expression for the compression block depth a_0 in torsion according to Collins and Mitchell (Ref. 5.11):

$$a_0 = \frac{A_{oh}}{p_h}\left[1 - \sqrt{1 - \frac{T_n p_h}{0.85 f'_c A_{oh}^2}\left(\tan\theta + \frac{1}{\tan\theta}\right)}\right] \tag{5.68}$$

where T_n is the nominal torsional moment strength at the limit state at failure.

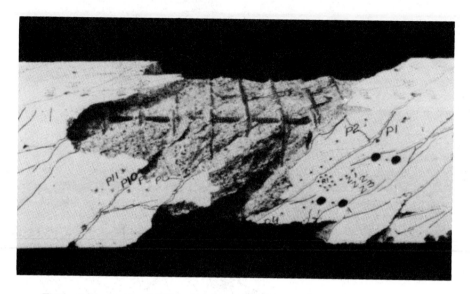

Figure 5.40 Torsional failure of web-reinforced beam after spalling of cover (Collins and Mitchell, Ref. 5.11).

For combined torsion and shear, the shearing stress at nominal strengths T_n and V_n in Equation 5.61 c becomes

$$\tau_n = \frac{T_n p_h}{A_{oh}^2} + \frac{V_n - V_p}{b_v d_v} \qquad (5.69)$$

where V_p = vertical component on the prestressing force.

b_v = minimum effective *web* width within shear depth d_v after spalling of cover. Subtract the diameters of ducts from the web width if ungrouted, or half the diameter of ducts for grouted tendons.

d_v = effective shear depth. This can be taken as the flexural lever arm, but *not less* than the vertical distance between the centers of bars or prestressing tendons at the corners of the stirrups.

The predicted values of the compressive strut inclination θ in Figure 5.38 range between 24° for pure torsion and 90° for pure flexure. Hence, the lower the value of θ selected for a given torque, the less is the transverse hoop steel needed and the more is the required area of longitudinal steel. Since transverse closed stirrups or ties are more expensive than longitudinal bars, a choice of lower values of θ is more economical in design.

The compression field theory assumes that the goemetrical properties of the designed section are chosen on the basis of yielding of the transverse web reinforcement and longitudinal steel *prior* to diagonal crushing of the concrete. Consequently, the transverse strain ϵ_t in Equations 5.64 and 5.66 a should be taken as the yield strain ϵ_{ty}.

The range of the compression strut angle θ in degrees can be evaluated from

$$10 + \frac{35(\tau_n/f_c')}{0.42 - 50\epsilon_l} < \theta < 80 - \frac{35(\tau_n/f_c')}{0.42 - 65\epsilon_{ty}} \qquad (5.70)$$

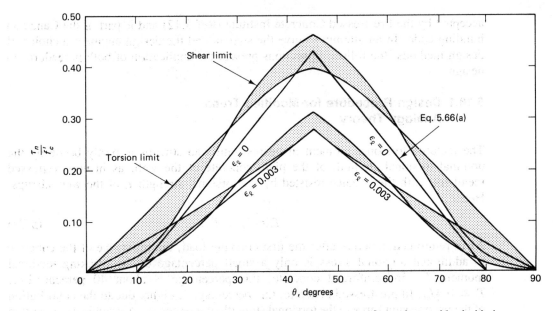

Figure 5.41 Design limits on angle θ of diagonal compression for torsion combined with shear (Ref. 5.11).

where the reinforcement is assumed to yield before crushing of the concrete. For $f_y = 60,000$ psi, Equation 5.70 can be simplified to

$$\theta_{\min} = 10 + 83.75\left(\frac{\tau_n}{f'_c}\right) \tag{5.71}$$

$$\theta_{\max} = 80 - 122.58\left(\frac{\tau_n}{f'_c}\right) \tag{5.72}$$

When $\theta_{\min} < \theta_{\max}$, diagonal crushing will not occur and the chosen section is adequate since the flexural stress effect is then more dominant than the torsional shear stress effect. Note that the limits of the angle θ conservatively take into consideration the flexure interaction that exists with the combined torsion and shear. The value of τ_n is calaculated from Equation 5.69; Equation 5.70 is plotted in Figure 5.41 (Ref. 5.11).

5.18 DESIGN OF PRESTRESSED CONCRETE BEAMS FOR TORSION

The ACI building code does not give direct procedures for the torsional design of prestressed concrete elements. Rather, the basic skew-bending theory of Section 5.17.1 or the basic space truss analogy of Section 5.17.2 can be modified to take into consideration the effects of the compressive prestressing force on the torsional behavior of beams, with web reinforcement producing what is termed here the modified truss analogy theory. The compression field theory presented in Section 5.17.3 can be readily used to evaluate the strength of prestressed sections in shear and torsion. It is accepted by the Prestressed Concrete Institute (Ref 5.12) and is part of the Canadian building code. In an attempt to give the student and the design engineer a choice of design methods, the following sections present the application of both procedures in design.

5.18.1 Design Procedure for Modified Truss Analogy Theory

The factored torsional moment in beams with web stirrups is partly borne by the nominal torsional strength of the plain concrete of the web, as in nonprestressed members, with the balance resisted by the torsional strength T_s of the web stirrups. Hence,

$$T_n = T_c + T_s \tag{5.73}$$

As in reinforced concrete after the first cracking load, the resistance of the concrete to additional torsional stress is only a small percentage of the cracking torsional moment T_{cr}. In reinforced concrete, this percentage is about 40 percent, i.e., $T_c \cong 0.4T_{cr}$. In prestressed concrete, this percentage is higher due to the contribution of the prestressing force to the torsional strength of the section. A graphical description of the difference in levels of torsional resistance after the torsional cracking moment T_{cr} develops is shown in Figure 5.42, as presented by Zia and McGee (Refs 5.13 and 5.14).

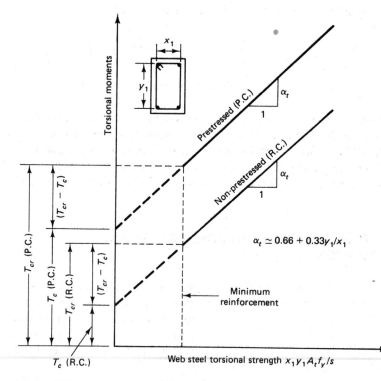

Figure 5.42 Relative torsional strength relationships in reinforced and prestressed concrete.

In the figure, the slope α_t relates the web steel torsional strength to the cracking torsional moment such that

$$T_s = \alpha_t \frac{x_1 y_1}{s} A_t f_y \tag{5.74}$$

where

$$\alpha_t = 0.66 + 0.33 \frac{y_1}{x_1} \le 1.50 \tag{5.75}$$

and where x_1 = shorter center-to-center dimension of closed stirrup
y_1 = longer center-to-center dimension of closed stirrup
s = spacing of torsional reinforcement
A_t = area of *one* leg of a closed stirrup resisting torsion within a distance s, in^2.

Equation 5.74 is equally applicable to prestressed and to reinforced concrete, and the torsional strength deduction $(T_{cr} - T_c)$ can also be assumed to be the same in both, as schematically shown in Figure 5.42.

When external torsion is accompanied by external shear, the same section is subjected to higher shearing stresses because of the combined effect of the two loading types as they interact with each other. A beam's resistance to combined torsion and shear is less than its resistance to either of these two parameters when acting alone. Consequently, an interaction relationship becomes necessary in a similar manner to that developed for combined bending and axial load discussed in Ref. 5.3. Figure 5.43 represents the following nondimensional interaction expressions relating torsion to shear:

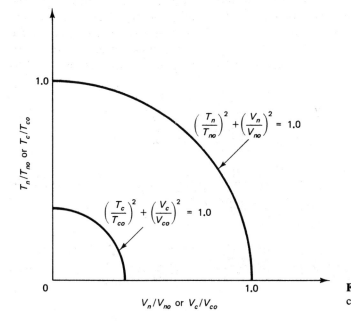

$$\left(\frac{T_n}{T_{no}}\right)^2 + \left(\frac{V_n}{V_{no}}\right)^2 = 1.0$$

$$\left(\frac{T_c}{T_{co}}\right)^2 + \left(\frac{V_c}{V_{co}}\right)^2 = 1.0$$

V_n/V_{no} or V_c/V_{co}

Figure 5.43 Interaction diagram for combined torsion and shear.

Sec. 5.18   Design of Prestressed Concrete Beams for Torsion

1. Concrete strength only:

$$\left(\frac{T_c}{T_{co}}\right)^2 + \left(\frac{V_c}{V_{co}}\right)^2 \le 1.0 \qquad (5.76\text{ a})$$

T_c and V_n are the nominal external torsion and shear when acting *simultaneously*. T_{co} and V_{co} are the nominal values for torsion and shear when each acts alone.

2. Nominal total strength at failure:

$$\left(\frac{T_n}{T_{no}}\right)^2 + \left(\frac{V_n}{V_{no}}\right)^2 \le 1.0 \qquad (5.76\text{ b})$$

T_n and V_n represent the nominal torsional and shear strengths to resist T_u and V_u when acting *simultaneously*. $T_{no} = T_c + T_s$ represents the nominal torsional resistance of the reinforced web when pure *torsion alone* acts on the section; $V_{no} = V_c + V_s$ represents the nominal shear resistance of the reinforced web when *shear alone* acts on the section.

Algebric transformation of Equation 5.76 gives the nominal torsion and shear carried by the plain concrete:

$$T_c = \frac{T_{co}}{\sqrt{1 + \left(\frac{T_{co}}{V_{co}}\right)^2 \left(\frac{V_c}{T_c}\right)^2}} \qquad (5.77\text{ a})$$

$$V_c = \frac{V_{co}}{\sqrt{1 + \left(\frac{V_{co}}{T_{co}}\right)^2 \left(\frac{T_c}{V_c}\right)^2}} \qquad (5.77\text{ b})$$

It is generally reasonable to assume in reinforced and prestressed members that the ratio of shear force to torsional moment V_c/T_c remains constant during the loading history. This ratio can be approximated by an equivalent ratio V_u/T_u, and hence by the required V_n/T_n. Zia and Hsu (Ref. 5.15) propose a modification to the format of Equations 5.77 as follows for prestressed beams and assuming that the ratio V_c/T_c is taken to be V_n/T_n:

$$T_c = \frac{T_{co}}{\sqrt{1 + \left(\frac{T_{co}}{V_{co}}\right)^2 \left(\frac{V_n}{T_n}\right)^2}} \qquad (5.78\text{ a})$$

$$V_c = \frac{V_{co}}{\sqrt{1 + \left(\frac{V_{co}}{T_{co}}\right)^2 \left(\frac{T_n}{V_n}\right)^2}} \qquad (5.78\text{ b})$$

where

$$T_{co} = 6\lambda\sqrt{f_c'}\left[\sqrt{1 + 10\bar{f_c}/f_c'} - k\right]\left(\sum \eta x^2 y\right) \qquad (5.79\text{ a})$$

and where $k = (1 - 0.133/\eta) \geq 0.60$

$\eta = 0.35/(0.75 + x/y)$

V_{co} = lesser of V_{ci} and V_{cw} from Equations 5.11 and 5.15, or V_c from Equation 5.16

$\bar{f}_c$ = compressive stress in concrete after prestress losses at the centroid (cgc) of the section.

The effect of torsion can be neglected if

$$T_u < 1.5\lambda\sqrt{f_c'}\left[\sqrt{1 + \frac{10\bar{f}_c}{f_c'}}\right]\left(\sum \eta x^2 y\right) \tag{5.79 b}$$

The nominal torsional strength at the limit state of failure becomes, for *combined* torsion and shear,

$$T_n = T_c + T_s = \frac{T_{co}}{\sqrt{1 + \left(\dfrac{T_{co}}{V_{co}}\right)^2\left(\dfrac{V_n}{T_v}\right)^2}} + \alpha_t\frac{x_1 y_1}{s}A_t f_y \tag{5.80 a}$$

and

$$V_n = \frac{\text{smaller of } V_{ci} \text{ or } V_{cw}}{\sqrt{1 + \left(\dfrac{V_{co}}{T_{co}}\right)^2\left(\dfrac{T_n}{V_n}\right)^2}} + \frac{A_v f_y d}{s} \tag{5.80 b}$$

The upper limits of T_n and V_n in Equations 5.80, as in Ref. 5.14, are

$$T_{n,\,max} = \frac{\gamma'\lambda\sqrt{f_c'}}{\sqrt{1 + \left(\dfrac{\gamma'}{10}\right)^2\left(\dfrac{V_n}{T_n}\right)^2\left(\dfrac{\sum \eta x^2 y}{b_w d}\right)^2}}\left(\sum \eta x^2 y\right) \tag{5.81 a}$$

$$V_{n,\,max} = \frac{10\lambda\sqrt{f_c'}}{\sqrt{1 + \left(\dfrac{10}{\gamma'}\right)^2\left(\dfrac{T_n}{V_n}\right)^2\left(\dfrac{b_w d}{\sum \eta x^2 y}\right)^2}}(b_w d) \tag{5.81 b}$$

where $\gamma' = \gamma\sqrt{1 + 10\bar{f}_c/f_c'}$ and $\gamma = 14 - 0.33(\bar{f}_c/f_c')$. If the member is subjected to torsional moment, only the upper limit of T_n is

$$T_{n,\,max} = \gamma\lambda\sqrt{f_c'}\sqrt{1 + 1 - \frac{\bar{f}_c}{f_c'}}\left(\sum \eta x^2 y\right) \tag{5.81 c}$$

5.18.1.1 Torsional web reinforcement. Meaningful additional torsional strength due to the addition of torsional reinforcement can be achieved only by using *both* stirrups and longitudinal bars. Ideally, equal volumes of steel in both the closed stirrups and the longitudinal bars should be used so that both participate equally in resisting the twisting moments. This principle is the basis of the ACI expressions for proportioning the torsional web steel. If s is the spacing of the stirrups, A_l is the total cross-sectional area of the longitudinal bars, and A_t is the cross section of one stirrup leg, where the dimensions of the stirrup are x_1 in the short direction and y_1 in the long direction, then

$$2A_t(x_1 + y_1) = A_l s \qquad (5.82\ a)$$

so that

$$2A_t = \frac{A_l s}{x_1 + y_1} \qquad (5.82\ b)$$

Hence, the total torsional web steel, including both the closed stirrups and the longitudinal bars for Equations 5.82 a and 5.82 b, becomes

$$A_{\text{total}} = 2A_t + \frac{A_l s}{x_1 + y_1} \qquad (5.83\ a)$$

But, from Equation 5.57,

$$A_t = \frac{T_s s}{\alpha_1 x_1 y_1 f_y} \qquad (5.83\ b)$$

where $\alpha_1 = 0.66 + 0.33 y_1/x_1 \le 1.5$ and T_s is the torsional resisting moment of the torsional web steel. If T_c is the nominal torsional resistance of the plain concrete in the web,

$$T_s = T_n - T_c \qquad (5.84)$$

From Equation 5.82 b, and using the ACI expression for A_t for combined torsion and shear, where

$$2A_t = \frac{200 \times s}{f_y} \frac{T_u}{T_u + V_u/3C_t} \qquad (5.85)$$

the longitudinal torsional reinforcement can be expressed as

$$A_l = \left(\frac{400 \times s}{f_y} \frac{T_u}{T_u + V_u/3C_t} - 2A_t \right) \frac{x_1 + y_1}{s} \qquad (5.86)$$

where $C_t = b_w d/\Sigma x^2 y$. The term $2A_t$ in Equation 5.86 cannot be less than $50 b_w s/f_y$, since this value is the minimum $2A_t$ for the torsional stirrups to be effective. A thorough discussion and detailed derivation of Equation 5.86 is presented in Ref. 5.9.

For combined torsion and shear in prestressed beams, the longitudinal reinforcement area A_l of Equation 5.82 b can be conveniently used in lieu of Equation 5.86 for simplification.

5.18.2 Design Procedure for Compression Field Theory

The design of web reinforcement for combined torsion and shear by the compression field theory is based on an interaction equation similar to the one described in Section 5.18.1, namely,

$$\left(\frac{M_{cr}}{M_{ocr}} \right)^2 + \left(\frac{V_{cr}}{V_{ocr}} \right)^2 + \left(\frac{T_{cr}}{T_{ocr}} \right)^2 = 1.0 \qquad (5.87)$$

where

$$M_{ocr} = \frac{I}{y_t}(7.5\lambda\sqrt{f_c'} + f_{ce})$$ (5.88 a)

or

$$M_{ocr} = S_b = 7.5\lambda\sqrt{f_c'} + f_{ce})$$ (5.88 b)

and where M_{cr} = cracking moment under combined loading
$\quad\quad\quad M_{ocr}$ = pure flexural cracking strength
$\quad\quad\quad V_{cr}$ = cracking shear under combined loading
$\quad\quad\quad V_{ocr}$ = pure shear cracking strength
$\quad\quad\quad y_t$ = distance from centroidal axis to extreme fibers in tension
$\quad\quad\quad T_{cr}$ = torsional cracking moment under combined loading
$\quad\quad\quad T_{ocr}$ = pure torsional cracking strength
$\quad\quad\quad f_{ce}$ = compressive stress in concrete due to effective prestress after losses at extreme fibers of section where tensile stress is caused by external load.

Also,

$$V_{ocr} = 4\lambda\sqrt{f_c'}\left[\sqrt{1 + \frac{\bar{f_c}}{4\sqrt{f_c'}}}\right](b_w d_p) + V_p$$ (5.89)

and

$$T_{ocr} = 4\lambda\sqrt{f_c'}\left[\sqrt{1 + \frac{\bar{f_c}}{(4\lambda\sqrt{f_c'})}}\right]\left(\frac{A_c^2}{p_c}\right)$$ (5.90)

where V_p = vertical component of prestressing force
$\quad\quad\quad \bar{f_c}$ = concrete compressive stress due to prestress at *centroid* cgc of section
$\quad\quad\quad A_c$ = area enclosed by perimeter of cross section p_c
$\quad\quad\quad p_c$ = outside perimeter of concrete section.

It can be assumed that $T_{cr}/V_{cr} \cong T_u/V_u \cong$ required T_n/V_n for all practical purposes. Similarly, $M_{cr}/V_{cr} \cong M_u/V_u \cong$ required M_n/V_n, but this ratio should not be taken to be less than the effective depth d_v.

The cracking shear V_{cr} required for combined loading to be obtained from Equation 5.87 *should not be less* than the service-load shear V_{SL} at the section under consideration in order to satisfy the service-load-level cracking requirement; otherwise a larger section has to be chosen.

5.18.2.1 Torsional web reinforcement. Torsional web reinforcement is needed if the factored torsional moment T_u based on uncracked stiffness is

$$T_u = \phi(0.25T_{ocr})$$ (5.91)

where T_{ocr} is the pure torsional cracking strength.

(a) *Transverse Shear Reinforcement:* When torsion has to be considered, the transverse reinforcing stirrups should be proportioned using the expression

$$V_n = \frac{A_v f_y}{s} \frac{d_v}{\tan \theta} + V_p \qquad \text{(5.92 a)}$$

or

$$A_n = \frac{(V_n - V_p)s \tan \theta}{f_y d_v} \qquad \text{(5.92 b)}$$

where A_v = area of all stirrups' legs at the section under consideration
d_v = vertical distance between longitudinal bars in corners of stirrups.

The maximum spacing is given by

$$s = \frac{d_v}{3 \tan \theta} \qquad \text{(5.92 c)}$$

(b) *Transverse Web Torsional Reinforcement:* Combined shear and torsion should be proportioned using the expression

$$T_n = \frac{A_t f_y}{s} \times \frac{2A_o}{\tan \theta} \qquad \text{(5.93 a)}$$

or

$$A_t = \frac{T_n s \tan \theta}{2f_y A_o} \qquad \text{(5.93 b)}$$

where A_t is the area of one stirrup leg, A_o is as given in Equation 5.65, and p_h is the closed stirrup centerline perimeter. The maximum stirrup spacing is

$$s = \frac{p_h}{8 \tan \theta} \qquad \text{(5.93 c)}$$

and the total area of stirrups for combined torsion and shear is

$$A_{tt} = A_v + 2A_t \qquad \text{(5.94)}$$

(c) *Longitudinal Reinforcement:* The longitudinal reinforcement is designed to resist the applied factored moment M_u, any axial load N_u, and an axial tension ΔN_u caused by the shear and torsion given by

$$\Delta N_u = \frac{1}{\tan \theta} \sqrt{(V_u - \phi V_p)^2 + \left(\frac{T_u p_o}{2A_o}\right)^2} \qquad \text{(5.95)}$$

where $p_o = p_h - 4a_o$ (the value of a_o is gotten from Equation 5.68.) Consequently, the longitudinal reinforcement requirement A_l is satisfied if

$$+M_n \geq \left(\frac{+M_u}{\phi_f}\right) + \left(\frac{d_v \Delta N_u}{2\phi_v}\right) \qquad \text{(5.96 a)}$$

and

$$-M_n \geq \left(\frac{d_v \Delta N_u}{2\phi_v}\right) - \left(\frac{M_u}{\phi_f}\right) \qquad \text{(5.96 b)}$$

Dauphin Island Bridge, Alabama, assembly of segmental bridge units. (*Courtesy,* Prestressed Concrete Institute.)

where ϕ_v = torsion strength reduction factor = 0.85
ϕ_f = flexure strength reduction factor = 0.90

If the section under consideration is closer than $d_v/\tan\theta$ from the inner edge of the bearing area for simply supported beams, then

$$-M_n \leq \frac{d_v T_u p_o}{2\tan\theta\,(\phi_v 2A_o)} - \frac{M_u}{\phi_f} \qquad (5.97\ a)$$

The lesser of the two values of the required $-M_n$ from Equations 5.96 b and 5.97 a is to be used for proportioning the longitudinal steel. For interior continuous supports,

$$-M_n \leq \frac{d_v T_u p_o}{2\tan\theta\,(\phi_v 2A_o)} + \frac{M_u}{\phi_f} \qquad (5.97\ b)$$

The nominal diameter of the bar or strand at each corner of the section should not be less than $5\tan\theta/16$, or $\frac{1}{2}$ in.

(d) *Web Crushing at the Bearing Points:* To avoid web crushing at the bearing points, the effective shear depth at the end of the member should be

$$d_{ve} \geq \frac{V_n/b_v}{0.012\theta' f'_c - T_n p_h/A_{oh}^2} \qquad (5.98)$$

where $\theta' = \theta - 10$ if $\theta \leq 45°$
$= 80 - \theta$ if $\theta > 45°$

and where A_{oh} = concrete area enclosed by the centerline of the closed stirrup dimensions.

Some effective shear widths b_v and shear depths d_v of typical prestressed sections are shown in Figure 5.44.

The areas A_o enclosed by the shear flow and the equivalent depth a_o of the compression block in torsion are shown in Figure 5.45.

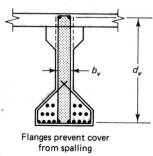

Flanges prevent cover
from spalling

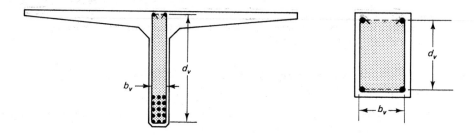

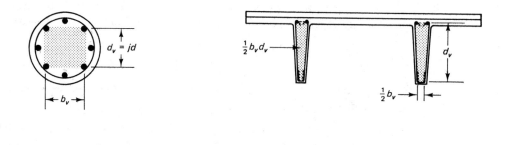

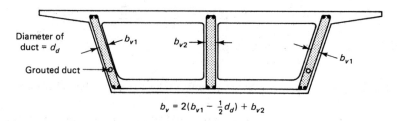

$$b_v = 2(b_{v1} - \tfrac{1}{2}d_d) + b_{v2}$$

Figure 5.44 Effective shear width and depth of typical prestressed concrete sections.

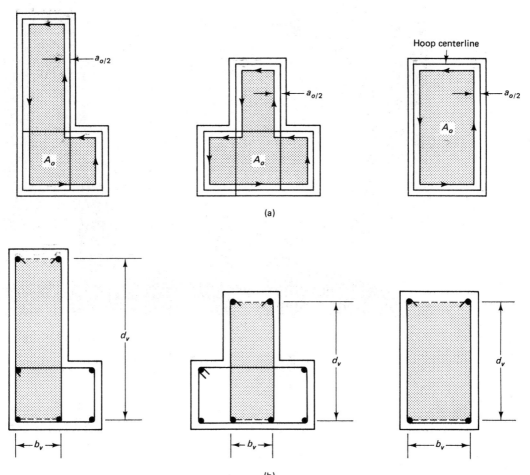

Figure 5.45 Shear-flow geometry and effective shear area. (a) Effective depth a_0 of compressive block in torsion with respect to shear-flow area A_0. (b) Effective shear width b_v (width less concrete spalling cover) and effective shear depth d_v.

5.19 STEP-BY-STEP PROCEDURE FOR DESIGNING WEB REINFORCEMENT FOR COMBINED SHEAR AND TORSION

5.19.1 Compression Field Method

1. Calculate T_u, V_u, T_{SL}, and V_{SL}, where the subscripts u and SL denote the factored load and service load, respectively. Then draw torsional moment, shear, and flexural moment diagrams along the span.

2. Determine the minimum flexure reinforcement requirement from the nominal moment strength M_n:

$$M_n \geq 1.2M_{ocr} = S_b(7.5\lambda\sqrt{f'_c} + f_{ce})$$

The Woodley Park Zoo station, Washington, D.C. (*Courtesy*, H. Wilden & Assoc.).

$$\bar{f}_c = \frac{P_e}{A_c} = \text{concrete stress at centroid}$$

$$f_{ce} = \frac{P_e}{A_c} + \frac{P_e e}{S_b}$$

$$a = \frac{A_{ps} f_{ps}}{0.85 f'_c b}$$

$$M_n = A_{ps} f_{ps} \left(\frac{d - a}{2} \right)$$

3. Verify whether torsional reinforcement is needed. Torsional moment can be neglected if the factored $T_u < 0.25 T_{ocr}$, where $T_{ocr} = 4\lambda \sqrt{f'_c} [\sqrt{1 + \bar{f}_c/(4\lambda \sqrt{f'_c})} \times (A_c^2/p_c)$, in which p_c is the outside perimeter of the concrete cross section.

4. Check the cross-section size and choose the torsion compression angle θ.

(a) Calculate the required nominal strengths T_n and V_n, and find the shear-flow area A_{oh} and the closed stirrups' centerline perimeter p_h.

(b) Find the nominal torsional shear stress

$$\tau_n = \frac{T_n p_h}{A_{oh}^2} + \frac{V_n - V_p}{b_v d_v}$$

where V_p is the vertical component of the prestressing force.

(c) The range of the torsion compression field angle θ in degrees is given by

$$\left[10 + \frac{35(\tau_n/f_c')}{0.42 - 50\epsilon_\ell}\right] < \theta < \left[80 - \frac{35(\tau_n/f_c')}{0.42 - 65\epsilon_{ty}}\right]$$

where ϵ_ℓ = yield strain in longitudinal steel
  ϵ_{ty} = yield strain in stirrups

For f_y = 60,000 psi,

$$\theta_{min} = 10 + 83.75\left(\frac{\tau_n}{f_c'}\right)$$

$$\theta_{max} = 80 - 122.58\left(\frac{\tau_n}{f_c'}\right)$$

Select a value of θ close to θ_{min} to minimize the stirrups area and maximize the longitudinal steel area.

5. Check whether the section satisfies the diagonal cracking requirement. Find the pure shear cracking strength:

(a)
$$V_{ocr} = 4\lambda\sqrt{f_c'}\left[\sqrt{1 + \frac{\bar{f_c}}{4\sqrt{f_c'}}}\right](b_w d_p) + V_p$$

(b)
$$T_{cr} = \frac{T_u}{V_u}(V_{cr})$$

and

$$M_{cr} = \frac{M_u}{V_u}(V_{cr})$$

V_{cr} is obtained from the interaction equation

$$\left(\frac{M_{cr}}{M_{ocr}}\right)^2 + \left(\frac{V_{cr}}{V_{ocr}}\right)^2 + \left(\frac{T_{cr}}{T_{ocr}}\right)^2 = 1.0$$

If $V_{cr} > V_{SL}$, the section satisfies the diagonal cracking criteria at service load; otherwise change the section.

6. Design the transverse reinforcement. Determine the required nominal shear and torsional moment strengths at a distance $d_v/2 \tan \theta$ from the face of the support.

(a) Find the required shear stirrups area

$$\frac{A_v}{s} = \frac{(V_n - V_p)\tan \theta}{f_y d_v}$$

per unit length per two legs, where d_v is the effective shear depth, approximately equal to the distance from the centers of the bars or prestressing tendons at the corners of the closed stirrups, and V_n is the factored V_u/ϕ.

(b) Find the required torsional stirrups area

$$\frac{A_t}{s} = \frac{T_n \tan \theta}{2 f_y A_o}$$

per unit length per one leg, where T_n is the factored T_u / ϕ, $A_o = (A_{oh} - a_o / 2 p_h)$, and

$$a_o = \frac{A_{oh}}{p_h} \left[1 - \sqrt{1 - \frac{T_n p_h}{0.85 f'_c A^2_{oh}} \left(\tan \theta + \frac{1}{\tan \theta} \right)} \right]$$

is the depth of the torsional compressive block.

(c) Find the total area of the closed stirrups or ties:

$$\frac{A_{tt}}{s} = \frac{A_v}{s} + 2 \frac{A_t}{s}$$

(d) Select the size of the stirrups and the cross-sectional area A_s. The spacing is given by

$$s = \frac{A_s}{A_{tt}/s}$$

The maximum allowable spacing is the smaller of $s = p_h / 8 \tan \theta$ and $s = d_o / 3 \tan \theta$, but not more than 12 in.

7. Design the longitudinal reinforcement.

(a) At the midspan section $(+M_n)$, find the axial tension caused by shear and torsion given by

$$\Delta N_u = \frac{1}{\tan \theta} \sqrt{(V_u - \phi V_p)^2 + \left(\frac{T_u p_o}{2 A_o} \right)^2}$$

where $p_o = p_h - 4 a_o$.

$$\text{Required } +M_{n, \min} = \frac{M_u}{\phi_f} + \frac{d_v \Delta N_u}{2 \phi_v}$$

where $\phi_f = 0.90$ and $\phi_v = 0.85$. Then check to make sure that M_n from step $2 > M_{n, \min}$.

(b) At the support section $(-M_n)$, ΔN_n is the same as in (a), and we then have

$$\text{Required } -M_{n, \min} = \frac{d_v \Delta N_u}{2 \phi_v} - \frac{M_u}{\phi_f}$$

Design the longitudinal steel to provide for the reinforcement to develop the smaller of the two values of moment obtained from (a) and (b). Then choose the size and number of longitudinal bars, and space them equally on both side faces of the beam. The maximum spacing $s = 12$ in.

8. Draw the beam section and show the details of all the reinforcement.

5.19.2 Modified Truss Analogy Method

1. Calculate T_u, and V_u, and verify whether torsional reinforcement is needed. Torsional moment can be neglected if the factored $T_u < 1.5\lambda\sqrt{f'_c} \times [\sqrt{1 + 10\bar{f}_c/f'_c}](\Sigma\eta x^2 y)$, where $\eta = 0.35/(0.75 + x/y)$, in which x is the smaller segment dimension and y the larger segment dimension.

2. Calculate the required T_n and V_n, and test whether the section is adequate.

(a) Find the required T_n and V_n at a section $d/2$ from the face of the support.

(b) Find the upper limits of the torsional and shear nominal strengths given by

$$T_{n,\max} = \frac{\gamma'\lambda\sqrt{f'_c}}{\sqrt{1 + \left(\dfrac{\gamma'}{10}\right)^2\left(\dfrac{V_n}{T_n}\right)^2\left(\dfrac{\Sigma\eta x^2 y}{b_w d}\right)^2}}\left(\Sigma\eta x^2 y\right)$$

and

$$V_{n,\max} = \frac{10\lambda\sqrt{f'_c}}{\sqrt{1 + \left(\dfrac{10}{\gamma'}\right)^2\left(\dfrac{T_n}{V_n}\right)^2\left(\dfrac{b_w d}{\Sigma\eta x^2 y}\right)^2}}(b_w d)$$

where $\gamma = 14 - 13.33\bar{f}_c/f'_c$ and $\gamma' = \gamma\sqrt{1 + 10\bar{f}_c/f'_c}$.

(c) If $T_n < T_{n,\max}$ and $V_n < V_{n,\max}$, do not enlarge the section.

3. Calculate the concrete nominal torsional strength T_c and the nominal shear strength V_c:

(a)
$$T_c = \frac{T_{co}}{\sqrt{1 + \left(\dfrac{T_{co}}{V_{co}}\right)^2\left(\dfrac{V_n}{T_n}\right)^2}}$$

where $T_{co} = 6\lambda\sqrt{f'_c}[\sqrt{1 + 10\bar{f}_c/f'_c} - k](\Sigma x^2 y)$ for $k = 1 - 0.33/\eta \geq 0.60$.

(b)
$$V_c = \frac{V_{co}}{\sqrt{1 + \left(\dfrac{V_{co}}{T_{co}}\right)^2\left(\dfrac{T_n}{V_n}\right)^2}}$$

where V_{co} is the lesser of V_{ci} or V_{cw} from Equations 5.11 and 5.15 or V_c from Equation 5.16. If $T_n = T_u/\phi < T_c$ and $V_n = V_u/\phi < V_c$, use minimum steel; otherwise go to step 4.

4. Design the transverse closed stirrups for combined torsion and shear.

(a) Find

$$T_s = \frac{T_u}{\phi} - T_c = T_n - T_c$$

$$V_s = \frac{V_u}{\phi} - V_c = V_n - V_c$$

(b)
$$T_s = \alpha_t \frac{x_1 y_1}{s} A_t f_y$$

where $\alpha_t = 0.66 + 0.33 y_1/x_1 \leq 1.50$, in which x_1 is the smaller dimension, center to center, of the stirrups and y_1 is the larger dimension, center to center, of the stirrups.

(c) Find the torsional stirrups area

$$\frac{A_t}{s} = \frac{T_s}{\alpha_t x_1 y_1 f_y}$$

per unit length per one leg.

(d) Find the shear stirrups area

$$\frac{A_v}{s} = \frac{V_s}{d f_y}$$

per unit length per two legs.

(e) The total area of the stirrups is

$$\frac{A_{tt}}{s} = \frac{A_v}{s} + 2\frac{A_t}{s}$$

(f) Select the size of the stirrups and the cross-sectional area A_s of stirrups in one transverse section where the spacing is

$$s = \frac{A_s}{A_{tt}/s}$$

5. Design the longitudinal steel for torsion strength.

(a) Find the total area of longitudinal bars

$$A_\ell = 2A_t\left(\frac{x_1 + y_1}{s}\right)$$

where $A_t = A_{tt}$ from step 4. Then select the size and area of longitudinal bars.

(b) $$\text{No. of bars} = \frac{A_\ell}{A_s}$$

Space the bars equally on both side faces of the beam with a maximum spacing of 12 in., and then draw the beam section and reinforcing details.

Figure 5.46 shows a flowchart for designing web steel for combined torsion and shear by the compression field theory. Figure 5.47 shows a flowchart for designing web reinforcement for combined shear and torsion by the modified truss analogy theory.

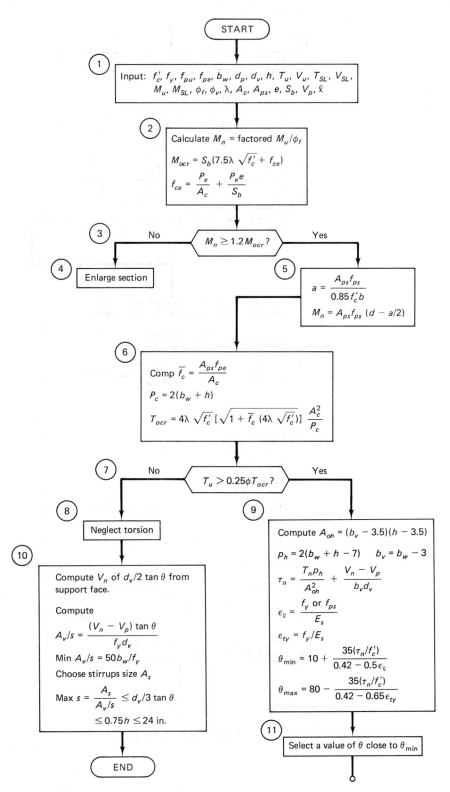

Figure 5.46 Flowchart for torsional design by compression field theory.

(12)

$$V_{ocr} = 4\lambda \sqrt{f_c'} \left[\sqrt{1 + \overline{f_c}/4\sqrt{f_c'}}\right] (b_w d_p) + V_p$$

$$T_{cr} = \frac{T_u}{V_u} (V_{cr})$$

$$M_{cr} = \frac{M_u}{T_u} (V_{cr})$$

$$\left(\frac{M_{cr}}{M_{ocr}}\right)^2 + \left(\frac{V_{cr}}{V_{ocr}}\right)^2 + \left(\frac{T_{cr}}{T_{ocr}}\right)^2 = 1 \text{ gives } V_{cr}$$

(14) Go to next section ◄── No ── **(13)** $V_{cr} > V_{SL}$? ── Yes ──►

(15)

Compute V_n and T_n at $d_v/2 \tan\theta$ from support

Calculate $A_v/s = \dfrac{(V_n - V_p)\tan\theta}{f_y d_v}$

$$a_o = \frac{A_{oh}}{p_h} \left[1 - \sqrt{1 - \frac{T_n p_h}{0.85 f_c' A_{oh}^2}} \right] (\tan\theta + 1/\tan\theta)$$

$$A_o = A_{oh} - \frac{a_o}{2} p_h$$

$$A_t/s = \frac{T_n \tan\theta}{2 f_y A_o}$$

$$A_{tt}/s = A_v/s + 2A_t/s$$

Choose stirrups size A_s

$$\text{Max } s = \frac{A_s}{A_{tt}/s} \leq \frac{p_h}{8\tan\theta} \leq \frac{d_v}{3\tan\theta} \leq 12 \text{ in.}$$

$$\Delta N_u = \frac{1}{\tan\theta} \sqrt{(V_u - \phi V_p)^2 + \left(\frac{T_u p_o}{2A_o}\right)^2}$$

$$p_o = p_h - 4a_o$$

$$+ M_{n,\min} = \frac{M_u}{\phi_f} + \frac{d_v \Delta N_u}{2\phi_v} \text{ for midspan } M_u$$

$$- M_{n,\min} = \frac{d_v \Delta N_u}{2\phi_v} - \frac{M_u}{\phi_f} \text{ for support moment}$$

Choose A_ℓ for smaller of $+M_n$ or $-M_n$
Space at $s \leq 12$ in. max.

END

Figure 5.46 (*continued*)

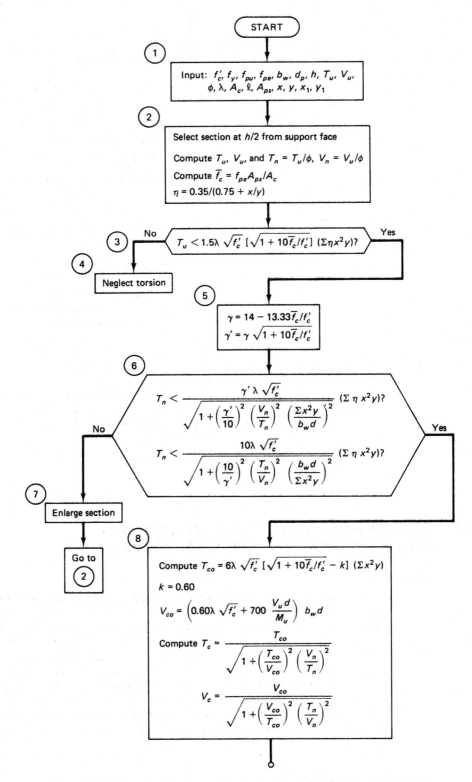

Figure 5.47 Flowchart for torsional design by the modified truss analogy theory.

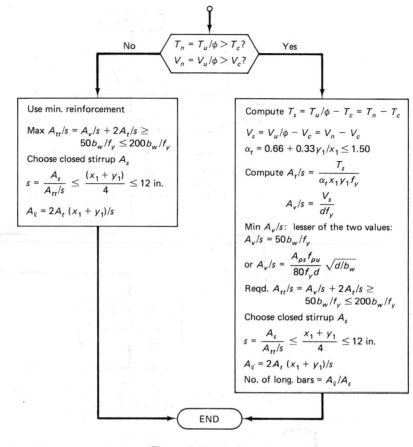

Figure 5.47 (*continued*)

5.20 DESIGN OF WEB REINFORCEMENT FOR COMBINED TORSION AND SHEAR IN PRESTRESSED BEAMS

Example 5.8

A parking garage floor for medium-size cars has the prestressed concrete flooring system shown in Figure 5.48. The floor panels are 36 ft × 54 ft (11 m × 16.5 m) on centers, and 54 ft (16.5 m) span precast double T's are supported by typical precast prestressed concrete spandrel L-beams spanning 36 ft (11 m) on centers (Figures 5.47(a) and (b)). The spandrel beams are torsionally restrained by their connections to the supporting columns. The floor is subjected to a service superimposed dead load due to the double T's of $W_{SD} = 77$ psf (3,687 Pa) and a service live load of $W_L = 50$ psf (2,394 Pa).

Design the spandrel beam web reinforcement to resist the combined torsion and shear to which it is subjected. Use (a) the compression field method and (b) the modified truss analogy method in the solution, and compare the results. Given data are the following:

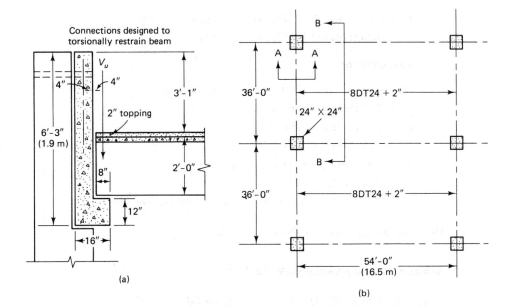

(a)

(b)

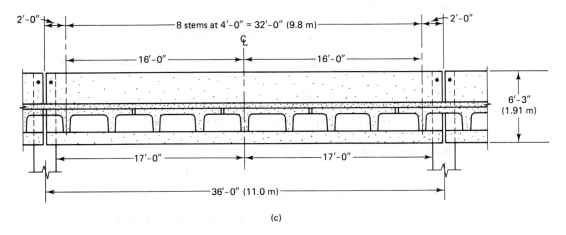

(c)

Figure 5.48 Geometrical details of structure in Example 5.8. (a) Section A-A. (b) Partial plan. (c) Section B-B.

Beam Properties

$A_c = 696$ in^2 (4,491 cm^2)

$I_c = 364{,}520$ in^4 (93.3 $\times$ 10^6 cm^4)

$c_b = 33.2$ in. (84.3 cm)

$c_t = 41.8$ in. (106.2 cm)

$S^t = 8{,}720$ in^3 (142,895 cm^3)

$S_b = 10{,}990$ in^3 (180,094 cm^3)

$W_D = 725$ plf (10.6 kN/m)

$f'_c = 5,000$ psi (34.5 MPa), normal-weight concrete

$f_y = 60,000$ psi (41.8 MPa) for stirrups

Prestressing

$A_{ps} = $ six $\frac{1}{2}$ in. dia, 270 K stress-relieved strands

$f_{pu} = 270,000$ psi (1,862 MPa)

$f_{ps} = 255,000$ psi (1,758 MPa)

$f_{pe} = 155,000$ psi (1,069 MPa)

$E_{ps} = 28 \times 10^6$ psi (193 $\times 10^3$ MPa)

$d_p = 68$ in. (172.7 cm)

$e = 68 - 41.8 = 26.2$ in. (66.5 cm), straight tendon

Disregard the effects of winds or earthquake.

Solution A: Compression Field Method

1. *Calculate T_u, V_u, M_u, T_{SL}, V_{SL} acting on L-beam*

(a) Service load

$$W_D = 725 \text{ plf (10.6 kN/m)}$$

$$W_{SD} = \frac{77 \times 54}{2} \times 4 \text{ ft} = 8,316 \text{ lb/stem (37.0 kN)}$$

$$W_L = \frac{50 \times 54}{2} \times 4 \text{ ft} = 5,400 \text{ lb/stem (24.0 kN)}$$

Total P_{SL} per stem $= 8,316 + 5,400 = 13,716$ lb (61.0 kN)

(b) Factored loads

$$W_{Du} = 1.4 \times 725 = 1,015 \text{ plf (14.8 kN/m)}$$

$$W_{SDu} = 1.4 \times 8,316 = 11,642 \text{ lb/stem (51.8 kN/m)}$$

$$W_{Lu} = 1.7 \times 5,400 = 9,180 \text{ lb/stem (40.8 kN)}$$

Total P_u per stem $= 11,642 + 9,180 = 20,822$ lb (92.6 kN)

T_u at face of support $= \frac{1}{2} P_u \times$ arm $\times$ no. of stems

$$= \frac{20,822 \times 8}{2} \times \frac{9}{12} = 62,466 \text{ ft-lb (84.7 kN-m)}$$

T_{SL} at face of support $= \dfrac{13,716}{20,822} \times 62,466 = 41,148$ ft-lb (55.8 kN-m)

V_u at face of support $= \frac{1}{2}(P_u \times$ no. of stems $+$ factored $W_D \times$ span$)$

$$= \frac{1}{2}(20,822 \times 9 + 1,015 \times 34) = 110,954 \text{ lb (494 kN)}$$

V_{SL} at face of support $= \frac{1}{2}(13,716 \times 9 + 725 \times 34) = 74,047$ lb (329 kN)

M_u at face of support $= 0$

Similarly, calculate the values of T_u, V_u, and M_u, and the corresponding service-load values at each transverse stem contact point along the span of the L-beam, and construct the torsion, shear, and moment diagrams as shown in Figure 5.49.

2. *Determine minimum flexure reinforcement requirement*

$$A_{ps} = 6 \times 0.153 = 0.918 \text{ in}^2$$

$$P_e = A_{ps}f_{pe} = 0.918 \times 155,000 = 142,290 \text{ lb (633 kN)}$$

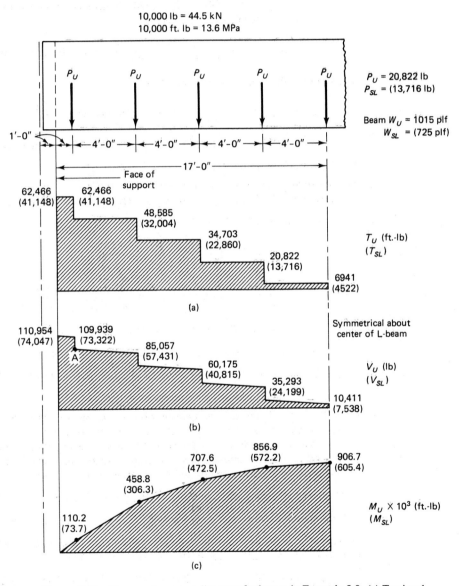

Figure 5.49 Force and moment diagrams for beams in Example 5.8. (a) Torsional moment. (b) Shear. (c) Flexural moment. Bracketed values are for service-load level.

$$\bar{f}_c = \frac{P_e}{A_c} = 142{,}290/696 = 204.4 \text{ psi } (1.4 \text{ MPa})$$

$$f_{ce} = \frac{P_e}{A_c} + \frac{P_e e}{S_b} = \frac{142{,}290}{696} + \frac{142{,}290 \times 26.2}{10{,}990} = 543.7 \text{ psi } (3.7 \text{ MPa})$$

From Equation 5.88 b, the pure flexural cracking moment is

$$M_{ocr} = S_b(7.5\lambda\sqrt{f_c'} + f_{ce}) = 10{,}990(7.5 \times 1.0\sqrt{5{,}000} + 543.7)$$

$$= 11{,}803{,}591 \text{ in.-lb } (1{,}334 \text{ kN-m}) \text{ since } \lambda = 1.0 \text{ for}$$
normal-weight concrete

$$a = \frac{A_{ps}f_{ps}}{0.85 f_c' b} = \frac{0.918 \times 255{,}000}{0.85 \times 5000 \times 8} = 6.9 \text{ in. } (17.5 \text{ cm})$$

The nominal flexural moment strength is

$$M_n = A_{ps}f_{ps}\left(d_p - \frac{a}{2}\right) = 0.918 \times 255{,}000\left(68 - \frac{6.9}{2}\right)$$

$$= 15{,}110{,}510 \text{ in.-lb} > 1.2 M_{ocr} = 14{,}164{,}306 \text{ } (1{,}601 \text{ kN-m})$$

Hence, minimum flexural reinforcement is satisfied.

3. *Verify whether torsional reinforcement is needed*
   From Equation 5.90 for the magnitude of pure torsional cracking strength,

$$p_c = 12 + 16 + 75 + 8 + 63 + 8 = 182 \text{ in.}$$

$$T_{ocr} = 4\lambda\sqrt{f_c'}\left[\sqrt{1 + \frac{\bar{f}_c}{(4\lambda\sqrt{f_c'})}\left(\frac{A_c^2}{p_c}\right)}\right]$$

$$= 4 \times 1.0\sqrt{5{,}000}\left[\sqrt{1 + \frac{204.4}{4\sqrt{5{,}000}} \times \frac{(696)^2}{182}}\right]$$

$$= 988{,}080 \text{ in.-lb } (112 \text{ kN-m})$$

$$\text{Max. } T_u = 62{,}466 \times 12 = 749{,}592 \text{ in.-lb } (84.7 \text{ kN-m})$$

$$0.25\phi T_{ocr} = 0.25 \times 0.85 \times 988{,}080 = 209{,}967 \text{ in.-lb } (23.7 \text{ kN-m})$$

$T_u > 0.25\phi T_{ocr}$, so torsional reinforcement is needed.

4. *Check cross-sectional size and choose torsion compression angle θ*
   In order to prevent crushing of the web concrete at the face of the support, a check of the cross-sectional bearing area has to be made. Assume that the load applied by the double-T stem nearest the column is directly transferred to the column such that no shear stresses are transferred by it to the spandrel beam. Then we have

   Min. required value of $T_n = T_u/\phi = 48{,}585 \times 12/0.85 = 685{,}906$ in.-lb

   Min. required value of $V_n = V_u/\phi = 85{,}057/0.85 = 100{,}067$ lb

Now, assume $b_v = 8 - 2 \times 1.5 = 5.0$ in. (12.7 cm) if the cover $= 1.5$ in. Also, assume $d_v = 75 - 2 \times 2.25 = 70.5$ in. (179 cm). Then

$$A_{oh} = 5.0 \times 70.5 + (12 - 2 \times 1.5)(8 - 1.5) = 411 \text{ in}^2 \text{ } (2{,}652 \text{ cm}^2)$$

$$p_h = 5 + 70.5 + 13 + 9 + 6.5 + 63 = 167 \text{ in. } (424 \text{ cm})$$

The torsional shear stress, from Equation 5.69, is

$$\tau_n = \frac{T_n p_h}{A_{oh}^2} + \frac{V_n - V_p}{b_v d_v}$$

where $V_p = 0$ since the tendon is straight. Thus, we have

$$\tau_n = \frac{685,906 \times 167}{(411)^2} + \frac{100,067 - 0}{5 \times 70.5} = 962 \text{ psi (6.6 MPa)}$$

$\tau_n / f_c' = 962/5,000 = 0.192$

From Equation 5.70,

$$\left[10 + \frac{35(\tau_n/f_c')}{0.42 - 50\epsilon_1}\right] < \theta < \left[80 - \frac{35(\tau_n/f_c')}{0.42 - 65\epsilon_{ty}}\right]$$

where $\epsilon_1 = 0.002$ in./in. and $\epsilon_{ty} = 0.002$ in./in.

Assuming #4 mild steel bars are used at each corner of the section, either Equation 5.70 can be used, or if $f_y = 60,000$ psi, Equation 5.71 or 5.72 can be used instead. We obtain

$$\left(10 + \frac{35 \times 0.192}{0.42 - 0.10}\right) < \theta < \left(80 - \frac{35 \times 0.192}{0.42 - 0.13}\right)$$

or $31.00° < \theta < 56.83°$. Since $\theta_{min} < \theta_{max}$, diagonal crushing of the compression struts or field will not occur, and the concrete section size is therefore adequate for the torsional requirement. Accordingly, choose $\theta = 32°$.

5. *Check whether section satisfies diagonal cracking requirement*

From Equation 5.89, the pure cracking shear is

$$V_{ocr} = 4\lambda\sqrt{f_c'} \times \sqrt{1 + \frac{\bar{f}_c}{4\sqrt{f_c'}}}\,(b_w d_p) + V_p$$

$$= 4 \times 1.0\sqrt{5,000} \times \sqrt{\frac{1 + 204.4}{4\sqrt{5,000}}}\,(8 \times 68) + 0$$

$$= 201,950 \text{ lb}$$

But

$$\frac{T_{cr}}{V_{cr}} \cong \frac{T_u}{V_u} \cong \text{req. } \frac{T_n}{V_n}$$

and

$$\frac{M_{cr}}{V_{cr}} \cong \frac{M_u}{V_u}$$

Then at the face of the support,

$$T_{cr} = \frac{T_u}{V_u}\,(V_{cr}) = \frac{62,466}{110,954}\,V_{cr}$$

$$= 0.56V_{cr} \text{ ft-lb} = 6.72V_{cr} \text{ in.-lb}$$

At the support,

$$M_{cr} = \frac{M_u}{V_u} (V_{cr}) = 0$$

From Equation 5.87 for combined loading,

$$\left(\frac{M_{cr}}{M_{ocr}}\right)^2 + \left(\frac{V_{cr}}{V_{ocr}}\right)^2 + \left(\frac{T_{cr}}{T_{ocr}}\right)^2 = 1.0$$

From step 3, $T_{ocr} = 988{,}080$ in.-lb, or

$$0 + \left(\frac{V_{cr}}{201{,}950}\right)^2 + \left(\frac{6.72 V_{cr}}{988{,}080}\right)^2 = 1$$

Hence,

$$V_{cr}^2 \left[\frac{1}{(201{,}950)^2} + \frac{1}{(147{,}036)^2}\right] = 1$$

So

$$V_{cr} = 118{,}868 \text{ lb } (529 \text{ kN}) > V_{SL} = 67{,}130 \text{ lb } (299 \text{ kN})$$

Thus, the section satisfies the diagonal cracking serviceability requirements.

6. *Design of transverse reinforcement*
Considering the region near the support for combined torsion and shear as critical, calculate V_u and T_u at a distance $d_v/2 \tan \theta$ from the face of the support. We have

d_v = effective shear depth $\cong$ distance between centers of bars or prestressing tendons at corners of the closed stirrups

$$= 75 \text{ in.} - (2 \times 1.5(\text{cover}) + 0.5(\text{stirrup dia}) + 0.5(\text{dia of bar})$$

$$= 71 \text{ in.}$$

So $d_v/2 \tan \theta = 71/2 \tan 32° = 56.81$ in. (144 cm). Now, from Figure 5.49(b), the factored shear at point A is

$$V_u = 109{,}939 - 20{,}822 = 89{,}117 \text{ lb}$$

Similarly, the factored shear at 56.81 in. from the face of the support is

$$V_u = 89{,}117 - \left[\left(\frac{56.81 - 12}{48}\right)\right](89{,}117 - 85{,}057)$$

$$= 85{,}327 \text{ lb } (380 \text{ kN})$$

So

$$V_n = V_u/\phi = 85{,}327/0.85 = 100{,}385 \text{ lb } (447 \text{ kN})$$

$$T_u = 48{,}585 \text{ ft-lb} = 583{,}020 \text{ in.-lb}$$

$$T_n = T_u/\phi = 583{,}020/0.85 = 685{,}906 \text{ in.-lb } (775 \text{ kN-m})$$

Next, calculate the torsional depth of the compression strut from Equation 5.68:

$$a_o = \frac{A_{oh}}{p_h}\left[1 - \sqrt{1 - \frac{T_n p_h}{0.85 f_c' A_{oh}^2}\left(\tan \theta + \frac{1}{\tan \theta}\right)}\right]$$

Shear and Torsional Strength Design   Chap. 5

From step 4, $A_{oh} = 411$ in^2 and $p_h = 167$ in. Thus,

$$a_o = \frac{411}{167}\left[1 - \sqrt{\frac{685,906 \times 167}{0.85 \times 5,000 \times (411)^2}\left(\tan 32° + \frac{1}{\tan 32°}\right)}\right]$$

$$= 0.48 \text{ in. } (1.22 \text{ cm})$$

From Equation 5.65, $A_o = A_{oh} - a_o \times (p_h/2) = 411 - (0.48/2) \times 167 = 371$ in^2 (2,394 cm^2). From Equation 5.92 b, the shear stirrups needed are given by

$$\frac{A_v}{s} = \frac{(V_n - V_p)\tan\theta}{f_y d_v} = \frac{(100,385 - 0)\tan 32°}{60,000 \times 71} = 0.0147 \text{ in}^2/\text{in. per two legs}$$

From Equation 5.93 b, the torsional stirrups needed are given by

$$\frac{A_t}{s} = \frac{T_n \tan\theta}{2f_y A_o} = \frac{685,906 \times \tan 32}{2 \times 60,000 \times 371} = 0.0096 \text{ in}^2/\text{in. per one leg}$$

From Equation 5.94 c, the total area of stirrups for combined torsion and shear is

$$\frac{A_{tt}}{s} = \frac{A_v}{s} + \frac{2A_t}{s} = 0.0147 + 2 \times 0.0096 = 0.0339 \text{ in}^2/\text{in. per two legs}$$

Trying #4 closed stirrups, we have $A_s = 2 \times 0.20 = 0.40$ in^2 and

$$s = \frac{A_s}{A_{tt}/s} = \frac{0.40}{0.0339} = 11.8 \text{ in. center to center}$$

$$\text{Max. allow. } s = \frac{p_h}{8\tan\theta} = \frac{167}{8\tan 32} = 33.4 \text{ in.} > 12.31 \text{ in.}$$

Also,

$$\text{Max. allow. } s = \frac{d_v}{3\tan\theta} = \frac{71}{3\tan 32} = 37.9 \text{ in.} > 12.31 \text{ in.}$$

Use #4 closed stirrups (12.7 mm dia) at 12 in. (30.5 cm) center to center.

7. *Design of longitudinal reinforcement*

(a) *Midspan section*
From step 6, $A_o = 371$ in^2. Also,

$$p_o = p_h - 4a_o = 167 - 4 \times 0.48 = 165 \text{ in.}$$

From Figure 5.49, $V_u = 10,411$ lb and $T_u = 6,941$ ft-lb $= 83,292$ in.-lb. Also, from Equation 5.95, the axial tension caused by shear and torsion is

$$\Delta N_u = \frac{1}{\tan\theta}\sqrt{(V_u - \phi V_p)^2 + \left(\frac{T_u p_o}{2A_o}\right)^2}$$

$$= \frac{1}{\tan 32}\sqrt{(10,411 - 0)^2 + \left(\frac{83,292 \times 165}{2 \times 371}\right)^2} = 34,003 \text{ lb } (151 \text{ kN})$$

From Equation 5.96 a,

$$\text{Min. rqd. } +M_n = \frac{M_u}{\phi_f} + \frac{d_v \Delta N_u}{2\phi_v}$$

$$= \frac{906,700 \times 12}{0.90} + \frac{71 \times 34,003}{2 \times 0.85}$$

$$= 13,509,459 \text{ in.-lb } (1,527 \text{ kN-m})$$

From step 2, the nominal moment strength supplied by the prestressing strands at midspan is

$$M_n = 15,110,510 \text{ in.-lb} > 13,599,459 \text{ in.-lb, O.K.}$$

Note that if the required min $+M_n$ exceeds the nominal moment strength of the strands, one should take into account the additional capacity supplied by the mild longitudinal steel at the corners, and even add extra longitudinal bars as necessary.

(b) *Section at the support*

From Figure 5.49, $M_u = 0$, $V_u = 110,954$ lb, and $T_u = 62,466$ ft-lb $=$ 749,592 in.-lb. From Equation 5.95,

$$\Delta N_u = \frac{1}{\tan \theta} \sqrt{(V_u - \phi V_p)^2 + \left(\frac{T_u p_o}{2 A_o}\right)^2}$$

$$= \frac{1}{\tan 32} \sqrt{(110,954 - 0)^2 + \left(\frac{749,592 \times 165}{2 \times 371}\right)^2}$$

$$= 320,450 \text{ lb } (1,425 \text{ kN})$$

Since the section is closer than $d_v/\tan \theta$ from the inner edge of the bearing area of the L-beam, use the lesser of the two values obtained from Equations 5.96 b and 5.97 a:

$$\text{Reg. min. } (-M_n) = \frac{d_v \Delta N_u}{2 \theta_v} - \frac{M_u}{\phi_f} = \frac{71 \times 320,450}{2 \times 0.85} - 0 = 13,383,500 \text{ in.-lb}$$

or

$$\text{Req. min. } (-M_n) = \frac{d_v T_u p_o}{2 \tan \theta (\phi_v 2 A_o)} - \frac{M_u}{\phi_f}$$

$$= \frac{71 \times 749,592 \times 165}{2 \tan 32 \times 0.85 \times 2 \times 371} - 0$$

$$= 11,140,963 \text{ in.-lb}$$

Req. min. $(-M_n)$ is the lesser of the two values; hence, $\min(-M_n) = 11,140,963$ in.-lb (1,259 kN-m), and reinforcing bars have to be provided to resist this nominal moment strength at the support section due to combined torsion and shear.

Since there are two #4 bars at the top corners, the moment M_n provided by these bars is

$$d = 75 - (1.5 + 0.5 + 0.25) \cong 72.5 \text{ in. } (184 \text{ cm})$$

$$b = 16 \text{ in.}$$

$$A_s = 2 \times 0.20 = 0.40 \text{ in}^2$$

$$a = \frac{A_s f_y}{0.85 f_c' b} = \frac{0.40 \times 60,000}{0.85 \times 5000 \times 16} = 0.35 \text{ in.}$$

$$M_n = A_s f_y \left(d - \frac{a}{2} \right) = 0.40 \times 60,000 \left(72.5 - \frac{0.35}{2} \right)$$

$$= 1,735,000 \text{ in.-lb (196 kN-m)}$$

The moment balance for which additional longitudinal mild steel has to be provided is

$$M_n = 11,140,963 - 1,735,800 = 9,405,163 \text{ in.-lb}$$

and the minimum allowable bar diameter is

$$\frac{s \tan \theta}{16} = \frac{12 \tan 32°}{16} = 0.47 \text{ in.}$$

$$= \tfrac{1}{2} \text{ in. dia bars, i.e., } \#4 \text{ bars}$$

Accordingly, distribute the #4 longitudinal bars along both vertical faces to provide the additional required $M_n = 9,405,163$ in.-lb.

Now assume that the #4 longitudinal bars are spaced at 6 inches center to center on both vertical faces of the 75 in. (191 cm) deep L-beam. Then calculate the M_n contribution of each row of two bars in the same manner that M_n for the two top corner bars was calculated, assuming that all longitudinal bars yield as the compression field theory postulates. The nominal moment strength of the additional 11 rows of #4 bars (22 bars) is $M_n \cong 9,450,000$ in.-lb. Hence, the design may be adopted. Figure 5.50 shows a cross section of the L-beam with the details of reinforcement.

For the reinforcing details to be complete, a design of the ledge and hanger reinforcement would be required, as well as the details of the anchorage of the longitudinal reinforcement at the supports. Chapter 10, on the design of connections, provides these details.

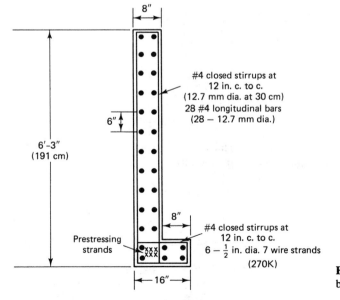

Figure 5.50 Reinforcement details of beam in Example 5.8.

Solution B: Modified Truss Analogy Method

1. *Calculate T_u and V_u, and verify whether torsional reinforcement is needed*
   From part A, the torsion and shear factored values at the face of the support are respectively

$$T_u = 62,466 \text{ ft-lb} = 749,592 \text{ in.-lb}$$

and

$$V_u = 110,954 \text{ lb}$$

Also,

$$\bar{f_c} = 204.4 \text{ psi}$$

Check whether torsion can be neglected: from Equation 5.79 b,

$$T_u = 1.5\lambda\sqrt{f_c'} \times \sqrt{1 + 10\bar{f_c}/f_c'} \; (\Sigma \, nx^2y)$$

$$n = \frac{0.35}{0.75 + x/y} = \frac{0.35}{0.75 + 8/75} = 0.409$$

$$\Sigma \, x^2y = 8^2 \times 75 + 8^2 \times 12 = 5,568 \text{ in}^3$$

$$n\Sigma \, x^2y = 0.409 \times 5,568 = 2,277 \text{ in}^3$$

$$T_u = 1.5 \times 1.0\sqrt{5,000} \times \sqrt{1 + \frac{10 \times 204.4}{5,000}} \, (2,277)$$

$$= 286,657 \text{ in.-lb} < 749,592 \text{ in.-lb}$$

Hence, design for torsion.

2. *Calculate the required T_n and V_n, and test whether the section is adequate*
   The upper limits for the torsional and shear nominal strengths, from Equations 5.81 a and b, are

$$T_{n,\,max} = \frac{\gamma'\lambda\sqrt{f_c'}}{\sqrt{1 + \left(\dfrac{\gamma'}{10}\right)^2 \left(\dfrac{V_n}{T_n}\right)^2 \left(\dfrac{\Sigma \, nx^2y}{b_wd}\right)^2}} (\Sigma \, nx^2y)$$

and

$$V_{n,\,max} = \frac{10\lambda\sqrt{f_c'}}{\sqrt{1 + \left(\dfrac{10}{\gamma'}\right)^2 \left(\dfrac{T_n}{V_n}\right)^2 \left(\dfrac{b_wd}{\Sigma \, nx^2y}\right)^2}} (b_wd)$$

Use these values of the required T_n and V_n at a distance $d/2$ from the face of the support (ACI Code stipulation for torsion only):

$$\frac{d}{2} = \frac{71}{2} = 35.5 \text{ in.}$$

From Figure 5.49,

$$T_u = 48,585 \text{ ft-lb} = 583,020 \text{ in.-lb}$$

$$\text{Req. } T_n = T_u/\phi = 583,020/0.85 = 685,906 \text{ in.-lb}$$

V_u at point A of Figure 5.49 = 109,939 − 20,822 = 89,117 lb

So the distance from point A to $d/2$ from the face of the support is $71/2 - 12 = 23.5$ in. Now,

$$V_u \text{ at } d/2 = 85,057 + \left(\frac{48 - 23.5}{48}\right)(89,117 - 85,057) = 87,129 \text{ lb}$$

$$\text{Req } V_n = V_u/\phi = 87,129/0.85 = 102,505 \text{ lb}$$

$$\gamma = 14 - 13.33\bar{f_c}/f_c' = 14 - 13.33 \times 204.4/5,000 = 13.46$$

$$\gamma' = \gamma\sqrt{1 + 10\bar{f_c}/f_c'} = 13.46\sqrt{1 + 10 \times 204.4/5,000} = 15.97$$

$$T_{n,\max} = \frac{15.97 \times 1.0\sqrt{5,000}}{\sqrt{1 + \left(\frac{15.97}{10}\right)^2\left(\frac{102,505}{685,906}\right)^2\left(\frac{2277}{8 \times 71}\right)^2}}(2,277)$$

$$= 2,571,301 \text{ in.-lb} > \text{Req. } T_n = 685,906 \text{ in.-lb, O.K.}$$

$$V_{n,\max} = \frac{10 \times 1.0\sqrt{5,000}}{\sqrt{1 + \left(\frac{10}{15.97}\right)^2\left(\frac{685,906}{102,505}\right)^2\left(\frac{8 \times 71}{2,277}\right)^2}}(8 \times 71)$$

$$= 277,656 \text{ lb} > \text{Req. } V_n = 102,505 \text{ lb, O.K.}$$

Hence, the section is adequate.

3. *Calculate the concrete nominal torsional strength T_c and nominal shear strength V_c*

From Equation 5.79 a, the nominal torsional strength when torsion is acting alone is

$$T_{co} = 6\lambda\sqrt{f_c'}[\sqrt{1 + 10\bar{f_c}/f_c'} - k](\Sigma nx^2y)$$

$$k = 1 - 0.33/n = 1 - 0.33/0.409 = 0.193 < 0.60$$

Accordingly, use $k = 0.60$. Then

$$T_{co} = 6 \times 1.0\sqrt{5,000}[\sqrt{1 + 10 \times 204.4/5,000} - 0.6](2,277)$$

$$= 567,002 \text{ in.-lb}$$

V_{co} is the lesser of V_{ci} and V_{cw} from Equations 5.11 and 5.15, or, alternatively, V_c from Equation 5.16. So use

$$V_{co} = \left(0.60\lambda\sqrt{f_c'} + 700\frac{V_u d}{M_u}\right)b_w d > 2\lambda\sqrt{f_c'}\,b_w d < 5\lambda\sqrt{f_c'}\,b_w d$$

from Equation 5.16.
From Figure 5.49,

$$M_u \text{ at } d/2 = 110,200 + \frac{(35.5 - 12)}{48}(458,800 - 110,200)$$

$$= 280,869 \text{ ft-lb} = 3,370,425 \text{ in.-lb}$$

$$V_u d/M_u = 87,129 \times 71/3,370,425 = 1.84 > 1.0$$

So use $V_u d / M_u = 1.0$. Then

$$V_{co} = (0.60 \times 1.0\sqrt{5,000} + 700 \times 1.0)(8 \times 71) = 421,698 \text{ lb}$$

$$2\lambda\sqrt{f'_c}\,b_w d = 2 \times 1.0\sqrt{5,000}\,(8 \times 71) = 80,327 \text{ lb}$$

$$5\lambda\sqrt{f'_c}\,b_w d = 5 \times 1.0\sqrt{5,000}\,(8 \times 71) = 200,818 \text{ lb} < 421,698 \text{ lb}$$

Use $V_{co} = 200,818$ lb.

Using Equations 5.78 a and b for combined torsion and shear, we obtain

$$T_c = \frac{T_{c0}}{\sqrt{1 + \left(\dfrac{T_{co}}{V_{co}}\right)^2 \left(\dfrac{V_n}{T_n}\right)^2}}$$

and

$$V_c = \frac{V_{c0}}{\sqrt{1 + \left(\dfrac{V_{co}}{T_{co}}\right)^2 \left(\dfrac{T_n}{V_n}\right)^2}}$$

So

$$\text{Req. } T_n = 685,906 \text{ in.-lb}$$

$$T_{co} = 567,002 \text{ in.-lb}$$

$$\text{Req. } V_n = 102,505 \text{ lb}$$

$$V_{co} = 200,818 \text{ lb}$$

$$T_c = \frac{566,504}{\sqrt{1 + \left(\dfrac{567,002}{200,818}\right)^2 \left(\dfrac{102,505}{685,906}\right)^2}} = 522,401 \text{ in.-lb (59.9 kN-m)}$$

$$V_c = \frac{200,818}{1 + \left(\dfrac{200,818}{567,002}\right)^2 \left(\dfrac{685,906}{102,505}\right)^2} = 78,070 \text{ lb (315 kN)}$$

4. *Design of transverse web steel for combined torsion and shear*

$$T_s = T_n - T_c = 685,906 - 522,401 = 163,505 \text{ in.-lb (18.5 kN-m)}$$

$$V_s = V_n - V_c = 102,505 - 78,070 = 24,435 \text{ lb (91 kN)}$$

From Equation 5.74,

$$T_s = \alpha_t \frac{x_1 y_1}{s} A_t f_y$$

where $\alpha_t = 0.66 + 0.33(y_1/x_1) \leq 1.50$. Since

$$x_1 = 8 - 2(1.5 + 0.50) = 4 \text{ in.}$$

$$y_1 = 75 - 2(1.5 + 0.5) = 71 \text{ in.}$$

we have

$$\alpha_t = 0.66 + 0.33 \times \frac{71}{4} = 6.51 > 1.50$$

So use $\alpha_t = 1.50$. Then

$$\frac{A_t}{s} = \frac{T_s}{\alpha_t x_1 y_1 f_y} = \frac{163,505}{1.5 \times 4 \times 71 \times 60,000} = 0.0064 \text{ in}^2/\text{in. per one leg}$$

From Equation 5.21 b

$$\text{Rqd.} \frac{A_v}{s} = \frac{V_s}{d f_y} = \frac{24,435}{71 \times 60,000} = 0.0057 \text{ in}^2/\text{in. per two legs}$$

$$\text{Rqd. total} \frac{A_{tt}}{s} = \frac{A_v}{s} + 2\frac{A_t}{s} = 0.0057 + 2 \times 0.0064$$
$$= 0.0185 \text{ in}^2/\text{in. per two legs}$$

Assuming # 4 closed stirrups, $A_s = 2 \times 0.2 = 0.4 \text{ in}^2$. Then

$$s = \frac{A_s}{A_{tt}/s} = \frac{0.40}{0.0185} = 21.6 \text{ in.}$$

$$\text{Max allow. } s = \frac{x_1 + y_1}{4} = \frac{4 + 71}{4} = 18.75 \text{ in.} \leq 12 \text{ in.}$$

$$\text{Min} \frac{A_v}{s} = \text{lesser of } \frac{A_v}{s} = \frac{50 b_w}{f_y}$$

or

$$\frac{A_v}{s} = \frac{A_{ps} f_{py}}{80 f_y d} \sqrt{\frac{d}{b_w}}$$

$$50 \frac{b_w}{f_y} = 50 \times \frac{8}{60,000} = 0.0067 \text{ in}^2/\text{in. per two legs}$$

$$\frac{A_{ps} f_{pu}}{80 f_y} \sqrt{\frac{d}{b_w}} = \frac{6 \times 0.153}{80} \times \frac{270,000}{60,000 \times 71} \sqrt{\frac{71}{8}}$$

$= 0.002 \text{ in}^2/\text{in. per two legs} < 0.0067$ for min. A_v/s; 0.002 in^2/in. per two legs controls. Required $A_v/s = 0.0057 > min. A_v/s = 0.002$. Hence, controls. So use $A_v/s = 0.0057$ in^2/in. per two legs. We have

$$\text{Min.} \frac{A_t}{s} = \frac{A_v}{2s} = 0.0033 \text{ in}^2/\text{in. per one leg}$$

$$\text{Total Min.} \frac{A_{tt}}{s} = \frac{A_v}{s} + 2\frac{A_t}{s} = 0.0134 \text{ in}^2/\text{in. per two legs}$$

$$< \text{Rqd.} \frac{A_{tt}}{s} = 0.0185 \text{ in}^2/\text{in. per two legs}$$

$$\text{Max} \frac{A_{tt}}{s} = 4(\text{Min.} \frac{A_{tt}}{s} \text{ per one leg}) = 2 \times 0.0134$$

$$= 0.0268 \text{ in}^2/\text{in. per two legs}$$

Hence, use #4 closed stirrups at 12 in. center to center for combined torsion and shear (12.7 mm at 30 cm).

5. *Design of longitudinal bars for torsion strength*

From Equation 5.82 b,

$$A_\ell = 2A_t \frac{(x_1 + y_1)}{s} = 2 \times 0.0185 \, (4 + 71) = 2.775 \text{ in}^2$$

Using #4 longitudinal bars, $A_s = 0.2$ in². So

$$\text{No. of bars} = \frac{A_\ell}{A_s} = \frac{2.775}{0.2} = 13.9 \text{ bars}$$

It is advisable to space the longitudinal bars at a spacing not exceeding 12 in. center to center. We then have

$$\text{No. of spaces} = \frac{71 - 0.5}{12} = 5.9$$

across the beam depth. So use fourteen #4 longitudinal bars, half on each side.

Comparison of Solutions A and B. The following table presents the various features of the two solutions to Example 5.8. The required nominal strengths at the face of the support for both solutions are $T_n = 881{,}873$ in.-lb and $V_n = 120{,}385$ lb.

Parameter	Solution A	Solution B
Critical section from face of support, in.	56.81	35.5
T_n (in.-lb)	685,906	685,906
V_n (lb)	100,385	102,505
T_{ocr} (in.-lb)	988,080	566,504
V_{ocr} (lb)	201,950	200,818
A_{tt}/s (in²/in. per two legs)	0.0339	0.0185
Stirrups size & spacing	#4 @ 12 in.	#4 @ 12 in.
Longitudinal steel A_ℓ	28 #4	14 #4

While the solution by the compression field theory (Method A) seems more conservative in terms of the longitudinal steel needed for torsion, the expressions used in the design are substantiated by exensive tests to failure of prestressed concrete beams subjected to combined torsion and shear. Hence, the reinforcement details given in Figure 5.50 are used.

REFERENCES

5.1 ACI Committee 318. *Building Code Requirements for Reinforced Concrete, 318–89, and Commentary to the Building Code Requirements for Reinforced Concrete, 318–89.* Detroit: American Concrete Institute, 1989.

5.2 Nawy, E. G. *Reinforced Concrete—A Fundamental Approach.* Englewood Cliffs, N.J.: Prentice Hall, 1985.

5.3 Mattock, A. H., Chen, K. C., and Soogswang, K. "The Behavior of Reinforced Concrete Corbels." *Journal of the Prestressed Concrete Institute* 21 (1976): 52–77.

5.4 Nawy, E. G. *Simplified Reinforced Concrete.* Englewood Cliffs, N.J.: Prentice Hall, 1986.

5.5 Sozen, M. A., Zwoyer, E. M., and Siess, C. P. Strength in Shear of Beams without Web Reinforcement. Urbana, Illinois: Bulletin No. 452, Engineering Experiment Station, University of Illinois, April 1959.

5.6 ACI Committee 318. *Building Code Requirements for Reinforced Concrete, 318–63,* and *Commentary on the Building Code Requirements for Reinforced Concrete, 318–63.* Detroit: American Concrete Institute, 1963.

5.7 Nawy, E. G., and Ukadike, M. M. "Shear Transfer in Concrete and Polymer Modified Concrete Members Subjected to Shear Load." *Journal of the American Society for Testing and Materials,* March 1983, pp. 83–97.

5.8 Prestressed Concrete Institute. *Manual of Design and Detailing of Precast and Prestressed Connections.* Chicago: Prestressed Concrete Institute, 1987.

5.9 Hsu, T. T. C. *Torsion in Reinforced Concrete.* New York: Van Nostrand Reinhold, 1983.

5.10 Hsu, T. T. C. "Torsion in Structural Concrete—Uniformly Prestressed Members without Web Reinforcement." *Journal of the Prestressed Concrete Institute* 13 (1968): 34–44.

5.11 Collins, M. P., and Mitchell, D. "Shear and Torsion Design of Prestressed and Non-Prestressed Concrete Beams." *Journal of the Prestressed Concrete Institute* 25 (1980): 32–100.

5.12 Prestressed Concrete Institute. *PCI Design Handbook.* Chicago: Prestressed Concrete Institute, 1985.

5.13 Zia, P., and McGee, W. D. "Torsion Design of Prestressed Concrete." *Journal of the Prestressed Concrete Institute* 19 (1974): 46–65.

5.14 McGee, W. D., and Zia, P. "Prestressed Concrete Members under Torsion, Shear and Bending." *Journal of the American Concrete Institute* 73 (1976): 26–32.

5.15 Zia, P., and Hsu, T. T. C. *Design for Torsion and Shear in Prestressed Concrete.* Chicago: ASCE Annual Convention, Reprint No. 3423, 1979.

5.16 Abeles, P. W., and Bardhan-Roy, B. K. *Prestressed Concrete Designer's Handbook.* 3d Ed. London: Viewpoint Publications, 1981.

PROBLEMS

5.1 A post-tensioned bonded prestressed beam has the cross section shown in Figure P5.1. It has a span of 75 ft (22.9 m) and is subjected to a service superimposed dead load

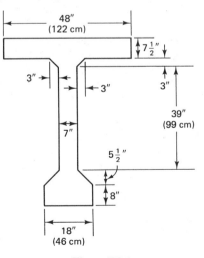

Figure P5.1.

W_{SD} = 450 plf (6.6 kN/m) and a superimposed service live load W_L = 2,300 plf (33.6 kN/m). Design the web reinforcement necessary to prevent shear cracking (a) by the detailed design method and (b) by the alternative method at a section 15 ft (4.6 m) from the face of the support. The profile of the prestressing tendon is parabolic. Use #3 stirrups in your design, and detail the section. The following data are given:

A_c = 876 in² (5,652 cm²)

I_c = 433,350 in⁴ (18.03 × 10⁶ cm⁴)

r^2 = 495 in² (3,194 cm²)

c_t = 25 in. (63.5 cm)

S^t = 17,300 in³ (2.83 × 10⁵ cm³)

c_b = 38 in. (96.5 cm)

S_b = 11,400 in³ (1.86 × 10⁵ cm³)

W_d = 910 plf (13.3 kN/m)

e_c = 32 in. (81.3 cm)

$e_e = 2$ in. (5 cm)

$f'_c = 5,000$ psi (44.5 MPa), normal-weight concrete

$f'_{ci} = 3,500$ psi (24.1 MPa)

f_y for stirrups $= 60,000$ psi (41.8 MPa)

$f_{pu} = 270,000$ psi (1,862 MPa) low-relaxation strands

$f_{ps} = 243,000$ psi (1,675 MPa)

$f_{pe} = 157,500$ psi (1,086 MPa)

$A_{ps} =$ twenty-four $\frac{1}{2}$ in. dia (12.7 mm dia) seven-wire strands

5.2 Find the shear strengths V_c, V_{ci}, and V_{cw} for the beam in Problem 5.1 at 1/10 span intervals along the entire span, and plot the variations in their values along the span in a manner similar to the plot in Figure 5.13.

5.3 Assume that a 4 in. (10 cm) topping of width $b = 8$ ft 6 in. (2.6 m) is situ cast on the precast section of Problem 5.1. If the top surface of the precast section is unroughened, design the necessary dowel reinforcement to ensure full composite action. Use the ACI coefficient of friction for determining the area and spacing of the shear-friction reinforcement, and use $f'_c = 3,000$ psi (20.7 MPa) for the topping. Compare the results with those obtained using the PCI coefficient of friction.

5.4 A 14 in. (35.6 cm) standard PCI double-T simply supported beam is shown in Figure P5.4. It has a span of 40 ft (12.2 m) and is subjected to a service dead load $W_{SD} = 25$ psf (1,197 Pa) plus self-weight $W_D = 31$ psf (1,484 Pa) and a service live load $W_L = 45$ psf (2,155 Pa). Design the web-shear reinforcement at $\frac{1}{2}d_p$ from the support

and at quarter span by (a) the detailed method and (b) the alternative method, and then compare the two designs. The tendon is harped at midspan. Given data are as follows:

f'_c (precast) $= 5,000$ psi, lightweight concrete

$f'_{ci} = 3,500$ psi

f'_c (topping) $= 3,000$ psi, normal weight

$f_{pu} = 270,000$ psi, low-relaxation strand

$f_{ps} = 189,000$ psi (1,303 MPa)

$f_{pe} = 156,000$ psi (1,076 MPa)

Stirrups $f_y = 60,000$ psi (41.8 MPa)

$A_{ps} =$ six $\frac{1}{2}$ in. (12.7 mm) dia seven-wire strands

$e_c = 8.01$ in. (20.3 cm)

$e_e = 4.51$ in. (11.5 cm)

Use $d_p = 10$ in. in the solution. The values of the section properties are as follows:

Section properties	Untopped	Topped
A_c	306 in²	
I_c	4,508 in⁴	7,173 in⁴
c_b	10.51 in.	12.40 in.
c_t	3.49 in.	3.60 in.
S_b	429 in³	578 in³
S^t	1,292 in³	1,992 in³
W_D	31 psf	56 psf

5.5 Design a bracket to support a concentrated factored load $V_u = 125,000$ lb (556 kN) acting at a lever arm $a = 4$ in. (101.6 mm) from the column face. The horizontal factored force $N_{uc} = 40,000$ lb (177.9 kN).

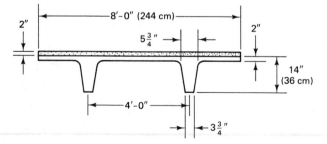

Figure P5.4.

Given data are:

$b = 14$ in. (355.6 mm)

$f'_c = 5,000$ psi (34.47 MPa), normal-
weight concrete

$f_y = 60,000$ psi (413.7 MPa)

Assume that the bracket was cast after the supporting column cured, and that the column surface at the bracket location was not roughened before casting the bracket. Detail the reinforcing arrangements for the bracket.

5.6 Solve Problem 5.5 if the structural system was made from monolithic sand-lightweight concrete in which the corbel or bracket was cast simultaneously with the supporting column.

5.7 Design the transverse and longitudinal reinforcement in Example 5.8 for combined torsion and shear assuming that the L-beam concrete is made of sand-lightweight concrete. Use both methods A and B in your design.

5.8 Design the web reinforcement for the beam in Example 5.8 for combined shear and torsion assuming that the centerline dimensions of the interior floor panels are 30 ft × 56 ft (9.1 m × 17.1 m). The floor is subjected to a service superimposed dead load due to the double T's of $W_{SD} = 77$ psf (3,687 Ma) and a service live load of 60 psf (2,873 MPa). Use both the compression field theory and the modified truss analogy theory, and compare the two solutions.

CHAPTER 6

Indeterminate Prestressed Concrete Structures

6.1 INTRODUCTION

As in reinforced concrete and other structural materials, continuity can be achieved at intermediate supports and knees of portal frames. The reduction of moments and stresses at midspans through the design of continuous systems results in shallower members that are stiffer than simply supported members of equal span and of comparable loading and are of lesser deflection.

Consquently, lighter structures with lighter foundations reduce the cost of materials and construction. In addition, the structural stability and resistance to longitudinal and lateral loads are usually improved. As a result, the span-to-depth ratio is also improved, depending on the type of continuous system being considered. For continous flat plates, a ratio of 40 to 45 is reasonable, while in box girders this ratio can be 25 to 30.

An additional advantage of continuity is the elimination of anchorages at intermediate supports through continuous post-tensioning over several spans, thereby reducing further the cost of materials and labor.

Continuous prestressed concrete is widely applied in the United States in the construction of flat plates for floors and roofs with continuity in one or both directions and with prestressing in one or both directions. Also, continuity is widely used in long-span prestressed concrete bridges, particularly situ-cast post-tensioned spans. Cantilevered box girder bridges, widely used in Europe as segmental bridges, are increasingly being used in the United States for very large spans, and cable-stayed bridges with prestressed decks are increasingly built as well.

Lincoln Executive Plaza, Arlington Heights, Illinois. (*Courtesy*, Prestressed Concrete Institute.)

The success of prestressed concrete construction is largely due to the economy of using precast elements, with the associated high quality control during fabrication. This desirable feature has been widely achieved by imposing continuity on the precast elements through placement of situ-cast reinforced concrete at the intermediate supports. The situ-cast concrete tends to resist the superimposed dead load and the live load that act on the spans after the concrete hardens. Note that forming, shoring, and reshoring can also be avoided in this type of construction, thereby reducing the costs further as compared with the costs of reinforced concrete.

6.2 DISADVANTAGES OF CONTINUITY IN PRESTRESSING

There are several disadvantages to having continuously prestressed elements:

1. Higher frictional losses due to the larger number of bends and longer tendons.
2. Concurrence of moment and shear at the support sections, which reduces the moment strength of those sections.
3. Excessive lateral forces and moments in the supporting columns, particularly if they are rigidly connected to the beams. These forces are caused by the elastic shortening of the long-span beams under prestress.
4. Effects of higher secondary stresses due to shrinkage, creep temperature variations, and settlement of the supports.
5. Secondary moments due to induced reactions at the supporting columns caused by the prestressing force (to be subsequently discussed).
6. Possible serious reversal of moments due to alternate loading of spans.
7. Moment values at the interior supports that require additional reinforcement at these supports, which might otherwise not be needed in simply supported beams.

All these factors can be accounted for through appropriate design and construction of the final system, including special provisions for bearings at the supporting columns.

6.3 TENDON LAYOUT FOR CONTINUOUS BEAMS

The construction system used, the length of the adjacent spans, and the engineering judgment and ingenuity of the design engineer determine the type of layout and method of framing to be used for achieving continuity. Basically, there are two categories of continuity in beams:

1. Monolithic continuity, where all the tendons are generally continuous throughout all or most of the spans and all tendons are prestressed at the site. Such prestressing is accomplished by post-tensioning.
2. Nonmonolithic continuity, where precast elements are used as simple beams on which continuity is imposed at the support sections through situ-cast reinforced

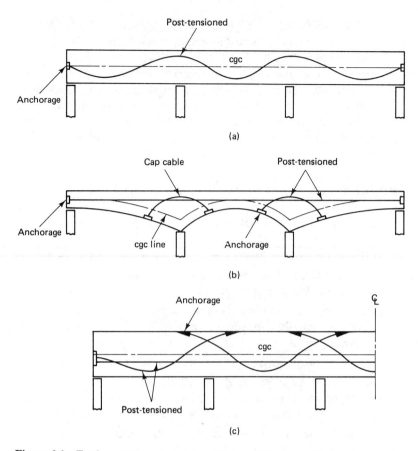

Figure 6.1 Tendon geometry in beams with monolithic continuity. (a) Beam with constant depth. (b) Nonprismatic beam with overlapping tendons. (c) Prismatic beam with overlapping tendons.

concrete which provides the desired level of continuity to resist the superimposed dead load and live load after the concrete hardens.

Figure 6.1 schematically demonstrates the various systems and combinations of systems to achieve monolithic continuity. Figure 6.2 illustrates how continuity is achieved in nonmonolithic construction. Figure 6.1(a) presents a simple continuity system in which all the spans are situ cast and post-tensioning is accomplished after the concrete hardens. Problems are encountered, however, in the accurate evaluation of frictional losses due to the large number of bends. The system shown in Figure 6.1(c), using variable-depth beams, namely, nonprismatic sections, can add to the cost of formwork.

Frictional losses in the post-tensioned straight cables are easier to evaluate accurately. Additional costs are incurred due to the necessity of several anchorages. The system shown in Figure 6.1(a) has the advantage on that of Figure 6.1(b) in that the cost of formwork will in general be less because the continuous beam has a constant depth, although architectural considerations sometimes require nonprismatic continuous sections.

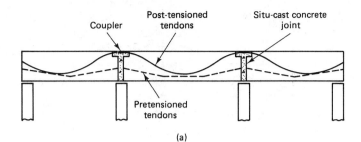

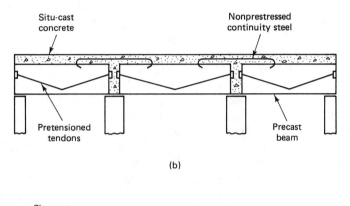

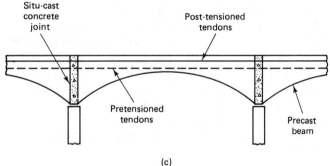

Figure 6.2 Continuity using precast pretensioned beams. (a) Post-tensioned continuity using couplers. (b) Continuity using nonprestressed steel. (c) Continuity in post-tensioning for nonprismatic beams.

Continuity achieved through the use of precast pretensioned beams with situ-cast concrete connecting joints can in many cases be easier to erect, and considerable savings may accrue since formwork and shoring at the site are generally not needed. Figures 6.2(a) and (c) are essentially comparable in the degree of their accuracy in estimating frictional and other losses.

The system illustrated in Figure 6.2(b) is probably the simplest for achieving continuity in prestressed concrete composite construction. The precast pretensioned elements are designed to carry the prestressing and self-weight moments, while the nonprestressed steel at the negative moment region at the support is designed to resist the additional superimposed dead load and the applied live load moments. If design for continuity due to the total dead load is to be achieved, the precast beams have to be shored before placing the composite concrete topping.

In general, precast elements, including those shown in Figures 6.2(a) and (c), are designed to resist their own weight as well as handling and transportation stresses by the strength provided in pretensioning. Post-tensioning or the use of nonprestressed steel at the supports provides the strength required to resist the live load and super-imposed load stresses, and no shoring is used in the construction process.

6.4 ELASTIC ANALYSIS FOR PRESTRESS CONTINUITY

6.4.1 Introduction

Reinforced concrete structures are usually statically indeterminate due to the continuity provided by monolithic construction. Advantageously, the bending moments are always smaller than those of comparable statically determinate beams, leading to shallower, more economical sections. Deformations due to axial loads are usually ignored except in very stiff members, and the settlement of the supports is also rarely considered since creep and shrinkage do not cause major stresses.

In prestressed concrete, continuity also leads to reduced bending moments. However, the bending moments due to the eccentric prestressing forces cause *secondary reactions* and secondary bending moments. These secondary forces and moments increase or decrease the primary effect of the eccentric prestressing forces. Also, the effects of elastic shortening, shrinkage, and creep become considerable as compared to those in reinforced concrete continuous structures.

Because prestressed elements, including those that are partially prestressed, have very limited flexural cracking as compared with reinforced concrete elements, the elastic theory for indeterminate structures can be applied with sufficient accuracy at the limit state of service load. In other words, the prestressed elements can be essentially considered homogeneous elastic material because of the limited cracking level, whereas in reinforced concrete it would not be rational to make such an assumption since flexural cracks start to generate at almost 10–15 percent of the failure load.

6.4.2 Support Displacement Method

Figure 6.3(a) shows a two-span continuous prestressed concrete beam. In part (b), the central support is assumed to have been removed. Because of the induced *secondary force or reaction R* at the internal support caused by the eccentric prestress, the original moments due to prestressing, namely, $M_1 = P_e e_1$, will be called *primary moments*, and the moments M_2 caused by the induced reactions will be called *secondary moments*. The effect of the secondary moment is to shift the location of the line of thrust, the C-line, at the intermediate supports of the continuous structure, and to return the beam section at the support to its original position before prestressing (see Figure 6.3(c)). The line of thrust is the center line of compressive force acting along the beam span. The secondary reaction R causes the camber Δ to be neutralized and the beam to be held down at the intermediate support by an equal but opposite reaction R,

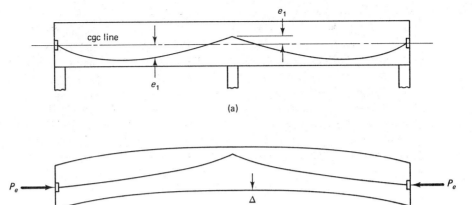

(a)

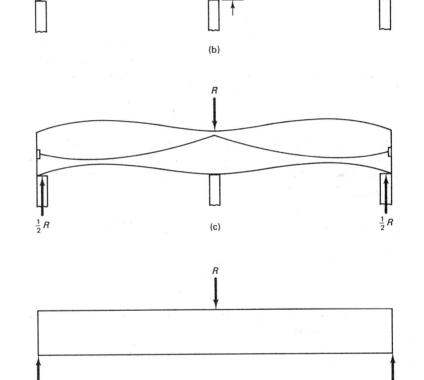

(b)

(c)

(d)

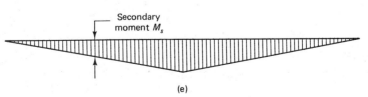

(e)

Figure 6.3 Secondary moments in continuous prestressed beams. (a) Tendon profile prior to prestressing. (b) Profile after prestressing if beam is not restrained by central support. (c) Secondary reaction to eliminate uplift or camber. (d) Reaction R on theoretically simply supported beam. (e) Secondary moment diagram due to R.

provided that the C-line at the intermediate support is above the cgs line. If the two lines coincide, the reaction R will be zero, as explained in Section 6.6.

The primary structure bending moment diagram M_1 due to the prestressing force is shown in Figure 6.4(a). If it is superimposed on the secondary moment diagram M_2 in Figure 6.4(b), a resulting moment diagram $M_3 = (M_1 + M_2)$ (Figure 6.4(c)) is generated due to the prestressing force for the condition where the beam lower fibers just touch the intermediate support, with the thrust line (C-line) moving a distance y from the tendon cgs profile, i.e., the T-line (Figure 6.4(d)). As a sign convention, the bending moments diagrams are drawn on the *tension* side of the beams. Such a convention can help eliminate errors in superposition in the analysis of portal frames and other systems whose vertical members are subjected to moments.

The deviation of the C-line from the cgs line is

$$y = \frac{M_2}{P_e} \qquad (6.1)$$

and the new location of the tendon profile cgs is determined from the net moment $M_3 = M_1 + M_2$ using the appropriate moment sign, positive $(+)$ above and negative $(-)$ below the base line. The resulting limit eccentricity of the C-line is

$$e' = e_3 = \frac{M_3}{P_e} \qquad (6.2)$$

where P_e is the effective prestressing force after losses. Note that e' is negative when the thrust line is *above* the neutral axis, as is the case for the intermediate support section. The concrete fiber stresses due to prestress only at an intermediate support become, from Equations 1.4 a and b,

$$f^t = -\frac{P_e}{A_c}\left(1 + \frac{e'_e c_t}{r^2}\right) \qquad (6.3\ a)$$

and

$$f_b = -\frac{P_e}{A_c}\left(1 - \frac{e'_e c_b}{r^2}\right) \qquad (6.3\ b)$$

The concrete fiber stresses at the support due to prestressing and the self-weight support moment are

$$f^t = -\frac{P_e}{A_c}\left(1 + \frac{e'_e c_t}{r^2}\right) + \frac{M_D}{S^t} \qquad (6.4\ a)$$

and

$$f_b = -\frac{P_e}{A_c}\left(1 - \frac{e'_e c_t}{r^2}\right) - \frac{M_D}{S_b} \qquad (6.4\ b)$$

Alternatively, using the M_3 moment values in Equtions 6.4a and b, the net moment at the section is $M_4 = M_3 - M_D$, and the concrete fiber stresses at the support where the tendons are above the neutral axis are evaluated from

$$f^t = -\frac{P_e}{A_c} - \frac{M_4}{S^t} \qquad (6.5\ a)$$

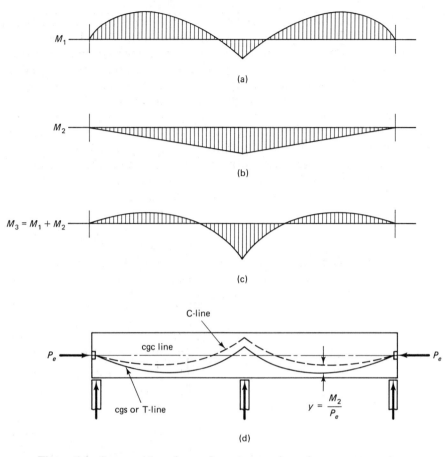

Figure 6.4 Superposition of secondary moments due only to prestress and transformation of the thrust C-line. (a) Primary moments M_1. (b) Secondary moments M_2. (c) Superposition of (b) on (a) to give resulting moment M_3. (d) Transformation of the C-line from the T-line.

and

$$f_b = -\frac{P_e}{A_c} + \frac{M_4}{S_b} \qquad (6.5\text{ b})$$

Both Equations 6.4 and 6.5 should give the same results whether applied to support, midspans sections, or any other sections along the span provided that the appropriate sign convention is maintained.

6.4.3 Equivalent Load Method

The equivalent load method is based on theoretically replacing the effects of the prestressing force by equivalent loads produced by the prestressing moments profile along the span due to the primary moment M_1 in Figure 6.5(b). If the shear diagram causing moments M_1 is constructed as in Figure 6.5(c), and the load producing this

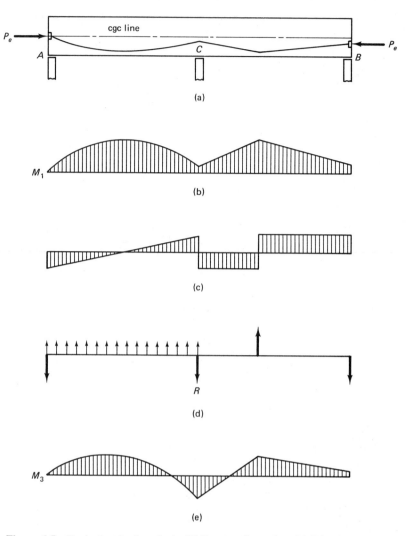

Figure 6.5 Equivalent load method of C-line transformation. (a) Primary structure after prestressing. (b) Primary moment M_1 due to prestressing. (c) Shear diagram for moments M_1. (d) Load causing moment in (b) and shear in (c). (e) Moment diagram for loads in (d) after moment distribution.

shear is evaluated as in Figure 6.5(d), the reaction R is the same as the displacement reaction R in the method described in Section 6.4.1. Calculation of the moment distribution due to the loading on the continuous beam in Figure 6.5(d) produces the moment diagram of moment M_3 in part (e) of the figure. This moment is the same as the net moment M_3 in Section 6.4.1, so that the resulting limit eccentricity of the cgs line will be $e_3 = M_3/P_e$. The prestress interior support reaction R is obtained from Figure 6.5(d) in order to determine the secondary moment M_2 caused by a load R acting at point c of a simple span AB. The deviation of the C-line from the cgs line is then $y = M_2/P_e$, as in the previous method.

6.5 EXAMPLES INVOLVING CONTINUITY

6.5.1 Effect of Continuity on Transformation of C-Line for Draped Tendons

Example 6.1

A bonded post-tensioned prestressed prismatic beam is continuous on three supports. It has two equal spans of 90 ft (27.4 m), and the tendon profile is shown in Figure 6.6. The effective prestressing force P_e after losses is 300,000 lb (1,334 kN). The beam overall dimensions are $b = 12$ in. (30 cm) and $h = 34$ in. (86 cm). Compute the primary and secondary moments due to prestressing, and find the concrete fiber stresses at the intermediate support C due to the prestressing force. Use both the support displacement method and the equivalent load method; assume that the variation in tension force along the beam can be neglected.

Solution (a)

Support Displacement Method. The primary moment M_1 due to prestressing causes upward *camber* or deflection at the intermediate support C. This camber, Δ_c, can be readily obtained from basic mechanics by the moment area method, taking the moments of areas AEC and ADC about point A in Figure 6.7(a) to get the tangential deviation of the elastic curve at c from the horizontal at A. From the figure,

$$EI\Delta_c = \left[(3.0 \times 10^6 + 1.05 \times 10^6)\frac{90 \times 2}{3}\right]\frac{90}{2} \times 144$$
$$- \left(\frac{2.1 \times 10^6 \times 90}{2}\right)\frac{90 \times 2}{3} \times 144$$
$$= 7.58 \times 10^{11} \text{ in}^3\text{-lb}$$

Similarly, from Figure 6.7(c),

$$EI\Delta_c = \frac{45R \times 12 \times 90}{2} \times \frac{90 \times 2}{3} \times 144 = 2.1 \times 10^8 R \text{ in}^3\text{-lb}$$

Equating the right sides of these equations to each other yields

$$7.58 \times 10^{11} = 2.1 \times 10^8 R$$

Then

$$R_c = \frac{7.58 \times 10^{11}}{2.1 \times 10^8} = 3,610 \text{ lb} \downarrow (16 \text{ kN})$$

$$R_A = R_B = 1,805 \text{ lb} \uparrow (8 \text{ kN})$$

The secondary moment M_2 due to concentrated load R_c varies linearly from interior support C to end supports A and B in Figures 6.6(d) and 6.7(c). From Figure 6.6(c),

$$M_2 = \frac{R_c}{2} \times 90 \times 12 = \frac{3,610}{2} \times 90 \times 12 = 1.95 \times 10^6 \text{ in.-lb}$$

The total moment M_3 at C due to prestress continuity is

$$M_1 + M_2 = 2.1 \times 10^6 + 1.95 \times 10^6 = 4.05 \times 10^6 \text{ in.-lb}$$

From Equation 6.1, the distance through which the C-line has to be transformed upwards

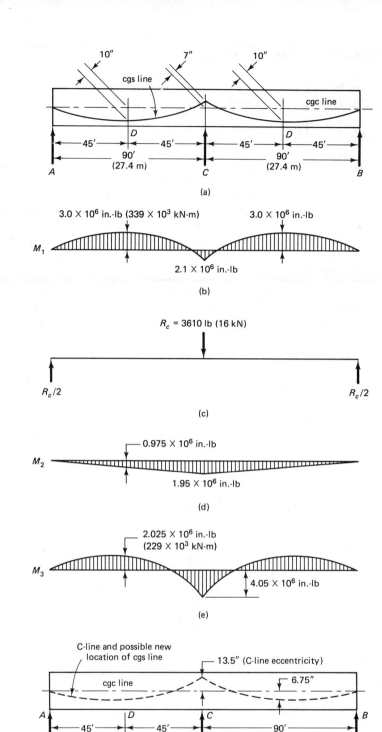

Figure 6.6 Transformation of thrust line in Example 6.1 due to continuity. (a) Tendon geometry: one possible location. (b) Primary moment M_1 due to prestress P_e. (c) Reaction R on theoretically simple beam. (d) Secondary moment M_2 due to R. (e) Final moment $M_3 = M_1 + M_2$. (f) New location of C-line and possible cgs line.

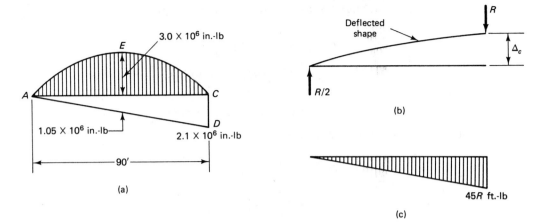

Figure 6.7 Camber Δ_c due to M_1. (a) Primary moment M_1. (b) Deflected shape due to R. (c) Secondary moment M_2 due to R.

at support C is

$$y_c = \frac{M_2}{P_e} = \frac{1.95 \times 10^6}{300,000} = 6.5 \text{ in. (16.5 cm)}$$

From Equation 6.2, the distance of the C-line above the cgc line, i.e., the eccentricity of the C-line above the cgc line at interior support C, is

$$e_c = \frac{M_3}{P_e} = \frac{4.05 \times 10^6}{300,000} = 13.5 \text{ in. (34.3 cm)}$$

The midspan total moment is

$$M_3 = 3.0 \times 10^6 - \frac{1}{2} \times 1.95 \times 10^6$$

$$= 2.025 \times 10^6 \text{ in.-lb}$$

and the C-line eccentricity is

$$e_D = \frac{2.025 \times 10^6}{300,000} = 6.75 \text{ in. (17.1 cm)}$$

Concrete Fiber Stresses at Interior Support C Due to Prestress Only

$$c_t = c_b = \frac{34}{2} = 17 \text{ in.}$$

$$e_c = 13.5 \text{ in.}$$

$$A_c = bh = 12 \times 34 = 408 \text{ in}^2$$

$$I_c = \frac{bh^3}{12} = \frac{12(34)^3}{12} = 39,304 \text{ in}^4$$

$$r^2 = \frac{I_c}{A_c} = \frac{39,304}{408} = 96.33 \text{ in}^2$$

From Equations 4.2 a and b, the top concrete fiber stress is

$$f^t = -\frac{P_e}{A_c}\left(1 + \frac{ec_t}{r^2}\right)$$

$$= -\frac{300{,}000}{408}\left(1 + \frac{13.5 \times 17}{96.33}\right)$$

$$= -2{,}487 \text{ psi } (C) \ (17.1 \text{ MPa})$$

and the bottom concrete fiber stress is

$$f_b = -\frac{P_e}{A_c}\left(1 - \frac{ec_b}{r^2}\right)$$

$$= -\frac{300{,}000}{408}\left(1 - \frac{13.5 \times 17}{96.33}\right)$$

$$= +1{,}016 \text{ psi } (T) \ (7.0 \text{ MPa})$$

Although the bottom fiber stress in tension is high and well beyond the maximum allowable, it is only due to prestress. Once self-weight is considered, it will diminish considerably.

Solution (b)

Equivalent Load Method.   From Equation 1.16 for load balancing,

$$W_b = \frac{8Pa}{\ell^2}$$

where a is the eccentricity of the tendon from the cgc line. So

$$W_b = \frac{8 \times 300{,}000 \times 13.5}{(90)^2 \times 12} = 333.3 \text{ lb/ft}$$

$$\text{FEM} = \frac{W_b \ell^2}{12} = \frac{333.3(90)^2}{12} = 224{,}978 \text{ ft-lb}$$

From the moment distribution operation in Figure 6.8, the final moment at the interior support C is $M_3 = M_1 + M_2 = 337{,}467$ ft-lb $= 4.05 \times 10^6$ in.-lb, which is the same value as solution (a).

Since the M_1 diagram is the primary moment diagram as in Figure 6.6(b), the diagram for the secondary moment M_2 can be constructed from $M_3 - M_1$ as shown in Figure 6.8(b), which is identical to Figure 6.6(d). Thereafter, all other steps for calculation of fiber stresses and location of both fiber stresses and the C-line are identical and give the same results as solution (a).

6.5.2 Effect of Continuity on Transformation of C-line for Harped Tendons

Example 6.2

Solve Example 6.1 assuming that the prestressing tendon is harped at the midspan of both adjacent spans. Use the support displacement method in your solution.

Solution   Construct the primary and secondary moment diagram shown in Figure 6.9.

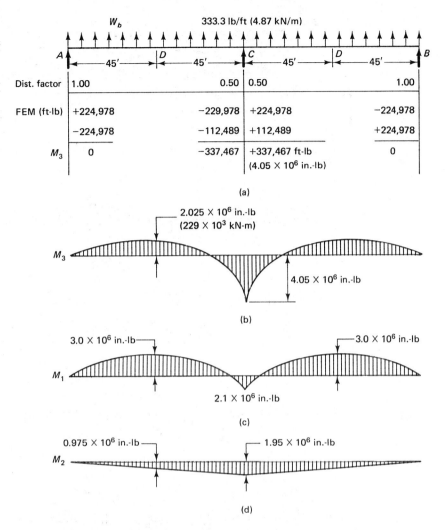

Figure 6.8 Equivalent load method for continuous beam analysis. (a) Equivalent load and moment distribution. (b) Total moment M_3. (c) Primary moment M_1. (d) Secondary moment M_2.

Then

$$EI\Delta_c = (3.0 \times 10^6 + 1.05 \times 10^6)90 \times \frac{1}{2} \times \frac{90}{2} \times 144 - \frac{2.1 \times 10^6 \times 90}{2}$$

$$\times \frac{90 \times 2}{3} \times 144 = 3.65 \times 10^{11} \text{ in}^3\text{-lb}$$

From Figure 6.7(c),

$$EI\Delta_c = 2.1 \times 10^8 R_c$$

$$R_c = \frac{3.65 \times 10^{11}}{2.1 \times 10^8} = 1,738 \text{ lb} \downarrow$$

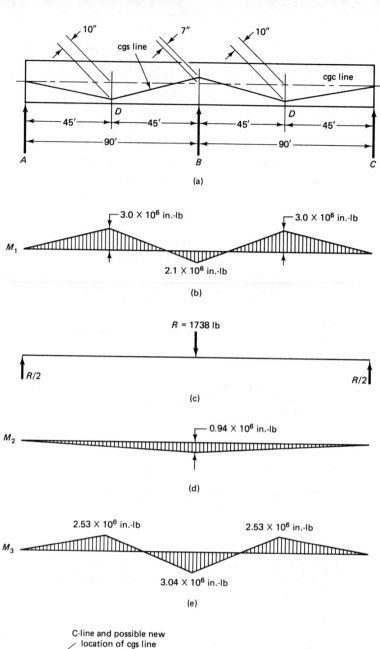

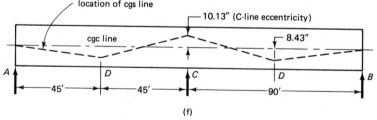

Figure 6.9 Transformation of thrust line in Example 6.2 due to continuity. (a) Tendon geometry: one possible location. (b) Primary moment M_1 due to prestress P_e. (c) Reaction R on theoretically simple beam. (d) Secondary moment M_2 due to R. (e) Final moment $M_3 = M_1 + M_2$. (f) New location of C-line and possible cgs line.

The secondary moment ordinate at interior support C is

$$M_2 = \frac{R}{2} \times 90 \times 12 = \frac{1,738}{2} \times 90 \times 12 = 0.94 \times 10^6 \text{ in.-lb}$$

Thus, the total moment $M_3 = 2.1 \times 10^6 + 0.94 = 3.04 \times 10^6$ in.-lb (344 × 10^3 kN-m).

From Equation 6.1, the distance through which the C-line has to be transformed upwards at support C is

$$y_c = \frac{M_2}{P_e} = \frac{0.94 \times 10^6}{300,000} = 3.13 \text{ in. (8 cm)}$$

From Equation 6.2, the distance of the C-line above the cgc line, i.e., the eccentricity of the C-line above the cgc line at interior support C, is

$$e_c = \frac{M_3}{P_e} = \frac{3.04 \times 10^6}{300,000} = 10.13 \text{ in. (25.7 cm)}$$

So the midspan total moment is

$$M_3 = 3.0 \times 10^6 - \frac{0.94 \times 10^6}{2} = 2.53 \times 10^6 \text{ in.-lb}$$

and the C-line eccentricity is

$$e_D = \frac{2.53 \times 10^6}{300,000} = 8.43 \text{ in. (21.4 cm)}$$

Concrete Fiber Stresses at Interior Support C Due to Prestress Only. $e_c = 10.13$ in. for the C-line or thrust line. So the top concrete fiber stress is

$$f^t = -\frac{P_e}{A_c}\left(1 + \frac{ec_t}{r^2}\right)$$

$$= -\frac{300,000}{408}\left(1 + \frac{10.13 \times 17}{96.33}\right)$$

$$= -2,049 \text{ psi } (C) \text{ (14.1 MPa)}$$

and the bottom concrete fiber stress is

$$f_b = -\frac{P_e}{A_c}\left(1 - \frac{ec_b}{r^2}\right)$$

$$= -\frac{300,000}{408}\left(1 - \frac{10.13 \times 17}{96.33}\right)$$

$$= +579 \text{ psi } (T) \text{ (4.0 MPa)}$$

Comparing the results of the harped tendon case of this example to the draped parabolic tendon of Example 6.1 reveals that smaller total continuity moments M_3 resulted at the intermediate support since the triangular area of the moment diagram is one-half the product of the span and the moment ordinate while the parabolic area is two-thirds of that same product. Consequently, the concrete fiber stresses are lower.

6.6 LINEAR TRANSFORMATION AND CONCORDANCE OF TENDONS

It can be recognized from the discussion in Section 6.4 that the profile of the line of thrust (the C-line) follows the profile of the prestressing tendon (the cgs line). This is to be expected since the C-line ordinates are the moment ordinates resulting from the product of the prestressing force P_e and the tendon eccentricity from the cgc line varying along the span. The deflection behavior of any beam is a function of the variation in moment along the span, and the shape of the moment diagram is a function of the type of load, namely, concentrated or distributed. Consequently, tendon profiles are often draped for distributed loads, while they are harped for concentrated loads, as illustrated in Chapters 1 and 4.

Examples 6.1 and 6.2 show that in a continuous beam, the deviation of the C-line from the cgc line is directly proportional to the secondary moment M_2. Since the M_2 diagram varies *linearly* with the distance from the support, it is possible to *linearly* transform the C-line by raising or lowering its position at the *interior* support while preferably maintaining its original position at the exterior simple supports. The profile of the C-line remains the same because of the linearity of the transformation. Consequently, it is possible to linearly transform the cgs line, i.e., the prestressing tendon profile along the beam span, without changing the profile positions of the C-line. This flexibility has major practical significance in the design of continuous prestressed concrete beams.

Example 6.3 points up the said flexibility. Compare the tendon profile it presents with that of Example 6.1, and note that the C-line in Figure 6.6(f) is the same as the C-line in Figures 6.10(f) and 6.11(f), although the cgs line locations along the span are not the same as those in Figures 6.6(a), 6.10(a), and 6.11(a). Also, note that in solution b, where the tendon profile coincides with the C-line profile, the reaction $R = 0$ and the secondary moment $M_2 = 0$. This means that when the cgs line *coincides* with C-line, the beam just touches the intermediate support and behaves like a simply supported beam. Such a beam is called a *concordant beam,* and the prestressing tendon is called a *concordant tendon.*

6.6.1 Verification of Tendon Linear Transformation Theorem

Example 6.3

The continuous beam of Example 6.1 has a new tendon profile as shown in

    (a) Figure 6.10(a) with eccentricities $e_C = 0$ and an eccentricity $e_D = 13.5$ in. (24.3 cm) at midspan similar to the intermediate support eccentricity in Example 6.1, and

    (b) Figure 6.11(a) with eccentricities $e_C = 13.5$ in. (24.3 cm) and $e_D = 6.75$ in. (17.1 cm) that are the same C-line eccentricities as in Example 6.1.

Verify that the profile and alignment of the C-line in (a) and (b) are the same as the C-line properties in Example 6.1 and Figure 6.6(f). The total moment at the intermediate

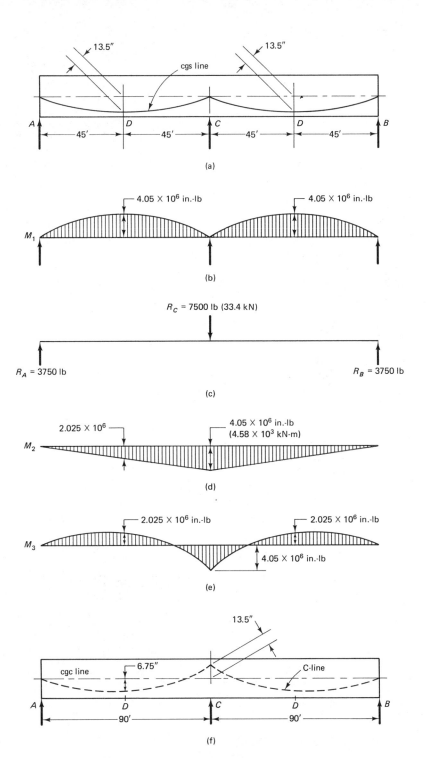

Figure 6.10 Tendon transformation in Example 6.3 (alternative a).

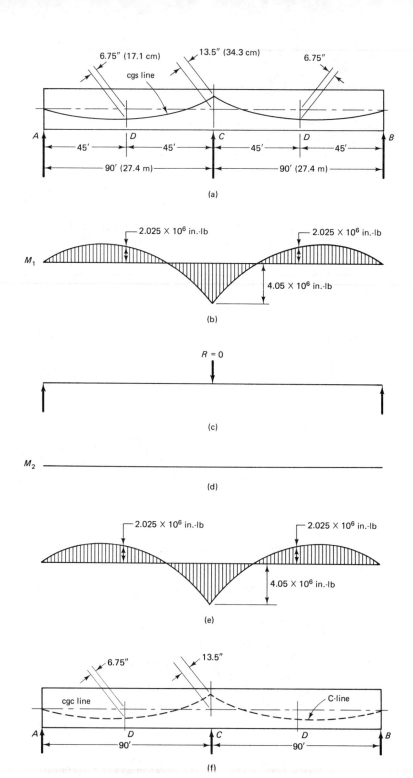

Figure 6.11 Tendon transformation in Example 6.3 (alternative b).

support C in both cases is $M_3 = 4.05 \times 10^6$ in.-lb, and at midspan D the moment is $M_3 = 2.025 \times 10^6$ in.-lb, as in Example 6.1.

Solution (a) The secondary moment is

$$M_2 = 4.05 \times 10^6 \text{ in.-lb } (458 \times 10^3 \text{ kN-m})$$

Also,

$$\frac{R_c}{2} \times 90 \times 12 = 4.05 \times 10^6$$

$$R_c = \frac{4.05 \times 10^6 \times 2}{90 \times 12} = 7{,}500 \text{ lb } (33.4 \text{ kN}) \downarrow$$

$$R_A = R_B = 3{,}750 \text{ lb } (16.7 \text{ kN}) \uparrow$$

Solution (b) Both $M_2 = 0$ and $R = 0$.

It can be seen from both solution (a) in Figure 6.10 and solution (b) in Figure 6.11 that the C-line coordinates are the same and close to the values obtained in Example 6.1. The fiber stresses are also the same, viz., $f^t = -2{,}487$ psi (C) (17.1 MPa) and $f_b = +1{,}016$ psi (T) (7.0 MPa).

6.6.2 Concordance Hypotheses

The following list summarizes the hypotheses defining the transformation and concordance of tendons in continuous prestressed beams:

1. Any tendon profile can be linearly transformed without affecting the C-line position.

2. A beam with a concordant tendon is a continuous beam whose C-line coincides with its cgs line.

3. A concordant tendon induces no reactions on intermediate supports.

4. The eccentricity of any concordant tendon measured from the cgc line produces a moment diagram representing a profile similar in form to the moment profile due to the superimposed load.

5. Any line of thrust (C-line) is a profile for a concordant tendon.

6. Superposition of several concordant tendons produces a concordant tendon, but superposition of concordant and nonconcordant tendons produces a nonconcordant tendon.

7. A change in eccentricity at one or both *end* supports results in a shift of the C-line, but a change in eccentricity at *intermediate* supports does not affect the position of the C-line.

8. The choice of concordance or nonconcordance is determined by concrete cover and efficiency in beam depth selection. Bending moments and shear diagrams for superposition of transverse loads on continuous beams is shown in Figure 6.12.

It is advisable to start a design assuming concordance in order to eliminate the need for calculating the secondary moment M_2. By trial and adjustment, one can arrive at the final beam section depth that fulfills the design requirements with the cgs location either concordant or nonconcordant, as the final design dictates.

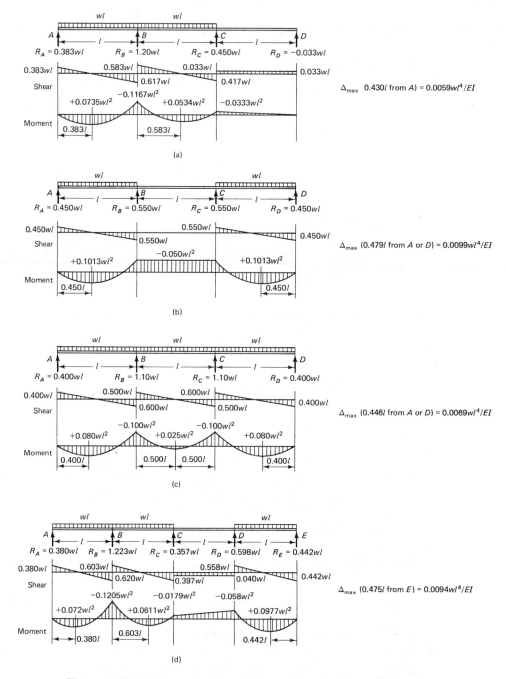

Figure 6.12 Bending moments and shear diagrams for continuous structures. (a) Continuous beam, three equal spans, one end span unloaded. (b) Continuous beam, three equal spans, end spans loaded. (c) Continuous beam, three equal spans, all spans loaded. (d) Continuous beam, four equal spans, third span unloaded. (e) Continuous beam, four equal spans, first and third spans loaded. (f) Continuous beam, four equal spans, all spans loaded. (g) Continuous beam, two equal spans, concentrated load at center of one span. (h) Continuous beam, two equal spans, concentrated load at any point.

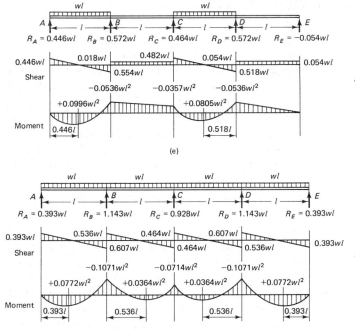

$$\Delta_{max}\ (0.477l \text{ from } A) = 0.0097wl^4/EI$$

(e)

$$\Delta_{max}\ (0.440l \text{ from } A \text{ and } E) = 0.0065wl^4/EI$$

(f)

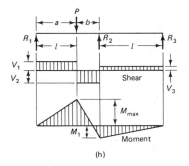

(g)

Total equivalent uniform load $= \dfrac{13}{8}P$

$R_1 = V_1 \qquad\qquad = \dfrac{13}{32}P$

$R_2 = V_2 + V_3 \qquad = \dfrac{11}{16}P$

$R_3 = V_3 \qquad\qquad = -\dfrac{3}{32}P$

$V_2 \qquad\qquad\qquad = \dfrac{19}{32}P$

M_{max} (at point of load) $= \dfrac{13}{64}Pl$

M_1 (at support R_2) $\quad -\dfrac{3}{32}Pl$

$\Delta_{max}\ (0.480l \text{ from } R_1) = 0.015Pl^3/EI$

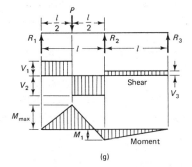

(h)

$R_1 = V_1 \qquad\qquad = \dfrac{Pb}{4l^3}[4l^2 - a(l+a)]$

$R_2 = V_2 + V_3 \qquad = \dfrac{Pa}{2l^3}[2l^2 + b(l+a)]$

$R_3 = V_3 \qquad\qquad = -\dfrac{Pab}{4l^3}(l+a)$

$V_2 \qquad\qquad\qquad = \dfrac{Pa}{4l^3}[4l^2 + b(l+a)]$

M_{max} (at point of load) $= \dfrac{Pab}{4l^3}[4l^2 - a(l+a)]$

M_1 (at support R_2) $\quad = \dfrac{F}{4l^2}$

Figure 6.12 *(continued)*

6.7 ULTIMATE STRENGTH AND LIMIT STATE AT FAILURE OF CONTINUOUS BEAMS

The service-load design of continuous prestressed beams assumes *elastic* behavior of the material up to the limit of allowable tensile stress in the concrete due to all loads. This limit in tensile stress is based on allowing some limited cracking beyond the first cracking load as determined by the modulus of rupture of concrete. As cracking becomes more effective during overload conditions, internal plastic deformation at the critical regions of maximum or peak moments and plastic redistribution of elastic moments from the negative to the positive moment regions are generated. At this stage of overload and beyond, up to the limit state at failure, plastic hinging develops at the most highly stressed regions in the continuous beam.

Total redistribution and full development of plastic hinges at the continuous supports of a fully bonded prestressed beam render the beam statically determinate, as if concordance of tendons were present with zero moments at the supports. Theoretically, in such cases the secondary moments M_2 can be disregarded beyond the first cracking load. However, such an assumption can result in an unsafe design unless a concordant tendon is assumed from the beginning and is executed in the final design with no overload conditions permitted. Otherwise, it is important to take into consideration the secondary moment M_2 due to prestressing up to the limit state of the failure load, but with a load factor of 1.0. Cosideration of the secondary moments is mandated by the fact that the elastic deformation caused by the nonconcordant tendons changes the amount of *inelastic* rotation required to obtain a given amount of redistribution. Conversely, for a beam with a given elastic rotational capacity, the amount by which the moment at the support may be varied is changed by an amount equal to the secondary moment at the support due to prestressing (Ref. 6.2).

In order to determine the moments to be used in the design, the following sequence of steps is recommended:

1. Determine the moments due to the dead and live loads at factored load level.
2. Establish the percentage of redistribution to be allowed, if any, and adjust the

Seven Mile Bridge, Florida Keys. (*Courtesy*, Post-Tensioning Institute.)

support moment values by increasing or decreasing them by the percentage established or a lower percentage if desirable.

3. Combine the adjusted factored moments from step 2 with the values of the elastic secondary moment M_2 due to prestressing. A positive secondary moment at the support caused by transforming a tendon downwards from a concordant profile will therefore *reduce* the negative moments near the supports and *increase* the positive moment in the midspan region. A tendon that is transformed upwards will have the opposite effect.

The ACI code permits an increase or decrease in negative moments calculated by the elastic theory for any assumed loading arrangements up to a limit defined by the percentage

$$p_d = 20 \left[1 - \frac{\omega_p + \dfrac{d}{d_p}(\omega - \omega')}{0.36\beta_1} \right] \tag{6.6}$$

where $\omega_p = \rho_p f_{ps}/f'_c$
$\quad\quad \omega' = \rho' f_y/f'_c$
$\quad\quad \omega = \rho f_y/f'_c$
$\quad$ for $\rho_p = A_p/bd_p$
$\quad$ and $\rho' = A'_s/bd$.

The modified negative moments are to be used for calculating moments at sections within spans for the same loading arrangements.

The ACI Code also requires that redistribution of moments be made only when the section at which the moment is reduced is so designed that whichever reinforcement index ω applies is $\leq 0.24\beta_1$, i.e., either

$$\omega_p \leq 0.24\beta_1 \tag{6.7 a}$$

or

$$\left[\omega_p + \frac{d}{d_p}(\omega - \omega') \right] \leq 0.24\beta_1 \tag{6.7 b}$$

or

$$\left[\omega_{pw} + \frac{d}{d_p}(\omega_w - \omega'_w) \right] \leq 0.24\beta_1 \tag{6.7 c}$$

6.8 TENDON PROFILE ENVELOPE AND MODIFICATIONS

The envelope for limiting tendon eccentricities for continuous beams can be constructed in the same manner as discussed in Section 4.4.3 for simply supported beams. A determination has to be made as to whether tension is to be allowed in the design in order to establish the limiting maximum and minimum ordinates of the upper and lower envelopes relative to the top and bottom kerns.

The tendon profile at the intermediate supports cannot have the same peak configuration as that of the negative bending moment diagram, due to both design and practical considerations. These considerations include the magnitude of frictional losses that increase with the decrease in the radius of curvature, the high level of compressive stress concentration in cases of abrupt changes in the tendon, and the additional difficulties that can be encountered in post-tensioning. Consequently, it is advisable to modify the tendon profile at the support so as to have a curvilinear transition at the support zone. Such modification has to be accounted for by modifying the primary moment M_1 diagram and the total moment M_3 diagram.

Typical tendon alternative profiles with equal upper and lower eccentricities at the peak moment sections are shown in Figure 6.13, where the chain-dotted line gives the average bending moments used in the service-load design. Selection of the tendon profile should be based on the following considerations:

1. The eccentricity should be as large as possible at the point where the largest bending moment develops at the limit state at failure.

2. Where possible, the total prestressing moment at any section should be sufficient to counteract the average service-load bending moment at that section. The test for this condition should be based on the resulting stress values rather than moment values, since the prestressing force causes axial load stress as well as bending moment stress.

3. A tendon profile alternative that produces the *least* frictional losses should be chosen in the design.

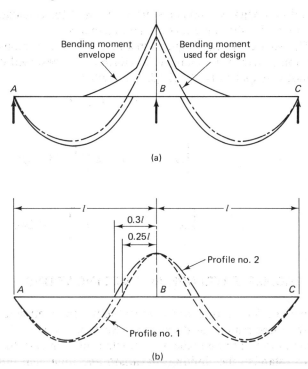

(a)

(b)

Figure 6.13 Tendon profile modification. (a) Bending moment diagram for continuous beam. (b) Tendon profile alternatives.

4. It is essential to consider the ultimate-load requirements when selecting the tendon profile; a tendon chosen only on the basis of linear transformation and concordance is not necessarily satisfactory, as it might neither totally fulfill the service-load stress requirements nor fully satisfy the ultimate-load requirements at both the midspan and interior supports.

5. A decrease in eccentricities at the intermediate supports through additional tendon transformation decreases the service-load compressive stresses and can result in allowing additional live-load moments, and hence more live load, *provided* that the midspan profile eccentricity and moments produce satisfactory concrete stresses.

6.9 TENDON AND C-LINE LOCATION IN CONTINUOUS BEAMS

Example 6.4

Design the tendon profile in the beam of Example 4.7, assuming the unshored beam to be continuous over three spans 64 ft (19.5 m) each, with only uniformly distributed live load $w_L = 1,514$ plf (22.1 kN/m). Consider the post-tensioned prestressing tendon to be continuous throughout the structure and fully grouted. Disregard tension force variation due to frictional losses in the bends, and assume that a maximum allowable concrete compressive fiber stress $f_c = 0.45f'_c = 2,250$ psi (15.5 MPa) is reached at the extreme top fibers of the composite section when the live load acts on the section. Use a modified effective width $b_m = 65$ in. (165 cm) for the compression flange to account for the modular ratio of the topping and precast concrete.

Solution

1. *Input data from example 4.7*

   (a) *Stress data*

$$\text{Precast} \quad f'_c = 5,000 \text{ psi (34.5 MPa), normal-weight concrete}$$
$$\text{Topping } f'_c = 3,000 \text{ psi (20.7 MPa), normal-weight concrete}$$
$$f_{ci} = 4,000 \text{ psi (27.6 MPa)}$$
$$f_{pu} = 270,000 \text{ psi (1,862 MPa)}$$
$$f_{ps} = 240,000 \text{ psi (1,655 MPa)}$$
$$f_{pi} = 189,000 \text{ psi (1,303 MPa)}$$
$$f_{pe} = 0.8(0.7f_{pu}) = 151,200 \text{ psi (1,043 MPa)}$$
$$\gamma = 0.8$$
$$\text{Midspan } f_t = 12\sqrt{f'_c} = 12\sqrt{5,000} = 849 \text{ psi (5.85 MPa)}$$
$$\text{Support } f_t = 6\sqrt{f'_c} = 425 \text{ psi (2.93 MPa)}$$

   (b) *Load data*

$$W_D = 583 \text{ plf (8.5 kN/m)}$$
$$W_{SD} = 1\tfrac{3}{4}/12 \times 7 \text{ ft} \times 150 = 153 \text{ plf (2.2 kN/m) for } 1\tfrac{3}{4} \text{ in. precast}$$
$$\text{slab formwork}$$

$$W_{CSD} = 4/12 \times 7 \text{ ft} \times 150 = 350 \text{ plf (5.1 kN/m) for 4 in. situ-cast topping}$$

$$W_L = 1,514 \text{ plf (22.1 kN/m) (obtained from } M_L = 9,300,000 \text{ in.-lb)}$$

$$\text{Span} = 64 \text{ ft (19.5 m)}$$

(c) *Section properties*

AASHTO Type III

Property	Precast	Composite
A_c, in²	560	994
I_c, in⁴	125,390	297,045
r^2, in²	223.9	318.1
c_b, in.	20.27	31.32
c_t, in.	24.73	13.86
S_b, in³	6,186	9,490
S^t, in³	5,070	21,714
S_{cbs}, in³ (bottom of slab)		19,251
S'_{cs}, in³ (top of slab)		15,288

$$n = E_c(\text{topping})/E_c(\text{precast}) = 0.77$$

Precast $h = 45$ in. (114.3 cm)

Transformed $b = 65$ in. (165 cm)

Situ-cast $h_f = 5.75$ in. (14.6 cm)

(d) *Prestressing steel*

Twenty-two $\frac{1}{2}$ in. dia (12.7 mm dia) seven-wire low-relaxation strands
$A_{ps} = 22 \times 0.153 = 3.366$ in² (21.7 cm²)

2. *Assume a trial tendon profile location as shown in Figure 6.14*
The prestressing force after losses is

$$P_e = 3.366 \times 151,200 = 508,940 \text{ lb (2,264 kN)}$$

while the primary moment at support B is

$$M_B = P_e \times e_B = 508,940 \times 15 = 7.63 \times 10^6 \text{ in.-lb (862} \times 10^3 \text{ kN-m)}$$

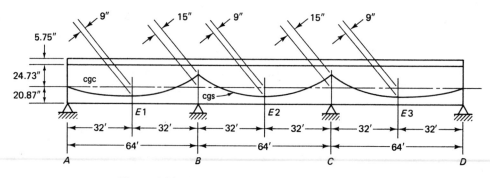

Figure 6.14 Trial profile of the prestressing tendon.

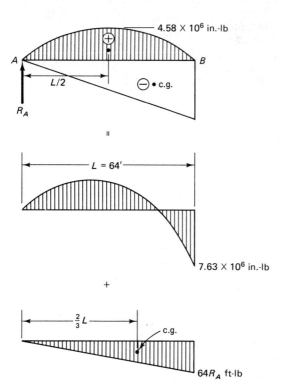

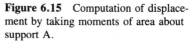

Figure 6.15 Computation of displacement by taking moments of area about support A.

The primary moment at midspan E_1 is

$$M_{E1} = 508,940 \times 9 = 4.58 \times 10^6 \text{ in.-lb}$$

Taking moments of areas about A in Figure 6.15 yields

$$E_c I_c \Delta_B = E_c I_c \Delta_c = \left[\left(\frac{7.63}{2} + 4.58 \right) 10^6 \times \frac{64 \times 2}{3} \right] \times \frac{64}{2} \times 144$$

$$- \left[\frac{7.63 \times 10^6 \times 64}{2} \right] \times \frac{64 \times 2}{3} \times 144$$

$$= 15.0 \times 10^{10} \text{ in.-lb}$$

Also,

$$E_c I_c \Delta_B = (R_A \times 64 \times 12) \left(\frac{64 \times 12}{2} \right) \left(\frac{64 \times 12 \times 2}{3} \right)$$

$$= 151 \times 10^6 R_B \text{ in-lb}$$

$$R_A = R_B = \frac{15.0 \times 10^{10}}{151 \times 10^6} = 994 \text{ lb} \downarrow$$

The secondary moment M_2 at support B is

$$R_B \times 64 \times 12 = 995 \times 64 \times 12 = 0.76 \times 10^6 \text{ in.-lb}$$

The total moment at support B due to prestressing only is

$$\text{Support } M_3 = (7.63 + 0.76) \times 10^6 = 8.39 \times 10^6 \text{ in.-lb (see Figure 6.16)}$$

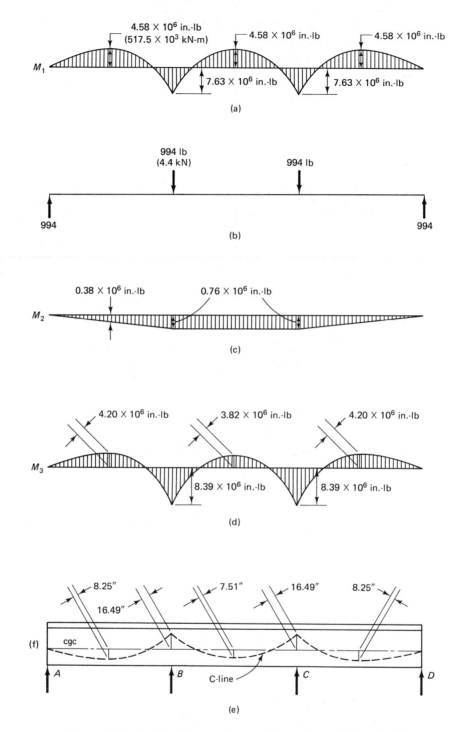

Figure 6.16 Tendon C-line profile in continuous beam of Example 6.4.

The C-line eccentricity is

$$e_B' = -\frac{8.39 \times 10^6}{508{,}940} = -16.49 \text{ in. (41.9 cm)}$$

(The eccentricity e' of the C-line when it is above the neutral axis is considered negative in order to conform to Equations 6.3 a and b.) Finally, the total moment at midspans E_1, E_2, and E_3 due to prestressing only is

$$\text{Midspan } M_3 = (4.58 - 0.38)10^6 = 4.20 \times 10^6 \text{ in.-lb}$$

and the C-line eccentricity is

$$e_{E1}' = +\frac{4.20 \times 10^6}{508{,}940} = +8.25 \text{ in. (21.0 cm)}$$

3. *Concrete fiber stresses due to prestress and self-weight (583 plf)*
   Using the moment factors and reaction factors from Figure 6.12, we have

$$M_D \text{ at } B = M_{B1} = 0.10 \times 583(64)^2 \times 12 = 2.87 \times 10^6 \text{ in.-lb}$$

$$\text{(see Figure 6.17)}$$

$$P_e = A_{ps} f_{pe} = 3.366 \times 151{,}200 = 508{,}940 \text{ lb}$$

M_D at $E_1 = 2.12 \times 10^6$ in.-lb (Max $+M$ is not at midspan; hence, calculate M_{E1} from the area of the shear diagram)

$$\text{Net } R_B = 1.1W\ell \uparrow - 994 = 1.1 \times 583 \times 64 - 994 = 40{,}048 \text{ lb}$$

The total support B moment due to prestressing and self-weight is, then,

$$M_4 = M_3 - 2.87 \times 10^6 = (8.39 - 2.87)10^6$$
$$= 5.52 \times 10^6 \text{ in.-lb (624} \times 10^3 \text{ kN-m)}$$

(i) *Support section B or C*
   The construction process in this stage involves mounting the precast I-beams and prestressing them. The fiber stresses due to prestressing and self-weight, from Equations 6.5 a and b, are as follows:

$$f_1^t = -\frac{P_e}{A_c} - \frac{M_4}{S^t} = -\frac{508{,}940}{560} - \frac{5.52 \times 10^6}{5{,}070} = -908.8 - 1{,}008.8$$

$$= -1{,}917.6 \text{ psi } (C)(13.2 \text{ MPa}) < 2{,}250 \text{ psi, O.K.}$$

$$f_{1b} = -\frac{P_e}{A_c} + \frac{M_4}{S_b} = -\frac{508{,}940}{560} + \frac{5.52 \times 10^6}{6186} = -908.8 + 892.3$$

$$= -16.5 \text{ psi } (C), \text{ no tension, O.K.}$$

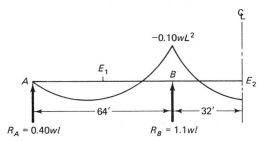

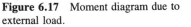

Figure 6.17 Moment diagram due to external load.

Alternatively, the concrete fiber stresses at the support can be computed from Equations 4.1 a and b or Equations 6.4 a and b using the C-line eccentricities $e'_e = -16.49$ in. We obtain

$$f'_1 = -\frac{P_e}{A_c}\left(1 + \frac{e'_e c_t}{r^2}\right) + \frac{M_D}{S^t}$$

$$= -\frac{508,940}{560}\left(1 + \frac{16.49 \times 24.73}{223.9}\right) + \frac{2.87 \times 10^6}{5,070}$$

$$f'_1 = -2,564.1 + 566.1 = -1,917.9 \text{ psi } (C) \text{ (13.8 MPa)}$$

$$f_{1b} = -\frac{P_e}{A_c}\left(1 - \frac{e'_e c_b}{r^2}\right) - \frac{M_D}{S_b}$$

$$= -\frac{508,940}{560}\left(1 - \frac{16.49 \times 20.27}{223.91}\right) - \frac{2.87 \times 10^6}{6,186}$$

$$= +447.9 - 463.9 = -16.0 \text{ psi } (C)$$

(ii) *Outer span midspan E_1*

The eccentricity at the midspan is $e'_c = +8.25$ in. Hence,

$$f'_1 = -\frac{P_e}{A_c} + \frac{M_4}{S^t}$$

$$f_{1b} = -\frac{P_e}{A_c} - \frac{M_4}{S_b}$$

Total moment M_4 at $E_1 = (4.20 - 2.12)10^6 = 2.08 \times 10^6$ in.-lb

$$f'_1 = -\frac{P_e}{A_c} + \frac{M_4}{S^t} = -\frac{508,940}{560} + \frac{2.08 \times 10^6}{5,070}$$

$$= -908.8 + 410.3 = -498.5 \text{ psi } (C) \text{ (3.4 MPa), O.K.}$$

$$f_{1b} = -908.8 - \frac{2.08 \times 10^6}{6,186} = -1,245 \text{ psi } (C) \text{ (8.6 MPa), O.K.}$$

Figure 6.16 gives the moments, eccentricities, and C-line geometry of the continuous beam in this example.

4. *Effect of adding the superimposed dead loads on support sections B and C*
At this stage of the construction process the $1\frac{3}{4}$ in. thick precast slabs (W_{SD}) are erected, followed by placing the 4 in. thick layer of wet concrete (W_{SD}). Thus, $W_{SD} = 153 + 350 = 503$ plf (7.3 kN/m). The support moment due to superimposed load is

$$M_{B2} = 0.1w\ell^2 = 0.1 \times 503(64)^2 \times 12 = -2.47 \times 10^6 \text{ in.-lb}$$

Also,

$$f'_2 = -\frac{P_e}{A_c} - \frac{M_5}{S^t}$$

$$f_{b2} = -\frac{P_e}{A_c} + \frac{M_5}{S_b}$$

$$M_5 = (M_4 - 2.47 \times 10^6) \text{ in.-lb} = 3.05 \times 10^6 \text{ in.-lb}$$

Hence,

$$f_2' = -908.8 - \frac{3.05 \times 10^6}{5,070} = 1,510.4 \text{ psi } (C)$$

$$f_{2b} = -908.8 + \frac{3.05 \times 10^6}{6,168} = -414.3 \text{ psi } (C)$$

5. *Effects of adding the superimposed live load on support sections B and C*
   After the concrete cures, resulting in full composite action for supporting the total service live load, the section moduli for the precast beam become

$$S_c^t = 21,714 \text{ in}^3$$

and

$$S_{bc} = 9,490 \text{ in}^3$$

The loading combination which causes the highest stress condition is when the load acts on only two *adjacent* spans AB and BC. We have

$$W_L = 1,514 \text{ plf } (22.1 \text{ kN/m})$$

including W_{CDL}. From Figure 6.12, the support moments due to live load on two adjacent spans is

$$M_{B3} = 0.1167W_L\ell^2 = 0.1167 \times 1,514(64)^2 \times 12 = -8.68 \times 10^6 \text{ in.-lb}$$

The live-load moment causes tension at the top, and compression at the bottom fibers of the support section. The resulting total fiber stresses in the precast section at the support due to all loads become

$$f_T^t = f_2' + \frac{M_{B3}}{S_c^t}$$

and

$$f_{bT} = f_{b2} - \frac{M_{B3}}{S_{bc}}$$

Hence,

$$f_T^t = -1,510.4 + \frac{8.68 \times 10^6}{21.714}$$

$$\cong -1,111 \text{ psi } (C) \ (7.7 \text{ MPa}) < 2,250 \text{ psi, O.K.}$$

$$f_{bT} = -414.3 - \frac{8.68 \times 10^6}{9,490} = -1,329 \text{ psi } (9.2 \text{ MPa}) < 2,250 \text{ psi, O.K.}$$

Alternative one-step solution using Equations 4.19 a and b

$$f_T^t = -\frac{P_e}{A_c}\left(1 - \frac{ec_t}{r^2}\right) - \frac{M_D + M_{SD}}{S^t} - \frac{M_{CSD} + M_L}{S_c^t}$$

$$f_{bT} = -\frac{P_e}{A_c}\left(1 + \frac{ec_b}{r^2}\right) + \frac{M_D + M_{SD}}{S_b} + \frac{M_{CSD} + M_L}{S_{cb}}$$

where $e = e_b' = M_3/P_e$ and M_{CSD} is the additional superimposed dead load at service after erection, assumed zero here.

From before, the C-line eccentricity is $e'_B = M_3/P_e = -8.39 \times 10^6/508,940 = -16.49$ in. (41.9 cm). Also, the support moments, using the appropriate signs, are

$$M_D = -2.87 \times 10^6 \text{ in.-lb}$$

$$M_{SD} = -2.47 \times 10^6 \text{ in.-lb}$$

and

$$M_L = -8.68 \times 10^6 \text{ in.-lb (including } M_{CDS})$$

Hence,

$$f'_T = -\frac{508,940}{560}\left[1 - \frac{(-16.49) \times 24.73}{223.91}\right]$$

$$+ \frac{(2.87 + 2.47) \times 10^6}{5,070} + \frac{8.68 \times 10^6}{21.714}$$

$$= -2,560 + 1053.2 + 399.7 = -1,111 \text{ psi } (C)$$

$$f_{bT} = -\frac{508,940}{560}\left[1 + \frac{(-16.49) \times 20.27}{223.91}\right]$$

$$- \frac{(2.87 + 2.47) \times 10^6}{6,186} - \frac{8.68 \times 10^6}{9,490}$$

$$= +447.9 - 863.2 - 914.6$$

$$= -1,330 \text{ psi } (C) \text{ (9.2 MPa)} < 2,250 \text{ psi, O.K.}$$

Stresses at top and bottom fibers of the situ-cast 4 in. slab

From Equations 4.20 a and b, the maximum fiber stresses at the top and bottom fibers of the composite slab at the support section are evaluated using section moduli S'_{CS} and S_{bCS}, where

$$f'_S = +\frac{M_L}{S'_{CS}} \times \text{ modular ratio } n = 0.77$$

$$f_{bS} = +\frac{M_L}{S_{bCS}} \times n$$

Hence,

$$f'_S = \frac{8.68 \times 10^6}{15,288} \times 0.77 \cong +437 \text{ psi } (T) \text{ (3 MPa)} < 849 \text{ psi, O.K.}$$

$$f_{bS} = +\frac{8.68 \times 10^6}{19,251} \times 0.77 \cong +347 \text{ psi } (T) \text{ (2.4 MPa)}$$

Superposition of the tensile stress at the bottom fibers of the slab and the compressive stress of $-1,111$ psi at the extreme top fibers of the precast section can result in a net compressive stress at the bottom of the slab $= -1,111 + 347 = -764$ psi.

The fiber stress at the extreme lower fibers of the precast section at support B or C is

$$f_{bT} = -414.3 - \frac{8.68 \times 10^6}{9,490} \cong 1329 \text{ psi } (9.2 \text{ MPa})$$

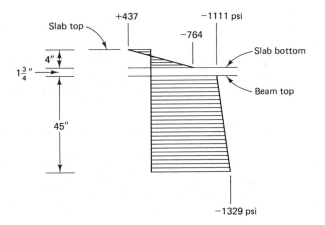

Figure 6.18 Stress distribution in the concrete at service load.

The final distribution of stress is shown in Figure 6.18.

Note that the concrete fiber stresses are considerably below the maximum allowable stresses at service load for the same live load and spans as the simply supported beam of Example 4.7. Consequently, the selected continuous tendon profile is not the most efficient in this example, since the superimposed live load is the same in both cases and the section is the same for the span lengths used. Example 6.5 demonstrates preferable modifications.

6. *Limit state at failure*

   (a) *Degree of ductility for moment redistribution*
   From Equations 6.6 and 6.7, the maximum redistribution percentage is

$$p_d = 20 \left[1 - \frac{\omega_p + \dfrac{d}{d_p}(\omega - \omega')}{0.36\beta_1} \right]$$

where

$$\omega_p + \frac{d}{d_p}(\omega - \omega') \le 0.24\beta_1$$

Also,

$$h = 45 + 5.75 = 50.75 \text{ in.}$$

$$d = 50.75 - (1.5 + 0.5 \text{ for stirrup} + 0.25) \cong 48.5 \text{ in.}$$

Support $d_p = c_b + e_e = 20.27 + 15$ from bottom fibers

$$= 35.27 \text{ in. (89.6 cm)}$$

Since the width of the compression flange at support is $b = 22$ in., try two #4 bars as compression steel and four #4 bars as tension steel. We obtain

$$\omega' = \frac{A_s' f_y}{bd f_c'} = \frac{2 \times 0.20}{22 \times 48.5} \times \frac{60,000}{5,000} = 0.0045$$

$$\omega = \frac{4 \times 0.2}{22 \times 48.5} \times \frac{60,000}{50,000} = 0.0090$$

$$\omega_p = \frac{A_{ps}\,f_{ps}}{bd_p\,f_c'} = \frac{3.366 \times 240{,}000}{22 \times 35.27 \times 5{,}000} = 0.2064$$

$$\beta_1 = 0.80 \text{ for 5,000 psi concrete}$$

$$\frac{d}{d_p}(\omega - \omega') = \frac{48.5}{35.27}(0.0090 - 0.0045) = 0.0062$$

$$\omega_p + \frac{d}{d_p}(\omega - \omega') = 0.2064 + 0.0062 = 0.2126$$

$$0.24\beta_1 = 0.24 \times 0.80 = 0.19 < 0.2126 \text{ at the support}$$

Hence, no redistribution of moment from support to midspan is permitted, i.e., the redistribution factor ρ_D in Equation 6.6 is zero. Thus, $0.36\beta_1 = 0.36 \times 0.80 = 0.45 > 0.2126$, O.K.

(b) *Flexural moments modifications*
It is advisable to apply the redistribution modifications separately to the dead and live loads since alternate span loading for live load has to be considered for worst loading conditions while dead load acts simultaneously on all spans. Since no redistribution occurs here, the elastic moments at support are

$$M_D + M_{SD} = (2.87 + 2.47)10^6 = 5.34 \times 10^6 \text{ in.-lb}$$

and

$$M_L = 8.68 \times 10.6 \text{ in.-lb}$$

(c) *Nominal moment strength*
The support factored M_u is

$$1.4 \times 5.34 \times 10^6 + 1.7 \times 8.68 \times 10^6 = 22.23 \times 10^6 \text{ in.-lb}$$

From step 3, the factored elastic secondary moment induced by reactions due to prestress, using a load factor of 1.0 as stipulated in the ACI code, is $M_2 = 0.76 \times 10^6$ in.-lb, and the total factored moment $M_u = (22.23 - 0.76) \times 10^6$ in.-lb $= 21.47 \times 10^6$ in.-lb (2.42 kN-m). Also, the required nominal moment strength $M_n = M_u/\phi = 21.47 \times 10^6/0.90 = 23.86 \times 10^6$ in.-lb (2.70 kN-m). If A_s' yielded, $f_y = f_s' = 60{,}000$ psi, width b at bottom $= 22$ in. So

$$a = \frac{A_{ps}f_{ps} + A_s f_y - A_s' f_y}{0.85 f_c' b}$$

$$= \frac{3.366 \times 240{,}000 + 0.8 \times 60{,}000 - 0.4 \times 60{,}000}{0.85 \times 5000 \times 22}$$

$$= 8.90 \text{ in. (22.6 cm)}$$

The depth of the flange up to the 7-in.-wide web section is

$$5.75 + 7 + \frac{4.5}{2} = 15 \text{ in.} > 8.90 \text{ in.}$$

The neutral axis is inside the flange, and the section behaves like a rectangular section.

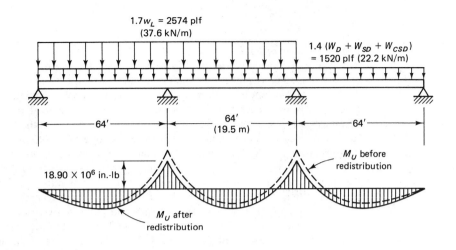

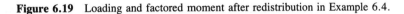

Figure 6.19 Loading and factored moment after redistribution in Example 6.4.

From Equation 4.29 a,

$$\text{Available } M_n = A_{ps}f_{ps}\left(d_p - \frac{a}{2}\right) + A_sf_y\left(d - \frac{a}{2}\right) + A'_sf_y\left(\frac{a}{2} - d'\right)$$

$$= 3.366 \times 240{,}000\left(35.27 - \frac{8.90}{2}\right)$$

$$+ 0.80 \times 60{,}000\left(48.5 - \frac{8.9}{2}\right) + 0.40 \times 60{,}000\left(\frac{8.9}{2} - 3\right)$$

$$= 27.05 \times 10^6 \text{ in.-lb } (3.1 \times 10^6 \text{ kN-m}) > 23.86 \times 10^6 \text{ in.-lb}$$

Hence, the section is safe but not efficient. Figure 6.19 shows the load and moment distributions.

In sum, the section is efficient neither at service load nor at ultimate load. It could be reduced in size by either changing the tendon eccentricities or enlarging the span or allowing a higher live load with new tendon eccentricities.

6.10 TENDON TRANSFORMATION TO UTILIZE ADVANTAGES OF CONTINUITY

Example 6.5

Linearly transform the prestressing tendon in the continuous prestressed beam in Example 6.4 such that the superimposed live load W_L on the 64 ft (19.8 m) spans can be increased by at least 50 percent.

Solution

1. *Transformation of tendon*

   The tendon eccentricity at support B in Example 6.4 produces compressive fiber stress at the support precast section due to all loads of 1,111 psi at the top, and 1,329 psi at the bottom, fibers. These are lower than the maximum allowable $f_c = -2,250$ psi. Therefore, the beam can sustain more load if the concrete compressive stress capacity is to be utilized. In order to allow the beam to carry more live load, more compressive stress at the midspan bottom fibers due to prestress needs to be developed through an increase in the tendon eccentricity.

   Accordingly, assume that the tendon is linearly transformed throughout all the spans such that $e_B = e_C = 11$ in. (27.9 cm), as shown in Figure 6.20. Then the transformation vertical distance $= 15 - 11 = 4$ in. (10 cm), the midspan eccentricity $e_{E1} = 9 + 4 = 13$ in. (33 cm), and the primary support B moment $M_1 = 508,940 \times 11 = 5.60 \times 10^6$ in.-lb (0.63×10^6 kN-m). Also, the primary midspan E_1 moment $M_1 = 508,940 \times 13 = 6.62 \times 10^6$ in.-lb (0.75×10^6 kN-m), and we have

   $$E_c I_c \Delta_B = \left[\left(\frac{5.60}{2} + 6.62 \right) 10^6 \times \frac{64 \times 2}{3} \right] \times \frac{64}{2} \times 144$$

   $$- \left[\frac{5.60 \times 10^6 \times 64}{2} \right] \times \frac{64 \times 2}{3} \times 144 = 75.1 \times 10^{10}$$

   Also,

   $$E_c I_c \Delta_c = \left[64 R_A \times 12 \times \frac{64 \times 12}{2} \right] \frac{64 \times 2}{3} \times 12$$

   $$= 15.1 \times 10^7 R_B \text{ in.-lb}$$

   $$R_A = R_B = \frac{75.1 \times 10^{10}}{15.1 \times 10^7} = 4,974 \text{ lb}$$

   $$M_2 = R_A \times 64 \times 12 = 4,974 \times 64 \times 12$$

   $$= 3.82 \times 10^6 \text{ in.-lb } (0.44 \times 10^6 \text{ kN-m})$$

   Support $M_3 = (5.60 + 3.82) \times 10^6 = 9.42 \times 10^6$ in.-lb (see Figure 6.20)

   Midspan $M_3 = \left(6.62 - \frac{1}{2} \times 3.82 \right) 10^6 = 4.71 \times 10^6$ in.-lb

   From Example 6.4, $P_e = 508,940$ lb. So calculate the final C-line eccentricities $= M_3/P_e$ as shown in Figure 6.20(f).

2. *Concrete fiber stresses due to prestress and self-weight (583 plf)*

   From step 3 of the solution to Example 6.4, the moment due to self-weight is 2.87×10^6 in.-lb. so the total moment $M_4 = M_3 - 2.87 \times 10^6 = (9.42 - 2.87)10^6 = 6.55 \times 10^6$ in.-lb.

   (i) *Support section B or C*

   $$f^t = -\frac{P_e}{A_c} - \frac{M_4}{S^t} = -\frac{508,940}{560} - \frac{6.55 \times 10^6}{5,070}$$

   $$= -908.8 - 1,291.9$$

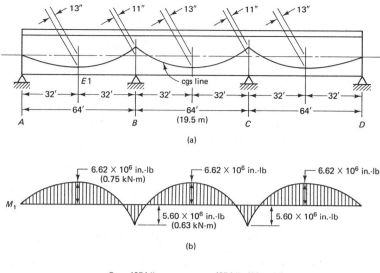

(a)

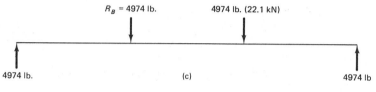

(b)

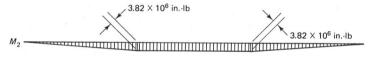

(c)

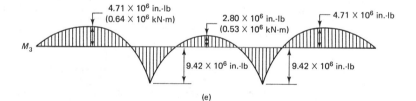

(d)

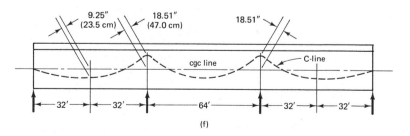

(e)

(f)

Figure 6.20 Tendon transformation in continuous beam of Example 6.5.

$$= -2,200.7 \text{ psi } (C) \ (15.2 \text{ MPa}) < 2,250 \text{ psi, O.K.}$$

$$f_b = -908.8 + \frac{6.55 \times 10^6}{6186} = -908.8 + 1,058.8$$

$$= +150.0 \text{ psi } (T), \text{ O.K.}$$

(ii) *Outer span midspan section E_1*

From Example 6.4, $M_D = 2.12 \times 10^6$ in.-lb. So the net moment $M_4 = M_3 - 2.12 \times 10^6 = (4.71 - 2.12)10^6 = 2.59 \times 10^6$ in.-lb $(0.4 \times 10^6$ kN-m), and

$$f^t = -908.8 + \frac{2.59 \times 10^6}{5,070} = -398.0 \text{ psi } (C) \ (2.7 \text{ MPa}), \text{ O.K.}$$

$$f_b = -908.8 - \frac{2.59 \times 10^6}{6,186}$$

$$= -1,327.5 \text{ psi } (C) \ (9.2 \text{ MPa}), \text{ no tension, O.K.}$$

3. *Determination of live-load intensity for new tendon profile for unshored construction*

$$f^t_T = -\frac{P_e}{A_c}\left(1 - \frac{e'_B c_t}{r^2}\right) - \frac{M_D + M_{SD}}{S^t} - \frac{M_{CSD} + M_L}{S^t_c}$$

$$f_{bT} = -\frac{P_e}{A_c}\left(1 + \frac{e'_B c_b}{r^2}\right) + \frac{M_D + M_{SD}}{S_b} + \frac{M_{CSD} + M_L}{S_{cb}}$$

(i) *Support section (at B or C)*

From Figure 6.20 (e), $M_3 = -9.42 \times 10^6$ in.-lb. Hence,

$$e'_B = -\frac{9.42 \times 10^6}{508,940} = -18.5 \text{ in. } (47.0 \text{ cm})$$

From Example 6.4, $M_D + M_{SD} = -(2.87 + 2.47)10^6 = -5.34 \times 10^6$ in.-lb. Also, from Figure 6.12, the live-load moment for a three-span beam with one span unloaded is

$$M_L = -0.1167 W_L \ell^2 = -0.1167 W_L (64)^2(12)$$

$$= -5,736 W_L \text{ in.-lb, including } M_{CSD}$$

The maximum allowable tensile stress $f_t = +849$ psi at midspan and $f_t = +425$ psi at support.

$$f^t_T = +425 = -\frac{508,940}{560}\left(1 - \frac{(-18.5) \times 24.73}{223.91}\right)$$

$$+ \frac{5.34 \times 10^6}{5,070} + \frac{5,736 W_L}{21,714}$$

giving $W_L = 8092$ plf.

(ii) *Midspan section (at E_1)*

Since $W_D = 583$ plf, by proportioning we obtain

$$M_D + M_{SD} = +(2.12 + 1.83)10^6 = +3.95 \times 10^6 \text{ in.-lb}$$

$$M_L = +2.12 \times 10^6 \times \frac{W_L}{583} = +3,636W_l \text{ in.-lb}$$

$$f'_T = -2,250 = -\frac{508,940}{560}\left(1 - \frac{9.25 \times 24.73}{223.91}\right)$$

$$-\frac{3.95 \times 10^6}{5,070} - \frac{3,636W_L}{21,714}$$

$$\frac{3,636W_L}{21,714} = 20 - 779 + 2,250$$

$$W_L = 8,904 \text{ plf}$$

$$f_{bT} = 849 = -\frac{508,940}{560}\left(1 + \frac{9.25 \times 20.27}{223.91}\right)$$

$$+\frac{3.95 \times 10^6}{6,186} + \frac{3,636W_L}{9,490}$$

$$\frac{3,636W_L}{9,490} = 849 + 1,669.8 - 538.5$$

$$W_L = 5,168 \text{ plf}$$

Hence, $W_L = 5,168$ plf controls for service load levels, to be verified by checking the ultimate moment strength available.

4. *Available nominal moment strength*

$$A_{ps} = 3.366 \text{ in}^2$$

$$\text{Support } d_p = 11 + 20.27 = 31.27 \text{ in.}$$

Assume that A_s at supports B and C is increased to four #8 bars in order to facilitate an increased live load. Then

$$A_s = 4 \times 0.79 = 3.16 \text{ in}^2$$

$$\omega = \frac{3.16}{22 \times 48.5} \times \frac{60,000}{5,000} = 0.0355$$

$$\omega_p + \frac{d}{d_p}(\omega - \omega') = 0.2064 + \frac{48.5}{31.27}(0.0355 - 0.0045)$$

$$= 0.2545 > 0.24\beta_1 > 0.19$$

Hence, there is no redistribution of moment from support to midspan.
Now, from before, the elastic $M_2 = 3.82 \times 10^6$ in.-lb and we have

$$M_u = 1.4(M_D + M_{SD}) + 1.7M_L - M_2$$

or

$$M_u = 1.4 \times 5.34 \times 10^6 + 1.7M_L - 3.82 \times 10^6$$

$$= 3.66 \times 10^6 + 1.7M_L$$

Next,

$$\text{Required } M_n = \frac{M_u}{\phi} = 4.07 \times 10^6 + 1.89 M_L$$

$$a = \frac{A_{ps}f_{ps} + A_s f_y - A'_s f_y}{0.85 f'_c b}$$

If A'_s yielded $f_y = f'_s = 60,000$ psi, then

$$a = \frac{3.366 \times 240,000 + (3.16 - 0.40)60,000}{0.85 \times 5,000 \times 22} = 10.41 \text{ in. } (26.4 \text{ cm})$$

which is less than the flange depth up to the 7-in.-wide web section. Hence, the neutral axis is inside the flange, and the section behaves like a rectangular section. Accordingly, we have

$$\text{Available } M_n = A_{ps}f_{ps}\left(d_p - \frac{a}{2}\right) + A_s f_y\left(d - \frac{a}{2}\right) + A'_s f_y\left(\frac{a}{2} - d'\right)$$

$$= 3.366 \times 240,000\left(31.27 - \frac{10.41}{2}\right) + 3.16 \times$$

$$60,000\left(48.5 - \frac{10.41}{2}\right) + 0.40 \times 60,000\left(\frac{10.41}{2} - 3\right)$$

$$= 28.91 \times 10^6 \text{ in.-lb} = 4.07 \times 10^6 + 1.89 M_L$$

Hence,

$$M_L = \frac{(28.91 - 4.07) \times 10^6}{1.89} = 13.14 \times 10^6 \text{ in.-lb}$$

If $M_L = 0.1167 W_L \ell^2$, then

$$13.14 \times 10^6 = 0.1167 W_L (64)^2 \times 12$$

So $W_L = 2,290$ plf $< W_L = 5,168$ plf from the service-load analysis. Hence, $W_L = 2,290$ plf controls, and the percent increase in live load is

$$\frac{(2,290 - 1,514)}{1,514} \times 100 = 51.3\% \cong 50\%, \text{ O.K.}$$

Thus, we can adopt the new profile of the tendon with four #8 bars at the support top fibers in the situ-cast slab and two #4 bars at the bottom precast section fibers.

6.11 DESIGN FOR CONTINUITY USING NONPRESTRESSED STEEL AT SUPPORT

Example 6.6

Design the beam in Example 6.5 such that the section and tendon profile of the AASHTO type III bridge beam used in Example 4.7 is made continuous through the use of nonprestressed mild steel reinforcement to carry the superimposed service dead load

$W_{SD} = 503$ plf and service live load $W_L = 2,290$ plf (33.4 kN/m). Assume that the tendon profile in the precast simply supported section is the same as the one in Example 4.7, namely, with midspan eccentricity $e_c = 16.27$ in. (41.3 cm) and end eccentricity $e_e = 10.0$ in. (25.4 cm). Sketch the prestressing tendon and other reinforcement details if the maximum allowable concrete compressive fiber stress at service load is $f_c = 2,250$ psi (15.5 MPa).

Solution

1. *Data for strength design at support*

   Because continuity is obtained in this case through the use of reinforced concrete at the supports, it is suggested that the topping concrete also be of $f'_c = 5,000$ psi compressive strength. Thus, we have

   $$f'_c = 5,000 \text{ psi (34.5 MPa), normal weight}$$
   $$f_y = 60,000 \text{ psi (413.7 MPa)}$$

   Design the continuity to resist the superimposed dead and live loads only, and *not* the self-weight:

   $$d = 50.75 - (1.5 \text{ in. cover} + 0.5 \text{ in. for stirrup} + 0.5 \text{ in. for half bar dia})$$
   $$= 48.25 \text{ in. (122.6 cm)}$$
   $$b_b = 22 \text{ in. at bottom}$$
   $$b_t = 84 \text{ in. at top (modified } b_m = 65 \text{ in.)}$$

2. *Nominal moment strength*

   From Example 6.5, the required moment strength due to $(W_{SD} + W_L)$ at support B is $M_n = 1.4 \times 2.47 \times 10^6 + 1.7 \times 13.14 \times 10^6 = 25.80 \times 10^6$ in.-lb. The bonded prestressed steel does not extend through the supports; hence, consider $\omega_p = 0$ for calculating the moment redistribution factor. Assume

   $$A_s = 8.5 \text{ in}^2$$

   $$\omega = \frac{8.5}{22 \times 48.25} \times \frac{60,000}{5,000} = 0.0961$$

   $$A'_s = \text{two #4} = 0.40 \text{ in}^2$$

   and

   $$\omega' = 0.0961 \times \frac{0.4}{8.5} = 0.0045$$

   Then from Example 6.4, $d_p = 35.87$ in., $d = 48.25$ in., $\omega_p + d/d_p(\omega - \omega') = 0 + 48.25/35.87(0.0961 - 0.0045) = 0.1232$, and $0.24\beta_1 = 0.192 > 0.1232$. Hence, the moment redistribution factor is

   $$\rho_d = 20\left[1 - \frac{\omega_p + d/d_p(\omega - \omega')}{0.36\beta_1}\right]$$

   $$= 2 - \left[1 - \frac{0.1232}{0.36 \times 0.80}\right] = 11.44\%$$

So use a distribution factor of 0.10:

$$\text{Rqd. } M_n = (1 - 0.10) \times 25.80 \times 10^6 = 23.22 \times 10^6 \text{ in.-lb}$$

$$(2.2 \times 10^6 \text{ kN-m})$$

$$M_n = A_s f_y \left(d - \frac{a}{2} \right)$$

Assume for the first trial that $d - a/2 \cong 0.9d$. Then

$$23.22 \times 10^6 = A_s \times 60{,}000 \times 0.9(48.25)$$

$$A_s = \frac{23.22 \times 10^6}{60{,}000 \times 0.9 \times 48.25} = 8.95 \text{ in}^2 \ (48.1 \text{ cm}^2)$$

$$a = \frac{A_s f_y}{0.85 f'_c b} = \frac{8.95 \times 60{,}000}{0.85 \times 5{,}000 \times 22} = 5.74 \text{ in.}$$

The depth of the flange to the web is $5.75 + 7 + 4.5/2$ for the AASHTO type 3 section $= 15$ in. > 7.19 in. The neutral axis is inside the flange, and the section behaves like a rectangular section. Therefore,

$$A_s = \frac{M_n}{f_y(d - a/2)} = \frac{23.22 \times 10^6}{60{,}000(48.25 - 5.74/2)} = 8.53 \text{ in}^2 \ (70.2 \text{ cm}^2)$$

Since the assumed $A_s = 8.5$ in^2, the moment distribution factor is satisfactory.

The area A_s is to be distributed over the total actual flange width of 84 in. A_s per ft width $= (8.53/84) \times 12 = 1.22$ in^2/12 in. Using #6 bars, A_s per bar $= 0.44$ in^2 (2.82 cm^2), and the spacing is

$$s = \frac{\text{Bar } A_s}{\text{Rqd. } A_s/12 \text{ in.}} = \frac{0.44}{1.22/12} = 4.32 \text{ in.}$$

Thus, use #6 bars at $4\frac{1}{4}$ in. center to center over the 84-in. width (19.1 mm dia. bars at 10.8 cm). The total number of bars over the 84-in.-width flange is

$$\frac{84 - (2 \times 1.5\text{-in. cover})}{4.25} + 1 \cong 20$$

so that we have

$$\text{Total } A_s = 20 \times 0.44 = 8.80 \text{ in}^2, \text{ O.K.}$$

Accordingly, adopt the design for flexure. Note that the complete design would involve dowel design for composite action, stirrup design for web shear, end block design, and design for serviceability requirements in deflection and crack control as detailed in earlier examples. Note also that continuity on three spans in this example using mild steel only at the supports allowed a 50% increase in the live-load intensity from 1,514 plf to 2,290 plf.

3. *Beam geometry schematic details*
   Figure 6.21 gives the reinforcing and tendon profile details of the continuous beam of this example. Note how the normal-weight concrete and mild steel provide continuity at the supports for the superimposed dead and live loads.

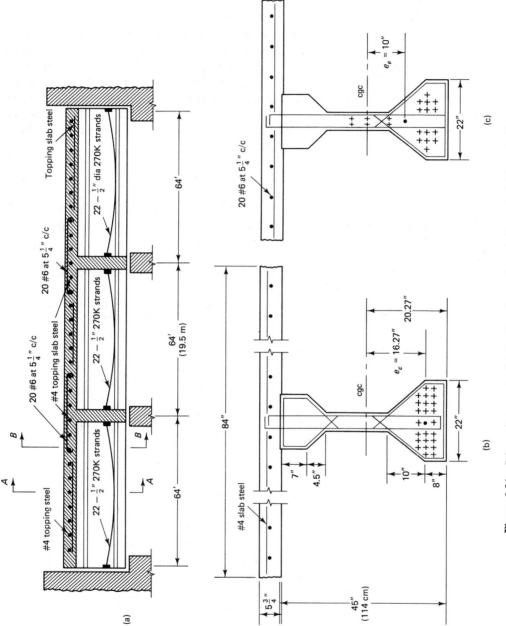

Figure 6.21 Schematic geometry details of continuous beam in Example 6.6 (see also Example 4.7). (a) Longitudinal section of bridge beam (not to scale). (b) Midspan section A-A. (c) Support section B-B.

6.12 INDETERMINATE FRAMES AND PORTALS

6.12.1 General Properties

Concrete frames are indeterminate structures consisting of horizontal, vertical, or inclined members joined in such a manner that the connection can withstand the stress and bending moments that act on it. The degree of indeterminacy depends upon the number of spans, the number of vertical members, and the type of end reactions. Typical frame configurations are shown in Figure 6.22. If n is the number of joints, b the number of members, r the number of reactions, and s the number of indeterminacies, the degree of indeterminacy is determined from the following inequalities:

$$3n + s > 3b + r \qquad \text{(unstable)} \qquad (6.8\,a)$$

$$3n + s = 3b + r \qquad \text{(statically determinate)} \qquad (6.8\,b)$$

$$3n + s < 3b + r \qquad \text{(statically indeterminate)} \qquad (6.8\,c)$$

The degree of indeterminacy is

$$s = 3b + r - 3n \qquad (6.8\,d)$$

where $3n$ equations of static equilibrium are always available and the total number of

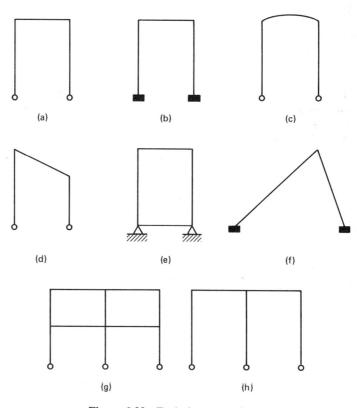

Figure 6.22 Typical structure frames.

unknowns is $3b + r$. As an example, the degree of indeterminancy of the frame in Figure 6.22(a) is

$$s = 3 \times 3 + 2 \times 2 - 3 \times 4 = 1$$

and for the frame in part (g) of the same figure it is

$$s = 3 \times 10 + 2 \times 3 - 3 \times 9 = 9$$

Note that in order for a frame to perform satisfactorily, the following conditions have to be satisfied:

1. The design must be based on the most unfavorable moment and shear combinations. If moment reversal is possible due to reversal of live-load direction, the highest values of positive and negative bending moments have to be considered in the design.
2. Proper foundation support for horizontal thrust has to be provided. If the frame is designed as hinged, an expensive construction procedure, an actual hinge system has to be provided.

6.12.2 Forces and Moments in Portal Frames

The behavior of concrete frames before cracking can be considered reasonably elastic, as was done in the case of a continuous beam at service-load and slight-overload conditions. Consequently, well before the development of plastic hinging, the bending moment diagrams shown in Figures 6.23 and 6.24 will be used in the design of

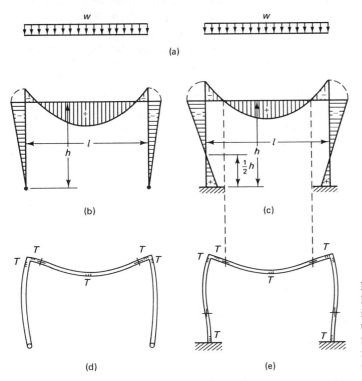

Figure 6.23 Right-angled portal frame loaded with gravity load intensity w (T indicates tension fibers). (a) Load intensity. (b) Bending moment (hinged-base frame). (c) Bending moment (fixed-base frame). (d) Deformation of frame in (b). (e) Deformation of frame in (c).

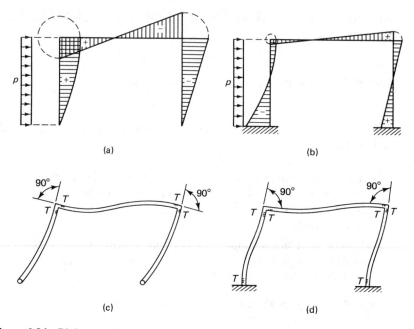

Figure 6.24 Right-angled portal frame loaded with wind load intensity p (T indicates tension fibers). (a) Bending moment (hinged-base frame). (b) Bending moment (fized-base frame). (c) Deformation of frame in (a). (d) Deformation of frame in (b).

indeterminate prestressed concrete frames. The usual methods of analyses of indeterminate structures including frames, such as virtual work, stiffness matrix, and flexibility matrix procedures, as well as the clapeyron three- or four-moment equations, are assumed in this text, so that only the minimum guidelines and simplifications are presented.

6.12.2.1. Uniform gravity loading on single-bay portal. Suppose that the moments of inertia I_c of the vertical columns and I_b of the horizontal beam of the portal in Figure 6.25(a) are not equal. The following values of the moments and thrusts can be inferred:

End Shear in Beam

$$V_B = V_C = \frac{1}{2} W\ell \qquad (6.9 \text{ a})$$

Horizontal Thrust

$$H = \frac{1}{h} C_1 w\ell \qquad (6.9 \text{ b})$$

where

$$C_1 = \frac{1}{12\left(\dfrac{2}{3}\dfrac{I_b}{I_c}\dfrac{h}{\ell} + 1\right)} \qquad (6.9 \text{ c})$$

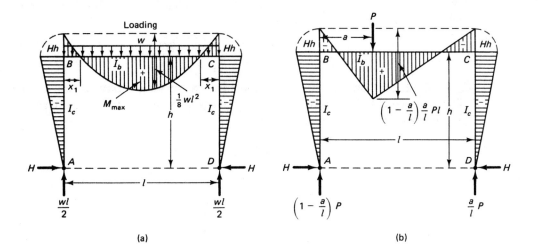

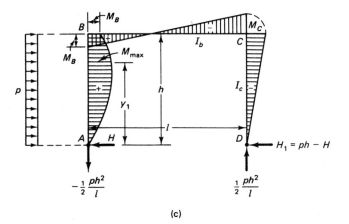

Figure 6.25 Bending moment ordinates in single-bay frame. (a) Uniform gravity loading. (b) Concentrated gravity loading. (c) Uniform horizontal pressure.

Maximum Negative Moment at Midspan

$$M_B = M_c = -Hh = -C_1 w\ell^2 \qquad (6.9\ d)$$

Maximum Positive Moment at Midspan

$$M_{max} = \frac{1}{8} w\ell^2 - Hh = \left(\frac{1}{8} - C_1\right)w\ell^2 \qquad (6.9\ e)$$

Bending Moments at Any Point x

$$M_x = \frac{1}{2} x(\ell - x)w - C_1 w\ell^2 \qquad (6.9\ f)$$

where the points of contraflexure from either corner of the portal are

$$x_1 = \frac{1}{2}\left(1 - \sqrt{1 - 8C_1}\right)\ell = C_2\ell \qquad (6.9\ g)$$

and

$$C_2 = \frac{1}{2}\left(1 - \sqrt{1 - 8C_1}\right) \qquad (6.9\ h)$$

6.12.2.2 Concentrated gravity loading on single-bay portal. Since the concentrated load P does not have to act at midspan, nonsymmetry of shears results. From Figure 6.25(b), the end shears are

$$V_B = \left(1 - \frac{a}{\ell}\right)P$$

and

$$V_c = \frac{a}{\ell}P \qquad (6.10\ a)$$

Horizontal Thrust

$$H = C_3 \frac{a}{\ell}\left(1 - \frac{a}{\ell}\right)P\frac{\ell}{h} \qquad (6.10\ b)$$

where

$$C_3 = \frac{1}{2\left(\dfrac{2}{3}\dfrac{I_b}{I_c}\dfrac{h}{\ell} + 1\right)} \qquad (6.10\ c)$$

Bending Moments at Corners

$$M_B = M_c = -Hh = -C_3\frac{a}{\ell}\left(1 - \frac{a}{\ell}\right)P\ell \qquad (6.10\ d)$$

Bending Moments at Any Point along BC. For $x < a$,

$$M_x = \left(1 - \frac{a}{\ell}\right)\left(\frac{x}{\ell} - \frac{a}{\ell}C_3\right)P\ell \qquad (6.10\ e)$$

For $x > a$,

$$M_x = \frac{a}{\ell}\left[1 - \frac{x}{\ell} - \left(1 - \frac{a}{\ell}\right)\right]C_3P\ell \qquad (6.10\ f)$$

Maximum Positive Moment at $x = a$

$$M_{max} = \frac{a}{\ell}\left(1 - \frac{a}{\ell}\right)P\ell - Hh = (1 - C_3)\frac{a}{\ell}\left(1 - \frac{a}{z}\right)P\ell \qquad (6.10\ g)$$

Horizontal Thrust for Several Concentrated Gravity Loads

$$H = \frac{1}{h}C_3\left[P_1\frac{a_1}{\ell}\left(1 - \frac{a_1}{\ell}\right) + P_2\frac{a_2}{\ell}\left(1 - \frac{a_2}{\ell}\right) + \cdots\right] \qquad (6.10\ h)$$

or

$$H = \frac{1}{h}C_3\sum P\frac{a}{\ell}\left(1 - \frac{a}{\ell}\right) \qquad (6.10\ i)$$

6.12.2.3 Uniform horizontal pressure on single-bay portal. From Figure 6.25(c), we have the following:

Vertical Reactions at Supports

$$R_A = -\frac{1}{2}ph\frac{h}{\ell}$$

and

$$R_D = +ph\frac{h}{\ell} \tag{6.11 a}$$

Horizontal Reactions. For windward hinge A,

$$H_A = \frac{1}{8}\frac{11\frac{I_b}{I_c}\frac{h}{\ell} + 18}{2\frac{I_b}{I_c}\frac{h}{\ell} + 3}ph = C_4ph \tag{6.11 b}$$

where

$$C_4 = \frac{1}{8}\frac{11\frac{I_b}{I_c}\frac{h}{\ell} + 18}{2\frac{I_b}{I_c}\frac{h}{\ell} + 3} \tag{6.11 c}$$

For leeward hinge D,

$$H_D = ph - H_A = (1 - C_4)ph \tag{6.11 d}$$

The bending moments at any point y along the column height due to horizontal pressure, with y being measured from the *bottom*, are

$$M_y = H_A y - \frac{1}{2}py^2 \tag{6.11 e}$$

Maximum Moment at Windward Column

$$M_{max} = \frac{1}{2}\left(\frac{1}{8}\frac{11\frac{I_b}{I_c}\frac{h}{\ell} + 18}{2\frac{I_b}{I_c}\frac{h}{\ell} + 3}\right)ph^2 = \frac{1}{2}C_4ph^2 \tag{6.11 f}$$

Point of Maximum Bending Moment above Support A

$$y_1 = \frac{1}{8}\left(\frac{11\frac{I_b}{I_c}\frac{h}{\ell} + 18}{2\frac{I_b}{I_c}\frac{h}{\ell} + 3}\right) = C_4h \tag{6.11 g}$$

$$M_B = H_A h - \frac{1}{2}ph^2 = \frac{3}{8}\frac{\dfrac{I_b}{I_c}\dfrac{h}{\ell} + 2}{2\dfrac{I_b}{I_c}\dfrac{h}{\ell} + 3}ph^2$$

$$= (C_4 - 0.5)ph^2 \tag{6.11 h}$$

$$M_c = -H_D h = -(1 - C_4)ph^2 \tag{6.11 i}$$

The constants C_1, C_2, C_3, and C_4 in Equations 6.9, 6.10, and 6.11 can be graphically represented as shown in Figure 6.26. Canned computer programs for the analysis of indeterminate beams and frames render the use of charts such as this unnecessary except for a quick check of numerical values.

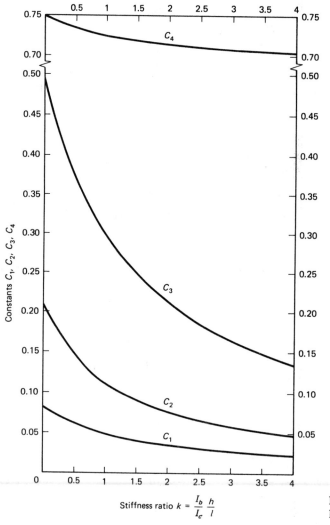

Figure 6.26 Constants C_1 through C_4 in Equations 6.9, 6.10, and 6.11.

6.12.3 Application to Prestressed Concrete Frames

As with continuous beams, a tendon profile has to be assumed at the start in order to determine the secondary bending moments M_2 for the portal frame horizontal beam and vertical legs. A concordant tendon is assumed for the horizontal beam for symmetrical gravity loading, and the vertical columns or legs are proportioned to resist the horizontal pressure and the extra moment caused by the shortening of the beam.

Longitudinal shortening of the horizontal beam caused by the prestressing force results in tensile stresses at the outside face of the frame columns. The prestressing vertical tendon should be designed to resist these stresses as well as others. The longitudinal shortening also results in horizontal reactions at the column's supports. Consequently, in order to obtain a prestressing force P in the longitudinal member, a force $P + \Delta P$ has to be applied to the frame. The incremental force ΔP can be evaluated by means of the following expressions.

Frame with Two Hinges at Supports

$$\Delta P = \frac{M_B}{h} = \frac{3}{2k + 3} \frac{E_c I_c}{h^2} \epsilon_{BC} \tag{6.12}$$

where $k = (I_b/I_c)(h/\ell)$ and ϵ_{BC} is the total strain due to elastic shortening and movement due to shrinkage and creep. The subscripts B and C denote the member extremities of the frame in Figures 6.25 and 6.27.

Frame with Fixed Supports

$$\Delta P = \frac{M_B - M_A}{h} = \frac{E_c I_c}{h} \left(\frac{3}{k + 2} + \frac{k + 3}{k(k + 2)} \right)$$

$$= \frac{3(2k + 1)}{k(k + 2)} \frac{EI_c}{h^2} \tag{6.13}$$

Figure 6.27 shows the axial deformation due to the strain $\epsilon_{BC} = \Delta \ell / \ell$ caused by shortening, creep, and shrinkage.

The *tributary* moments M_A and M_B due to the logitudinal shortening of member BC in Figure 6.27 are as follows.

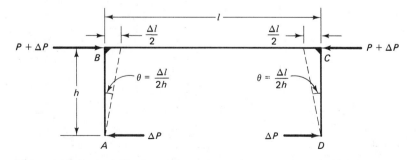

Figure 6.27 Longitudinal deformation of beam BC due to elastic shortening, creep, and shrinkage.

Frame with Two Hinges at Supports

$$M_B = \frac{6}{2k+3} \frac{E_c I_b \theta}{\ell} = \frac{3}{2k+3} \frac{E_c I_c}{h} \epsilon_{BC} \tag{6.14}$$

as $k \longrightarrow 0$, $M_B \longrightarrow \frac{EI_c}{h} \epsilon_{BC}$.

Frame with Fixed Supports

$$M_B = \frac{6}{k+2} \frac{E_c I_c \theta}{h} = \frac{3}{k+2} \frac{E_c I_c}{h} \epsilon_{BC} \tag{6.15}$$

as $k \longrightarrow 0$, $M_B \longrightarrow \frac{1.5 E_c I_c}{h} \epsilon_{BC}$ and $M_A \longrightarrow \infty$.

The reason for the drastic change in moment values M_B and M_A is that as k approaches zero, ΔP approaches infinity. In such a case the stiffness of the vertical members relative to the horizontal member approaches infinity, and the horizontal member becomes very flexible, as shown in Figure 6.28. The effects of the horizontal reactions on the prestressing force are schematically shown in Figure 6.29 for both hinged-base and fixed-base frames.

Note that the discussion here and in the previous section applies equally to

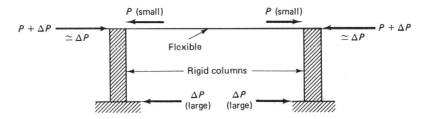

Figure 6.28 Effect of tributary moments due to elastic shortening.

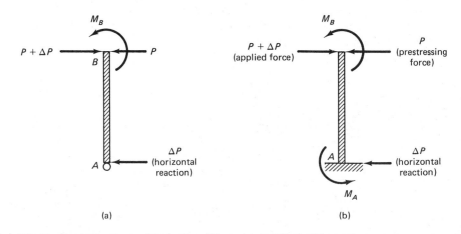

(a)

(b)

Figure 6.29 Horizontal reaction effect on prestressing force. (a) Hinged-base frame. (b) Fixed-base frame.

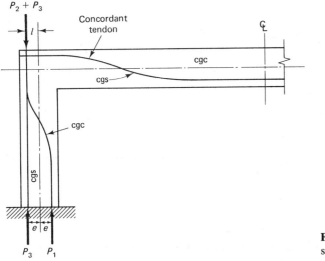

Figure 6.30 Tendon profile in a pre-stressed frame.

continuously cast and precast prestressed composite frames. Continuity at the corner of the frames has to be accomplished in the construction process. A typical prestressing tendon profile for a frame is shown in Figure 6.30. The prestressing force P_1 is assumed to be less than P_2 in order to allow for the frictional losses in prestress.

6.12.4 Design of Prestressed Concrete Bonded Frame

Example 6.7

A warehouse structure is made of a prestressed single-bay hinged-base portal frame made of standard double-T sections for both the horizontal beam and the two vertical columns. The units are 8 ft. wide (2.44 m) The frame has a clear span of 80 ft (24.4 m) and is subjected to a uniform gravity live-load intensity $W_L = 240$ plf (3.5 kN/m) and a uniform horizontal wind pressure of intensity $p_w = 65$ plf (0.95 kN/m) at the windward side and a suction of intensity $p_L = 40$ plf (0.58 kN/m) at the leeward side, as shown in Figure 6.31. Design the frame, the profile, and the location of the prestressing tendons for service-load and ultimate load conditions given the following data:

$f_{pu} = 270,000$ psi (1,862 MPa) for low-relaxation tendons
$f_{ps} = 235,000$ psi (1,620 MPa)
$f_{pi} = 189,000$ psi (1,303 MPa)
Total losses $= 21\%$, losses after one month of prestressing $= 17\%$
f_{pe} (one month) $= (1 - 0.17)189,000 = 156,870$ psi (1,082 MPa)
f_{pe} (final) $= (1 - 0.21)189,000 = 149,310$ psi (1,029 MPa)
$f'_c = 5,000$ psi (34.5 MPa)
$f_c = 0.45f'_c = 2,250$ psi (15.5 MPa)
$f'_{ci} \cong 0.70f'_c = 3,500$ psi (24.1 MPa)
$f_{ci} = 0.6f'_{ci} = 2,100$ psi (14.5 MPa)
f_{ti} (midspan) $= 3\sqrt{f'_{ci}} = 177$ psi
f_{ti} (support) $= 6\sqrt{f'_{ci}} = 355$ psi
$f_t = 6\sqrt{f'_c}$ to $12\sqrt{f'_c} = 849$ psi (5.85 Mpa)

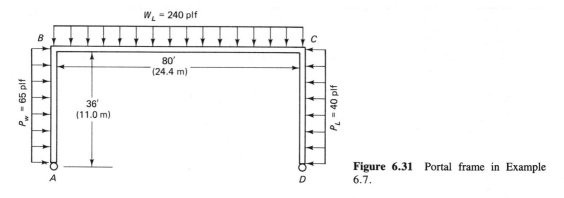

Figure 6.31 Portal frame in Example 6.7.

Solution

Frame Horizontal Beam BC

Preliminary Analysis. Assume that self-weight $W_D = 600$ plf (8.8 N/m). Then from Equation 6.9 c, the stiffness coefficient is

$$C_1 = \frac{1}{12\left(\dfrac{2}{3}k + 1\right)}$$

where

$$k = \frac{I_b}{I_c}\frac{h}{\ell}$$

Assume at this stage that $I_b = I_c$, where I_b is the moment of inertia of the beam BC and I_c is the moment of inertia of the column AB or DC. Then

$$k = \frac{h}{\ell} \times 1.0 = \frac{36}{80} = 0.45$$

From the chart for C_1 in Figure 6.26, $C_1 = 0.064$. So given total losses = 21%, it follows that $\gamma = 0.79$.

Now, assume 2 in. of concrete topping ($f'_c = 3,000$ psi lightweight), 5 psf insulation, and a waterproofing width of the segment = 8 ft. Then

$$W_{SD} = \left(\frac{2}{12} \times 110 + 5\right)8 \text{ ft.} = 187 \text{ plf}$$

Beam BC is to be designed as a concordant cable, i.e., it is to behave as a simply supported beam *for self-weight* W_D. But it would be considered continuous for the superimposed dead load W_{SD} and live load W_L as part of the rigid portal frame. The C-line would then coincide with the cgs line due to concordance. Accordingly, if $W_D \cong 600$ plf, then from Equation 6.9 e, the *midspan* moment is

$$M = \left(\frac{1}{8} - C_1\right)w\ell^2$$

so that

$$M_D = \frac{w\ell^2}{8} = \frac{600(80)^2}{8} \times 12 = 5.760 \times 10^6 \text{ in.-lb}$$

$$M_{SD} = \left(\frac{1}{8} - 0.064\right)187(80)^2 \times 12 = 0.876 \times 10^6 \text{ in.-lb}$$

$$M_L = \left(\frac{1}{8} - 0.064\right)240(80)^2 \times 12 = 1.124 \times 10^6 \text{ in.-lb}$$

Assume $f_t = 0$. Then from Equation 4.4 b, the minimum section modulus at the bottom fibers for an efficient section is given by

$$S_b \geq \frac{(1 - \gamma)M_D + M_{SD} + M_L}{f_t - \gamma f_{ci}}$$

or

$$S_b = \frac{(1 - 0.79)5.760 \times 10^6 + 0.876 \times 10^6 + 1.124 \times 10^6}{0 - 0.79(-2,100)} = 1,935 \text{ in}^3$$

The closest section is PCI 8DT32 + 2 Double-T type 168-DI with 2 in. concrete topping (Ref. 6.15).

Properties of Preliminary Section

Property	Untopped	Topped
A_c (in^2)	567	759
I_c (in^4)	55,464	71,886
$r^2 = I_c/A_c$ (in^2)	97.8	94.7
c_b (in.)	21.21	23.66
c_t (in.)	10.79	10.34
S^t (in^3)	5,140	6,952
S_b (in^3)	2,615	3,038
W_D (plf)	591	738

$e_e = 8.21$ in. (20.9 cm)
$e_c = 17.46$ in. (44.3 cm)
sixteen $\frac{1}{2}$ in. dia (12.7 mm dia) 270 K, stress-relieved strands
$A_{ps} = 16 \times 0.153 = 2.45\text{in}^2$ (15.8 cm^2)

Analysis of Section at Transfer

(a) **Midspan Section** ($e_c = 17.46$ in.)
   Adjust

$$M_D = 5.760 \times 10^6 \times \frac{591}{600} = 5.673 \times 10^6 \text{ in.-lb}$$

where

$$591 = \frac{567}{12 \times 12} \times 150 \text{ plf}$$

From Equation 4.1 a,

$$f^t = -\frac{P_i}{A_c}\left(1 - \frac{ec_t}{r^2}\right) - \frac{M_D}{S^t} \leq f_{ti}$$

$$P_i = A_{ps}f_{pi} = 2.45 \times 189{,}000 = 463{,}050 \text{ lb}$$

$$f^t = -\frac{463{,}050}{567}\left(1 - \frac{17.46 \times 10.79}{97.8}\right) - \frac{5.673 \times 10^6}{5{,}140}$$

$$= -347 \text{ psi } (C), \text{ no tension, O.K.}$$

So provide nonprestressed steel at the top fibers at midspan to account for any possible tensile stresses. Then, from Equation 4.1 b,

$$f_b = -\frac{P_i}{A_c}\left(1 + \frac{ec_b}{r^2}\right) + \frac{M_D}{S_b}$$

$$= -\frac{463{,}050}{567}\left(1 + \frac{17.46 \times 21.21}{97.8}\right) + \frac{5.673 \times 10^6}{2{,}615}$$

$$= -1{,}740 \text{ psi } (C) < f_{ci} = 2{,}100 \text{ psi, O.K.}$$

(b) *Support Section* $(e_e = 8.21 \text{ in.})$

$$f^t = -\frac{463{,}050}{567}\left(1 - \frac{8.21 \times 10.79}{97.8}\right) - 0$$

$$= -76.9 \text{ psi } (C), \text{ no tension, O.K.}$$

$$f_b = -\frac{463{,}050}{567}\left(1 + \frac{8.21 \times 21.21}{97.8}\right) + 0$$

$$= -2{,}271 \text{ psi } (C) \ (15.7 \text{ MPa}) > f_{ci} = 2{,}100 \text{ psi, unsatisfactory}$$

Hence, lower the magnitude of the prestressing force by *debonding* some strands over a length of 15 percent of span from the support face, or change the eccentricity of the tendon. If the former technique is employed, debond four strands over a length $= 0.15 \times 80 \text{ ft} = 12 \text{ ft} \ (366 \text{ m})$ from the support, releasing the anchorage of the four grouted strands. We obtain

$$A_{ps} = (16 - 4)0.153 = 1.836 \text{ in}^2$$

$$P_i = 1.836 \times 189{,}000 = 347{,}004 \text{ lb} \ (1{,}543 \text{ kN})$$

$$f_b = -\frac{347{,}004}{567}\left(1 + \frac{8.21 \times 21.21}{97.8}\right) + 0$$

$$= -1{,}701.6 \text{ psi} < 2{,}100 \text{ psi, O.K.}$$

Frame Vertical Column Analysis. Choose a double-T as walls for the frame and suppose that e_b and S_b refer to the outer face and that e_t and S_t refer to the inner face of the vertical T-section. Since it was assumed, in calculating the stiffness coefficient k in the previous section, that $I_b = I_c$, choose also 8DT32, hinged at the base. This vertical member will act as a compression member subject to large axial load and bending. The bending moments are caused by wind load and moments from the frame horizontal beam BC. In such a case it is preferable to spread the tendon across the section, as shown in Figure 6.32 comparing the beam section and the column section.

Assume that the center of gravity of the prestressing strands coincides with the cgc line, and design the distribution of the strands according to

$$e_c = e_e = 0$$

Try using twenty $\frac{1}{2}$ in. dia 270 K low-relaxation strands:

$$A_{ps} = 20 \times 0.153 = 3.06 \text{ in}^2$$

$$P_i = A_{ps} f_{pi} = 3.06 \times 189,000 = 578,340 \text{ lb}$$

$$f' = f_b = -\frac{P_i}{A_c} \pm 0 = -\frac{578,340}{567} = -1,020 \text{ psi } (C) < f_{ci} = 2,100 \text{ psi, O.K.}$$

Frame Moments and Reactions at Service-Load Level

Horizontal Portal Beam BC

Free Support W_D Stage. Assume that the length of precast beams is $80 - 1.3 = 78.7$ ft. Then the midspan moment $M_E = w\ell^2/8 = 591(78.7)^2/8 \times 12 = 5.491 \times 10^6$ in.-lb. and the reaction at the column-wall bracket support is $R_D = 591 \times 78.7/2 = 23,256$ lb.

Composite Topping W_{SD} Stage. From before, the midspan moment is $M_{SD} = 0.876 \times 10^6$ in.-lb. The support moment is then

$$M_B = M_c = \frac{w\ell^2}{8} - 0.876 \times 10^6$$

$$= \frac{187(80)^2 \times 12}{8} - 0.876 \times 10^6 = 0.919 \times 10^6 \text{ in.-lb}$$

Redistribution of Moments

From Equations 6.6 and 6.7c, the maximum distribution percentage is

$$p_d = 20\left[1 - \frac{\omega_p + \dfrac{d}{d_p}(\omega - \omega')}{0.36\beta_1}\right]$$

$$d = 32 + 2 - 2.5 \cong 31.5 \text{ in.}$$

$$\text{compression side } b = 2 \times 4.75 = 9.50 \text{ in.}$$

Use $d_p = 0.8h = 0.8 \times 32 = 25.6$ in., and assume two #5 bars per rib at the compression side and two #7 bars per rib at the tension side of both the horizontal roof beams and the vertical wall beams. We obtain

$$A'_s = 4 \times 0.305 = 1.22 \text{ in}^2$$

$$\omega' = \frac{A'_s}{bd} \frac{f_y}{f'_c} = \frac{1.22}{9.5 \times 31.5} \times \frac{60,000}{5,000} = 0.0489$$

$$A_s = 4 \times 0.60 = 2.40 \text{ in}^2$$

$$\omega = \frac{A_s}{bd} \frac{f_y}{f'_c} = \frac{2.40}{9.5 \times 31.5} \times \frac{60,000}{5000} = 0.0962$$

Use $\omega_p = (A_{ps}/bd_p)(f_{ps}/f'_c) = 0$ since the prestressing steel is not continuous over corners of the portal frame. Then $\omega_p + (d/d_p)(\omega - \omega') = 0 + (31.5/25.6) \times (0.0962 - 0.0489) = 0.0582$. Also, $0.24\beta_1 = 0.24 \times 0.80 = 0.192 > 0.0582$. Hence, moment redistribution is permissible and we have

$$\text{distribution factor } \rho_D = 20\left[1 - \frac{\omega_p + (d/d_p)(\omega - \omega')}{0.36\beta_1}\right]$$

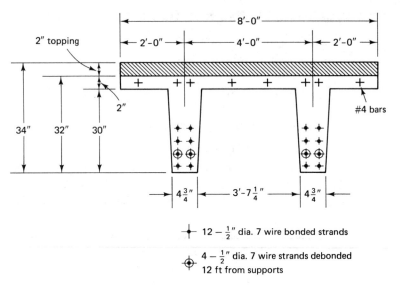

(a)

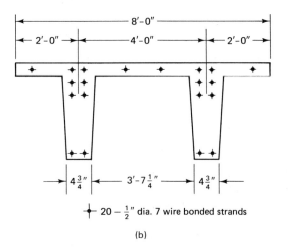

(b)

Figure 6.32 Details of beam and wall double-T's in Example 6.7. (a) Horizontal beam standard PCI section 8DT+2 (168–D1). (b) Vertical wall section 8DT32 with twenty 1/2 in. dia strands with $e_c = e_e = 0$.

$$= 20\left(1 - \frac{0 + 0.0582}{0.36 \times 0.80}\right) = 15.95\%$$

Accordingly, use a moment distribution factor of 0.12 for transferring 12 percent of the moment from the frame corners B and C to midspan BC. Also, rigid connecting steel plates should be used at the portal upper joints and be so designed to provide a moment connection capable of transferring at least 12 percent of the support moment to the midspan. Then the adjusted $M_B = M_c = (1 - 0.12)0.919 \times 10^6 = 0.809 \times$

10^6 in.-lb, the adjusted midspan $M_E = 0.876 \times 10^6 \times 1.12 = 0.981 \times 10^6$ in.-lb, and the superimposed dead-load reaction R_{SD} at the support $= (187 \times 80)/2 = 7,480$ lb.

Live-load W_L Stage. From before, the midspan moment is $M_L = 1.124 \times 10^6$ in.-lb. So the support moment is

$$M_B = M_c = \frac{240\,(80)^2 \times 12}{8} - 1.124 \times 10^6$$

$$= 1.180 \times 10^6 \text{ in.-lb}$$

The adjusted $M_B = M_c = (1 - 0.12) \times 1.180 \times 10^6 = 1.038 \times 10^6$ in.-lb, and the adjusted midspan $M_L = 1.124 \times 10^6 \times 1.12 = 1.259 \times 10^6$ in.-lb. The live-load reaction at the vertical support is

$$R_L = \frac{240 \times 80}{2} = 9,600 \text{ lb}$$

Wind Pressure Stage. From Equations 6.11 h and i,

$$M_B = (C_4 - 0.5)ph^2$$

$$M_c = -(1 - C_4)ph^2$$

From before,

$$k = \frac{I_b}{I_c}\frac{h}{\ell} = 0.45 \text{ for } I_b = I_c$$

From the chart for C_4 in Figure 6.26, $C_4 = 0.73$.
Windward side moment M_B

$$M_{B1} = (0.73 - 0.5)65(36)^2 \times 12 = 232,502 \text{ in.-lb}$$

$$M_{B2} = (1 - 0.73)40(36)^2 \times 12 = 167,961 \text{ in.-lb}$$

$$\text{Total } M_B = 232,502 + 167,961 = 400,463 \text{ in.-lb}$$

Leeward side moment M_c

$$M_{c1} = -(1 - 0.73)65(36)^2 \times 12 = -272,938 \text{ in.-lb}$$

$$M_{c2} = -(0.73 - 0.5)40(36)^2 \times 12 = -143,078 \text{ in.-lb}$$

$$\text{Total } M_c = -272,938 - 143,078 = -416,016 \text{ in.-lb}$$

The controlling wind moment $M_W = 416,016$ in.-lb, since wind can blow from either the left or the right.

From Equation 6.11 a, the vertical reactions at A and D due to wind are

$$R_{WA} = -\tfrac{1}{2}ph\frac{h}{\ell} = -\frac{(65 + 40)(36)^2}{2 \times 80} = -851 \text{ lb}$$

$$R_{WD} = +\tfrac{1}{2}ph\frac{h}{\ell} = +851 \text{ lb}$$

Loads and Moments Due to Long-Term Effects

Moments to Restrain End Rotations at B and C Due to Long-Term Prestress Losses. Figure 6.33 shows the moment distributions on horizontal member BC. One month after prestressing we have:

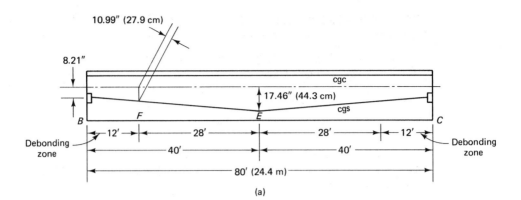

(a)

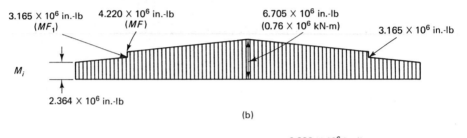

(b)

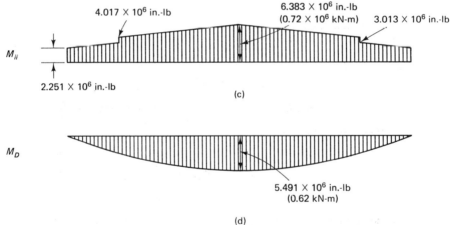

(c)

(d)

Figure 6.33 Bending moment diagrams for primary and self-weight moments for beam BC. (a) Tendon profile. (b) Prestressing moments one month after initial prestress. (c) Effective prestressing moment after all losses. (d) Beam BC self-weight moments.

$$f_{pe1} = 156{,}870 \text{ psi}$$

$$\text{Midspan } P_e = 16 \times 0.153 \times 156{,}870 = 384{,}018 \text{ lb}$$

$$\text{Support } P_e = (16 - 4) \times 0.153 \times 156{,}870 = 288{,}013 \text{ lb}$$

$$\text{Midspan moment } M_E = 384{,}018 \times 17.46 = 6.705 \times 10^6 \text{ in.-lb}$$

$$\text{Support moment } M_B = 288{,}013 \times 8.21 = 2.364 \times 10^6 \text{ in.-lb}$$

The eccentricity at section F, where four strands were debonded, is

$$e = (17.46 - 8.21)\frac{12}{40} + 8.21 = 10.99 \text{ in.}$$

and the moment at section F is

$$M_F = 384{,}018 \times 10.99 = 4.220 \times 10^6 \text{ in.-lb.}$$

The reduced M_F due to debonding is $288{,}013 \times 10.99 = 3.165 \times 10^6$ in.-lb, and $M_{F1} = 3.165 \times 10^6$ in.-lb.

The service load after all losses have occurred is as follows:

$$f_{pe} = 149{,}310 \text{ psi}$$

$$f_{pe}/f_{pe1} = \frac{149{,}310}{156{,}870} = 0.952$$

$$M_E = 6.705 \times 10^6 \times 0.952 = 6.383 \times 10^6 \text{ in.-lb}$$

$$M_F = 4.220 \times 10^6 \times 0.952 = 4.017 \times 10^6 \text{ in.-lb}$$

$$M_{F1} = 3.165 \times 10^6 \times 0.952 = 3.013 \times 10^6 \text{ in.-lb}$$

$$M_B = M_c = 2.364 \times 10^6 \times 0.952 = 2.251 \times 10^6 \text{ in.-lb}$$

Slopes at B and C at Beam Erection One Month After Prestressing

$$\text{Slope } \theta = \frac{1}{E_c I_b}[M\ell]$$

To find the areas of the moment diagrams for half the span due to symmetry, (i) add half of Figure 6.33(b) to half of Figure 6.33(d), and (ii) add half of Figure 6.33(c) to half of Figure 6.33(d). Then subtract (ii) from (i) to get the rotation of the beam at B or C that would have to be restrained by a welded connection to develop continuity at the portal frame corners B and C. We have:

(i)

$$\theta E_c I_b \times 10^{-6} = M_{(i)}\ell \text{ at beam erection}$$

$$= 2.364 \times 12 \times 12 + (3.165 - 2.364) \times 12 \times 12 \times \frac{1}{2}$$

$$+ 4.220 \times 28 \times 12 + (6.705) - 4.220) \times 28 \times 12 \times \frac{1}{2}$$

$$- 5.491 \times 40 \times 12 \times \frac{2}{3} = 2233.49 - 1757.12 = 476.47$$

(ii)

$$\theta E_c I_b \times 10^{-6} = M_{(ii)}\ell \text{ at service load}$$

$$= 2.251 \times 12 \times 12 + (3.013 - 2.251) \times 12 \times 12 \times \frac{1}{2}$$

$$+ 4.017 \times 28 \times 12 + (6.383 - 4.017) \times 28 \times 12 \times \frac{1}{2}$$

$$- 5.491 \times 40 \times 12 \times \frac{2}{3} = 2{,}126.21 - 1{,}757.12 = 369.09$$

The rotational angle θ at B or C caused by the reduction in the prestressing force due to long-term losses is

$$\frac{1}{E_c I_b}(476.47 - 369.09)10^{-6} = \frac{107.4 \times 10^6}{E_c I_b}$$

If M_r is the resisting moment at the connection weld to restrain the member against this rotation,

$$\text{Slope } \theta = \frac{M_r \ell/2}{E_c I_b} = \frac{M_r \times 480}{E_c I_b}$$

Equating the right sides of the preceding equations yields

$$\frac{107.4 \times 10^6}{E_c I_b} = \frac{480 M_r}{E_c I_b}$$

$$M_r = \frac{107.4 \times 10^6}{480} = 0.224 \times 10^6 \text{ in.-lb}$$

Moments Resulting from Creep and Shrinkage Long-Term Losses

(a) Creep

$$P_i = 463,050 \text{ lb}$$

$$P_e \text{ at erection} = 384,018 \text{ lb}$$

It is reasonable to take the creep force as the average of P_i and P_e. Thus,

$$\epsilon_{CR} = \frac{1}{A_c E_c}\left[\left(\frac{P_i + P_e}{2}\right)C_u\right]$$

Use the creep coefficient $C_u = 2.25$:

$$E_c = 57,000\sqrt{f'_c} = 57,000\sqrt{5,000} = 4.03 \times 10^6 \text{ psi}$$

$$\epsilon_{CR} = \frac{1}{567 \times 4.03 \times 10^6}\left(\frac{463,050 + 384,018}{2} \times 2.25\right)$$

$$= 417 \times 10^{-6} \text{ in./in.}$$

(b) Shrinkage

From Equation 3.15, the shrinkage strain from the time of erection (30 days after prestressing) to one year later is

$$\epsilon_{SH} = 8.2 \times 10^{-6} K_{SH}\left(1 - 0.06\frac{V}{S}\right)(100 - RH)$$

Now, $V/S = 1.79$, and if we assume that $RH = 75$ percent, then, from Table 3.4, which states that after 30 days $K_{SH} = 0.45$, we have

$$\epsilon_{SH} = 8.2 \times 10^{-6} \times 0.45(1 - 0.06 \times 1.79)(100 - 75)$$

$$= 82.3 \times 10^{-6} \text{ in./in.}$$

So the total deformation due to creep and shrinkage is $(417 + 82.3)10^{-6} = 499.3 \times 10^{-6}$ in./in.

Indeterminate Prestressed Concrete Structures   Chap. 6

From Equation 6.12,

$$M_B = M_c = \frac{3}{(2k + 3)} \frac{E_c I_c}{h} \epsilon_{BC}$$

From before, the stiffness coefficient $k = 0.45$, and $E_c = 4.03 \times 10^6$ psi. Also, precast $I_c = 55,464$ in.^{4}. Consequently,

$$M_B = M_C = \frac{3}{(2 \times 0.45 + 3)} \times \frac{4.03 \times 10^6 \times 55,464}{36 \times 12} \times 499.3 \times 10^{-6}$$

$$= 198,724 \text{ in.-lb} = 0.199 \times 10^6 \text{ in.-lb}$$

These moments due to long-term effects will produce tensile stresses at the inside face of the vertical member and bottom face of the horizontal member.

Final Moments and Stresses in the Horizontal Beam BC

Midspan Section ($e_c = 17.46$ in.)

$$M_D = 5.491 \times 10^6 \text{ in.-lb } (0.62 \times 10^6 \text{ kN-m})$$

$$M_{SD} = 0.981 \times 10^6 \text{ in.-lb}$$

$$M_L = 1.259 \times 10^6 \text{ in.-lb}$$

$$M_r = 0.224 \times 10^6 \text{ in.-lb}$$

$$M_{CR+SH} = 0.199 \times 10^6 \text{ in.-lb}$$

P_e after all losses $= 2.45 \times 149,310 = 365,810$ lb

The total superimposed moments are

$$M_T = M_{SD} + M_L + M_r + M_{CR+SH}$$

$$= (0.981 + 1.259 + 0.244 + 0.199) \times 10^6$$

$$= 2.663 \times 10^6 \text{ in.-lb}$$

$$f_b = -\frac{P_e}{A_c}\left(1 - \frac{e_c c_b}{r^2}\right) - \frac{M_D}{S^b} + \frac{M_T}{S_{bc}}$$

$$= -\frac{365,810}{567}\left(1 + \frac{17.46 \times 21.21}{97.8}\right) + \frac{5.491 \times 10^6}{2,615} + \frac{2.663 \times 10^6}{3,038}$$

$$= -111.8 \text{ psi } (C), \text{ no tension, O.K.}$$

$$f^t = -\frac{P_e}{A_c}\left(1 - \frac{e_c c_t}{r^2}\right) - \frac{M_D}{S^t} - \frac{M_T}{S_c^t}$$

$$= -\frac{365,810}{567}\left(1 - \frac{17.46 \times 10.79}{97.8}\right) - \frac{5.491 \times 10^6}{5,140} - \frac{2.663 \times 10^6}{6,952}$$

$$= -853.7 \text{ psi } (C) < f_c = 2,250 \text{ psi, O.K.}$$

Support Section ($e_e = 8.21$ in.)

$$M_D = 0$$

$$M_{SD} = 0.809 \times 10^6 \text{ in.-lb}$$

$$M_L = 1.038 \times 10^6 \text{ in.-lb}$$

$$M_W = 0.416 \times 10^6 \text{ in.-lb}$$

Not including the relief moments due to rotation, creep, and shrinkage, which cause compressive stresses, the total negative moments at supports B or C are

$$-M_T = (0.809 + 1.038 + 0.416)10^6 = 2.26 \times 10^6 \text{ in.-lb}$$

The sections at supports B and C are virtually reinforced concrete. P_e for 12 strands at either support = 288,013 lb, and

$$f^t = -\frac{288,013}{567}\left(1 - \frac{8.21 \times 10.79}{97.8}\right) + 0 + \frac{2.26 \times 10^6}{6,952}$$

$$= +278.7 \text{ psi } (T) < f_t = 12\sqrt{f_c'} = 849 \text{ psi, O.K.}$$

Provide nonprestressed steel to accommodate all the tensile stress. Also,

$$f_b = -\frac{288,013}{567}\left(1 + \frac{8.21 \times 21.21}{97.8}\right) - 0 - \frac{2.26 \times 10^6}{3,038}$$

$$= -2,159.6 \text{ psi } (C) < f_c = 2,250 \text{ psi, O.K.}$$

$$M_u = 1.4 \times 0.809 \times 10^6 + 1.7(1.038 \times 10^6 + 0.416 \times 10^6)$$

$$= 3.61 \times 10^6 \text{ in.-lb}$$

Rqd. $M_n = \dfrac{M_u}{\phi} = \dfrac{3.61 \times 10^6}{0.90} = 4.01 \times 10^6 \text{ in.-lb}$

$$M_n = A_s f_y\left(d - \frac{a}{2}\right)$$

Assume a moment arm $d - a/2 \cong 0.9d = 0.9 \times 31.5 = 28.35$ in. Then

$$4.02 \times 10^6 = A_s \times 60,000 \times 28.35$$

$$A_s = \frac{4.01 \times 10^6}{60,000 \times 28.35} = 2.36 \text{ in}^2 \ (15.2 \text{ cm}^2)$$

$$a = \frac{A_s f_y}{0.85 f_c' b} = \frac{2.36 \times 60,000}{0.85 \times 5,000 \times 96}$$

$$= 0.35 \text{ in. } (0.89 \text{ cm}) < h_f = 4 \text{ in.}$$

Hence, treat as a rectangular section:

$$d - \frac{a}{2} = 31.5 - \frac{0.35}{2} = 31.3 \text{ in. } (79.5 \text{ cm})$$

$$A_s = \frac{4.01 \times 10^6}{60,000 \times 31.3} = 2.14 \text{ in}^2 \ (13.8 \text{ cm}^2)$$

Use two #7 bars (22 mm dia) in each rib. Then

$$A_s = 4 \times 0.60 = 2.40 \text{ in}^2 > 2.14 \text{ in}^2, \text{ O.K.}$$

Final Moments and Stresses in the Vertical Column Walls AB and DC. The direct load on the column is

$$R_D + R_{SD} + R_L + R_W = 23,256 + 7,480 + 9,600 + 851 = 41,187 \text{ lb } (183 \text{ kN})$$

at, say, 15-in. eccentricity. Then

$$M_D = 41,187 \times 15 = 0.617 \times 10^6 \text{ in.-lb}$$

$$M_{SD} = 0.809 \times 10^6$$

$$M_L = 1.038 \times 10^6$$

$$M_W = 0.416 \times 10^6$$

The total moment is

$$M_T = (0.617 + 0.809 + 1.038 + 0.416)10^6$$

$$= 2.880 \times 10^6 \text{ in.-lb } (0.33 \times 10^6 \text{ kN-m})$$

For 20 strands in the wall units,

$$P_e = A_{ps}f_{pe} = 3.06 \times 149,310 = 456,887 \text{ lb } (2,032 \text{ kN})$$

$$f_b(\text{outer face}) = -\frac{P_e}{A_c} - \frac{P}{A_c} + \frac{M_T}{S_b}$$

$$= -\frac{456,887}{567} - \frac{41,187}{567} + \frac{2.880 \times 10^6}{2,615}$$

$$= +223.0 \text{ psi } (T) < 849 \text{ psi, O.K.}$$

$$f_t(\text{innerface}) = -\frac{P_e}{A_c} - \frac{P}{A_c} - \frac{M_T}{S^t}$$

$$= -\frac{456,887}{567} - \frac{41.187}{567} - \frac{2.880 \times 10^6}{5,140}$$

$$= -1,438.8 \text{ psi } (C) \ (9.86 \text{ MPa}) < f_c = 2,250 \text{ psi } (15.5 \text{ MPa})$$

Consequently, adopt the double-T section 8DT32 for the walls with twenty $\frac{1}{2}$ in. dia seven-wire 270 K low-relaxation strands arranged as shown in Figure 6.32. Also, adopt the double-T section 8DT32 + 2(168 − D1) for the horizontal top beam BC with sixteen $\frac{1}{2}$ in. dia seven-wire 270 K low-relaxation strands with four strands debonded 12 ft (3.66 m) from the face of the supports.

Figure 6.34 gives a schematic of the configuration details of the prestressed concrete portal frame. The total design would involve designing the vertical wall brackets, shear strength, flexural strength, and serviceability checks as well as detailing the welded connections between the horizontal beam and the supporting wall columns.

6.13 LIMIT DESIGN (ANALYSIS) OF INDETERMINATE BEAMS AND FRAMES

The discussions presented so far deal with the proportioning of the controlling sections in the design process, such as the midspan and support sections, with redistribution factors p_c for continuity empirically provided by the code. The continuity factors assume that adequate longitudinal reinforcement is provided at the critical continuity zones to properly control the cracking levels of those zones.

Such a procedure does not necessarily give the most efficient solution to a statically indeterminate continuous beam or frame, since full redistribution at ultimate

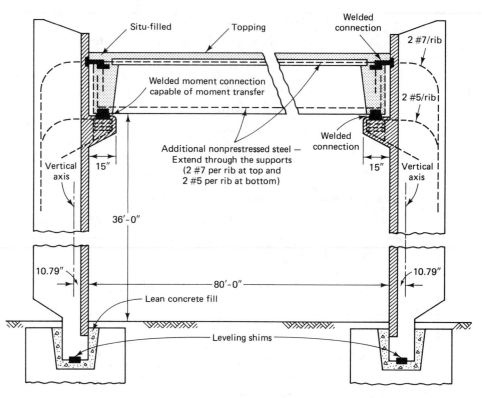

Figure 6.34 Sectional elevation and connection details of frame in Example 6.7.

load is not considered. As the applied load is gradually increased until the structure as a whole reaches its limit capacity, the critical sections, such as the supports or corners of frames, develop severe cracking, and the rotation becomes so large that, for all practical purposes, rotating *plastic hinges* have developed. If the number of plastic hinges that develop equals the number of indeterminacies, the structure becomes determinate, as *full redistribution* of moments would have taken place throughout it. With the development of an additional hinge, the structure becomes a mechanism tending toward collapse.

Analysis of the structure at *full* moment redistribution is termed as *plastic* or *limit* analysis. Since concrete cracks severely at high overloads, it is possible for the designer to *impose* the desirable locations of the plastic hinges by making the concrete member fail or making it adequately strong at any section by decreasing or increasing the reinforcement percentage without appreciably altering the stiffness of the member. This flexibility in proportioning is not available in the plastic design of steel structures, where the resulting locations of the plastic hinges are obtained from mechanisms determined by upper and lower bound solutions. Details of Baker's *theory of imposed rotations* are presented in Refs. 6.5, 6.6, and 6.7.

6.13.1 Method of Imposed Rotations

The imposed locations of the plastic hinges coincide with the locations of the maximum elastic moments for combined gravity loads and horizontal wind loads. These locations occur at the intermediate supports of continuous beams and beam-column corners of frames, as seen in the portal frame of Figure 6.35. By superposing part (a) on part (b), one plainly sees that the maximum elastic moment occurs at corner C. Since plastic moments are a magnification of the elastic moments, the natural location for the development of a plastic hinge is at that corner.

Because the structure is indeterminate to the first degree, only one hinge develops, resulting in a *basic* frame ABC, which is the fundamental frame for the imposed hinges seen in Figure 6.35(e), numbered in the order in which they are expected to form.

The structure in Figure 6.35(e) has nine indeterminacies; hence, nine plastic hinges are formed. A tenth hinge reduces the structure to a mechanism resulting in

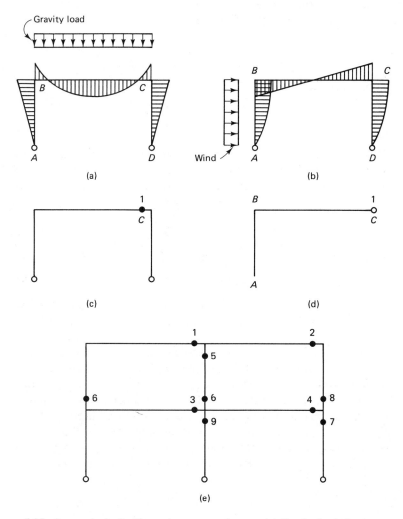

Figure 6.35 Imposed plastic hinges in concrete frames. (a) Gravity-load elastic moment. (b) Wind-load intensity moments. (c) Hinge 1 at C reducing frame to statically determinate. (d) Basic plastic frame. (e) Succession of plastic hinges in two-span, two-level frame.

collapse. Note that no plastic hinges are permitted to form at midspan of the horizontal members. The plastic moments resulting in hinges 1, 2, 3, . . . , n are denoted $\bar{X}_1$, $\bar{X}_2$, $\bar{X}_3$, . . . , $\bar{X}_n$ and are assumed to remain constant throughout the progressive deformation of the structure. Hence, the derivative of the total strain energy U with respect to the assumed plastic moments $\bar{X}_i$ at any hinge i is set equal to the plastic rotation at the hinge, i.e.,

$$\frac{\delta U}{\delta \bar{X}_i} = -\theta_i \tag{6.16}$$

If δ_{ik} is assumed to represent the relative rotation of the ith hinge due to a unit moment at the kth hinge, $\delta_{ik} = \delta_{ki}$ from Maxwell's reciprocal theorem. The coefficients δ_{ik} are called *influence coefficients*, because they represent the displacement or rotation at a particular section due to a unit moment at *another* section, i.e., $\delta_{ik} = -\theta_i$.

From the principle of virtual work,

$$\delta_{ik} = \sum \int_0^\ell \frac{M_i M_k}{E_c I} ds \tag{6.17}$$

Consequently,

$$\sum \int_0^\ell \frac{M_i M_k}{E_c I} ds = -\theta_i \tag{6.18}$$

The left-hand side of Equation 6.18 represents the integration of the products of the areas of the M_i diagrams and the ordinates of M_k diagrams at their centroids along the horizontal distance s along the span. Substituting δ_{i0} and δ_{ik} for M_k, we obtain

$$\delta_{i0} + \sum_{k=1}^{k=n} \delta_{ik} \bar{X}_k = -\theta_i \tag{6.19}$$

This is a structure having n plastic hinges to reduce it to statically determinate:

$$\delta_{10} + \delta_{11} \bar{X}_1 + \delta_{12} \bar{X}_2 + \ldots + \delta_{1n} \bar{X}_n = -\theta_1$$
$$\delta_{20} + \delta_{21} \bar{X}_1 + \delta_{22} \bar{X}_2 + \ldots + \delta_{2n} \bar{X}_n = -\theta_2 \tag{6.20}$$
$$\delta_{n0} + \delta_{n1} \bar{X}_1 + \delta_{n2} \bar{X}_2 + \ldots + \delta_{nn} \bar{X}_n = -\theta_n$$

The number of equations is equal to the number of redundancies or indeterminacies. By trial and adjustment of the redundant plastic moments $\bar{X}_1$, . . . , $\bar{X}_n$ in the solution of Equations 6.20 for controlled maximum allowable rotation of the largest rotating hinge θ_1, the plastic moments at the beam supports and column ends are obtained for the plastic design of the concrete structure. The arbitrary plastic moment values $\bar{X}_1$, $\bar{X}_2$, . . . , $\bar{X}_n$ are chosen to result in plastic rotations θ_1, θ_2, . . . , θ_n that give *full redistribution* of moments throughout the structure.

It can be proven that the influence coefficient δ_{ik} in Equations 6.20 is

$$\delta_{ik} = \frac{A_i}{E_I} \eta \tag{6.21}$$

where A_i is the area under the primary M_i bending moment diagram and η is the ordinate of the M_k moment diagram under the centroid of the M_i diagram (Ref. 6.5).

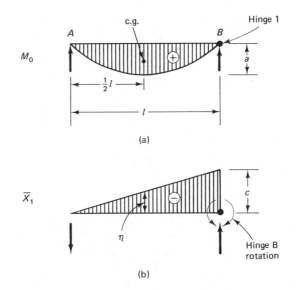

Figure 6.36 Influence coefficient determination from superposing M_0 and X_1. (a) Primary structure moment. (b) Redundant structure moment.

As an example, in Figure 6.36 the influence coefficient δ_{01} is obtained by superposing the moment diagram M_0 of the primary structure on the diagram $\overline{X}_1$ of the redundant structure created by the development of hinge 1. We have

$$A_i = \frac{2}{3}\, \ell a$$

and η under the centroid of the M_0 diagram $= c/2$, resulting in

$$\delta_{01} = -\frac{1}{EI}\left(\frac{2}{3}\,\ell a\right)\left(\frac{c}{2}\right) = \frac{1}{3EI}\,\ell a c$$

δ_{11} is obtained by superposing the redundant structure $\overline{X}_1$ on itself:

$$A_i = \frac{1}{2}\,\ell a,$$

$$\eta = \frac{2}{3}c$$

$$\delta_{11} = -\frac{1}{EI}\left(\frac{1}{2}\,\ell a \times \frac{2}{3}c\right) = -\frac{1}{3EI}\,\ell a c$$

Table 6.1 gives the values $\int M_i M_k\, ds$ for evaluating the influence coefficient values δ_{ik} for various combinations of primary and redundant moment diagrams. It can aid the designer in easily forming and solving sets of equations 6.20 for any indeterminate structural system.

TABLE 6.1 PRODUCT INTEGRAL VALUES $\int M_i M_k\, ds$ FOR VARIOUS MOMENT COMBINATIONS $EI\delta_{ik}$

M_k \ M_i	rect. a	a (dec. tri.)	a (inc. tri.)	Parabolic a	a (tri. peak)	d, b (trap.)
rect. c	lac	$\frac{1}{2}lac$	$\frac{1}{2}lac$	$\frac{2}{3}lac$	$\frac{1}{2}lac$	$\frac{1}{2}l(a+b)c$
c (dec. tri.)	$\frac{1}{2}lac$	$\frac{1}{3}lac$	$\frac{1}{6}lac$	$\frac{1}{3}lac$	$\frac{1}{4}lac$	$\frac{1}{6}l(2a+b)c$
c (inc. tri.)	$\frac{1}{2}lac$	$\frac{1}{6}lac$	$\frac{1}{3}lac$	$\frac{1}{3}lac$	$\frac{1}{4}lac$	$\frac{1}{6}l(a+2b)c$
Parabolic c	$\frac{2}{3}lac$	$\frac{1}{3}lac$	$\frac{1}{3}lac$	$\frac{8}{15}lac$	$\frac{5}{12}lac$	$\frac{1}{3}l(a+b)c$
c (tri. peak)	$\frac{1}{2}lac$	$\frac{1}{4}lac$	$\frac{1}{4}lac$	$\frac{5}{12}lac$	$\frac{1}{3}lac$	$\frac{1}{4}l(a+b)c$
c, d (trap.)	$\frac{1}{2}la(c+d)$	$\frac{1}{6}la(2c+d)$	$\frac{1}{6}la(c+2d)$	$\frac{1}{3}la(c+d)$	$\frac{1}{4}la(c+d)$	$\frac{1}{6}l[a(2c+d)+b(2d+c)]$

6.13.2 Determination of Plastic Hinge Rotations in Continuous Beams

Example 6.8

Determine the required plastic hinge rotation in the four-span beam of Figure 6.37. The beam is subjected to simple-span plastic moment M_0 so that the midspan moment is equal to the support moment $= \frac{1}{2}M_0$ before full rotation of the hinges and full moment redistribution take place.

Solution The structure is statically indeterminate to the third degree, so that three hinges will develop at the plastic limit. Assume the maximum ordinate c of the redundant moment at hinge location to be unity. Then, from Table 6.1 and Figure 6.38,

$$EI\delta_{10} = -\frac{2}{3}M_0\ell$$

$$\delta_{11} = \frac{2}{3}\ell$$

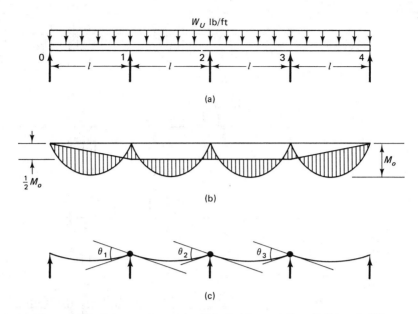

Figure 6.37 Primary moments and plastic hinge rotations in Example 6.8.

$$\delta_{12} = \frac{1}{6}\ell$$

$$\delta_{13} = 0$$

From Equation 6.16,

$$-\theta_1 = \delta_{10} + \delta_{11}\,\overline{X}_2 + \delta_{12}\,\overline{X}_2 + \delta_{13}\,\overline{X}_3$$

$$-EI\theta_1 = -\frac{2}{3}M_0\ell + 0.5M_0\left(\frac{2\ell}{3}\right) + 0.5M_0\left(\frac{\ell}{6}\right) + 0 = -\frac{M_0\ell}{4}$$

Also, again from Table 6.1 and Figure 6.38,

$$EI\delta_{20} = \frac{2}{3}M_0\ell\left(-\frac{1}{2}\right) + \frac{2}{3}M_0\ell\left(-\frac{1}{2}\right) = -\frac{2}{3}M_0\ell$$

$$EI\delta_{21} = \left(-\frac{\ell}{2}\right)\left(-\frac{1}{3}\right) = +\frac{\ell}{6}$$

$$EI\delta_{22} = 2\left(-\frac{\ell}{2}\right)\left(-\frac{2}{3}\right) = +\frac{2\ell}{3}$$

$$EI\delta_{23} = \left(-\frac{\ell}{2}\right)\left(-\frac{1}{3}\right) = +\frac{\ell}{6}$$

From Equation 6.16,

$$-\theta_2 = \delta_{20} + \delta_{21}\,\overline{X}_1 + \delta_{22}\,\overline{X}_2 + \delta_{23}\,\overline{X}_3$$

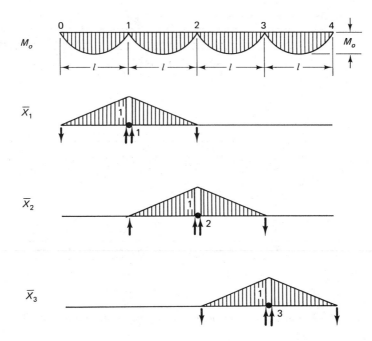

Figure 6.38 Primary and redundant moments in Example 6.8.

$$-EI\theta_2 = -\frac{2}{3}M_0\ell + 0.5M_0\left(+\frac{\ell}{6}\right) + 0.5M_0\left(+\frac{2}{3}\ell\right) + 0.5M_0\left(+\frac{\ell}{6}\right) = -\frac{M_0\ell}{6}$$

From symmetry, $\theta_3 = \theta_1$. Therefore, the required plastic hinge rotations at the support are

$$\theta_1 = \frac{M_0\ell}{4EI} = \theta_3$$

and

$$\theta_2 = \frac{M_0\ell}{6EI}$$

Since $\theta_2 < \theta_1$, the first hinge to develop, and the controlling one in the design, is $\theta_1 = M_0\ell/4EI$.

Note that the procedure used in Example 6.8 can be used in the limit design of any continuous beam or multistory frame. Also, it is important to maintain the correct sign convention by drawing all moments at the *tension* side of the member, as noted earlier.

The preceding discussion gives the basic *imposed rotations approach* embodied in Baker's theory. Other modified approches have been proposed by Cohn (Ref. 6.17), Sawyer (Ref. 6.18), and Furlong (Ref. 6.19). Cohn's method is based on the requirements of limit equilibrium and serviceability, with a subsequent check of rotational compatibility. Sawyer's method is based on the simultaneous requirements of limit equilibrium and rotational compatibility, with a subsequent check of serviceability.

Furlong's method is based on assigning ultimate moments for various loading patterns on the continuous spans that would satisfy serviceability and limit equilibrium for the worst case. The sections are reinforced in such a manner that the ultimate

Pretensioned T-beam with rectangular confining reinforcement at failure (Nawy, Potyondy).

moment strengths for each span are equal to or greater than the *product* of the maximum ultimate moment M_o in the span when the ends are free to rotate and a moment coefficient k_1 for various boundary conditions as listed in Table 6.2.

6.13.3 Rotational Capacity of Plastic Hinges

Rotation is the *total* change in slope along the short plasticity length concentrated at the hinge zone. It can also be described as the angle of discontinuity between the plastic parts of the member on either side of the plastic hinge. As Figure 6.39 shows, there are two types of hinges—tensile and compressive. In order that the first hinge that develops in the structure, usually the critical hinge, can rotate without rupture until the nth hinge develops, the concrete section at the first hinge has to be made ductile enough through section core confinement to be able to sustain the necessary rotation. This is equally applicable to both tension and compression hinges, where confinement of the concrete core is obtained through concentration of closed stirrups at the supports and column ends. A typical plot showing increase in rotation through increase in confining reinforcement is shown in Figure 6.40 (Ref. 6.14).

The plasticity length ℓ_p determines the extent of severe cracking and the mag-

TABLE 6.2 BEAM MOMENT COEFFICIENTS FOR ASSIGNED MOMENTS

Boundary condition	Moment type	Beam loaded by one concentrated load at midspan	All other beams
Span with ends restrained	Negative	0.37	0.50
	Positive	0.42	0.33
Span with one end restrained	Negative	0.56	0.75
	Positive	0.50	0.46

nitude of rotation of the hinge. Therefore, it is important to limit the magnitude of ℓ_p through the use of *closely spaced* ties or closed stirrups. In this manner, the strain capacity of the concrete at the confined section can be significantly increased, as

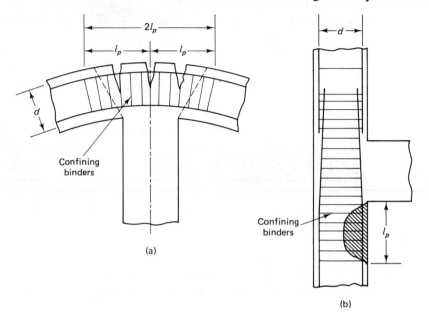

Figure 6.39 Plasticity zones ℓ_p in plastic hinges. (a) Tensile hinge. (b) Compressive hinge.

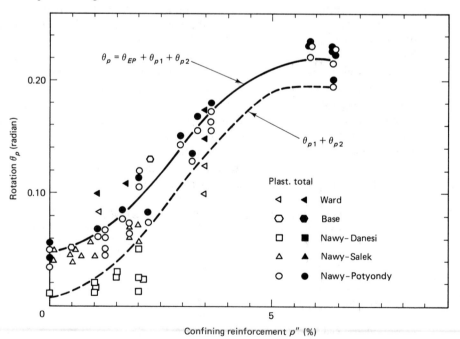

Figure 6.40 Comparison of plastic rotation with results of other authors.

experimentally demonstrated by several investigators, including Nawy (Refs. 6.12, 6.13, and 6.14). Several empirical expressions have been developed; see, for example, Baker (Ref. 6.5), Corley (Ref. 6.11), Nawy (Ref. 6.14), Sawyer (Ref. 6.18), and Mattock (Ref. 6.20). Two of them, for the plasticity length ℓ_p and the concrete strain ϵ_c (Ref. 6.20), are

$$\ell_p = 0.5d + 0.5Z \tag{6.22}$$

and

$$\epsilon_c = 0.003 + 0.02\frac{b}{Z} + 0.2\rho_s \tag{6.23}$$

where d = effective depth of the beam (in.)
$\quad Z$ = distance from the critical section to the point of contraflexure
$\quad \rho_s$ = ratio of volume of confining binder steel (including the compression steel) to the volume of the concrete core
$\quad \ell_p$ = *half* the plasticity length on each side of the centerline of plastic hinge.

Equation 6.22 can be more conservative for high values of ρ_s.

Once the concrete strain ϵ_c is determined, the angle of rotation of the plastic hinge is readily determined from the expression

$$\theta_p = \left(\frac{\epsilon_c}{c} - \frac{\epsilon_{ce}}{kd}\right)\ell_p \tag{6.24}$$

where c = neutral axis depth at the limit state at failure
$\quad \epsilon_{ce}$ = strain in the concrete at the extreme compression fibers when the yield curvature is reached
$\quad kd$ = neutral axis depth corresponding to ϵ_{ce}
$\quad \epsilon_c$ = concrete compressive strain at the end of the inelastic range or at the limit state at failure.

The strain ϵ_{ce} can usually be taken at the load level when the strain in the tension reinforcement reaches the yield strain $\epsilon_y = f_y/E_s$. It can be taken to be approximately 0.001 in./in. or higher, depending on whether the tension steel yields before the concrete crushes at the extreme compression fibers in cases of overreinforced beams, as is sometimes the case in prestressed beams. If concrete crushes first, the value of ϵ_{ce} will have to be higher than 0.001 in./in. A limit of allowable $\epsilon_c = 1.0$ percent for confined concrete is recommended in determining the maximum allowable plastic rotation θ_p, although strains of confined concrete as high as 13 percent could be obtained, as shown in Ref. 6.14.

The discussion in this entire section (6.13) is equally applicable to reinforced and prestressed concrete indeterminate structures at the plastic loading range where full redistribution of moments has taken place. As the prestressed concrete section is cracked and decompression in the prestressing steel has taken place, the structural system gradually starts to behave similarly to a reinforced concrete system. As the load reaches the limit state at failure, the flexural behavior of the prestressed concrete elements is expected to closely resemble that of reinforced concrete elements.

Sec. 6.13   Limit Design (Analysis) of Indeterminate Beams and Frames   **383**

6.13.4 Calculation of Available Rotational Capacity

Example 6.9

Determine the required and available rotational capacities of the critical plastic hinges in the continuous prestressed concrete beam in Example 6.8 for both confined and unconfined concrete. Given data are as follows:

$$M_u = \tfrac{1}{2}M_0 = 400bd^2$$

$$c = 0.28d$$

$$kd = 0.375d$$

$$\epsilon_{ce} = 0.001 \text{ in./in. at end of the elastic range}$$

$$\epsilon_{ce} = 0.004 \text{ in./in. at end of the inelastic range for unconfined sections}$$

$$\text{Max. allowable } \epsilon_c = 0.01 \text{ in/in. for confined sections}$$

$$E_c/I_c = 150,000bd^3 \text{ in}^3\text{-lb}$$

$$\frac{Z}{d} = 5.5$$

$$f'_c = 5,000 \text{ psi}$$

$$f_y = 60,000 \text{ psi for the mild steel}$$

Also, calculate the maximum allowable span-to-depth ratio ℓ/d for the beam if full redistribution of moments is to occur at the limit state at failure.

Solution

$$M_0 = 2 \times 400bd^2 = 800bd^2$$

From Example 6.8,

$$\text{Required } \theta_1 = \theta_3 = \frac{M_0\ell}{4E_c I_c} = \frac{800bd^2\ell}{4 \times 150,000bd^3} = \frac{1}{750}\frac{\ell}{d} \text{ radian}$$

$$\text{Required } \theta_2 = \frac{M_0\ell}{6EI} = \frac{800bd^2\ell}{6 \times 150,000bd^3} = \frac{1}{1,125}\frac{\ell}{d} \text{ radian}$$

From Equation 6.22,

$$\ell_p = 0.5d + 0.05Z = 0.5d + 0.05 \times 5.5d = 0.775d$$

The total plasticity length on both sides of the hinge centerline is $2 \times 0.775d = 1.55d$.

Unconfined Section. From Equation 6.24,

$$\text{Available } \theta_p = \left(\frac{\epsilon_c}{c} - \frac{\epsilon_{ce}}{kd}\right)\ell_p = \left(\frac{0.004}{0.28d} - \frac{0.001}{0.375d}\right)1.55d = 0.018 \text{ radian}$$

For full moment redistribution,

$$\frac{1}{750}\frac{\ell}{d} \leq 0.018 \qquad \text{and} \qquad \frac{1}{1,125}\frac{\ell}{d} \leq 0.018$$

or

$$\frac{\ell}{d} \leq 13.5 \qquad \text{and} \qquad \frac{\ell}{d} \leq 20.3$$

Confined Sections

Max. allow $\epsilon_c = 0.01$ in./in.

$$\text{Available } \theta_p = \left(\frac{0.01}{0.28d} - \frac{0.001}{0.375d}\right)1.55d = 0.51 \text{ radian}$$

For full moment redistribution,

$$\frac{1}{750}\frac{\ell}{d} \le 0.051 \qquad \text{and} \qquad \frac{1}{1,125}\frac{\ell}{d} \le 0.051$$

or

$$\frac{\ell}{d} \le 38.3 \qquad \text{and} \qquad \frac{\ell}{d} = 57.4$$

Comparing the results of the unconfined sections in the first case to the confined sections in the second case, one sees that confinement of the concrete at the plastic hinging zone permits more slender sections for full plasticity and, hence, a more economical indeterminate structural system.

6.13.5 Check for Plastic Rotation Serviceability

Example 6.10

If closed-stirrup binders are used in Example 6.9 with binder ratio $\rho_s = 0.025$ and $\ell/d = 35$ with c at failure $= 0.25d$, verify whether the continuous beam satisfies the rotation serviceability criteria given that $b = \frac{1}{2}d$.

Solution

$$\frac{Z}{d} = 5.5$$

Hence,

$$\frac{b}{Z} = \frac{1}{11}$$

Also,

$$\text{Available } \epsilon_c = 0.003 + 0.2\frac{b}{Z} + 0.2\rho_s$$

$$= 0.003 + 0.2 \times \frac{1}{11} + 0.2 \times 0.025 = 0.026 \text{ in./in.}$$

The maximum allowable to be utilized is $\epsilon_c = 0.01$ in./in. So use, for $\epsilon_c = 0.01$, the corresponding available plastic rotation:

$$\theta_p = \left(\frac{0.01}{0.25d} - \frac{0.001}{0.375d}\right)1.55d = 0.058 \text{ radian}$$

$$\text{Rqd. } \theta_1 = \frac{1}{750}\frac{\ell}{d} = \frac{35}{750} = 0.046 \text{ radian}$$

$$\text{Rqd. } \theta_2 = \frac{1}{1,125}\frac{\ell}{d} = \frac{35}{1,125} = 0.031 \text{ radian}$$

Available $\theta_p = 0.058$ radian $>$ reqd. $\theta = 0.046$ radian. Thus, the beam satisfies the serviceability criteria for plastic rotation.

The foregoing discussion for the limit design of reinforced and prestressed concrete indeterminate beams and frames permits the design engineeer to provide ductile connections at beam-column supports and generate full moment redistribution throughout the structure, resulting in full utilization of the strength of the prestressed system. Also, continuity in both pretensioned and post-tensioned systems to withstand seismic loading can be effectively utilized through the appropriate confinement of the connecting zones by means of the procedures presented in this section.

REFERENCES

6.1 ACI Committee 318. *Building Code Requirements for Reinforced Concrete, ACI Standard 318-89*. Detroit: American Concrete Institute, 1989.

6.2 ACI Committee 318. *Commentary on Building Code Requirements for Reinforced Concrete, ACI 318 R-89*. Detroit: American Concrete Institute, 1989.

6.3 Taylor, F. W., Thompson, S. E., and Smulski, E. *Concrete Plain and Reinforced*, vol. 2. New York: Wiley, 1947.

6.4 Abeles, P. W., and Bardhan Roy, B. K. *Prestressed Concrete Designer's Handbook*. 3d ed. London: Viewpoint Publications, 1981.

6.5 Baker, A. L. L. *The Ultimate Load Theory Applied to the Design of Reinforced and Prestressed Concrete Frames*. London: Concrete Publications Ltd., 1956.

6.6 Baker, A. L. L. *Limit State Design of Reinforced Concrete*. London: Cement and Concrete Association, 1970.

6.7 Ramakrishnan, V., and Arthur, P. D. *Ultimate Strength Design of Structural Concrete*. London: Wheeler, 1977.

6.8 Nawy, E. G. *Reinforced Concrete—A Fundamental Approach*. Englewood, Cliffs, N.J., Prentice Hall, 1985.

6.9 Lin, T. Y., and Thornton, K. "Secondary Moments and Moment Redistribution in Continuous Prestressed Concrete Beams." *Journal of the Prestressed Concrete Institute*, January–February 1972, pp. 8–20.

6.10 Nilson, A. H. *Design of Prestressed Concrete*. New York: Wiley, 1987.

6.11 Corley, W. G. "Rotational Capacity of Reinforced Concrete Beams." *Journal of the Structural Division*, ASCE 92 (1966): 121–146.

6.12 Nawy, E. G., and Salek, F. "Moment-Rotation Relationships of Non-Bonded Prestressed Flanged Sections Confined with Rectangular Spirals." *Journal of the Prestressed Concrete Institute*, August 1968, pp. 40–55.

6.13 Nawy, E. G., Danesi, R., and Grosco, J. "Rectangular Spiral Binders Effect on the Rotation Capacity of Plastic Hinges in Reinforced Concrete Beams." *Journal of the American Concrete Institute*, December 1968, pp. 1001–1010.

6.14 Nawy, E. G., and Potyondy, J. G. "Moment Rotation, Cracking and Deflection of Spirally Bound Pretensioned Prestressed Concrete Beams." *Engineering Research Bulletin No. 51*, New Brunswick, N.J.: Bureau of Engineering Research, Rutgers University, 1970, pp. 1–97.

6.15 Prestressed Concrete Institute. *PCI Design Handbook*. Chicago: Prestressed Concrete Institute, 1985.

6.16 Park, R., and Paulay, J. *Reinforced Concrete Structures*. New York: Wiley, 1975.

6.17 Cohn, M. Z., "Rotational Compatibility in the Limit Design of Reinforced Concrete Beams." *Proceedings of the International Symposium on the Flexural Mechanics of Reinforced Concrete*, ASCE-ACI. Miami: Nov. 1964, pp. 359–382.

6.18 Sawyer, H. A. "Design of Concrete Frames for Two Failure Stages." *Proceedings of the International Symposium on the Flexural Mechanics of Reinforced Concrete*, ASCE-ACI. Miami: Nov. 1964, pp. 405–431.

6.19 Furlong, R. W. "Design of Concrete Frames by Assigned Limit Moments." *Journal of the American Concrete Institute* 67 (1970): 341–353.

6.20 Mattock, A. H. "Discussion of Rotational Capacity of Reinforced Concrete Beams by W. G. Corley." *Journal of the Structural Division*, ASCE 93 (1967): 519–522.

PROBLEMS

6.1 A two-span continuous beam has a parabolic tendon profile shown in Figure P6.1. The prestressing force P_e after all losses is 450,000 lb (2,002 kN). The beam has a rectangular section 15 in. (38.1 cm) wide.
  (a) Find the final profile of the thrust C-line and the beam reactions at all supports.
  (b) Design the beam depth such that the concrete fiber stresses due only to prestressing do not exceed the maximum allowable for normal-weight concrete having cylinder strength $f'_c = 6,000$ psi (41.4 MPa).
  (c) Determine the shape of the concordant tendon, and draw a beam elevation of the tendon profile.

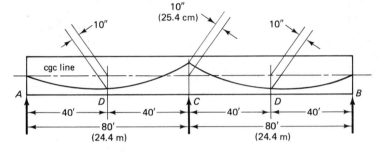

Figure P6.1.

6.2 Solve Problem 6.1 for a tendon profile harped at midspan points D, but having the same eccentricities. Compare the results with those of Problem 6.1.

6.3 Solve problem 6.1 for a tendon profile which has eccentricities $e_A = e_B = 3$ in. (7.6 cm) at the exterior supports above the cgc line.

6.4 Develop the tendon profile for the continuous beam in Example 6.4 if the beam is continuous over two equal spans of 64 ft (19.4 m).

6.5 Solve Problem 6.4 if the beam is continuous over four equal spans of 64 ft (19.4 m).

6.6 Design, for service loading, the frame in Example 6.7 using the same loading conditions if the span of the horizontal beam is 90 ft (27.4 m) and the height of the portal is 25 ft (7.6 m).

6.7 Design the portal frame of an aircraft hangar having the dimensions and the loading shown in Figure P6.7. Detail the connections and the configuration of the prestressing tendons of the horizontal member. Use the same allowable stresses as in Example 6.7.

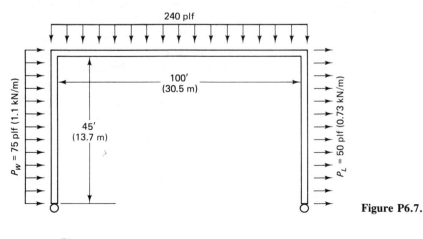

Figure P6.7.

CHAPTER 7

Camber, Deflection, and Crack Control

7.1 INTRODUCTION

Serviceability of prestressed concrete members in their deflection and cracking behavior is at least as important a criterion in design as serviceability of reinforced concrete elements. The fact that prestressed concrete elements are more slender than their counterparts in reinforced concrete, and their behavior more affected by flexural cracking, makes it more critical to control their deflection and cracking. The primary design involves proportioning the structural member for the limit state of flexural stresses at service load and for limit states of failure in flexure, shear, and torsion, including anchorage development strength. Such a design can only become complete if the magnitudes of long-term deflection, camber (reverse deflection), and crack width are determined to be within allowable serviceability values.

Prestressed concrete members are continuously subjected to sustained eccentric compression due to the prestressing force, which seriously affects their long-term creep deformation performance. Failure to predict and control such deformations can lead to high reverse deflection, i.e., camber, which can produce convex surfaces detrimental to proper drainage of roofs of buildings, to uncomfortable ride characteristics in bridges and aqueducts, and to cracking of partitions in apartment buildings, including misalignment of windows and doors.

The difficulty of predicting very accurately the total long-term prestress losses makes it more difficult to give a precise estimate of the magnitude of expected camber. Accuracy is even more difficult in partially prestressed concrete systems, where

Transamerica Pyramid, San Francisco, California.

limited cracking is allowed through the use of additional nonprestressed reinforcement. Creep strain in the concrete increases camber, as it causes a negative increase in curvature which is usually more dominant than the decrease produced by the decrease in prestress losses due to creep, shrinkage, and stress relaxation. A best estimate of camber increase should be based on accumulated experience, span-to-depth ratio code limitations, and a correct choice of the modulus E_c of the concrete. Calculation of the moment-curvature relationships at the major incremental stages of loading up to the limit state at failure would also assist in giving a more accurate evaluation of the stress-related load deflection of the structural element.

The cracking aspect of serviceability behavior in prestressed concrete is also critical. Allowance for limited cracking in "partial prestressing" through the additional use of nonprestressed steel is prevalent. Because of the high stress levels in the prestressing steel, corrosion due to cracking can become detrimental to the service life of the structure. Therefore, limitations on the magnitudes of crack widths and their spacing have to be placed, and proper crack width evaluation procedures used. The following discussion of the state of the art emphasizes the extensive work of the author on cracking in pretensioned and post-tensioned prestressed beams.

7.2 BASIC ASSUMPTIONS IN DEFLECTION CALCULATIONS

Deflection calculations can be made either from the moment diagrams of the prestressing force and the external transverse loading, or from the moment-curvature relationships. In either case, the following basic assumptions have to be made:

1. The concrete gross cross-sectional area is accurate enough to compute the moment of inertia except when refined computations are necessary.
2. The modulus of concrete $E_c = 33w^{1.5}\sqrt{f_c'}$, where the value of f_c' corresponds to the cylinder compressive strength of concrete at the age at which E_c is to be evaluated.
3. The principle of superposition applies in calculating deflections due to transverse load and camber due to prestressing.
4. All computations of deflection can be based on the center of gravity of the prestressing strands (cgs), where the strands are treated as a single tendon.
5. Deflections due to shear deformations are disregarded.
6. Sections can be treated as *totally elastic* up to the decompression load. Thereafter, the cracked moment of inertia I_{cr} can give a more accurate determination of deflection and camber.

7.3 SHORT-TERM (INSTANTANEOUS) DEFLECTION OF UNCRACKED AND CRACKED MEMBERS

7.3.1 Load-deflection Relationship

Short-term deflections in prestressed concrete members are calculated on the assumption that the sections are homogeneous, isotropic, and elastic. Such an assumption is

Supporting base of the Transamerica Pyramid, San Francisco, California.

an approximation of actual behavior, particularly that the modulus E_c of concrete varies with the age of the concrete and the moment of inertia varies with the stage of loading, i.e., whether the section is uncracked or cracked.

Ideally, the load-deflection relationship is trilinear, as shown in Figure 7.1. The three regions prior to rupture are:

Region I. Precracking stage, where a structural member is crack free.

Region II. Postcracking stage, where the structural member develops acceptable controlled cracking in both distribution and width.

Region III. Postserviceability cracking stage, where the stress in the tension reinforcement reaches the limit state of yielding.

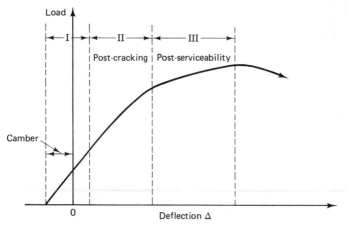

Figure 7.1 Beam load-deflection relationship. Region I, precracking stage; region II, postcracking stage; region III, postserviceability stage.

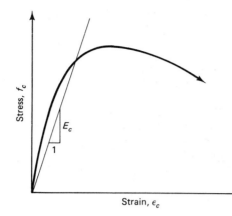

Figure 7.2 Stress-strain diagram of concrete.

7.3.1.1 Precracking stage: region I.

The precracking segment of the load-deflection curve is essentially a straight line defining full elastic behavior, as in Figure 7.1. The maximum tensile stress in the beam in this region is less than its tensile strength in flexure, i.e., it is less than the modulus of rupture f_r of concrete. The flexural stiffness EI of the beam can be estimated using Young's modulus E_c of concrete and the moment of inertia of the uncracked concrete cross section. The load-deflection behavior is significantly dependent on the stress-strain relationship of the concrete. A typical stress-strain diagram of concrete is shown in Figure 7.2.

The value of E_c can be estimated using the ACI empirical expression given in Chapter 2, viz.,

$$E_c = 33w^{1.5}\sqrt{f_c'} \tag{7.1 a}$$

or

$$E_c = 57{,}000\sqrt{f_c'} \qquad \text{for normal-weight concrete}$$

The precracking region stops at the initiation of the first flexural crack, when the concrete stress reaches its modulus of rupture strength f_r. Similarly to the direct tensile splitting strength, the modulus of rupture of concrete is proportional to the square root of its compressive strength. For design purposes, the value of the modulus for concrete may be taken as

$$f_r = 7.5\lambda\sqrt{f_c'} \tag{7.1 b}$$

where $\lambda = 1.0$ for normal-weight concrete. If all-lightweight concrete is used, then $\lambda = 0.75$, and if sand-lightweight concrete is used, $\lambda = 0.85$.

If the distance of the extreme *tension* fibers of concrete from the center of gravity of the concrete section is y_t, then the cracking moment is given by

$$M_{cr} = \frac{I_g}{y_t}\left[-\frac{P_e}{A_c}\left(1 + \frac{ec_b}{r^2}\right) + 7.5\lambda\sqrt{f_c'}\right] \tag{7.2 a}$$

or

$$M_{cr} = S_b\left[7.5\lambda\sqrt{f_c'} - \frac{P_e}{A_c}\left(1 + \frac{ec_b}{r^2}\right)\right] \tag{7.2 b}$$

More conservatively, from Equation 5.12, the cracking moment can be expressed as

$$M_{cr} = S_b[6\lambda \sqrt{f'_c} + f_{ce} - f_d] \qquad (7.3)$$

7.3.1.2 Calculation of cracking moment M_{cr}

Example 7.1

Compute the cracking moment M_{cr} for a prestressed rectangular beam section having a width $b = 12$ in. (305 mm) and a total depth $h = 24$ in. (610 mm), given that $f'_c = 4,000$ psi (27.6 MPa). The concrete stress f_b due to eccentric prestressing is 1,850 psi (12.8 MPa) in compression.

Solution The modulus of rupture $f_r = 7.5\sqrt{f'_c} = 7.5\sqrt{4,000} = 474.3$ psi (3.27 MPa). Also, $I_g = bh^3/12 = 12(24)^3/12 = 13,824$ in^4 (575,470 cm^2); $y_t = 24/2 = 12$ in. (305 mm) to the tension fibers; and $S_b = I_g/y_t = 13,824/12 = 1,152$ in^3 (18,878 cm^3).

From Equation 7.2 b,

$$M_{cr} = S_b\left[7.5\lambda \sqrt{f'_c} - \frac{P_e}{A_c}\left(1 + \frac{ec_b}{r^2}\right)\right] = 1,152[474.3 - (-1,850)]$$

$$= 2.68 \times 10^6 \text{ in.-lb (302.9 kN-m)}$$

If the beam were not prestressed, the moment would be $M_{cr} = f_r I_g/y_t = 473.3 \times 13,824/12 = 0.546 \times 10^6$ in.-lb (61.7 kN-m).

7.3.1.3 Postcracking service-load stage: region II.
The precracking region ends at the initiation of the first crack and moves into region II of the load-deflection diagram of Figure 7.1. Most beams lie in this region at service loads. A beam undergoes varying degrees of cracking along the span corresponding to the stress and deflection levels at each section. Hence, cracks are wider and deeper at midspan, whereas only narrow, minor cracks develop near the supports in a simple beam.

When flexural cracking develops, the contribution of the concrete in the tension area diminishes substantially. Hence, the flexural rigidity of the section is reduced, making the load-deflection curve less steep in this region than in the precracking stage segment. As the magnitude of cracking increases, stiffness continues to decrease, reaching a lower bound value corresponding to the reduced moment of inertia of the cracked section. The moment of inertia I_{cr} of the cracked section can be calculated from the basic principles of mechanics.

7.3.1.4 Postserviceability cracking stage and limit state of deflection behavior at failure: region III.
The load-deflection diagram of Figure 7.1 is considerably flatter in region III than in the preceding regions. This is due to substantial loss in stiffness of the section because of extensive cracking and considerable widening of the stabilized cracks throughout the span. As the load continues to increase, the strain ϵ_s in the steel at the tension side continues to increase beyond the yield strain ϵ_y with no additional stress. The beam is considered at this stage to have structurally failed by initial yielding of the tension steel. It continues to deflect without additional loading, the cracks continue to open, and the neutral axis continues to rise toward the outer compression fibers. Finally, a secondary compression failure develops, leading to total crushing of the concrete in the maximum moment region followed by rupture.

7.3.2 Uncracked Sections

7.3.2.1 Deflection calculations. Deflection calculations for uncracked prestressed sections tend to be more accurate than those for cracked sections since the assumptions of elastic behavior are more applicable. The use of the moment of inertia of the gross section rather than the transformed section does not appreciably affect the accuracy sought in the calculations.

Suppose a beam is prestressed with a constant eccentricity tendon as shown in Figure 7.3. Use the sign convention of plotting the primary moment diagram on the tension side of the beam, and employ the elastic weight method by converting the moment diagram ordinates to elastic weights $M_1/(E_c I_c)$ on a beam span ℓ. Then the moment of the weight intensity $(Pe)/E_c I_c$ of the half-span AC in Figure 7.3(c) about the midspan point C gives

$$\delta_c = \frac{Pe\ell}{2E_c I_c}\left(\frac{\ell}{2}\right) - \frac{Pe}{E_c I_c}\left(\frac{\ell}{2} \times \frac{\ell}{4}\right) = \frac{Pe\ell^2}{E_c I_c 8} \qquad (7.4)$$

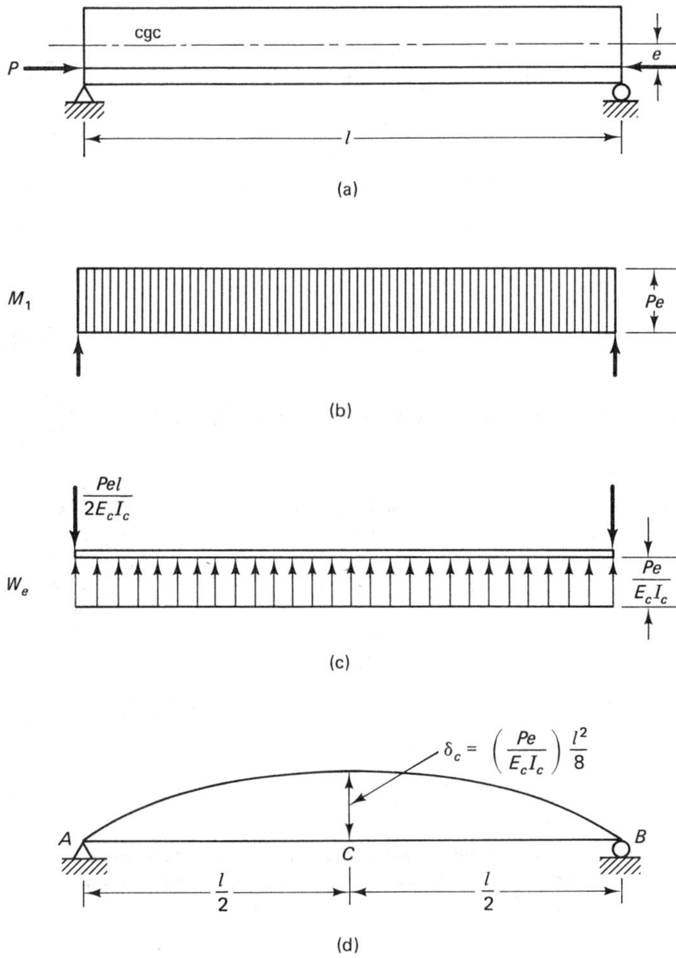

M_1

(a)

(b)

$\frac{Pe l}{2E_c I_c}$

W_e

$\frac{Pe}{E_c I_c}$

(c)

$\frac{Pe}{E_c I_c} \cdot \frac{\ell}{2} \cdot \frac{\ell}{4} = \frac{Pe\ell^2}{E_c I_c \cdot 8}$

$\delta_c = \left(\frac{Pe}{E_c I_c}\right)\frac{l^2}{8}$

A

C

B

$\frac{l}{2}$

$\frac{l}{2}$

(d)

Figure 7.3 Calculation of deflection by elastic weight or moment-area method. (a) Prestressing force. (b) Primary moment M_1. (c) Elastic weight $W_e = M/E_c I_c$. (d) Deflection.

Notice that the deflection diagram in Figure 7.3(d) is drawn *above* the base line, as the beam cambers upwards due to prestressing.

Similar computations can be performed for any tendon profile and any type of transverse loading regardless of whether the tendon geometry or loading is symmetrical or not. The final camber or deflection is the superposition of the deflections due to prestressing on the deflections due to external loads.

7.3.2.2 Strain and curvature elevation.

The distribution of strain across the depth of the section at the controlling stages of loading is linear, as is shown in Figure 7.4, with the angle of curvature dependent on the top and bottom concrete extreme fiber strains ϵ_{ct} and ϵ_{cb}. From the strain distributions, the curvature at the various stages of loading can be expressed as follows:

(1) Initial prestress:

$$\phi_i = \frac{\epsilon_{cbi} - \epsilon_{cti}}{h} \qquad (7.5\ a)$$

(2) Effective prestress after losses:

$$\phi_e = \frac{\epsilon_{cbe} - \epsilon_{cte}}{h} \qquad (7.5\ b)$$

(3) Service load:

$$\phi = \frac{\epsilon_{ct} - \epsilon_{cb}}{h} \qquad (7.5\ c)$$

(4) Failure:

$$\phi = \frac{\epsilon_{cr}}{c} \qquad (7.5\ d)$$

Use a *plus* sign for tensile strain and a *minus* sign for compressive strain.

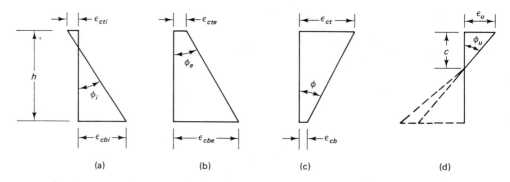

Figure 7.4 Strain distribution and curvature at controlling loading stages. (a) Initial prestress, $\phi_i = (\epsilon_{cbi} - \epsilon_{cti})/h$. (b) Effective prestress after losses, $\phi_e = (\epsilon_{cbe} - \epsilon_{cte})/h$. (c) Service load, $\phi = (\epsilon_{cb} - \epsilon_{ct})/h$. (d) Failure, $\phi_u = \epsilon_u/h$.

Priest Point Park Bridge in Olympia, Washington, a cast-in-place prestressed concrete structure. (*Courtesy*, Arvid Grant and Associates, Inc.)

The effective curvature ϕ_e in Figure 7.4(b) after losses is the sum, using the appropriate sign, of the initial curvature ϕ_1, the change in curvature $d\phi_1$ due to loss of prestress from creep, relaxation, and shrinkage, and the change in curvature $d\phi_2$ due to creep of concrete under sustained prestressing force, i.e.,

$$\phi_e = \phi_i + d\phi_1 + d\phi_2 \tag{7.6}$$

where, from the basic mechanics of materials,

$$\phi = \frac{M}{E_c I_c} \tag{7.7 a}$$

For the primary moment, $M_1 = P_e e$, so that

$$\phi = \frac{P_e e}{E_c I_c} \tag{7.7 b}$$

Substituting into Equation 7.4 for simply supported beams with constant-eccentricity tendons yields

$$\delta_c = \frac{\phi \ell^2}{8} \tag{7.8}$$

The general expression for deflection in terms of curvature as proposed by Tadros in Ref. 7.3 gives

$$\delta = \phi_c \frac{\ell^2}{8} - (\phi_e - \phi_c)\frac{a^2}{6} \tag{7.9}$$

where ϕ_c = curvature at midspan
ϕ_e = curvature at the support
a = length parameter as a function of the tendon profile.

7.3.2.3 Instantaneous deflection of simply supported beam prestressed with parabolic tendon

Example 7.2

Find the immediate midspan deflection of the beam shown in Figure 7.5 prestressed by a parabolic tendon with maximum eccentricity e at midspan and effective prestressing force P_e. Use both the elastic weight method and the equivalent weight method. The span of the beam is ℓ ft, and its stiffness is $E_c I_c$.

Solution

Elastic weight method. From Figure 7.5(b),

$$R'_e = \frac{1}{2}\left(\frac{P_e e \ell}{E_c I_c} \times \frac{2}{3}\right) = \frac{P_e e \ell}{3E_c I_c}$$

The moment due to the elastic weight W_e about the midspan point C is

$$M_C = R'_e\left(\frac{\ell}{2}\right) - \left[\frac{P_e e \ell}{E_c I_c} \times \frac{2}{6}\left(\frac{3}{8} \times \frac{\ell}{2}\right)\right]$$

$$= \frac{1}{E_c I_c}\left(\frac{P_e e \ell^2}{6} - \frac{3P_e e \ell^2}{48}\right) = \frac{5P_e e \ell^2}{48E_c I_c}$$

Then

$$\delta_c = \frac{5}{48}\frac{P_e e \ell^2}{E_c I_c} \tag{a}$$

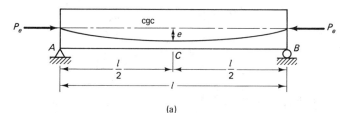

(a)

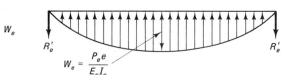

(b)

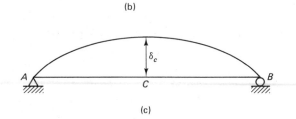

(c)

Figure 7.5 Deflection of beam in Example 7.2. (a) Tendon profile. (b) Elastic weight $M/E_c I_c$. (c) Deflection.

Camber, Deflection, and Crack Control   Chap. 7

Equivalent weight method. From Chapter 1, the equivalent balancing load intensity w resulting from the pressure of the parabolic tendon on the concrete is

$$w = \frac{8P_e e}{\ell^2}$$

Also, from the basic mechanics of materials, the midspan deflection of a uniformly loaded simply supported beam is

$$\delta_c = \frac{5}{384} \frac{w \ell^4}{E_c I_c} \tag{b}$$

Substituting for the load intensity w from the previous equation into this one yields

$$\delta_c = \frac{5}{48} \frac{P_e e \ell^2}{E_c I_c} \tag{c}$$

As expected, Equation (c) is identical to Equation (a) for the midspan deflection of the beam.

Figure 7.6 shows typical midspan deflection expressions for simply supported beams, complementing the shear and moment expressions for continuous beams given earlier in Figure 6.12.

7.3.3 Cracked Sections

7.3.3.1 Effective-moment-of-inertia computation method. As the prestressed element is overloaded, or in the case of partial prestressing where limited controlled cracking is allowed, the use of the gross moment of inertia I_g underestimates the camber or deflection of the prestressed beam. Theoretically, the cracked moment of inertia I_{cr} should be used for the section across which the cracks develop while the gross moment of inertia I_g should be used for the beam sections between the cracks. However, such refinement in the numerical summation of the deflection increases along the beam span is sometimes unwarranted because of the accuracy difficulty of deflection evaluation. Consequently, an effective moment of inertia I_e can be used as an average value along the span of a simply supported bonded tendon beam, a method developed by Branson in Refs. 7.4 and 7.5. According to this method,

$$I_e = I_{cr} + \left(\frac{M_{cr}}{M_a}\right)^3 (I_g - I_{cr}) \leq I_g \tag{7.10}$$

Equation 7.10 can also be written in the form

$$I_e = \left(\frac{M_{cr}}{M_a}\right)^3 I_g + \left[1 - \left(\frac{M_{cr}}{M_a}\right)^3\right] I_{cr} \leq I_g \tag{7.11}$$

where I_{cr} = moment of inertia of the cracked section, from Equation 7.13 to follow

I_g = gross moment of inertia

M_{cr} = cracking moment of section of maximum moment M_a

M_a = maximum moment M_{max} acting on the span at the stage for which deflection is calculated.

The effective moment of inertia I_e in Equations 7.10 and 7.11 thus depends on the

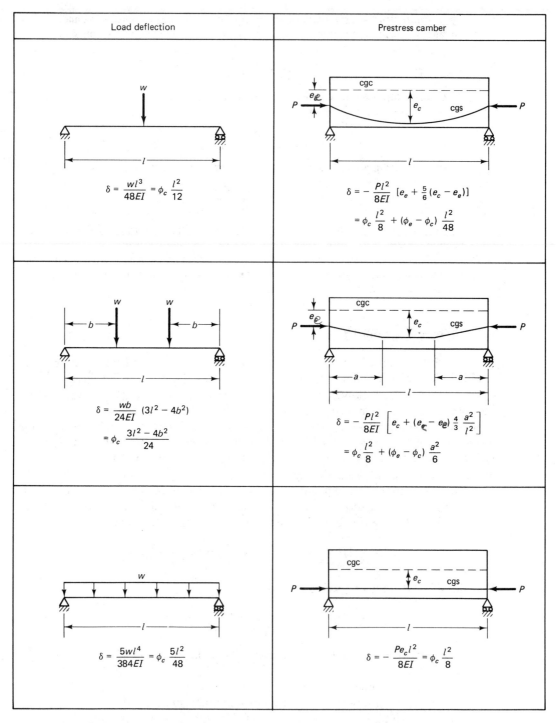

Load deflection	Prestress camber

$$\delta = \frac{wl^3}{48EI} = \phi_c \frac{l^2}{12}$$

$$\delta = -\frac{Pl^2}{8EI}\left[e_e + \frac{5}{6}(e_c - e_e)\right]$$
$$= \phi_c \frac{l^2}{8} + (\phi_e - \phi_c)\frac{l^2}{48}$$

$$\delta = \frac{wb}{24EI}(3l^2 - 4b^2)$$
$$= \phi_c \frac{3l^2 - 4b^2}{24}$$

$$\delta = -\frac{Pl^2}{8EI}\left[e_c + (e_c - e_e)\frac{4}{3}\frac{a^2}{l^2}\right]$$
$$= \phi_c \frac{l^2}{8} + (\phi_e - \phi_c)\frac{a^2}{6}$$

$$\delta = \frac{5wl^4}{384EI} = \phi_c \frac{5l^2}{48}$$

$$\delta = -\frac{Pe_c l^2}{8EI} = \phi_c \frac{l^2}{8}$$

Figure 7.6 Short-term deflection in prestressed beams. Subscirpt c indicates midspan, subscript e support.

Camber, Deflection, and Crack Control   Chap. 7

maximum moment M_a along the span in relation to the cracking moment capacity M_{cr} of the section.

In the case of continuous beams with both ends continuous,

$$\text{Avg. } I_e = 0.70 I_m + 0.15(I_{e1} + I_{e2}) \tag{7.12 a}$$

and for continuous beams with one end continuous,

$$\text{Avg. } I_e = 0.85 I_m + 0.15(I_{\text{cont.end}}) \tag{7.12 b}$$

where I_m is the midspan section moment of inertia and I_{e1} and I_{e2} are the end-section moments of inertia.

7.3.3.2 Bilinear computation method.

In graphical form, the bilinear moment-deflection relationship follows stages I and II described in Section 7.3.1 in accordance with ACI code. The idealized diagram for the I_g and I_{cr} zones is shown in Figure 7.7. Branson's effective I_e gives the average total *immediate* deflection $\delta_{\text{tot}} = \delta_e + \delta_{cr}$ described in the previous section.

The ACI code requires that computation of deflection in the cracked zone in the bonded tendon beams be based on the transformed section whenever the tensile stress f_t in the concrete exceeds $6\sqrt{f_c'}$. Hence, δ_{cr} in Figure 7.6 is evaluated using the transformed I_{cr} utilizing the contribution of the reinforcement in the bilinear method of deflection computation. The cracking moment of inertia can be calculated by the PCI approach (Ref. 7.7) for fully prestressed members by means of the equation

without mild steel

$$I_{cr} = n_p A_{ps} d_p^2 (1 - 1.67 \sqrt{n_p \rho_p}) \tag{7.13}$$

where $n_p = E_{ps}/E_c$. If nonprestressed reinforcement is used to carry tensile stresses, namely, in "partial prestressing," Equation 7.13 can be modified to give

$$I_{cr} = (n_p A_{ps} d_p^2 + n_s A_s d^2)(1 - 1.67 \sqrt{n_p \rho_p + n_s \rho}) \tag{7.14}$$

where $n_s = E_s/E_c$ for the nonprestressed steel d = effective depth to center of mild steel or nonprestressed strand steel.

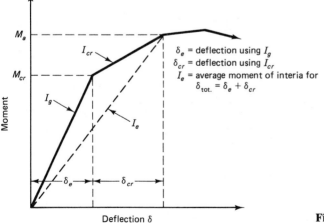

δ_e = deflection using I_g
δ_{cr} = deflection using I_{cr}
I_e = average moment of interia for
$\delta_{\text{tot.}} = \delta_e + \delta_{cr}$

Figure 7.7 Moment-deflection relationship.

7.3.3.3 Incremental moment-curvature method. The cracked moment of inertia can be calculated more accurately from the moment-curvature relationship along the beam span and from the stress and, consequently, strain distribution across the depth of the critical sections. As shown in Figure 7.4(d) for strain ϵ_{cr} at first cracking,

$$\phi_{cr} = \frac{\epsilon_{cr}}{c} = \frac{M}{E_c I_{cr}} \tag{7.14}$$

where ϵ_{cr} is the strain at the extreme concrete compression fibers and M is the total moment, including the prestressing primary moment M_1, about the centroid cgc of the section under consideration. Equation 7.14 can be rewritten to give

$$I_{cr} = \frac{Mc}{E_c \epsilon_{cr}} = \frac{Mc}{f} \tag{7.15}$$

where f is the concrete stress at the extreme compressive fibers of the section.

A flowchart for instantaneous deflection calculation and construction of the moment-curvature diagram in step-by-step increments is given in Figure 7.8.

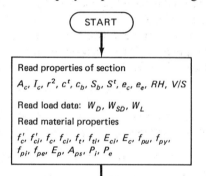

Figure 7.8 Flowchart for immediate moment-curvature camber and deflection.

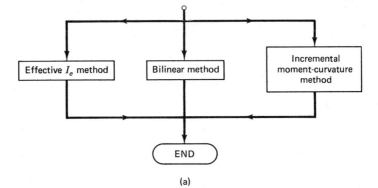

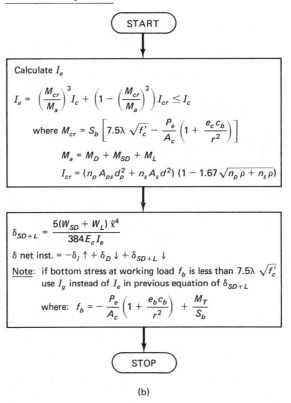

(a)

Subroutine for I_e method

START

Calculate I_e

$$I_e = \left(\frac{M_{cr}}{M_a}\right)^3 I_c + \left(1 - \left(\frac{M_{cr}}{M_a}\right)^3\right) I_{cr} \leq I_c$$

where $M_{cr} = S_b \left[7.5\lambda \sqrt{f_c'} - \frac{P_e}{A_c}\left(1 + \frac{e_c c_b}{r^2}\right)\right]$

$M_a = M_D + M_{SD} + M_L$

$I_{cr} = (n_p A_{ps} d_p^2 + n_s A_s d^2)(1 - 1.67\sqrt{n_p\rho + n_s\rho})$

$$\delta_{SD+L} = \frac{5(W_{SD} + W_L)\ell^4}{384 E_c I_e}$$

δ net inst. $= -\delta_i \uparrow + \delta_D \downarrow + \delta_{SD+L} \downarrow$

Note: if bottom stress at working load f_b is less than $7.5\lambda \sqrt{f_c'}$ use I_g instead of I_e in previous equation of δ_{SD+L}

where: $f_b = -\frac{P_e}{A_c}\left(1 + \frac{e_b c_b}{r^2}\right) + \frac{M_T}{S_b}$

STOP

(b)

Subroutine for bilinear method

START

Check bottom stress at midsection for working load

$$f_{bc} = -\frac{P_e}{A_c}\left(1 + \frac{e_b c_b}{r^2}\right) + \frac{M_T}{S_b}$$

Figure 7.8   (continued)

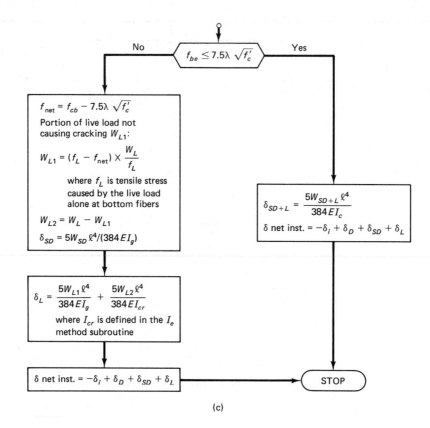

$$f_{be} \leq 7.5\lambda \sqrt{f'_c}$$

No   Yes

$$f_{net} = f_{cb} - 7.5\lambda \sqrt{f'_c}$$

Portion of live load not causing cracking W_{L1}:

$$W_{L1} = (f_L - f_{net}) \times \frac{W_L}{f_L}$$

        where f_L is tensile stress
        caused by the live load
        alone at bottom fibers

$$W_{L2} = W_L - W_{L1}$$

$$\delta_{SD} = 5W_{SD} \ell^4/(384 E I_g)$$

$$\delta_{SD+L} = \frac{5W_{SD+L} \ell^4}{384 E I_c}$$

δ net inst. $= -\delta_i + \delta_D + \delta_{SD} + \delta_L$

$$\delta_L = \frac{5W_{L1} \ell^4}{384 E I_g} + \frac{5W_{L2} \ell^4}{384 E I_{cr}}$$

where I_{cr} is defined in the I_e method subroutine

δ net inst. $= -\delta_i + \delta_D + \delta_{SD} + \delta_L$

STOP

(c)

Subroutine for incremental moment-curvature method

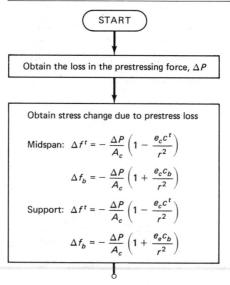

START

Obtain the loss in the prestressing force, ΔP

Obtain stress change due to prestress loss

Midspan: $\Delta f^t = -\dfrac{\Delta P}{A_c} \left(1 - \dfrac{e_c c^t}{r^2}\right)$

$\Delta f_b = -\dfrac{\Delta P}{A_c} \left(1 + \dfrac{e_c c_b}{r^2}\right)$

Support: $\Delta f^t = -\dfrac{\Delta P}{A_c} \left(1 - \dfrac{e_c c^t}{r^2}\right)$

$\Delta f_b = -\dfrac{\Delta P}{A_c} \left(1 + \dfrac{e_e c_b}{r^2}\right)$

Figure 7.8   (*continued*)

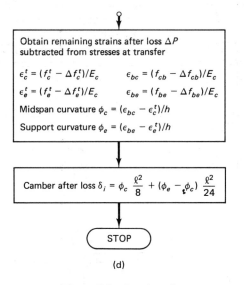

Obtain remaining strains after loss ΔP subtracted from stresses at transfer

$$\epsilon_c^t = (f_c^t - \Delta f_c^t)/E_c \qquad \epsilon_{bc} = (f_{cb} - \Delta f_{cb})/E_c$$

$$\epsilon_e^t = (f_e^t - \Delta f_e^t)/E_c \qquad \epsilon_{be} = (f_{be} - \Delta f_{be})/E_c$$

Midspan curvature $\phi_c = (\epsilon_{bc} - \epsilon_c^t)/h$

Support curvature $\phi_e = (\epsilon_{be} - \epsilon_e^t)/h$

Camber after loss $\delta_i = \phi_c \dfrac{\ell^2}{8} + (\phi_e - \phi_c) \dfrac{\ell^2}{24}$

STOP

(d)

Figure 7.8 (*continued*)

7.4 COMPUTATION OF INITIAL CAMBER

Example 7.3

Evaluate the initial camber of the bonded T-beam in Example 4.1 at the time of prestress transfer, assuming that fourteen $\frac{1}{2}$ in. dia (12.7 mm dia) seven-wire 270 K stress-relieved strands are used for prestressing.

Solution

1. *Data*

(a) *Elevation*
Figure 7.9 shows the geometry of the bonded T-beam.

(b) *Geometrical Properties*

$$A_c = 782 \text{ in}^2 \ (5,045 \text{ cm}^2)$$

$$I_c = 169,020 \text{ in}^4 \ (7.04 \times 10^6 \text{ cm}^4)$$

$$S_b = 4,803 \text{ in}^3 \ (6.69 \times 10^4 \text{ cm}^3)$$

$$S^t = 13,194 \text{ in}^3$$

$$W_D = 815 \text{ plf}$$

$$W_{SD} = 100 \text{ plf} \ (1.46 \text{ KN/m})$$

$$W_L = 1,100 \text{ plf} \ (16.05 \text{ KN/m})$$

$$e_c = 33.14 \text{ in.}$$

$$e_e = 20.00 \text{ in.}$$

$$c_b = 35.19 \text{ in.}$$

$$c_t = 12.81 \text{ in.}$$

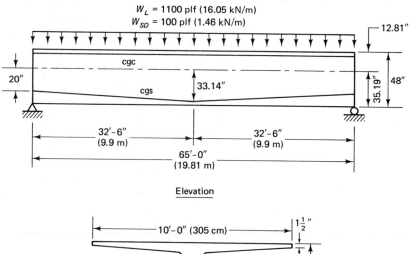

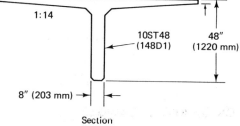

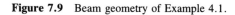

Figure 7.9 Beam geometry of Example 4.1.

$V/S = 2.33$ in.

$RH = 70\%$

$f'_c = 5,000$ psi

$f'_{ci} = 3,750$ psi

$f_{ci} = 2,250$ psi

$f_c = 2,250$ psi

$f_{ti} = 184$ psi

$f_t = 849$ psi

$f_{pu} = 270,000$ psi (1,861.7 MPa)

$f_{pi} = 189,000$ psi (1,303.2 MPa)

$f_{pe} = 154,980$ psi (1,068.6 MPa)

$f_{py} = 230,000$ psi

$E_{ps} = 27.5 \times 10^6$ psi (189.6 $\times 10^6$ MPa)

$A_{ps} = 14 \times 0.153 = 2.142$ in^2 (13.82 cm^2)

$P_i = 404,838$ ib (1,801 kN)

$P_e = 331,967$ lb (1,477 kN)

2. *Midspan Section Stresses*

$$e_c = 33.14 \text{ in. (872 mm.)}$$

From Example 4.1, $M_D = 5,386,875$ in.-lb.

(a) *At Transfer*

From Equation 4.1 a,

$$f^t = -\frac{P_i}{A_c}\left(1 - \frac{e_c c_t}{r^2}\right) - \frac{M_D}{S^t}$$

$$= -\frac{404,838}{782}\left(1 - \frac{33.14 \times 12.81}{216.14}\right) - \frac{5,386,875}{13,194}$$

$$= +499.1 - 408.3 \cong 91 \text{ psi } (T) < 184 \text{ psi, O.K.}$$

From Equation 4.1 b,

$$f_b = -\frac{P_i}{A_c}\left(1 + \frac{e_c c_b}{r^2}\right) + \frac{M_D}{S_b}$$

$$= -\frac{404,838}{782}\left(1 + \frac{33.14 \times 35.19}{216.14}\right) + \frac{5,386,875}{4,803}$$

$$= -3,311.0 + 1,121.6 \cong -2,190 \text{ psi } < 2,250 \text{ psi, O.K.}$$

(b) *At Service Load*

$$M_{SD} = \frac{100(65)^2 12}{8} = 633,750 \text{ in.-lb (72 kN-m)}$$

$$M_L = \frac{1,100(65)^2 12}{8} = 6,971,250 \text{ in.-lb (788 kN-m)}$$

$$\text{Live-load } f^t = \frac{6,971,250}{13,194} = -528.4 \text{ psi } (C)$$

$$\text{Live-load } f_b = \frac{6,971,250}{4,803} = 1,451.4 \text{ psi } (T)$$

$$\text{Total moment } M_T = M_D + M_{SD} + M_L = 5,386,875 + 7,605,000$$

$$= 12,991,875 \text{ in.-lb (1,468 kN-m)}$$

From Equation 4.3 a,

$$f^t = -\frac{P_e}{A_c}\left(1 - \frac{e c_t}{r^2}\right) - \frac{M_T}{S^t}$$

$$= -\frac{331,967}{782}\left(1 - \frac{33.14 \times 12.81}{216.14}\right) - \frac{12,991,875}{13,194}$$

$$= +409.3 - 987.4 = -575.4 \text{ psi } < f_c = -2,250 \text{ psi, O.K.}$$

From Equation 4.3 b,

$$f_b = -\frac{P_e}{A_c}\left(1 + \frac{e c_b}{r^2}\right) + \frac{M_T}{S_b}$$

$$= -\frac{331,967}{782}\left(1 + \frac{33.14 \times 35.19}{216.14}\right) + \frac{12,991,875}{4,803}$$

$$= -2,715 + 2,705 = +10 \text{ psi } (T) < f_t = +846 \text{ psi, O.K.}$$

3. *Support Section Stresses*
From Example 4.1,

$$e_e = 20.0 \text{ in. (508 mm)}$$
$$f_{ti} = 6\sqrt{f'_{ci}} = 6\sqrt{3,750} = 367.4 \text{ psi}$$
$$f_t = 6\sqrt{f'_c} = 6\sqrt{2,250} = 284.6 \text{ psi}$$

(a) *At Transfer*

$$f_b = -\frac{404,838}{782}\left(1 + \frac{20.0 \times 35.19}{216.14}\right) + 0 = -2,203.4 \text{ psi } (C)$$

$$< 2,250 \text{ psi, O.K.}$$

$$f^t = -\frac{404,838}{782}\left(1 - \frac{20.0 \times 12.81}{216.14}\right) - 0 = +96 \text{ psi, O.K.}$$

(b) *At Service Load*

$$f^t = -\frac{331,967}{782}\left(1 - \frac{20.0 \times 12.81}{216.14}\right) - 0 = +78.7 \text{ psi } (T)$$

$$< f_t = 846 \text{ psi, O.K.}$$

$$f_b = -\frac{331,967}{782}\left(1 + \frac{20.0 \times 35.91}{216.14}\right) + 0 = -1,835.1 \text{ psi } (C)$$

$$< f_c = -2,250 \text{ psi, O.K.}$$

Hence, fourteen $\frac{1}{2}$ in. dia strands are adequate.

(c) *Summary of Fiber Stresses (psi)*

	Midspan		Support	
	f^t	f_b	f^t	f_b
Prestress P_i only	+499	−3,311	+96	−2,203
At transfer & W_D	+91	−2,190	+96	−2,203
At service Load	−575	+10	+79	−1,835
Live load only	−528	+1,451	0	0

(1 psi = 6.895 MPa)

4. *Deflection and Camber Calculation*

From basic mechanics or from Figure 7.6, for $a = \dfrac{\ell}{2}$, the camber at midspan due to a single harp or depression of the prestressing tendon is

$$\delta \uparrow = \frac{P e_c \ell^2}{8EI} + \frac{P(e_e - e_c)\ell^2}{24EI}$$

So

$$E_{ci} = 57,000\sqrt{f'_{ci}} = 57,000\sqrt{3,750} = 3.49 \times 10^6 \text{ psi (24.1 MPa)}$$
$$E_c = 57,000\sqrt{f'_c} = 57,000\sqrt{5,000} = 4.03 \times 10^6 \text{ psi (27.8 MPa)}$$

$$\delta_{pi} \uparrow = -\frac{404,838 \times 33.14 \times (65 \times 12)^2}{8 \times 3.49 \times 10^6 \times 169,020}$$

$$+ \frac{-404,838(20 - 33.14)(65 \times 12)^2}{24 \times 3.49 \times 10^6 \times 169,020}$$

$$= -1.73 + 0.23 = -1.50 \text{ in. } \uparrow \ (38 \text{ mm})$$

due to prestress only. The self-weight per inch is $815/12 = \text{lb/in.}$, and the deflection caused by self-weight is $\delta_D \downarrow = 5wl^4/384EI$, or

$$\delta_{pi} \downarrow = \frac{5(67.9)(65 \times 12)^4}{384 \times 3.49 \times 10^6 \times 169,020} = 0.55 \text{ in. } \downarrow \ (14 \text{ mm})$$

Thus, the net camber at transfer is $-1.50 \uparrow + 0.55 \downarrow = -0.95$ in. $\uparrow$ (24 mm).

7.5 TOTAL IMMEDIATE DEFLECTION AT SERVICE LOAD

Example 7.4

Evaluate the total immediate elastic deflection of the beam in Example 7.3 using (a) the effective-moment-of-inertia computation method, (b) the bilinear computation method, and (c) the incremental moment-curvature method. The beam is bonded post-tensioned, with $E_c = 57,000\sqrt{5,000} = 4.03 \times 10^6$ psi, $A_{ps} = $ fourteen $\frac{1}{2}$ in. dia strands $= 2.142$ in². Disregard the contribution of the nonprestressed steel in calculating the cracked moment of inertia, and assume that the value of the effective prestress $P_e = 331,967$ lb occurs at the first load application 30 days after erection and does not include all the time-dependent losses.

Solution (a): Effective I_e Method   From Equation 7.11,

$$I_e = \left(\frac{M_{cr}}{M_a}\right)^3 I_g + \left[1 - \left(\frac{M_{cr}}{M_a}\right)^3\right] I_{cr} \le I_g$$

The fiber stresses (psi) due to the superimposed load $W_{SD} + W_L$ at midspan are given by

$$\text{Live load: } f'_{(L)} = -528 \text{ psi } (C), f_{b(L)} = +1,451 \text{ psi } (T)$$

$$\text{All loads: } f' = -575 \text{ psi } (C), f_b = +10 \text{ psi } (T)$$

The modulus of rupture $f_r = 7.5\sqrt{f'_c} = 7.5\sqrt{5,000} = 530$ psi. From Equation 5.12 or 7.3,

$$M_{cr} = \frac{I_c}{y_t}(6\lambda\sqrt{f'_c} + f_{ce} - f_d)$$

$$= S_b(6\lambda\sqrt{f'_c} + f_{ce} - f_d)$$

From Example 4.1, the stresses at the bottom fibers are

$$f_{ce} = -2,715 \text{ psi } (C)$$

and

$$f_d = \frac{M_D}{S_b} = \frac{5,386,875}{4,803} = +1,122 \text{ psi } (T)$$

so that

$$M_{cr} = 4,803(6 \times 1.0\sqrt{5,000} + 2,715 - 1,122)$$

$$= 9.69 \times 10^6 \text{ in.-lb } (1.09 \times 10^6 \text{ kN-m})$$

$$M_a = 12.99 \times 10^6 \text{ in.-lb } (1.47 \times 10^6 \text{ kN-m}) \text{ (from Example 4.1)}$$

$$\left(\frac{M_{cr}}{M_a}\right)^3 = \left(\frac{9.69 \times 10^6}{12.99 \times 10^6}\right)^3 = 0.42$$

$$I_g = 169,020 \text{ in}^4$$

From Equation 7.12,

$$I_{cr} = n_p A_{ps} d_p^2 (1 - 1.67\sqrt{n_p \rho_p})$$

$$n_p = \frac{E_{ps}}{E_c} = \frac{28 \times 10^6}{4.03 \times 10^6} = 6.95$$

$$d_p = e_c + c_t = 33.14 + 12.81 = 45.95 \text{ in.} > 0.8h$$

Use $d_p = 45.95$ in. and $A_{ps} = 2.142$ in^2. Then

$$\rho_p = \frac{A_{ps}}{A_c} = \frac{2.142}{782.0} = 0.003$$

$$I_{cr} = 6.95 \times 2.142(45.95)^2(1 - 1.67\sqrt{6.95 \times 0.003})$$

$$= 23,853 \text{ in}^4 \ (10.0 \times 10^5 \text{ cm}^4)$$

$$I_e = 0.42 \times 169,020 + (1 - 0.42)23,853 \leq 169,020$$

$$= 84,823 \text{ in}^4 \ (35.31 \times 10^5 \text{ cm}^4)$$

However, since at service load $f_b = 10$ psi $(T) < f_r = 530$ psi, the section *did not crack* under short-term loading, and the gross I_g should be used, i.e.,

$$I_e = I_g = 169,020 \text{ in}^4 \ (70.36 \times 10^5 \text{ cm}^4)$$

The superimposed load $W_{SD} + W_L$ per inch of span is $(100 + 1,100)/12 = 100$ lb per inch, and the deflection due to the superimposed load is

$$\delta_{SD+L} \downarrow = \frac{5w\ell^4}{384E_c I_e} = \frac{5 \times 100(65 \times 12)^4}{384 \times 4.03 \times 10^6 \times 169,020}$$

$$= +0.71 \text{ in.} \downarrow \ (18 \text{ mm})$$

From Example 7.3, the net camber is $\delta = -0.95$ in. $\uparrow$ (24 mm), and the net short-term deflection due to all loads is

$$\delta_{net} = -0.95 \uparrow + 0.71 \downarrow = -0.24 \text{ in. } (6.1 \text{ mm})$$

very small indeed for a 65-ft span.

Solution (b): Bilinear Method From solution (a),

$$I_{cr} = 23,853 \text{ in}^4$$

and

$$f_r = +530 \text{ psi } (T)$$

The net tensile stress beyond the first cracking load at the modulus of rupture is

$$f_{net} = f_b - f_r = +10 - 530 = -443 \text{ psi } (C) \ (3,054 \text{ MPa})$$

where f_b is the total fiber stress at the tension side due to all loads. Since there is no residual tension in this case at the bottom fiber with all loads acting, the gross moment of inertia I_g has to be used for the uncracked section, and $\delta_{(I_{cr})} = 0$ in the bilinear method.

(Example 7.5, next, deals with the development of residual tension, where the bilinear calculation becomes applicable, with a portion or percentage of the live load contributing to the cracked section deflection.

From solution (a), $W_{L+SD} = 100$ lb/in., and the uncracked deflection is

$$\delta_{(t_g)} = \frac{5w\ell^4}{384E_cI_g} = \frac{5 \times 100(65 \times 12)^4}{384 \times 4.03 \times 10^6 \times 169,020} = +0.71 \text{ in.} \downarrow$$

and

$$\delta_{(t_{cr})} = 0$$

The total deflection at service in the bilinear method is

$$\delta_{(t_g)} + \delta_{(t_{cr})} = +0.71 \text{ in.} \downarrow \quad (18 \text{ mm})$$

and the net instantaneous deflection is

$$\delta_{net} = -0.95 \uparrow +0.71 \downarrow = -0.24 \text{ in.} \uparrow \quad (6.1 \text{ mm}) \text{ camber}$$

Solutions (a) and (b) are both expected to give the same values in this case since the section did not crack at service.

Solution (c): Incremental Moment-Curvature Method From the statement of the problem, 30 days' P_e in this example is considered equivalent to 331,967 lb. So

$$30 \text{ days' prestress loss } \Delta P = P_i - P_e = 404,838 - 331,967$$
$$= 72,871 \text{ lb } (324 \text{ kN})$$

Strains at transfer due to prestressing

$$E_c \text{ at 7 days} = 3.49 \times 10^6 \text{ psi}$$

(i) *Due to prestressing force* (P_i)

Midspan: $f^t = +499$ psi

$$f_b = -3,311 \text{ psi}$$

$$\epsilon_c^t = \frac{+499}{3.49 \times 10^6} = +143 \times 10^{-6} \text{ in./in.}$$

$$\epsilon_{cb} = -949 \times 10^{-6} \text{ in./in.}$$

Support: $f^t = +96$ psi

$$f_b = -2,203 \text{ psi}$$

$$\epsilon_e^t = +28 \times 10^{-6} \text{ in./in.}$$

$$\epsilon_{eb} = -631 \times 10^{-6} \text{ in./in.}$$

(1 psi = 6.895 MPa)

(ii) *Due to prestressing and self-weight* $(P_i + W_D)$

Midspan: $f^t = +91$ psi $\epsilon_c^t = +26 \times 10^{-6}$ in./in.

$f_b = -2,190$ psi $\epsilon_{cb} = -628$ in./in.

Support: same as in (i).

Strain change due to prestress loss

$$-\Delta P = 72{,}871 \text{ lb}$$

$$E_c = 4.03 \times 10^{-6} \text{ psi}$$

Midspan Section

$$\Delta f^t = -\frac{(-\Delta P)}{A_c}\left(1 - \frac{ec_t}{r^2}\right) = +\frac{72{,}871}{782}\left(1 - \frac{33.14 \times 12.81}{216.14}\right)$$

$$= -90 \text{ psi } (C)$$

$$\Delta\epsilon_c^t = \frac{-90}{4.03 \times 10^6} = -22 \times 10^{-6} \text{ in./in.}$$

$$\Delta f_b = -\frac{(-\Delta P)}{A_c}\left(1 + \frac{ec_b}{r^2}\right) = \frac{72{,}871}{782}\left(1 + \frac{33.14 \times 35.19}{216.14}\right) = 596 \text{ psi } (T)$$

$$\Delta\epsilon_{cb} = \frac{596}{4.03 \times 10^6} = +148 \times 10^{-6} \text{ in./in.}$$

Support Section

$$\Delta f^t = -\frac{-\Delta P}{A_c}\left(1 - \frac{ec_t}{r^2}\right) = +\frac{72{,}871}{782}\left(1 - \frac{20 \times 12.81}{216.14}\right)$$

$$= -17.3 \text{ psi } (C)$$

$$\Delta\epsilon_e^t = \frac{-17.3}{4.03 \times 10^6} = -4 \times 10^{-6} \text{ in./in.}$$

$$\Delta f_b = -\frac{-\Delta P}{A_s}\left(1 - \frac{ec_b}{r^2}\right) = +\frac{72{,}871}{782}\left(1 + \frac{20 \times 35.19}{216.14}\right)$$

$$= 396.6 \text{ psi } (T)$$

$$\Delta\epsilon_{eb} = \frac{+396.6}{4.03 \times 10^6} = +89 \text{ in./in.}$$

Superimposing the strain at transfer on the strain due to prestress loss gives the stress distributions at *service load* due to *prestress only*, as shown in Figure 7.10. From Figure 7.10,

$$\text{Midspan curvature } \phi_c = \frac{-801 - 121}{48} \times 10^{-6} = -19.21 \times 10^{-6} \text{ rad/in.}$$

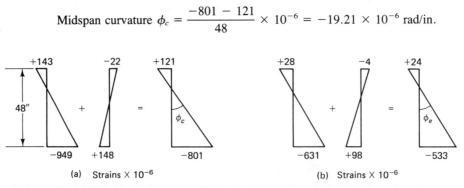

(a) Strains $\times 10^{-6}$ (b) Strains $\times 10^{-6}$

Figure 7.10 Strain distribution across section depth at prestress transfer in Example 7.4. (a) Midspan section. (b) Support section.

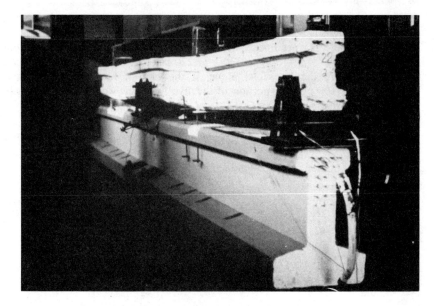

Deflection of continuous beam (Nawy et al.).

$$\text{Support curvature } \phi_e = \frac{-533 - 24}{48} \times 10^{-6} = -11.60 \times 10^{-6} \text{ rad/in.}$$

From Figure 7.6, for $a = \ell/2$, the beam camber after losses due only to P_e is

$$\delta_e \uparrow = \phi_c\left(\frac{\ell^2}{8}\right) + (\phi_e - \phi_c)\frac{\ell^2}{24}$$

$$= -19.20 \times 10^{-6}\frac{(65 \times 12)^2}{8} + (-11.60 + 19.21)$$

$$\times 10^{-6}\frac{(65 \times 12)^2}{24} = \frac{(65 \times 12)^2}{24} \times 10^{-6}(-19.21 \times 3 + 7.61)$$

$$= -1.27 \text{ in } \uparrow \quad (32 \text{ mm}) \text{ (camber)}$$

Deflection due to gravity loads $W_D + W_{SD} + W_L$

From solution (a), $\delta_g = 0.55 + 0.71 = +1.26$ in. $\downarrow$ (32 mm). So the net deflection $\delta_{net} = -1.27 \uparrow + 1.26 \downarrow = -0.01$ in. $\uparrow$ (camber). We compare this value with $-0.24 \uparrow$ in. in the other two solutions. From Table 7.2, allow $\delta_L = 1/180 = (65 \times 12)/180 = 4.33$ in. (110 mm) $\gg 0.71$ in.; hence, this value is satisfactory.

If the tensile stress at the midspan lower fibers at service load were to exceed the modulus of rupture (see next example) the net deflection would increase considerably.

7.6 IMMEDIATE DEFLECTION OF PRESTRESSED BEAM BY BILINEAR METHOD

Example 7.5

Solve Example 7.4 by the bilinear method for a condition of tensile stress of 750 psi at the midspan bottom fibers at service load, i.e., the tensile stress exceeds the modulus of rupture $f_r = 7.5\sqrt{f_c'} = 530$ psi. Assume that the net beam camber due to prestress and self-weight is $\delta = 0.95$ in.

Solution The net tensile stress beyond the first cracking load at the modulus of rupture is $f_{net} = f_b - f_r = 750 - 530 = +220$ psi (T). From Example 7.4, the tensile stress caused by the live load alone at the bottom fibers is $+1,451$ psi. Now, since $W_L = 1,100$ plf, the portion of the live load that would result in tensile stress at the bottom fibers is

$$W_1 = \frac{(1,451 - 220)}{1,451} \times 1,110 = 933 \text{ plf} = \frac{933}{12} = 78 \text{ lb/in.}$$

The deflection determined by the uncracked I_g is

$$\delta_g = \frac{5W_1\ell^4}{384E_cI_g} = \frac{5 \times 78(65 \times 12)^4}{384 \times 4.03 \times 10^6 \times 169,020} = 0.55 \text{ in. } \downarrow \text{ (14 mm)}$$

So

$$W_2 = \frac{(1,100 - 933)}{12} = 14 \text{ lb/in.}$$

and

$$I_{cr} = 23,853 \text{ in}^4$$

The deflection determined by the cracked I_{cr} is

$$\delta_{cr} = \frac{5W_2\ell^4}{384E_cI_c} = \frac{5 \times 14(65 \times 12)^4}{384 \times 4.03 \times 10^6 \times 23,853} = 0.70 \text{ in. } \downarrow \text{ (15.5mm)}$$

Thus, the total deflection $\delta_{net} \downarrow = 0.55 + 0.70 = +1.25$ in $\downarrow$ (295 mm), which compares with 0.71 in $\downarrow$ in Example 7.4, and the net deflection $\delta_{net} = -0.95 \uparrow + 1.25 \downarrow = 0.30$ in. $\downarrow$ (5 mm).

Notice that the cracked section changed the net deflection from 0.24 in. (6.1 mm) upward camber to 0.30 in. (7.6 mm) downward deflection, an increase in deflection of 0.54 in. (13.7 mm).

7.7 CONSTRUCTION OF MOMENT-CURVATURE DIAGRAM

Example 7.6

Construct the moment-curvature diagram for the midspan section of the bonded single-T beam in Example 7.3 for the following incremental strain steps:

1. Strain at transfer $f_{pi} = 189,000$ psi due to P_i only.
2. Strain at $f_{pe} = 154,980$ psi prior to gravity loads.
3. Decompression at tendon cgs level.
4. Cracked section, strain ϵ_{c1} at top $= 0.001$ in./in.
5. Cracked section, strain ϵ_{c2} at top $= 0.003$ in./in.

Note the strain distribution in Figure 7.11 due to the prestressing force P_e. Use the stress-strain diagram of Figure 7.12 for the prestressing steel and that of Figure 7.13 for the concrete to determine the actual stresses through strain compatibility.

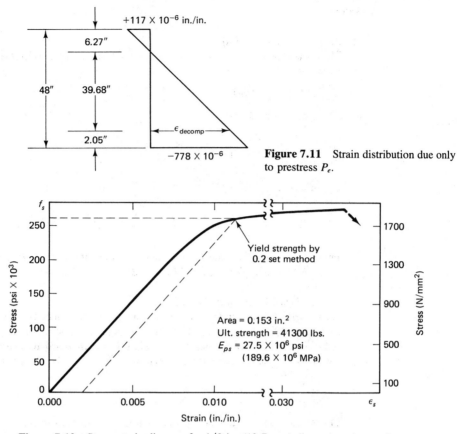

Figure 7.11 Strain distribution due only to prestress P_e.

Figure 7.12 Stress-strain diagram for 1/2 in. (12.7 mm) dia prestressing tendons.

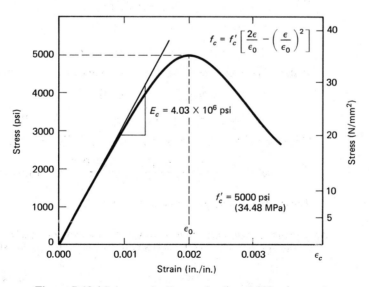

Figure 7.13 Stress-strain diagram for $f'_c = 5,000$ psi concrete.

Solution

1. *Prestress Transfer Stage*

   From the data for Example 7.3, the midspan stresses due only to prestress P_i are as follows:

   $$f^t = +499 \text{ psi}$$

   $$f_b = -3{,}311 \text{ psi}$$

   $$\epsilon_c^t = \frac{499}{3.49 \times 10^6} = +143 \times 10^{-6} \text{ in./in.}$$

   $$\epsilon_{cb} = \frac{-3{,}311}{3.99 \times 10^6} = -949 \times 10^{-6} \text{ in./in.}$$

   $$\phi_i = \frac{(\epsilon_{cd} - \epsilon_c^t)}{h} = \frac{(-949 - 143)}{48} \times 10^{-6} = -22.75 \times 10^{-6} \text{ rad/in.}$$

   From Example 7.4, the corresponding moments due to $P_i + M_D$ are $M_i = -404{,}938 \times 33.14 + 5{,}386{,}875 = -8.03 \times 10^{-6}$ in.-lb.

2. *Service-Load Stage after Losses*

   At the service-load stage, only the strains due to the effective prestress P_e have to be considered. This is because in the subsequent decompression stage a moment value M_g due to gravity loads has to be found which would reduce the stress in the prestressing steel to zero. From Example 4.1, $P_e = 331{,}967$ lb. Hence,

   $$\frac{P_e}{P_i} = \frac{331{,}967}{404{,}838} = 0.82$$

   The stresses and strains at midspan at transfer prestress P_i are

   $$f_c^t = +499 \text{ psi}$$

   $$f_{cb} = -3{,}311 \text{ psi}$$

   $$\epsilon_c^t = +143 \times 10^{-6} \text{ in./in.}$$

   $$\epsilon_{cb} = -949 \times 10^{-6} \text{ in./in.}$$

   Reduce the strains up to the P_e stage as follows:

   $$\epsilon_c^t = 0.82(+143 \times 10^{-6}) = +117 \times 10^{-6} \text{ in./in.}$$

   $$\epsilon_{cb} = 0.82(-949 \times 10^{-6}) = -778 \times 10^{-6} \text{ in./in.}$$

   The strain distribution becomes, as shown in Figure 7.11,

   $$\phi_2 = \frac{(\epsilon_{ce} - \epsilon_c^t)}{h} = \frac{(-778 - 117)10^{-6}}{48} = -18.64 \times 10^{-6} \text{ rad/in.}$$

   The corresponding gravity-load moment $M_g = 0$.

3. *Decompression Stage with Zero Concrete Stress at Tendon cgs*

   From Figure 7.11, the decompression strain at the cgs level is

   $$\epsilon_{\text{decomp.}} = -778 \times 10^{-6} \times \frac{39.68}{39.68 + 2.05} = 740 \times 10^{-6} \text{ in./in.}$$

and

$$\epsilon_{pe} = \frac{f_{pe}}{E_{ps}} = \frac{154{,}980}{27.5 \times 10^6} = 5{,}636 \times 10^{-6} \text{ in./in.}$$

Compatibility of strain requires that the prestressing tendon in the *bonded* beam undergo the same change in strain as the surrounding concrete, *increasing* the tensile strain in the tendon in order to reduce the compressive stress in the concrete at the cgs level to zero. Thus,

Total $\epsilon_{pe} = 5{,}636 \times 10^{-6} + 740 \times 10^{-6} = 6{,}376 \times 10^{-6}$ in./in.

From the stress-strain diagram in Figure 7.12 the corresponding stress $f_{pe} = 177{,}000$ psi. Consequently, we have

Adjusted $P_e = 177{,}000 \times 14 \times 0.153 = 379{,}134$ lb

Adjusted $f^t = -\dfrac{379{,}134}{782}\left(1 - \dfrac{33.14 \times 12.81}{216.14}\right) \cong +467$ psi (T)

$$\epsilon_c^t = -\frac{+467}{4.03 \times 10^6} = 116 \times 10^{-6} \text{ in./in.}$$

Adjusted $f_b = -\dfrac{379{,}134}{782}\left(1 + \dfrac{33.14 \times 35.19}{216.14}\right) \cong -3{,}101$ psi (C)

$$\epsilon_{cb} = \frac{-3{,}101}{4.03 \times 10^6} = 769 \times 10^{-6} \text{ in./in.}$$

$$f_{\text{decomp.}} = \frac{M_{\text{decomp.}} \times y}{I_c} = \frac{M_{\text{decomp.}} \times 33.14}{169{,}020}$$

$$M_{\text{decomp.}} = \frac{2{,}949 \times 169{,}020}{33.14} = 15.04 \times 10^6 \text{ in.-lb } (1.70 \times 10^6 \text{ N-m})$$

$$f_t = \frac{M_{\text{decomp.}}}{S^t} = \frac{15.04 \times 10^6}{13{,}194} = -1{,}140 \text{ psi } (C) \ (7.86 \text{ MPa})$$

Net stress $f^t = -1{,}140 + 467 = -673$ psi (C) (4.64 MPa)

$$\epsilon_c^t = \frac{-673}{4.03 \times 10^6} = -167.0 \times 10^{-6} \text{ in./in.}$$

$$f_b = \frac{15.04 \times 10^6}{S_b} = \frac{15.04 \times 10^6}{4{,}803} = +3{,}131 \text{ psi } (T)$$

Net $f_b = +3{,}131 - 3{,}101 = +30$ psi (T)

$$\epsilon_{cb} = \frac{+30}{4.03 \times 10^6} = +7.4 \times 10^{-6} \text{ in./in.}$$

$$\phi_{\text{decomp.}} = \frac{(\epsilon_{cb} - \epsilon_c^t)}{h} = \frac{(7.4 + 167.0)}{48} \times 10^{-6} = +3.63 \times 10^{-6} \text{ rad/in.}$$

$$= +3.63 \times 10^{-6} \text{ rad/in.}$$

Corresponding $M_g = 15.04 \times 10^6$ in.-lb

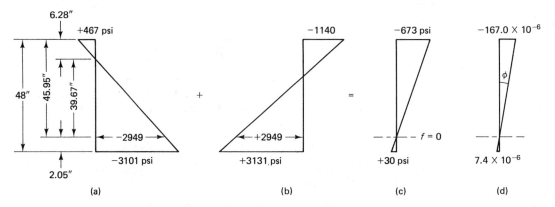

Figure 7.14 Strain distribution at decompression in Example 7.6. (a) Loading stress. (b) Decompression stress. (c) Final stress. (d) Unit strain.

Figure 7.14 gives the stress and strain distributions in this beam at the decompression state.

4. *Cracked-Section Stage,* $\epsilon_c = 0.001$ in./in.
From before, $\epsilon_{pe} = 6,376 \times 10^{-6} = 0.0064$ in./in. By trial and adjustment, assume a neutral axis depth $c = 1.5$ in. below the top fibers of the flange. Then $\Delta\epsilon_{ps}$ is the additional strain in bonded prestressing strands due to $\epsilon_c = 0.001$ in./in. at the top fibers, and from similar triangles in Figure 7.15,

$$\Delta\epsilon_{ps} = \frac{(45.95 - 1.5)}{1.5} \times 0.001 = 0.0303 \text{ in./in.}$$

So the total $\epsilon_{ps} = 0.0303 + 0.0064 = 0.0367$ in./in.
From the stress-strain diagram of the prestressing steel in Figure 7.11, the corresponding stress is

$$f_{ps} = 261,000 \text{ psi}$$

and

$$A_{ps} = 14 \times 0.153 = 2.142 \text{ in}^2$$

Thus, the tensile force $T_p = 261,000 \times 2.142 = 559,062$ lb.
From Figure 7.13, $f_c = 3,000$ psi corresponds to $\epsilon_c = 0.001$ in./in. The compressive force is then $C_c = (10 \text{ ft} \times 12 \times 1.5)3,000 = 540,000$ lb $< T = 559,062$ lb. Hence, the neutral axis depth should be increased.

Second Trial

Assume $c = 1.55$ in. Then

$$\Delta\epsilon_{ps} = \frac{45.95 - 1.55}{1.55} \times 0.001 = 0.0286 \text{ in./in.}$$

and

$$\text{Total } \epsilon_{ps} = 0.0286 + 0.0063 = 0.0349 \text{ in./in.}$$

From Figure 7.12, $f_{ps} \cong 260,500$ psi, $T_p = 260,500 \times 2.142 = $

Camber, Deflection, and Crack Control   Chap. 7

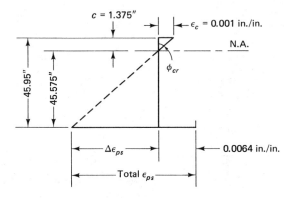

Figure 7.15 Strain distribution at $\epsilon_c = 0.001$ in./in. in Example 7.6.

557,991 psi, and $C_c = (10 \times 12 \times 1.55)3,000 = 558,000$ psi $\cong T_p$. Hence, assumed $c = 1.55$ in. O.K.

$$\text{Moment } M = 557,991\left(45.95 - \frac{1.55}{3}\right) = 25.4 \times 10^6 \text{ in./lb}$$

and from Equation 7.5 d,

$$\phi_u = \frac{\epsilon_u}{c} = \frac{0.001}{1.55} = 645 \times 10^{-6} \text{ rad/in.}$$

5. *Cracked-Section Stage,* $\epsilon_c = 0.003$ in./in
   $\epsilon_c = 0.003$ in./in is the maximum unit strain allowed by the ACI Code at ultimate load. Assume $f_{ps} = 263,000$ psi. Then

$$a = \frac{A_{ps}f_{ps}}{0.85 f'_c b} = \frac{2.142 \times 263,000}{0.85 \times 5,000 \times 120} = 1.1 \text{ in.}$$

$$c = \frac{a}{\beta_1} = \frac{1.1}{0.80} = 1.375 \text{ in.}$$

From Figure 7.15,

$$\epsilon_{ps} = \frac{45.95 - 1.375}{1.375} \times 0.003 = 0.0973 \text{ in./in.}$$

Total $\epsilon_{ps} = 0.0973 + 0.0063 = 0.1036$ in./in.

From the stress diagram in Figure 7.12, $f_{ps} \cong f_{pu} = 270,000$ psi. So use $a \cong 1.13$ in., giving

$$M_n = A_{ps}f_{ps}\left(d_p - \frac{a}{2}\right) = 2.142 \times 270,000\left(45.95 - \frac{1.13}{2}\right)$$

$$= 26.2 \times 10^6 \text{ in.-lb}$$

Use $c \cong 1.4$ in

$$\phi_u = \frac{\epsilon_u}{c} = \frac{0.003}{1.4} = 2,142 \times 10^{-6} \text{ rad/in.}$$

A schematic plot of the moment-curvature diagram is shown in Figure 7.16. The load-deflection diagram has the same form and can be inferred from the moment-curvature diagram.

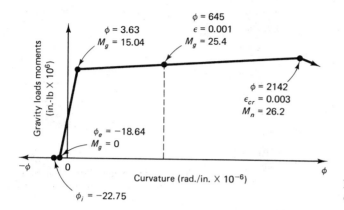

Figure 7.16 Moment-curvature diagram of Example 7.6.

7.8 LONG-TERM EFFECTS ON DEFLECTION AND CAMBER

7.8.1 PCI Multipliers Method

The ACI Code provides the following equation for estimating the time-dependent factor for deflection of nonprestressed concrete members:

$$\lambda = \frac{\xi}{1 + 50\rho'} \tag{7.16}$$

where ξ = time-dependent factor for sustained load
ρ' = compressive reinforcement ratio
λ = multiplier for *additional* long-term deflection.

In a similar manner, the PCI multipliers method provides a multiplier C_1 which takes account of long-term effects in prestressed concrete members. C_1 differs from λ in Equation 7.16, because the determination of long-term cambers and deflections in prestressed members is more complex due to the following factors:

1. The long-term effect of the prestressing force and the prestress losses.
2. The increase in strength of the concrete after release of prestress due to losses.
3. The camber and deflection effect during erection.

Because of these factors, Equation 7.16 cannot be readily used.

Table 7.1, based on Refs. 7.7 and 7.8, can provide reasonable multipliers of immediate deflection and camber provided that the upward and downward components of the initial calculated camber are separated in order to take into account the effects of loss of prestress, *which only apply to the upward component.*

Shaikh and Branson, in Ref. 7.6, propose that substantial reduction can be achieved in long-term camber by the addition of nonprestressed steel. In that case, a reduced multiplier C_2 can be used given by

$$C_2 = \frac{C_1 + A_s/A_{ps}}{1 + A_s/A_{ps}} \tag{7.17}$$

TABLE 7.1 C_1 MULTIPLIERS FOR LONG-TERM CAMBER AND DEFLECTION

	Without composite topping	With composite topping
At erection:		
(1) Deflection (downward) component—apply to the elastic deflection due to the member weight at release of prestress	1.85	1.85
(2) Camber (upward) component—apply to the elastic camber due to prestress at the time of release of prestress	1.80	1.80
Final:		
(3) Deflection (downward) component—apply to the elastic deflection due to the member weight at release of prestress	2.70	2.40
(4) Camber (upward) component—apply to the elastic camber due to prestress at the time of release of prestress	2.45	2.20
(5) Deflection (downward)—apply to the elastic deflection due to the superimposed dead load only	3.00	3.00
(6) Deflection (downward)—apply to the elastic deflection caused by the composite topping	—	2.30

where C_1 = multiplier from Table 7.1
A_s = area of nonprestressed reinforcement
A_{ps} = area of prestressed strands.

7.8.2 Incremental Time-Steps Method

The incremental time-steps method is based on combining the computations of deflections with those of prestress losses due to time-dependent creep, shrinkage, and relaxation. The design life of the structure is divided into several time intervals selected on the basis of specific concrete strain limits, such as unit strain levels $\epsilon_{c_1} = 0.001$ and 0.002 in./in., and ultimate allowable strain $\epsilon_c = 0.003$ in./in. The strain distributions, curvatures, and prestressing forces are calculated for each interval together with the incremental shrinkage, creep, and relaxation strain losses during the particular time interval. The procedure is repeated for all subsequent incremental intervals, and an integration or summation of the incremental steps is made to give the total time-dependent deflection at the particular section along the span. These calculations should be made for a sufficient number of points along the span, such as midspan and quarterspan points, to be able to determine with sufficient accuracy the form of the moment-curvature diagram.

The general expression for the total rotation at the end of a time interval can be expressed as

$$\phi_t = -\frac{P_i e_x}{E_c I_c} + \sum_0^t \left(P_{n-1} - P_n\right) \frac{e_x}{E_c I_c} - \sum_0^t (C_n - C_{n-1}) P_{n-1} \frac{e_x}{E_c I_c} \qquad (7.17)$$

where P_i = initial prestress before losses

$$e_x = \text{eccentricity of tendon at any section along the span}$$

Subscript $n - 1$ = beginning of a particular time step

Subscript n = end of the aforementioned time step

C_{n-1}, C_n = creep coefficients at beginning and end, respectively, of a particular time step

$P_n - P_{n-1}$ = prestress loss at a particular time interval from all causes.

Obviously, this elaborate procedure is justified only in the evaluation of deflection and camber of very large-span bridge systems such as segmental bridges, where the erection and assembly of the segments require a relatively accurate estimate of deflections. From Equation 7.17, the total deflection at a particular section is

$$\delta_x = \phi_t k \ell^2 \tag{7.18}$$

Now, suppose that the following strains from subsequent Example 7.8 are used to illustrate calculation of incremental and total rotations:

ϵ^t_{n-1} = gross strain due only to prestress at the top fibers, e.g., $\epsilon^t_c = +143 \times 10^{-6}$ in./in. (Figure 7.19)

$\epsilon_{b,n-1}$ = gross strain due only to prestress at the bottom fibers, e.g., $\epsilon_{cb} = -949 \times 10^{-6}$ in./in. (Figure 7.19)

$\Delta\epsilon^t_{CR,n}$ = gross creep incremental strain at the top fibers, e.g., $\Delta\epsilon^t_{CRc} = +127 \times 10^{-6}$ in./in. (Figure 7.20)

$\Delta\epsilon_{CRb,n}$ = gross creep incremental strain at the bottom fibers, e.g., $\Delta\epsilon_{CRcb} = -841 \times 10^{-6}$ in./in. (Figure 7.20)

$\Delta\epsilon_{ps,n}$ = strain reduction due to prestress loss caused by creep force $\Delta P, n$ (such as 162×10^{-6} in./in., as seen from Figure 7.20)

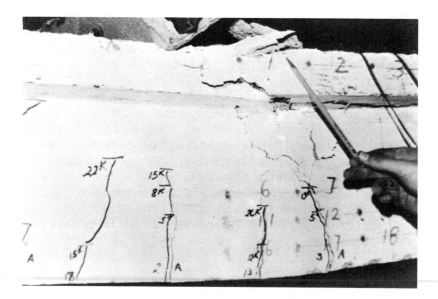

Prestressed I-beam at the limit state of failure load (Nawy et al.).

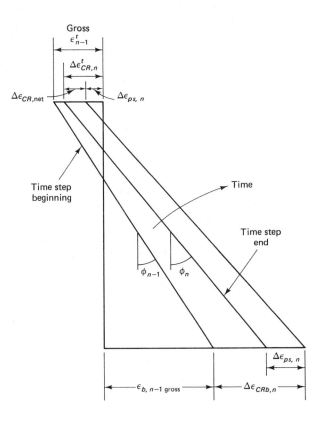

Figure 7.17 Strain changes and rotations at step n.

Then the net incremental creep strain that will result in incremental rotation ϕ_n is

$$\Delta\epsilon_{CR,\text{net}}^t = (\Delta\epsilon_{CR,n}^t - \Delta\epsilon_{ps,n}^t) \qquad (7.19\ a)$$

for the top fibers and

$$\Delta\epsilon_{CRb,\text{net}} = (\Delta\epsilon_{CRb,n} - \Delta\epsilon_{psb,n}) \qquad (7.19\ b)$$

for the bottom fibers.

The incremental rotation is, then,

$$\Delta\phi_n = \frac{\Delta\epsilon_{CR,\text{net}}^t - \Delta\epsilon_{CRb,\text{net}}}{h} \qquad (7.19\ c)$$

and the total rotation becomes

$$\phi_T = \phi_{n-1} + \Delta\phi_n \qquad (7.20)$$

A schematic of the changes in strains and rotations from time step $n - 1$ to time step n is shown in Figure 7.17.

The selection of the time intervals depends on the refinement level desired in the computation of cambers. For each time step, the incremental creep and shrinkage strains and relaxation loss in prestress are computed as shown in Example 7.8 to give a curvature increment $\Delta\phi$. Thereafter, new values of stress, strain, and curvature are obtained at the end of the time interval, adding the curvature increment $\Delta\phi_n$ to the total curvature ϕ_{n-1} at the beginning of the desired intervals, as given in Equation 7.17.

Clearly, the incremental time-step procedure is lengthy and, hence, justified only in evaluation and assembly of segments requiring relatively accurate estimates of deformations.

The total camber (↑) or deflection (↓) due to the prestressing force can be obtained from Equation 7.20 as

$$\delta_T = \phi_T k \ell^2 \qquad (7.21)$$

where k is a function of the span and geometry of the section and the prestressing tendon.

Several investigators have proposed different formats for estimating the additional time-dependent deflection $\Delta\delta$ from the moment-curvature relationship ϕ modified for creep. Both Tadros and Dilger recommend integrating the modified curvature along the beam span, while Naaman expresses the long-term deflection in terms of midspan and support curvatures at a time interval t (Refs. 7.10, 7.11, 7.12). As an example, Naaman's expression gives, for a parabolic tendon,

$$\Delta\delta(t) = \phi_1(t)\frac{\ell^2}{8} + [\phi_2(t) - \phi_1(t)]\frac{\ell^2}{48}$$

where $\phi_1(t)$ = midspan curvature at time t
$\phi_2(t)$ = support curvature at time t

in which

$$\phi(t) = \frac{M}{E_{ce}(t)I_c}$$

where $E_{ce}(t)$ = time adjusted modulus

$$= \frac{E_c(t_1)}{1 + KC_c(t)}$$

in which $E_c(t_1)$ = modulus of concrete at start of interval
$C_c(t)$ = creep coefficient at end of time interval.

7.8.3 Approximate Time-Steps Method

The approximate time-steps method is based on a simplified form of summation of constituent deflections due to the various time-dependent factors. If C_u is the long-term creep coefficient, the curvature at effective prestress P_e can be defined as

$$\phi_e = -\frac{P_i e_x}{E_c I_c} + (P_i - P_e)\frac{e_x}{E_c I_c} - \left(\frac{P_i + P_e}{2}\right)\frac{e_x}{E_c I_c}C_u \qquad (7.22)$$

The final deflection under P_e is

$$\delta_{et} = -\delta_i + (\delta_i - \delta_e) - \left(\frac{\delta_i + \delta_e}{2}\right)C_u \qquad (7.23\ a)$$

or

$$\delta_{et} = \delta_e - \left(\frac{\delta_i + \delta_e}{2}\right)C_u \qquad (7.23\ b)$$

Adding the deflection due to self-weight δ_D and superimposed dead load δ_{SD}, which are affected by creep, together with the deflection due to live load δ_L gives the final time-dependent *increase* in deflection due to prestressing and sustained loads, given by

$$\Delta\delta = -\left(\frac{\delta_i + \delta_c}{2}\right)C_u + (\delta_D + \delta_{SD})(1 + C_u) \qquad (7.24\ a)$$

and the final total *net* deflection is

$$\delta_T = -\delta_e - \left(\frac{\delta_i + \delta_c}{2}\right)C_u + (\delta_D + \delta_{SD})(1 + C_u) + \delta_L \qquad (7.24\ b)$$

Intermediate deflections are found by substituting C_t for C_u in Equations 7.24 a and b, where

$$C_t = \frac{t^{0.60}}{10 + t^{0.60}}C_u \qquad (7.25)$$

in which $t^{0.60}/(10 + t^{0.60})$ is the creep ratio α.

Branson et al., in Refs. 7.4–7.6, proposed the following expression for predicting the time-dependent increase in deflection $\Delta\delta$ of Equation 7.24 a:

$$\Delta\delta = \left[\eta + \frac{(1 + \eta)}{2}k_r C_t\right]\delta_{i(P_i)} + k_r C_t \delta_{i(D)} + K_a k_r C_t \delta_{i(SD)} \qquad (7.26)$$

where $\eta = P_e/P_i$
c_t = creep coefficient at time t
K_a = factor corresponding to age of concrete at superimposed load application
   = $1.25t^{-0.118}$ for moist-cured concrete
   = $1.13t^{-0.095}$ for steam-cured concrete
t = age, in days, at loading
k_r = $1/(1 + A_s/A_{ps})$ when $A_s/A_{ps} \leq 2.0$
   $\cong 1$ for all practical purposes.

For the final deflection increment, C_u is used in place of C_t in Equation 7.26.

For noncomposite beams, the total deflection $\delta_{T,t}$ becomes (Ref. 7.9)

$$\delta_{T,t} = -\delta_{pi}\left[1 - \frac{\Delta P}{P_0} + \lambda(k_r C_t)\right] + \delta_D[1 + k_r C_t]$$

$$+\delta_{SD}[1 + K_a k_r C_t] + \delta_L \qquad (7.27)$$

where δ_p = deflection due to prestressing
ΔP = total loss of prestress excluding initial elastic loss
λ = $1 - \Delta P/2P_0$
in which P_0 = prestress force at transfer after elastic loss
   = P_i less elastic loss.

For composite beams, the total deflection is

$$\delta_T = -\delta_{pi}\left[1 - \frac{\Delta P}{P_0} + K_a k_r C_u \lambda\right] + \delta_D + [1 + K_a k_r C_u]$$

$$+ \delta_{pi} \frac{I_e}{I_{\text{comp.}}} \left[1 - \frac{\Delta P - \Delta P_c}{P_0} + k_r C_u (\lambda - \alpha \lambda') \right]$$

$$+ (1 - \alpha) k_r C_u \delta_D \frac{I_c}{I_{\text{comp.}}} + \delta_s \left[1 + \alpha k_r C_u \frac{I_c}{I_{\text{comp.}}} \right]$$

$$+ \delta_{df} + \delta_L \qquad (7.28)$$

where $\lambda' = 1 - (\Delta P_c / 2P_0)$

ΔP_c = loss of prestress at time composite topping slab is cast, excluding initial elastic loss

$I_{\text{comp.}}$ = moment of inertia of composite section

δ_{df} = deflection due to differential shrinkage and creep between precast section and composite topping slab

= $F y_{cs} \ell^2 / 8 E_{cc} I_{\text{comp.}}$ for simply supported beams (for continuous beams, use the appropriate factor in the denominator)

y_{cs} = distance from centroid of composite section to centroid of slab topping

F = force resulting from differential shrinkage and creep

E_{cc} = modulus of composite section

α = creep strain at time t divided by ultimate creep strain

= $t^{0.60} / (10 + t^{0.60})$.

In sum, comparing the relative rigor involved in applying the three methods of Sections 7.8.1, 7.8.2, and 7.8.3, it is important to recognize that the degree of spread can be very large. Engineering judgment has to be exercised in determining a reasonably accurate concrete modulus E_c value at the various loading stages and in achieving values of creep coefficients that are neither under- nor overestimated.

7.8.4 Deflection of Composite Beams

Computation of deflections for composite prestressed beams is similar to that for noncomposite sections. The process becomes more rigorous if the incremental time-steps method is used. The additional steps at the various construction stages of the precast element and the situ-cast top slab require consideration of the changes in the moments of inertia from the precast to the composite values at the appropriate stages. Moreover, the difference in the shrinkage characteristics and time-step increments due to the difference in shrinkage values of the precast section and the added concrete topping increase the rigor of the computational process. Fortunately, the use of computer programs facilitates speedy evaluation of camber and deflection in composite elements.

7.9 PERMISSIBLE LIMITS OF CALCULATED DEFLECTION

The ACI Code requires that the calculated deflection has to satisfy the serviceability requirement of minimum permissible deflection for the various structural conditions listed in Table 7.2. Note that long-term effects cause measurable increases in deflection and camber with time and result in excessive overstress in the concrete and the reinforcement, *requiring* computation of deflection and camber.

AASHTO permissible deflection requirements, shown in Table 7.3, are more rigorous because of the dynamic impact of moving loads on bridge spans.

TABLE 7.2 ACI MINIMUM PERMISSIBLE RATIOS OF SPAN (ℓ) TO DEFLECTION (δ) (ℓ = LONGER SPAN)

Type of member	Deflection δ to be considered	$(\ell/\delta)_{min}$
Flat roofs not supporting and not attached to nonstructural elements likely to be damaged by large deflections	Immediate deflection due to live load L	180[a]
Floors not supporting and not attached to nonstructural elements likely to be damaged by large deflections	Immediate deflection due to live load L	360
Roof or floor construction supporting or attached to nonstructural elements likely to be damaged by large deflections	That part of total deflection occurring after attachment of nonstructural elements; sum of long-term deflection due to all sustained loads (dead load plus any sustained portion of live load) and immediate deflection due to any additional live load[b]	480[c]
Roof or floor construction supporting or attached to nonstructural elements not likely to be damaged by large deflections		240[c]

[a] Limit not intended to safeguard against ponding. Ponding should be checked by suitably calculating deflection, including added deflections due to ponded water, and considering long-term effects of all sustained loads, camber, construction tolerances, and reliability of provisions for drainage.

[b] Long-term deflection has to be determined, but may be reduced by the amount of deflection calculated to occur before attachment of nonstructural elements. This reduction is made on the basis of accepted engineering data relating to time–deflection characteristics of members similar to those being considered.

[c] Ratio limit may be lower if adequate measures are taken to prevent damage to supported or attached elements, but should not be lower than tolerance of nonstructural elements.

TABLE 7.3 AASHTO MAXIMUM PERMISSIBLE DEFLECTION (ℓ = LONGER SPAN)

Type of member	Deflection considered	Maximum permissible deflection	
		Vehicular traffic only	Vehicular and pedestrian traffic
Simple or continuous spans	Instantaneous due to service live load plus impact	$\dfrac{\ell}{800}$	$\dfrac{\ell}{1000}$
Cantilever arms		$\dfrac{\ell}{300}$	$\dfrac{\ell}{375}$

Following is a step-by-step procedure for computing deflection:

1. Determine the properties of the concrete, including the concrete modulus E_c, concrete creeps, and the shrinkage and prestress relation at the various loading stages.

2. Choose the time increments to be used in the deflection calculations.

3. Compute the concrete fiber stresses at the top and bottom extreme fibers due to all loads.

4. Compute the initial strains e_{ci} at the top and bottom fibers and the corresponding rotations, as well as subsequent strains and rotations. Use the equations

$$\phi_i = \frac{\epsilon_{chi} - \epsilon_{cti}}{h}$$

$$\phi_e = \frac{\epsilon_{cbe} - \epsilon_{cte}}{h}$$

$$\phi = \frac{\epsilon_{ct} - \epsilon_{cb}}{h}$$

$$\phi_u = \frac{\epsilon_u}{c}$$

Also, compute the strains at the cgs line and compute the relaxation of the strands during the first time interval.

5. Compute the total change of stress in the prestressing steel due to creep, shrinkage, and relaxation acting as a force F at the cgs. Then compute the concrete fiber and cgs stresses due to F.

6. Add the result of step 5 to the result of step 3.

7. Repeat the same procedure for all time intervals, and add the effect of superimposed dead loads.

8. Add the deflections due to live load to get the total deflection δ_T.

9. Verify whether the computed δ_T is within the permissible limits. If not, change the section.

Figure 7.18 presents a flowchart for computation of deflection by the approximate time-step method.

7.10 LONG-TERM CAMBER AND DEFLECTION CALCULATION BY THE PCI MULTIPLIERS METHOD

Example 7.7

Given $f_{pi} = 189,000$ psi, evaluate the long-term camber and deflection of the bonded T-beam in Examples 7.3, 7.4, and 7.5 by the PCI multipliers method, and verify whether the deflection values satisfy the ACI permissible limits. If the beam were to be post-tensioned, assume that f_{pi} would be equal to 189,000 psi after anchorage losses and after

$$\text{START}$$

1. Input section properties: A_c, I_g, r^2, c^t, c_b, S_b, S^t, e_c, e_e, RH, V/S
   Input load data W_D, W_{SD}, W_L
   Input material properties f'_c, f'_{ci}, f_c, f_{ci}, f_t, f_{ti}, E_{ci}, E_c, f_{pu}, f_{py}, f_{pi}, f_{pe}, E_{ps}, A_{ps}, P_i, P_o, P_e, C_u
   Time intervals t, prestress loss ΔP, k_r, λ, F, y_{cs}

2. Calculate fiber stresses at midspan and support section at transfer

$$f^t_c = -\frac{P_i}{A_c}\left(1 - \frac{e_c c^t}{r^2}\right) \qquad f_{cb} = -\frac{P_i}{A_c}\left(1 + \frac{e_c c_b}{r^2}\right)$$

$$\phi_{ci} = \frac{(f_{cb} - f^t_c)}{E_{ci}h}$$

$$f^t_e = -\frac{P_i}{A_c}\left(1 - \frac{e_e c^t}{r^2}\right) \qquad f_{eb} = -\frac{P_i}{A_c}\left(1 + \frac{e_e c_b}{r^2}\right)$$

$$\phi_{ei} = \frac{(f_{eb} - f^t_e)}{E_{ci}h}$$

3. Compute

Prestress camber $\delta_{pi} \uparrow = k_1 \dfrac{P_i e_c \ell^2}{8 E_{ci} I_g} + k_2 \dfrac{P_i (e_e - e_c) \ell^2}{8 E_{ci} I_g}$

Self-weight deflection $\delta_D \downarrow = \dfrac{5 W_D \ell^4}{384 E_{ci}}$

$\delta_{net} = -\delta_i + \delta_D$ where k_1, k_2: a function for tendon as in Table 7.1

4. Compute $I_e = \left(\dfrac{M_{cr}}{M_a}\right)^3 I_g + \left[1 - \left(\dfrac{M_{cr}}{M_a}\right)^3\right] I_{cr} \le I_g$

   where $M_{cr} = S_b\left[7.5\lambda \sqrt{f'_c} - \dfrac{P_e}{A_c}\left(1 + \dfrac{e_c c_b}{r^2}\right)\right]$

$I_{cr} = (n_p A_{ps} d^2_p + n_s A_s d^2)(1 - 1.67\sqrt{n_p \rho_p + n_s \rho})$

5. Compute $\delta_{SD} = \dfrac{5 W_{SD} \ell^4}{384 E_c I_e}$ $\qquad \delta_L = \dfrac{5 W_L \ell^4}{384 E_c I_e}$

6. Compute time-dependent factors for each time interval

$C_t = \dfrac{t^{0.60}}{10 + t^{0.60}} C_u$

$K_a = 1.25 t^{-0.118}$ for moist-cured
$\quad = 1.13 t^{-0.095}$ for steam-cured concrete

Figure 7.18 Flowchart for computation of deflection.

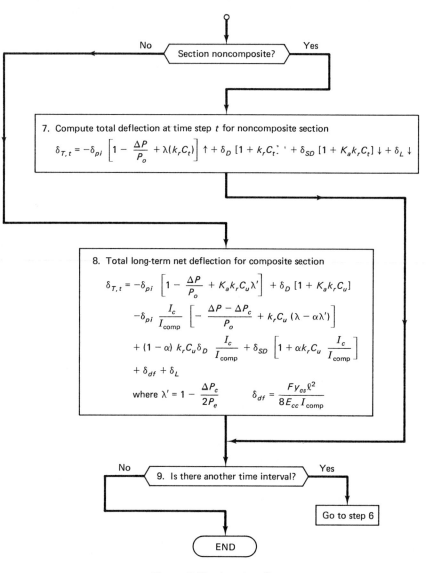

Figure 7.18 (*continued*)

eliminating frictional losses by jacking from both beam ends and then rejacking so as to maintain the net f_{pi} = 189,000 psi prior to erection. Also, assume that the nonstructural elements attached to the structure will not be damaged by deflections and that live load is transient. Use E_c = 4.03 × 10^6 psi for all loads in the solution.

Solution

$$I_g = 169,020 \text{ in}^4$$

$$W_D = 815 \text{ plf} = 67.9 \text{ lb/in.}$$

$$\delta_D = \frac{5W\ell^4}{384E_cI_g} = \frac{5 \times 67.9(65 \times 12)^4}{384 \times 3.49 \times 10^6 \times 169,020} = 0.55 \text{ in. } \downarrow \text{ (14 mm)}$$

$$W_{SD} = 100 \text{ plf} = 8.3 \text{ lb/in.}$$

TABLE 7.4 LONG-TERM CAMBER AND DEFLECTION VALUES IN EXAMPLE 7.7 BY PCI MULTIPLIERS METHOD

Load	Transfer δ_p in.	Multiplier	Erection in.	Multiplier	Final δ_{net} in.
	(1)		(2)		(3)
Prestress	1.50 ↑	1.80	2.70 ↑	2.45	3.68 ↑
W_D net	0.55 ↓	1.85	1.02 ↓	2.70	1.49 ↓
	0.95 ↑		1.68 ↑		2.19 ↑
W_{SD} net	0.06 ↓		0.06 ↓	3.0	0.18 ↓
			1.62 ↑		2.01 ↑
W_L net	0.65 ↓				0.65 ↓
					1.36 ↑

$$\delta_{SD} = \frac{5 \times 8.3(65 \times 12)^4}{384 \times 4.03 \times 10^6 \times 169,020} = 0.06 \text{ in. } \downarrow \ (1.5 \text{ mm})$$

From the solution in Example 7.4 for the effective moment of inertia of the cracked section,

$$W_L = 1,100 \text{ plf} = 91.7 \text{ lb/in.}$$

$$I_e = 86,812 \text{ in}^4$$

but since the section did not crack,

$$I_e = I_g = 169,020 \text{ in}^4 \ (\text{max. } f_t < f_r = 530 \text{ psi})$$

$$\delta_L = \frac{5 \times 91.7(65 \times 12)^4}{384 \times 4.03 \times 10^6 \times 169,020} = 0.65 \text{ in. } \downarrow \ (20 \text{ mm})$$

If the section is composite after erection, $I_{comp.}$ has to be used in calculating δ_L and δ_{SD} if the beam is shored during placement of concrete topping.

The values of long-term camber using the PCI multipliers are given in Table 7.4. From the table, the camber after erection and installation of the superimposed dead load at 30 days = 2.19 in ↑ (56 mm). Also, the final net camber after five years = 0.75 in. ↑ (19 mm), the live-load deflection = 0.65 in ↓ (20 mm), and the allowable deflection = $\ell/240 = (65 \times 12)/240 = 3.25$ in. (82.5 mm) > 2.29 in. camber if the live load is assumed all transient in this case, which is satisfactory.

7.11 LONG-TERM CAMBER AND DEFLECTION CALCULATION BY THE INCREMENTAL TIME-STEPS METHOD

Example 7.8

Solve Example 7.7 by the incremental time-steps method assuming that $f_{pi} = 189,000$ psi and that prestress losses are incrementally evaluated at prestressing (seven days after casting), 30 days after transfer (completion of erection and application of the superimposed dead load), 90 days, and 5 years. Assume that the ultimate creep coefficient $C_u = 2.35$ for the concrete and $f_{py} = 230,000$ psi for the prestressing steel used in the beam. Plot the camber-time and deflection-time relationships for the beam,

using $E_c = 4.03 \times 10^6$ psi for all incremental steps in this solution, except at transfer, where $f'_{ci} = 3,750$ psi.
Solution

Instantaneous Stresses, Strains, and Deflections

$$E_{ci} = 57,000\sqrt{3,750} = 3.49 \times 10^6 \text{ psi}$$

From Example 7.3 and Figure 7.9, the initial fiber stresses (psi) and strains (in./in.) for the beam at transfer due to prestress force P_i and $P_i + W_D$ are as follows:

(a) Prestress Force P_i

$$\textit{Midspan: } f^t = +499 \text{ psi (3.1 MPa)}$$

$$f_b = -3,311 \text{ psi (22.8 MPa)}$$

$$\epsilon'_c = \frac{+499}{3.49 \times 10^6} = +143 \times 10^{-6} \text{ in./in.}$$

$$\epsilon_{cb} = -949 \times 10^{-6}$$

$$\textit{Support: } f^t = +96 \text{ (0.7 MPa)}$$

$$f_b = -2,203 \text{ (15.2 MPa)}$$

$$\epsilon'_e = +28 \times 10^{-6}$$

$$\epsilon_{eb} = -631 \times 10^{-6}$$

Note that unless otherwise stated, the modulus E_c of concrete should be calculated for the time change at each incremental step.

Continuing, we have

$$\text{Midspan } \phi_{ci} = \frac{\epsilon_{cb} - e^t_c}{h} = \frac{-949 - 143}{48} \times 10^{-6} = 22.75 \times 10^{-6} \text{ rad/in.}$$

$$\text{Support } \phi_{ei} = \frac{-631 - 28}{48} \times 10^{-6} = -13.73 \times 10^{-6}$$

Typical deflection prior to limit state at failure (Nawy et al.).

From Figure 7.6,

$$\delta_i \uparrow = \phi_c \frac{\ell^2}{8} + (\phi_e - \phi_c)\frac{\ell^2}{24}$$

$$\delta_i \uparrow = -22.75 \times 10^{-6} \times \frac{(65 \times 12)^2}{8} + (-13.73 + 22.75) \times 10^{-6}$$

$$\times \frac{(65 \times 12)^2}{24} = \frac{(65 \times 12)^2}{24} \times 10^{-6}(-22.75 \times 3 + 9.02)$$

$$= -1.50 \text{ in. } \uparrow \ (38 \text{ mm})$$

Notice that this value is the same as that obtained by the moment expression in Example 7.3.

Finally,

$$\text{Self-wt. } \delta_D = +\frac{5w\ell^4}{384E_c I_g} = \frac{5 \times \left(\dfrac{815}{12}\right)(65 \times 12)^4}{384 \times 3.49 \times 10^6 \times 169{,}020} = +0.55 \text{ in} \downarrow \ (14 \text{ mm})$$

$$\text{Net camber at transfer} = -1.50 \uparrow \ + 0.55 \downarrow \ = -0.95 \text{ in. } \uparrow \ (24 \text{ mm})$$

Time-Dependent Factors

(a) *Creep*. From Equation 3.10,

$$\epsilon_{CR} = \frac{C_t}{E_c}(f_{cs}) = C_t \epsilon_{cs}$$

where f_{cs} = concrete stress at cgs level
  ϵ_{cs} = concete strain at cgs level
  ϵ'_{CR} = unit creep strain per unit stress at ultimate creep = C_u/E_c
    = $2.35/4.03 \times 10^6 = 0.583 \times 10^{-6}$ in./in. per unit stress.

Note that creep strain has to be calculated at the centroid of the reinforcement in order to calculate the creep loss in prestress.

From Equation 3.9 b, the creep coefficient at any time, in days, is

$$C_t = \frac{t^{0.60}}{10 + t^{0.60}}C_u$$

As an example, at 30 days after transfer

$$\epsilon'_{CR,t} = \epsilon'_{CR}\left(\frac{t^{0.60}}{10 + t^{0.60}}\right) = 0.583 \times 10^{-6}\left(\frac{30^{0.60}}{10 + 30^{0.60}}\right)$$

$$= 0.254 \times 10^{-6} \text{ in./in. per unit stress}$$

Creep strains at other time intervals are similarly computed.

(b) *Shrinkage of Concrete*. From Equation 3.15 a for moist-cured concrete,

$$\epsilon_{SH,t} = \frac{t}{t + 35}\epsilon_{SH}$$

where $\epsilon_{SH} = 800 \times 10^{-6}$ in./in. for moist-cured concrete.

Thirty days after transfer, the shrinkage time $t = 30$ days if the beam is post-

TABLE 7.5 TIME-DEPENDENT INCREMENTAL PRESTRESS LOSS FACTORS IN EXAMPLE 7.8

Time	Creep $\times 10^{-6}$		Shrinkage $\times 10^{-6}$		Relaxation	
Days	$\epsilon'_{CR,t}$	$\Delta\epsilon'_{CR}$	$\epsilon_{SH,t}$	$\Delta\epsilon_{SH}$	R	ΔR
(1)	(2)	(3)	(4)	(5)	(6)	(7)
-7	—	—	0	—	—	—
P/S	0	—	133	133	0	—
30	0.254	0.254	369	236	0.0776	0.0776
90	0.349	0.095	576	207	0.0906	0.0130
365	0.452	0.103	730	154	0.1071	0.0165
5 yrs	0.525	0.073	785	55	0.1261	0.0190

Δ = Incremental increase. Columns 2, 4, and 6 give cumulative values.

tensioned and $t = 30 + 7 = 37$ days if it is pretensioned. Hence,

$$\epsilon_{SH,30} = \frac{30}{30 + 35} \times 800 \times 10^{-6} = 369 \times 10^{-6} \text{ in./in.}$$

In a similar manner, ϵ_{SH} may be calculated for all other steps tabulated in Table 7.5.

(c) Relaxation of Strands. From Equation 3.6,

$$\frac{f_{pR}}{f_{pi}} = 1 - \left(\frac{\log t_2 - \log t_1}{10}\right)\left(\frac{f_{pi}}{f_{py}} - 0.55\right)$$

where $\log t$, in hours, is to the base 10, f_{pi}/f_{py} exceeds 0.55, and f_{pR} is the remaining stress in the steel after relaxation for 30 days $= 720$ hr after prestressing. If the relaxation loss ratio is

$$R = 1 - \frac{f_{pR}}{f_{pi}} = \left(\frac{\log 720 - 0}{10}\right)\left(\frac{189,000}{230,000} - 0.55\right) = 0.078$$

we must find the R values for all time steps using $t_1 = 0$ as a base.

Table 7.5 gives the incremental time-dependent parameters for prestress loss factors in this example for time steps 7, 30, 90, and 365 days, and 5 years after prestressing.

Transfer to Erection (Step end = 30 days)

(a) Concrete Fiber Stresses at the cgs Level for Calculation of Creep
The tendon eccentricities are $e_c = 33.14$ in. and $e_e = 20.0$ in. Figure 7.19 gives the instantaneous stresses and the corresponding *gross* strains before losses due to prestress and self-weight W_D. From the figure,

$$f_{bcc} = -3,311 \times \frac{39.66}{39.66 + 2.05} = -3,148 \text{ psi } (C) \text{ (21.7 MPa)}$$

$$\epsilon_{bcc} = \frac{-3,148}{4.03 \times 10^6} = -781 \times 10^{-6} \text{ in./in.}$$

$$f_{bec} = -2,203 \times \frac{30.81}{30.81 + 15.19} = -1,476 \text{ psi } (C) \text{ (10.2 MPa)}$$

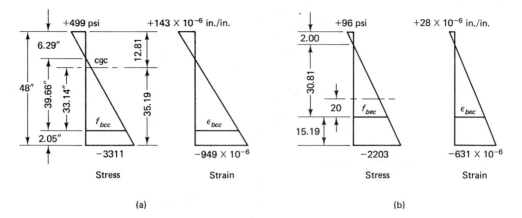

Figure 7.19 Stress and strain at transfer due only to prestress before losses in Example 7.8. (a) Midspan section. (b) Support section.

$$\epsilon_{bec} = \frac{-1,476}{4.03 \times 10^6} = -366 \times 10^{-6} \text{ in./in.}$$

Creep Incremental Strain

$$\Delta\epsilon_{CR} = \Delta\epsilon'_{CR} \times \text{stress } f \text{ at cgs}$$

From Table 7.5, $\Delta\epsilon'_{CR} = 0.254 \times 10^{-6}$ in./in. per unit stress. So

Midspan $\Delta\epsilon_{CR} = \Delta\epsilon'_{CR}f_{bcc} = 0.254 \times 10^{-6}(-3,148) = -800 \times 10^{-6}$ in./in.

Support $\Delta\epsilon_{CR} = \Delta\epsilon'_{CR}f_{bec} = 0.254 \times 10^{-6}(-1,476) = -375 \times 10^{-6}$ in./in.

Shrinkage Incremental Strain

$$\Delta\epsilon_{SH} = -236 \times 10^{-6} \text{ in./in.}$$

Relaxation Stress Loss

$$\Delta P_{R30} = 0.0776 \times 189,000 = 14,666 \text{ psi (101.1 MPa)}$$

Total Steel Stress Losses

$$\Delta f_T = (\Delta\epsilon_{CR} + \Delta\epsilon_{SH})E_{ps} + \Delta f_R$$

Midspan $\Delta f_{T30} = (800 + 236) \times 10^{-6} \times 27.5 \times 10^6 + 14,666 = 43,156$ psi

Support $\Delta f_{T30} = (375 + 236)10^{-6} \times 27.5 \times 10^6 + 14,666 = 31,468$ psi

Hence, use an average $\Delta f_{T30} = \frac{1}{2}(43,156 + 31,468) = 37,312$ psi (257.3 MPa).

(b) Corresponding Change in Concrete Fiber Stresses and Strains

Prestress force loss $\Delta P_{30} = \Delta f_{T30}A_{ps} = 37,312 \times 2.142 = 79,922$ lb. (355.5 kN)

(i) *Midspan section* (1 psi = 6.895 MPa)

$$\Delta f' = -\frac{\Delta P_{30}}{A_c}\left(1 - \frac{ec_t}{r^2}\right) = -\frac{79,922}{782}\left(1 - \frac{33.14 \times 12.81}{216.14}\right)$$

$$= +98 \text{ psi } (T)$$

$$\Delta \epsilon_c' = \frac{+98}{4.03 \times 10^6} = +24 \times 10^{-6} \text{ in./in.}$$

$$\Delta f_b = -\frac{\Delta P_{30}}{A_c}\left(1 + \frac{ec_b}{r^2}\right) = -\frac{79,922}{782}\left(1 + \frac{33.14 \times 35.19}{216.14}\right)$$

$$= -653 \text{ psi } (C)$$

$$\Delta \epsilon_{cb} = \frac{-653}{4.03 \times 10^6} = -162 \times 10^{-6} \text{ in./in.}$$

(ii) *Support section*

$$\Delta f^t = -\frac{79,922}{782}\left(1 - \frac{20 \times 12.81}{216.14}\right) = +19 \text{ psi } (T)$$

$$\Delta \epsilon_e^t = \frac{+19}{4.03 \times 10^6} = +5 \times 10^{-6} \text{ in./in.}$$

$$\Delta f_b = -\frac{79,922}{782}\left(1 + \frac{20 \times 35.19}{216.14}\right) = -435 \text{ psi } (C)$$

$$\Delta \epsilon_{eb} = \frac{-435}{4.03 \times 10^6} = -108 \text{ in./in.}$$

(c) Net Strains, Resulting Curvatures, and Camber

(i) *Net creep strain (in./in.)*

Fiber gross strain

Midspan

$$\Delta \epsilon_{CRc}' = f_7^t \times \Delta \epsilon_{CR}' = +499 \times 0.254 \times 10^{-6} = +127 \times 10^{-6} \text{ in./in.}$$

$$\Delta \epsilon_{CRcb} = f_{7b} \times \Delta \epsilon_{CR}' = -3,311 \times 0.254 \times 10^{-6} = -841 \times 10^{-6} \text{ in./in.}$$

Support

$$\Delta \epsilon_{CRe}' = +96 \times 0.254 \times 10^{-6} = +24 \times 10^{-6} \text{ in./in.}$$

$$\Delta \epsilon_{CReb} = -2,203 \times 0.254 \times 10^{-6} = -560 \times 10^{-6} \text{ in./in.}$$

Net creep strains (in./in.)

$$\Delta_{\text{net}} \epsilon_{CR} = \Delta \epsilon_{CR} - \Delta \epsilon_{ps}$$

where $\Delta \epsilon_{ps}$ is the strain loss due to prestress loss Δf in part (b) of the solution. From Figure 7.20, we have the following:

Midspan

$$\Delta \epsilon_{CRc, \text{net}}' = \Delta \epsilon_{CRc}' - \Delta \epsilon_{pscb}' = (+127 - 24)10^{-6} = +103 \times 10^{-6}$$

$$\Delta \epsilon_{CRcb, \text{net}} = \Delta \epsilon_{CRcb} - \Delta \epsilon_{pscb} = (-841 + 162)10^{-6} = -679 \times 10^{-6}$$

Support

$$\Delta \epsilon_{CReb, \text{net}}' = \Delta \epsilon_{CRe}' - \Delta \epsilon_{pse}' = (+24 - 5)10^{-6} = +19 \times 10^{-6}$$

$$\Delta \epsilon_{CReb, \text{net}} = \Delta \epsilon_{CReb} - \Delta \epsilon_{pseb} = (-560 + 108)10^{-6} = -452 \times 10^{-6}$$

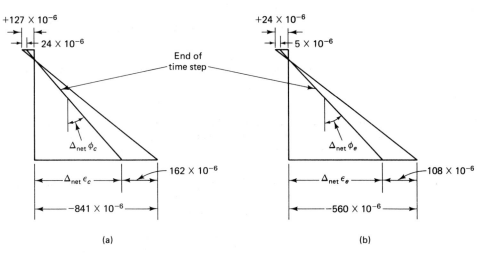

+127 × 10⁻⁶

Figure 7.20 Creep incremental strains at 30 days in Example 7.8.

(ii) *Curvatures (rad/in.)*

$\Delta\phi_{30}$ is the added curvature due to losses at the end of 30 days after transfer based on the adjusted net strains, in other words the curvature increment for this step.

Midspan

$$\Delta\phi_{c30} = \frac{\Delta\epsilon_{CRcb,\,net} - \Delta\epsilon'_{CRc,\,net}}{h} = \frac{(-679 - 103)10^{-6}}{48}$$

$$= -16.29 \times 10^{-6} \ (0.64 \times 10^{-6} \ \text{rad/mm})$$

Support

$$\Delta\phi_{e30} = \frac{\Delta\epsilon_{CReb,\,net} - \Delta\epsilon'_{CRe,\,net}}{h} = \frac{(-452 - 19)10^{-6}}{48}$$

$$= -9.81 \times 10^{-6} \ (0.34 \times 10^{-6} \ \text{rad/mm})$$

(iii) *Total curvature (rad/in.) and camber*

From before, $\phi_{ci} = -22.75 \times 10^{-6}$ rad/mm and $\phi_{ei} = -13.73 \times 10^{-6}$ rad/mm. So the total curvature at 30 days after transfer is

$$\phi_T = \phi_i + \Delta\phi_{30}$$

Midspan

$$\phi_{Tc30} = (-22.75 - 16.29) \times 10^{-6} = -39.04 \times 10^{-6} \ (1.54 \times 10^{-6} \ \text{rad/mm})$$

Support

$$\phi_{Te30} = (-13.73 - 9.81) \times 10^{-6} = -23.54 \times 10^{-6} \ (0.93 \times 10^{-6} \ \text{rad/mm})$$

From Figure 7.6, the camber due to prestress at the end of 30 days is

$$\delta_{P30} \uparrow = \phi_{Tc30}\left(\frac{\ell^2}{8}\right) + (\phi_{Te30} - \phi_{Tc30})\frac{\ell^2}{24}$$

$$= -39.04 \times 10^{-6} \times \frac{(65 \times 12)^2}{8}$$

$$+ (-23.54 + 39.04)10^{-6} \times \frac{(65 \times 12)^2}{24}$$

$$\delta_{30\,(camb.)} \uparrow = \frac{(65 \times 12)^2}{24}(-39.04 \times 3 + 15.50) \times 10^{-6}$$

$$= -2.58 \text{ in.} \uparrow \ (65.5 \text{ mm})$$

(d) Long-term Deflections Due to Gravity Loads at 30 Days after Transfer. Assume $E_c = 4.03 \times 10^{-6}$ psi as a reasonable value for the modulus of concrete for the rest of the example. We have $W_D = 815$ plf $= 67.9$ lb/in. Also,

$$\text{Self-weight deflection } \delta_D = \frac{5W\ell^4}{384E_cI_g} = +0.55 \text{ in.} \downarrow \ (14 \text{ mm}) \text{ from before}$$

$$W_{SD} = 100 \text{ plf}$$

$$\delta_{SD} = \frac{5 \times \dfrac{100}{12}(65 \times 12)^4}{384 \times 4.03 \times 10^{-6} \times 169{,}020}$$

$$= +0.06 \text{ in.} \downarrow \ (1.5 \text{ mm})$$

$$C_t = \frac{t^{0.06}}{10 + t^{0.60}} \quad C_u = \frac{30^{0.60}}{10 + 30^{0.60}} \times 2.35 = 1.02$$

(C_t for W_{SD} at ~15 days = 0.80)

$$\delta_{D30} = 0.55(1 + 1.02) = +1.11 \text{ in} \downarrow \ (28 \text{ mm})$$

$$\delta_{SD30} = 0.06(1 + 0.80) = +0.11 \text{ in} \downarrow \ (2.8 \text{ mm})$$

$$\delta_L = 0 \text{ (building occupied at 90 days)}$$

Total gravity load deflections $= 1.11 + 0.11 + 0 = +1.22$ in. $\downarrow$ (31 mm)

$$\delta_{net\,30} = -2.58 \uparrow + 1.22 \downarrow$$

$$= -1.36 \text{ in.} \uparrow \ (34.5 \text{ mm}) \text{ (camber)}$$

Service-Load Step—90 days after Transfer

(a) New Reduced Concrete Fiber Stresses and Strains Due to Prestress Losses in the Previous Stage

Prestress force P_i

(i) *Midspan*

$$f^t = +499 - 98 = +401 \text{ psi}$$

$$f_b = -3{,}311 + 653 = -2{,}658 \text{ psi}$$

$$\epsilon_c^t = \frac{+401}{4.03 \times 10^6} = +100 \times 10^{-6} \text{ in./in.}$$

$$\epsilon_{cb} = -660 \times 10^{-6} \text{ in./in.}$$

(ii) *Support*

$$f^t = +96 - 19 = +77 \text{ psi}$$

$$f_b = -2,203 + 435 = -1,811 \text{ psi}$$

$$\epsilon_e^t = +20 \times 10^{-6} \text{ in./in.}$$

$$\epsilon_{eb} = -449 \times 10^{-6} \text{ in./in.}$$

From Figure 7.21,

$$f_{bcc} = -2,658 \times \frac{39.66}{39.66 + 2.05} = -2,527 \text{ psi}$$

$$\epsilon_{bcc} = -627 \times 10^{-6} \text{ in./in.}$$

$$f_{bec} = -1,811 \times \frac{30.81}{30.81 + 15.19} = -1,213 \text{ psi}$$

$$\epsilon_{bec} = -301 \times 10^{-6} \text{ in./in.}$$

Creep incremental strain

$$\Delta\epsilon_{CR} = 0.095 \times 10^{-6} \text{ in./in. per unit stress (Table 7.5)}$$

$$\text{Midspan } \Delta\epsilon_{CR} = \Delta\epsilon_{CR}' f_{bcc} = 0.095 \times 10^{-6}(-2,527) = -240 \times 10^{-6} \text{ in./in.}$$

$$\text{Support } \Delta\epsilon_{CR} = \Delta\epsilon_{CR}' f_{bec} = 0.095 \times 10^{-6}(-1,213) = -115 \times 10^{-6} \text{ in./in.}$$

Shrinkage incremental strain

$$\Delta\epsilon_{SH} = -207 \times 10^{-6} \text{ in./in.}$$

Relaxation stress loss

$$\Delta P_R = 0.0130(189,000 - 37,312) = 1,972 \text{ psi}$$

Total steel stress losses

$$\Delta f_T = (\Delta\epsilon_{CR} + \Delta\epsilon_{SH})E_{ps} + \Delta f_R$$

$$\text{Midspan } \Delta f_{T90} = (240 + 207) \times 10^{-6} \times 27.5 \times 10^6 + 1,972 = 14,265 \text{ psi}$$

$$\text{Support } \Delta f_{T90} = (115 + 207) \times 10^{-6} \times 27.5 \times 10^6 + 1,972 = 10,827 \text{ psi}$$

$$\text{Average } \Delta f_{T90} = \frac{1}{2}(14,265 + 10,827) = 12,546 \text{ psi}$$

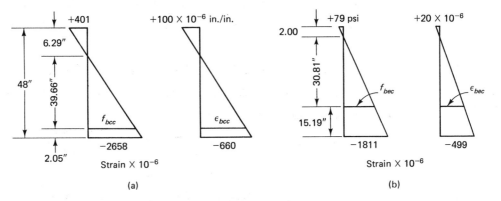

Figure 7.21 Adjusted stress and strain at 30 days due only to prestress in Example 7.8. (a) Midspan section. (b) Support section.

(b) Corresponding Change in Concrete Fiber Stresses and Strains. Prestress force loss $\Delta P_{90} = \Delta f_{T90} A_{ps} = 12,546 \times 2.142 = 26,648$ lb (118.5 kN)

(i) *Midspan section* (1 psi = 6.895 MPa)

$$\Delta f^t = +98 \times \frac{26,648}{79,922} = +33 \text{ psi}$$

$$\Delta f_b = -218 \text{ psi}$$

$$\Delta \epsilon_c^t = +8 \times 10^{-6} \text{ in./in.}$$

$$\Delta \epsilon_{cb} = -54 \times 10^{-6} \text{ in./in.}$$

(ii) *Support section*

$$\Delta f^t = +6 \text{ psi}$$

$$\Delta f_b = -145 \text{ psi}$$

$$\Delta \epsilon_e^t = +2 \times 10^{-6} \text{ in./in.}$$

$$\Delta \epsilon_{eb} = -36 \times 10^{-6} \text{ in./in.}$$

(c) Net Strains, Resulting Curvatures, and Camber
(i) *Net creep strain (in./in.)*

Fiber gross strain

Midspan

$$\Delta \epsilon_{CRc}^t = f_{30}^t \times \Delta \epsilon_{CR}^t = 401 \times 0.095 \times 10^{-6} = +38 \times 10^{-6} \text{ in./in.}$$

$$\Delta \epsilon_{CRcb} = f_{30b} \times \Delta \epsilon_{CR}^t = -2,658 \times 0.095 \times 10^{-6} = -253 \times 10^{-6}$$

Support

$$\Delta \epsilon_{CRe}^t = +79 \times 0.095 \times 10^{-6} = +8 \times 10^{-6}$$

$$\Delta \epsilon_{CReb} = -1,811 \times 0.095 \times 10^{-6} = -172 \times 10^{-6}$$

Net creep strains (in./in.)

Midspan

$$\Delta \epsilon_{CRc, \text{net}}^t = \Delta \epsilon_{CRc}^t - \Delta \epsilon_{psc}^t = (+38 - 8) \times 10^{-6} = +30 \times 10^{-6}$$

$$\Delta \epsilon_{CRcb, \text{net}} = \Delta \epsilon_{CRcb} - \Delta \epsilon_{pscb} = (-253 + 54) \times 10^{-6} = -199 \times 10^{-6}$$

Support

$$\Delta \epsilon_{CRe, \text{net}}^t = \Delta \epsilon_{CRe}^t - \Delta \epsilon_{pse}^t = (+8 - 2) \times 10^{-6} = +6 \times 10^{-6}$$

$$\Delta \epsilon_{CReb, \text{net}} = \Delta \epsilon_{CReb} - \Delta \epsilon_{pseb} = (-172 + 36) \times 10^{-6} = -136 \times 10^{-6}$$

(ii) *Curvature (rad./in.)*

$$\text{Midspan } \Delta \phi_{c90} = \frac{\Delta \epsilon_{CRc, \text{net}}^t - \Delta \epsilon_{CRcb, \text{net}}}{h} = \frac{(-199 - 30) \times 10^{-6}}{48}$$

$$= -4.77 \times 10^{-6} \text{ rad/in.}$$

$$\text{Support } \Delta \phi_{e90} = \frac{\Delta \epsilon_{CRe, \text{net}}^t - \Delta \epsilon_{CReb, \text{net}}}{h} = \frac{(-136 - 6) \times 10^{-6}}{48}$$

$$= -2.96 \times 10^{-6} \text{ rad/in.}$$

(iii) *Total curvature (rad/in.) and camber*

From before, $\phi_{c30} = -39.04 \times 10^{-6}$ and $\phi_{e30} = -23.54 \times 10^{-6}$. So the total curvature is

$$\phi_T = \phi_{30} + \Delta\phi_{90}$$

and we also have all of the following:

Midspan $\phi_{Tc90} = (-39.04 - 4.77) \times 10^{-6} = -43.81 \times 10^{-6}$ (1.72 rad/mm)

Support $\phi_{Te90} = (-23.54 - 2.96) \times 10^{-6} = -26.50 \times 10^{-6}$ (1.04 rad/mm)

$$\delta_{p90}(\text{camber}) \uparrow = \phi_{Tc90}\left(\frac{\ell^2}{8}\right) + (\phi_{Te90} - \phi_{Tc90})\frac{\ell^2}{24}$$

$$= -43.81 \times 10^{-6} \times \frac{(65 \times 12)^2}{8}$$

$$+ (-26.50 + 43.81) \times 10^{-6}\frac{(65 \times 12)^2}{24}$$

$$= \frac{(65 \times 12)^2}{24}(-43.81 \times 3 + 17.31) \times 10^{-6}$$

$$= -2.89 \text{ in.} \uparrow \quad (73 \text{ mm})$$

(d) *Long-term Deflections Due to Gravity Loads at 90 Days after Transfer*

$$t = 90 \text{ days}$$

$$C_t = \frac{t^{0.60}}{10 + t^{0.60}}C_u = \frac{90^{0.60}}{10 + 90^{0.60}} \times 2.35 = 1.41$$

$$(C_t = 1.27 \text{ for } t = 60 \text{ days})$$

$$\delta_{D90} = \delta_i(1 + C_t) = 0.55(1 + 1.41) = +1.33 \text{ in.} \downarrow$$

$$\delta_{SD90} = 0.06(1 + 1.27) = +0.14 \text{ in.} \downarrow$$

$$\delta_L \text{ (from Example 7.7)} = +0.65 \text{ in.} \downarrow$$

So the total deflection at 90 days due to gravity loads is

$$\delta_g = +1.33 + 0.14 + 0.65 = +2.12 \text{ in.} \downarrow \quad (69 \text{ mm})$$

and the net deflection is

$$\text{Net } \delta_{\text{net}, 90} = -2.89 \uparrow +2.12 \downarrow = -0.77 \text{ in.} \uparrow \quad (20 \text{ mm}) \text{ (camber)}$$

Service-Load Deflection at 5 Years. The same steps as the previous give the results tabulated in Tables 7.6 and 7.7, while a plot of the cambers and deflections as a function of time is shown in Figure 7.22. The deflection-time relationship becomes almost asymptotic.

From Table 7.7, the final net deflection (camber) at five years is 0.87 in., which is much less than the maximum allowable deflection or camber from Table 7.2 ($\ell/240 = (65 \times 12)/240 = 3.25$ in. $\geqslant 0.87$ in.). Hence, the beam satisfies the serviceability requirements for time-dependent deflection. Note that long-term creep losses can be considerably reduced by the addition of nonprestressed reinforcement to the section.

TABLE 7.6 LONG-TERM STRESS AND STRAIN CHANGES IN EXAMPLE 7.8

Time at step end	Gross fiber stress in concrete at start of time increment				Creep strain increment $\times 10^{-6}$ in./in. per unit stress	Shrinkage strain increment $\times 10^{-6}$ in./in.	Steel relaxation increment psi	Total steel stress loss increment psi	Concrete stress change due to losses psi				Concrete strain change due to losses $\times 10^{-6}$ in./in.			
	Midspan		End						Midspan		End		Midspan		End	
Days	f^t	f_b	f^t	f_b	$\Delta\varepsilon_{CRc}$ / $\Delta\varepsilon_{CRb}$	$\Delta\varepsilon_{SH}$	Δf_R	$\Delta f_{TR,t}$	Δf^t	Δf_b	Δf^t	Δf_b	$\Delta\varepsilon_c^t$	$\Delta\varepsilon_{cb}$	$\Delta\varepsilon_e^t$	$\Delta\varepsilon_{eb}$
(1)	(2)		(3)		(4)	(5)	(6)	(7)	(8)		(9)		(10)		(11)	
*P/S	0	0	0	0	0	133	—	—	—	—	—	—	—	—	—	—
30	+499	−3,311	+96	−2,203	−800 / −375	236	14,666	37,312	+98	−653	+19	−435	+24	−162	+5	−108
90	+401	−2,658	+79	−1,811	−240 / −115	207	1,972	12,546	+33	−218	+6	−145	+8	−54	+2	−36
365	+368	−2,440	+73	−1,666	−239 / −115	154	2,296	11,399	+30	−199	+6	−133	+7	−50	+1	−33
5 yr	+338	−2,241	+67	−1,533	−155 / −75	55	2,427	7,112	+19	−124	+4	−83	+5	−31	+1	−21

*P/S = step at transfer of prestress at 7 days after concrete is cast; 1,000 psi = 6,895 MPa.

TABLE 7.7 LONG-TERM CURVATURES AND DEFLECTIONS IN EXAMPLE 7.8

Time at step end	Net creep strain increment × 10^{-6} in./in.		Curvature increment × 10^{-6} rad/in.		Total cumulative curvature × 10^{-6} rad/in.		Deflection		
							Prestress camber force P in.	Gravity loads $W_D + W_{SD} + W_L$ in.	Net in.
Days	Midspan $\Delta\epsilon^t_{CR,net}$ / $\Delta\epsilon_{CRb,net}$	End $\Delta\epsilon^t_{CR,net}$ / $\Delta\epsilon_{CRb,net}$	Midspan $\Delta\phi_c$	End $\Delta\phi_e$	Midspan $\Sigma\,\phi_c$	End $\Sigma\,\phi_e$	$\delta_{p,t}$	δ_g	$\delta_{net,t}$
(1)	(2)	(3)	(4)	(5)	(6)	(7)	(8)	(9)	(10)
P/S*	0	0	−22.75	−13.73	−22.75	−13.73	−1.50	+0.55	−0.95 ↑
30	+103 / −679	+19 / −452	−16.29	−9.81	−39.04	−23.54	−2.58	+1.22	−1.36 ↑
90	+30 / −199	+6 / −136	−4.77	−2.96	−43.81	−26.50	−2.89	+2.12	−0.77 ↑
365	+31 / −201	+7 / −139	−4.83	−3.04	−48.64	−29.54	−3.21	+2.37	−0.84 ↑
5 yr	+20 / −133	+4 / −99	−3.19	−2.15	−51.83	−31.69	−3.43	+2.56	−0.87 ↑

*P/S = step at transfer of prestress at 7 days after concrete is cast; 1 in. = 25.4 mm.

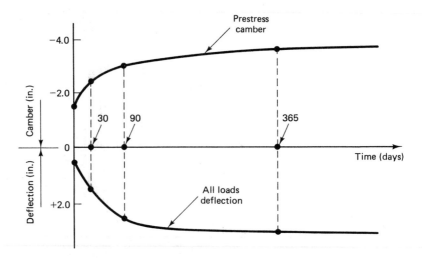

Figure 7.22 Prestressed camber and load deflections vs. time in Example 7.8.

7.12 LONG-TERM CAMBER AND DEFLECTION CALCULATION BY THE APPROXIMATE TIME-STEPS METHOD

Example 7.9

Solve Example 7.7 by the approximate time-steps method using the same allowable steel and concrete stresses. Compare this solution with those of Examples 7.7 and 7.8.

Solution

Data for This Alternative Solution. From Example 7.8,

$$P_i = 404,838 \text{ lb}$$
$$P_e = 331,967 \text{ lb}$$
$$A_s = 2.40 \text{ in}^2$$
$$A_{ps} = 2.142 \text{ in}^2$$
$$C_u = 2.35$$
$$C_{t30} = 1.02$$
$$C_{t90} = 1.41$$
$$C_{t365} = 1.82$$
$$C_{t5yr} = 2.12$$

Use the same C_t value for δ_{SD} as for δ_D, and consider it accurate enough. Then

$$K_a = 1.25t^{-0.118} \text{ for moist-cured concrete}$$

$$t = \text{age at loading, in days} = 30, 90, 365, 5 \text{ yr}$$

$$k_r = \frac{1}{1 + A_s/A_{ps}}$$

where $A_s/A_{ps} \leq 2.0$.

Deflection at failure of prestressed T-beam with confining reinforcement (Nawy, Potyondy).

Use $k_r \cong 1$ as accurate enough for practical purposes, since usually $A_s/A_{ps} \geqslant 2$.

Instantaneous Camber and Deflections. From Example 7.3,

$\delta_{i(P_i)} = \delta_{pi}$ = instantaneous initial prestress camber = -1.05 in. $\uparrow$

$\delta_{i(D)}$ = instantaneous dead-load deflection = $+0.55$ in. $\downarrow$

$\delta_{i(SD)}$ = instantaneous superimposed dead-load deflection = $+0.06$ in. $\downarrow$

δ_L = live-load deflection = $+0.65$ in. $\downarrow$

From Equation 7.26, the incremental time-step *net* deflection is

$$\Delta\delta_T = -\left[\eta + \left(\frac{1 + \eta}{2}\right)k_r C_u\right]\delta_{p(P_i)} \uparrow + k_r C_u \delta_{i(D)} \downarrow + K_a k_r C_u \delta_{i(SD)} \uparrow$$

δ_p is the deflection (camber) due to prestressing = $\delta_{i(P_i)}$ and $\eta = P_e/P_i$. From Equation 7.27, the total *net* deflection due to loads is

$$\delta_T = -\delta_p\left[1 - \frac{\Delta P}{P_e} + \lambda(k_r C_t)\right] + \delta_D[1 + k_r C_t] \downarrow + \delta_{SD}[1 + K_a k_r C_t] \downarrow + \delta_L \downarrow$$

where $\Delta P = (P_i - P_e)$ is the total loss of prestress excluding any initial elastic loss, $\lambda = 1 - \Delta P/2P_0$, and δ_p and η are as before.

Transfer to Erection (30 Days). Assume $P_0 \cong P_i$. Then $\Delta P = P_i - P_e = 404{,}838 - 331{,}967 = 72{,}871$ lb, and $\Delta P/P_i = 72{,}871/404{,}838 = 0.18$. So

$$\lambda = 1 - \frac{72{,}871}{2 \times 404{,}838} = 0.91$$

$$k_r = 1$$

$$K_a = 1.25(15)^{-0.118} = 0.91$$

$$C_t = 1.02$$

$$\delta_{T30} = -1.50(1 - 0.18 + 0.91 \times 1.02) \uparrow + 0.55(1 + 1.02) \downarrow$$
$$+ 0.06(1 + 0.91 \times 1.02) \downarrow + 0$$
$$= -2.62 \uparrow + 1.23 \downarrow = -1.39 \text{ in. } \uparrow \ (35 \text{ mm})$$

Erection to Service-Load Deflection (90 Days). The total interval from transfer is $t = 90$ days. So we have

$$K_a = 1.25(90)^{-0.118} = 0.74$$

$$C_t = 1.41$$

$$\delta_L = +0.65 \text{ in. } \downarrow$$

$$\delta_{T90} = -1.50(1 - 0.19 + 0.91 \times 1.41) \uparrow + 0.55(1 + 1.41) \downarrow$$
$$+ 0.06(1 + 0.74 \times 1.41) \downarrow + 0.65 \downarrow$$
$$= -3.14 \uparrow + 2.10 \downarrow = -1.04 \text{ in. } \uparrow \ (11 \text{ mm})$$

Service-Load Deflection at 365 Days

$$K_a = 1.25(365)^{-0.118} = 0.52$$

$$C_t = 1.82$$

$$\delta_{T365} = -1.50(1 - 0.19 + 0.91 \times 1.82) \uparrow + 0.55(1 + 1.82) \downarrow$$
$$+ 0.06(1 + 0.62 \times 1.82) \downarrow + 0.65 \downarrow$$
$$= -3.70 \uparrow + 2.33 \downarrow = -1.37 \text{ in. } \uparrow \ (19 \text{ mm})$$

Service-Load Deflections at 5 Years

$$K_a = 1.25(1,825)^{-0.118} = 0.52$$

$$C_t = 2.12$$

$$\delta_{T5yr} = -1.50(1 - 0.19 + 0.91 \times 2.12) \uparrow + 0.55(1 + 2.12) \uparrow$$
$$+ 0.06(1 + 0.52 \times 2.12) \downarrow + 0.65 \downarrow$$
$$= -4.11 \uparrow + 2.49 \downarrow = -1.62 \text{ in. } \uparrow \ (25.5 \text{ mm})$$

Comparison of Deflection Calculations by the Three Methods. Table 7.8 gives the calculated values of camber and deflection using the three methods in Examples 7.7, 7.8, and 7.9. The minus sign $(-)$ indicates upward camber $\uparrow$, and the plus sign $(+)$ indicates downward deflection $\downarrow$. The camber is the upward deflection due to the prestress force *less* the reduction in deflection due to self-weight.

Comparison of the net deflections shows that the multipliers method and the approximate time-steps method give essentially comparable results, while the incremental time-steps method gives slightly lower camber values (approximately $\frac{3}{4}$ in. difference). This variation is expected because incremental prestress losses are determined at each step rather than as a single lump-sum loss taken at the final stage. The incremental step method is time consuming, and use of computers is necessary to justify its use. A large number of incremental time steps need to be investigated in large-span major structures such as segmental or cable-stayed bridges where accuracy of deflection and computations of camber are a major concern.

TABLE 7.8 CAMBER AND DEFLECTION COMPARISONS (δ INCH)

Time at step end, days	Methods								
	PCI multipliers			Incremental time step			Approximate time step		
	Camber	δ_g	δ_{net}	Camber	δ_g	δ_{net}	Camber	δ_g	δ_{net}
(1)	(2)	(3)	(4)	(5)	(6)	(7)	(8)	(9)	(10)
7	−1.50	+0.55	−0.95	−1.50	+0.55	−0.95	−1.50	+0.55	−0.95
30	−2.70	+1.08	−1.62	−2.58	+1.22	−1.36	−2.62	+1.23	−1.39
90				−2.89	+2.12	−0.77	−3.14	−2.10	−1.04
365				−3.21	+2.37	−0.84	−3.70	+2.33	−1.37
5 yrs	−3.68	+2.32	−1.36	−3.40	+2.56	−0.87	−4.10	+2.49	−1.62

Typical cracking propagation in prestressed concrete beams (Nawy et al.).

7.13 CRACKING BEHAVIOR AND CRACK CONTROL IN PRESTRESSED BEAMS

7.13.1 Introduction

The increased use of partial prestressing, allowing limited tensile stresses in the concrete under service-load and overload conditions while allowing nonprestressed steel to carry the tensile stresses, is becoming prevalent due to practicality and economy. Consequently, an evaluation of the flexural crack widths and spacing and control of their development become essential. Work in this area is relatively limited because of the various factors affecting crack width development in prestressed concrete. However, experimental investigations support the hypothesis that the major controlling parameter is the reinforcement stress change beyond the decompression stage. Nawy, *et al.* have undertaken extensive research since the 1960s on the cracking behavior of prestressed pretensioned and post-tensioned beams and slabs because of the great vulnerability of the highly stressed prestressing steel to corrosion and other environmental effects and the resulting premature loss of prestress (Refs. 7.13–7.17). Serviceability behavior under service and overload conditions can be controlled by the design engineer through the application of the criteria presented in this section.

7.13.2 Mathematical Model Formulation for Serviceability Evaluation

Crack Spacing. Primary cracks form in the region of maximum bending moment when the external load reaches the cracking load. As loading is increased, additional cracks will form and the number of cracks will be stabilized when the stress in the concrete no longer exceeds its tensile strength at further locations regardless of load increase. This condition essentially produces the absolute minimum crack spacing that can occur at high steel stresses, here termed *stabilized minimum crack spacing*. The maximum possible crack spacing under this stabilized condition is twice the minimum and is termed the *stabilized maximum crack spacing*. Hence, the stabilized mean crack spacing a_{cs} is the mean value of the two extremes.

The total tensile force T transferred from the steel to the concrete over the stabilized mean crack spacing can be defined as

$$T = \gamma a_{cs} \mu \sum o \qquad (7.29)$$

where γ is a factor reflecting the distribution of bond stress, μ is the maximum bond stress which is a function of $\sqrt{f_c'}$, and $\sum o$ is the sum of the reinforcing elements' circumferences. The resistance R of the concrete area in tension A_t can be defined (see Figure 7.23) as

$$R = A_t f_t' \qquad (7.30)$$

By equating Equations 7.29 and 7.30, the following expression for a_{cs} is obtained, where c is a constant to be developed from the tests:

$$a_{cs} = c \frac{A_t f_t'}{\sum o \sqrt{f_c'}} \qquad (7.31)$$

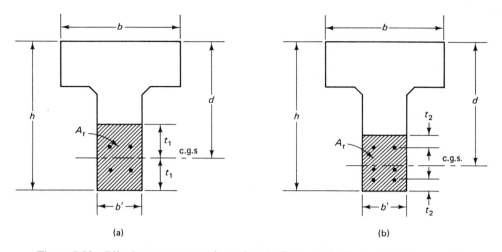

Figure 7.23 Effective concrete area in tension. (a) For even distribution of reinforcement in concrete. (b) For noneven distribution of reinforcement in concrete.

From extensive tests (see Refs. 7.13, 7.14, and 7.15), $cf'_t/\sqrt{f'_c}$ is found to have an average value of 1.2 for pretensioned, and 1.54 for post-tensioned prestressed beams.

Crack Width. If Δf_s is the net stress in the prestressed tendon or the magnitude of the tensile stress in the normal steel at any crack width load level in which the decompression load (decompression here means $f_c = 0$ at the level of the reinforcing steel) is taken as the reference point, then for the prestressed tendon

$$\Delta f_s = f_{nt} - f_d \quad \text{ksi} \ (= 1,000 \ \text{psi}) \tag{7.32}$$

where f_{nt} is the stress in the prestressing steel at any load level beyond the decompression load and f_d is the stress in the prestressing steel corresponding to the decompression load.

The unit strain $\epsilon_s = \Delta f_s/E_s$. Because it is logical to disregard as insignificant the unit strains in the concrete due to the effects of temperature, shrinkage, and elastic shortening, the maximum crack width can be defined as

$$w_{\max} = ka_{cs}\epsilon_s^{\alpha} \tag{7.33}$$

or

$$w_{\max} = k'\alpha_{cs} (\Delta f_s)^a \tag{7.34}$$

where k and α are constants to be established by tests.

7.13.3 Expressions for Pretensioned Beams

Equation 7.34 is rewritten in terms of Δf_s so that the following expression at the reinforcement level is obtained based on large numbers of tests:

$$w_{\max} = 1.4 \times 10^{-5} a_{cs} (\Delta f_s)^{1.31} \tag{7.35}$$

A 25 percent band of scatter envelops all the data for the expression in Equation 7.35 for $\Delta f_s = 20$ to 80 ksi.

Sec. 7.13   Cracking Behavior and Crack Control in Prestressed Beams   **447**

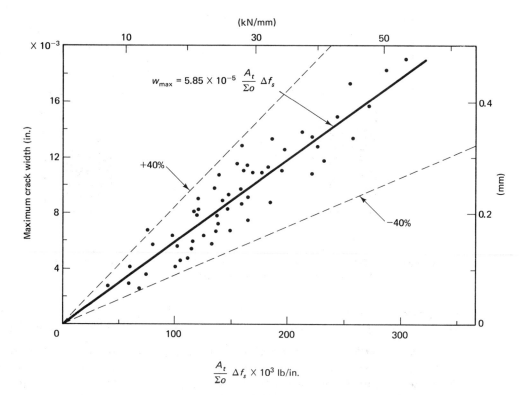

Figure 7.24 Linearized maximum crack width versus $(A_t/\Sigma_o)\Delta f_s$, pretensioned beams.

Linearizing Equation 7.35 for easier use by the design engineer leads to the simplified expression

$$w_{max} = 5.85 \times 10^{-5}\frac{A_t}{\Sigma o}(\Delta f_s) \qquad (7.36\ a)$$

of maximum crack width at the reinforcing steel level, and a maximum crack width (in.) at the tensile face of the concrete of

$$w'_{max} = 5.85 \times 10^{-5}R_i\frac{A_t}{\Sigma o}(\Delta f_s) \qquad (7.36\ b)$$

where R_i is the ratio of distance from neutral axis to tension face to the distance from neutral axis to centroid of reinforcement.

A plot of the data and the best-fit expression for Equation 7.36 a is given in Figure 7.24 with a 40 percent spread, which is reasonable in view of the randomness of crack development and the linearization of the original expression (Equation 7.35).

7.13.4 Expressions for Post-Tensioned Beams

The expression developed for the crack width in post-tensioned bonded beams which contain mild steel reinforcement is

$$w_{max} = 6.51 \times 10^{-5}\frac{A_t}{\Sigma o}(\Delta f_s) \qquad (7.37\ a)$$

for the width at the reinforcement level closest to the tensile face, and

$$w_{\max} = 6.51 \times 10^{-5} R_i \frac{A_t}{\Sigma o} (\Delta f_s) \tag{7.37 b}$$

at the tensile face of the concrete lower fibers.

For nonbonded beams, the factor 6.51 in Equations 7.37 a and 7.37 b becomes 6.83. Figure 7.25 gives a regression plot of Equation 7.37 a that shows a scatter band

Flexural cracking propagation in pretensioned prestressed T-beam (Nawy et al.).

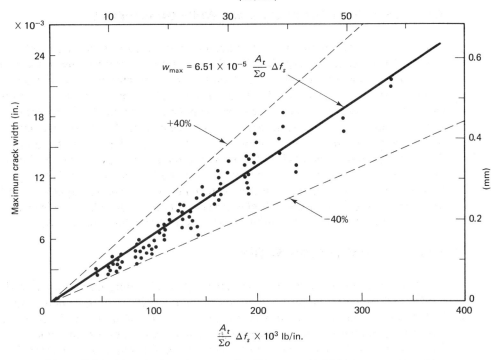

Figure 7.25 Linearized maximum crack width versus $(A_t/\Sigma_o)\Delta f_s$, post-tensioned beams.

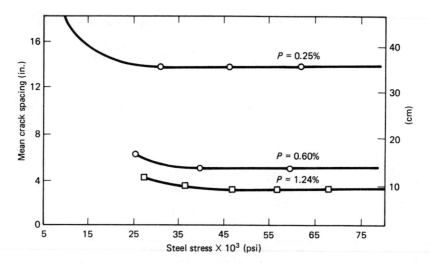

Figure 7.26 Reinforcement percentage effect on the relationship between crack spacing and incremental reinforcement stress.

of ±40 percent, which is not unexpected in flexural cracking behavior. The crack spacing stabilizes itself beyond an incremental stress Δf_s of 30,000 psi to 35,000 psi, as shown in Figure 7.26, depending on the *total* reinforcement percent ρ_T of both the prestressed and the nonprestressed steel.

7.13.5 Long-Term Effects on Crack Width Development

Limited studies on crack width development and increase with time show that both sustained and cyclic loadings increase the amount of microcracking in the concrete. Also, microcracks formed at service-load levels in partially prestressed beams do not seem to have a recognizable effect on the strength or serviceability of the concrete element. Macroscopic cracks, however, do have a detrimental effect, particularly in terms of corrosion of the reinforcement and appearance. Hence, an increase of crack width due to sustained loading significantly affects the durability of the prestressed member regardless of whether prestressing is circular, such as in tanks, or linear, such as in beams. Information obtained from sustained load tests of up to two years and fatigue tests of up to one million cycles indicates that a *doubling* of crack width with time can be expected. Therefore, engineering judgment has to be exercised as regards the extent of tolerable crack width under long-term loading conditions.

7.13.6 Permissible Crack Widths

The maximum crack width that a structural element should be permitted to develop depends on the particular function of the element and the environmental conditions to which the structure is liable to be subjected. Table 7.9 from the ACI Committee 224 report on cracking serves as a reasonable guide on the permissible crack widths in concrete structures under the various environmental conditions encountered.

TABLE 7.9 MAXIMUM PERMISSIBLE FLEXURAL CRACK WIDTHS

	Crack width	
Exposure condition	in.	mm
Dry air or protective membrane	0.016	0.41
Humidity, moist air, soil	0.012	0.30
Deicing chemicals	0.007	0.18
Seawater and seawater spray; wetting and drying	0.006	0.15
Water-retaining structures (excluding nonpressure pipes)	0.004	0.10

7.14 CRACK WIDTH AND SPACING EVALUATION IN PRETENSIONED T-BEAM WITHOUT MILD STEEL

Example 7.10

A pretensioned prestressed concrete beam has a T section as shown in Figure 7.27. It is prestressed with fifteen $\frac{7}{16}$-in. dia seven-wire strand 270-K grade. The locations of the neutral axis and center of gravity of steel are shown in the figure. $f'_s = 5,000$ psi, $E_c = 57,000\sqrt{f'_c}$, and $E_c = 28 \times 10^6$ psi. Find the mean stabilized crack spacing and the crack widths at the steel level as well as at the tensile face of the beam at $\Delta f_s = 30 \times 10^3$ psi. Assume that no failure in shear or bond takes place.

Solution

$$\Delta f_s = 30,000 \text{ psi} = 30 \text{ ksi}$$

Mean Stabilized Crack Spacing

$$A_t = 7 \times 14 = 98 \text{ sq in.}$$

$$\sum o = 15\pi D = 15\pi\left(\frac{7}{16}\right) = 20.62 \text{ in.}$$

$$a_{cs} = 1.2\left(\frac{A_t}{\sum o}\right) = 1.2\left(\frac{98}{20.62}\right) = 5.7 \text{ in (145 mm)}$$

Maximum Crack Width at Steel Level

$$w_{max} = 5.85 \times 10^{-5}\left(\frac{A_t}{\sum o}\right)\Delta f_s = 5.85 \times 10^{-5}\left(\frac{98}{20.62}\right)30$$

$$= 834.1 \times 10^{-5} \text{ in.} \cong 0.0083 \text{ in. } (0.21 \text{ mm})$$

Maximum Crack Width at Tensile Face of Beam

$$R_i = \frac{25 - 10.36}{25 - 10.36 - 3.5} = 1.31$$

$$w'_{max} = w_{max} R_i = 0.0083 \times 1.31 = 0.011 \text{ in } (0.28 \text{ mm})$$

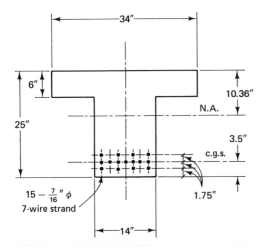

Figure 7.27 Beam cross-section in Example 7.10.

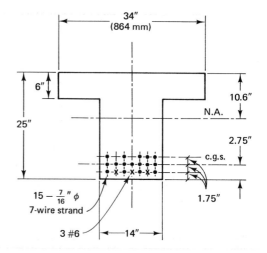

Figure 7.28 Beam cross-section in Example 7.11.

7.15 CRACK WIDTH AND SPACING EVALUATION IN PRETENSIONED T-BEAM CONTAINING NONPRESTRESSED STEEL

Example 7.11

The beam in Example 7.10 also contains three #6 nonprestressed mild steel as shown in Figure 7.28. Find the crack spacing and width for an incremental steel stress $\Delta f_s = 30,000$ psi $= 30$ ksi (207 MPa)

Solution

Mean Stabilized Crack Spacing

$$A_t = 14(3 \times 1.75 + \tfrac{1}{2} \times \tfrac{7}{16} + 1\tfrac{3}{8}) = 14 \times 6.84 = 95.8 \text{ in}^2$$

$$\sum o = 20.62 + 3 \times 2.36 = 27.70 \text{ in.}$$

$$a_{cs} = 1.2\left(\frac{A_t}{\sum o}\right) = 1.2\left(\frac{95.8}{27.7}\right) = 4.15 \text{ in. (105 mm)}$$

Maximum Crack Width at Steel Level

$$w_{max} = 5.85 \times 10^{-5}\left(\frac{A_t}{\sum o}\right)\Delta f_s = 5.85 \times 10^{-5}\left(\frac{95.8}{27.7}\right)30$$

$$= 606.9 \times 10^{-5} \cong 0.0061 \text{ in (0.15 mm)}$$

Maximum Crack Width at Tensile Face of Beam

$$R_i = \frac{25 - 10.6}{25 - 10.6 - 2.75} = 1.24$$

$$w'_{max} = w_{max} R_i = 0.0061 \times 1.24 = 0.007 \text{ in. (0.18 mm)}$$

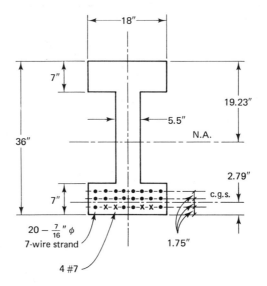

Figure 7.29 Beam cross-section in Example 7.12.

7.16 CRACK WIDTH AND SPACING EVALUATION IN PRETENSIONED I-BEAM CONTAINING NONPRESTRESSED MILD STEEL

Example 7.12

A pretensioned prestressed concrete I-beam has the geometry shown in Figure 7.29. It is prestressed with twenty $\frac{7}{16}$ in. dia seven-wire 270 K grade low-relaxation strands and four #7 mild steel bars having yield strength $f_y = 60,000$ psi. Find the mean stabilized crack spacing and the crack widths at the steel level as well as at the tensile face of the beam at incremental steel stress $\Delta f_s = 20,000$ psi (138 MPa). Assume that no failure in shear or bond takes place, and check whether the crack widths that develop satisfy the crack control criteria for deicing chemicals.

Solution $\Delta f_s = 20,000$ psi = 20 ksi (138 MPa)

Mean Stabilized Crack Spacing

$$A_t = 18 \times (3 \times 1.75 + \tfrac{1}{2} \times \tfrac{7}{16} + 1\tfrac{5}{16}) = 122.06 \text{ in}^2$$

$$\textstyle\sum o = 20\pi D + 4 \times 2.75 = 20\pi \times \tfrac{7}{16} + 4 \times 2.75 = 38.49 \text{ in.}$$

$$a_{cs} = 1.2\left(\frac{A_t}{\sum o}\right) = 1.2\left(\frac{122.06}{38.49}\right) = 3.8 \text{ in. (97 mm)}$$

Maximum Crack Width at Steel Level

$$w_{max} = 5.85 \times 10^{-5}\left(\frac{A_t}{\sum o}\right)\Delta f_s = 5.85 \times 10^{-5}\left(\frac{122.06}{38.49}\right)20$$

$$= 371.0 \times 10^{-5} \cong 0.0037 \text{ in. (0.1 mm)}$$

Maximum Crack Width at Tensile Face of Beam

$$R_i = \frac{36 - 19.23}{36 - 19.23 - 2.79} = 1.2$$

$$w'_{max} = w_{max} R_i = 0.0037 \times 1.2 = 0.004 \text{ in. (0.1 mm)}$$

Maximum Allowable Crack Width for Deicing. From Table 7.9, the maximum allowable crack width for deicing is $W_{max} = 0.007 > 0.004$ in. (0.1 mm). Hence, serviceability requirement is satisfied.

7.17 CRACK WIDTH AND SPACING EVALUATION FOR POST-TENSIONED T-BEAM CONTAINING NONPRESTRESSED STEEL

Example 7.13

A post-tensioned prestressed concrete beam has a T section as shown in Figure 7.30. It is prestressed with twelve $\frac{7}{16}$ in. dia seven-wire strands of 270 K grade and additionally reinforced with four #6 nonprestressed steel bars. The locations of the neutral axis and center of gravity of steel are shown in the figure. Assume that $f'_c = 5,000$ psi, $E_c = 57,000\sqrt{f'_c}$ psi, and $E_s = 28,000$ psi. Find the mean stabilized crack spacing and the crack widths at the steel level as well as at the tensile face of the beam at $\Delta f_s = 30,000$ psi, assuming there is no failure in shear or bond. Then determine whether the beam satisfies the serviceability criteria for crack control for humidity and moist air.
Solution

$$\Delta f_s = 30,000 \text{ psi} = 30 \text{ ksi}$$

Mean Stabilized Crack Spacing

$$A_t = 8 \times 12 = 95 \text{ in}^2$$

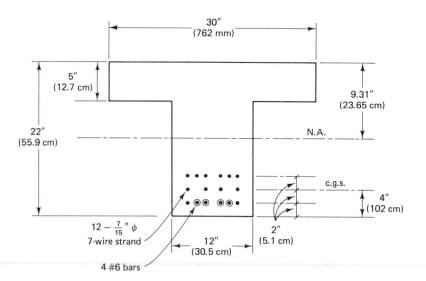

Figure 7.30 Beam cross-section in Example 7.13.

$$\sum o = 12 \times \tfrac{7}{16} + 4 \times 2.36 = 25.93 \text{ in.}$$

$$a_{cs} = 1.54 \frac{A_t}{\sum o} = 1.54 \times \frac{96}{25.93} = 5.70 \text{ in. (145 mm)}$$

Maximum Crack Width at Steel Level

$$w_{max} = 6.51 \times 10^{-5} \frac{A_t}{\sum o} (\Delta f_s) = 6.51 \times 10^{-5} \times \frac{96}{25.93} \times 30$$

$$= 0.0072 \text{ in. (0.18 mm)}$$

Maximum Crack Width at Tensile Face of Beam

$$R_i = \frac{22 - 9.31}{22 - 9.31 - 4} = 1.46$$

$$w'_{max} = w_{mas} R_i = 0.007 \times 1.46 = 0.0105 \text{ in (0.27 mm)}$$

Maximum Allowable Crack Width for Humidity. From Table 7.9, the maximum allowable crack width for the stated humidity condition is 0.012 in. (0.3 mm) > 0.0105 in. (0.27 mm), which is satisfactory.

REFERENCES

7.1 ACI Committee 318. *Building Code Requirements for Reinforced Concrete 318–89*. Detroit: American Concrete Institute, 1989.

7.2 ACI Committee 318. *Commentary on Building Code Requirements for Reinforced Concrete 318R–89*. Detroit: American Concrete Institute, 1989.

7.3 Tadros, M. K. "Designing for Deflection." Paper presented at CIP Seminar on Advanced Design Concepts in Precast Prestressed Concrete, Prestressed Concrete Institute Convention, Dallas, October 1979.

7.4 Branson, D. E. *Deformation of Concrete Structures*, New York: McGraw Hill, 1977.

7.5 Branson, D. E. "The Deformation of Non-Composite and Composite Prestressed Concrete Members." In *Deflection of Concrete Structures*, ACI Special Publication SP-43. Detroit: American Concrete Institute, 1974, pp. 83–127.

7.6 Shaikh, A. F., and Branson, D. E. "Non-Tensioned Steel in Prestressed Concrete Beams. *Journal of the Prestressed Concrete Institute* 15 (1970): 14–36.

7.7 Prestressed Concrete Institute. *PCI Design Handbook*. 3d ed. Chicago: Prestressed Concrete Institute, 1985.

7.8 Martin, L. D., "A Rational Method of Estimating Camber and Deflections of Precast, Pre-

stressed Concrete Members." *Journal of the Prestressed Concrete Institute* 22 (1977): 100–108.

7.9 Nilson, A. H. *Design of Prestressed Concrete*. New York: Wiley, 1987.

7.10 Tadros, M. K., Ghali, A., and Dilger, W. H. "Effect of Non-Prestressed Steel on Prestress Loss and Deflection." *Journal of the Prestressed Concrete Institute* 22 (1977): 50–63.

7.11 Tadros, M. K. "Expedient Service Load Analysis of Cracked Prestressed Concrete Sections. *Journal of the Prestressed Concrete Institute* 27 (1982): 86–111; also Discussions and Author's Closure, 28 (1983): 137–158.

7.12 Naaman, A. E. "Partially Prestressed Concrete: Review and Recommendations. *Journal of the Prestressed Concrete Institute* 30 (1985): 30–71.

7.13 Nawy, E. G., and Potyondy, J. G. "Flexural Cracking Behavior of Pretensioned Prestressed Concrete I- and T-Beams." *Journal of the American Concrete Institute* 65 (1971): 335–360.

7.14 Nawy, E. G., and Huang, P. T. "Crack and Deflection Control of Pretensioned Prestressed Beams." *Journal of the Prestressed Concrete Institute* 22 (1977): 30–47.

7.15 Nawy, E. G., and Chiang, J. Y. "Serviceability Behavior of Post-Tensioned Beams." *Journal of the Prestressed Concrete Institute* 25 (1980): 74–95.

7.16 Cohn, M. Z. "Partial Prestressing From Theory to Practice." *NATO–ASI Applied Science Series,* Vols. I and II, Publ. Martinus Nijhoff Publishers, in Cooperation with NATO Scientific Affairs Division, Dordrecht, 1986, Vol. 1, p. 405; Vol. II, p. 425.

7.17 Nawy, E. G. "Flexural Cracking Behavior of Pretensioned and Post-Tensioned Beams: The State of the Art." *Journal of the American Concrete Institute,* December 1985, pp. 890–900.

7.18 Nawy, E. G. "Flexural Cracking Behavior and Crack Control of Pretensioned and Post-

7.19 Neville, A. M. *Properties of Concrete.* 3d ed. London: Pitman, 1981.

7.20 Bazant, Z. P. "Prediction of Creep Effects Using Age Adjusted Effective Modulus Method." *Journal of the American Concrete Institute,* April 1972, pp. 212–217.

7.21 Libby, J. R. *Modern Prestressed Concrete.* New York: Van Nostrand Reinhold, 1984.

Tensioned Prestressed Beams." *Proceedings of the NATO–NSF Advanced Research Workshop,* vol. 2. Dordrecht-Boston: Martinus Nijhoff, 1986, pp. 137–156.

PROBLEMS

7.1 Calculate the instantaneous and long-term cambers and deflections of the AASHTO beam of Example 4.2 for 7, 30, 180, and 365 days, and 5 years by (a) the PCI multipliers method, (b) the incremental time-steps method, and (c) the approximate time-steps method. Then tabulate and compare the results. Are the deflections within the AASHTO permissible limits on deflection?

7.2 A 72-ft (21.9-m) span simply supported lightweight concrete double-T beam is subjected to a superimposed topping load $W_{SD} = 250$ plf (3.65 kN/m) and a service live load $W_L = 300$ plf (4.38 kN/m). Calculate the immediate camber and deflection of this beam by the bilinear method and the time-dependent deflections at intervals of 7, 30, 90, and 365 days using (a) the PCI multipliers method, (b) the incremental time-

steps method, and (c) the approximate time-steps method. Then tabulate and compare the results, and verify whether they are within the permissible ACI limits on deflection for the conditions where non-structural elements are not likely to be damaged by large deflections. Use Figure P7.2 and the data that follow.

	Noncomposite	Composite
A_c, in^2	615 (3,968 cm^2)	855
I_c, in^4	59,720 (24.9 × 10^5 cm^4)	77,118
r^2, in^2	97	90
c_b, in.	21.98 (558 mm)	24.54
c_t, in.	10.02 (255 mm)	9.46
S_b, in^3	2,717 (4.5 × 10^4)	3142
S^t, in^3	5,960 (9.8 × 10^4)	8152
W_d, plf	641 (9.34 kN/m)	891

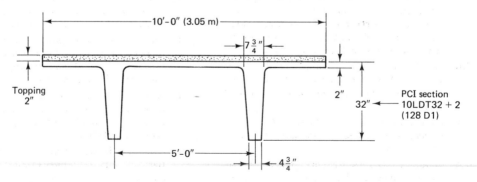

Figure P7.2

$V/S = 615/364 = 1.69$ in. (4.3 mm)

$RH = 75\%$

$e_c = 18.73$ in. (476 mm)

$e_e = 12.81$ in. (325 mm)

$f'_c = 5,000$ psi (34.5 MPa)

$f'_{ci} = 3,750$ psi (25.7 MPa)

f_t at bottom fibers $= 12\sqrt{f'_c} = 849$ psi
(5.6 MPa)

$A_{ps} = $ twelve $\frac{1}{2}$ in. dia low-relaxation, stress-relieved steel depressed at midspan only

$f_{pu} = 270,000$ psi (1,862 MPa), stress relieved

$f_{pi} = 189,000$ psi (1,303 MPa)

$f_{py} = 260,000$ psi (1,793 MPa)

7.3 Determine the crack width and stabilized mean crack spacing in the double-T beam of Problem 7.2 for an incremental stress of 15,000 psi (103 MPa) beyond the decompression state. Also, determine whether the maximum crack width obtained satisfies the serviceability requirement for crack control for a humid and moist environment.

7.4 A simply supported bonded single-T beam has a 68-ft span and is subjected to a uniform live load of 1,000 plf and a superimposed dead load of 200 plf. It has the shape and geometrical dimensions shown in Figure P7.4, and its geometrical properties and maximum allowable stresses are as follows:

$A_c = 782$ in^2

$I_c = 169,020$ in^4

$S_b = 4803$ in^3

$S^t = 13,194$ in^4

$W_D = 815$ plf

$V/S = 2.33$ in.

$f'_c = 5,000$ psi

$f_c = 2,250$ psi

$f'_{ci} = 3,750$ psi

$f_{ti} = 184$ psi

$f_{pu} = 270,000$ psi

$f_{py} = 235,000$ psi

$f_{pi} = 195,000$ psi

$E_{ps} = 28 \times 10^6$ psi

$c_b = 35.19$ in.

$c_t = 12.81$ in.

$RH = 70\%$

$f_{ci} = 2,250$ psi

$f_t = 849$ psi

$f_{pe} = 150,000$ psi

$f_y = 60,000$ psi

$A_{ps} = $ sixteen $\frac{1}{2}$ in. dia seven-wire low-relaxation strands

(a) Find the eccentricities e_c and e_e that would result in a tensile stress $f_t = 750$ psi at the lower fiber at midspan *at service load*, and tensile stresses within the allowable limits at the support section both at initial prestress and at service load. Use *nonprestressed* reinforcement where necessary.

(b) Find the long-term camber and deflection of the beam by the *approximate*

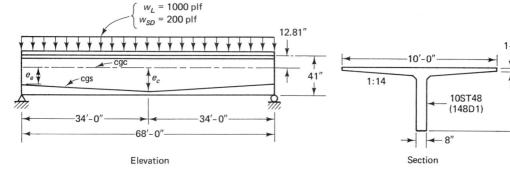

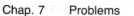

Elevation

Section

Figure P7.4.

time-step procedure for $t = 7$ days and $t = 180$ days, assuming that the ultimate creep coefficient $C_u = 2.35$. Use the moment-curvature approach to determine the initial camber at transfer. Are the values within the allowable ACI limits?

(c) Calculate the flexural crack width at service load for a stress increment $\Delta f_s = 15,000$ psi beyond the decompression stress.

7.5 Find the long-term camber and deflection in Example 7.8 by the incremental time-steps method, assuming that twelve $\frac{1}{2}$ in. dia seven-wire 270 K stress-relieved strands are used for prestressing the beam section. Calculate the flexural crack width at service load for a stress increment $\Delta f_s = 15,000$ psi beyond the decompression stress.

CHAPTER 8

Prestressed Compression and Tension Members

8.1 INTRODUCTION

Although prestressing is predominantly used in flexural members such as beams and slabs, it is also used in axially loaded members such as long columns (compression members) and ties for arches and truss elements (tension members). Yet another use is in pretensioned and post-tensioned prestressed piles and masts.

The theory, analysis, and design of prestrssed compression members are similar to those of reinforced concrete members. The internal axial prestressing force in the *bonded* tendon produces no column action; hence, no buckling can result as long as the prestressing steel and the surrounding concrete are in direct contact along the total length of the element. As such, the bending tendency of the concrete at midlength is neutralized by the stretching effect of the axially embedded prestressing strands.

Columns are normally subjected to bending in addition to axial load, since external loads are rarely concentric. As a result, the concrete section is subjected to tension at the side farthest from the line of action of the longitudinal load. Cracking develops, but it can be prevented through the use of prestress in the columns. If the applied load is concentric, prestressing is inconsequential if not altogether disadvantageous, as the compressive stress on the concrete section is needlessly increased.

A compression member can be considered fully prestressed throughout its length if no loss in development of prestressing stress occurs at its ends. If partial loss occurs, the reinforcement segment in the development zone is considered nonprestressed and the section at the end zone is treated as a reinforced concrete eccentrically loaded section.

The George Moscone Convention Center, San Francisco, California, design by T.Y. Lin International. (*Courtesy,* Post-Tensioning Institute.)

Tension members are normally subjected to direct tension only. These elements are mostly linear, as for example, railroad ties, restraining ties for arch bridges, tension members in trusses, and foundation anchorages for earthwork retaining structures. Tension members can also be circular or parabolic in shape, as witness prestressed circular containers or catenary-shaped bridge elements. The fundamental function of the pure tension members is to prevent their cracking at service load and to enble them to sustain all the necessary deformation needed to develop full resistance to the external service loads and overloads. Being crack free would fully protect the reinforcement from corrosion and other environmental conditions.

8.2 PRESTRESSED COMPRESSION MEMBERS: LOAD–MOMENT INTERACTION IN COLUMNS AND PILES

In order to evaluate the nominal strength of a column at various eccentric load levels, it is necessary to evaluate all possible combinations of ultimate nominal loads P_n and ultimate nominal moments M_n given by

$$M_n = P_n e_i \tag{8.1}$$

where e_i is the eccentricity of the load at the various load-moment combinations. A plot of the relationship between P_n and M_n is shown in the interaction diagram of Figure 8.1 for both nonslender columns (material failure) and slender columns (stability failure). In the nonslender column, failure occurs as the load reaches the value A along path OA, and the concrete arches at the compression side. In the slender column, the maximum load is reached at B along the path OBC, which intersects the interaction diagram at C. Instability occurs once the load is reached. A quantitative definition of slender and nonslender columns is given later.

The basic assumptions made in regard to prestressed concrete, similar to those made in regard to reinforced concrete columns, are as follows:

1. The strain distribution in the concrete varies linearly with depth.
2. The stress distribution in the compression zone is parabolic and is replaced in analysis and design by an equivalent rectangular block.

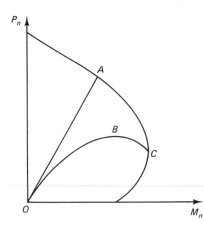

Figure 8.1 Basic interaction diagram for columns. (a) Path OA for material failure in nonslender column. (b) Path OBC for buckling failure of slender column.

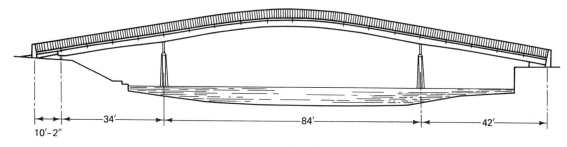

Elevation

(a)

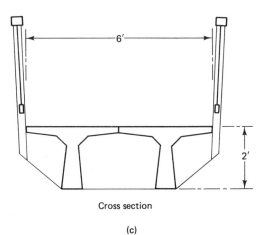

(b)

Cross section

(c)

Eel Pie Island, Twickenham, England, prestressed concrete footbridge.

3. The stress-strain diagrams of the concrete and the prestressing steel are known.

4. The crushing strain of the conrete in combined bending and axial load at the extreme fibers is $\epsilon_c = 0.003$ in./in., and the average crushing strain at mid-depth of a concrete section subjected mainly to axial load is $\epsilon_0 = 0.002$ in./in.

5. The section is considered to have failed when the strain in the concrete at the extreme compression fibers reaches $\epsilon_c = 0.003$ in./in. or $\epsilon_0 = 0.002$ in./in. at mid-depth. Note that $\epsilon_c = 0.003$ is the value used in the ACI Code, whereas other codes use a higher value of 0.0035 or 0.0038.

6. Compatibility of strain is postulated between the concrete and the prestressing steel.

The modes of failure are also similar to those of reinforced concrete columns:

1. *Initial compression failure, small eccentricity*. This failure mode develops when the strain in the concrete at the loaded side reaches $\epsilon_{cu} = 0.003$ in./in. while the strain in the prestressing steel at the far side is below the yield strain. The eccentricity e of the axial load is smaller than the balanced eccentricity e_b.

2. *Initial tension failure, large eccentricity*. This failure mode is the reverse of the preceding one. The steel at the far side yields prior to the crushing of the conrete at the loaded side. The eccentricity e of the axial load is larger than the balanced eccentricity e_b.

3. *Balanced failure, balanced eccentricity*. This failure mode defines the simultaneous tension yielding of the prestressing steel at the far side and crushing of the concrete at the loaded side. The eccentricity of the axial load is defined as the balanced eccentricity e_b.

The three major controlling points on the interaction diagram are:

1. $M_u = 0$, corresponding to $\epsilon_0 = 0.002$ in./in. at failure due to the concentric load P_u. The neutral axis position is at infinity.

2. No tension at the extreme concrete tensile fibers and $\epsilon_{cu} = 0.003$in./in. at the extreme concrete compression fibers. The neutral axis position is at the extreme tension fibers.

3. $P_u = 0$ and $\epsilon_{cu} = 0.003$ in./in. at the extreme compression fibers. The neutral axis is inside the section and is determined by trial and adjustment, assuming a depth c and then testing the assumption.

Figure 8.2 shows the strain and stress distribution for these three cases.

The remaining points on the interaction diagram are for cases that lie between stages (a), (b), and (c) of Figure 8.2, namely, from concentric loading to pure bending. In the case of columns, pure bending defines the state where the ratio of the factored axial load P_u to the factored flexural moment M_u is negligible. The parabolic distribution of stress for cases (b) and (c) is replaced by the equivalent rectanguar block, where the block depth $a = \beta_1 c$, as is done in the case of flexural beams.

The typical case of a compression member lies between stages (b) and (c) of Figure 8.2 The strains, stresses, and forces for such a cases are shown in Figure 8.3 for the critical section at the limit state of ultimate load by material failure. Cutting the free-body diagram at the column midheight above section 1-1, the cross-section of the member is shown in part (b) of the figure, and the strain and stress at failure in parts (c) and (d), respectively. The strain ϵ_{ce} is the *uniform* strain in the concrete

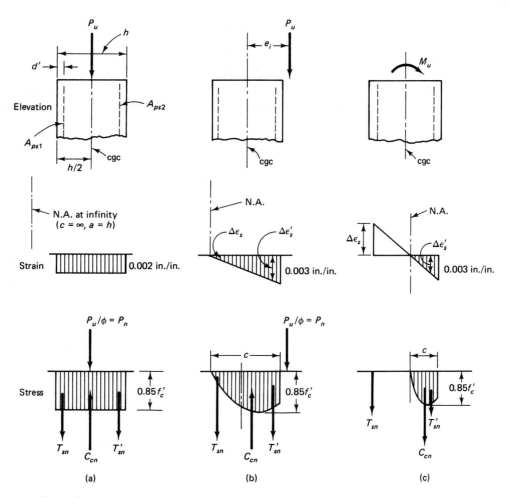

Figure 8.2 Strain and stress distribution alternatives across concrete section depth. (a) $M_u = 0$. (b) $\epsilon_c = 0.003$ in./in., and zero tension at the extreme tensile fibers. (c) $P_u = 0$ and $\epsilon_c = 0.003$ in./in. at the extreme compressive fibers.

under effective prestress after creep, shrinkage, and relaxation losses given respectively by

$$C_{cn} = 0.85 f_c' b a \qquad (8.1\ a)$$

$$T_{s'n} = A_{ps}' f_{ps}' \qquad (8.1\ b)$$

and

$$T_{sn} = f_{ps} A_{ps2} \qquad (8.1\ c)$$

Equilibrium of forces then gives

$$P_n = C_{cn} - T_{sn}' - T_{sn} \qquad (8.2)$$

If the effective prestressing force after all losses is P_e, the corresponding strain in the tendons prior to the application of the external load is

$$\epsilon_{pe} = \frac{f_{pe}}{E_{ps}} = \frac{P_e}{(A_{ps} + A'_{ps})E_{ps}} \tag{8.3}$$

The change in strain in the prestressing steel area A'_{ps} as the compression member passes from the effective prestressing stage to the ultimate load can be defined as

$$\Delta\epsilon'_{ps} = \epsilon_{cu}\left(\frac{c - d'}{c}\right) - \epsilon_{ce} \tag{8.4 a}$$

$$\Delta\epsilon_{ps} = \epsilon_{cu}\left(\frac{d - c}{c}\right) + \epsilon_{ce} \tag{8.4 b}$$

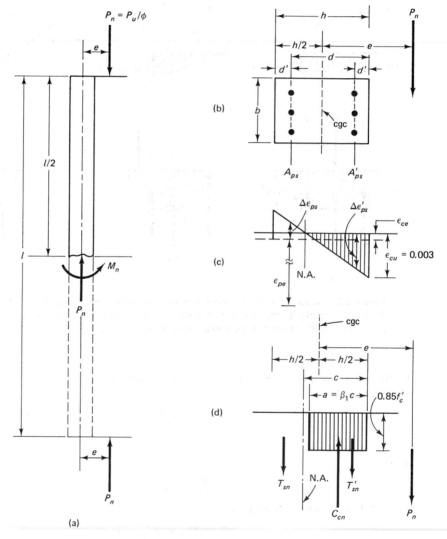

Figure 8.3 Stresses and forces in typical eccentrically loaded nonslender column. (a) Elevation. (b) Cross section. (c) Strain distribution. (d) Stresses and forces.

Cable-stayed Sunshine Skyway Bridge, Tampa, Florida. (*Courtesy,* Figg and Muller Engineers, Inc.)

$$T'_{sn} = A'_{ps}f'_{ps} = A'_{ps}E_{ps}(\epsilon_{pe} - \Delta\epsilon'_{ps})$$

or

$$T'_{sn} = A'_{ps}E_{ps}\left[\epsilon_{pe} - \epsilon_{cu}\left(\frac{c - d'}{c}\right) + \epsilon_{ce}\right] \qquad (8.5)$$

Similarly,

$$T_{sn} = A_{ps}f_{ps} = A_{ps}E_{ps}(\epsilon_{pe} + \Delta\epsilon'_{ps})$$

or

$$T_{sn} = A_{ps}E_{ps}\left[\epsilon_{pe} + \epsilon_{cu}\left(\frac{d - c}{c}\right) + \epsilon_{ce}\right] \qquad (8.6)$$

Taking moments about the plastic centroid cgc of the section gives

$$M_n = P_n \overset{\curvearrowright}{e} = C_{cn}\left(\frac{h}{2} - \frac{a}{2}\right)\overset{\curvearrowright}{} - T'_{sn}\left(\frac{h}{2} - d'\right)\overset{\curvearrowright}{} + T_{sn}\left(d - \frac{h}{2}\right)\overset{\curvearrowright}{} \qquad (8.7)$$

From Equations 8.1 a, 8.5, 8.6, and 8.7, the nominal strengths P_n and M_n for several eccentricities e_i can be evaluated in order to construct the P-M interaction diagram for any section or develop nondimensional series of P-M interaction diagrams for various concrete strength levels. The design strengths are evaluated from the nominal strength values as

$$\text{Design } P_u = \phi P_n$$

and

$$\text{Design } M_u = \phi M_n = \phi P_n e$$

where ϕ is the strength reduction factor for compression members. Note that the *design* P_u and M_u have to have a value close to, but not less than, the *factored* values P_u and M_u. The load-moment interaction diagram for the controlling eccentricities is shown in Figure 8.4.

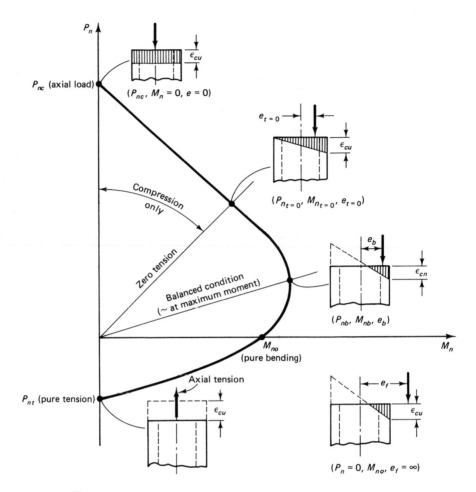

Figure 8.4 Load-moment interaction diagram controlling eccentricities.

8.3 STRENGTH REDUCTION FACTOR ϕ

For members subject to flexure and relatively small axial loads, failure is initiated by yielding of the tension reinforcement and takes place in an increasingly ductile manner. Hence, for small axial loads it is reasonable to permit an increase in the ϕ factor from that required for pure compression members. When the axial load vanishes, the member is subjected to pure flexure, and the strength reduction factor ϕ becomes 0.90. Figure 8.5 shows the zone in which the value of ϕ can be increased from 0.7 to 0.9 for tied columns and 0.75 to 0.9 for spiral columns. In part (a), as the factored design compression load ϕP_n decreases beyond $0.1A_g f_c'$, the ϕ factor increases from 0.7 to 0.9 for tied columns and 0.75 to 0.9 for spiral columns. For those cases where the value of P_{nb} is less than $0.1A_g f_c'$, ϕ values increase when the load $P_u < P_{ub}$, or $\phi P_n < \phi P_{nb}$, as in part (b) of the figure.

The value of $0.10 f_c' A_g$ is chosen by the ACI Code as the design axial load value

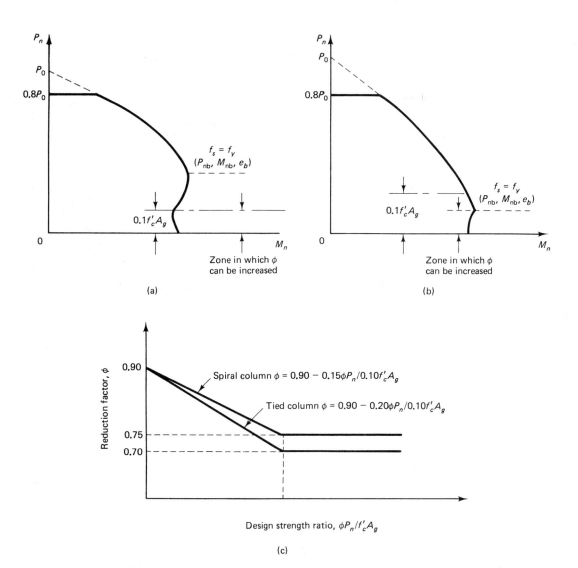

Figure 8.5 Controlling zones for modification of reduction factor in columns. (a) $0.1f_c'A_s < P_{nb}$. (b) $0.1f_c'A_s > P_{nb}$. (c) Variation of ϕ for symmetrically reinforced compression members (P_n is the nominal axial strength at given eccentricity).

ϕP_n below which the ϕ factor can safely be increased for most compression members. In summary, if initial failure is in compression, the strength reduction factor ϕ is always 0.70 for tied columns and 0.75 for spirally reinforced columns.

The following expressions give variations in the value of ϕ for symmetrically reinforced compression members. The columns should have an effective depth not less than 70 percent of the total depth. For tied columns,

$$\phi = 0.90 - \frac{0.20\phi P_n}{0.1f_c'A_g} \geq 0.70 \tag{8.8}$$

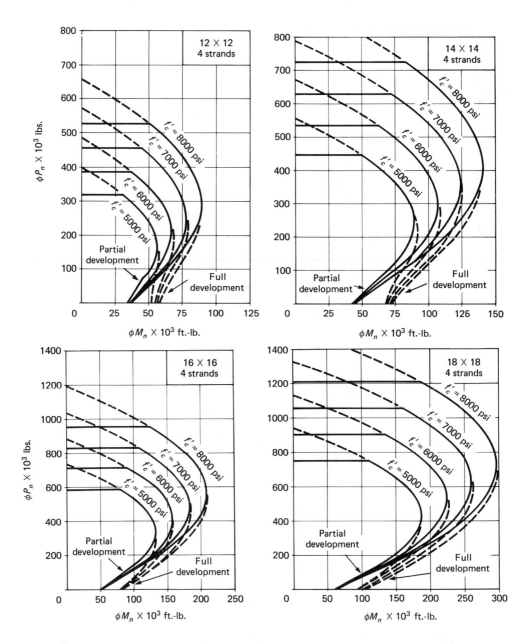

Figure 8.6 Typical design load-moment interaction plots for columns 12 in. × 12 in., 14 in. × 14 in., 16 in. × 16 in., and 24 in. × 24 in. (Ref. 8.3).

In both of these equations, if ϕP_{nb} is less than $0.1 f'_c A_g$, then ϕP_{nb} should be substituted for $0.1 f'_c A_g$ in the denominator, using $0.7 P_{nb}$ for tied, and $0.75 P_{nb}$ for spirally reinforced, columns. Figure 8.5(c) presents the graphical representation of Equations 8.8 and 8.9.

Prestressed Compression and Tension Members   Chap. 8

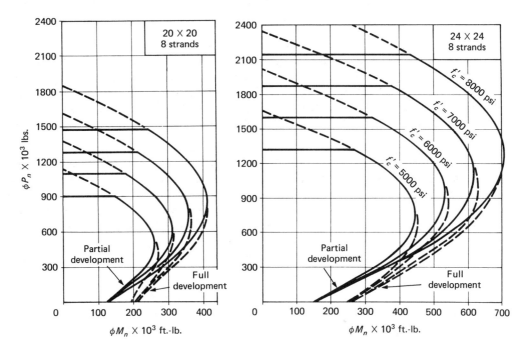

For spirally reinforced columns,

$$\phi = 0.90 + \frac{0.15\phi P_n}{0.1f_c' A_g} \geq 0.75 \qquad (8.9)$$

It should be noted that the balanced condition in *prestressed* compression members is highly indeterminate. A reasonable approximation can be made by calculating by trial and adjustment the maximum moment value coordinate on the interaction diagram as the balanced momemt M_{nb}, or by assuming that the tensile strain in the tension layers equals a strain increment $\Delta \epsilon_{ps} = 0.0012$ to 0.0020 in./in. beyond the service-load level and computing the compressive stress block depth a of the concrete accordingly. This assumption has to be verified and the M_{nb} value adjusted after the interaction diagram is plotted in order to determine the maximum moment ordinate in the diagram for which the neutral axis depth c_b will be used in the computations.

Typical load-moment interaction diagrams are given in Figure 8.6 (see Ref. 8.3).

8.4 OPERATIONAL PROCEDURE FOR THE DESIGN OF NONSLENDER PRESTRESSED COMPRESSION MEMBERS

The following steps can be carried out for the design of nonslender (short) columns where the behavior is controlled by material failure:

1. Evaluate the factored external axial load P_u and the factored moment M_u. Calculate the applied eccentricity $e = M_u/P_u$.

2. Assume a cross section and type of lateral reinforcement to be used, namely, tied or spiral. Avoid fractional quantities in selecting sectional dimensions.

3. Assume the number and size of strands.

4. Calculate M_{nb} for the assumed section as the balanced moment and the corresponding P_{nb} and $e_b = M_{nb}/P_{nb}$ through evaluation of the maximum moment coordinate in the interaction diagram by trial and adjustment. Or, alternatively, assume that the strain in the extreme tensile fibers is equal to the yield strain ϵ_{py} of the prestressing steel, and then proceed to calculate the balanced axial load P_{nb} and the balanced moment M_{nb}. This step also enables one to verify the value of the strength reduction factor.

5. Assume a neutral axis depth c, and find the corresponding P_n and M_n. Then check for the adequacy of the assumed section, i.e., whether $\phi P_n >$ the factored P_u and $\phi M_n >$ the factored M_u. If the section cannot support the factored load or is oversized and hence uneconomical, revise the cross section and the reinforcement through trial and adjustment by repeating steps 4 and 5 as necessary, including the construction of an interaction in the diagram.

6. Design the lateral reinforcement.

Figure 8.7 presents a flowchart of the trial-and-adjustment sequences of the design or analysis procedure.

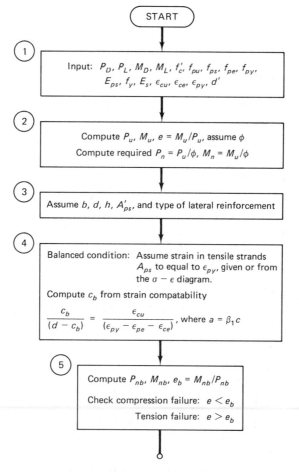

START

① Input: P_D, P_L, M_D, M_L, f'_c, f_{pu}, f_{ps}, f_{pe}, f_{py}, E_{ps}, f_y, E_s, ϵ_{cu}, ϵ_{ce}, ϵ_{py}, d'

② Compute P_u, M_u, $e = M_u/P_u$, assume ϕ
Compute required $P_n = P_u/\phi$, $M_n = M_u/\phi$

③ Assume b, d, h, A'_{ps}, and type of lateral reinforcement

④ Balanced condition: Assume strain in tensile strands A_{ps} to equal to ϵ_{py}, given or from the $\sigma - \epsilon$ diagram.

Compute c_b from strain compatability

$$\frac{c_b}{(d - c_b)} = \frac{\epsilon_{cu}}{(\epsilon_{py} - \epsilon_{pe} - \epsilon_{ce})}, \text{ where } a = \beta_1 c$$

⑤ Compute P_{nb}, M_{nb}, $e_b = M_{nb}/P_{nb}$

Check compression failure: $e < e_b$

Tension failure: $e > e_b$

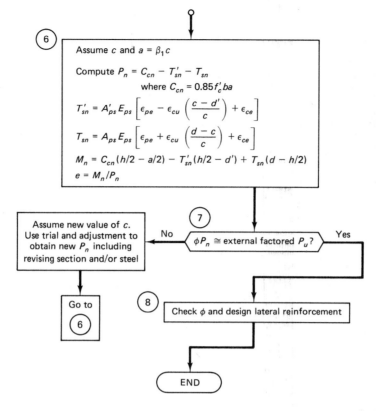

Figure 8.7 Flowchart for design or analysis of prestressed concrete nonslender compression members.

8.5 CONSTRUCTION OF NORMAL LOAD–MOMENT (P_n–M_n) AND DESIGN (P_u–M_u) INTERACTION DIAGRAMS

Example 8.1

Construct the nominal load–moment interaction diagram for a prestressed concrete compression member 14 in. (356 mm) wide and 14 in. deep. The member is reinforced with eight $\frac{1}{2}$ in. (12.7 mm) dia seven-wire stress-relieved 270 K strands, half on each side of the two faces parallel to the neutral axis as shown in Figure 8.8. The stress-strain diagram of the strands is shown in Figure 8.9. The effective prestress after all losses is $f_{pe} = 150,000$ psi (1,034 MPa). Additionally, draw the design interaction diagram using the appropriate strength reduction factor values. Consider the strands fully developed throughout the length of the member. Given data are as follows:

$f'_c = 6,000$ psi (47.5 MPa), normal-weight concrete

$E_{ps} = 29 \times 10^6$ psi (200×10^3 MPa)

$f_{ps} = 240,000$ psi (1,655 MPa)

$\epsilon_{cu} = 0.003$ in./in. at failure

$\epsilon_{ce} = 0.005$ in./in. when P_e acts on the section

$\epsilon_{py} = $ strand yield strain $\cong 0.012$ in./in. from Figure 8.9

Interior of the George Moscone Convention Center, San Francisco, California, design by T.Y. Lin International. (*Courtesy*, Post-Tensioning Institute.)

Solution

Nominal Strength P_n–M_n Diagram

1. *Axial Compression: $M_u = 0$, $c = \infty$*

   The compressive block depth $a = 14$ in. (356 mm), and the effective depth $d = 14 - 2 = 12$ in. (305 mm). So we have $C_{cn} = 0.85 f_c' ba = 0.85 \times 6000 \times 14 \times 14 = 999{,}600$ lb (4,446 kN).

   From Equation 8.5,

   $$T_{sn}' = A_{ps}' E_{ps}\left[\epsilon_{pe} - \epsilon_{cu}\left(\frac{c - d'}{c}\right) + \epsilon_{ce}\right]$$

   $$A_{ps}' = 4 \times 0.153 = 0.612 \text{ in}^2 \text{ (3.95 cm}^2)$$

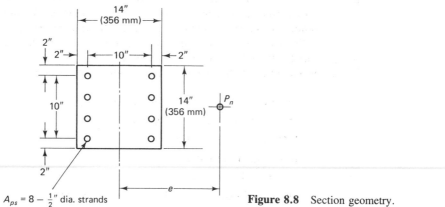

$A_{ps} = 8 - \frac{1}{2}''$ dia. strands

Figure 8.8 Section geometry.

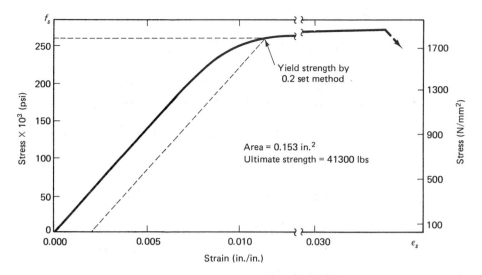

Figure 8.9 Stress-strain diagram for $\frac{1}{2}$ in. (12.7 mm) dia 270 K prestressing tendon.

From Figure 8.9, for $E_{ps} = 29 \times 10^6$ psi (200×10^3 MPa), $\epsilon_{pe} = 0.0052$ in./in. and $\epsilon_{cu} = 0.003$ in./in. Thus,

$$T'_{sn} = 0.612 \times 29 \times 10^6\left[0.0052 - 0.003\left(\frac{\infty - 2}{\infty}\right) + 0.0005\right]$$

$$= 0.612 \times 29 \times 10^6(0.0052 - 0.003 + 0.0005)$$

$$= 47,920 \text{ lb } (213.1 \text{ kN})$$

From Equation 8.6,

$$T_{sn} = A_{ps}E_{ps}\left[\epsilon_{pe} + \epsilon_{cu}\left(\frac{d - c}{c}\right) + \epsilon_{ce}\right]$$

$$= 0.612 \times 29 \times 10^6\left[0.0052 + 0.003\left(\frac{12 - \infty}{\infty}\right) + 0.0005\right]$$

$$= 47,920 \text{ lb } (213.1 \text{ kN})$$

From Equation 8.2,

$$P_n = C_{cn} - T'_{sn} - T_{sn} = 999,600 - 47,920 - 47,920$$

$$= 903,760 \text{ lb } (4,020 \text{ kN})$$

From Equation 8.7,

$$M_n = C_{cn}\left(\frac{h}{2} - \frac{a}{2}\right) - T'_{sn}\left(\frac{h}{2} - d'\right) + T_{sn}\left(d - \frac{h}{2}\right)$$

$$= 999,600\left(\frac{14}{2} - \frac{14}{2}\right) - 47,920\left(\frac{14}{2} - 2\right) + 47,920\left(12 - \frac{14}{2}\right)$$

$$= 0$$

$$e_1 = \frac{M_n}{P_n} = 0$$

2. *Zero Tension at Tension Face, c = 14 in.*

$$\beta_1 = 0.85 - \frac{0.05(f'_c - 4{,}000)}{1{,}000} = 0.75$$

$$a = \beta_1 c = 0.75 \times 14 = 10.5 \text{ in. } (267 \text{ mm})$$

$$C_{cn} = 0.85 \times 6{,}000 \times 14 \times 10.5 = 749{,}700 \text{ lb } (3{,}335 \text{ kN})$$

$$T'_{sn} = 0.612 \times 29 \times 10^6 \left[0.0052 - 0.003\left(\frac{14 - 2}{14}\right) + 0.0005 \right]$$

$$= 55{,}526 \text{ lb } (247.0 \text{ kN})$$

$$T_{sn} = 0.612 \times 29 \times 10^6 \left[0.0052 + 0.003\left(\frac{12 - 14}{14}\right) + 0.0005 \right]$$

$$= 93{,}557 \text{ lb } (416.1 \text{ kN})$$

$$P_n = C_{cn} - T'_{sn} - T_{sn} = 749{,}700 - 55{,}526 - 93{,}557$$

$$= 600{,}617 \text{ lb } (2{,}672 \text{ kN})$$

$$M_n = 749{,}700\left(\frac{14}{2} - \frac{10.5}{2}\right) - 55{,}526\left(\frac{14}{2} - 2\right) + 93{,}557\left(12 - \frac{14}{2}\right)$$

$$= 1{,}502{,}130 \text{ in.-lb } (169.7 \text{ kN-m})$$

$$e_2 = \frac{1{,}502{,}130}{600{,}617} = 2.50 \text{ in. } (63.5 \text{ mm})$$

3. *Pure Bending: $P_u = 0$*

Neglecting the effect of the compression steel A'_{ps}, we have

$$a = \frac{A_{ps}f_{ps}}{0.85 f'_c b} = \frac{0.612 \times 240{,}000}{0.85 \times 6{,}000 \times 14} = 2.06 \text{ in. } (52.3 \text{ mm})$$

$$c = \frac{2.06}{0.75} = 2.75 \text{ in. } (69.9 \text{ mm})$$

$$M_n = A_{ps}f_{ps}\left(d - \frac{a}{2}\right) = 0.612 \times 240{,}000\left(12 - \frac{2.06}{2}\right)$$

$$= 1{,}611{,}274 \text{ in.-lb}$$

$$e_3 = \frac{1{,}611{,}274}{0} = \infty$$

4. *Balanced Condition: P_{nb}, M_{nb}, e_b*

Assume the strain in the tensile strands A_{ps} to be equal to the incremental strain $\Delta\epsilon_{py}$ beyond the service-load level P_e. Taking a value $\Delta\epsilon_{py} \cong 0.0014$ to be modified by trial and adjustment, and from Figure 8.10, similar triangles give

$$\frac{c}{(d - c)} = \frac{\epsilon_{cu}}{\Delta_{py}} = \frac{0.003}{0.0014}$$

Hence $c = 8.15$ in. (207 mm). So

$$a_b = \beta_1 c = 0.75 \times 8.15 = 6.11 \text{ in } (155.2 \text{ mm})$$

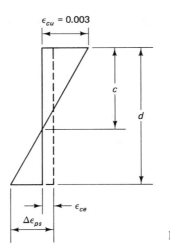

$\epsilon_{cu} = 0.003$

c

d

ϵ_{ce}

$\Delta\epsilon_{ps}$

Figure 8.10 Strain distribution.

$$C_{cn} = 0.85 \times 6{,}000 \times 6.10 \times 14 = 435{,}540 \text{ lb } (1{,}937 \text{ kN})$$

$$T'_{sn} = 0.612 \times 29 \times 10^6 \left[0.0052 - 0.003 \left(\frac{8.13 - 2}{8.13} \right) + 0.0005 \right]$$

$$= 61{,}018 \text{ lb } (271.4 \text{ kN})$$

$$T_{sn} = 0.612 \times 29 \times 10^6 \left[0.0052 + 0.003 \left(\frac{12 - 8.13}{8.13} \right) + 0.0005 \right]$$

$$= 126{,}509 \text{ lb } (562.7 \text{ kN})$$

$$P_{nb} = 435{,}540 - 61{,}018 - 126{,}509 = 248{,}013 \text{ lb } (1{,}103 \text{ kN})$$

$$M_{nb} = 435{,}540 \left(\frac{14}{2} - \frac{6.10}{2} \right) - 61{,}018 \left(\frac{14}{2} - 2 \right) + 126{,}509 \left(12 - \frac{14}{2} \right)$$

$$= 2{,}047{,}838 \text{ in.-lb } (272.1 \text{ kN-m})$$

$$e_4 = e_b = \frac{2{,}047{,}838}{248{,}013} = 8.26 \text{ in. } (210 \text{ mm})$$

The coordinates for the preceding four cases are the controlling points on the P_n–M_n interaction diagram. Other points need to be computed as well, in order to develop an accurate diagram to cover the entire loading range. For example, additional points between the coordinates of the second and third cases have to be determined, assuming additional values of the neutral axis depth c and calculating P_n, M_n, and e for the c-values assumed. Table 8.1 summarizes the values of the coordinates used for plotting the P_n–M_n interaction diagram as well as the P_u–M_u design diagram. From the diagram, it is seen that the maximum moment ordinate seems to have a value close to $M_n = 2{,}047{,}838$ in.-lb. Hence, an assumption of $c_b = 8.15$ in. is verified.

Design Load–Moment (P_u–M_u) Diagram   From Section 8.3, $0.1A_g f'_c = 0.1 \times 14 \times 14 \times 6{,}000 = 117{,}600$ lb (522 kN), and $P_{nb} = 248{,}013$ lb $> 117{,}600$ lb; hence, Figure 8.5a applies for the zone in which ϕ can be increased beyond the 0.70 value for tied columns. To increase ϕ for M_{n7}, we have, from Equation 8.8,

$$\phi = 0.90 - \frac{0.20\phi P_n}{0.1 f'_c A_g}$$

Point	c in.	a in.	$P_n \times 10^3$ lb	$M_n \times 10^3$ in.-lb	ϕ	$P_u \times 10^3$ lb	$M_u \times 10^3$ in.-lb	e in.
1	∞	14	903.8	0	0.7	632.7*	0	0
5	18	13.5	826.6	388.9	0.7	578.7*	272.2	0.5
2	14	10.5	600.6	1,502.1	0.7	420.4	1,051.5	2.5
6	10	7.5	365.1	2,005.6	0.7	255.6	1,403.9	5.5
4	8.15	6.1	248	2,047.9	0.7	173.6	1,433.5	8.3
7	6	4.5	101.2	1,969.9	0.77	77.9	1,516.8	19.5
3	2.75	2.1	0	1,611.3	0.90	0	1,450.1	∞

* Max P_u allowed by the code for tied columns = $0.80\phi P_n$ = 498,906 lb (2,219 kN). Also,

1,000 lb = 4.448 kN

1,000 in.-lb = 0.1130 kN-m

1 in. = 25.4 mm.

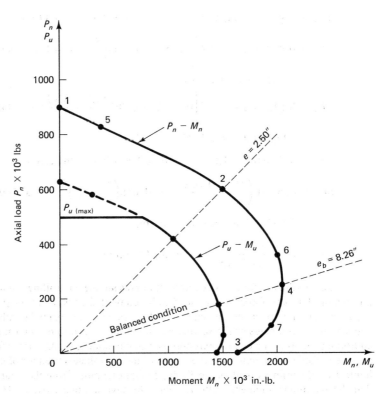

Figure 8.11 Load-moment interaction plots in Example 8.1.

Assume $\phi = 0.77$. Then

$$\phi = 0.90 - \frac{0.20 \times 0.77 \times 101,224}{117,600} \cong 0.77$$

$$P_{u7} = \phi P_{n7} = 0.77 \times 101,224 = 77,638 \text{ lb } (345 \text{ kN})$$

$$M_{u7} = \phi M_{n7} = 0.77 \times 1{,}969{,}875 = 1{,}516{,}804 \text{ in.-lb (171.4 kN-m)}$$

$$M_{u3} \text{ for pure bending} = \phi M_{n3} = 0.90 \times 1{,}611{,}274$$

$$= 1{,}450{,}147 \text{ in.-lb (163.9 kN-m)}$$

$$P_{u1} = \phi P_n = 0.70 \times 903{,}760 = 623{,}632 \text{ lb (not usable)}$$

The ACI Code requires that the maximum design axial load strength ϕP_n for tied prestressed columns should not exceed $0.80\phi P_n$, and for spirally reinforced prestressed columns should not exceed $0.85\phi P_n$. We thus have

$$\text{Max } P_u = 0.8\phi P_n = 0.8 \times 623{,}632$$

$$= 498{,}906 \text{ lb (2,219 kN)}$$

$$P_{u5} = 0.7 \times 826{,}648 = 578{,}654 \text{ (2,574 kN)}$$

The remaining values of the coordinates are summarized in Table 8.1. Plots of the interaction diagrams for the nominal strength $(P_m\text{–}M_n)$ and the design strength $(P_u\text{–}M_u)$ are shown in Figure 8.11.

8.6 LIMIT STATE AT BUCKLING FAILURE OF SLENDER (LONG) PRESTRESSED COLUMNS

Considerable literature exists on the behavior of columns subjected to stability testing. If the column slenderness ratio exceeds the limits for short columns, the compression member will buckle prior to reaching its limit state of material failure. The strain in the compression face of the concrete at buckling load will then be less than the 0.003 in./in. shown in Figure 8.12. Such a column would be a slender member subjected

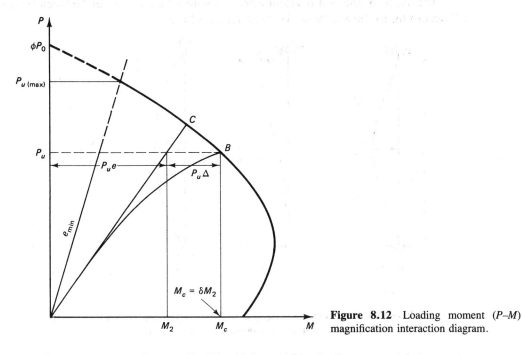

Figure 8.12 Loading moment $(P\text{–}M)$ magnification interaction diagram.

to combined axial load and bending, deforming laterally and developing additional moment due to the $P\Delta$ effect, where P is the axial load and Δ is the deflection of the column's buckled shape at the section being considered.

Consider a slender column subjected to axial load P_u at an eccentricity e. The buckling effect produces an additional moment of $P_u\Delta$. This moment reduces the load capacity from point C to point B in the interaction diagram of Figure 8.12. The total moment $P_u e + P_u\Delta$ is represented by point B in the diagram, and the column can be designed for a larger or magnified moment M_c as a nonslender column.

In such a case, the load P_u is assumed to act at an eccentricity $e + \Delta$ to produce moment M_c. The ratio M_c/M_2 is termed the *magnification factor* δ. If kl_u/r is the slenderness ratio, then the lower limit for disregarding the slenderness ratio effect, i.e., disregarding the need for stability analysis as specified by the ACI Code, is

$$\frac{kl_u}{r} < 34 - 12\frac{M_1}{M_2} \qquad \text{(for braced frames)} \qquad (8.10\ a)$$

$$\frac{kl_u}{r} < 22 \qquad \text{(for unbraced frames)} \qquad (8.10\ b)$$

where k is the column length factor, as shown in Figure 8.13, and M_1 and M_2 are the moments at the opposite ends of the compression member. M_2 is always larger than M_1, and the ratio M_1/M_2 is taken as positive for single curvature and negative for double curvature, as shown in Figure 8.14(a).

The effective length kl_u is used as the modified length of the column to account for end restraints other than being pinned. kl_u represents the length of an auxiliary pin-ended column which has an Euler buckling load equal to that of the column under consideration. Alternatively, it is the distance between the points of contraflexure of the member in its buckled form.

The value of the end restraint effective length factor k varies between 0.5 and 2.0 according to the nature of the restraint as follows:

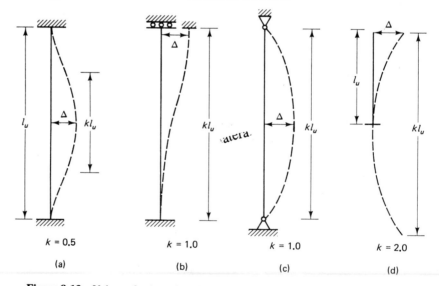

Figure 8.13 Values of column length factor k for typical end conditions. (a) Fixed-fixed. (b) Fixed-fixed with lateral motion. (c) Pinned. (d) Fixed-free.

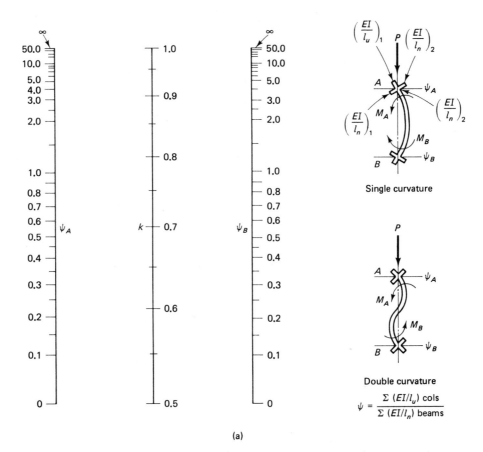

Figure 8.14 Effective length factor k for (a) braced and (b) unbraced frames.

Both column ends pinned, no lateral motion	$k = 1.0$
Both column ends fixed	$k = 0.5$
One end fixed, other end free	$k = 2.0$
Both column ends fixed, lateral motion exists	$k = 1.0$

Typical cases illustrating the buckled shape of the column for several end conditions and the corresponding length factors k are shown in Figure 8.13.

For members in a structural frame, the end restraint lies between the hinged and fixed conditions. The actual k value can be determined from the Jackson and Moreland alignment charts in Figure 8.14. In lieu of these charts, the following equations suggested in the ACI Code commentary can also be used for calculating k:

1. *Braced compression members.* An upper bound to the effective length factor may be taken as the smaller of the expressions

$$k = 0.7 + 0.05(\psi_A + \psi_B) \leq 1.0 \qquad (8.11\ a)$$

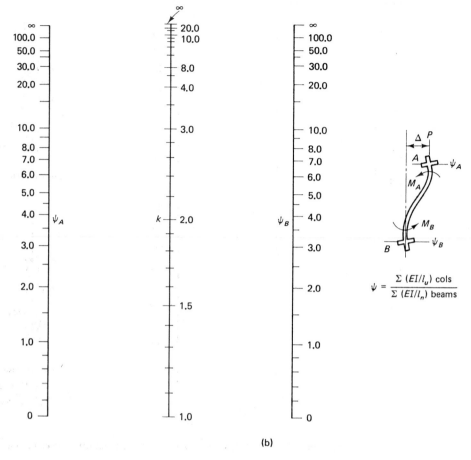

(b)

Figure 8.14 (*continued*)

and

$$k = 0.85 + 0.05\psi_{min} \le 1.0 \qquad (8.11\ b)$$

where ψ_A and ψ_B are the values of ψ at the two ends of the column and ψ_{min} is the smaller of the two values. ψ is the ratio of the stiffness of all compression members to the stiffness of all flexural members in a plane at one end of the column. That is,

$$\psi = \frac{\Sigma\ EI/l_u\ \text{columns}}{\Sigma\ EI/l_n\ \text{beams}} \qquad (8.12)$$

where l_u is the unsupported length of the column and l_n is the clear beam span.

2. *Unbraced compression members restrained at both ends.* The effective length may be taken as follows:

For $\psi_m < 2$,

$$k = \frac{20 - \psi_m}{20}\sqrt{1 + \psi_m} \qquad (8.13\ a)$$

For $\psi_m \geq 2$,

$$k = 0.9\sqrt{1 + \psi_m} \qquad (8.13 \text{ b})$$

where ψ_m is the average of the ψ values at the two ends of the compression member.

3. *Unbraced compression members hinged at one end.* The effective length factor may be taken as

$$k = 2.0 + 0.3\psi \qquad (8.14)$$

where ψ is the value at the restrained end.

The radius of gyration $r = \sqrt{I_g/A_g}$ can be taken as $r = 0.3h$ for rectangular sections, where h is the column dimension perpendicular to the axis of bending. For circular sections, r is taken as $0.25h$.

If the value of kl_u/r is larger than that obtained from Equations 8.10, two methods of stability analysis having general acceptance are recommended:

1. The *moment magnification method,* where the design of the member is based on a magnified moment

$$M_c = \delta M_2 = \delta_b M_{2b} + \delta_s M_{2s} \qquad (8.15)$$

Lateral loads tend to increase or magnify the moments more than gravity loads. To account for the difference between the lateral and gravity loads, the magnifying factor δ is broken into two components, δ_b and δ_s, where δ_b is the magnification factor for the predominant gravity-load moment M_{2b}, which is defined as the larger factored end moment on a column due to loads that produce no appreciable side sway, that is, only gravity-load moments.

δ_s is the magnification factor applied to the larger end moment M_{2s} due to loads that result in appreciable side sway, such as wind-load moments. In the case of frames braced with shear walls against side sway, the entire moment acting on the column is considered to be M_{2b}, and δ_s is therefore assumed to be zero. Normally, if the lateral deflection on the building is less than span $l_n/1,500$, the frame is considered a braced frame.

2. A *second-order analysis,* taking into consideration the effect of deflections. This analysis must be used in cases where $kl_u/r > 100$.

Note that all columns have to be designed for a minimum eccentricity of at least $(0.6 + 0.03h)$ in.

8.7 MOMENT MAGNIFICATION METHOD

The moment magnifiers δ_b and δ_s are dependent on the slenderness of the member, the stiffness of the entire frame, the restraint or applied moment at its ends, and the design cross section such that

$$\delta_b = \frac{C_m}{1 - P_u/\phi P_c} \geq 1 \qquad (8.16 \text{ a})$$

$$\delta_s = \frac{1}{1 - \Sigma P_u/\phi\Sigma P_c} \geq 1 \qquad (8.16\ b)$$

where P_c = Euler buckling load = $\pi^2 EI/(kl_u)^2$
$\qquad kl_u$ = effective length (between points of inflection)
$\Sigma P_u, \Sigma P_c$ = summations for all columns in a story
$\qquad l_u$ = unsupported length of column
$\qquad C_m$ = a factor relating the actual moment diagram to an equivalent uniform moment diagram for braced members subject to end loads only.

Note that some differences might result in choosing the prestressed concrete column section as compared with choosing a reinforced concrete section when the magnification method is applied in the design of slender prestressed concrete compression members. However, these results are inconclusive.

The factor

$$C_m = 0.6 + 0.4\frac{M_1}{M_2} \geq 0.4 \qquad (8.17)$$

where $M_1 \leq M_2$ and $M_1/M_2 > 0$ if no point of inflection exists between the column ends (single curvature; see Figure 8.14(a)). For other conditions, $C_m = 1.0$

When the computed end eccentricities are less than $(0.6 + 0.03h)$ in., one may use the computed end moments to evaluate M_1/M_2 in Equation 8.17. If computations show that there is essentially no moment at both ends of a compression member, the ratio M_1/M_2 should be taken equal to 1.

An estimate of EI must include the effects of cracking and creep under long-term loading. For all members,

$$EI = \frac{(E_c I_g/5) + E_s I_s}{1 + \beta_d} \qquad (8.18\ a)$$

For lightly reinforced members ($\rho_g \leq 3\%$), this may be simplified to

$$EI = \frac{E_c I_g/2.5}{1 + \beta_d} \qquad (8.18\ b)$$

where

$$\beta_d = \frac{\text{factored dead-load moment}}{\text{factored total moment}} \leq 1$$

8.8 SECOND-ORDER ANALYSIS

The rigorous mathematical approach of second-order analysis is needed if the slenderness ratio kl_u/r exceeds 100. The effect of deflection has to be taken into account, and an appropriate reduced tangent modulus for concrete has to be used. The designer, with the aid of computers, can solve with limited efforts the set of simultaneous equations needed to determine the size of the reinforced concrete slender column. Charts can also be developed for the various eccentricity ratios and reinforcement and concrete strength combinations. The majority of columns in concrete building frames do not necessitate a second-order analysis since the slenderness ratio kl_u/r is, in most cases, below 100.

8.9 OPERATIONAL PROCEDURE FOR THE DESIGN OF SLENDER (LONG) COLUMNS

1. Determine whether the frame has an appreciable side sway. If it does, use the magnification factors δ_b and δ_s. If the side sway is negligible, assume that $\delta_s = 0$. Then assume a cross section, calculate the eccentricity using the greater of the end moments, and check whether it is more than the minimum allowable eccentricity, that is,

$$\frac{M_2}{P_u} \geq (0.6 + 0.03h) \text{ in.}$$

If the given eccentricity is less than the specified minimum, use the minimum value.

2. Calculate ψ_A and ψ_B using Equation 8.12, and then obtain k using Figure 8.14 or Equations 8.13. Calculate kl_u/r, and determine whether the column is a short or long column. If the column is slender and kl_u/r is less than 100, calculate the magnified moment M_c. Then, using the value obtained, calculate the equivalent eccentricity to be used if the column is to be designed as a short column. If kl_u/r is greater than 100, perform a second-order analysis.

3. Design the equivalent nonslender column. The flowchart in Figure 8.15 presents the sequence of calculations. The necessary equations are provided in Section 8.2 and in the flowchart.

8.10 DESIGN OF SLENDER (LONG) PRESTRESSED COLUMN

Example 8.2

A square tied prestressed bonded column is part of a 5 × 3 bays frame building subjected to uniaxial bending. Its clear height is $\ell_u = 15$ ft (4.54 in.), and it is not braced against sidesway. The factored external load $P_u = 300,000$ lb (1,334 kN), and the factored end moments are $M_1 = 425,000$ in.-lb (48.0 kN-m) and $M_2 = 750,000$ in.-lb (84.8 kN-m). Design the column section and the reinforcement necessary for the following two conditions:

1. Consider gravity loads only, assuming negligible lateral sidesway due to wind.
2. Suppose sidesway wind effects cause a factored $P_u = 24,000$ lb (106.8 kN) and a factored $M_u = 220,000$ in.-lb (24.9 kN-m). The loads per floor of all columns at that level are $\Sigma P_u = 4.5 \times 10^6$ lb (20 × 10³ kN) and $\Sigma P_c = 31.0 \times 10^6$ lb (138 × 10³ kN).

Use $\frac{1}{2}$ in. dia 270 K stress-relieved prestressing strands. Given data are as follows:

$\beta_d = 0.4$
$\psi_A = 1.0$

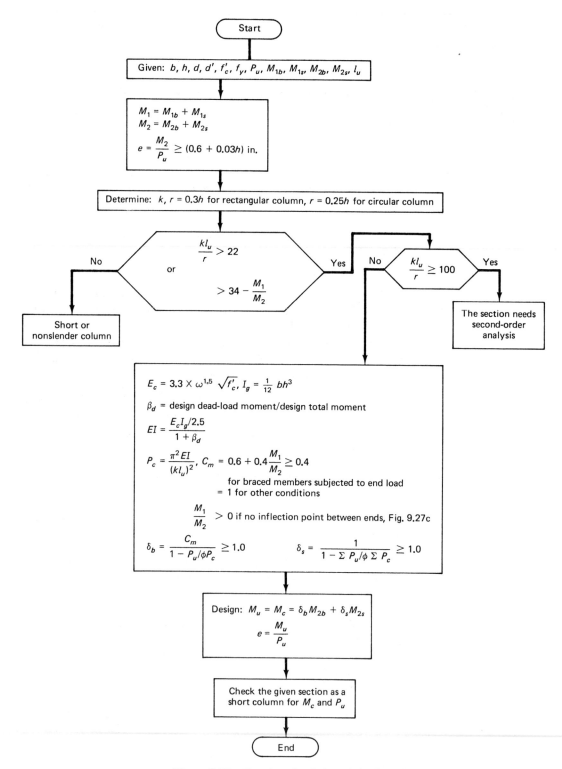

Figure 8.15 Flowchart for design of slender columns.

$\psi_B = 2.0$

$f'_c = 6,000$ psi (41.4 MPa)

$f_{pu} = 270,000$ psi (1,862 MPa)

$f_{ps} = 240,000$ psi (1,655 MPa)

$f_{pe} = 150,000$ psi (1,034 MPa)

$E_{ps} = 28 \times 10^6$ psi $(200 \times 10^3$ MPa)

ϵ_{cu} at failure = 0.003 in./in.

$\epsilon_{ce} = 0.0005$ in./in. when P_e acts on the section

$d' = 2$ in. (50.8 mm)

Ties $f_y = 60,000$ psi (414 MPa)

The stress-strain diagram of the prestressing steel is as in Figure 8.9.

Solution $\epsilon_{pe} = 0.0052$ in./in. from the stress-strain diagram of Figure 8.9, corresponding to $f_{pe} = 150,000$ psi. Similarly, $\epsilon_{py} \cong 0.012$ in./in. from the same figure, corresponding to $f_{py} = 260,000$ psi.

Gravity Loads Only

Check for No Sidesway and Minimum Eccentricity (Step 1). Since the frame has no appreciable sidesway, the entire M_2 is taken to be M_{2b}, and the magnification factor for sidesway, δ_s, is taken to be equal to zero in Equation 8.15. By trial and adjustment, a column section is assumed and analyzed. Accordingly, we try a section 15 in. × 15 in. (381 mm × 381 mm) as shown in Figure 8.16(a) and obtain

$$\text{Actual eccentricity} = \frac{M_{2b}}{P_u} = \frac{750,000}{300,000} = 2.50 \text{ in. (63.5 mm)}$$

$$\text{Minimum allowable eccentricity} = 0.6 + 0.03h = 0.6 + 0.03 \times 15$$

$$= 1.05 \text{ in. (2.67 mm)} < 2.50 \text{ in.}$$

Hence, use $M_{2b} = 750,000$ in.-lb as the larger of the moments M_1 and M_2 on the column.

Calculate the Eccentricity to Be Used for Equivalent Short Column (Step 2). From

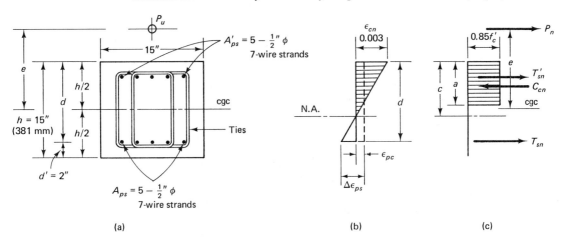

Figure 8.16 Proposed column section geometry in Example 8.2. (a) Cross-sectional details. (b) Strain distribution. (c) Stress block and forces.

the chart in Figure 8.14(b), $k = 1.45$ and the slenderness ratio is

$$\frac{k\ell_u}{r} = \frac{1.45 \times 15 \times 12}{0.3 \times 15} = 58.0$$

Since $58.0 > 22$ and < 100, use the moment magnification method. We obtain

$$E_c = 33w^{1.5}\sqrt{f'_c} = 33 \times 145^{1.5}\sqrt{6{,}000} = 4.46 \times 10^6 \text{ psi } (32 \times 10^3 \text{ MPa})$$

$$I_g = \frac{15(15)^3}{12} = 4{,}218.8 \text{ in}^4$$

$$EI = \frac{E_cI_g/2.5}{1 + \beta_d} = \frac{4.46 \times 10^6 \times 4{,}218.3}{2.5} \times \frac{1}{1 + 0.4}$$

$$= 5.34 \times 10^9 \text{ lb.-in}^2$$

$$(k\ell_u)^2 = (1.45 \times 15 \times 12)^2 = 68.1 \times 10^3 \text{ in}^2$$

Hence,

$$P_c = \text{Euler buckling load} = \frac{\pi^2 EI}{(k\ell_u)^2} = \frac{\pi^2 \times 5.34 \times 10^9}{68.1 \times 10^3}$$

$$= 773{,}132 \text{ lb} = 773.1 \text{ Kips } (3{,}439 \text{ kN})$$

Now, $C_m = 1.0$ for a nonbraced column. Assume $\phi = 0.7$. Then we have

$$\text{Moment magnifier } \delta_b = \frac{C_m}{1 - P_u/\phi P_c} = \frac{1.0}{1 - \dfrac{300{,}000}{0.7 \times 773{,}132}} = 2.24$$

$$\text{Design moment } M_c = \delta_b M_{2b} = 2.24 \times 750{,}000$$

$$= 1{,}680{,}000 \text{ in.-lb } (189.8 \text{ kN-m})$$

$$\text{Required } P_n = \frac{P_u}{\phi} = \frac{300{,}000}{0.7} = 428{,}571 \text{ lb } (1{,}906 \text{ kN})$$

$$\text{Required } M_n = \frac{1{,}680{,}000}{0.7} = 2{,}400{,}000 \text{ in.-lb } (271.2 \text{ kN-m})$$

$$\text{Eccentricity } e = \frac{2{,}400{,}000}{428{,}571} = 5.60 \text{ in. } (142 \text{ mm})$$

Design of an Equivalent Nonslender Column (Step 3). The equivalent column has to carry a minimum nominal axial load $P_n = 428{,}571$ lb and a minimum nominal uniaxial moment $M_n = 2{,}400{,}000$ in.-lb.

To design the equivalent nonslender column, we analyze the assumed 15 in. × 15 in. column section assuming five $\frac{1}{2}$ in. dia seven-wire stress-relieved strands on each of the two faces parallel to the neutral axis, as in Example 8.1. Then

$$A_{ps} = A'_{ps} = 5 \times 0.153 = 0.765 \text{ in}^2 \text{ (4.94 cm}^2\text{)}$$

Balanced Failure Condition

$$d = h - 2 = 15 - 2 = 13 \text{ in. } (330 \text{ mm})$$

Comparing with Example 8.1 and using trial and adjustment, a reasonable assumption of the neutral axis depth for the balanced condition would be a value of $c_b = 8.3$ in. (211 mm). Then $a_b = \beta_1 \times c_b = 0.75 \times 8.3 = 6.23$ in. (158 mm).

Next, from Figure 8.3,

$$C_{cn} = 0.85 \times 6{,}000 \times 15 \times 6.23 = 476{,}595 \text{ lb (2,119 kN)}$$

From Equation 8.5,

$$T'_{sn} = 0.765 \times 28 \times 10^6 \left[0.0052 - 0.003 \left(\frac{8.3 - 2}{8.3} \right) + 0.0005 \right]$$

$$= 73{,}318 \text{ lb (385 kN)}$$

From Equation 8.6,

$$T_{sn} = 0.765 \times 28 \times 10^6 \left[0.0052 + 0.003 \left(\frac{13 - 8.3}{8.3} \right) + 0.0005 \right]$$

$$= 158{,}482 \text{ lb (704 kN)}$$

From Equation 8.2,

$$P_{nb} = C_{cn} - T'_{sn} - T_{sn}$$

$$= 476{,}595 - 73{,}318 - 158{,}482$$

$$= 229{,}310 \text{ lb (1,020 kN)}$$

From Equation 8.7,

$$M_n = 476{,}595 \left(\frac{15}{2} - \frac{6.23}{2} \right) - 73{,}318 \left(\frac{15}{2} - 2 \right) + 158{,}482 \left(13 - \frac{15}{2} \right)$$

$$= 2{,}103{,}124 \text{ in.-lb (237.7 kN-m)}$$

$$e_b = \frac{M_{nb}}{P_{nb}} = \frac{2{,}103{,}124}{229{,}310} = 9.17 \text{ in. (233 mm)} > \text{actual } e = 5.60 \text{ in.}$$

The prestressed column load has small eccentricity, and initial failure would be in compression. Also, $\phi = 0.7$, as assumed.

Assume Neutral Axis Depth c = 12 in.

$$a = \beta_1 c = 0.75 \times 12 = 9.0 \text{ in.}$$

From Equation 8.1 a,

$$C_{cn} = 0.85 f'_c ba = 0.85 \times 6{,}000 \times 15 \times 9 = 688{,}500 \text{ lb}$$

From Equation 8.5,

$$T'_{sn} = A'_{ps} E_{ps} \left[\epsilon_{pe} - \epsilon_{cu} \left(\frac{c - d'}{c} \right) + \epsilon_{ce} \right]$$

$$= 0.765 \times 28 \times 10^6 \left[0.0052 - 0.003 \left(\frac{12 - 2}{12} \right) + 0.0005 \right]$$

$$= 70{,}993 \text{ lb}$$

From Equation 8.6,

$$T_{sn} = A_{ps} E_{ps} \left[\epsilon_{pe} + \epsilon_{cu} \left(\frac{d - c}{c} \right) + \epsilon_{ce} \right]$$

$$= 0.765 \times 28 \times 10^6 \left[0.0052 + 0.003 \left(\frac{13 - 12}{12} \right) + 0.0005 \right]$$

$$= 127{,}449 \text{ lb.}$$

From Equation 8.2,

$$P_n = C_{cn} - T'_{sn} - T_{sn}$$

$$\text{Available } P_n = 688{,}500 - 70{,}993 - 127{,}449$$

$$= 490{,}058 \text{ lb} > \text{required } P_n = 428{,}571 \text{ lb}$$

Accordingly, we go on to a second trial-and-adjustment cycle.

Assume Neutral Axis Depth c = 11.2 in.

$$a = \beta_1 c = 0.75 \times 11.2 = 8.4 \text{ in.}$$

$$C_{cn} = 0.85 f'_c ba = 0.85 \times 6{,}000 \times 15 \times 8.4 = 642{,}600 \text{ lb}$$

$$T'_{sn} = 0.765 \times 28 \times 10^6 \left[0.0052 - 0.003 \left(\frac{11.2 - 2}{11.2} \right) + 0.0005 \right]$$

$$= 69{,}309 \text{ lb}$$

$$T_{sn} = 0.765 \times 28 \times 10^6 \left[0.0052 + 0.003 \left(\frac{13 - 11.2}{11.2} \right) + 0.0005 \right]$$

$$= 132{,}421 \text{ lb}$$

$$\text{Available } P_n = 642{,}600 - 69{,}309 - 132{,}421$$

$$= 440{,}870 \text{ lb} \cong \text{ required } P_n = 428{,}571 \text{ lb, O.K.}$$

From Equation 8.7,

$$M_n = C_{cn} \left(\frac{h}{2} - \frac{a}{2} \right) - T'_{sn} \left(\frac{h}{2} - d' \right) + T_{sn} \left(d - \frac{h}{2} \right)$$

$$= 642{,}600 \left(\frac{15}{2} - \frac{8.4}{2} \right) - 69{,}309 \left(\frac{15}{2} - 2 \right) + 132{,}421 \left(13 - \frac{15}{2} \right)$$

$$= 2{,}467{,}696 \text{ in.-lb} > 2{,}400{,}000 \text{ in.-lb (278.8 kN-m} > 271.2 \text{ kN-m), O.K.}$$

Consequently, adopt a section 15 in. × 15 in. with five $\frac{1}{2}$ in. dia. seven-wire stress-relieved 270 K strands at each of the two faces parallel to the neutral axis. Then design the necessary transverse ties.

Gravity and Wind Loading (Sidesway). From part 1, $P_c = 773{,}132$ lb and $U = 0.75(1.4D + 1.7L + 1.7W)$. Also, $U = 0.9D + 1.3W$ (did not control), $P_u = 0.75(300{,}000 + 24{,}000) = 243{,}000$ lb, $M_{2b} = 0.75 \times 750{,}000$ in.-lb $= 562{,}500$ in.-lb, and $M_{2s} = 0.75 \times 220{,}000 = 165{,}000$ in.-lb.

From Equation 8.16(a),

$$\delta_b = \frac{1.0}{1 - \dfrac{P_u}{\phi P_c}} = \frac{1.0}{1 - \dfrac{243{,}000}{0.7 \times 773{,}132}} = 1.81$$

From Equation 8.16 b,

$$\delta_s = \frac{1.0}{1 - \dfrac{\Sigma P_u}{\phi \Sigma P_c}} = \frac{1.0}{1 - \dfrac{4.5 \times 10^6}{0.7 \times 31.0 \times 10^6}} = 1.26$$

From Equation 8.15,

$$M_c = \delta_c M_{2b} + \delta_s M_{2s} = 1.81 \times 562{,}500 + 1.26 \times 165{,}000$$

$$= 1{,}226{,}025 \text{ in.-lb}$$

$$\text{Required } P_n = \frac{243{,}000}{0.7} = 347{,}143 \text{ lb}$$

$$\text{Required } M_n = \frac{1{,}226{,}025}{0.7} = 1{,}751{,}464 \text{ in.-lb.}$$

$$\text{Eccentricity } e = \frac{1{,}751{,}464}{347{,}143} = 5.05 \text{ in.} < e_b = 9.17 \text{ in.}$$

Hence, initial compression failure occurs and the assumed $\phi = 0.7$ is correct. We then have $P_n = 347{,}143 < P_n = 428{,}571$ lb in case 1.

The conditions for case 2 with sidesway do not control, since failure is still in compression and the required P_n is *less* than that for case 1. Accordingly, we adopt the same 15 in. × 15 in. section of case 1, with five $\frac{1}{2}$ in. dia 270 K stress-relieved prestressing strands on each of the two faces parallel to the neutral axis.

One of the earliest prestressed concrete footbridges in England at the Festival of Britain, London, 1951.

8.11 COMPRESSION MEMBERS IN BIAXIAL BENDING

8.11.1 Exact Method of Analysis

Columns in corners of buildings are compression members subjected to biaxial bending about both the x and the y axes as shown in Figure 8.17. Also, biaxial bending occurs due to imbalance of loads in adjacent spans and almost always in bridge piers. Such columns are subjected to moment M_{xx} about the x axis creating a load eccentricity e_y, and a moment M_{yy} about the y axis creating a load eccentricity e_x. Thus, the neutral axis is inclined at an angle θ to the horizontal.

The angle θ depends on the interaction of the bending moments about both axes and the magnitude of the total P_u. The compressive area in the column section can have

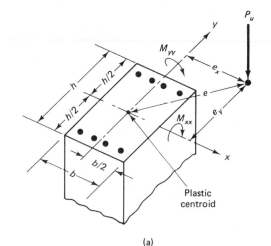

(a)

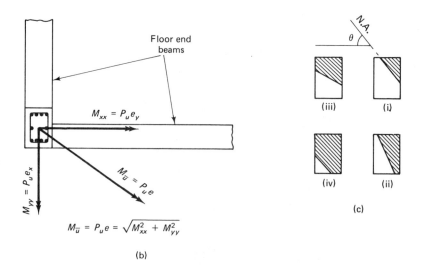

(b)

Figure 8.17 Corner column subjected to axial load. (a) Biaxially stressed column cross section. (b) Vector moments M_{xx} and M_{yy} in column plan. (c) Neutral axis direction.

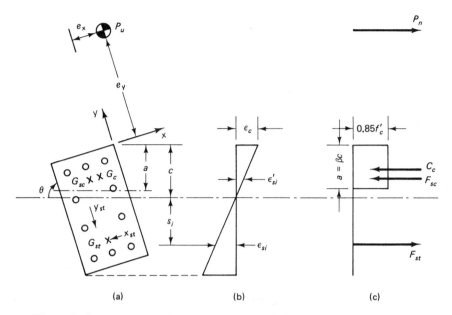

Figure 8.18 Strain compatibility and forces in biaxially loaded rectangular columns. (a) Cross section. (b) Strain. (c) Forces.

one of the alternative shapes shown in Figure 8.17(c). Since such a column has to be designed from first principles, the trial-and-adjustment procedure has to be followed where compatibility of strain has to be maintained at all levels of the reinforcing bars. Additional computational effort is also needed, because of the position of the inclined neutral-axis plane and the four different possible forms of the concrete compression area.

Figure 8.18 shows the strain distribution and forces on a biaxially loaded rectangular column cross section. G_c is the center of gravity of the concrete compression area, having coordinates x_c and y_c from the neutral axis in the x and y directions, respectively. G_{sc} is the resultant position of the steel forces in the compression area having coordinates x_{sc} and y_{sc} from the neutral axis in the x and y directions, respectively. G_{st} is the resultant position of the steel forces in the tension area having coordinates x_{st} and y_{st} from the neutral axis in the x and y directions, respectively. From equilibrium of internal and external forces,

$$P_n = 0.85f'_c A_c + F_{sc} - F_{st} \tag{8.19}$$

where A_c = area of the compression zone covered by the rectangular stress block
F_{sc} = resultant steel compressive forces ($\Sigma A'_s f_{sc}$)
F_{st} = resultant steel tensile force ($\Sigma A_s f_{st}$)

Also, from equilibrium of internal and external moments,

$$P_n e_x = 0.85f'_c A_c x_c + F_{sc} x_{sc} + F_{st} x_{st} \tag{8.20 a}$$

$$P_n e_y = 0.85f'_c A_c y_c + F_{sc} y_{sc} + F_{st} y_{st} \tag{8.20 b}$$

The position of the neutral axis has to be assumed in each trial and the stress calculated in *each* bar using

$$f_{si} = E_s \epsilon_{si} = E_c \epsilon_c \frac{s_i}{c} < f_y \qquad (8.21)$$

8.11.2 Load Contour Method of Analysis

One method of arriving at a rapid solution is to design the column for the vector sum of M_{xx} and M_{yy} and use a circular reinforcing cage in a square section for the corner column. However, such a procedure cannot be economically justified in most cases. Another design approach well proven by experimental verification is to transform the biaxial moments into an equivalent uniaxial moment and an equivalent uniaxial eccentricity. The section can then be designed for uniaxial bending, as previously discussed in this chapter, to resist the actual factored biaxial bending moments.

Such a method considers a failure surface instead of failure planes and is generally termed the *Bresler–Parme contour method*. The method involves cutting the three-dimensional failure surfaces in Figure 8.19 at a constant value P_n to give an interaction plane relating M_{nx} and M_{ny}. In other words, the contour surface S can be viewed as a curvilinear surface which includes a family of curves, termed the *load contours*.

The general nondimensional equation for the load contour at a constant load P_n may be expressed as

$$\left(\frac{M_{nx}}{M_{ox}}\right)^{\alpha_1} + \left(\frac{M_{ny}}{M_{oy}}\right)^{\alpha_2} = 1.0 \qquad (8.22)$$

where $M_{nx} = P_n e_y$
$\quad\quad M_{ny} = P_n e_x$

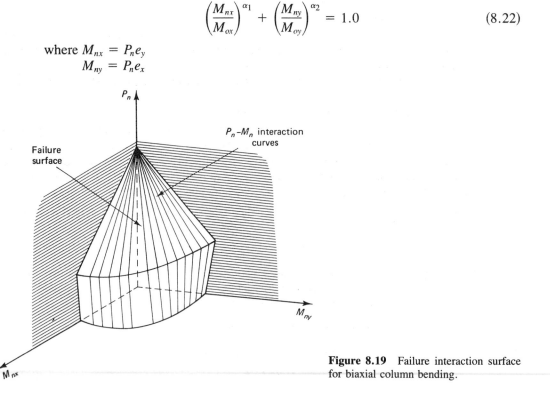

Figure 8.19 Failure interaction surface for biaxial column bending.

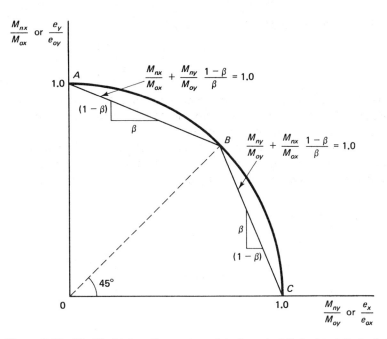

Figure 8.20 Modified interaction contour plot of constant P_n for biaxially loaded column.

$M_{ox} = M_{nx}$ at an axial load P_n such that M_{ny} or $e_x = 0$

$M_{oy} = M_{ny}$ at an axial load P_n such that M_{nx} or $e_y = 0$

The moments M_{ox} and M_{oy} are the *required* equivalent resisting moment strengths about the x and y axes, respectively, while α_1 and α_2 are exponents that depend on the cross-sectional geometry and the steel percentage and its location and material stress f'_c and f_y.

Equation 8.22 can be simplified using a common exponent and introducing a factor β for one particular axial load value P_n such that the ratio M_{nx}/M_{ny} would have the same value as the ratio M_{ox}/M_{oy} as detailed by Parme and associates. Such simplification leads to

$$\left(\frac{M_{nx}}{M_{ox}}\right)^{\alpha} + \left(\frac{M_{ny}}{M_{oy}}\right)^{\alpha} = 1.0 \tag{8.23}$$

where $\alpha = \log 0.5 / \log \beta$. Figure 8.20 gives a contour plot *ABC* from Equation 8.23.

For design purposes, the contour is approximated by two straight lines *BA* and *BC*, and Equation 8.23 can be simplified to two conditions:

1. For *AB* when $M_{ny}/M_{oy} < M_{nx}/M_{ox}$,

$$\frac{M_{nx}}{M_{ox}} + \frac{M_{ny}}{M_{oy}}\left[\frac{1-\beta}{\beta}\right] = 1.0 \tag{8.24 a}$$

2. For *BC* when $M_{ny}/M_{oy} > M_{nx}/M_{ox}$,

$$\frac{M_{ny}}{M_{oy}} + \frac{M_{nx}}{M_{ox}}\left[\frac{1-\beta}{\beta}\right] = 1.0 \qquad (8.24\ b)$$

In both of these equations, the *actual* controlling equivalent uniaxial moment strength M_{oxn} or M_{oyn} should be at least equivalent to the *required* controlling moment strength M_{ox} or M_{oy} of the chosen column section.

For rectangular sections where the reinforcement is evenly distributed along all the column faces, the ratio M_{oy}/M_{ox} can be taken to be approximately equal to b/h. In that case, Equations 8.24 can be modified as follows:

1. For $\dfrac{M_{ny}}{M_{nx}} > b/h$,

$$M_{ny} + M_{nx}\frac{b}{h}\frac{1-\beta}{\beta} \cong M_{oy} \qquad (8.25\ a)$$

2. For $\dfrac{M_{ny}}{M_{nx}} \leq b/h$,

$$M_{nx} = M_{ny}\frac{h}{b}\frac{1-\beta}{\beta} \cong M_{ox} \qquad (8.25\ b)$$

The controlling required moment strength M_{ox} or M_{oy} for designing the section is the larger of the two values as determined from Equations 8.25.

Plots like those of Figure 8.21 are used in the selection of β in the analysis and design of the columns just described. In effect, the modified load-contour method can be summarized in Equation 8.23 as a method for finding equivalent required moment strengths M_{ox} and M_{oy} for designing the columns as if they were uniaxially loaded.

8.11.3 Step-by-Step Operational Procedure for the Design of Biaxially Loaded Columns

The following steps can be used as a guideline for the design of columns subjected to bending in both the *x* and *y* directions. The procedure assumes an equal area of reinforcement on all four faces.

1. Calculate the uniaxial bending moments assuming an equal number of bars on each column face. Assume a value of an interaction contour β factor between 0.50 and 0.70 and a ratio of h/b. This ratio can be approximated to M_{nx}/M_{ny}. Using Equations 8.25 determine the equivalent required uniaxial moment M_{ox} or M_{oy}. If M_{nx} is larger than M_{ny}, use M_{ox} for the design and vice versa.

2. Assume a cross section for the column and a reinforcement ratio $\rho = \rho' \cong 0.01$ to 0.02 on each of the two faces parallel to the axis of bending of the larger equivalent moment. Then make a preliminary selection of the steel bars, and verify the capacity P_n of the assumed column cross section. In the completed design, the same amount of longitudinal steel should be used on all four faces.

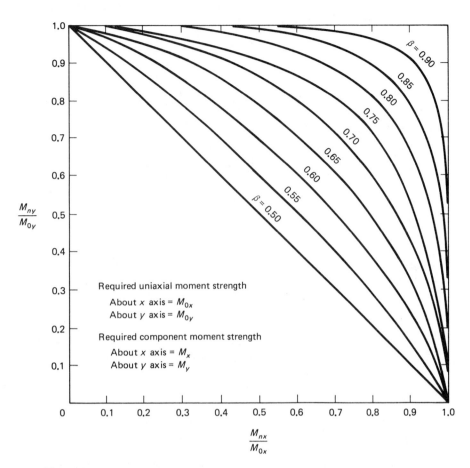

Figure 8.21 Contour β-factor chart for rectangular columns in biaxial bending.

3. Calculate the *actual* nominal moment strength M_{oxn} for equivalent uniaxial bending about the x axis when $M_{ox} = 0$. Its value has to be at least equivalent to the *required* moment strength M_{ox}.

4. Calculate the actual nominal moment strength M_{oyn} for the equivalent uniaxial bending moment about the y axis when $M_{oy} = 0$.

5. Find M_{ny} by entering M_{nx}/M_{oxn} and the trial β value into the β factor contour plots of Figure 8.21.

6. Make a second trial and adjustment, increasing the assumed β value if the M_{ny} value obtained from entering the chart is less than the required M_{ny}. Repeat this step until the two values of M_{ny} converge, either through changing β or changing the section.

7. Design the lateral ties and detail the section.

A flowchart for the primary steps in evaluating the controlling moment values in biaxially loaded columns is given in in Figure 8.22. A detailed computational example, together with a discussion of biaxially loaded columns, is presented in Ref. 8.2.

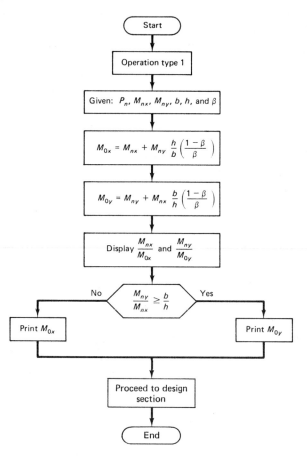

Figure 8.22 Flowchart for evaluation of controlling moment values in biaxially loaded columns.

8.12 PRACTICAL DESIGN CONSIDERATIONS

The following guidelines are presented for the design and arrangement of reinforcement to arrive at a practical design.

8.12.1 Longitudinal or Main Reinforcement

The average effective prestress in the concrete in prestressed compression members should not be less than 225 psi (1.55 MPa). This code requirement sets a minimum reinforcement ratio such that compressive members with lower prestress values will have a minimum nonprestressed reinforcement ratio of one percent.

8.12.2 Lateral Reinforcement for Columns

8.12.2.1 Lateral ties. Lateral reinforcement is required to prevent spalling of the concrete cover or local buckling of the longitudinal bars. This reinforcement could be in the form of ties evenly distributed along the height of the column at specified intervals. Longitudinal bars spaced more than 6 in. apart should be supported by lateral ties, as shown in Figure 8.23.

The following guidelines are to be followed for the selection of the size and spacing of ties.

1. The size of the tie should not be less than a #3 (9.5 mm) bar.
2. The vertical spacing of the ties must not exceed
   (a) Forty-eight times the diameter of the tie
   (b) Sixteen times the diameter of the longitudinal bar
   (c) The least lateral dimension of the column

Figure 8.23 shows a typical arrangement of ties for four, six, and eight longitudinal bars in a column cross section.

8.12.2.2 Spirals. The other type of lateral reinforcement is spirals or helical lateral reinforcement, as shown in Figure 8.24. Spirals are particularly useful in increasing ductility or member toughness, and hence are mandatory in high-earthquake-risk regions. Normally, concrete outside the confined core of the spirally

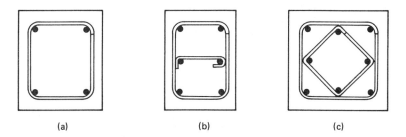

(a)   (b)   (c)

Figure 8.23 Typical arrangement of ties for four, six, and eight longitudinal bars in a column. (a) One tie. (b) Two ties. (c) Two ties.

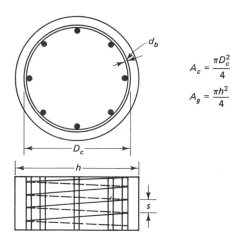

$$A_c = \frac{\pi D_c^2}{4}$$

$$A_g = \frac{\pi h^2}{4}$$

Figure 8.24 Helical or spiral reinforcement for columns.

reinforced column can totally spall under unusual and sudden lateral forces such as earthquake-induced forces. The columns have to be able to sustain most of the load even after the spalling of the cover in order to prevent the collapse of the building. Hence, the spacing and size of spirals are designed to maintain most of the load-carrying capacity of the column, even under such severe load conditions.

Closely spaced spiral reinforcement increases the ultimate-load capacity of columns. The spacing or pitch of the spiral is so chosen that the load capacity due to the confining spiral action compensates for the loss due to spalling of the concrete cover.

Equating the increase in strength due to confinement to the loss of capacity in spalling, and incorporating a safety factor of 1.2, we obtain the minimum spiral reinforcement ratio

$$\rho_s = 0.45\left(\frac{A_g}{A_c} - 1\right)\frac{f'_c}{f_{sy}}$$

(8.26)

where $\rho_s = \dfrac{\text{volume of the spiral steel per one revolution}}{\text{volume of concrete core contained in one revolution}}$

$$A_c = \frac{\pi D_c^2}{4}$$

(8.27 a)

$$A_g = \frac{\pi h^2}{4}$$

(8.27 b)

h = diameter of the column
a_s = cross-sectional area of the spiral
d_b = nominal diameter of the spiral wire
D_c = diameter of the concrete core out-to-out of the spiral
and f_{sy} = yield strength of the spiral reinforcement

To determine the pitch s of the spiral, calculate ρ_s using Equation 8.26, choose a bar diameter d_b for the spiral, calculate a_s, and then obtain pitch b using Equation 8.29 b below.

The spiral reinforcement ratio ρ_s can be written

$$\rho_s = \frac{a_s \pi (D_c - d_b)}{(\pi/4)D_c^2 s}$$

(8.28)

Therefore, the pitch is given by

$$s = \frac{a_s \pi (D_c - d_b)}{(\pi/4)D_c^2 \rho_s}$$

(8.29 a)

or

$$s = \frac{4a_s(D_c - d_b)}{D_c^2 \rho_s}$$

(8.29 b)

The spacing or pitch of spirals is limited to a range of 1 to 3 in. (25.4 to 76.2 mm), and the diameter should be at least $\frac{3}{8}$ in. (9.53 mm). The spiral should be well anchored by providing at least $1\frac{1}{2}$ extra turns when splicing of spirals rather than welding is used.

8.13 DESIGN OF SPIRAL LATERAL REINFORCEMENT

Example 8.3

Design the lateral spiral reinforcement for a circular prestressed concrete column $h = 20$ in. (508 mm) and clear cover $d_c = 1.5$ in. (38.1 mm) given that $f_y = 60,000$ psi (413.7 MPa).

Solution   Using Equation 8.26,

$$\text{required } \rho_s = 0.45\left(\frac{A_g}{A_c} - 1\right)\frac{f'_c}{f_{sy}}$$

Using #3 spirals with a yield strength $f_y = 60,000$ psi, we obtain

Clear concrete cover $d_c = 1.5$ in. (38.1 mm)

$$f_{sy} = 60,000 \text{ psi}$$

$$D_c = h - 2d_c = 20.0 - 2 \times 15 = 17.0 \text{ in. (431.8 mm)}$$

$$A_c = \frac{\pi(17.0)^2}{4} = 226.98 \text{ in}^2$$

$$A_g = 314.0 \text{ in}^2$$

$$\rho_s = 0.45\left(\frac{314.0}{226.98} - 1\right)\frac{4,000}{60,000} = 0.0115$$

For #3 spirals, $a_s = 0.11$ in^2. So using Equation 8.29 b, we get

$$\text{pitch } s = \frac{4a_s(D_c - d_b)}{D_c^2\rho_s} = \frac{4 \times 0.11(17.0 - 0.375)}{(17.0)^2 \times 0.0115} = 2.20 \text{ in. (55.9 mm)}$$

Accordingly, provide #3 spirals at $2\frac{1}{4}$ in. pitch (9.53 mm dia spiral at 54.0 mm pitch).

8.14 PRESTRESSED TENSION MEMBERS

8.14.1 Service-Load Stresses

Tension elements and systems such as railroad ties, bridge truss tension members, foundation anchors for retaining walls, and ties in walls of liquid-retaining tanks combine the high strength of the prestressing strands with the stiffness of the concrete. As such, they provide tensile resistance and reduced deformations that could not be provided if the member were made of an all-steel section to carry the same load. The limited and controlled deformation of the prestressed tension member in spite of its slenderness makes it especially useful as a tie or as part of an overall structural system.

Figure 8.25 compares the elongation of a prestressed concrete member in direct tension with a structural steel member of similar capacity. The elongation of the tension tie results from application of the external force F, while the elongation of the unstressed tendon in part (a) due to force F is, from basic mechanics,

$$\Delta L_{ps} = \frac{FL}{A_{ps}E_{ps}} \tag{8.30}$$

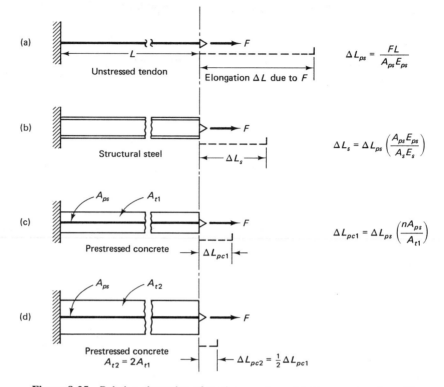

Figure 8.25 Relative elongation of tension members. (a) Unstressed tendon. (b) Structural steel. (c) Prestressed concrete, area $A_{t1} = A_g + (n-1)A_{ps}$. (d) Prestressed concrete, area $A_{t2} = 2A_{t1}$.

If the tendon is replaced by a rolled structural member, the change in the properties of the section results in a deformation

$$\Delta L_s = \Delta L_{ps}\left(\frac{A_{ps}E_{ps}}{A_sE_s}\right) \tag{8.31}$$

where A_s is considerably larger than A_{ps}. Hence, a considerably reduced elongation is seen as shown in Figure 8.25(b). The transformed area of concrete in Figure 8.25(c) is

$$A_{t1} = A_g + (n-1)A_{ps} \tag{8.32}$$

and if the change in stress in the concrete is

$$\Delta f_c = \frac{F}{A_{t1}} \tag{8.33}$$

the corresponding change in stress in the prestressing steel is

$$\Delta f_{ps} = \frac{n_p F}{A_{t1}} \tag{8.34}$$

where n_p is the modular ratio E_{ps}/E_c. Therefore,

$$\Delta L_{pc1} = \Delta L_{ps}\left(\frac{n_p A_{ps}}{A_{t1}}\right) \tag{8.35}$$

or

$$\Delta L = \frac{n_p FL}{A_t E_{ps}} = \frac{FL}{A_t E_c} \qquad (8.36)$$

Equation 8.36 gives the elongation of the tension tie due to the external load F at service load. It is readily seen that if the area of concrete is doubled, the elongation is halved for the same tensile force F. In sum, through a judicious choice of the geometry of the section, it is possible to reduce the elongation of prestressed tension members considerably.

Shortening of the tension tie occurs due to both the prestressing force and the long-term effects. The stress in the concrete due to initial prestress after transfer is

$$f_{ci} = -\frac{P_i}{A_c} \qquad (8.37)$$

where A_c is the net area of concrete in the section. The reduced stress in the concrete that results in an effective prestressing force P_e after all time-dependent losses have occurred is

$$f_{ce} = \frac{P_e}{A_c} \qquad (8.38\ a)$$

while the effective stress in the prestressing steel is

$$f_{pe} = -\frac{P_e}{A_{ps}} \qquad (8.38\ b)$$

Superimposing Equation 8.33 on Equation 8.38 a for the total effect of the external tensile force F and the prestressing force P_e, we have, for the total stress in the concrete,

$$f_c = -\frac{P_e}{A_c} + \frac{F}{A_t} \qquad (8.39)$$

Laboratory prestressing bed. (*Courtesy*, Building Research Establishment, Garston, Watford, England.)

The corresponding stress in the tendon is

$$f_{ps} = f_{pe} + \frac{nF}{A_t}$$ (8.40)

If cracking is not allowed in the tension member, the change in length ΔL_{pc} (elongation) at service load, from Equation 8.36, is essentially equivalent to the reduction in length Δp due to the effective prestressing force P_e (see next).

8.14.2 Deformation Behavior

Evaluation of the deformation of the tension member is critical to the overall design of the entire structure. Changes in the length of the member can induce severe stresses in the adjoining members, which could lead to structural failure. If the member is post-tensioned and fully bonded, then the reduction in length due to the initial prestress P_i alone is

$$\Delta_i = -\frac{P_i L}{A_c E_c}$$ (8.41 a)

and the elastic reduction in length after losses is

$$\Delta p = -\frac{P_e L}{A_c E_c}$$ (8.41 b)

Due to creep, the initial prestressing force P_i reduces to the effective force P_e. So using a creep coefficient C_u and assuming, as in Chapter 7, an average force $(P_i + P_e)/2$ as sufficiently accurate for evaluating creep loss, the change in length due to creep is

$$\Delta_{CR} = -\frac{L}{A_c E_c}\left[C_u\left(\frac{P_i + P_e}{2}\right)\right]$$ (8.42)

To account for shrinkage, the change in length is

$$\Delta_{SH} = \epsilon_{SH} L$$ (8.43)

so that the total effective reduction in length becomes

$$\Delta_e = -\left\{\frac{L}{A_c E_c}\left[P_e + C_u\left(\frac{P_i + P_e}{2}\right)\right] + \epsilon_{SH} L\right\}$$ (8.44)

Conversely, the loss in tension due to creep and shrinkage alone, from Equations 8.42 and 8.43, is

$$\Delta P = \frac{\Delta_{CR} + \Delta_{SH}}{L} E_{ps} A_{ps}$$ (8.45)

8.14.3 Decompression and Cracking

In considering decompression and cracking, it has to be assumed that no cracking is allowed at service load. If the probability of cracking exists due to a possible overload, an additional prestressing force and the addition of nonprestressed reinforcement become necessary to control cracking.

Assuming that only prestressed steel is provided, the tensile stress in the concrete at the *first cracking load* F_{cr} should not exceed the direct tensile strength of the concrete, ranging between $f'_t = 3\lambda \sqrt{f'_c}$ and $f'_t = 5\lambda \sqrt{f'_c}$, where $\lambda = 1, 0.85$, and 0.75 for normal-weight, sand-lightweight and all-lightweight concrete, respectively. The cracking load F_{cr} can be evaluated from Equation 8.39 using the appropriate value of f'_t, viz.,

$$f'_t = -\frac{P_e}{A_c} + \frac{F_{cr}}{A_t} \tag{8.46}$$

Any overload beyond F_{cr} is expected to cause a dynamic increase in cracking such that all the applied load starts to be carried by the prestressing steel alone, with a consequent failure of the member. Therefore, provision of an adequate level of residual compressive stress in the concrete becomes necessary in major tension members.

A *decompression load,* namely, the load at $f'_t = 0$ in Equation 8.46, should be the maximum allowable service load to which the member can be subjected. Equation 8.46 then becomes

$$f'_t = 0 = -\frac{P_e}{A_c} + \frac{F_{dec}}{A_t} \tag{8.47}$$

8.14.4 Limit State at Failure and Safety Factors

After cracking, the entire tension in the member is assumed to be carried by the prestressing tendon. Consequently, the nominal strength of the linear tension member is

$$F_n = A_{ps} f_{pu} \tag{8.48}$$

and the design ultimate load is

$$F_u = \phi_n F_n = \phi A_{ps} f_{pu} \tag{8.49}$$

It is important to provide a minimum safety factor of 1.5 for the decompression load F_{dec}. The safety factor level is determined by the importance of the tension member in the structure, the importance of the structure itself, and the negative effects of long-term reductions in length of the tension member on the integrity of the overall structure. It is not unreasonable under certain conditions to use a safety factor of 2.0 or more in a particular design.

8.15 SUGGESTED STEP-BY-STEP PROCEDURE FOR THE DESIGN OF TENSION MEMBERS

1. Determine the factored load F_u and the corresponding required nominal strength $F_n = F_u/\phi$.

2. Choose a uniform concrete compressive stress value at service load due to P_e to range between $f_c = 0.20 f'_c$ and $f_c = 0.30 f'_c$. Select the area A_{ps} and the size of the prestressing strands, and then compute the net concrete area $A_c = A_g - A_{duct}$ and the transformed area $A_t = A_g + (n - 1)A_{ps}$.

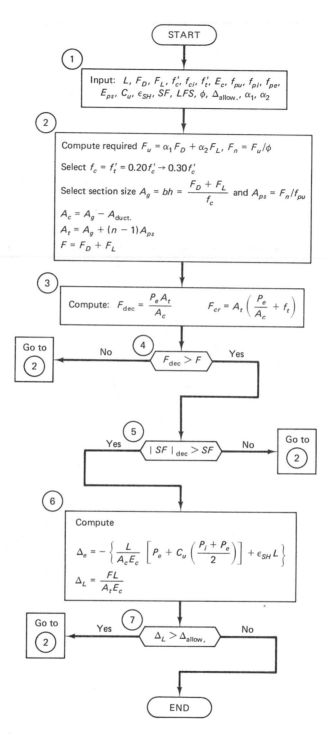

START

①

Input: L, F_D, F_L, f'_c, f_{ci}, f'_t, E_c, f_{pu}, f_{pi}, f_{pe}, E_{ps}, C_u, ϵ_{SH}, SF, LFS, ϕ, $\Delta_{allow.}$, α_1, α_2

② Compute required $F_u = \alpha_1 F_D + \alpha_2 F_L$, $F_n = F_u/\phi$

Select $f_c = f'_t = 0.20 f'_c \rightarrow 0.30 f'_c$

Select section size $A_g = bh = \dfrac{F_D + F_L}{f_c}$ and $A_{ps} = F_n/f_{pu}$

$A_c = A_g - A_{duct.}$
$A_t = A_g + (n - 1) A_{ps}$
$F = F_D + F_L$

③ Compute: $F_{dec} = \dfrac{P_e A_t}{A_c}$ $F_{cr} = A_t \left(\dfrac{P_e}{A_c} + f_t \right)$

Go to ② ← No —— ④ $F_{dec} > F$ —— Yes

⑤ Yes —— $|SF|_{dec} > SF$ —— No → Go to ②

⑥ Compute

$$\Delta_e = - \left\{ \dfrac{L}{A_c E_c} \left[P_e + C_u \left(\dfrac{P_i + P_e}{2} \right) \right] + \epsilon_{SH} L \right\}$$

$$\Delta_L = \dfrac{FL}{A_t E_c}$$

Go to ② ← Yes —— ⑦ $\Delta_L > \Delta_{allow.}$ —— No

END

Figure 8.26 Flowchart for the design (analysis) of linear prestressed tension members.

Prestressed Compression and Tension Members   Chap. 8

3. Compute the maximum allowable external force F based on decompression stress $f'_t = 0$, i.e., find F_{dec}. Then compute the first cracking load F_{cr}.

4. Find the factors of safety due to forces F, F_{dec}, and F_{cr} to verify that they exceed the value 1.5.

5. Compute the length-shortening deformations due to creep and shrinkage (Equation 8.44):

$$\Delta_e = -\left\{ \frac{L}{A_c E_c}\left[P_e + C_u\left(\frac{P_i + P_e}{2} \right) \right] + \epsilon_{SH} L \right\}$$

Then check whether the value obtained causes excessive stress in adjoining members. Next, compute the elongation

$$\Delta_L = \frac{FL}{A_t E_c}$$

due to external load (Equation 8.36) and verify whether the value obtained is within the specified limits of the design.

6. Adopt the design if all requirements are satisfied; otherwise, proceed through another trial-and-adjustment cycle.

A flowchart for the step-by-step trial-and-adjustment procedure that can be used for the design and/or analysis of linear tension members is presented in Figure 8.26.

8.16 DESIGN OF LINEAR TENSION MEMBERS

Example 8.4

A linear post-tensioned fully grouted direct-tension tie for an underground shelter shown in Figure 8.27 has a length $L = 130$ ft (39.6 m). The tie is to be designed for a net horizontal roof and earthfill dead load of thrust $F_d = 115,000$ lb (512 kN) and live load of thrust $F_L = 55,000$ lb (245 kN). Design the tie with a safety factor not less than 1.5 against cracking, using $\frac{1}{2}$ in. (12.7 mm) dia 270 K stress-relieved prestressing strands. The maximum allowable elongation due to external load is $\frac{1}{2}$ in., and given data are as follows:

$f'_c = 6,000$ psi, (41.4 MPa) normal weight

$f'_{ci} = 4,000$ psi (31.0 MPa)

$f'_t = 4\sqrt{f'_c} = 310$ psi (2.14 MPa)

$E_{ci} = 4.06 \times 10^6$ psi (28 MPa)

$E_c = 4.69 \times 10^6$ psi (32.2 MPa)

$f_{pu} = 270,000$ psi (1,862 MPa)

$f_{pi} = 190,000$ psi (1,310 MPa)

$f_{pe} = 150,000$ psi (1,034 MPa)

$E_{ps} = 29 \times 10^6$ psi (200 $\times 10^3$ MPa)

$n_p = E_{ps}/E_c = 29 \times 10^6/4.69 \times 10^6 = 6.18$

$C_u = 2.35$

$\epsilon_{SH} = 700 \times 10^{-6}$ in./in.

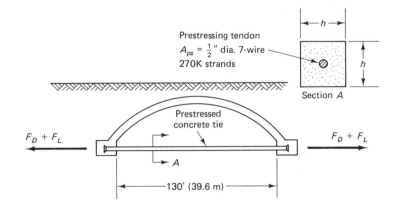

Figure 8.27 Prestressed concrete tie in Example 8.3.

The stress-strain diagram of the strands is given in Figure 8.9. Use a load factor of 1.4 for F_D and 1.7 for F_L.

Solution

Factored Loads and Choice of Strands (Steps 1 and 2). The factored load is $F_u = 1.4F_D + 1.7F_L = 1.4 \times 115,000 + 1.7 \times 55,000 = 254,000$ lb (1,132 kN), and the required nominal strength is $F_n = F_u/\phi = 254,500/0.9 = 282,778$ lb (1,258 kN). So $A_{ps} = F_n/f_{pu} = 282,778/270,000 = 1.05$ in^2 (6.77 cm^2). Hence, trying $\frac{1}{2}$ in. (12.7 mm) dia seven-wire strands, we find that the number of strands $= 1.05/0.153 = 6.86$. So we use seven strands and obtain $A_{ps} = 7 \times 0.153 = 1.07$ in^2 (6.9 cm^2). Then the available nominal strength is $F_n = 1.07 \times 270,000 = 288,900$ lb (1,285 kN).

Now, assume that the uniform compressive stress that is caused by the prestressing force P_e is $f_c = 0.25f'_c$. Then $P_e = 1.07 \times 150,000 = 160,500$ lb and $A_g = 160,500/(0.25 \times 6,000) = 107$ in^2. Accordingly, we try a tie rod section 11 in. × 11 in. (279 mm × 279 mm) and obtain $A_g = 121$ in^2. Then, assuming the tendon duct has a 2 in. diameter,

$$A_c = A_g - \frac{\pi(2)^2}{4} = 117.86 \text{ in}^2 \ (760.4 \text{ cm}^2)$$

From Equation 8.32, the transformed concrete area is

$$A_t = A_g + (n - 1)A_{ps} = 121 + (6.18 - 1)1.07 = 126.54 \text{ in}^2 \ (816.4 \text{ cm}^2)$$

Check of Forces F, F_{dec}, and F_{cr} (Steps 3 and 4). From Equation 8.47,

$$f'_t = 0 = -\frac{P_e}{A_c} + \frac{F_{dec}}{A_t}$$

or

$$0 = -\frac{160,500}{117.86} + \frac{F_{dec}}{126.54}$$

So the maximum allowable F is

$$\frac{160,500}{117.86} \times 126.54 = 172,320 \text{ lb } (766 \text{ kN})$$

and the actual $F = F_D + F_L = 115{,}000 + 55{,}000 = 170{,}000$ lb $< 172{,}320$ lb, which is satisfactory.

The first cracking load F_{cr} is obtained from Equation 8.46. Assume that the maximum tensile stress in the concrete at the first cracking load is $4\lambda\sqrt{f_c'} = 4 \times 1.0 \times \sqrt{6{,}000} = 310$ psi. Then

$$f_t' = -\frac{P_e}{A_c} + \frac{F_{cr}}{A_t}$$

or

$$310 = -\frac{160{,}500}{117.86} + \frac{F_{cr}}{126.54}$$

$$F_{cr} = 211{,}548 \text{ lb}$$

From before, the available nominal strength is $F_n = 288{,}900$ lb. So the design strength is $F_u = \phi F_n = 0.9 \times 288{,}900 = 260{,}010$ lb $>$ required $F_u = 254{,}500$ lb, which is satisfactory. The cracking SF is $F_u/F_{cr} = 260{,}010/211{,}548 = 1.23$. This is close to 1.0, indicating that once the member is cracked it almost reaches its limit state at failure. Finally, the decompression SF is $F_u/F_{dec} = 260{,}010/172{,}320 = 1.51$ and the actual SF is $\phi F_n/F = 260{,}010/170{,}000 = 1.53$. Thus, the available SF is greater than the required SF $= 1.5$, which is satisfactory.

Deformation Check (Step 5). We know that $P_i = f_{pi}A_{ps} = 190{,}000 \times 1.07 = 203{,}300$ lb (904 kN). From Equation 8.41 a, the initial shortening at prestress transfer is

$$\Delta_i = -\frac{P_i L}{A_c E_c} = -\frac{203{,}300 \times 130 \times 12}{117.86 \times 4.69 \times 10^6} = 0.57 \text{ in. (14.5 mm)}$$

From Equation 8.44, the effective prestress shortening and long-term shortening due to creep and shrinkage is

$$\Delta_e = -\left\{\frac{L}{A_c E_c}\left[P_e + C_u\left(\frac{P_i + P_e}{2}\right)\right] + \epsilon_{SH}L\right\}$$

$$= -\left\{\frac{130 \times 12 \times 10^3}{117.86 \times 4.69 \times 10^6}\left[160.5 + 2.35\left(\frac{203.3 + 160.5}{2}\right)\right]\right.$$

$$\left. + 700 \times 10^{-6}(130 \times 12)\right\}$$

$$= 2.75 \text{ in. (69.9 mm)}$$

From Equation 8.36, the elongation due to external load is

$$\Delta_L = \frac{FL}{A_t E_c} = \frac{170{,}000(130 \times 12)}{126.54 \times 4.69 \times 10^6} = 0.45 \text{ in. (11.4 mm)}$$

which is satisfactory since the allowable Δ is $\frac{1}{2}$ in. So the net shortening is $-2.75 + 0.45 = -2.30$ in. (58.4 mm). Thus, a deformation of this magnitude has to be imposed on the adjoining elements in order to determine whether the resulting stresses are within the allowable limits. Assuming they are, we can adopt the design of an 11 in. × 11 in. tie prestressed with seven $\frac{1}{2}$ in. seven-wire 270 K stress-relieved strands in one tendon.

Sec. 8.16   Design of Linear Tension Members

REFERENCES

8.1 ACI Committee 318. *Building Code Requirements for Reinforced Concrete, ACI Standard 318–89*. Detroit: American Concrete Institute, 1989; and *Commentary on Building Code Requirements for Reinforced Concrete 318R-89*. Detroit: American Concrete Institute, 1989.

8.2 Nawy, E. G., *Reinforced Concrete—A Fundamental Approach*. Englewood Cliffs, N.J.: Prentice Hall, 1985.

8.3 Prestressed Concrete Institute. *PCI Design Handbook* 3d ed. Chicago: Prestressed Concrete Institute, 1985.

8.4 Post-Tensioning Institute. *Post-Tensioning Manual*. 4th ed. Phoenix: Post-Tensioning Institute, 1985.

8.5 Lin, T. Y., and Lakhwara, T. R. "Ultimate Strength of Eccentrically Loaded Partially Prestressed Columns." *PCI Journal* 11 (1986): 37–49.

8.6 Gerwick, B. C., Jr. *Construction of Prestressed Concrete Structures*. New York: Wiley-Interscience, 1971.

8.7 Zia, P., and Moriadith, F. L. "Ultimate Load Capacity of Prestressed Concrete Columns." *Journal of the American Concrete Institute* 63 (1986): 767–788.

8.8 Gerwick, B. C., Jr. "Prestressed Concrete Developments in Japan." *Journal of the Prestressed Concrete Institute* 23 (1978): 66–76.

8.9 Wheen, R. J. "Prestressed Concrete Members in Direct Tension." *Journal of the Structural Division, American Society of Civil Engineers,* 105 (1979): 1471–1487.

8.10 Nawy, E. G. *Simplified Reinforced Concrete*. Englewood Cliffs, N.J.: Prentice Hall, 1986.

8.11 Wilhelm, W. J., and Zia, P. "Effects of Creep and Shrinkage on Prestressed Concrete Columns." *Journal of the Structural Division, American Society of Civil Engineers* 96 (1970): 2103–2123.

PROBLEMS

8.1 Compute the nominal strengths P_n and M_n of the precast prestressed tied concrete non-slender column having the cross section shown in Figure P8.1 and an eccentricity $e = 9$ in. The column is prestressed with six $\frac{1}{2}$ in. dia seven-wire 270 K stress-relieved prestressing strands having the stress-strain properties shown in Figure 8.9. Determine the type of initial failure of the column, and design the size and spacing of the necessary ties. Given data are as follows:

$f'_c = 7,000$ psi (48.6 MPa), normal-weight concrete

$f'_{ci} = 4,900$ psi (33.8 MPa)

$f_{pu} = 270,000$ psi (1,862 MPa)

$f_{py} = 255,000$ psi (1,758 MPa)

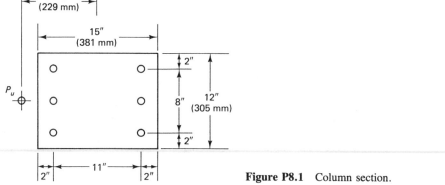

Figure P8.1 Column section.

$f_{pe} = 150,000$ psi (1,034 MPa)

$\epsilon_{cu} = 0.0038$ in./in.

$\epsilon_{ce} = 0.0008$ in./in.

$E_{ps} = 29 \times 10^6$ psi (200×10^3 MPa)

$d' = 2$ in. (50.8 mm)

8.2 Construct the nominal load–moment P_n–M_n and the design P_u–M_u diagrams for the prestressed concrete columns in Example 8.1 if the overall sectional dimensions of the column are 20 in. (508 mm) × 20 in. (508 mm) and the prestressing reinforcement is twelve $\frac{1}{2}$ in. dia seven-wire 270 K stress-relieved strands, half of which are placed at each face parallel to the neutral axis.

8.3 Design a square tied prestressed bonded slender column having a clear height $\ell_u = 20$ ft (6.10 mm) and which is not braced against sidesway. The column is loaded with the same magnitudes of loads and moments and is constructed of the same properties as the materials used in Example 8.2. The design should cover the following two loading conditions:

1. Gravity loading only, assuming that lateral sidesway due to wind is negligible.
2. Sidesway of the force and moment magnitudes of Example 8.2.

Compare the size needed in Problem 8.2 for $\ell_u = 20$ ft to the size of the column in Example 8.2 with $\ell_u = 15$ ft (4.57 m).

8.4 Design the prestressed tension member in Example 8.3 if its length $L = 90$ ft (27.4 m), and compare the section obtained with the section of the 130-ft- (39.6-m)-long tie in the example.

CHAPTER 9

Two-Way Prestressed Concrete Floor Systems

9.1 INTRODUCTION: REVIEW OF METHODS

Supported floor systems are usually constructed of reinforced concrete cast in place. Two-way slabs and plates are those panels in which the dimensional ratio of length to width is less than 2. The analysis and design of framed floor slab systems represented in Figure 9.1 encompasses more than one aspect of such systems. The present state of knowledge permits reasonable evaluation of (1) the moment capacity, (2) the slab–column shear capacity, and (3) serviceability behavior, as determined by deflection control and crack control. Note that flat plates are slabs supported directly on columns without beams, as shown in part (a) of the figure, compared to part (b) for slabs on beams, and part (c) for waffle slab floors.

Essentially the same principles are used in the analysis of continuous two-way prestressed concrete flat plate systems as in the analysis of reinforced concrete plate systems. The techniques of construction differ, however, and it is often unlikely that economic considerations alone can justify using prestressed two-way floor systems of the types shown in Figures 9.1(b) and (c). The prestressing is normally post-tensioned after the two-way plate is cast. Sometimes, on-site precast two-way slabs, called *lift slabs,* are used as a distinct structural system that is fast to construct and perhaps more economical than cast-in-place prestressed two-way slabs. However, the construction technique in lift slabs and the absence of the expertise required for such construction can create hazardous conditions which may result in loss of stability and structural collapse.

The technique for producing lift slabs involves casting a ground-level slab which can double as a casting bed over which all the other floor slabs are cast and stacked,

University of Wyoming, addition. (*Courtesy,* Prestressed Concrete Institute.)

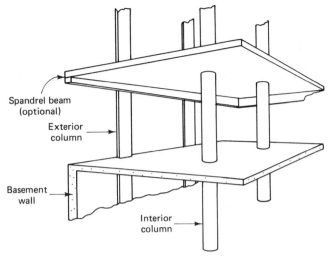

Spandrel beam
(optional)

Exterior
column

Basement
wall

Interior
column

(a)

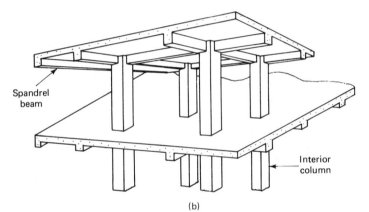

Spandrel
beam

Interior
column

(b)

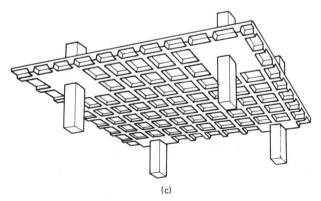

(c)

Figure 9.1 Two-way-action floor systems. (a) Two-way flat-plate floor. (b) Two-way slab floor on beams. (c) Waffle slab floor.

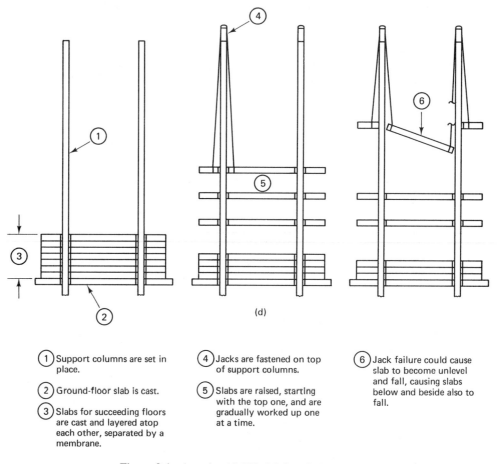

① Support columns are set in place.

② Ground-floor slab is cast.

③ Slabs for succeeding floors are cast and layered atop each other, separated by a membrane.

④ Jacks are fastened on top of support columns.

⑤ Slabs are raised, starting with the top one, and are gradually worked up one at a time.

⑥ Jack failure could cause slab to become unlevel and fall, causing slabs below and beside also to fall.

(d)

Figure 9.1 (*cont.*) (d) Lift-slab installation technique.

separated by a membrane or sprayed parting agent. The columns, which can be steel or concrete, are built prior to the casting of the basic bottom slab extending to the building's height. All the other slabs are cast around the columns, with steel collars having sufficient clearance to permit *lifting* (jacking) of the slab or plate to the appropriate floor level, as shown in Figure 9.1(d). Lifting is accomplished through the use of jacks placed on top of the columns and connected to threaded rods extending down the faces of the columns to the lifting collars embedded in the slabs. Simultaneous activation of all the jacks is essential in order to maintain the slab in a totally horizontal state to avoid imbalance.

The analysis of slab behavior in flexure up to the 1940s and early 1950s followed the classical theory of elasticity, particularly in the United States. The small-deflections theory of plates, assuming the material to be homogeneous and isotropic, formed the basis of ACI Code recommendations with moment coefficient tables. The work, principally by Westergaard, which empirically allowed limited moment redistribution, guided the thinking of the code writers. Hence, the elastic solutions, complicated even for simple shapes and boundary conditions when no computers were available, made it mandatory to idealize, and sometimes render empirical, conditions beyond economic bounds.

In 1943, Johansen presented his yield-line theory for evaluating the collapse capacity of slabs. Since that time, extensive research into the ultimate behavior of reinforced concrete slabs has been undertaken. Studies by many investigators, such as those of Ockleston, Mansfield, Rzhanitsyn, Powell, Wood, Sawczuk, Gamble-Sozen-Siess, and Park, contributed immensely to a further understanding of the limit-state behavior of slabs and plates at failure as well as at serviceable load levels.

The various methods that are used for the analysis and design of two-way action slabs and plates are summarized in the following subsections.

9.1.1 The Semielastic ACI Code Approach

The ACI approach gives two alternatives for the analysis and design of a framed two-way action slab or plate system: the direct design method and the equivalent frame method. Both methods are discussed in more detail in Section 9.3. The equivalent frame method will be used in the design and analysis of prestressed slabs and plates.

9.1.2 The Yield-Line Theory

Whereas the semielastic code approach applies to standard cases and shapes and has an inherent excessively large safety factor with respect to capacity, the yield-line theory is a plastic theory easy to apply to irregular shapes and boundary conditions. Provided that serviceability constraints are applied, Johansen's yield-line theory represents the true behavior of concrete slabs and plates, permitting evaluation of the bending moments from an assumed collapse mechanism which is a function of the type of external load and the shape of the floor panel. This topic will be discussed in more detail in Section 9.14.

9.1.3 The Limit Theory of Plates

The interest in developing a limit solution became necessary due to the possibility of finding a variation in the collapse field which could give a lower failure load. Hence, an upper bound solution requiring a valid mechanism when supplying the work equation was sought, as well as a lower bound solution requiring that the stress field satisfy everywhere the differential equation of equilibrium, that is,

$$\frac{\partial^2 M_x}{\partial x^2} - 2\frac{\partial^2 M_{xy}}{\partial x\,\partial y} + \frac{\partial^2 M_y}{\partial y^2} = -w \tag{9.1}$$

where M_x, M_y, and M_{xy} are the bending moments and w is the unit intensity of load. Variable reinforcement permits the lower bound solution still to be valid. Wood, Park, and other researchers have given more accurate semiexact predictions of the collapse load.

For limit-state solutions, the slab is assumed to be completely rigid until collapse. Further work at Rutgers by Nawy incorporated the deflection effect at high load levels as well as the compressive membrane force effects in predicting the collapse load.

9.1.4 The Strip Method

The strip method was proposed by Hillerborg, in attempting to fit the reinforcement to the strip fields. Since practical considerations require the reinforcement to be placed in orthogonal directions, Hillerborg set twisting moments equal to zero and transformed the slab into intersecting beam strips, hence the name "strip method."

Except for Johansen's yield-line theory, most of the other solutions are lower bound. Johansen's upper bound solution can give the highest collapse load as long as a valid failure mechanism is used in predicting the collapse load.

9.1.5 Summary

The equivalent frame method will be the chief method discussed, because of the limitations of the use of the direct design method in its applicability to two-way prestressed floor systems and the need for more refined determination of the stiffnesses at the column-slab joints in the design process. The yield-line theory for the limit state at failure evaluation of slabs and plates will also be concisely presented.

9.2 FLEXURAL BEHAVIOR OF TWO-WAY SLABS AND PLATES

9.2.1 Two-Way Action

Consider a single rectangular panel supported on all four sides by unyielding supports such as shear walls or stiff beams. We seek to visualize the physical behavior of the panel under gravity load. The panel will deflect in a dishlike form under the external load, and its corners will lift if it is not monolithically cast with the supports. The contours shown in Fig. 9.2(a) indicate that the curvatures and consequently the moments at the central area C are more severe in the shorter direction y with its steep contours than in the longer direction x.

Evaluation of the division of moments in the x and y directions is extremely complex, as the behavior is highly statically indeterminate. The simple case of the panel in part (a) of Figure 9.2 is expounded by taking strips AB and DE at midspan, as in part (b), such that the deflection of both strips at the central point C is the same.

The deflection of a simply supported uniformly loaded beam is $5wl^4/384EI$, i.e., $\Delta = kwl^4$, where k is a constant. If the thickness of the two strips is the same, the deflection of strip AB is $kw_{AB}L^4$ and the deflection of strip DE is $kw_{DE}S^4$, where w_{AB} and w_{DE} are the portions of the total load intensity w transferred to strips AB and DE, respectively, i.e., $w = w_{AB} + w_{DE}$. Equating the deflections of the two strips at the central point C, we get

$$w_{AB} = \frac{wS^4}{L^4 + S^4} \tag{9.2 a}$$

and

$$w_{DE} = \frac{wL^4}{L^4 + S^4} \tag{9.2 b}$$

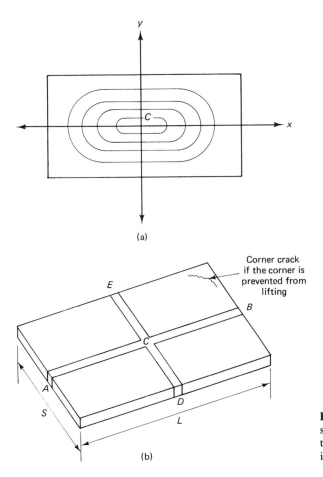

(a)

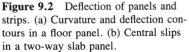

Corner crack
if the corner is
prevented from
lifting

(b)

Figure 9.2 Deflection of panels and strips. (a) Curvature and deflection contours in a floor panel. (b) Central slips in a two-way slab panel.

It is seen from these equations that the shorter span S of strip DE carries the heavier portion of the load. Hence, the shorter span of such a slab panel on unyielding supports is subjected to the larger moment, supporting the foregoing discussion of the steepness of the curvature contours in Figure 9.2(a).

9.2.2 Relative Stiffness Effects

Alternatively, one has to consider a slab panel supported by flexible supports such as beams and columns, or flat plates supported by a grid of columns. In either case, the distribution of moments in the short and long directions is considerably more complex. This complexity arises from the fact that the degree of stiffness of the yielding supports determines the intensity of steepness of the curvature contours in Figure 9.2(a) in both the x and y directions and the redistribution of moments.

The ratio of the stiffness of the beam supports to the slab stiffness can result in curvatures and moments in the long direction larger than those in the short direction, as the total floor behaves as an orthotropic *plate* supported on a grid of columns without beams. If the long span L in such floor systems of slab panels without beams is considerably larger than the short span S, the maximum moment at the center of a plate panel would approximate the moment at the middle of a uniformly loaded strip of span L that is clamped at both ends.

In sum, as slabs are flexible and highly underreinforced, redistribution of moments in both the long and short directions depends on the relative stiffnesses of the supports and the supported panels. Overstress in one region is reduced by such redistribution of moments to the lesser stressed regions.

9.3 THE EQUIVALENT FRAME METHOD

9.3.1 Introduction

The following discussion of the equivalent frame method of analysis for two-way systems summarizes the ACI Code approach to the evaluation and distribution of the total moments in a two-way slab panel. The Code assumes that vertical planes cut through an entire rectangularly planned multistory building along lines AB and CD in Figure 9.3 midway between columns. A rigid frame results in the x direction. Similarly, vertical planes EF and HG result in a rigid frame in the y direction. A solution of such an idealized frame consisting of horizontal beams or equivalent slabs and supporting columns enables the design of the slab as the beam part of the frame. The equivalent frame method thus treats the idealized frame in a manner similar to an actual frame, and hence is more exact and has fewer limitations than the direct design method. Basically, it involves a full moment distribution of many cycles as compared

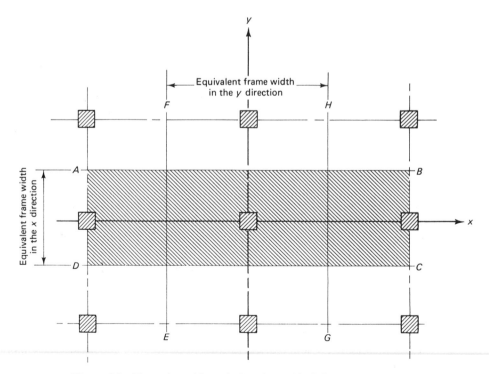

Figure 9.3 Floor plan with equivalent frame (shaded area in x direction).

to the direct design method, which involves a one-cycle-moment distribution approximation.

9.3.2 Limitations of the Direct Design Method

The following are the limitations of the direct design method:

1. There is a minimum of three continuous spans in each direction.
2. The ratio of the longer to the shorter span within a panel should not exceed 2.0.
3. Successive span lengths in each direction should not differ by more than one-third the length of the longer span.
4. Columns may be offset a maximum of 10 percent of the span in the direction of the offset from either axis between the center lines of successive columns.
5. All loads shall be due to gravity only and uniformly distributed over the entire panel. The live load shall not exceed three times the dead load.
6. If the panel is supported by beams on all sides, the relative stiffness of the beams in two perpendicular directions shall not be less than 0.2 or greater than 5.0.

Given these limitations, for prestressed concrete floor slabs, it is necessary to use the equivalent frame method.

9.3.3 Determination of the Statical Moment M_0

There are basically four major steps in the design of the floor panels:

1. Determination of the total statical moment in each of the two perpendicular directions.
2. Distribution of the total moment for the design of sections for negative and positive moment.
3. Distribution of the negative and positive moments to the column and middle strips and to the panel beams, if any. A column strip is a width that is 25 percent of the equivalent frame width on each side of the column center line, and whose middle strip is the balance of the equivalent frame width.
4. Proportioning of the size and distribution of the reinforcement in the two perpendicular directions.

Given the above, correct determination of the values of the distributed moments becomes a principal objective. Consider typical interior panels having center line dimensions l_1 in the direction of the moments being considered and dimensions l_2 in the direction perpendicular to l_1, as shown in Figure 9.4. The clear span l_n extends from face to face of columns, capitals, or walls. Its value should not be less than $0.65l_1$, and circular supports shall be treated as square supports having the same cross-sectional area. The total statical moment of a uniformly loaded simply supported beam as a one-dimensional member is $M_0 = wl^2/8$. In a two-way slab panel as a two-dimensional member, the idealization of the structure through conversion to an

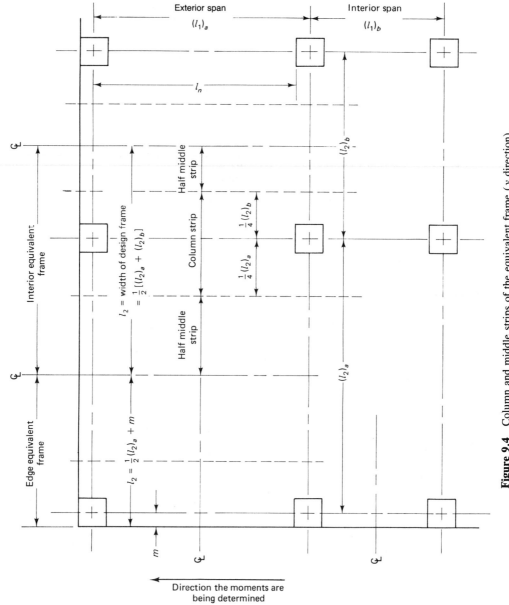

Figure 9.4 Column and middle strips of the equivalent frame (y direction).

Sydney Opera House during construction.

equivalent frame makes it possible to calculate M_0 once in the x direction and again in the orthogonal y direction. If one takes as a free-body diagram the typical interior panel shown in Figure 9.5(a), symmetry reduces the shears and twisting moments to zero along the edges of the cut segment. If no restraint existed at ends A and B, the panel would be considered simply supported in the span l_n direction. If one cuts at midspan, as in Figure 9.5(b), and considers half the panel as a free-body diagram, the moment M_0 at midspan would be

$$M_0 = \frac{wl_2 l_{n1}}{2} \frac{l_{n1}}{2} - \frac{wl_2 l_{n1}}{2} \frac{l_{n1}}{4}$$

or

$$M_0 = \frac{wl_2(l_{n1})^2}{8} \tag{9.3}$$

Due to the existence of restraint at the supports, M_0 in the x direction would be distributed to the supports and midspan such that

$$M_0 = M_C + \tfrac{1}{2}(M_A + M_B) \tag{9.4 a}$$

The distribution would depend on the degree of stiffness of the support. In a similar manner, M_0 in the y direction would be the sum of the moments at midspan and the average of the moments at the supports in that direction.

In the orthogonal direction, Equation 9.4 a becomes

$$M_0' = M_C' + \tfrac{1}{2}(M_A' + M_B') \tag{9.4 b}$$

where M_0', M_A', M_B', and M_C' are at 90 degrees to M_0, M_A, M_B, and M_C, respectively. Also, in an analogous manner to equation 9.3,

$$M_0' = \frac{W \ell_1 (\ell_{n2})^2}{8} \tag{9.5}$$

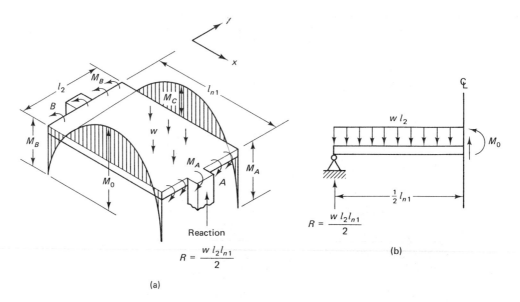

Figure 9.5 Simple moment M_0 acting on an interior two-way slab panel in x direction. (a) Moment on panel. (b) Free-body diagram.

The load intensity W at service load in the prestressed concrete slab would be W_w per unit area.

9.3.4 Equivalent Frame Analysis

The structure, divided into continuous frames as shown in Figure 9.6 for frames in both orthogonal directions, would have the row of columns and a wide continuous beam (slab) ABCDE for gravity loading. Each floor is analyzed separately, whereby the columns are assumed fixed at the floors above and below. To satisfy statical and equilibrium considerations, each equivalent frame must carry the total applied load; alternate span loading has to be used for the worst live-load condition.

It is necessary to account for the rotational resistance of the column at the joint when running a moment relaxation or distribution, except when the columns are so slender as to have very small rigidity compared to the rigidity of the slab at the joint. In such cases, as, for example, in lift slab construction, only a continuous beam is necessary. A schematic illustration of the constituent elements of the equivalent frame is given in Figure 9.7. The slab strips are assumed to be supported by transverse slabs. The column provides a resisting torque M_T equivalent to the applied torsional moment intensity m_t. The exterior ends of the slab strip rotate more than the central section because of torsional deformation. In order to account for this rotation and deformation, the actual column and the transverse slab strip are conceptually replaced by an equivalent column such that the flexibility of the equivalent column is *equal* to the *sum* of the flexibilities of the actual column and the slab strip. This assumption is represented by the equation

$$\frac{1}{K_{ec}} = \frac{1}{\Sigma K_c} + \frac{1}{K_t} \tag{9.6}$$

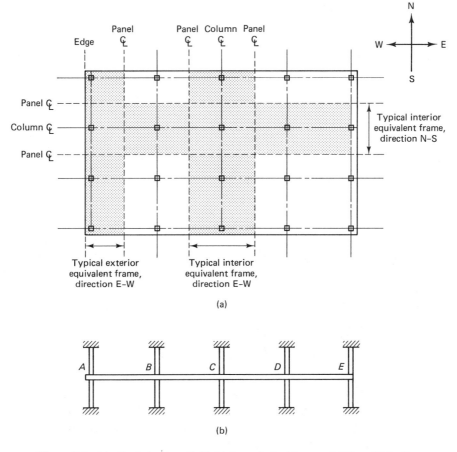

Figure 9.6 Idealized structure divided into equivalent frames. (a) Plan. (b) Section in E–W direction.

where K_{ec} = flexural stiffness of the equivalent column, moment per unit rotation
ΣK_c = sum of flexural stiffnesses of the upper and lower columns at the joint, moment per unit rotation
K_t = torsional stiffness of the torsional beam, moment per unit rotation.

Alternatively, Equation 9.6 can be written as the stiffness equation

$$K_{ec} = \frac{\Sigma K_c}{1 + \dfrac{\Sigma K_c}{K_t}} \tag{9.7}$$

and the column stiffness for an equivalent frame (Ref. 9.9) can be defined as

$$K_c = \frac{EI}{\ell'}\left[1 + 3\left(\frac{L}{L'}\right)^2\right] \tag{9.8}$$

where I is the column moment of inertia, L is the centerline span, and L' is the clear span of the equivalent beam. The carryover factors are approximated by

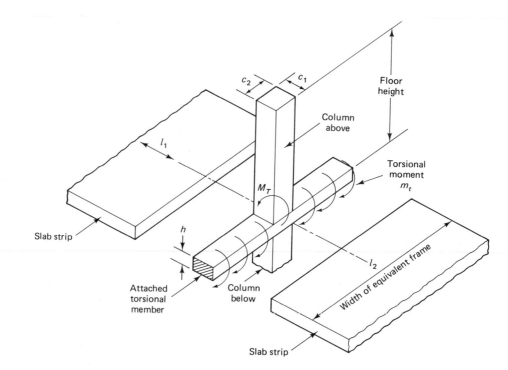

Figure 9.7 Constituent elements of the equivalent frame.

$-\frac{1}{2}(1 + 3h/L)$. An exact computation of the carry-over factor can be made by the column-analogy method using the slab as an analogous column.

A simpler expression for K_c (Ref 9.10) gives results within five percent of the more refined values from Equation 9.8, viz.,

$$K_c = \frac{4EI}{L - 2h} \qquad (9.9)$$

where h is the slab thickness. The torsional stiffness of the slab in the column line is

$$K_t = \sum \frac{9E_{cs}C}{L_2\left(1 - \dfrac{c_2}{L_2}\right)^3} \qquad (9.10\ a)$$

where L_2 = band width
L_n = span
c_2 = column dimensions in the direction parallel to the torsional beam

and the torsional constant is

$$C = \sum(1 - 0.63\ x/y)x^3y/3 \qquad (9.10\ b)$$

in which x = shorter dimension of the rectangular part of the cross-section at the column junction (such as the slab depth)

y = longer dimension of the rectangular part of the cross-section at the column junction (such as the column width).

The slab stiffness is given by the equation

$$K_s = \frac{4E_{cs}I_s}{L_n - c_1/2} \qquad (9.11)$$

As the effective stiffness K_{ec} of the column and the slab stiffness K_s are established, the analysis of the equivalent frame can be performed by any applicable methods, such as relaxation or moment distribution.

The distribution factor for the fixed-end moment *FEM* is

$$DF = \frac{K_s}{\Sigma K} \qquad (9.12)$$

where $\Sigma K = K_{ec} + K_{s(left)} + K_{s(right)}$. A carryover factor of COF $\cong \frac{1}{2}$ can be used without loss of accuracy since the nonprismatic section causes only very small effects on fixed-end moments and carryover factors. The fixed-end moment *FEM* for a uniformly distributed load is $w\ell_2(\ell_n)^2/12$ at the supports, such that after moment redistribution the sum of the negative distributed moment at the support and the midspan moment is always equal to the static moment $M_0 = w\ell_2(\ell_n)^2/8$.

9.3.5 Pattern Loading of Spans

Loading all spans simultaneously does not necessarily produce the maximum positive and negative flexural stresses. Consequently, it is advisable to analyze the multispan frame using alternative span loading patterns for the *live* load. For a three-span frame, the suggested patterns for the live load are shown in Figure 9.8. The ACI Code, however, permits the full factored live load to be used on the entire slab system if the live load is less than 75 percent of the dead load.

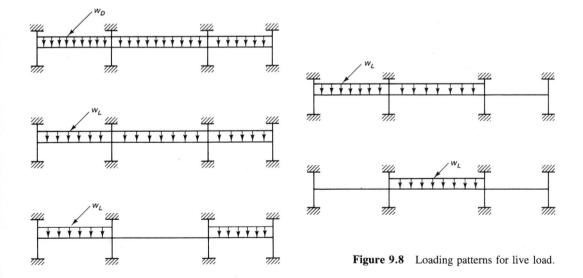

Figure 9.8 Loading patterns for live load.

9.4 TWO-DIRECTIONAL LOAD BALANCING

As mentioned in Chapter 1, load balancing represents the forces counteracting the external gravity load. These forces are produced by the transverse component of the longitudinal prestressing force in a parabolic or harped tendon. The load w in Equations 9.3–9.5 represents the *downward* external transverse load intensity, which can be either the working load intensity w_w or the factored load intensity w_u. The *upward* load intensity in the slab field due to the transverse component of the prestressing force, also mentioned in Chapter 1, would reduce the effect of w_w and can be chosen so as to *exactly balance* a particular downward load intensity. Under such a condition, the two-way slab would be subjected to neither bending nor twisting moments, and the analysis would be considerably simplified.

Two-directional load balancing in two-way slabs is different from one-directional load balancing in beams. The balancing load produced by the tendons in one direction either increases or decreases the balancing load produced by the tendons in the perpendicular direction. Hence, the prestressing forces and tendon profiles in the two orthogonal directions are totally interrelated, fully maintaining the basic principles of statics. The ultimate benefit of load balancing is in the creation of a design of a structural prestressed floor such that the upward component of the prestressing force results in a load intensity distribution in each direction that is equivalent to the downward external load intensity. Such a design is called a pure *balanced design*. Any deviation from the balanced condition should be analyzed as a load acting on the slab without being affected by the transverse upward component of prestress.

If a two-way slab on rigid supports such as walls is prestressed in both orthogonal directions having spans L_S and L_L in the short and long directions, respectively, as shown in Figure 9.9, the intensity of the upward balancing load required to produce balanced design loads is given, from Equation 1.15 a, by the equations

$$W_{bal(S)} = \frac{8P_S e_S}{L_S^2} \qquad (9.13\ a)$$

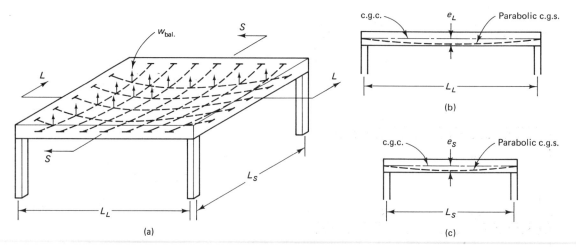

Figure 9.9 Balancing loads in two-way prestressed panel. (a) Three-dimensional view. (b) Section L–L in the long direction. (c) Section S–S in the short direction.

and

$$W_{bal(L)} = \frac{8P_L e_L}{L_L^2}$$ (9.13 b)

where P_S and P_L are the effective prestressing forces after losses in the short and long directions L_S and L_L, respectively, *per unit width of slab,* and e_S and e_L are the corresponding maximum eccentricities of the prestressing tendons. The total balancing load per unit width then becomes

$$W_{bal} = W_{bal(S)} + W_{bal(L)} = \frac{8P_S e_S}{L_S^2} + \frac{8P_L e_L}{L_L^2}$$ (9.14)

The designer would select the level of W_{bal} and determine the values of the prestressing forces P_S and P_L accordingly. Many combinations of P_S and P_L can satisfy the statics equation 9.5. If the slab panels were supported on beams, or if simple panels were supported on a wall, the most economical design would be to carry the load W only in the short direction, or to carry $\frac{1}{2}W$ in each direction in the case of a square slab panel. The slab panel loaded by W_{bal} and stressed by prestressing forces P_S and P_L would be subjected to a uniform stress distribution P_S/h and P_L/h in the respective directions, h being the slab thickness. The slab panel would be totally level, with no deflection or cambers. Any deviation of the applied load from W_{bal} would require the use of the usual elastic theory for the analysis of two-way plates.

As prestressed post-tensioned two-way slabs are usually flat plates supported directly on columns, all of the load has to be carried in both directions using either uniformly distributed prestressing tendons or banded tendons, with a concentration of tendons at the columns strips of the two-way plate panels.

Uniform stress distribution and zero deflection/camber are not mandatory for proportioning the floor system. If they were, load balancing would not necessarily be the most economical way of determining the prestressing forces. Instead, the designer would often use a partial balancing load $W_{bal} < W_D + W_L$ for a multipanel floor system, as will be seen in Example 9.2. If the load intensity $W_w = W_D + W_L$ is larger than the balanced load W_{bal} from Equation 9.5, then unit moments M_S and M_L will result in the directions S and L, respectively.

The unit stress in the concrete in the short and long directions due to the unbalanced loading is obtained by superimposing the uniform compression due to balanced loading on the flexural stress in the concrete caused by the bending moments M_S and M_L that result from the unbalanced load $W_w - W_{bal}$. The resulting concrete stresses at the top and bottom fibers in each direction are given as follows:

Short direction

$$f^t = -\frac{P_S}{bh} - \frac{M_S c}{I_s}$$ (9.15 a)

$$f_b = -\frac{P_S}{bh} + \frac{M_S c}{I_s}$$ (9.15 b)

Long direction

$$f^t = -\frac{P_L}{bh} - \frac{M_L c}{I_L}$$ (9.16 a)

Kishwaukee River segmental bridge during construction, Winnebago County, Illinois. (*Courtesy,* H. Wilden & Associates, Macungie, Pennsylvania and the Prestressed Concrete Institute.)

$$f_b = -\frac{P_L}{bh} + \frac{M_L c}{I_L} \qquad (9.16\ b)$$

In these equations, upscript t represents the top of the slab, subscript b represents the bottom of the slab, $c = \frac{1}{2}h$, width $b = 12$ in., and

$$P_s = \frac{\text{total } P_S}{L}$$

and

$$P_L = \frac{\text{total } P_L}{S}$$

are the unit prestress forces. The service-load moment coefficient for evaluating M_S and M_L can be obtained from the chart in Figure 9.10 for any boundary condition (Ref. 9.13–CP 114). The bending moment coefficients there are for the maximum positive and negative bending moments, where βx_2 and $\beta x_2'$ apply to $+M$ and $-M$, respectively, on the short span L_x. Similarly, βy_2 and $\beta y_2'$ apply to the maximum positive and negative bending moments, respectively, on the long span L_y. In an analogous fashion, the charts in Figure 9.11 give a rapid method for the evaluation of the ultimate bending moment coefficients (Ref. 9.13–CP 110) in continuous two-way-action concrete plates.

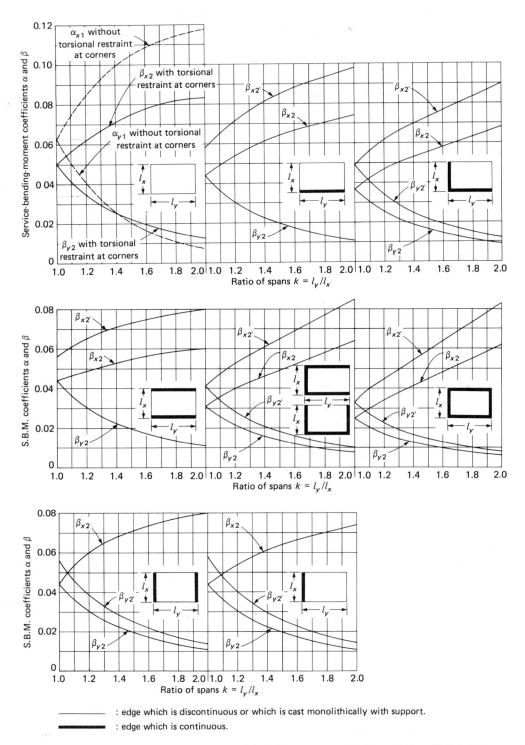

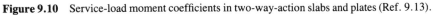

——————— : edge which is discontinuous or which is cast monolithically with support.

▬▬▬▬▬ : edge which is continuous.

Figure 9.10 Service-load moment coefficients in two-way-action slabs and plates (Ref. 9.13).

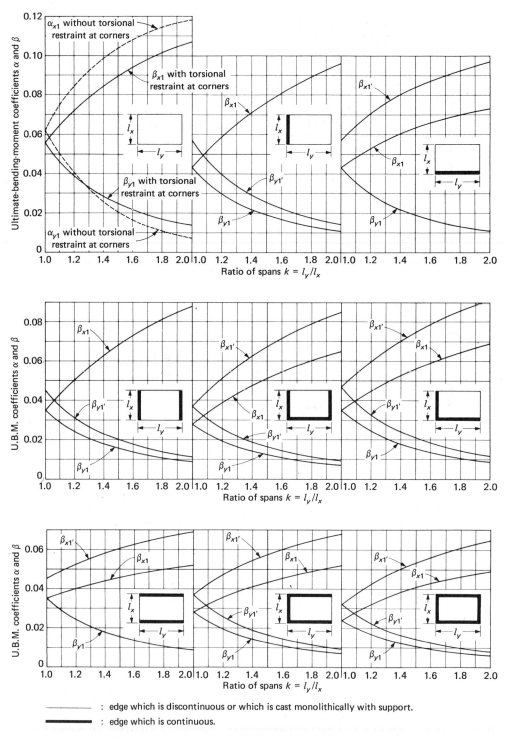

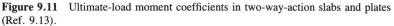

——————— : edge which is discontinuous or which is cast monolithically with support.

▬▬▬▬▬ : edge which is continuous.

Figure 9.11 Ultimate-load moment coefficients in two-way-action slabs and plates (Ref. 9.13).

9.5 FLEXURAL STRENGTH OF PRESTRESSED PLATES

9.5.1 Design Moments M_u

The design moments for statically indeterminate prestressed bonded members are determined by combining the frame distributed moments M_u due to the *factored* dead plus live loads with the secondary moments M_s induced into the frame by the tendons. The load balancing approach discussed in the previous section directly includes both the primary moments M_1 and the secondary moments M_s. Hence, for service-load intensity values, only the net loads W_{net} need to be considered in the calculation of the fixed-end factored moments, while W_{bal} needs to be considered for flexural strength analysis.

Fixed-end M_u for moment distribution

If $M_1 = P_e e = Fe$ is the primary moment, M_{bal} is the balanced moment due to W_{bal}, M_S = distributed $M_{bal} - M_1$ is the secondary moment, and $\overline{M}_u$ is the fixed-end factored moment due to the factored load intensity W_u, then the design ultimate moment would be at least

$$\text{Design } M_u = \text{distributed } \overline{M}_u - M_S \qquad (9.17\ a)$$

and the available moment strength would be

$$M_n = \frac{M_u}{\phi} \qquad (9.17\ b)$$

Inelastic redistribution of moments due to continuity would be applied to the available moment strength M_n at the support towards the required M_n at midspan.

Where bonded reinforcement is provided at the supports with minimum non-prestressed steel provided in accordance with Equations 9.19 and 9.20, negative moments calculated by the elastic theory for any assumed loading arrangement may

Bridge deck prestressing reinforcement.

be increased or decreased by not more than the percentage given by the inelastic moment redistribution factor

$$\rho_D = 20\left[1 - \frac{\omega_p + (d/d_p)(\omega - \omega')}{0.36\beta_1}\right] \text{ percent} \qquad (9.18)$$

The modified negative moment should be used for calculating moments at sections within spans, viz., the positive moments, for the same loading arrangement. Inelastic moment redistribution of the negative moments can be made only when the section at which the moment is reduced is so designed that ω_p or $\omega_p + (d/d_p)(\omega - \omega')$ is not greater than $0.24\beta_1$.

Example 9.2 below illustrates in detail the equivalent frame analysis procedure both for service-load and ultimate-load conditions, and the inelastic moment redistribution due to continuity that is utilized in the strength analysis.

9.6 BENDING OF PRESTRESSING TENDONS AND LIMITING CONCRETE STRESSES

9.6.1 Distribution of Prestressing Tendons

Each plate panel is assumed to be supported continuously along the transverse column center lines. The assumption is also made, as previously stated, that the slab panel behaves as an orthogonal panel of two wide beams whose width is equal to the panel width and which is supported along the column centerlines. Consequently, 100 percent of the load to be balanced is considered to be supported by the wide beam in *each* of the two perpendicular directions.

It is also known that the lateral distribution of moments is *not uniform* across the width of the panel, but tends to be more concentrated in the column strips. Consequently, it is not unreasonable to concentrate a large percentage of the tendons in the column strip, as defined in Figure 9.4, and to spread the remaining tendons in the middle strip. For continuous spans, 65–75 percent of the moment in each direction is carried by the respective column strip, while the total area and number of tendons required by the total applied moment are maintained.

The width of half the column strip on either side of the column is one-fourth of the *smaller* of the two panel dimensions. The middle strip is the slab band bound by the two column strips. Accordingly, the distribution or banding of the prestressing tendons follows the percentage distribution of the moment between the column and middle strips. Consequently, if 70 percent of the tendons are concentrated in the column strip, it is reasonable to expect that the column strip will carry *essentially* 70 percent, and the middle strip will carry the remaining 30 percent, of the total moment.

Since tendons exert downward loads at the high points which join adjacent parabolic profiles, the accompanying downward reactions should, as nearly as practicable, be resisted by columns, walls and/or upward tendon loads to achieve minimum deflections and maximum shear capacity. Consequently, a logical statical distribution of tendons can be one in which all tendons in one direction are placed through or immediately adjacent to the columns, and the tendons in the perpendicular direction are spaced uniformly across the bay width.

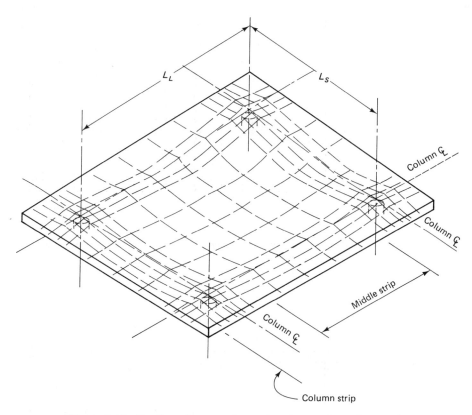

Figure 9.12 Bending of prestressing tendons in two-way-action panels.

Figure 9.12 shows the resulting distribution of prestressing tendons in the two orthogonal directions. As a general guideline, the recommended spacing of the tendons in the column strip is about three to four times the slab thickness, while the maximum spacing of the tendons in the middle strip should not exceed six times the slab thickness. An average compressive stress in the concrete in each direction should be at least about 125 psi (0.90 MPa).

Investigations have verified (Ref. 9.4), through testing of the four-panel prestressed plates shown in Figure 9.13, that variation of distribution of prestressing strands does not alter the deflection behavior or magnitude and capacity for the *same* total prestressing steel percentage. Banding of the prestressing tendons as shown in part (b) of the figure, with about 65–75 percent of the tendons in the column strip, seems to be the most effective, particularly in enhancing the shear-moment transfer capacity at the column support section of the two-way slab.

9.6.2 Limiting Concrete Tensile Stresses at Service Load (f_t, psi)

9.6.2.1 Flexure. The ACI 318 Code limits the tensile stresses in the concrete for prestressed elements in order to control flexural cracking development. The following tabulated values give the maximum permissible tensile stress in the prestressed elements for the various moment regions.

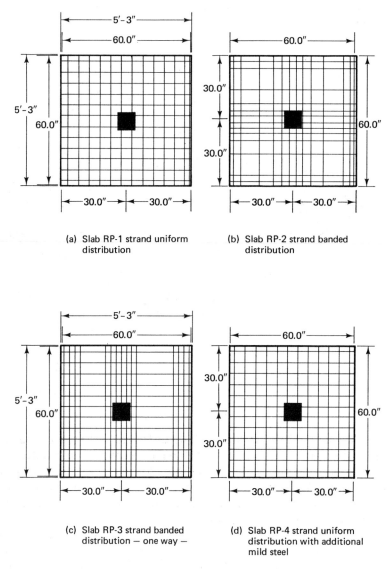

(a) Slab RP-1 strand uniform distribution

(b) Slab RP-2 strand banded distribution

(c) Slab RP-3 strand banded distribution — one way —

(d) Slab RP-4 strand uniform distribution with additional mild steel

Figure 9.13 Various alternatives of tendon distribution (Ref. 9.4).

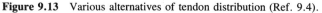

1. Negative moment area with the addition of nonprestressed reinforcement $\qquad$ $6\sqrt{f_c'}$

2. Negative moment area without the addition of nonprestressed reinforcement $\qquad$ 0

3. Positive moment area with the addition of nonprestressed reinforcement $\qquad$ $2\sqrt{f_c'}$

4. Positive moment area without the addition of nonprestressed reinforcement $\qquad$ 0

5. Compressive stress in the concrete $\qquad$ $f_c = 0.45 f_c'$

9.6.2.2 Reinforcement. The minimum area of bonded reinforcement, except as required by Equation 9.20 below, is

$$A_s = 0.004A \qquad (9.19\ a)$$

where A is the area in square inches of that part of the cross section between the flexural tension face and the center of gravity of the gross section. In positive-moment areas where the computed tensile stress in the concrete at service load exceeds $2\sqrt{f'_c}$, the minimum area of bonded reinforcement has to be computed from

$$A_s = \frac{N_c}{0.5f_y} \qquad (9.19\ b)$$

where N_c is the tensile force in the concrete due to the unfactored dead plus live load, and $f_y = 60,000$ psi. In negative-moment areas at column supports, the minimum area of bonded reinforcement in each direction has to be determined from

$$A_s = 0.00075hL \qquad (9.20)$$

where L = span length in the direction parallel to the reinforcement being determined
h = slab thickness.

The reinforcement obtained from Equation 9.20 has to be distributed within a slab band width between the lines that are $1.5h$ outside opposite faces of the column support. At least four bars or wires should be provided in both directions.

The minimum length of bonded reinforcement in the *positive* area should be *one-third* the clear span, centered in the positive-moment area. The minimum length of bonded reinforcement in the *negative* area should extend *one-sixth* the clear span on each side of the support, placed at the *top* fibers. The stress f_{ps} in the prestressing reinforcement at nominal strength, as required by the ACI 318 Code, is governed by the following requirements.

Bonded Tendons. For bonded tendons,

$$f_{ps} = f_{pu}\left(1 - \frac{\gamma_p}{\beta_1}\left[\rho_p \frac{f_{pu}}{f'_c} + \frac{d}{d_p}(\omega - \omega')\right]\right) \qquad (9.21)$$

where $\omega' = \rho' = \dfrac{f_y}{f'_c}$
and $\gamma_p = 0.40$ for $f_{py}/f_{pu} \geq 0.85$
$= 0.28$ *for* $f_{py}/f_{pu} \geq 0.90$.

If any compression reinforcement is considered, the term $[\rho_p f_{pu}/f'_c + (d/d_p)(\omega - \omega')]$ in Equation 9.21 shall be taken not less than 0.17, and d' shall be no greater than $0.15d_p$.

Nonbonded Tendons. For nonbonded tendons with a slab span-to-depth ratio ≤ 35,

$$f_{ps} = f_{se} + 10,000 + \frac{f'_c}{100\rho_p} \qquad (9.22)$$

where $f_{ps} \leq f_{py} \leq f_{pe} + 60,000$.

For nonbonded tendons with a slab span-to-depth ratio > 35,

$$f_{ps} = f_{pe} + 10,000 + \frac{f'_c}{300\rho_p} \qquad (9.23)$$

where $f_{ps} \leq f_{py} \leq f_{pe} + 30,000$.

9.6.2.3 Shear

Column Support Sections in Flat Plates. The nominal shear strength provided by the concrete at the column junctions of two-way prestressed slabs is given by

$$V_c = (\beta_p \sqrt{f'_c} + 0.3\bar{f}_c)b_o d + V_p \qquad (9.24\ a)$$

or the nominal unit shearing strength is

$$v_c = \beta_p \sqrt{f'_c} + 0.3\bar{f}_c + \frac{V_p}{b_o d} \qquad (9.24\ b)$$

where b_o = perimeter of the critical shear section at distance $d/2$ from face of support

$\bar{f}_c$ = average value of the effective compressive stress in the concrete due to externally applied load for the two orthogonal directions computed at the section centroid after all prestress losses (termed f_{pc} by the ACI Code)

V_p = vertical component of all effective prestressing forces crossing the critical section

β_p = the smaller of the two values of 3.5 or $(\alpha_s d/b_o + 1.5)$, where α_s is 40 for interior columns, 30 for edge columns, and 20 for corner columns

In slabs with distributed tendons, the term V_p can be conservatively disregarded; otherwise it becomes necessary to use the actual reverse curvature tendon geometry in the calculations in order to assess the shear carried by the tendons crossing the critical section. According to the ACI 318 Code, no portion of the column cross section shall be closer to a discontinuous edge than four times the slab thickness, f'_c in Equations 9.24 shall not exceed 5,000 psi, and $\bar{f}_c$ in each direction shall be neither less than 125 psi nor greater than 500 psi.

If these requirements are not satisfied, V_c shall be computed by the smaller of the values obtained from the expressions

$$\text{(i)} \quad V_c = \left(2 + \frac{4}{\beta_c}\right)\sqrt{f'_c}\,b_o d \qquad (9.25\ a)$$

$$\text{(ii)} \quad V_c = \left(\frac{\alpha_s d}{b_o} + 2\right)\sqrt{f'_c}\,b_o d \qquad (9.25\ b)$$

$$\text{(iii)} \quad V_c = 4\sqrt{f'_c}\,b_o d \qquad (9.25\ c)$$

where β_c = ratio of long to short side of the column or concentrated load area.

Continuous Edge Support. For distributed load and continuous edge support such as supporting beams or wall support, if the effective prestress is not less than 40 percent of the tensile strength of the reinforcement, the maximum allowable shear stress is

$$V_c = \left[0.60\sqrt{f_c'} + 700\frac{V_u d}{M_u} \right] b_w d_p \geq 2\sqrt{f_c'}\, b_w d \qquad (9.26)$$

$$< 5\sqrt{f_c'}\, b_w d$$

where b_w is taken as the strip width and $V_u d/M_u$ at a distance $d_p/2$ from the face of the support. $d_p \geq 0.80\, h$.

Values of $\sqrt{f_c'}$ in all the above equations are to be multiplied by a factor $\lambda = 1.0$ for normal-weight concrete, $\lambda = 0.85$ for sand-lightweight concrete, and $\lambda = 0.75$ for all-lightweight concrete.

Shear Force Coefficients. The maximum shear force at the edge of a two-way slab panel carrying uniformly distributed load and supported along the length of its perimeters can be approximated as follows

$$V = \tfrac{1}{3}wL_s \text{ (short edge)} \qquad (9.27\ a)$$

$$V = kwL_s/(2k + 1) \text{ (long edge)} \qquad (9.27\ b)$$

where k is the ratio of the long span L_L to the short span L_S. The same values are applicable to a panel that is fixed or continuous along all four edges. For other conditions, the distribution of shearing forces, the stresses due to which are rarely critical, have to be adjusted on the basis that the shearing force is slightly larger at a continuous edge than at a simply supported edge.

The ACI Code allows a 15 percent increase in the shear force at the first interior continuous support for one-way action.

9.7 LOAD-BALANCING DESIGN OF A SINGLE-PANEL TWO-WAY FLOOR SLAB

Example 9.1

A two-way single-panel prestressed warehouse lift slab 20 ft × 24 ft (6.10 m × 7.32 m) has the plan shown in Figure 9.14. It is supported on masonry walls on all sides, with negligible rotational restraint at these boundaries. The slab has to carry a superimposed service dead load of 15 psf (0.72 kPa) in addition to its self-weight and a service live load of 75 psf (3.59 kPa). No deflection is allowed under full dead load.

Design the slab as a post-tensioned nonbonded prestressed two-way floor using $\frac{1}{2}$ in. dia seven-wire 270 K stress-relieved strands (12.7 mm dia strands). Given data are as follows:

$f_c' = 5,000$ psi (34.47 MPa), normal weight

$f_{ci}' = 3,750$ psi (25.86 MPa)

E–W max. f_c due to net prestress after losses = 200 psi

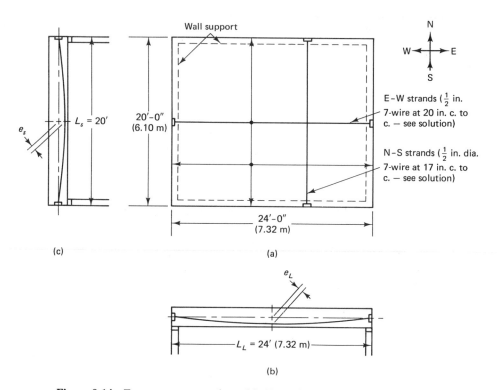

Figure 9.14 Two-way prestressed panel in Example 9.1. (a) Plan. (b) Section in longitudinal E–W direction. (c) Section in transverse N–S direction.

N–S max. f_c due to net prestress after losses not to exceed 350 psi (ACI allows up to 500 psi)

Max. f_c due to combined stresses $= 0.45f_c'$

$E_c = 57,000\sqrt{f_c} = 4.03 \times 10^6$ psi (27.8 × 10⁶ MPa)

$f_{ps} \leq 0.70f_{pu} = 189,000$ psi (1,303 MPa), as required by the ACI Code

$f_{py} = 240,000$ psi (1,655 MPa)

$f_{pe} = 159,000$ psi (1,096 MPa)

$E_{ps} = 29 \times 10^6$ psi (200 × 10³ MPa)

$f_y = 60,000$ psi (413.7 MPa)

$E_s = 29 \times 10^6$ psi (200 × 10³ MPa)

Solution

$$e_S = e_L = \frac{6}{2} - 1 = 2.0 \text{ in. (51 mm)}$$

Choose a trial slab thickness on the basis of a span-to-depth ratio $\cong 45$:

$$h = \frac{(20 + 24) \times 12}{2} \times \frac{1}{45} = 5.87 \text{ in.}$$

So try a 6-in. (152-mm)-thick slab assuming a duct diameter $\cong 0.5$ in. and an effective depth $d = 6.0 - (0.5/2 + \frac{3}{4}) = 5.0$ in. (127 mm).

Balancing Load

$$W_D = 15 \text{ psf} + \frac{6}{12} \times 150 = 90 \text{ psf} \ (4.31 \text{ kPa})$$

Since a balancing load is required for zero deflection or camber due to dead load, assume that $W_{bal} = W_D = 90 \text{ psf} \ (4.31 \text{ kPa})$. Also, since f_c due to prestressing $= 200 \text{ psi}$ (given), assume this to be the stress in the E–W direction. Then the effective prestressing force in the E–W direction is $P_L = 200 \times 6 \times 12 = 14{,}400 \text{ lb per strip}$, and from Equation 9.13 b,

$$W_{bal(L)} = \frac{8 P_L e_L}{L_L^2} = \frac{8 \times 14{,}400 \times 2}{(24)^2 \times 12} \cong 33 \text{ psf} \ (1.58 \text{ kPa})$$

The uplift to be provided by the tendons in the short direction (preferably the load is to be carried by the short-direction span) becomes $W_{bal(S)} = W_D - W_{bal(L)} = 90 - 33 = 57 \text{ psf} \ (2.73 \text{ kPa})$. Then, from Equation 9.13 a,

$$P_S = \frac{W_{bal(S)} L_S^2}{8 e_S} = \frac{57 \times (20)^2 \times 12}{8 \times 2}$$

$= 17{,}100 \text{ lb/ft} \ (249.7 \text{ kN/m})$ after losses, and the compression in the concrete after losses in prestress in the N–S direction is

$$f_c = \frac{P_S}{bh} = \frac{17{,}100}{12 \times 6} = 238 \text{ psi} < 350 \text{ psi}$$

which is satisfactory. Hence, use $\frac{1}{2}$ in. dia seven-wire 270 K stress-relieved strands with effective prestressing force $P_e = 159{,}000 \times 0.153 = 24{,}327 \text{ lb} \ (108.2 \text{ kN})$.

Required Spacing in the N–S Direction. The required spacing in the N–S direction is

$$s_S = \frac{24{,}327}{17{,}100} = 1.42 \text{ ft} = 17 \text{ in.} \ (432 \text{ mm})$$

Required Spacing in the E–W Direction. The required spacing in the E–W direction is

$$S_L = \frac{24{,}327}{14{,}400} = 1.69 \text{ ft} \cong 20 \text{ in.} \ (508 \text{ mm})$$

Note that both spacings correspond to the recommended spacing of 3 to 5 times the slab thickness. As an additional measure, to prevent splitting of the concrete in the anchorage zones at the wall, add two #4 nonprestressed mild steel bars (12.7 mm dia) along the anchorage line on the slab perimeter.

Service-load Stresses. The service live load $W_L = 75 \text{ psf} \ (3.59 \text{ kPa})$, and the aspect ratio $k = L_L/L_S = 24/20 = 1.20$. From Figure 9.10, the moment coefficients for the maximum midspan moments in the short and long directions are $\alpha_{N-S} = 0.062$ and $\alpha_{E-W} = 0.035$, respectively, assuming that the corners of the two-way slab are held down, i.e., torsionally restrained.

We assume an effective $L_S = 19.5 \text{ ft}$ and $L_L = 23.5 \text{ ft}$.

Live-load Moments. The live-load moments are

$$M_S = 0.062 \times 75 \times (19.5)^2 \times 12 = 21{,}218 \text{ in.-lb/ft}$$

and

$$M_L = 0.035 \times 75 \times (23.5)^2 \times 12 = 17,396 \text{ in.-lb/ft}$$

The moment of inertia is

$$I_s = \frac{12(6)^3}{12} = 216 \text{ in}^4$$

Concrete Stresses Due to Live Load. In the short direction, we have

$$f = \frac{Mc}{I_s} = \frac{21,218 \times 3}{216} = 295 \text{ psi (2.03 MPa)}$$

while in the long direction,

$$f = \frac{17,396 \times 3}{216} = 242 \text{ psi (1.67 MPa)}$$

The combined axial stresses due to the balanced load and the flexural stresses due to the live load, from Equations 9.15 and 9.16, become, in the short direction (N-S),

$$f^t = -\frac{P_s}{bh} - \frac{M_S c}{I_s} = -238 - 295 = -533 \text{ psi } (C) \text{ (3.68 MPa)}$$

and

$$f_b = -238 + 295 = +57 \text{ psi } (T) \text{ (very small and, hence, negligible)}$$

and in the long direction (E-W),

$$f^t = -200 - 242 = -442 \text{ psi } (C) \text{ (3.05 MPa)}$$

and

$$f_b = -200 + 242 = +42 \text{ psi } (T) \text{ (again negligible)}$$

The compressive stresses are $f_c = 0.45 \times 5,000 = 2,250$ psi, which is well below the ACI allowable stresses and, hence, satisfactory. These low stress levels can justify making the slab thinner than 6 in., provided that the deflection due to live load is acceptable. Note that the slab develops zero deflection or camber under dead load in this example, due to the load balancing.

Deflection Check. Only the deflection due to live load is checked. From principles of mechanics, we have

$$\Delta = \frac{5}{48} \frac{ML^2}{E_c I_s}$$

$$I_s = 216 \text{ in}^4$$

$$E_c = 4.03 \times 10^6 \text{ psi}$$

$$\Delta_{\text{E-W}} = \frac{5}{48} \frac{17,396 (24 \times 12)^2}{4.03 \times 10^6 \times 216} = 0.17 \text{ in.}$$

$$\Delta_{\text{N-S}} = \frac{5}{48} \frac{21,218 (20 \times 12)^2}{4.03 \times 10^6 \times 216} = 0.15 \text{ in.}$$

Avg. midspan deflection $\Delta = \dfrac{0.17 + 0.15}{2} = 0.16$ in. (4.1 mm)

$$\text{Acceptable deflection} = \frac{L_s}{360} = \frac{20 \times 12}{360} = 0.67 \text{ in. } (17 \text{ mm}) \gg 0.16 \text{ in.}$$

Consequently, a second cycle reducing the slab thickness to $5\frac{1}{2}$ in. could also be used, provided that the resulting slab nominal moment strength were adequate to carry the load. In this case, $h = 5\frac{1}{2}$ in. is not adequate, as will be shown in the next part of the solution.

Nominal Moment Strength

$$W_u = 1.4 \times 90 + 1.7 \times 75 = 254 \text{ psf } (12.2 \text{ kPa})$$

Also,

$$\text{Effective } L_S = 19.5 \text{ ft}$$

$$L_L = 23.5 \text{ ft}$$

From Figure 9.11, the moment coefficients for maximum factored moment are

$$\alpha_{\text{N-S}} = 0.072$$

and

$$\alpha_{\text{E-W}} = 0.038$$

N-S Direction. In the N-S direction, we have

$$\text{Factored } M_u = 0.072 \times 254(19.5)^2 \times 12 = 83{,}448 \text{ in.-lb/ft}$$

$$\text{Required } M_n = \frac{M_u}{\phi} = \frac{83{,}448}{0.9} = 92{,}720 \text{ in.-lb/ft}$$

Note that prestressing forces in this structure do not produce any secondary moments M_s since there is no continuity at the slab boundaries. We have $A_{ps} = 0.153$ in^2 on 1.42 ft center to center (from before), and $A_{ps}/\text{ft} = 0.153/1.42 = 0.11$ in^2/ft. Also, the effective $f_{pe} = 159{,}000$ psi. So in cases where the A_{ps} used exceeds $P_e/$ initial A_{ps}, reduce f_{pe} accordingly.

We have, further,

$$\rho_{\text{N-S}} = \frac{0.11}{12 \times 5} = 0.0018$$

$$\text{Slab span-to-depth ratio} = \frac{20 \times 12}{6} = 40$$

From Equation 9.23 b,

$$f_{ps} = f_{pe} + 10{,}000 + \frac{f'_c}{300\rho_p} \leq f_{py} \leq f_{pe} + 30{,}000$$

$$f_{ps} = 159{,}000 + 10{,}000 + \frac{5{,}000}{300 \times 0.0018} = 178{,}259 \text{ psi}$$

$$< f_{py} = 240{,}000 \text{ psi} < f_{pe} + 30{,}000 = 189{,}000 \text{ psi}$$

$$< \text{limit } f_{ps} = 189{,}000 \text{ psi, O.K.}$$

$$a = \frac{A_{ps}f_{ps}}{0.85f'_c b} = \frac{0.11 \times 178{,}259}{0.85 \times 5{,}000 \times 12} = 0.38 \text{ in.}$$

$$\text{Available } M_n = A_{ps}f_{ps}\left(d - \frac{a}{2}\right) = 0.11 \times 178,259\left(5 - \frac{0.38}{2}\right)$$

$$= 94,316 \text{ in.-lb/ft} > \text{req } M_n = 92,720 \text{ in.-lb, O.K.}$$

E-W Direction. In the E-W direction, we have

$$\text{Factored } M_u = 0.038 \times 254(235)^2 \times 12 = 63,964 \text{ in.-lb/ft}$$

$$\text{Required } M_n = \frac{M_u}{\phi} = \frac{63,964}{0.90} = 71,071 \text{ in.-lb/ft}$$

$$A_{ps} = 0.153 \text{ in}^2 \text{ per 1.69 ft center to center (from before)}$$

$$A_{ps}/\text{ft} = \frac{0.153}{1.69} = 0.09 \text{ in}^2/\text{ft}$$

$$\rho_{\text{E-W}} = \frac{0.09}{12 \times 5} = 0.0015$$

$$f_{ps} = 159,000 + 10,000 + \frac{5,000}{300 \times 0.0015} = 180,111 \text{ psi, O.K.}$$

$$a = \frac{0.09 \times 180,111}{0.85 \times 5,000 \times 12} = 0.32 \text{ in.}$$

$$\text{Available } M_n = 0.09 \times 180,111\left(5 - \frac{0.32}{2}\right)$$

$$= 78,456 \text{ in.-lb/ft} > \text{Req. } M_n \text{ of } 71,071 \text{ in.-lb/ft, O.K.}$$

$$(29.1 \text{ kN-m/m} > 26.3 \text{ kN-m/m})$$

Shear Strength. From before, the aspect ratio $k = 1.2$, and from Equations 4.27,

$$V_u = \tfrac{1}{3}w_uL_S = \tfrac{1}{3} \times 254 \times 19.5 = 1,651 \text{ lb/ft} \qquad \text{(N-S)}$$

$$V_u = \frac{kw_uL_S}{2k + 1} \qquad \text{(E-W)}$$

$$= 1.2 \times 254 \times \frac{19.5}{2 \times 1.2 + 1} = 1,748 \text{ lb/ft (25.5 kN/m)}$$

From Equation 9.26,

$$2\sqrt{f_c'}b_wd_p \le V_c = \left(0.6\sqrt{f_c'} + 700\frac{V_ud}{M_u}\right)b_wd_p \le 5\sqrt{f_c'}b_wd_p$$

Consider $M_u = 0$ at the boundaries of the single-panel wall-supported slab in this example. In such a case,

$$V_c = 0.6\sqrt{5,000} \times 12 \times 5 = 2,546 \text{ lb/ft (37.2 kN/m)} \gg 1,748 \text{ lb/ft}$$

which is satisfactory. Hence, adopt the following design:

$$h = 6 \text{ in. (152 mm)}$$

$$d_p = 5 \text{ in. (127 mm)}$$

Use $\tfrac{1}{2}$ in. dia seven-wire stress-relieved 270 K strands spaced 17 in. (432 mm) center to center in the N–S direction, and 20 in. (508 mm) center to center in the E–W direction. Also, use two #4 bars (12.7 mm dia) along the anchorage zone line around all the slab perimeter.

9.8 ONE-WAY SLAB SYSTEMS

One-way prestressed slabs behave similarly to beams regardless of whether they are simply supported or continuous over several supports. A one-way slab is therefore designed as a 12-in.-wide beam. The main prestressing strands are placed in the direction of the slab length, namely, spanning the continuous spans. The aspect ratio, the ratio of the span to the width of the slab band, has to have a value of 2 or greater in order for the slab to be considered one-way.

The same design and analysis procedures and examples as in Chapter 6 are to be used for the analysis and design of continuous one-way-action prestressed floor systems.

9.9 SHEAR-MOMENT TRANSFER TO COLUMNS SUPPORTING FLAT PLATES

9.9.1 Shear Strength

The shear behavior of two-way slabs and plates is a three-dimensional stress problem. The critical shear failure plane follows the perimeter of the loaded area and is located at a distance that gives a minimum shear perimeter b_0. Based on extensive analytical and experimental verification, the shear plane should not be closer than a distance $d/2$ from the concentrated load or reaction area.

If no special shear reinforcement is provided, the nominal shear strength V_c, as required by the ACI, is defined in Equations 9.24, 9.25, and 9.26. The coefficients for evaluating the factored external shear force V_u in a continuous two-way slab supported all along the panel perimeters can be evaluated approximately from Equations 9.27.

9.9.2 Shear-Moment Transfer

The unbalanced moment at the column face support of a slab without beams is one of the more critical design considerations in proportioning a flat plate or a flat slab. To ensure adequate shear strength requires moment transfer to the column by flexure across the perimeter of the column and by eccentric shearing stress such that approximately 60 percent is transferred by flexure and 40 percent by shear.

The fraction γ_v of the moment transferred by eccentricity of the shear stress decreases as the width of the face of the critical section resisting the moment increases in such a manner that

$$\gamma_v = 1 - \frac{1}{1 + \frac{2}{3}\sqrt{\frac{b_1}{b_2}}} \tag{9.28}$$

where $b_2 = c_2 + d$ is the width of the face of the critical section resisting the moment and $b_1 = c_1 + d$ is the width of the face at right angles to b_2.

The remaining portion γ_f of the unbalanced moment transferred by flexure is given by and acting on an effective slab width between lines that are $1\frac{1}{2}$ times the total slab thickness h on both sides of the column support.

$$\gamma_f = \frac{1}{1 + \frac{2}{3}\sqrt{\frac{b_1}{b_2}}} = 1 - \gamma_v \qquad (9.29)$$

For exterior columns, $b_1 = c_1 + \frac{1}{2}d$.

The distribution of shear stresses around the column edges is as shown in Figure 9.15. It is considered to vary linearly about the centroid of the critical section. The factored shear force V_u and the unbalanced factored moment M_u, both assumed to be acting at the column face, have to be transferred to the centroidal axis $c-c$ of the critical section. Thus, the axis position has to be located, thereby obtaining the shear force arm g (the distance from the column face to the centroidal axis plane) of the critical section $c-c$ for the shear moment transfer.

For calculating the maximum shear stress sustained by the plate in the edge column region, the ACI Code requires using the full nominal moment strength M_n provided by the column strip in Equations 9.30 to follow as the unbalanced moment, multiplied by the transfer fraction factor γ_v. This unbalanced moment $M_n \geq M_{ue}/\phi$ is composed of two parts: the negative end panel moment $M_{ne} = M_e/\phi$ at the face of the column, and the moment $(V_u/\phi)g$ due to the eccentric factored perimetric shear force V_u. The limiting value of the shear stress intensity is expressed as

$$\frac{v_{u(AB)}}{\phi} = \frac{V_u}{\phi A_c} + \frac{\gamma_v M_{ue} c_{AB}}{\phi J_c} \qquad (9.30\ a)$$

$$\frac{v_{u(CD)}}{\phi} = \frac{V_u}{\phi A_c} - \frac{\gamma_v M_{ue} c_{CD}}{\phi J_c} \qquad (9.30\ b)$$

where the nominal shear strength intensity is

$$v_n = \frac{v_u}{\phi} \qquad (9.30\ c)$$

and where A_c = area of concrete of assumed critical section
$\qquad = 2d(c_1 + c_2 + 2d)$ for an interior column
and J_c = property of assumed critical section analogous to polar moment of inertia.

The value of J_c for an interior column is

$$J_c = \frac{d(c_1 + d)^3}{6} + \frac{d^3(c_1 + d)}{6} + \frac{d(c_2 + d)(c_1 + d)^2}{2}$$

and the value of J_c for an edge column with bending parallel to the edge is

$$J_c = \frac{(c_1 + d/2)(d)^3}{6} + \frac{2(d)}{3}(c^3_{AB} + c^3_{CD}) + (c_2 + d)(d)(c_{AB})^2$$

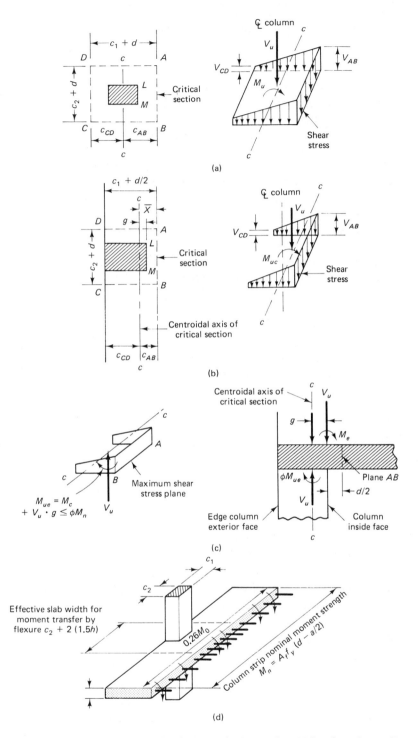

Figure 9.15 Shear stress distribution around column edge. (a) Interior column. (b) End column. (c) Critical surface. (d) Transfer nominal moment strength M_n.

Sec. 9.9   Shear-Moment Transfer to Columns Supporting Flat Plates

From basic principles of the mechanics of materials, the shearing stress is

$$v_u = \frac{V_u}{A_c} + \gamma_v \frac{Mc}{J}$$

where the second term on the right-hand side is the shearing stress resulting from the torsional moment at the column face.

If the nominal moment strength M_n of the shear moment transfer zone after the design of the reinforcement results in a larger value than M_{ue}/ϕ, this value of M_n should be used in Equations 9.30 a and b in lieu of M_{ue}/ϕ. In such a case, where the moment strength value $M_n = M_{ne} + (V_u/\phi)g$ is increased because of the use of flexural reinforcement in excess of what is needed to resist M_{ue}/ϕ, the slab stiffness is raised, thereby increasing the transferred shear stress v_u calculated from Equations 9.30 a and b for development of full moment transfer. Consequently, it is advisable to maintain a design moment M_{ue} with a value close to the factored moment value M_{ue} if an increase in the shear stress due to additional moment transfer needs to be avoided and a resulting need for additional increase in the plate design thickness prevented.

Example 9.2 illustrates the procedure for calculating the limit perimeter shear stress in the plate at the edge column region.

A higher perimetric shearing stress v_u can occur than that evaluated by Equation 9.30 a or b when adjoining spans are unequal or unequally loaded in the case of an interior column. The ACI Code stipulates, in the slab section pertaining to factored moments in columns and walls, that the supporting element such as a column or a wall has to resist an unbalanced moment

$$M' = 0.07[(w_{nd} + 0.5w_{nl})l_2 l_{n2} - w'_{nd}l'_2(l'_n)^2] \tag{9.31}$$

where w'_{nd}, l'_2, and l'_n refer to the shorter span. Hence, an additional term is added to Equation 9.30 a or b in such cases so that

$$v_u = \frac{V_u}{A_c} + \frac{\gamma_v M_u c_{AB}}{J_c} + \frac{\gamma_v M' c}{J'_c} \tag{9.32}$$

where J'_c is the polar moment of inertia with moment areas taken in a direction perpendicular to that used for J_c.

9.9.3 Deflection Requirements for Minimum Thickness— An Indirect Approach

For a preliminary estimate of the two-way slab thickness, it is necessary to use some approximate guidelines in order to expeditiously select a trial depth. Span-to-depth ratios in prestressed slabs are expected to be higher than those in reinforced concrete slabs if the advantages of prestressing are not to be economically lost.

Service live loads rather than total dead plus live loads should be used in evaluating deflection. Load balancing from the transverse component of the prestressing force would have to be used to neutralize the dead-load deflection or even produce camber if the live load is excessively high. An approximate guideline of a

TABLE 9.1 MINIMUM PERMISSIBLE RATIOS OF SPAN (*l*) TO DEFLECTION (*a*)

(*l* = longer span)

Type of member	Deflection *a* to be considered	$(l/a)_{min}$
Flat roofs not supporting and not attached to nonstructural elements likely to be damaged by large deflections	Immediate deflection due to live load *L*	180[a]
Floors not supporting and not attached to nonstructural elements likely to be damaged by large deflections	Immediate deflection due to live load *L*	360
Roof or floor construction supporting or attached to nonstructural elements likely to be damaged by large deflections	That part of total deflection occurring after attachment of nonstructural elements; sum of long-time deflection due to all sustained loads (dead load plus any sustained portion of live load) and immediate deflection due to any additional live load[b]	480[c]
Roof or floor construction supporting or attached to nonstructural elements not likely to be damaged by large deflections		240[c]

[a] Limit not intended to safeguard against ponding. Ponding should be checked by suitable calculations of deflection, including added deflections due to ponded water, and considering long-term effects of all sustained loads, camber, construction tolerances, and reliability of provisions for drainage.

[b] Long-term deflection has to be determined, but may be reduced by the amount of deflection calculated to occur before attachment of nonstructural elements. This reduction is made on the basis of accepted engineering data relating to time-deflection characteristics of members similar to those being considered.

[c] Ratio limit may be lower if adequate measures are taken to prevent damage to supported or attached elements, but should not be lower than tolerance of nonstructural elements.

span-to-depth ratio of 16 to 25 for solid cantilever slabs and 40 to 50 for two-way continuous slabs is not unreasonable to use. For waffle slabs, a lower value of 35 to 40 is recommended. For simply supported spans and for single- and double-T's, use 90 percent of these values for the first trial.

The ACI requires that the minimum span-to-deflection ratio be restricted depending on the type of loading and conditions of use. This limitation is obviated by the need to prevent cracking of plaster in the ceilings and damage to sensitive supported equipment as well as cracking of partitions and ponding of water in roofs. Table 9.1 gives recommended values of span-to-deflection ratios for deflection control.

A more accurate evaluation of the deflection/camber of reinforced concrete and prestressed concrete two-way-action slabs and plates is presented in Section 9.12. This approach utilizes the stiffnesses of the interconnecting members using the equivalent frame method in the deflection analysis. It is both logical and expedient to use, particularly since the stiffness factors of the various elements have already been computed in the flexural analysis of the equivalent continuous frames.

9.10 STEP-BY-STEP TRIAL-AND-ADJUSTMENT PROCEDURE FOR THE DESIGN OF A TWO-WAY PRESTRESSED SLAB AND PLATE SYSTEM

The following sequence of steps is suggested for the design or analysis of two-way action prestressed slabs:

1. Determine whether the slab geometry and loading require a two-way analysis by the equivalent frame method.

2. Select a trial slab thickness for the maximum of either the longitudinal $h = L/45$ or the transverse $h = L/45$. Compute the total service dead and live loads and the factored loads.

3. Assume a tendon profile across the continuous spans in both the E–W and N–S directions, and determine the prestressing force F, the concrete stress $f_c = F/A_c$, and the number of strands in a span. Calculate the balancing load intensity $W_{bal} = 8Fa/L^2$ and the net load $W_{net\downarrow} = W_{w\downarrow} - W_{bal\uparrow}$

4. Determine, by the equivalent frame method, the equivalent frame characteristics and the flexural and torsional stiffnesses of the slab given by

$$K_c \cong \frac{4EI}{L - 2h}$$

and

$$K_t = \sum \frac{9E_{cs}C}{L_2\left(1 - \dfrac{c_2}{L_2}\right)^3}$$

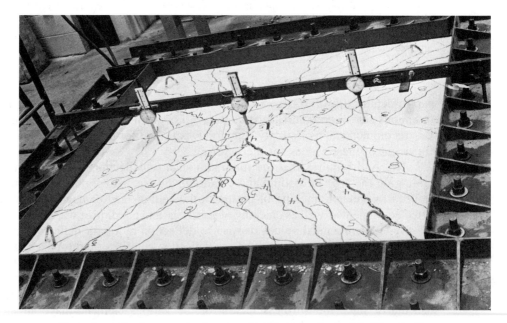

Flexural cracking in a restrained one-panel reinforced concrete slab. (Tests by Nawy et al.)

Two-Way Prestressed Concrete Floor Systems   Chap. 9

where $C = \Sigma(1 - 0.63x/y)x^3y/3$. Then determine

$$K_{ec} = \left(\frac{1}{K_c} + \frac{1}{K_t}\right)^{-1}$$

for the exterior and interior columns, and the slab stiffness

$$K_s \cong \frac{4EI}{L_1 - c_1/2}$$

where L_1 is the centerline span and c_1 is the column depth for each slab-column joint.

5. From the values of K_{ec} and K_s obtained at each joint, determine the moment distribution factors

$$DF = \frac{K_s}{\Sigma K}$$

for the slabs, where $\Sigma K = K_{ec} + K_{S(\text{left})} + K_{S(\text{right})}$. Then calculate the fixed-end moments FEM at the joints for the net loads given by $FEM = WL^2/12$ for a distributed load.

6. Run a moment distribution for the net load moment M_{net}, and adjust the redistributed moments to obtain the net moment values at the face of the supports; the equation is $M_n = $ centerline $M_n - V_c/3$. Then verify that the concrete stresses

$$f_t = -\frac{P}{A} + \frac{M_{\text{net}}}{S}$$

resulting from these moments are below the maximum allowable $f_t = 6\sqrt{f_c'}$ for support sections and $f_t = 2\sqrt{f_c'}$ for midspan sections.

7. Compute the balanced service-load fixed-end moments

$$FEM_{\text{bal}} = \frac{W_{\text{bal}}L^2}{12}$$

and run a moment distribution of the balanced load moments M_{bal}. Then find the primary moment $M_1 = P_e e$ and the secondary moment $M_s = $ Distributed $(M_{\text{bal}} - M_1)$.

8. Compute the fixed-end factored load moments $FEM_u^- = \dfrac{W_u L^2}{12}$ and run a moment distribution of the factored moments. Then determine the required design moment $M_u = $ Distributed $(M_u^- - M_s)$ for the slabs at all joints and at maximum positive M_u along the spans.

9. Determine the required nominal moment strengths $M_n = M_u/\phi$ for the negative support moments $-M_u$ and the positive span moments $+M_u$. Then check whether the available $-M_n$ and $+M_n$ for the slab and the prestressing steel are adequate. Next, determine the *inelastic* moment redistribution ΔM_R from

$$\rho_D = 20\left[1 - \frac{\omega_p + (d/d_p)(\omega - \omega')}{0.36\beta_1}\right] \text{ percent}$$

where $\Delta M_R = \rho_D(\text{support } M_u)$. Add mild steel to the support and midspan where necessary, recalling that the minimum nonprestressed steel $A_s = 0.00075hL$.

11. Check the nominal shear strength of the slab at the exterior and interior supports, and calculate the shear-moment transfer and the flexure-moment transfer to the columns. The moment shear factor is

$$\gamma_v = 1 - \frac{1}{1 + \frac{2}{3}\sqrt{b_1/b_2}}$$

and the moment flexure factor is

$$\gamma_f = \frac{1}{1 + \frac{2}{3}\sqrt{b_1/b_2}}$$

where $b_1 = c_1 + d/2$ for an exterior column
$\quad\quad b_1 = c_1 + d$ for an interior column
$\quad\quad b_2 = c_2 + d$

Then calculate c_{AB} and c_{CD} for exterior columns, as well as the total nominal unbalanced moment strength $M_n = M_{ue} + V_u g$.

12. Calculate the shear ultimate stress due to perimeter shear and effect of $\gamma_v M_n$:

$$v_n = \frac{V_u}{\phi_v A_c} + \frac{\gamma_v c_{AB} M_n}{J_c} \leq \text{max. allowable } v_c \text{ where}$$

$$\text{max. allowable shear stress } v_c = \beta_p \sqrt{f_c'} + 0.3\bar{f}_c + \frac{V_p}{b_o d}$$

$$\beta_p = \text{the smaller of the two values of 3.5 or } (\alpha_s d/b_o + 1.5)$$

where α_s is 40 for interior columns, 30 for edge columns, and 20 for corner columns.

Column section shall be at least 4 in. from the face of the discontinuous edge, and f' shall not exceed 5000 psi and $\bar{f}$ shall be 125 psi min. and 500 psi max.; otherwise v_c shall be computed from the smaller of the values obtained from the following expressions:

$$v_c = (2 + 4/\beta_c)\sqrt{f_c'} \quad \text{or} \quad v_c = (\alpha_s d/b_o)\sqrt{f_c'} \quad \text{or} \quad v_c = 4\sqrt{f_c'}$$

13. Calculate the factored moment value $\gamma_f M_n$, and check the available moment strength M_n of the section, concentrating the steel in the column band $[c + 2(1.5h)]$.

14. Check the deflection and camber serviceability behavior of the critical slab panels.

15. Adopt the design if it satisfies all the preceding criteria. Then perform the operations for the E–W and N–S directions of the floor system.

Figure 9.16 gives a flowchart for the design or analysis of two-way action prestressed concrete floor slabs and plates.

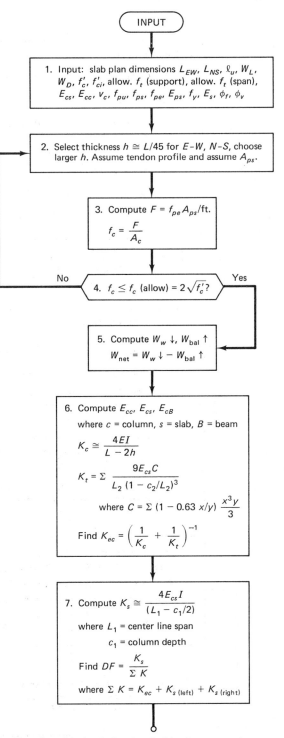

Figure 9.16 Flowchart for the design (analysis) of prestressed concrete two-way slabs and plates in flexure and shear.

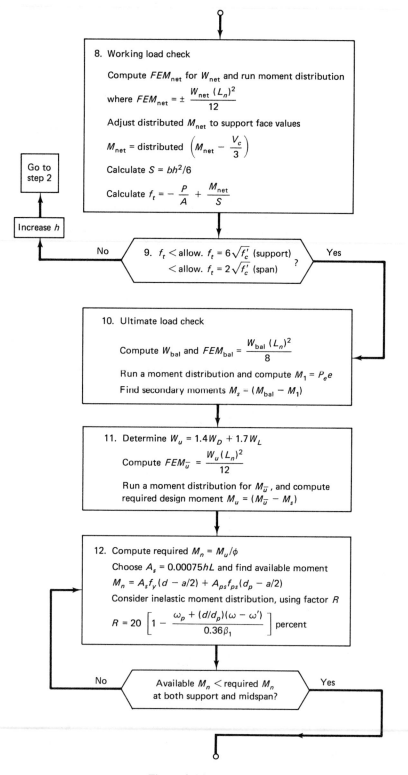

8. Working load check

Compute FEM_{net} for W_{net} and run moment distribution

where $FEM_{net} = \pm \dfrac{W_{net}(L_n)^2}{12}$

Adjust distributed M_{net} to support face values

$M_{net} = \text{distributed}\ \left(M_{net} - \dfrac{V_c}{3}\right)$

Calculate $S = bh^2/6$

Calculate $f_t = -\dfrac{P}{A} + \dfrac{M_{net}}{S}$

Go to step 2

Increase h

No

9. $f_t < \text{allow.}\ f_t = 6\sqrt{f_c'}$ (support)
$< \text{allow.}\ f_t = 2\sqrt{f_c'}$ (span) ?

Yes

10. Ultimate load check

Compute W_{bal} and $FEM_{bal} = \dfrac{W_{bal}(L_n)^2}{8}$

Run a moment distribution and compute $M_1 = P_e e$
Find secondary moments $M_s = (M_{bal} - M_1)$

11. Determine $W_u = 1.4W_D + 1.7W_L$

Compute $FEM_{\bar{u}} = \dfrac{W_u(L_n)^2}{12}$

Run a moment distribution for $M_{\bar{u}}$, and compute
required design moment $M_u = (M_{\bar{u}} - M_s)$

12. Compute required $M_n = M_u/\phi$

Choose $A_s = 0.00075hL$ and find available moment
$M_n = A_s f_y(d - a/2) + A_{ps}f_{ps}(d_p - a/2)$
Consider inelastic moment distribution, using factor R

$R = 20\left[1 - \dfrac{\omega_p + (d/d_p)(\omega - \omega')}{0.36\beta_1}\right]$ percent

No

Available $M_n <$ required M_n
at both support and midspan?

Yes

Figure 9.16 (*cont.*)

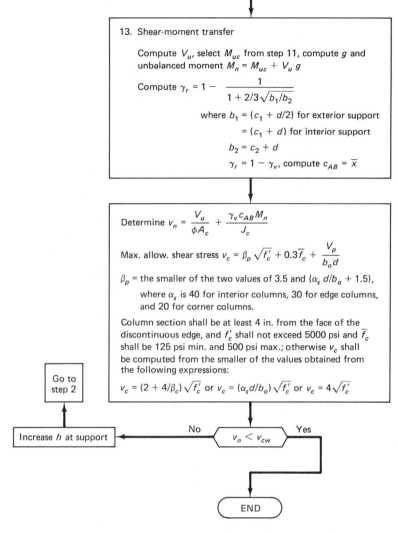

13. Shear-moment transfer

Compute V_u, select M_{uc} from step 11, compute g and unbalanced moment $M_n = M_{uc} + V_u\, g$

Compute $\gamma_r = 1 - \dfrac{1}{1 + 2/3\sqrt{b_1/b_2}}$

where $b_1 = (c_1 + d/2)$ for exterior support

$= (c_1 + d)$ for interior support

$b_2 = c_2 + d$

$\gamma_r = 1 - \gamma_v$, compute $c_{AB} = \overline{x}$

Determine $v_n = \dfrac{V_u}{\phi A_c} + \dfrac{\gamma_v c_{AB} M_n}{J_c}$

Max. allow. shear stress $v_c = \beta_p \sqrt{f_c'} + 0.3\overline{f}_c + \dfrac{V_p}{b_o d}$

β_p = the smaller of the two values of 3.5 and $(\alpha_s\, d/b_o + 1.5)$,

where α_s is 40 for interior columns, 30 for edge columns, and 20 for corner columns.

Column section shall be at least 4 in. from the face of the discontinuous edge, and f_c' shall not exceed 5000 psi and $\overline{f}_c$ shall be 125 psi min. and 500 psi max.; otherwise v_c shall be computed from the smaller of the values obtained from the following expressions:

$v_c = (2 + 4/\beta_c)\sqrt{f_c'}$ or $v_c = (\alpha_s d/b_o)\sqrt{f_c'}$ or $v_c = 4\sqrt{f_c'}$

Go to step 2

Increase h at support ← No — $v_n < v_{cw}$ — Yes

END

Figure 9.16 *(cont.)*

9.11 Design of Prestressed Post-Tensioned Flat-Plate Floor System

Example 9.2

A post-tensioned prestressed nonbonded flat-plate floor system for an apartment complex is shown in Figure 9.17. The end-panel centerline dimensions are 17 ft, 6 in. × 20 ft, 0 in. (5.33 m × 6.10 m), and the interior panel dimensions are 24 ft, 0 in. × 20 ft, 0 in. (7.32 m × 6.10 m). The heights ℓ_u of the intermediate floors are typically 8 ft, 9 in. (2.67 m). Design a typical floor panel to withstand a working live load $W_L = 40$ psf (1.92 kPa) and a superimposed dead load $W_D = 20$ psf (0.96 kPa) due to partitions and flooring. Assume in your solution that all panels are simultaneously loaded

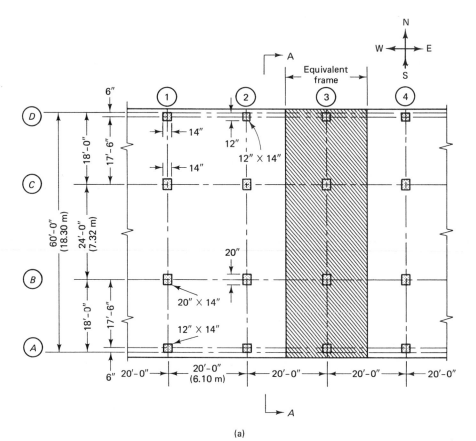

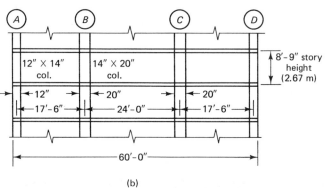

Figure 9.17 Flat-plate apartment structure in Example 9.1. (a) Plan. (b) Section A–A, N–S.

by the live load, and verify the shear-moment transfer capacity of the floor at the column supports. Use $\frac{1}{2}$ in. dia seven-wire 270 K stress-relieved prestressing strands and the equivalent frame method to arrive at your solution. Given data are as follows:

$f_c' = 4,000$ psi (27.58 MPa), normal weight

$f_{ci}' = 3,000$ psi (20.68 MPa)

Support $f_t = 6\sqrt{f_c'} = 380$ psi (2.62 MPa)

Midspan $f_t = 2\sqrt{f_c'} = 127$ psi (0.88 MPa)

Max. v_c required by the ACI Code

$f_{pu} = 270,000$ psi (1,862 MPa)

f_{ps} not to exceed 185,000 psi (1,276 MPa)

$f_{py} = 243,000$ psi (1,675 MPa)

$f_{pe} = 159,000$ psi (1,096 MPa)

$E_{ps} = 29 \times 10^6$ psi (200×10^3 MPa)

$f_y = 60,000$ psi (413.7 MPa)

Solution

N-S Direction

Service Load Analysis

1. *Loads*

   For deflection control, assume that the slab thickness $h \cong L/45$. Then the longitudinal $h = 20 \times 12/45 = 5.33$ in. and the transverse $h = 24 \times 12/45 = 6.40$ in. So try $h = 6\frac{1}{2}$ in. (165 mm) slabs $= 81$ psf. The superimposed dead load $= 20$ psf, and we have

$$\text{Total } W_D = 101 \text{ psf}$$

$$W_L = 40 \text{ psf}$$

$$\text{Total } W_w = W_{D+L} = 141 \text{ psf (6.75 kPa)}$$

$$W_u = 1.4W_D + 1.7W_L = 1.4 \times 101 + 1.7 \times 40 \cong 210 \text{ psf (10.05 kPa)}$$

L_n = bay span (N–S for this part of the solution)

L_2 = band width (E–W direction) = 20 ft (240 in.)

2. *Load Balancing and Tendon Profile*

   In order to make a preliminary estimate of the balanced load, assume an average intensity of compressive stress on the concrete due to load balancing of $f_c = 170$ psi (1.17 MPa). Then the unit $F = 170 \times 6.5 \times 12 = 13,260$ lb/ft (193.6 kN/m). So, trying $\frac{1}{2}$ in. dia 270 K seven-wire strands, we find that the effective force P_e per strand $= A_{ps}f_{pe} = 0.153 \times 159,000 = 24,327$ lb. For the $L = 20$-ft bay along the longitudinal direction of the structure, the total force is $F_e = FL = 13,260 \times 20 = 265,200$ lb (1,180 kN).

   The number of strands per bay is $F_e/P_e = 265,200/24,327 \cong 11$ strands, and the total $P_e = F_e = 24,327 \times 11 = 267,597$ lb. The actual unit force $F = 267,597/20 = 13,380$ lb/ft (195.3 kN/m), and the actual $f_c = F/A = 13,380/(6.5 \times 12) \cong 172$ psi $\cong 170$ psi, which is satisfactory. Consequently, use $f_c = 172$ psi due to load balancing, and assume a parabolic tendon profile as shown in Figure 9.18.

   Outside Spans AB or CD at Midspan

$$a_1 = a_3 = \frac{3.25 + 5.50}{2} - 1.75 = 2.625 \text{ in.}$$

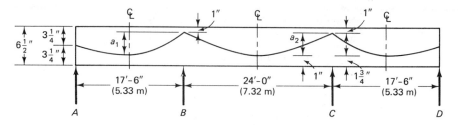

Figure 9.18 Tendon profile in N–S direction in Example 9.2.

From Equation 1.16 for a parabolic tendon,

$$W = \frac{8Fa}{L_n^2}$$

$$W_{bal} = \frac{8 \times 13{,}380 \times 2.625/12}{(18)^2} \cong 72 \text{ psf}$$

The net load intensity producing bending is

$$W_{net} = W_w - W_{bal} = 141 - 72 = 69 \text{ psf } (3.30 \text{ kPa})$$

Interior Span BC

$$a_2 = 6.5 - 1 - 1 = 4.5 \text{ in.}$$

$$W_{bal} = \frac{8Fa}{L_n^2} = \frac{8 \times 13{,}380 \times 4.5/12}{(24)^2} \cong 70 \text{ psf}$$

$$W_{net} = 141 - 70 = 71 \text{ psf } (3.40 \text{ kPa})$$

3. *Equivalent Frame Characteristics*

Take the equivalent frame in the N–S direction whose plan is shown in the shaded portion of Figure 9.17. The approximate flexural stiffness of the column above and below the floor joint (the moment per unit rotation), from Ref. 9.10 and Equation 9.9, is

$$K_c = \frac{4E_c I_c}{L_n - 2h}$$

where $L_n = \ell_u = 8$ ft, 9 in. = 105 in.

(a) *Exterior column (14 in. × 12 in.) stiffness*

For the exterior columns, $b = 14$ in., so $I_c = 14(12)^3/12 = 2{,}016$ in⁴. Assume that $E_{col}/E_{slab} = E_{cc}/E_{cs} = 1.0$, and use $E_{cc} = E_{cs} = 1.0$ in the calculations as E_{cs} drops out in the equation for K_{ec}. Then we obtain

$$\text{Total } K_c = \frac{4 \times 1 \times 2{,}016}{105 - (2 \times 6.5)} \times 2 \text{ (for top and bottom columns)}$$

$$= 175.3 \text{ in.-lb/rad}$$

From Equation 9.10 b, the torsional constant is

$$C = \sum \left(1 - 0.63\frac{x}{y}\right)\frac{x^3 y}{3}$$

$$= \left(1 - 0.63 \times \frac{6.5}{12}\right) 6.5^3 \times \frac{12}{3} = 724$$

The torsional stiffness of the slab at the column line is

$$K_t = \sum \frac{9E_{cs}C}{L_2\left(1 - \dfrac{c_2}{L_2}\right)^3}$$

$$= \frac{9 \times 1 \times 724}{20 \times 12(1 - 14/(12 \times 20))^3} + \frac{9 \times 1 \times 724}{20 \times 12(1 - 14/(12 \times 20))^3}$$

From Equation 9.7, the equivalent column stiffness is $K_{ec} = (1/K_c + 1/K_t)^{-1} = (1/175.3 + 1/65)^{-1} = 47$ in.-lb/rad/E_{cc}.

(b) *Interior column (14 in. $\times$ 20 in.) stiffness*
For the interior columns, $b = 14$ in., so $I = 14(20)^3/12 = 9,333$ in^4. Hence, we have

$$\text{Total } K_c = \frac{4 \times 1 \times 9,333}{105 - 2 \times 6.5} \times 2 = 812 \text{ in.-lb./rad/}E_{cc}$$

$$C = \left(1 - 0.63 \times \frac{6.5}{20}\right) \times (6.5)^3 \times \frac{20}{3} = 1,456$$

$$K_t = \frac{9 \times 1,456}{20 \times 12(1 - 14/(12 \times 20))^3}$$

$$+ \frac{9 \times 1,456}{20 \times 12(1 - 14/(12 \times 20))^3}$$

$$= 131 \text{ in.-lb/rad/}E_{cs}$$

$$K_{ec} = (1/812 + 1/131)^{-1} = 113 \text{ in.-lb/rad/}E_{cc}$$

(c) *Slab stiffness*
From Equation 9.9 and Ref. 9.10,

$$K_s = \frac{4E_{cs}I_s}{L_n - \dfrac{c_1}{2}}$$

where L_n is the centerline span and c_1 the column depth. The slab band width in the E–W direction is $20/2 + 20/2 = 20$ ft. Thus, $I_s = 20 \times 12(6.5)^3/12 = 5,493$ in^4, and for the slab at the right of exterior column A,

$$K_s = \frac{4 \times 1 \times 20(6.5)^3}{12 \times 17.5 - 12/2} = 108 \text{ in.-lb/rad/}E_{cs}$$

while for the slab at the left of interior column B,

$$K_s = \frac{4 \times 1 \times 20(6.5)^3}{12 \times 17.5 - 20/2} = 110 \text{ in.-lb/rad/}E_{cs}$$

and for the slab at the right of interior column B,

$$K_s = \frac{4 \times 1 \times 20(6.5)^3}{12 \times 24 - 20/2} = 79 \text{ in.-lb/rad/}E_{cs}$$

TABLE 9.2 MOMENT DISTRIBUTION OF NET LOAD MOMENTS M_{net}

	(A)		(B)	₵	(C)	
DF	0.697	0.364	0.262	⋮	0.262	0.364
COF	0.5	0.5	0.5	⋮	0.5	0.5
FEM_net × 10³ in.-lb	−21.1	21.1	40.9	⋮	40.9	−21.1
Dist.	+14.71	7.21	5.19	⋮	−5.19	
CO Dist.	3.61 / −2.52	7.36 / −1.73	−2.60 / −1.25	⋮		
Final M_net × 10³ per ft	−5.30	33.94	−39.56	⋮		

From Equation 9.12, the slab distribution factor at the joints is $DF = K_s/\Sigma K$, where $\Sigma K = K_{ec} + K_{s(left)} + K_{s(right)}$. So for the outer joint A slab, $DF = 108/(47 + 108) = 0.697$; for the left joint B slab, $DF = 110/(113 + 110 + 79) = 0.364$; and for the right joint B slab, $DF = 79/(113 + 110 + 79) = 0.262$.

4. *Design Service-load Moments and Stresses*
   Design net load moments
   For the exterior spans AB and CD, $W_{net} = 69$ psf. So the fixed-end moment is

$$FEM = \frac{WL_n^2}{12} = \frac{69 \times (17.5)^2}{12} \times 12 = 21.1 \times 10^3 \text{ in.-lb}$$

Similarly, for the interior span BC, $W_{net} = 71$ psf. So the fixed-end moment is

$$FEM = \frac{71(24)^2}{12} \times 12 = 40.9 \times 10^3 \text{ in.-lb}$$

By running a moment distribution analysis as shown in Table 9.2, a carryover factor $COF = \frac{1}{2}$ can be used for all spans. Such an assumption is justified, as the effect of nonprismatic sections would be negligible on the fixed-end moments and carryover factors. It can also be assumed in multispan frames that the frame at a joint two spans away from the left joint (joint C) can be considered fixed in the distribution of the moments.

Slab concrete tensile stress at support
The net moment at the interior face of column B is the difference of the centerline moment and $V_c/3$, i.e.,

$$M_{net, max} = 39.56 \times 10^3 - \frac{20}{3}\left(\frac{71 \times 24}{2}\right) = 33,880 \text{ in.-lb/ft}$$

The slab section modulus $S = bh^2/6 = 12(6.5)^2/6 = 84.5$ in³., and we have, for the support concrete stress,

$$f_t = -\frac{P}{A} + \frac{M}{S} = -172 + \frac{33,880}{84.5} = 229 \text{ psi (1.63 MPa)}$$

So the allowable $f_t = 6\sqrt{f_c'} = 380$ psi > 229 psi, which is satisfactory.
Slab concrete tensile stress at midspan
The net midspan maximum moment is $WL^2/8 - 39.56 \times 10^3$, or

$$M_{\text{net, max}} = \frac{71(24)^2}{8} \times 12 - 39.56 \times 10^3 = 21,784 \text{ in.-lb/ft (7.85 kN/m)}$$

Also,

$$\text{Midspan } f_t = -\frac{P}{A} + \frac{M}{S} = -172 + \frac{21,784}{84.5} = 86 \text{ psi (0.545 MPa)}$$

So the allowable $f_t = 2\sqrt{f_c'} = 127$ psi > 86 psi, which is satisfactory.
If f_t were to exceed the allowable f_t, the entire tensile force would have to be taken by mild steel reinforcement at a stress $f_s = \frac{1}{2}f_y$.

Ultimate Flexural Strength Analysis
Design Moments M_u

1. *Balanced moments M_{bal}*
   The secondary moment is given by $M_s = M_{\text{bal}} - M_1$, where M_{bal} is the balanced moment and M_1 is the primary moment $= P_e e = Fe$. For the span AB or CD,

$$FEM_{\text{bal}} = \frac{72(17.5)^2}{12} \times 12 = 22,050 \text{ in.-lb/ft}$$

and for the span BC,

$$FEM_{\text{bal}} = \frac{70(24)^2}{12} \times 12 = 40,320 \text{ in.-lb/ft}$$

Running a moment distribution as in Table 9.3 will determine the maximum M_{bal} for the exterior column joints.

2. *Secondary moments M_s and fatored load moment M_u^-*
   Span AB
   From the tendon profile of Figure 9.18, $e = 0$. So we have:

$$\text{Primary moment } M_1/\text{ft at } A = P_e e = 0$$

$$M_{\text{bal}} = 5,670 \text{ in.-lb/ft (from Table 9.3)}$$

$$M_s = M_{\text{bal}} - M_1 = 5,670 - 0 = 5.67 \times 10^3 = 5,670 \text{ in.-lb/ft}$$

$$\text{Factored load } FEM_u = \frac{W_u \ell^2}{12} = \frac{210(17.5)^2}{12} \times 12 = 64,313 \text{ in.-lb/ft}$$

Span BA
From the tendon profile in Figure 9.18, $e = 6.5/2 - 1 = 2.25$ in. So we have:

$$M_1 = 13,380 \times 2.25 = 30,105 \text{ in.-lb/ft (11.16 kN-m)}$$

$$M_{\text{bal}} = 34,460 \text{ in.-lb/ft (from Table 9.3)}$$

TABLE 9.3 MOMENT DISTRIBUTION OF BALANCED LOAD MOMENTS M_{bal}

	(A)		(B)	¢	(C)
DF	0.697	0.364	0.262	0.262	0.364
COF	0.5	0.5	0.5	0.5	0.5
FEM_{bal} $\times 10^3$ in.-lb per ft	−22.05	22.05	−40.32	40.3	−22.05
Dist.	+15.37	+6.65	+4.79	−4.79	−6.65
CO dist.	3.33 / −2.32	7.69 / −1.93	−2.40 / −1.39		
Final M_{bal} $\times 10^3$ per ft	−5.67	34.46	−39.32		

$$M_s = 34,460 - 30,105$$

$$= 4,355 \text{ in.-lb/ft } (1.61 \text{ kN-m/m})$$

Factored load $FEM_u = 64,313$ in.-lb/ft (23.84 kN-m/m)

Span BC

$$e = 2.25 \text{ in.}$$

$$M_1 = 30,105 \text{ in.-lb/ft}$$

$$M_{bal} = 39,320 \text{ in.-lb/ft (from Table 9.3)}$$

$$M_s = 39,320 - 30,105 = 9,215 \text{ in.-lb/ft } (3.42 \text{ kN-m/m})$$

$$\text{Factored load } FEM_u = \frac{210(24)^2}{12} \times 12 = 120,960 \text{ in.-lb/ft } (44.84 \text{ kN-m/m})$$

Run a moment distribution for the factored moments as in Table 9.4. Analysis of pattern loading of alternate spans should also be made to determine the worst conditions of service-load and factored-load moments.

3. *Design moment M_u*

The design moments M_u are the difference of the factored-load moments M_u^- and the secondary moments M_s, i.e., $M_u = M_u^- - M_s$ (from Equation 9.17).
Joint A (span AB) moment $-M_u$
For the joint A (span AB) moment, $M_s = 5,670$ in.-lb/ft (from before), and so the centerline $M_u = 16,370 - 5,670 = 10,700$ in.-lb/ft. The moment reduction to the column face of support A = $Vc/3$. Thus,

$$V_{AB} = \frac{W_u L}{2} - \frac{M_{u@B}^- - M_{u@A}^-}{L_n} = \frac{210 \times 175}{2} - \frac{10^3(101.88 - 16.37)}{17.5 \times 12}$$

$$= 1,837.5 - 407.2 = 1,430.3 \text{ lb/ft}$$

$$c = 12 \text{ in.}$$

TABLE 9.4 MOMENT DISTRIBUTION OF FACTORED LOADS

	(A)		(B)	¢	(C)
DF	0.697	0.364	0.262	0.262	0.364
COF	0.5	0.5	0.5	0.5	0.5
FEM_u^- $\times 10^3$ in.-lb per ft	−64.31	64.31	−120.96	120.96	−64.31
Dist.	+44.82	+20.62	+14.84	−14.84	−20.62
CO Dist.	10.31 / −7.19	22.41 / −5.46	−7.42 / −3.93		
Final $M_u^- \times 10^3$ per ft	−16.37	101.88	−117.47		

$$\text{Req column face } M_u = 10{,}700 - \frac{1{,}430.3 \times 12}{3}$$

$$= 10{,}700 - 5{,}720 = 4{,}980 \text{ in.-lb/ft } (1.85 \text{ kN-m/m})$$

$$\text{Req } -M_n = \frac{M_u}{\phi} = \frac{4{,}980}{0.9} = 5{,}533 \text{ in.-lb/ft } (2.05 \text{ kN-m/m})$$

Joint B (span BA) moment $-M_u$

For the joint B (span BA) moment, $M_s = 4{,}355$ in.-lb/ft (from before), and so the centerline $M_u = 101{,}880 - 4{,}355 = 97{,}525$ in.-lb/ft. Thus,

$$V_{BA} = 1{,}837.5 + 407.2 = 2{,}244.7 \text{ lb/ft.}$$

$$c = 20 \text{ in.}$$

$$\text{Req. column face } M_u = 97{,}525 - \frac{2{,}244.7 \times 20}{3}$$

$$= 97{,}525 - 14{,}965 = 82{,}560 \text{ in.-lb/ft } (30.61 \text{ kN-m/m})$$

$$\text{Req. } -M_n = \frac{M_u}{\phi} = \frac{82{,}560}{0.9} = 91{,}733 \text{ in.-lb/ft } (34 \text{ kN-m/m})$$

Joint B (span BC) moment $-M_u$

For the joint B (span BC) moment, $M_s = 9{,}215$ in.-lb/ft, and so the centerline $M_u = 117{,}470 - 9{,}215 = 108{,}255$ in.-lb/ft. Thus,

$$V_{BC} = \frac{210 \times 24}{2} = 2{,}520 \text{ lb/ft}$$

$$\text{Req column face } -M_u = 108{,}255 - \frac{2{,}520 \times 20}{3} = 108{,}255 - 16{,}800$$

$$= 91{,}455 \text{ in.-lb/ft } (37.32 \text{ kN-m/m})$$

$$\text{Req} -M_n = \frac{M_u}{\phi} = \frac{91,455}{0.9}$$

$$= 101,617 \text{ in.-lb/ft } (41.47 \text{ kN-m/m})$$

Span AB maximum positive moment $+M_u$

Assume that the point of zero shear and maximum moment is x ft from face A. Then $x = V_{AB}/W_u = 1,430/210 = 6.81$ ft. Also, from Table 9.4, the end M_u^- at A $= 16,370$ in.-lb/ft., and from before, $M_s = \frac{1}{2}(5,670 + 4,355) = 5,013$ in.-lb/ft. So we have

$$\text{Max. } +M_u = V_{AB}x - \frac{W_u x^2}{2} - M_u^- + M_s$$

$$= 1,430.3 \times 6.81 \times 12 - \frac{210(6.81)^2}{2} \times 12 - 16,370 + 5,013$$

$$= 116,884 - 58,434 - 16,370 + 5,013$$

$$= 47,093 \text{ in.-lb/ft } (17.46 \text{ kN-m/m}) \text{ at } 6.81 \text{ ft from A}$$

$$\text{Req} +M_n = \frac{M_u}{\phi} = \frac{47,093}{0.9} = 52,326 \text{ in.-lb/ft } (19.4 \text{ kN-m/m})$$

Span BC maximum positive moment $+M_u$

From before, $V_{BC} = 2,520$ lb/ft and $x = L_n/2 = 24/2 = 12$ ft. The simple span midspan moment is, then,

$$M = V_{BC} \times \frac{L_n}{2} - \left(W_u \times \frac{L}{2}\right)(L_4)$$

$$= 2,520 \times \frac{24}{2} - \frac{210(24)^2}{8} = 15,120 \text{ ft.-lb/ft} = 181,440 \text{ in.-lb/ft}$$

Alternatively, the simple span moment is

$$M = \frac{W_u L^2}{8} = \frac{210(24)^2}{8} \times 12$$

$$= 181,440 \text{ in.-lb/ft}$$

Now, $+M_u = M - M_u^- + M_s$. From Table 9.4, $M_u^- = -117,470$ in.-lb/ft, and $M_s = 9,215$ in.-lb/ft. So the required maximum $+M_u = 181,440 - 117,470 + 9,215 = 73,185$ in.-lb/ft (27.13 kN-m/m) at midspan. And the required $+M_n = M_u/\phi = 73,185/0.9 = 81,317$ in.-lb/ft (30.14 kN-m/m).

Figure 9.19 gives a plot of the required design moments M_u across the continuous spans and the peak values of the moments.

Flexural Strength M_n (Nominal Moment Strength). The ACI Code requires a minimum amount of nonprestressed reinforcement. From Equation 9.20,

$$A_s = 0.00075 h L_n$$

1. *Interior support section at B*

For the interior support section at B, the controlling required $M_n =$

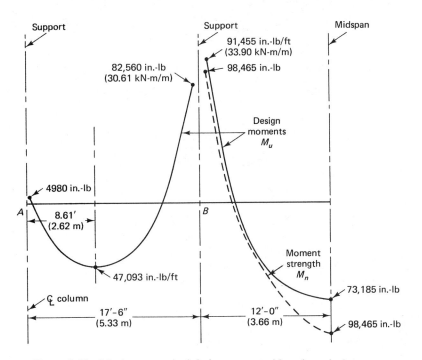

Figure 9.19 Maximum required design moments M_u and nominal moment strengths M_n in Example 9.2.

101,617 in.-lb/ft. We then have

$$A_s = 0.00075 \times 65 \left(\frac{18 + 24}{2}\right) \times 12 = 1.23 \text{ in}^2 \ (7.93 \text{ cm}^2)$$

Hence, try six #4 bars of 11-ft length, and space the bars at a maximum of 6 in. (152 mm) center to center so that they are concentrated over the column on a band width equal to the column width plus $1\frac{1}{2}$ slab thicknesses on each side of the column. Then

$$A_s = 6 \times 0.20 = 1.20 \text{ in}^2 \cong \text{required } 1.23 \text{ in}^2, \text{ O.K.}$$

Panel width = 20 ft

$$A_s/\text{ft} = \frac{1.2}{20} = 0.06 \text{ in}^2$$

From Equation 9.23 b, the design stress in the tendon is

$$f_{ps} = f_{pe} + \frac{f'_c}{300\rho_p} + 10,000 \text{ psi}$$

and

$$\rho_p = \frac{A_{ps}}{bd} = \frac{11 \times 0.153}{(20 \times 12)5.5} = 0.0013$$

$$f_{pe} = 159,000 \text{ psi}$$

$$f_{ps} = 159,000 + \frac{4,000}{300 \times 0.0013} + 10,000 = 179,256 \text{ psi } (1,236 \text{ MPa})$$

$$F_{ps} = \frac{179{,}256 \times 0.153 \times 11}{20} = 15{,}084 \text{ lb/ft}$$

$$F_s = 60{,}000 \times A_s/\text{ft} = 60{,}000 \times 0.06 = 3{,}600 \text{ lb/ft}$$

Then the total force $F/\text{ft} = F_{ps} + F_s = 15{,}084 + 3{,}600 = 18{,}684 \text{ lb/ft}$, and we also have

$$\text{Compression block depth } a = \frac{A_s f_y + A_{ps} f_{ps}}{0.85 f'_c b}$$

$$= \frac{18{,}684}{0.85 \times 4000 \times 12} = 0.46 \text{ in. (11.7 mm)}$$

The bars and tendons are to be placed at the same level $d = 6.5 - 1 = 5.5$ in. Also, $M_n = (A_s f_y + A_{ps} f_{ps})(d - a/2)$, the available $-M_n = 18{,}684 \times (5.5 - 0.46/2) = 98{,}465$ in.-lb/ft (36.5 MPa), and the required $M_n = 101{,}617$ in.-lb/ft $> 98{,}465$, which is unsatisfactory. Thus, additional moment strength $\Delta M_n = 101{,}617 - 98{,}465 = 3{,}152$ in.-lb/ft is needed.

We next check the allowable inelastic moment redistribution to midspan.

2. *Midspan section at span BC*
From before, $F_{ps} = A_{ps} f_{ps} = 15{,}084$ lb/ft, and

$$a = \frac{A_{ps} f_{ps}}{0.85 f'_c b} = \frac{15{,}084}{0.85 \times 4{,}000 \times 12} = 0.37 \text{ in.}$$

So the available $-M_n = A_{ps} f_{ps}(d - a/2) = 15{,}084(5.5 - 0.37/2) = 80{,}171$ in.-/ft, and the required $M_n = 81{,}317$ in.-lb/ft $> 80{,}171$ in-lb/ft, and hence is unsatisfactory. Accordingly, add two bars at midspan over a 20-ft width to get

$$A_s = 2 \times 0.20 = 0.40 \text{ in}^2$$

$$A_s f_y = \frac{0.40 \times 60{,}000}{20} = 1{,}200 \text{ lb/ft}$$

$$a = \frac{(15{,}084 + 1{,}200)}{0.85 \times 4{,}000 \times 12} = 0.40 \text{ in.}$$

$$\text{Available } +M_n = (A_s f_y + A_{ps} f_{pf})\left(d - \frac{a}{2}\right)$$

$$= (15{,}084 + 1{,}200)\left(5.5 - \frac{0.4}{2}\right) = 86{,}305 \text{ in.-lb/ft}$$

$$> \text{req } +M_n = 81{,}317 \text{ in.-lb/ft, O.K.}$$

Thus, the available ΔM_R to accommodate moment redistribution from the support is $86{,}305 - 81{,}317 = 4{,}988$ in.-lb/ft.

3. *Allowable inelastic moment redistribution ΔM_R at support junction toward midspan*
From Equation 9.18,

$$\rho_D = 20\left[1 - \frac{\omega_p + (d/d_p)(\omega - \omega')}{0.36\beta_1}\right] \text{ percent}$$

$$\omega_p = \frac{\rho_p f_{ps}}{f_c'} = 0.0013 \times \frac{179{,}256}{4{,}000} = 0.0583$$

$$d = d_p = 5.5 \text{ in.}$$

$$\omega = \omega' = 0$$

$$\beta_1 = 0.85$$

$$\rho_D = 20\left(1 - \frac{0.0583 + 0}{0.36 \times 0.85}\right) = 16.2 \text{ percent}$$

So use a 12 percent redistribution factor.

The inelastic redistribution from the support to midspan is

$$\Delta M_R = 0.120 \times 98{,}465 = 11{,}816 \text{ in.-lb/ft}$$

$$> \text{available } \Delta M_R = 4{,}988 \text{ in.-lb/ft}$$

which is unsatisfactory. Thus, add four #4 bars to make a total of six #4 bars at midspan. Then

$$\sum A_s f_y = \frac{6 \times 0.2 \times 60{,}000}{20} = 3{,}600 \text{ lb/ft}$$

$$a = \frac{15{,}084 + 3{,}600}{0.85 \times 4{,}000 \times 12} = 0.46 \text{ in. (11.7 mm)}$$

$$\text{Available } +M_n = (15{,}084 + 3{,}600)\left(5.5 - \frac{0.46}{2}\right)$$

$$= 98{,}465 \text{ in.-lb/ft}$$

$$\text{Available } \Delta M_R \text{ at midspan} = 98{,}465 - 81{,}317 = 17{,}148 \text{ in.-lb/ft}$$

$$> \Delta M_R = 11{,}816$$

which can be elastically redistributed from the support. Hence, the design is satisfactory.

Summary

The required $-M_n = 101{,}617 - 17{,}148 = 84{,}469$ in.-lb/ft, which is less than the support section nominal $-M_n = 98{,}465$ in.-lb/ft and, hence, satisfactory. The required $+M_n = 81{,}317 + 17{,}148 = 98{,}465$ in.-lb/ft, which is approximately equal to the available $+M_n = 98{,}465$ in.-lb/ft and, hence, satisfactory. Accordingly, use six #4 (12.7 mm dia) nonprestressed mild steel bars at the bottom fibers at midspan in addition to the continuous prestressing tendon in the 20-ft segment. Also, use six #4 nonprestressed mild steel bars at the top fibers at the support, centered through the column at 6 in. center-to-center spacing (six 12.7 mm dia bars at 152 mm center to center).

The midspan sections of spans AB and CD would have more than adequate positive nominal moment strength to resist the positive factored moments. The nominal negative moment strength of the sections at the exterior supports A and D are governed by the moment-shear transfer stresses.

4. *Banding the reinforcement at the column region*

There are eleven $\frac{1}{2}$ in. dia strands, and the width of a column strip $= 2(\frac{1}{4} \times 20 \times 12) = 120$ in. Assume that 70 percent of the tendons are concentrated in the column strip. Then the number of strands = $0.7 \times 11 = 7.7$. Accordingly, concentrate seven strands in the column strip, three of which are to pass through and be centered on the column section.

There are $11 - 7 = 4$ strands in the middle strip. On this basis, it can be reasonably assumed that the percentage distribution of moments between the column strip and the middle strip would be approximately as follows:

Column strip moment factor = $7/11 = 0.64$

Middle strip moment factor = 0.36

Max total $-M$ at *column face* B = 33,880 in.-lb/ft (see Table 9.2)

Max total $+M$ at midspan = 21,784 in.-lb/ft

Consequently, distribute the prestressing tendons between the column strips and middle strips as shown subsequently.

Nominal Shear Strength

1. *Exterior columns A and D*

(a) *Geometry and external load*

From before, $V_{AB} = 1430.3$ lb/ft, and the total shear is $V_B = 1430.3 \times 20 = 28,606$ lb. Assume an exterior wall and glass averaging a load of 500 plf:

$$\text{Wall } V_u = 1.4 \times 500 \times 20 = 14,000 \text{ lb}$$

$$\text{Slab } V_u = 28,606 \text{ lb}$$

$$\text{Total factored } V_{uA} = 42,606 \text{ lb (189.5 kN)}$$

The critical shear section is taken at $d'/2$ from the face of the column, as shown in Figure 9.20. We have

$$d = 6.5 - 1.0 = 5.5 \text{ in.}$$

$$\text{Max } d_p = d_v = 0.8h = 0.8 \times 6.5 = 5.2 \text{ in.(132 mm)}$$

$$c_1 = 12 \text{ in.}$$

$$c_2 = 14 \text{ in.}$$

$$b_1 = c_1 + \frac{d}{2} = 12 + \frac{5.2}{2} = 14.6 \text{ in.}$$

$$b_2 = c_2 + d = 14 + 5.2 = 19.2 \text{ in.}$$

$$A_c = b_0 d = 5.2(2 \times 14.6 + 19.2) = 252 \text{ in}^2$$

From the figure,

$$d(2c_1 + c_2 + 2d)\bar{x} = d\left(c_1 + \frac{d}{2}\right)^2$$

or

$$5.2(2 \times 12 + 14 + 2 \times 5.2)\bar{x} = 5.2(14.6)^2$$

$$\bar{x} = c_{AB} = 4.40 \text{ in.}$$

$$g = \bar{x} - \frac{d}{2} = 4.4 - \frac{5.2}{2} = 1.8 \text{ in.}$$

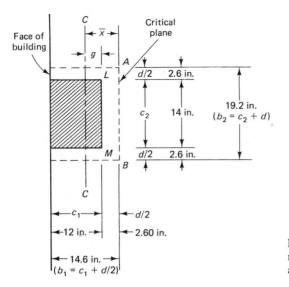

Figure 9.20 Critical planes for shear moment transfer in end column of Example 9.2 (line A, Figure 9.17).

Alternatively,

$$c_{AB} = \frac{b_1^2 d}{A_c} = \frac{(1.6)^2 \times 5.2}{252} = 4.40$$

$$c_{CD} = b_1 - c_{AB} = 14.6 - 4.4 = 10.2 \text{ in.}$$

From the geometrical properties of the exterior column shown in Figure 9.20, and from Equations 9.28 and 9.29,

$$\gamma_v = 1 - \frac{1}{1 + \frac{2}{3}\sqrt{b_1/b_2}} = 1 - \frac{1}{1 + \frac{2}{3}\sqrt{14.6/19.2}}$$

$$= 1 - 0.63 = 0.37$$

$$\gamma_f = \frac{1}{1 + \frac{2}{3}\sqrt{b_1/b_2}} = 0.63$$

Using d_v for d, the polar moment of inertia is

$$J_c = \frac{(c_1 + d/2)d^3}{6} + \frac{2d}{3}(c_{AB}^3 + c_{CD}^3) + (c_2 + d)(d)(c_{AB})^2$$

$$= \frac{14.6(5.2)^3}{6} + \frac{2 \times 5.2}{3}(4.4^3 + 10.2^3) + 19.2 \times 5.2(4.4)^2$$

$$= 342 + 3{,}974 + 1{,}933 = 6{,}249 \text{ in}^4$$

From before, the unit $-M_u = 10{,}700$ in.-lb/ft at the column centerline. So the total bay moment at the column centerline is $-M_c = 10{,}700 \times 20 = 214{,}000$ in.-lb. Now assume that the resultant V_u acts at the face of the column for shear-moment transfer. Then the shear moment transferred by eccentricity is $V_u g = -28{,}606 \times 1.8 = 51{,}491$ in.-lb, the total external factored moment $M_{ue} = -(214{,}000 + 51{,}491) = -265{,}491$ in.-lb, and the total required unbalanced moment strength $M_n = M_{ue}/\phi = 265{,}491/0.9 = 294{,}990$ in.-lb.

Placing concrete in a post-tensioned prestressed concrete slab.

(b) *Shear-moment transfer*

The fraction of the nominal moment strength to be transferred by shear is $\gamma_f M_n = 0.37 \times 294,990 = 109,146$ in.-lb. From Equation 9.30 a, the shearing stress due to perimeter shear, the effect of $\gamma_v M_n$ and the weight of the wall, is

$$v_n - \frac{V_u}{\phi A_c} + \frac{\gamma_v c_{AB} M_n}{J_c}$$

$$= \frac{42,606}{0.85 \times 252} + \frac{0.37 \times 4.4 \times 294,990}{6,249}$$

$$= 198.9 + 76.9 \cong 276 \text{ psi}$$

From the load balancing part of the solution, the average compressive stress in the concrete at the cross-section centroid due to externally applied load P_e is $\bar{f}_c = P_e/A_c = 172$ psi.

From equations 9.24 and 9.25 and disregarding the effect of the vertical component V_p of the prestressing force, the maximum allowable shear strength becomes

$$v_c = \beta_p \sqrt{f'_c} + 0.3\bar{f}_c$$

where the factor β_p is the smaller of $(\alpha_s d/b_0 + 1.5)$ and 3.5, and $\alpha_s = 30$ for end column support. From Figure 9.20, $b_0 = 2 \times 14.6 + 19.2 = 48.2$ in., and

$$\frac{\alpha_s d}{b_0} + 1.5 = \frac{30 \times 5.5}{48.2} + 1.5 = 4.92 > 3.5.$$

Hence use $\beta_p = 3.5$.

Max allowable $v_c = 3.5\sqrt{4,000} + 0.3 \times 172$

$$= 221 + 52 = 273 \text{ psi} \cong \text{actual } v_n = 276 \text{ psi, O.K.}$$

If V_p were accounted for, the maximum allowable v_c would have been higher than 273 psi.

(c) *Flexure moment transfer*

The fraction of nominal moment strength to be transferred by flexure is $M_n = 0.63 \times 294{,}990 = 185{,}844$ in.-lb. From Equation 9.20, Min $A_s = 0.00075hL = 0.00075 \times 6.5 \times 17.5 \times 12 = 1.02$ in². So use six #4 bars $\times$ 6 ft, including the standard hook, yielding $A_s = 6 \times 0.2 = 1.2$ in². The stress in the tendon strands is calculated from Equation 9.23 assuming that three strands pass the column at the exterior support at $e = 0$. We have $d = 6.5/2 = 3.25$ in., and the effective concrete width $b = c_2 + 2(1.5 + h) = 14 + 2(1.5 + 6.5) = 30$ in. Also,

$$\rho_p = \frac{A_{ps}}{bd_p} = \frac{3 \times 0.153}{30 \times 3.25} = 0.0047$$

$$f_{ps} = f_{pe} + 10{,}000 + \frac{f'_c}{300\rho_p}$$

$$= 159{,}000 + 10{,}000 + \frac{4{,}000}{300 \times 0.0047}$$

$$= 171{,}837 \text{ psi}$$

$$A_{ps} = 3 \times 0.153 = 0.459 \text{ in.}$$

$$a = \frac{A_s f_y + A_{ps} f_{ps}}{0.85 f'_c b} = \frac{1.20 \times 60{,}000 + 0.459 \times 171{,}837}{0.85 \times 4{,}000 \times 30}$$

$$= 1.48 \text{ in.}$$

$$\text{Available } M_n = A_s f_y\left(d - \frac{a}{2}\right) + A_{ps} f_{ps}\left(d_p - \frac{a}{2}\right)$$

$$= 1.2 \times 60{,}000\left(5.5 - \frac{1.48}{2}\right) + 0.459 \times 171{,}837\left(3.25 - \frac{1.48}{2}\right)$$

$$= 342{,}720 + 197{,}972 = 540{,}692 \text{ in.-lb}$$

$$\gg \gamma_f M_n = 185{,}844 \text{ in.-lb}$$

The available nominal moment strength is thus considerably larger than the moment being transferred by flexure. Figure 9.21 shows one scheme for banding both the prestressed and nonprestressed reinforcement to provide for shear-moment transfer at the exterior column zone.

2. *Interior columns B and C*

(a) *Geometry and external load*

From before, $V_{BA} + V_{BC} = 2{,}244.7 + 2{,}520 \cong 4765$ plf. The total shear is $V_{uB} = 4{,}765 \times 20 = 95{,}300$ lb (423.9 kN), and also, $c_1 = 20$ in., $c_2 = 14$ in., and $d = 6.5 - 1 = 5.5$ in.

Assume that $d_v = 0.8h \cong 5.2$ in.; calculate $g = \frac{1}{2}c_1 = 20/2 = 10$ in.

$$b_1 = c_1 + d = 20 + 5.2 = 25.2 \text{ in.}$$

$$b_2 = c_2 + d = 14 + 5.2 = 19.2 \text{ in.}$$

$$A_c = b_0 d = 2(25.2 \times 5.2 + 19.2 \times 5.2) = 462 \text{ in}^2$$

Using d_v for d, the polar moment of inertia is

$$J_c = \frac{d(c_1 + d)^3}{6} + \frac{d^3(c_1 + d)}{6} + \frac{d(c_2 + d)(c_1 + d)^2}{2}$$

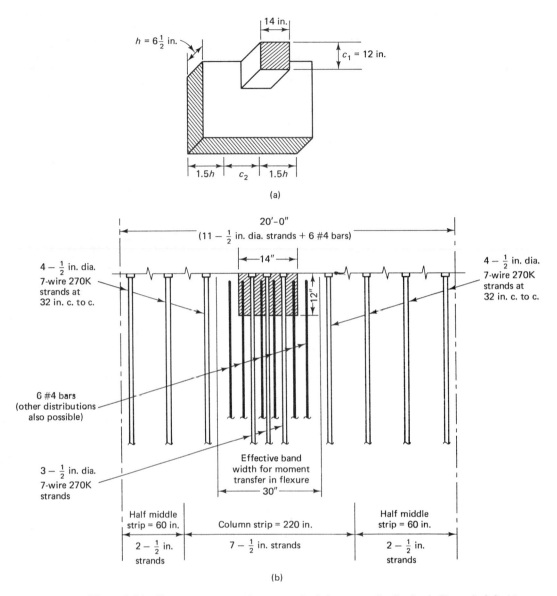

Figure 9.21 Shear-moment transfer zone and reinforcement distribution in Example 9.2. (a) Column zone band (33.5 in. wide). (b) Reinforcement distribution plan.

$$= \frac{5.2(25.2)^3}{6} + \frac{(5.2)^3(25.2)}{6} + \frac{5.2(19.2)(25.2)^2}{2}$$

$$= 46{,}161 \text{ in}^4$$

Figure 9.22 shows the geometrical properties of the *interior columns*:

$$\gamma_v = 1 - \frac{1}{1 + \frac{2}{3}\sqrt{25.2/19.2}} = 0.433$$

$$\gamma_f = 1 - 0.433 = 0.567$$

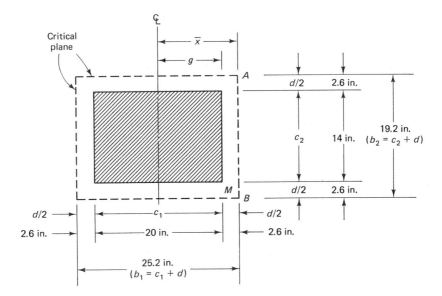

Figure 9.22 Critical plane for shear transfer in interior column of Example 9.2 (line B or C, Figure 9.17).

The moment $M_{ue} = M_e$ for each interior column, and the net unit moment $+M_e = 101{,}617 - 91{,}733 = 9{,}884$ in.-lb. The unbalanced shear moment is equal to the net $V_u \times g = 10(2{,}520 - 2{,}244.7) = 2{,}753$ in.-lb. Finally, the total moment $M_{ue} = 9{,}884 \times 20 + 2{,}753 = 200{,}433$ in.-lb, and the total required unbalanced moment strength is $M_n = M_{ue}/\phi = 200{,}433/0.9 = 222{,}703$ in.-lb.

(b) *Shear-moment transfer*

The fraction of nominal moment strength to be transferred by shear is $\gamma_v M_n = 0.433 \times 222{,}703 = 96{,}403$ in.-lb, and $c_{AB} = \frac{1}{2}(c_1 + d) = \frac{1}{2}b_1 = 25.2/2 = 12.6$ in.

From Equation 9.30 a, the shear stress due to perimeter shear and the effect of M_n is

$$v_n = \frac{V_u}{\phi A_c} + \frac{\gamma_v c_{AB} M_n}{J_c}$$

$$= \frac{95{,}300}{0.85 \times 462} + \frac{96{,}403 \times 12.6}{46{,}161} = 242.7 + 26.3 = 269 \text{ psi (185 MPa)}$$

$$< \text{allowable } v_c = 273 \text{ psi, O.K.}$$

(c) *Flexure moment transfer*

The fraction of nominal moment strength to be transferred by flexure is $\gamma_f M_n = 0.567 \times 222{,}703 = 126{,}273$ in.-lb, and $b = c_2 + 2(1.5 + h) = 14 + 2(1.5 + 6.5) = 30$ in., the same as for exterior column A. Assume, as in the case of exterior columns, that three strands pass the interior columns B and C. We have

$$d_p = 6.5 - 1 = 5.5$$

$$\rho_p = \frac{A_{ps}}{b d_p} = \frac{3 \times 0.153}{30 \times 5.5} = 0.0028$$

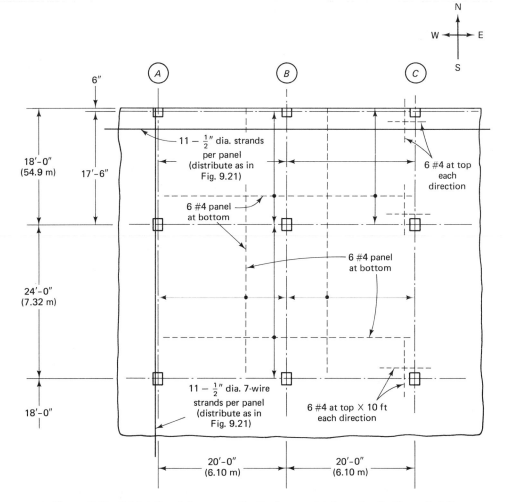

Figure 9.23 Schematic reinforcement distribution, partial floor plan for Example 9.2.

$$f_{ps} = f_{pe} + 10,000 + \frac{f'_c}{300\rho_p}$$

$$= 159,000 + 10,000 + \frac{4,000}{300 \times 0.0028} = 173,762 \text{ psi}$$

which is very close to f_{ps} for column A. Accordingly, using six #4 bars × 12 ft as minimum mild steel, as for the exterior columns, $a \cong 1.48$ in. and the available $M_n = 1.2 \times 60,000(5.5 - 1.48/2) + 0.459 \times 173,766(5.5 - 1.48/2) = 722,362$ in.-lb $\gg$ required $M_n = 126,273$ in.-lb, and hence satisfactory.

Figure 9.23 shows a schematic layout of the reinforcement in the continuous flat plate. The three $\frac{1}{2}$ in. dia strands in each direction should pass through the critical shear perimeter of the supporting columns. Of course, serviceability requirements for deflection should be checked, as in Section 9.13.

From the analysis made, we adopt the design and use the same pattern of reinforcement for both the N–S and E–W directions of the floor system, as the spans dimensions in both directions are very close in value.

9.12 DIRECT METHOD OF DEFLECTION EVALUATION

9.12.1 The Equivalent Frame Approach

As in the equivalent frame method for flexural analysis discussed in detail in the preceding sections, the structure is divided into continuous frames centered on the column lines in each of the two perpendicular directions. Each frame is composed of a row of columns and a *broad* band of slab together with column line beams, if any, between panel centerlines.

By the requirement of statics, the applied load must be accounted for in each of the two perpendicular (orthogonal) directions. In order to account for the torsional deformations of the support beams, an *equivalent* column is used whose flexibility is the *sum* of the flexibilities of the actual column and the torsional flexibility of the transverse beam or slab strips (stiffness is the inverse of flexibility). In other words,

$$\frac{1}{K_{ec}} = \frac{1}{\Sigma K_c} + \frac{1}{K_t} \tag{9.33}$$

where K_{ec} = flexural stiffness of the equivalent column, bending moment per unit rotation

ΣK_c = sum of flexural stiffnesses of upper and lower columns, bending moment per unit rotation

K_t = torsional stiffness of the transverse beam or slab strip, torsional moment per unit rotation.

The value of K_{ec} would thus have to be known in order to calculate the deflection by this procedure.

The slab–beam strips are considered supported *not* on the columns, but on *transverse* slab–beam strips on the column centerlines. Figure 9.24(a) illustrates this

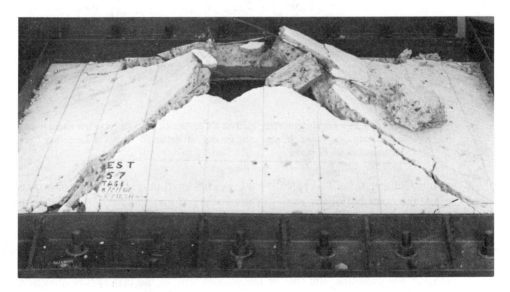

Rectangular concrete slab at rupture. (Tests by Nawy et al.)

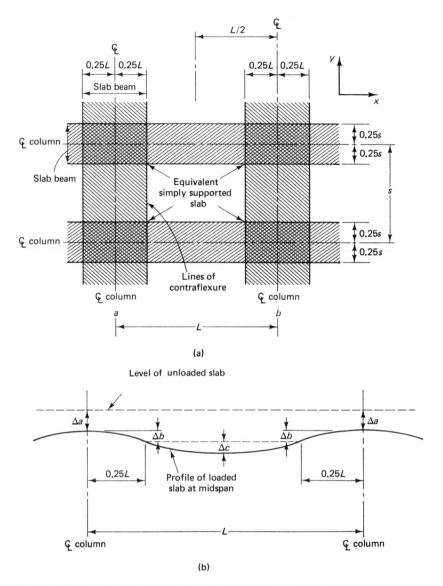

Figure 9.24 Equivalent frame method for deflection analysis. (a) Plate panel transferred into equivalent frames. (b) Profile of deflected shape at centerline. (c) Deflected shape of panel.

point. Deformation of a typical panel is considered in *one direction at a time*. Thereafter, the contributions in each of the two directions, x and y, are added to obtain the total deflection at any point in the slab or plate.

First, the deflection due to bending in the x direction is computed (Figure 9.24(b)). Then the deflection due to bending in the y direction is found. The midpanel deflection can now be obtained as the sum of the center-span deflections of the column strip in one direction and that of the middle strip in the orthogonal direction (Figure 9.24(c)).

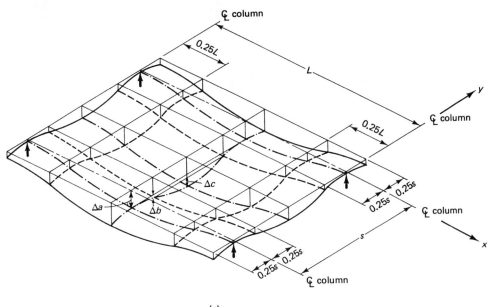

(c)

Figure 9.24 (*cont.*)

9.12.2 Column and Middle Strip Deflections

The deflection of each panel can be considered the sum of three components:

1. The basic midspan deflection of the panel, assumed fixed at both ends, given by

$$\delta' = \frac{wl^4}{384 E_c I_{\text{frame}}} \tag{9.34}$$

This has to be proportioned to separate deflections δ_c of the column strip and δ_s of the middle strip, such that

$$\delta_c = \delta' \frac{M_{\text{col strip}}}{M_{\text{frame}}} \frac{E_c I_{cs}}{E_c I_c} \tag{9.35 a}$$

and

$$\delta_s = \delta' \frac{M_{\text{slab strip}}}{M_{\text{frame}}} \frac{E_c I_{cs}}{E_c I_s} \tag{9.35 b}$$

where I_{cs} is the moment of inertia of the total frame, I_c the moment of inertia of the column strip, and I_s the moment of inertia of the middle slab strip.

2. The center deflection, $\delta''_{\theta L} = \frac{1}{8}\theta L$, due to rotation at the left end while the right end is considered fixed, where θL is the left M_{net}/K_{ec} and K_{ec} is the flexural stiffness of the equivalent column (moment per unit rotation).

3. The center deflection, $\delta''_{\theta R} = \frac{1}{8}\theta L$ due to rotation at the right end while the left end is considered fixed, where θL is the right M_{net}/K_{ec}. Hence,

$$\delta_{cx} \text{ or } \delta_{cy} = \delta_c + \delta''_{\theta L} + \delta''_{\theta R} \qquad (9.36 \text{ a})$$

$$\delta_{sx} \text{ or } \delta_{sy} = \delta_s + \delta''_{\theta L} + \delta''_{\theta R} \qquad (9.36 \text{ b})$$

In Equations 9.36 a and 9.36 b, use the values of δ_c, $\delta''_{\theta L}$, and $\delta''_{\theta R}$ which correspond to the applicable span directions. From Figures 9.24(b) and (c), the total deflection is

$$\Delta = \delta_{sx} + \delta_{cy} = \delta_{sy} + \delta_{cx} \qquad (9.37)$$

9.13 DEFLECTION EVALUATION OF TWO-WAY PRESTRESSED CONCRETE FLOOR SLABS

Example 9.3

Compute the central deflection of the exterior panels of the two-way post-tensioned prestressed concrete floor designed in Example 9.2 for both short-term and long-term loading. Assume that the maximum allowable deflection is 1/480 of the span.

Solution

Structural Data. From Example 9.2, we have the following data:

Plate thickness $h = 6\frac{1}{2}$ in. (165 mm)

Loads: $W_d = 100$ psf (4.84 kPa)

$\qquad W_L = 40$ psf (1.92 kPa)

$\qquad$ Span AB $W_{\text{bal}} = 72$ psf (3.45 kPa)

$\qquad W_{\text{net}} = W_D + W_L - W_{\text{bal}} = 101 + 40 - 72 = 69$ psf (3.3 kPa)

$\qquad$ Span BC $W_{\text{bal}} = 70$ psf (3.35 kPa)

$\qquad W_{\text{net}} = 141 - 70 = 71$ psf (3.4 kPa)

The floor plan is shown in Figure 9.25, and the overall details and vertical section of the building are presented in Figure 9.17. The distributed bending moments in the N-S direction taken from the flexural analysis for W_{net} in Table 9.2 are shown in Figure 9.26.

Stiffness Factors and Strip Moments

N-S Direction (Span 18 ft). The column stiffness factor K_{ec} values were computed in Example 9.2, with the following results:

Exterior column A: $K_{ec} = 47E_c$ in.-lb/rad

Interior column B: $K_{ec} = 113E_c$ in.-lb/rad

Net Frame $M_A = 5.30 \times 10^3$ in.-lb/ft

Net Frame $M_B = (39.56 - 33.94)10^3 = 5.62 \times 10^3$ in.-lb/ft

As discussed in Example 9.2, the column strip takes 64 percent of the moment and the middle strip takes 36 percent of the moment. The frame total $I_{cs} = bh^3/12 = 20 \times 12(6.5)^3/12 = 5,493$ in⁴, while the column strip $I_c =$ the middle strip $I_c = 5,493/2 = 2,747$ in⁴.

From Equation 9.34, the basic midspan deflection in the N-S direction at central

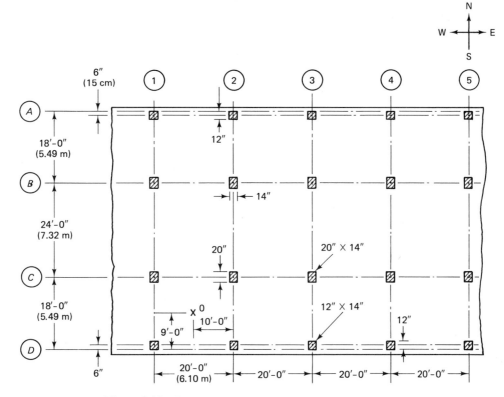

Figure 9.25 Two-way post-tensioned floor plan in Example 9.3.

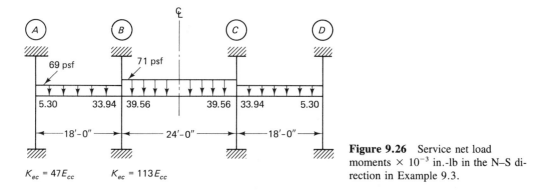

Figure 9.26 Service net load moments × 10⁻³ in.-lb in the N–S direction in Example 9.3.

point O in Figure 9.27, assuming both ends of the panel fixed, is

$$\delta' = \frac{WL^4}{384 E_c I_{cs}} = \frac{69 \times 20(18)^4(12)^3}{384 \times 4.03 \times 10^6 \times 5{,}493} = 0.029 \text{ in.}$$

This deflection has to be proportioned to separate deflections δ_c of the column strip and δ_s of the middle strip:

$$\delta_c = \delta' \frac{M_{\text{col. strip}}}{M_{\text{frame}}} \frac{E_c I_{cs}}{E_c I_c}$$

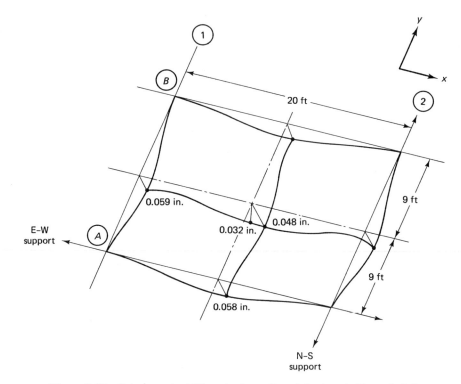

Figure 9.27 Column and middle strips immediate deflections in Example 9.3.

from Example 9.2, $M_{\text{col. strip}}/M_{\text{frame}} = 0.64$, so N-S $\delta_c = 0.029 \times 0.64 \times 2 = 0.037$ in., N-S $\delta_s = 0.029 \times 0.36 \times 2 = 0.021$ in., and the rotation at end A is

$$\theta_A = \frac{M_A}{K_{ec}} = \frac{5.30 \times 10^3 \times 20}{47 \times 4.03 \times 10^6} = 5.6 \times 10^{-4} \text{ rad}$$

$$\theta_B = \frac{M_B}{K_{ec}} = \frac{5.62 \times 10^3 \times 20}{113 \times 4.03 \times 10^6} = 2.5 \times 10^{-4} \text{ rad}$$

$$\delta'' = \frac{\theta\ell}{8} = \frac{(5.6 + 2.5)10^{-4}(18 \times 12)}{8} = 0.022 \text{ in.}$$

Therefore, N-S net $\delta_{cy} = 0.037 + 0.022 = 0.059$ in. and N-S net $\delta_{sy} = 0.021 + 0.022 = 0.043$ in.

E-W Direction (Span 20 ft). For the E-W direction, the width b of an equivalent frame $= \frac{1}{2}(18 + 24) = 21.0$ ft. The frame total I_{cs} is

$$\frac{bh^3}{12} = \frac{21 \times 12(6.5)^3}{12} = 5{,}767 \text{ in}^4$$

and the column strip $I_c = $ the middle strip $I_s = 5{,}767/2 = 2{,}884$ in⁴.

From Equation 9.34, the fixed-end central deflection at O is

$$\delta' = \frac{WL^4}{384E_cI_{cs}} = \frac{69 \times 21(20)^4(12)^3}{384 \times 4.03 \times 10^6 \times 5{,}767} = 0.045 \text{ in.}$$

If the same distribution of moments is assumed to exist between the column and middle strips, then E-W $\delta_c = 0.045 \times 0.64 \times 2 = 0.058$ in. and E-W $\delta_s = 0.045 \times 0.36 \times 2 = 0.032$ in.

For the case of all panels loaded in this example, the net moments at each column due to the difference in negative moments from the spans to the left and to the right of the column are zero. Hence, consider the net rotation $\theta = 0$, and use E-W net $\delta_{cx} = 0.058$ in. and E-W net $\delta_{sx} = 0.032$ in.

Figure 9.27 gives the column and middle strip deflections in both the N-S and E-W directions.

Total Immediate Central Deflection. The total central deflection $\Delta = \delta_{sx} + \delta_{cy} = \delta_{sy} + \delta_{cx}$; therefore $\Delta_{\text{N-S}} = \delta_{sy} + \delta_{cx} = 0.043 + 0.058 = 0.101$ in. and $\Delta_{\text{E-W}} = \delta_{sx} + \delta_{cy} = 0.032 + 0.059 = 0.091$ in. Hence, the average immediate deflection due to the net load is $W_{\text{net}} = \frac{1}{2}(\Delta_{\text{N-S}} + \Delta_{\text{E-W}}) = \frac{1}{2}(0.101 + 0.091) = 0.096$ in. (2.44 mm).

Long-Term Deflection. For the long-term deflection, $W_{\text{net}} = 69$ psf and the live load $W_L = 40$ psf. Assuming that 65 percent of the live load is sustained, the total sustained load intensity is $W_{\text{sust.}} = (69 - 40) + 0.65 \times 40 = 55$ psf. Assuming further a total creep factor of 2, we have

$$\text{Long-term deflection} = \frac{55}{69} \times 0.096 \times 2 = 0.153 \text{ in. (4.09 mm)}$$

$$\text{Total deflection} = 0.096 + 0.153 = 0.249 \text{ in. (6.33 mm)}$$

The maximum allowable deflection in this structure is

$$\Delta_{\text{allow.}} = \frac{L}{480} = \frac{20 \times 12}{480} = 0.50 \text{ in. (12.7 mm)} > \text{actual } \Delta = 0.249 \text{ in., O.K.}$$

Testing setup of four-panel prestressed concrete floor. (Tests by Nawy et al.)

9.14 YIELD-LINE THEORY FOR TWO-WAY-ACTION PLATES

A study of the hinge-field mechanism in a slab or plate at loads close to failure aids the engineering student in developing a feel for the two-way-action behavior of plates. Hinge fields are successions of hinge bands which are idealized by lines; hence the name *yield-line theory* by K. W. Johansen.

To do justice to this subject, an extensive discussion over several chapters or a whole textbook is necessary. The intention of this chapter is only to introduce the reader to the fundamentals of the yield-line theory and its application.

The yield-line theory is an upper-bound solution to the plate problem. This means that the predicted moment capacity of the slab has the highest expected value in comparison with test results. Additionally, the theory assumes a totally rigid-plastic behavior, namely, that the plate stays plane at collapse. Consequently, deflection is not accounted for, nor are the compressive membrane forces that will act in the plane of the slab or plate considered. The plates are assumed to be considerably under-reinforced, in such manner that the maximum reinforcement percentage ρ does not exceed $\frac{1}{2}$ percent of the section bd.

Since the solutions are upper bound, the slab thickness obtained by this process is in many instances thinner than what is obtained by the lower bound solutions, such as the equivalent frame method. Consequently, it is important to apply rigorously the serviceability requirements for deflection control and for crack control in conjunction with the use of the yield-line theory.

One distinct advantage of this theory is that solutions are possible for any shape of plate, whereas most other approaches are applicable only to the rectangular shapes with rigorous computations for boundary effects. The engineer can, with ease, find the moment capacity for a triangular, trapezoidal, rectangular, circular, or any other conceivable shape, provided that the failure mechanism is known or predictable. Since most failure patterns are presently identifiable, solutions can be readily obtained.

9.14.1 Fundamental Concepts of Hinge-Field Failure Mechanisms in Flexure

Under action of a two-dimensional system of bending moments, yielding of a rigid-plastic plate occurs when the principal moments satisfy Johansen's square yield

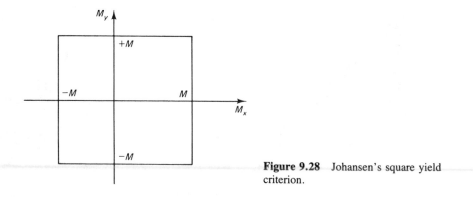

Figure 9.28 Johansen's square yield criterion.

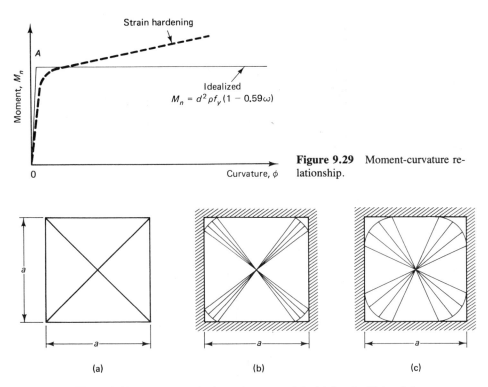

Figure 9.29 Moment-curvature relationship.

Figure 9.30 Failure mechanism of a square slab. (a) $i = 0$. (b) $i = 0.5$. (c) $i = 1.0$.

criterion as shown in Figure 9.28. According to this criterion, yielding is considered to have occurred when the numerically greater of the principal moments reaches the value of $\pm M$ at the yield-line cracks. The directions of the principal curvature rates are considered to coincide with the curvatures of the principal moments. The idealized moment–curvature relationship is shown as the solid line in Figure 9.29. Line OA is considered almost vertical at point O, and strain hardening is neglected.

If one considers the simplest case of a square slab with supports, with degree of fixity i varying from $i = 0$ for simply supported to $i = 1.0$ for fully restrained on all four sides, the failure mechanism would be as shown in Figure 9.30 when a uniformly distributed load is applied.

Consider the simply supported case (a). The yield-line moments along the yield lines are the principal moments. Hence, the twisting moments are zero in the yield lines and in most cases the shearing forces are also zero. Consequently, only the moment M per unit length of the yield line acts about the lines AD and BE in Figure 9.31. The total moments can be represented by a vector in the direction of the yield line whose value is the product of M and the length of the yield line, that is, $M(a/2 \cos \theta)$ in Figure 9.31(c). The virtual work of the yield moments of the shaded triangular segment ABO is the scalar product of the two moment vectors $Ma/2 \cos \theta$ on fracture lines AO and BO and rotation θ. In other words, the internal work is

$$E_I = \sum \overline{M}\,\overline{\theta}$$

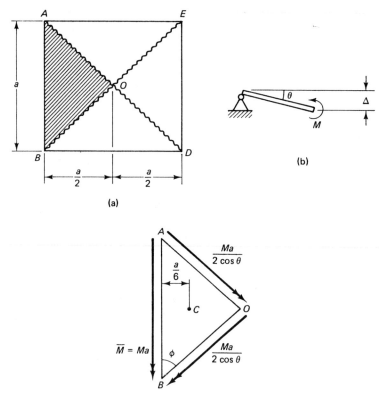

(a)

(b)

(c)

Figure 9.31 Vector moments on slab segment at failure.

If the displacement of the shaded segment at its center of gravity c is δ, the external work is

$$E_E = \text{force} \times \text{displacement} = \sum \iint w_u \, dx \, dy \, \delta$$

where w_u is the intensity of external load per unit area. But $E_I = E_E$; hence,

$$\sum \overline{M} \, \overline{\theta} = \sum \iint w_u \, dx \, dy \, \delta \qquad (9.38)$$

Applying Equation 9.38 to the case under discussion gives us

$$\overline{M} \, \overline{\theta} = Ma \, \frac{\Delta}{a/2}$$

since angle θ in Figure 9.31(b) is small ($\theta = \Delta/(a/2)$).

The work per triangular segment is

$$E_I = \overline{M} \, \overline{\theta} = 2M\Delta$$

$$E_E = \frac{w_u a^2}{4} \times \frac{\Delta}{3}$$

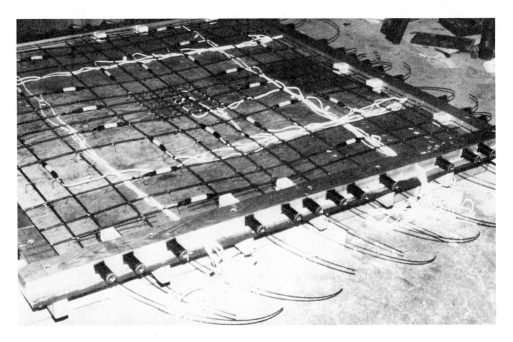

Preparing prestressing tendons in the forms for a four-panel continuous prestressed two-way-action plate (Nawy, Chakrabarti, et al.).

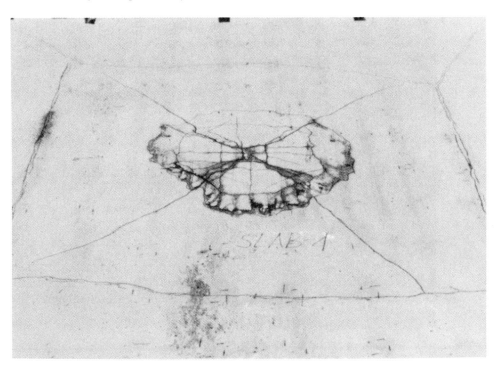

Yield-line pattern at failure at column reaction and panel boundaries of a two-way multipanel floor. (Tests by Nawy, Chakrabarti, et al.)

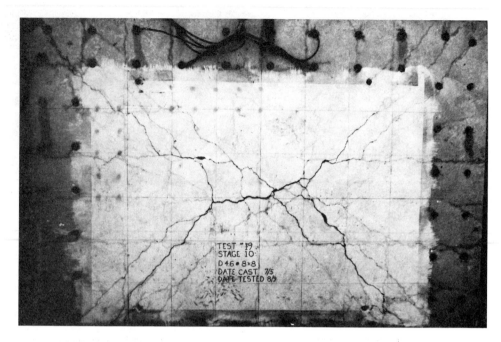

Yield-line patterns at failure at tensions face of rectangular restrained panel. (Tests by Nawy et al.)

where the deflection at the center of gravity of the triangle is $\Delta/3$. Therefore,

$$4(2M\Delta) = 4\left(\frac{w_u a^2}{12}\Delta\right)$$

and

$$\text{unit } M = \frac{w_u a^2}{24} \tag{9.39}$$

If the square slab is fully fixed on all four sides, $E_I = 4(4M\Delta)$ since fracture lines develop around not only the diagonals but also the four edges, as shown in Figure 9.30(c). Hence, for a fully fixed square slab,

$$\text{unit } M = \frac{w_u a^2}{48} \tag{9.40}$$

Observe that a lower bound solution as proposed by Mansfield's failure pattern in Figure 9.30(c) gives a value $M = w_u a^2/42.88$. Hence, for a uniformly loaded square slab with load intensity w_u per unit area and degree of support fixity i on all sides,

$$w_u a^2 = M[24(1 + i)] \tag{9.41}$$

The general equation for the yield-line moment capacity of a rectangular iso-tropic slab on beams and having dimensions $a \times b$ as shown in Figure 9.32, with side

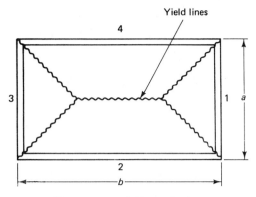

Figure 9.32 Rectangular slab. Note sequence of side numbers.

(Note sequence of side numbers)

a being the shorter dimension, is

$$\text{unit } M \frac{\text{ft-lb}}{\text{ft}} = \frac{w_u a_r^2}{24}\left[\sqrt{3 + \left(\frac{a_r}{b_r}\right)^2} - \frac{a_r}{b_r}\right]^2 \tag{9.42}$$

where $a_r = \dfrac{2a}{\sqrt{1 + i_2} + \sqrt{1 + i_4}}$

$b_r = \dfrac{2b}{\sqrt{1 + i_1} + \sqrt{1 + i_3}}$

i = degree of restraint, depending on stiffness ratios as discussed in Section 9.2.

Note that Equation 9.42 reduces to the simplified form of Equation 9.40 or 9.41 for the case of a square slab restrained on all four sides ($i = 1.0$).

Affine Slabs. Slabs that are reinforced differently in the two perpendicular directions are called *orthotropic slabs* (or *plates*). The moment in the *x* direction equals M and in the *y* direction equals μM, where μ is a measure of the degree of orthotropy, or the ratio

$$\frac{M_y}{M_x} = \frac{(A_s)_y}{(A_s)_x}$$

To simplify the analysis, the slab should be converted to an *affine* (isotropic) slab, where the strength and reinforcement area in both the *x* and *y* directions are the same. Such conversion can be made as follows:

1. *Divide* the linear dimension in the M direction by $\sqrt{\mu}$ for a slab to be reinforced for a moment M in both directions using the same unit load intensity w_u per unit area.

2. In the case of concentrated loads or total loads, also divide such loads by $\sqrt{\mu}$.

3. In the case of line loads, the line load has to be divided by $\sqrt{\mu \cos^2\theta + \sin^2\theta}$, where θ is the angle between the line load and the M direction.

If the slab is to be analyzed as an affine slab with the moment μM in both directions, the dimension in the μM direction has to be *multiplied* by $\sqrt{\mu}$. In either case, the result is of course the same.

9.14.2 Failure Mechanisms and Moment Capacities of Slabs of Various Shapes Subjected to Distributed or Concentrated Loads

The preceding concise introduction to the virtual-work method of yield-line moment evaluation should facilitate a good understanding of the mathematical procedures of most standard rectangular shapes subjected to uniform loading. More complicated slab shapes and other types of symmetrical and nonsymmetrical loading require more advanced knowledge of the subject. Also, the assumed failure shape and minimization energy principles can give values for particular cases that differ slightly from one experimenter to another depending on the mathematical assumptions made with respect to the failure shape.

The following summary of failure patterns and the respective moment capacities in terms of load, many of them due to Mansfield (Ref. 9.14), should give the reader adequate coverage of solutions to most cases expected in today's and tomorrow's structures.

1. Point load to corner of rectangular cantilever plates:

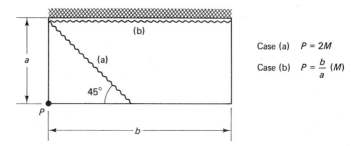

Case (a) $P = 2M$

Case (b) $P = \dfrac{b}{a}\,(M)$

2. Square plate centrally loaded and having boundaries simply supported against both downward and upward movements:

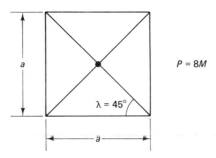

$P = 8M$

3. Regular *n*-sided plate with simply supported edges and centrally loaded ($n > 4$):

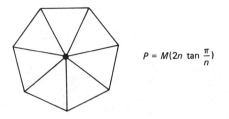

$$P = M(2n \tan \frac{\pi}{n})$$

4. Square plates centrally loaded and having boundaries simply supported against downward movement, but free for upward movement:

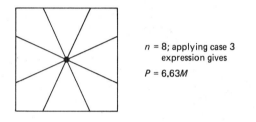

$n = 8$; applying case 3
expression gives

$P = 6.63M$

5. Circular centrally loaded plate simply supported along the edges:

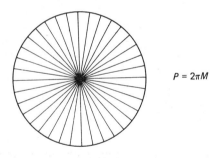

$P = 2\pi M$

6. Circular plate with fully restrained edges and centrally loaded by point load *P*:

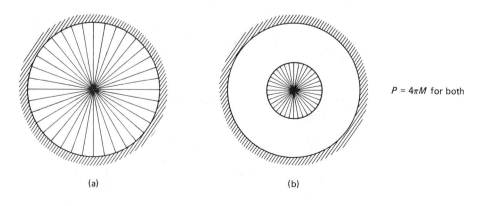

(a)   (b)

$P = 4\pi M$ for both

7. Point load P applied anywhere in arbitrarily shaped plate fully restrained on all boundaries:

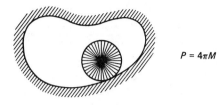

$$P = 4\pi M$$

8. Equilateral triangular plate with simply supported edges and centrally loaded by point load P:

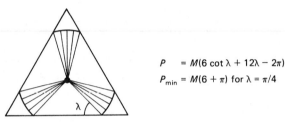

$$P = M(6 \cot \lambda + 12\lambda - 2\pi)$$
$$P_{min} = M(6 + \pi) \text{ for } \lambda = \pi/4$$

9. Acute-angled triangular plate on simply supported edges loaded with point load P at the center of the inscribed circle:

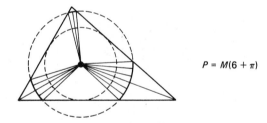

$$P = M(6 + \pi)$$

10. Obtuse-angled triangular plate with simply supported edges and load P at the center of the inscribed circle:

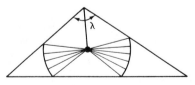

$$P = M(4 + 2\lambda + 2 \cot 1/2\lambda),$$
where ϕ is in radians

As λ approaches π, the plate degenerates into case 11

11. Long strip simply supported along the edges and load with point P midway between the edges:

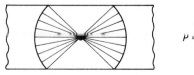

$$P = M(4 + 2\pi)$$

12. Simply supported strip with equal loads P between the edges:

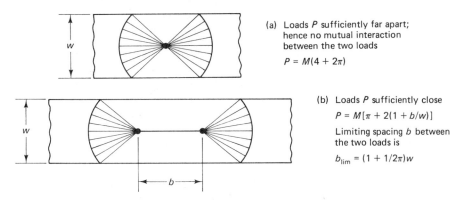

(a) Loads P sufficiently far apart; hence no mutual interaction between the two loads

$$P = M(4 + 2\pi)$$

(b) Loads P sufficiently close

$$P = M[\pi + 2(1 + b/w)]$$

Limiting spacing b between the two loads is

$$b_{lim} = (1 + 1/2\pi)w$$

13. Simply supported strip with unequal loads P and kP midway between the edges, where $k < 1.0$ and the loads are sufficiently apart:

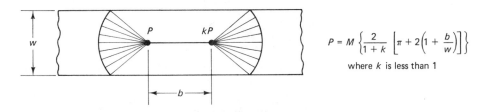

$$P = M\left\{ \frac{2}{1 + k} \left[\pi + 2\left(1 + \frac{b}{w}\right) \right] \right\}$$

where k is less than 1

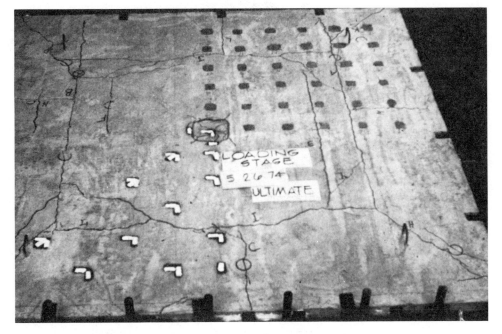

Four-panel slab at failure showing the yield-line patterns at the negative compression face of the supports. (Tests by Nawy and Chakrabarti.)

14. Uniformly loaded square slab with degree of fixity i varying between zero and 1.0:

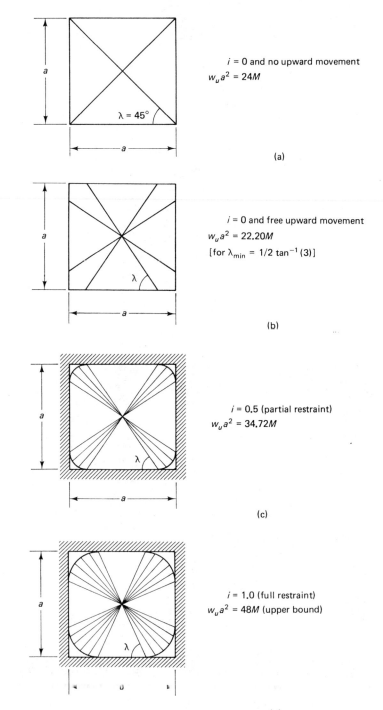

$i = 0$ and no upward movement
$w_u a^2 = 24M$

(a)

$i = 0$ and free upward movement
$w_u a^2 = 22.20M$
[for $\lambda_{min} = 1/2 \tan^{-1}(3)$]

(b)

$i = 0.5$ (partial restraint)
$w_u a^2 = 34.72M$

(c)

$i = 1.0$ (full restraint)
$w_u a^2 = 48M$ (upper bound)

(d)

15. Equilateral triangular plate ($\lambda = 60°$) uniformly loaded:

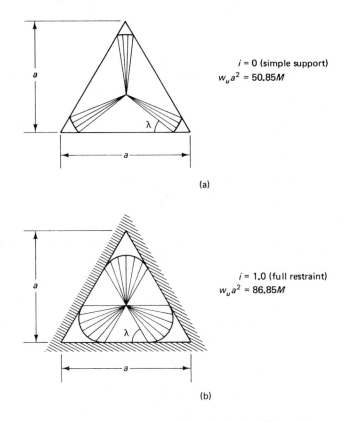

$i = 0$ (simple support)

$w_u a^2 = 50.85M$

(a)

$i = 1.0$ (full restraint)

$w_u a^2 = 86.85M$

(b)

16. Rectangular slab uniformly loaded with unit load of intensity w_u supported on all four sides with degree of restraint i varying from zero to 1.0 (note the sequence of numbers assigned to the panel sides):

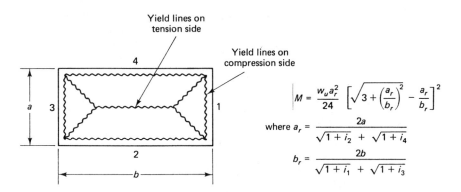

Yield lines on tension side

Yield lines on compression side

$$M = \frac{w_u a_r^2}{24} \left[\sqrt{3 + \left(\frac{a_r}{b_r}\right)^2} - \frac{a_r}{b_r} \right]^2$$

where $a_r = \dfrac{2a}{\sqrt{1 + i_2} + \sqrt{1 + i_4}}$

$b_r = \dfrac{2b}{\sqrt{1 + i_1} + \sqrt{1 + i_3}}$

As a general note, in the foregoing equations relating the load P to the moment M, load P is assumed to act at a point. To adjust for the fact that P acts on a finite area, assume that it acts over a circular area of radius ρ. For a slab fully restrained on all boundaries, the hinge field would be bound by a circle touching the slab boundary (circle radius = r). In such a case,

$$M + M' = \frac{P}{2\pi}\left(1 - \frac{2\rho}{3r}\right) \tag{9.43}$$

where M is the positive unit moment and M' the negative unit moment.

The reaction of columns supporting flat plates can be similarly considered for analyzing the flexural local capacity of the plate in the column area. For rectangular supports, an approximation to an equivalent circular support can be made in the use of Equation 9.43.

9.15 YIELD-LINE MOMENT STRENGTH OF A TWO-WAY PRESTRESSED CONCRETE PLATE

Example 9.4

Find the nominal moment strength of the two-way prestressed concrete plate in Example 9.2, assuming that the prestressing strands are bonded.

Solution

Loads. The total load intensity at the limit state of failure, from Example 9.2, is $W_u = 1.4W_D + 1.7W_L = 210$ psf. Assuming that the column reaction is an inverted concentrated load in the continuous plate field, the required moment strength M_n can be defined from case 7 of subsection 9.14.2 as follows:

$$P_A = 4\pi M_n$$

$$\text{Factored } P_u = 210 \times 20\left(\frac{24 + 18}{2}\right) = 88,200 \text{ lb (393 kN)}$$

$$\text{(the column weight is negligible)}$$

$$\text{Req } P_n = \frac{P_u}{\phi} = \frac{88,200}{0.9} = 98,000 \text{ lb (436 kN)}$$

$$\text{Req Unit } M_n \text{ for point load} = \frac{P_n}{4\pi} = \frac{98,000}{4 \times 3.14} = 7,799 \text{ lb (34.7 kN)}$$

$$\text{or } 7,779 \text{ in.-lb/in. (34.7 kN-m/m)}$$

$$\text{Equivalent } \rho = \frac{20 \times 14}{\pi^2} = 28.4 \text{ in.}$$

Assume $r \cong 17.5$ ft $= 210$ in. and $M' = M$. Then

$$M'_n = \frac{P_n}{4\pi}\left(1 - \frac{2\rho}{3r}\right) = 7,799\left(1 - \frac{2 \times 28.4}{210}\right) = 5,690 \text{ lb (25.3 kN)}$$

Available Moment Strength M_n. The available column area slab reinforcement is determined as follows.

Prestressing Steel

$$A_{ps} = \text{three } \tfrac{1}{2} \text{ in. dia seven-wire 270 K strands} = 3 \times 0.153 = 0.459 \text{ in}^2$$

$$f_{py} = 243,000 \text{ psi } (1,675 \text{ MPa})$$

$$f_{ps} = 179{,}256 \text{ psi at interior column}$$

$$f'_c = 4{,}000 \text{ psi } (27.58 \text{ MPa})$$

Accordingly, use f_{py} at the limit state of failure.

Nonprestressed Steel

$$A_s = \text{six } \#4 \text{ bars} = 6 \times 0.2 = 1.20 \text{ in}^2$$

$$f_y = 60{,}000 \text{ psi}$$

Moment Strength M_n

$$d_p = d = 6.5 - 1 = 5.5 \text{ in.}$$

$$b = 33.5 \text{ in. (from Example 9.2)}$$

$$a = \frac{A_s f_y + A_{ps} f_{py}}{0.85 f'_c b} = \frac{1.2 \times 60{,}000 + 0.459 \times 243{,}000}{0.85 \times 4{,}000 \times 33.5} = 1.61 \text{ in. } (40.9 \text{ mm})$$

$$\text{Available } M_n = A_s f_y \left(d - \frac{a}{2} \right) + A_{ps} f_{py} \left(d_p - \frac{a}{2} \right)$$

$$= 1.2 \times 60{,}000 \left(5.5 - \frac{1.61}{2} \right) + 0.459 \times 243{,}000 \left(5.5 - \frac{1.61}{2} \right)$$

$$= 338{,}040 + 523{,}666 = 861{,}706 \text{ in.-lb}$$

$$\text{Unit } M_n = \frac{861{,}706 \text{ in.-lb}}{33.5 \text{ in.}} = 25{,}723 \text{ in.-lb/in.} = 25{,}723 \text{ lb}$$

Check M_n For the Entire Panel Width

$$\text{N-S slab band width} = \frac{18 + 24}{2} = 21 \text{ ft } (6.4 \text{ m})$$

$$\text{E-W slab band with} = 20 \text{ ft}$$

So use $b = 21$ ft $= 252$ in. Then the total $A_{ps} =$ eleven $\frac{1}{2}$ in. (12.7 mm) dia seven-wire strands. Since the top nonprestressed steel is only in the column zone, disregarding it would be on the safe side. We then have

$$\text{Unit } A_{ps} = \frac{11 \times 0.153}{25.2} = 0.0067 \text{ in}^2/\text{in.}$$

$$a = \frac{0.0067 \times 243{,}000}{0.86 \times 4{,}000 \times 1} = 0.48 \text{ in.}$$

$$\text{Unit } M_n = 0.0067 \times 243{,}000 \left(5.5 - \frac{0.48}{2} \right) = 8{,}564 \text{ in.-lb/in.}$$

$$= 8{,}564 \text{ lb } (38.09 \text{ kN}), \text{ use}$$

$$\text{Req } M_n = 5{,}690 \text{ lb} < \text{available } M_n = 8{,}564 \text{ lb, O.K.}$$

Plainly, from this *limit theory* solution, quick analysis of a prestressed plate can be performed. Such an analysis, however, should also include an evaluation of yield-line shear strength at the support (Ref. 9.15) and serviceability checks for crack control and deflection control. The designer can easily choose the moment values for the applicable failure mechanism as presented in subsection 9.14.2. A serviceability check for crack control can be easily made using the criteria based on the extensive research reported in Refs. 9.19–9.21 and the discussion in Section 11.9 in this text on crack control in walls of large prestressed concrete tanks.

REFERENCES

9.1 Post-Tensioning Institute. *Post-Tensioning Manual*. 4th ed. Phoenix: Post-Tensioning Institute, 1985.

9.2 ACI Committee 318. *Building Code Requirements for Reinforced Concrete 318–89, and Commentary on Building Code Requirements for Concrete Institute 318R–89*. Detroit: American Concrete Institute, 1989.

9.3 Nawy, E. G. *Reinforced Concrete—A Fundamental Approach*. Englewood Cliffs, N.J.: Prentice Hall, 1985.

9.4 Nawy, E. G., and Chakrabarti, P. "Deflection of Prestressed Concrete Flat Plates." *Journal of the Prestressed Concrete Institute* 21 (1976): 86–102.

9.5 Lin, T. Y. "Load-Balancing Method for Design and Analysis of Prestressed Concrete Structures." *Journal of the American Concrete Institute* 60 (1963): 719–742.

9.6 Burns, N. H., and Hemabom, R. "Test of Scale Model Post-Tensioned Flat Plate." *Journal of the Structural Division, American Society of Civil Engineers* 103 (1977): 1237–1255.

9.7 Nilson, A. H. *Design of Prestressed Concrete*. New York: Wiley, 1987.

9.8 Scordelis, A. C., Lin, T. Y., and Itaya, R. "Behavior of a Continuous Slab Prestressed in Two Directions. *Journal of the American Concrete Institute* 56 (1959): 441–459.

9.9 Cross, H., and Morgan, N. *Continuous Frames of Reinforced Concrete*. New York: Wiley, 1954.

9.10 Rice, P. F., Hoffman, E. S., Gustafson, D. P., and Gouwens, A. J. *Structural Design Guide to the ACI Building Code*. 3d ed. New York: Van Nostrand Reinhold, 1985.

9.11 Nawy, E. G. "Strength, Serviceability, and Ductility." In *Handbook of Structural Concrete*, pp. 12-1 to 12-88. London and New York: McGraw Hill, 1983.

9.12 Lin, T. Y., and Burns, N. H. *Design of Prestressed Concrete Structures*. 3d ed. New York: Wiley, 1981.

9.13 Reynolds, C. E., and Steedman, J. C. *Reinforced Concrete Designer's Handbook*. 9th ed. London: Viewpoint Publications, 1981.

9.14 Mansfield, E. H. "Studies in Collapse Analysis of Rigid-Plastic Plates with a Square Yield Diagram." *Proceedings of the Royal Society* 241 (1957): 225–261.

9.15 Gesund, Hans, and Dikshit, O. P., "Yield Line Analysis of Punching Problem at Slab Column Intersections." In *International Symposium on Cracking, Deflection, and Ultimate Load of Concrete Slab Systems*, pp. 177–203. Detroit: American Concrete Institute, 1971.

9.16 Hung, T. Y., and Nawy, E. G. "Limit Strength and Serviceability Factors in Uniformly Loaded, Isotropically Reinforced Two-Way Slabs." In *International Symposium on Cracking, Deflection, and Ultimate Load of Concrete Slab Systems*, pp. 1–41 Detroit: American Concrete Institute, 1971.

9.17 Nilson, A. H., and Walters, D. B. "Deflection of Two-Way Floor Systems by the Equivalent Frame Method." *Journal of the American Concrete Institute* 72 (1975): 210–218.

9.18 Nawy, E. G., and Chakrabarti, P., "Serviceability Deflection Behavior of Two-Way Action Prestressed Concrete Plates." In *International Conference on Prestressed Concrete*, Sydney, Australia: Concrete Institute of Australia, 1976, pp 1–10.

9.19 Nawy, E. G. "Crack Control through Reinforcement Distribution in Two-Way Acting Slabs and Plates." *Journal of the American Concrete Institute* 69 (1972): 217–219.

9.20 Nawy, E. G., and Blair, K. *Further Studies of Flexural Crack Control in Structural Slab Systems*. Detroit: American Concrete Institute, 1971.

9.21 Vessey, J. V., and Preston, R. L. *A Critical Review of Code Requirements for Prestressed Concrete Reservoirs*. Paris: F. I. P., 1978.

9.22 Cohn, M. Z. "Partial Prestressing, From Theory to Practice." In *NATO—ASI Applied Science Series*, Vol. 1, p. 405, and Vol. 2, p. 425. Dordrecht, Netherlands: Martinus Nijhoff, in Cooperation with NATO Scientific Affairs Division, 1986.

PROBLEMS

9.1 Design the two-way prestressed floor in Example 9.2 by the equivalent frame method if the spacing of the columns in the E–W direction is changed to 24 ft (7.32 m) center to center. Analyze the floor for the worst condition of pattern live loading, and find the maximum long-term deflection of both the central and the end floor panels and compare it with the maximum allowable deflection if the floor carries equipment that is sensitive to excessive deflection.

9.2 Design the flat plate in Problem 9.1 considering it as a lift slab supported on steel columns as shown in Figure P9.2. Assume that no negative moments are transferred from the slab panels to the supporting columns, and check for deflections accordingly.

9.3 Analyze the flat plate in Problem 9.1 by the yield-line theory, and compare the design results with the equivalent frame design used in Problem 9.1.

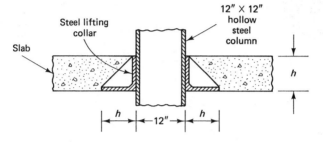

Figure P9.2 Lift slab at column support.

CHAPTER 10

Connections for Prestressed Concrete Elements

10.1 INTRODUCTION

The function of a connection is to economically transmit loads and stresses from one part of a structure to an adjoining part and provide stability to the structural system. The forces acting at the connection or joint are produced not only by gravity loads but by winds, seismic effects, volumetric changes due to long-term creep and shrinkage, differential movement of panels, and temperature effects.

Since a connection is the weakest link in the overall structural system, it has to be designed for nominal strength higher than the elements it connects. An additional load factor of at least 1.3 should be used in the design of connections, except in the case of insensitive connections, such as pads for column bases, where such an additional load factor is not necessary. All connections should be designed for a minimum horizontal tensile force of 0.2 times the vertical dead load, unless properly designed bearing pads are used.

The factors that have to be considered in design of a connection for strength are as follows:

1. The load transfer mechanism
2. Load factors
3. Volumetric changes
4. Ductility
5. Durability
6. Fire resistance

Gulf Life Center, Jacksonville, Florida. (*Courtesy*, Prestressed Concrete Institute.)

7. Required tolerances and clearances

8. Erection-related considerations

9. Considerations regarding hot weather and cold weather

10. The economics of the details of the connection

10.2 TOLERANCES

Clearances between elements must be realistically assessed. Where large tolerances are allowed in a supporting structure, or where no tolerances are specified, the clearances have to be increased to account for these factors. The following are recommended tolerances from Ref. 10.2 for deviations from idealized dimensions in beams, columns, and spandrel panels:

1. Variation in plan from specified location in plan: $\pm \frac{1}{2}$ in., any column or beam, any locations.

2. Deviation in plan from straight lines parallel to specified linear building lines: $\frac{1}{40}$ in. per ft, any beam less than 20 ft, or adjacent columns less than 20 ft apart; $\frac{1}{2}$ in., adjacent columns 20 ft or more apart.

3. Difference in relative position of adjacent columns from specified relative position: $\frac{1}{2}$ in. at any deck level.

4. Deviation from plumb: $\frac{1}{4}$ in. for every 10 ft of height; 1 in. maximum for the entire height.

5. Variation in elevation of bearing surfaces from specified elevation: $\pm \frac{1}{2}$ in., any column or beam, any location.

6. Deviation of top of spandrel from specified elevation: $\frac{1}{2}$ in., any spandrel.

7. Deviation in elevation of bearing surfaces from lines parallel to specified grade lines: $\frac{1}{40}$ in. per ft, any beam less than 20 ft or adjacent columns less than 20 ft apart; $\frac{1}{2}$ in. maximum, any beam 20 ft or more in length or adjacent columns 20 ft or more apart.

8. Variation from specified bearing length on support: $\pm \frac{3}{4}$ in.

9. Variation from specified bearing wdith on support: $\pm \frac{1}{2}$ in.

10. Jog in alignment of matching edges: $\frac{1}{4}$ in.

Table 10.1 gives tolerances applicable to connections.

10.3 COMPOSITE MEMBERS

As discussed in detail in Chapter 5 Sections 5.7 through 5.11, full transfer of horizontal shear forces must be assured at the interface of the precast member and the situ-cast topping. Figure 5.14 presents the interacting forces, and the flow chart of Section 5.8.2 gives the operational step-by-step design procedure and the applicable design equations. Figure 5.18 of Example 5.3 and the accompanying design give the size and

TABLE 10.1 TOLERANCES FOR CONNECTIONS

Item	Recommended tolerances* in.
Field-placed anchor bolts (transit or template)	$\pm \frac{1}{2}$
Elevation of field cast footings and piers	± 1
Structural Precast Concrete	
Position of plates	± 1
Location of inserts	$\pm \frac{1}{2}$
Location of bearing plates	$\pm \frac{3}{4}$
Location of blockouts	$\pm \frac{1}{2}$
Length	$\pm \frac{3}{4}$
Overall depth	$\pm \frac{1}{4}$
Width of stem	$\pm \frac{1}{8}$
Overall width	$\pm \frac{1}{4}$
Horizontal deviation of ends from square	$\pm \frac{1}{2}$
Vertical deviation of ends from square	$\pm \frac{1}{8}$ per ft of height
Bearing deviation from plane	$\pm \frac{3}{16}$
Position of post-tensioning ducts in precast members	$\pm \frac{1}{2}$
Architectural Precast Concrete	
Length or width	$\pm \frac{1}{16}$ per 10 ft, but not less than $\pm \frac{1}{8}$
Thickness	$+ \frac{1}{4} - \frac{1}{8}$
Location of blackouts	$\pm \frac{1}{2}$
Location of anchors and inserts	$\pm \frac{3}{8}$
Warpage or squareness	$\pm \frac{1}{8}$ in 6 ft
Joint widths —specified	$\frac{3}{8} - \frac{5}{8}$
—min. and max. dimensions	$\frac{1}{4}$ and $\frac{3}{4}$

*Other construction materials may control tolerances selected.

spacing of the dowels necessary to effect the full transfer of the horizontal shear forces between the interconnected elements.

10.4 REINFORCED CONCRETE BEARING IN COMPOSITE MEMBERS

A typical composite-action dowel reinforcement is shown in Figure 10.1. In order to prevent the concrete that is in direct bearing contact in such reinforcements from crushing due to excess direct compressive load, the external load has to be applied to an adequate bearing area size such that the resulting limit-state stresses do not exceed the compressive strength of concrete. The nominal bearing strength of *plain* concrete can be defined as

$$V_n = C_r(0.85f'_c A_1)\sqrt{A_2/A_1} \le 1.2f'_c A_1 \qquad (10.1)$$

where $C_r = 1.0$ when reinforcement is provided in the direction of the horizontal frictional force N_u shown in Figure 10.2 or when N_u is taken to be zero. C_r can be defined as $(S \times W/200)^{N_u/V_u}$, where the area $S \times W$, shown in Figure 10.3, should not exceed 9.0 in^2

$A_1 =$ direct bearing area

$A_2 =$ maximum area of the portion of the supporting surface that is geometrically similar to and concentric with the loaded area shown in Figure 10.3.

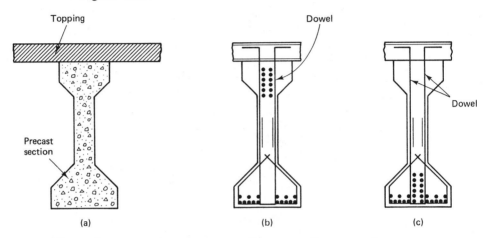

Figure 10.1 Typical dowel-action reinforcement. (a) Situ-cast topping on precast section. (b) Support section. (c) Midspan section.

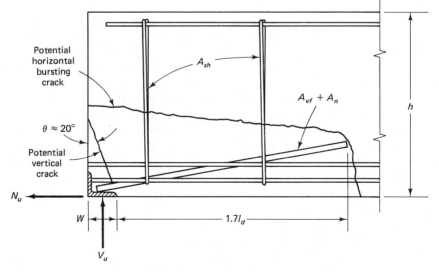

Figure 10.2 Reinforced bearing end in beam.

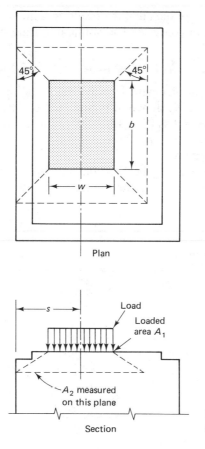

Plan

Section

Figure 10.3 Bearing area on a concrete pad.

The design bearing strength is

$$V_u = \phi V_n$$

where $\phi = 0.70$. In order to avoid accidental cracking or spalling at the ends of thin-stemmed members, a minimum reinforcement equal to $N_u / \phi f_y$, but not less than one #3 bar (9.52 mm dia), is recommended when the bearing area is less than 2 in^2 (12.9 cm^2).

If the applied factored load V_u exceeds the *design* bearing strength $V_u = \phi V_n$ as calculated from Equation 10.1, reinforcement is required in the bearing area. This reinforcement can be designed by the shear-friction theory presented in Chapter 5. *All* precast members ought to be designed for reinforced bearing except solid and hollow-core slabs, as recommended in Ref. 10.2, in order to prevent horizontal and vertical cracks from forming at the beam's extreme ties at the supports. The inclination of the end crack can be safely assumed to be approximately 20 degrees, as in Figure 10.2. Also, if V_u is equal to the applied factored shear force, in pounds, parallel to the assumed crack plane, it should be limited by the values given in Table 10.2 for the indicated maximum effective shear-friction coefficients μ_e.

The reinforcement area nominally perpendicular to the assumed crack plane can

TABLE 10.2 MAXIMUM APPLIED FACTORED FORCE V_u, LB

Crack interface condition	Recommended μ	Maximum μ_e	Maximum V_u, lb
1. Concrete to concrete, cast monolithically	1.4λ	3.4	$0.30\lambda^2 f'_c A_{cr} \leq 1,000\lambda^2 A_{cr}$
2. Concrete to hardened concrete with roughened surface	1.0λ	2.9	$0.25\lambda^2 f'_c A_{cr} \leq 1,000\lambda^2 A_{cr}$
3. Concrete to concrete	0.6λ	2.2	$0.20\lambda^2 f'_c A_{cr} \leq 800\lambda^2 A_{cr}$
4. Concrete to steel	0.7λ	2.4	$0.20\lambda^2 f'_c A_{cr} \leq 800\lambda^2 A_{cr}$

be found from

$$A_{vf} = \frac{V_{up}}{\phi \mu_e f_y} \tag{10.2}$$

where V_u/ϕ = nominal strength V_n
f_y = yield strength of A_{vf}, psi
V_{up} = applied factored shear force, limited by the values given in Table 10.2

and

$$\mu_e = \frac{1,000\lambda A_{cr}\mu}{V_{up}} \tag{10.3}$$

in which λ = 1.0 for normal-weight, 0.85 for sand-lightweight, and 0.75 for all-lightweight concrete

A_{cr} = area of the crack plane interface (in^2), which can be taken as $1.7\ell_d b$, where ℓ_d is the development length of the A_{vf} bars (in.) and b is the average member width (in.). A_{cr} should be taken as the lesser of $b \times w/\sin\theta$ and $b \times h/\cos\theta$, where θ is the angle of inclination of the end crack in Figure 10.2, h is the depth of the beam, and $b \times w$ is the bearing area.

Table 10.3 gives the development length ℓ_d for various bar sizes. The vertical reinforcement A_{sh} across potential horizontal cracks can be determined from

$$A_{sh} = \frac{(A_{vf} + A_n)f_y}{\mu'_e f_{ys}} \tag{10.4}$$

where

$$\mu'_e = \frac{1,000\mu A_{cr}\mu}{(A_{vf} + A_n)f_y} \tag{10.5}$$

and f_{ys} = yield strength of A_{sh}, psi
A_n = area of reinforcement to resist axial tension N_u in Figure 10.2, defined as

$$A_n = N_u/(\phi f_y) \tag{10.6}$$

in which N_u = factored applied horizontal tensile force nominally perpendicular to the assumed crack plane
ϕ = strength reduction factor = 0.85.

TABLE 10.3 DEVELOPMENT LENGTHS FOR GRADE 60 MILD STEEL BARS

Development and lap lengths in inches

Bar size	$f'_c = 3,000$ psi Tension ℓ_d	$1.3\ell_d$	$1.7\ell_d$	Compression ℓ_d	$f'_c = 4,000$ psi* Tension ℓ_d	$1.3\ell_d$	$1.7\ell_d$	Compression ℓ_d	$f'_c = 5,000$ psi Tension ℓ_d	$1.3\ell_d$	$1.7\ell_d$	Compression ℓ_d	Min. Comp. Splice
3	12	12	15	8	12	12	15	8	12	12	15	8	12
4	12	16	20	11	12	16	20	9	12	16	20	9	15
5	15	20	26	14	15	20	26	12	15	20	26	11	19
6	19	25	33	16	18	23	31	14	18	23	31	14	23
7	26	34	45	19	23	30	39	17	21	27	36	16	26
8	35	45	59	22	30	39	51	19	27	35	46	18	30
9	44	57	74	25	38	49	65	21	34	44	58	20	34
10	56	72	95	28	48	63	82	24	43	56	73	23	38
11	68	89	116	31	59	77	101	27	53	69	90	25	42

Bar size	$f'_c = 6,000$ psi Tension ℓ_d	$1.3\ell_d$	$1.7\ell_d$	Compression ℓ_d	$f'_c = 7,000$ psi* Tension ℓ_d	$1.3\ell_d$	$1.7\ell_d$	Compression ℓ_d	$f'_c = 8,000$ psi Tension ℓ_d	$1.3\ell_d$	$1.7\ell_d$	Compression ℓ_d	Min. Comp. Splice
3	12	12	15	8	12	12	15	8	12	12	15	8	12
4	12	16	20	9	12	16	20	9	12	16	20	9	15
5	15	20	26	11	15	20	26	11	15	20	26	11	19
6	18	23	31	14	18	23	31	14	18	23	31	14	23
7	21	27	36	16	21	27	36	16	21	27	36	16	26
8	24	32	42	18	24	31	41	18	24	31	41	18	30
9	31	40	53	20	29	37	49	20	27	35	46	20	34
10	39	51	67	23	36	47	62	23	34	44	58	23	38
11	48	63	82	26	45	58	76	26	42	54	71	26	42

Notes:

Tension

$\ell_d = 2{,}400\, A_b/\sqrt{f'_c}$; min. $24d_b$ or 12 in.

Compression development length

$\ell_d = 1{,}200 d_b/\sqrt{f'_c}$; min. $18d_b$ or 8 in.

Compression splice length

compression ℓ_d; min. $30d_b$ or 12 in.

where

A_b = area of individual bar, in^2

d_b = diameter of bar, in

For limitations, see ACI 318 Code

Multiply table values by

1.4 for top reinforcement
1.33 for all-lightweight concrete
1.18 for sand-lightweight concrete
0.8 for bar spacing 6 in. or more
  (3 in. from member face)

$\dfrac{A_d \text{ req'd}}{A_d \text{ prov'd}}$ for excess reinforcement

*for Grade 40 bars, required lengths are two-thirds of the table values, but not less than the required minimum lengths

Charlotte-Mecklenburg Government Center Parking Structure. (*Courtesy*, Prestressed Concrete Institute.)

Note that all reinforcement on either side of the assumed crack plane should be properly anchored by development length or welding to angles, plates, or hooks in order to develop the calculated resisting force.

10.4.1 Reinforced Bearing Design

Example 10.1

A PCI standard 16RB28 rectangular prestressed beam is subjected to a vertical factored end shear force $V_u = 90,000$ lb (400 kN) and a horizontal tensile force $N_u = 21,000$ lb (93.4 kN). The beam is supported on a teflon pad of size 4 in. $\times$ 4 in. (10 cm $\times$ 10 cm). Design the end reinforcement in the beam that can prevent the development of vertical or horizontal bearing cracks given the following data:

$f'_c = 5,000$ psi (34.47 MPa), normal-weight concrete

$f_y = 60,000$ psi for all mild reinforcement (413.7 MPa)

$\theta = 20$ degrees

Solution

Horizontal Reinforcement ($A_{vf} + {}_nn$). For the determination of the horizontal reinforcement, we have

$$\text{Beam depth } h = 28 \text{ in.}$$

$$\frac{b \times w}{\sin \theta} = \frac{16 \times 4}{\sin 20} = 187 \text{ in}^2 \ (1,207 \text{ cm}^2)$$

Dallas Municipal Center, Dallas, Texas. (*Courtesy*, Post-Tensioning Institute.)

$$\frac{b \times h}{\cos \theta} = \frac{16 \times 28}{\cos 20} = 477 \text{ in}^2 \ (3{,}078 \text{ cm}^2)$$

Use A_{cr} as the *lesser* of the two values, i.e., $A_{cr} = 187 \text{ in}^2$.
From Table 10.2 $\mu = 1.4$, and from Equation 10.3,

$$\mu_e = \frac{1{,}000\lambda A_{cr}\mu}{V_{up}} = \frac{1{,}000 \times 1.0 \times 187 \times 1.4}{90{,}000} = 2.91 < 3.4$$

Thus, use $\mu_e = 2.91$.
From Equation 10.2,

$$A_{vf} = \frac{V_{up}}{\phi f_y \mu_e} = \frac{90{,}000}{0.85 \times 60{,}000 \times 2.91} = 0.61 \text{ in}^2 \ (3.94 \text{ cm}^2)$$

$$N_u = 21{,}000 \text{ lb}$$

$$\frac{N_u}{V_u} = \frac{21{,}000}{90{,}000} = 0.23 > \text{minimum } 0.20$$

Hence, use $N_u = 21{,}000$ lb.
From Equation 10.6, $A_n = N_u/\phi f_y = 21{,}000/(0.85 \times 60{,}000) = 0.41 \text{ in}^2$ (2.65 cm²)

Total Steel

$$A_s = A_{vf} + A_n = 0.61 + 0.41 = 1.02 \text{ in}^2 \ (6.58 \text{ cm}^2)$$

So use three #6 bars $= 1.32 \text{ in}^2$ (8.52 cm²).

Vertical Reinforcement (A_{sh}). From Table 10.3, $\ell_d =$ development length of #6 bars $= 18$ in. (46 cm) and $A_{cr} = 1.7\ell_d b = 1.7 \times 18 \times 16 = 489.6 \text{ in}^2$

Walt Disney World Monorail, Orlando, Florida, a series of hollow precast prestressed concrete 100-box girders that are individually post-tensioned to provide a six-span continuous structure, design by ABAM Engineers. (*Courtesy,* Walt Disney World Co.)

(3,159 cm²). From Equation 10.5,

$$\mu_e' = \frac{1,000\lambda A_{cr}\mu}{(A_{vf} + A_n)f_y} = \frac{1,000 \times 1.0 \times 489.6 \times 1.4}{1.02 \times 60,000} = 11.20 > \text{allowable } \mu_e = 3.4$$

So use $\mu_e' = 3.4$. Then, from Equation 10.4,

$$A_{sh} = \frac{(A_{vf} + A_n)f_y}{\mu_e'f_{ys}} = \frac{1.02 \times 60,000}{3.4 \times 60,000} = 0.30 \text{ in}^2 \ (1.94 \text{ cm}^2)$$

Accordingly, use three #3 stirrups = 0.33 in² (2.13 cm²).

10.5 DAPPED-END BEAM CONNECTIONS

A dapped-end beam is a structural element with abruptly reduced depth at its ends in order to provide the necessary seating or bearing on corbels or brackets without loss of clear height between floors. A typical dapped end in a prestressed beam is shown in Figure 10.4. Two types of cracks can develop: crack 2 is a direct shear crack, while cracks 3, 4, and 5 are diagonal tension cracks caused by flexure and axial tension in the extended reduced depth and the stress concentration at the reentrant corner. Therefore, the following types of reinforcement, as shown in the figure, have to be provided:

1. Flexural reinforcement A_f plus axial tension reinforcement A_n, where $A_s = A_f + A_n$, to resist the cantilever bending stresses.

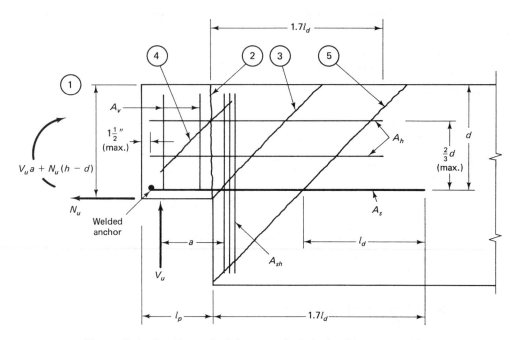

Figure 10.4 Cracking and reinforcement in dapped-end beam connections.

2. Shear-friction reinforcement $A_f + A_h$, plus axial tension reinforcement A_n, to resist the direct vertical shear force at the junction of the dapped and the undapped portion of the beam causing crack 2.

3. Shear reinforcement A_{sh}, to resist the diagonal tension generated at the reentrant corner causing crack 3.

4. Diagonal tension reinforcement $A_h + A_v$, to resist the potential diagonal tension crack 4 in the extended dapped portion of the beam.

5. Development length of $A_s = A_f + A_n$, to resist the potential diagonal tension crack 5 in the undapped portion of the beam.

10.5.1 Determination of Reinforcement to Resist Failure

10.5.1.1 Flexure and axial tension. For moment equilibrium in Figure 10.4, the total factored moment acting on the cantilever dapped portion at the plane of A_s is

$$M_u = V_u a + N_u(h - d) \qquad (10.7\ \text{a})$$

where h = depth of member above the dap

d = effective depth of the dap to center of reinforcement A_s

a = shear span.

M_u has to be resisted by a nominal moment strength $M_n = M_u/\phi$, or

$$M_n = \frac{V_u a + N_u(h - d)}{\phi} \qquad (10.7\ \text{b})$$

Assuming that the moment arm $jd \cong 0.9d$,

$$F_n = \frac{V_u a + N_u(h - d)}{0.9\phi d} \tag{10.8}$$

where $\phi = 0.90$ for flexure. Since $0.9\ \phi \cong 0.81$, for simplification use a value of $\phi = 0.85$ in Equation 10.8 to obtain

$$F_n = \frac{V_u a + N_u(h - d)}{\phi d} \tag{10.9 a}$$

or

$$F_n = \frac{V_u}{\phi}\left(\frac{a}{d}\right) + \frac{N_u}{\phi}\left(\frac{h - d}{d}\right) \tag{10.9 b}$$

The flexural reinforcement is then

$$A_f = \frac{F_n}{f_y} = \frac{V_u a + N_u(h - d)}{\phi f_y d} \tag{10.10}$$

and the direct tension reinforcement due to the tensile force N_u is

$$A_n = \frac{N_u}{\phi f_y} \tag{10.11}$$

The total area of the flexural and direct tension reinforcement then becomes, from Equations 10.10 and 10.11,

$$A_s = A_f + A_n = \frac{1}{\phi f_y}\left[V_u\left(\frac{a}{d}\right) + N_u\left(\frac{h}{d}\right)\right] \tag{10.12}$$

where, again, the adjusted $\phi = 0.85$.

10.5.1.2 Direct vertical shear. The potential direct shear crack 2 is resisted by the combination of reinforcements A_s and A_h in Figure 10.4. The horizontal reinforcement A_h needed to resist the direct shear can be evaluated as

$$A_h = 0.5(A_s - A_n) \tag{10.13}$$

where

$$A_s = \frac{2V_u}{3\phi f_y \mu_e} + A_n \tag{10.14 a}$$

$$A_n = \frac{N_u}{\phi F_y} \tag{10.14 b}$$

$$\mu_e = \frac{1,000\lambda bh\mu}{V_u} \tag{10.14 c}$$

with $\phi = 0.85$ and μ_e the same as in Equation 10.3. Hence,

$$A_s = \frac{1}{\phi f_y}\left(\frac{2V_u}{3\mu_e} + N_u\right) \tag{10.15}$$

Jesse H. Jones Memorial Bridge, Houston, Texas. (*Courtesy*, Post-Tensioning Institute.)

The value of A_s used in Equation 10.13 should be the greater of the two values obtained from Equations 10.12 and 10.15.

The reinforcement A_s should be extended a minimum of $1.7\ell_d$ past the end of the dap, or ℓ_d past crack 5, and anchored at the end of the beam by welding to cross bars, angles, or plates. Horizontal bars A_h should be similarly extended, and vertical bars A_{sh} and vertical or inclined bars A_v should be well anchored by hooks as required by the ACI Code.

The nominal shear strength of the dap end is limited to

$$V_n \leq 0.30 f'_c bd \leq 1,000bd \tag{10.16 a}$$

for normal-weight concrete, and

$$V_n \leq \left(0.20 - \frac{0.07a}{d}\right) f'_c bd \tag{10.16 b}$$

or

$$V_n \leq \left(800 - \frac{280a}{d}\right) bd \tag{10.16 c}$$

whichever is smaller, for sand-lightweight or all-lightweight concrete, where a is the shear span and d the effective depth of the beam.

10.5.1.3 Diagonal tension at reentrant corner. The reinforcement needed to resist the inclined diagonal tension cracking propagating from the center of stress concentration at the reentrant corner towards the undapped portion can be obtained from the expression

$$A_{sh} = \frac{V_u}{\phi f_y} \tag{10.17}$$

where $\phi = 0.85$ and f_y is the yield strength of the A_{sh} reinforcement.

10.5.1.4 Diagonal tension in the dapped end. In order to resist the potential diagonal crack 4 in the dapped end, additional reinforcement A_v has to be provided such that the total nominal shear strength V_n satisfies the equation

$$V_n = \frac{V_u}{\phi} = A_v f_y + A_h f_y + 2\lambda b d \sqrt{f_c'} \tag{10.18}$$

At least half of the reinforcement in this area has to be placed vertically, so that Equation 10.18 gives

$$\text{Min } A_v = \frac{1}{2f_y}\left(\frac{V_u}{\phi} - 2\lambda b d \sqrt{f_c'}\right) \tag{10.19}$$

Note that performance considerations require the following:

1. The depth of the dapped end should be at least one-half the beam depth, unless the beam is significantly deeper than required by design for other than structural considerations.

2. If the flexural stress calculated for the full depth of the section using factored loads and gross section properties exceeds $6\sqrt{f_c'}$ immediately beyond the dap, additional longitudinal reinforcement should be placed in the beam in order to develop the required flexural strength.

3. The diagonal tension reinforcement A_{sh} should be placed as closely as practicable to the reentrant corner. This reinforcement is in addition to the design shear reinforcement required for the full-depth beam section.

10.5.2 Dapped-End Beam Connection Design

Example 10.2

A PCI standard 16RB28 prestressed beam dapped at the end for bearing on a column corbel is subjected to a factored gravity end shear $V_u = 110,000$ lb (489 kN) and a horizontal axial tension $N_u = 20,000$ lb (97.9 kN). Design the flexural, direct shear, and diagonal tension reinforcements A_s, A_h, A_{sh}, and A_v required to prevent potential cracking due to dapping the beam ends. Given data are $f_c' = 5,000$ psi (34.47 MPa), normal weight, and $f_y = 60,000$ psi (413.7 MPa). Use bars for the reinforcement.
Solution Assume that the shear span $a = 6$ in. (152 mm), the dapped-end effective $d = 16$ in. (40.6 mm), and $h = 18$ in. (457 mm).

Flexure and Axial Tension Reinforcement A_s

$$\frac{N_u}{V_u} = \frac{20,000}{110,000} = 0.18 < 0.20$$

Hence, $N_u = 0.20 \times 110,000 = 22,000$ lb (97.9 kN). Thus,

$$A_s = \frac{1}{\phi f_y}\left[V_u\left(\frac{a}{d}\right) + N_u\left(\frac{h}{d}\right)\right]$$

$$= \frac{1}{0.85 \times 60,000}\left[110,000 \times \frac{6}{16} + 22,000 \times \frac{18}{16}\right] = 1.29 \text{ in}^2$$

Direct Shear Reinforcement A_s and A_h. From Table 10.2 $\mu = 1.4\lambda$ where $\lambda = 1.0$. Then, from Equation 10.14 c, where b for the 16RB28 section is 16 in.,

$$\mu_e = \frac{1,000\lambda bh\mu}{V_u} = \frac{1,000 \times 1.0 \times 16 \times 18 \times 1.4}{110,000} = 3.67$$

$$> \text{max allowable } \mu_e = 3.4$$

Thus, use $\mu_e = 3.4$. Then, from Equation 10.5,

$$A_s = \frac{1}{\phi f_y}\left(\frac{2V_u}{3\mu_e} + N_u\right) = \frac{1}{0.85 \times 60,000}\left(\frac{2 \times 110,000}{3 \times 3.4} + 22,000\right)$$

$$= 0.85 \text{ in.} < A_s = 1.29 \text{ from before}$$

Hence, use $A_s = 1.29$ in^2 (8.32 cm^2). Then three #6 bars $= 1.32$ in^2, which is satisfactory.

From Equation 10.4 b,

$$A_n = \frac{N_u}{\phi f_y} = \frac{22,000}{0.85 \times 60,000} = 0.37 \text{ in}^2 \text{ (2.39 cm}^2\text{)}$$

From Equation 10.13, the horizontal shear reinforcement across the depth of the beam is $A_h = 0.5(A_s - A_n) = 0.5(1.29 - 0.37) = 0.46$ in^2 (2.97 cm^2). So try two #3 U bars $= 2(2 \times 0.11) = 0.44$ in^2 (2.84 cm^2), which will be verified subsequently. As a check, the nominal shear strength, from Equation 10.16 a, is

$$\text{Available } V_n = 800bd = 800 \times 16 \times 16 = 204,800 \text{ lb}$$

$$\text{Required } V_n = \frac{V_u}{\phi} = \frac{110,000}{0.85} = 129,412 \text{ lb} < 204,800 \text{ lb, O.K.}$$

Diagonal Tension Vertical Reinforcement at Reentrant Corner. From Equation 10.17,

$$A_{sh} = \frac{V_u}{\phi f_y} = \frac{110,000}{0.85 \times 60,000} = 2.16 \text{ in}^2 \text{ (13.94 cm}^2\text{)}$$

So trying #4 closed ties, $A_s = 2 \times 0.20 = 0.40$ in^2. The number of ties $= 2.16/0.4 = 5.4$; hence, use six #4 ties, concentrated close to the reentrant corner.

Diagonal Tension Reinforcement A_v in the Dapped End. From Equation 10.19,

$$A_v = \frac{1}{2f_y}\left(\frac{V_u}{\phi} - 2\lambda bd\sqrt{f_c'}\right)$$

The nominal shear strength of the *plain concrete* is

$$2\lambda bd\sqrt{f_c'} = 2 \times 1.0 \times 16 \times 16\sqrt{5,000} = 36,034 \text{ lb}$$

Then

$$A_v = \frac{1}{2 \times 60,000}\left(\frac{110,000}{0.85} - 36,034\right) = 0.78 \text{ in}^2$$

Try two #4 U stirrups $= 2(2 \times 0.20) = 0.80$ in^2. From before, $A_h = 0.44$ in^2. So the total nominal shear strength of the section, from Equation 10.18, is

$$\text{Available } V_n = A_v f_y + A_h f_y + 2\lambda bd \sqrt{f_c'}$$

$$= 0.80 \times 60,000 + 0.44 \times 60,000 + 36,034$$

$$= 110,434 \text{ lb} < \frac{V_u}{\phi} = 129,412 \text{ lb}$$

which is unsatisfactory. Consequently, try increasing A_h by changing from #3 to #4 U stirrups:

$$A_h = 2(2 \times 0.2) = 0.80$$

$$\text{Revised available } V_n = 0.80 \times 60,000 + 0.80 \times 60,000 + 36,034$$

$$= 132,034 > \text{required } \frac{V_u}{\phi} = 129,412 \text{ lb, O.K.}$$

Check Development Length Requirements for Anchorage. The reinforcement A_s is three #6 bars. From Table 10.3 for #6 bars, $f_c' = 5,000$ psi and $\ell_d = 18$ in. past the 45° diagonal crack propagating from the reentrant corner. Also, the undapped beam depth = 2 ft, 4 in. = 28 in., and the total development length = $28 - d + \ell_d = 28 - 16 + 18 = 30$ in. Since the minimum $1.7\ell_d = 31$ in., use $\ell_d = 32$ in. = 2 ft, 8 in. (81 cm).

The reinforcement A_h is two #4 U bars. So from Table 10.3, $1.7\ell_d = 20$ in. (51 cm) beyond the beam dap. Figure 10.5 gives the reinforcement details for the dapped beam connection.

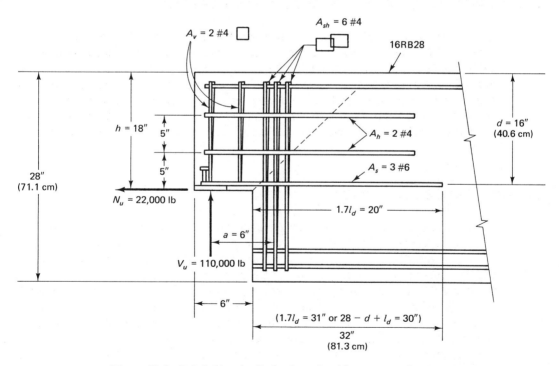

Figure 10.5 Reinforcing details for dapped-end beam connection in Example 10.2.

Sec. 10.5   Dapped-End Beam Connections

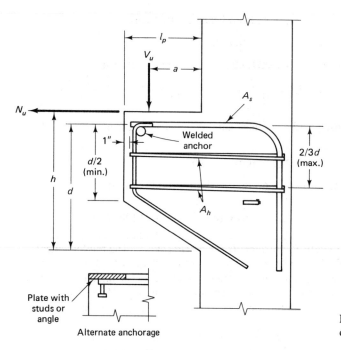

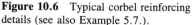

Figure 10.6 Typical corbel reinforcing details (see also Example 5.7.).

10.6 REINFORCED CONCRETE BRACKETS AND CORBELS

Corbels are short cantilevers whose shear span-to-depth ratio a/d does not exceed a value of 1.0. They are subjected to a direct shear V_u and a horizontal tension N_u. Section 5.14 in Chapter 5, the design flow chart in Sec 5.14.4 and the detailed design example 5.7 give a comprehensive discussion and application of the shear-friction theory in the design of corbels. Working out the details of reinforcement of the connection is of major significance in the success of the design of a corbel as regards its ability to resist applied loads. Typical corbel reinforcing details are shown in Figure 10.6.

10.7 CONCRETE BEAM LEDGES

Beam ledges are used to support transverse precast prestressed beam-end concentrated loads in a manner similar to the way corbels operate. Direct shear acting on the ledge can cause vertical cracks as shown in Figure 10.7. If the load is noncontinuous and comes from one side, the ledge beam in L-shaped form acts like a spandrel beam and is subjected to torsional moment in addition to direct shear. The design of the ledge beam itself follows the procedures and examples in Chapter 5. The discussion presented here covers the design of the shear reinforcement for the cantilevering ledge, which often has a shear span-to-depth ratio ℓ_p/d of $\frac{1}{2}$ or less.

The nominal shear strength of the ledge at the reentrant corner can be determined by the lesser of the two values obtained from the following expressions under the given conditions:

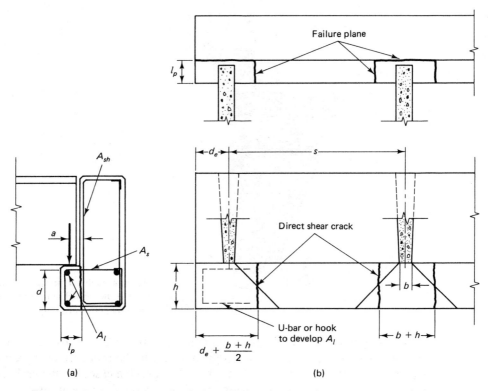

Figure 10.7 Connection design for beam ledges. (a) Ledge beam cross section. (b) Plan and elevation.

1. $s > b + h$

$$V_n = 3h\lambda\sqrt{f'_c}\,(2\ell_p + b + h) \qquad (10.20\ \text{a})$$

$$V_n = h\lambda\sqrt{f'_c}\,(2\ell_p + b + h + 2d_e) \qquad (10.20\ \text{b})$$

2. $s < b + h$, *and equal concentrated loads*

$$V_n = 1.5h\lambda\sqrt{f'_c}\,(2\ell_p + b + h + s) \qquad (10.21\ \text{a})$$

$$V_n = h\lambda\sqrt{f'_c}\left(\ell_p + \frac{b + h}{2} + d_e + s\right) \qquad (10.21\ \text{b})$$

where ℓ_p = ledge projection, in.
b = width of bearing area, in.
h = depth of beam ledge, in.
s = spacing of concentrated loads, in.
d_e = distance from center of load to end of beam, in.

If the ledge supports a continuous load or closely spaced concentrated loads, the nominal shear strength of the ledge section has to be evaluated from

$$V_n = 24h\lambda\sqrt{f'_c} \qquad (10.22)$$

where V_n is in lb per ft of length. The design strength V_u has to be at least equal to the factored force $V_u = \phi V_n$ for $\phi = 0.85$. If the applied factored load V_u exceeds the

Hoisting double-T prestressed roof element at the Civil Engineering Laboratory, Rutgers University.

design strength as determined from Equation 10.20, 10.21, or 10.22, special reinforcement has to be provided in a design similarly to the reinforcement required in a dapped beam end as discussed in Section 10.5. In such a case, the flexural reinforcement A_s is determined from Equation 10.12, the vertical diagonal tension "hanger" reinforcement A_{sh} from Equation 10.17, and the longitudinal reinforcement A_1 placed at the top and bottom fibers of the ledge from

$$A_\ell = \frac{200\ell_p d}{f_y} \qquad (10.23)$$

where A_ℓ is the area of the longitudinal reinforcement in the ledge. The reinforcement A_{sh} can be uniformly spaced over a width $6h$ on either side of the bearing, but not to exceed half the distance to the next load. The bar spacing should not exceed the ledge depth h or 18 in., and the A_{sh} designed for the ledge need not be additive to the shear and torsional reinforcement of the total ledge beam.

10.7.1 Design of Ledge Beam Connection

Example 10.3

A garage floor structure is composed of 10-ft-wide double-T's supported at the exterior end by standard L-beam sections. The layout of the double-T's is such that a stem can be placed at any point on the ledge. The vertical factored end shear $V_u = 24,000$ lb (107 kN) per stem, and the horizontal tensile force $N_u = 5,000$ lb (22.4 kN) per stem. Compute the nominal shear strength of the ledge and design the reinforcement if necessary, given that

$b = 4$ in.

$h = 12$ in.

Mariners Island Office Building, San Mateo, California (*Courtesy,* Robert Englekirk Consulting Structural Engineers and W2MH Group Architects, Los Angeles, California. Photo by Dixie Carillo.).

$d = 105$ in.

$\ell_p = 6$ in. (15 cm)

$s = 48$ in. (122 cm)

$f'_c = 5,000$ psi (34.47 MPa), normal weight

$f_y = 60,000$ psi (413.7 MPa)

Solution

$$V_u = 24,000 \text{ lb}$$

$$N_u = 5,000 \text{ lb}$$

$$s = 48 \text{ in.}$$

$$b + h = 4 + 14 = 18 \text{ in.}$$

$$\text{Min. } d_e = \tfrac{1}{2}b = 2 \text{ in.}$$

$$2\ell_p + b + h = 2 \times 6 + 4 + 14 = 30 \text{ in.}$$

Since $s > b + h$, and $d_e < 2\ell_p + b + h$, Equation 10.20 b applies, and the available $V_n = h\lambda\sqrt{f'_c}(2\ell_p + b + h + 2d_e) = 12 \times 1.0\sqrt{5,000}(2 \times 6 + 4 + 12 + 2 \times 2) = 27,153$ lb (120.8 kN). So the design $V_u = \phi V_n = 0.85 \times 27,153 = 23,080$ lb $<$ Factored $V_u = 24,000$ lb, and we can use the dapped section reinforcement design.

    Flexural Reinforcement A_s.   The shear span $a \cong 3\ell_p/4 + 1.5 = 3 \times 6/4 + 1.5 = 6$ in. (15 cm). Since $N_u/V_u = 5,000/24,000 = 0.20 > 0.21$, use $N_u = 5,000$ lb. Then, from Equation 10.12,

$$A_s = \frac{1}{\phi f_y}\left[V_u\left(\frac{a}{d}\right) + N_u\left(\frac{h}{d}\right)\right]$$

$$= \frac{1}{0.85 \times 60,000} \left[24,000 \frac{6}{10.5} + 5,000 \frac{12}{10.5} \right] = 0.38 \text{ in}^2 \ (2.45 \text{ cm}^2)$$

Since $6h = 6 \times 12 > s/2 = 24$ in., distribute the reinforcement $s/2 = 24$ in. on each side of the load.

The width of the band for placement of flexural reinforcement $A_s = 2 \times 24 = 48$ in., and the maximum bar spacing $= h = 12$ in. So use four #3 bars in each 48-in. band width $= 0.44$ in$^2 >$ required 0.38 in^2. Accordingly, place two additional bars at the beam end in order to provide equivalent reinforcement for the stem placed near the end.

Diagonal Tension Vertical Reinforcement A_{sh}. From Equation 10.17,

$$A_{sh} = \frac{V_u}{\phi f_y} = \frac{24,000}{0.85 \times 60,000} = 0.47 \text{ in}^2 \ (3.03 \text{ cm}^2)$$

over a 48-in.-width band. Thus, $A_{sh}/\text{ft} = 0.47/4 = 0.12$ in^2/ft, or #3 bars @ 11 in. Consequently, use five #3 ⌐ bars in each 48-in. band width $= 0.55$ in$^2 >$ required 0.47 in^2. Then, for practical considerations, use the same number and spacing for both the A_s and A_{sh} steel, namely, five #3 closed hoops. Note that only one leg of the hanger steel hoop A_{sh} is accounted for in the selection of the five #3 bars in order to provide for the required concentration of the steel near the reentrant corner.

Longitudinal Reinforcement A_ℓ. From Equation 10.23,

$$A_\ell = \frac{200 \ell_p - d}{f_y} = \frac{200 \times 6 \times 10.5}{60,000} = 0.21 \text{ in}^2$$

For practical field considerations, use #4 bars, one on each corner of the ledge, giving four #4 bars $= 0.80$ in^2 (12.7 mm dia) > 0.21 in^2, O.K.

The complete design of the ledge beam, of course, would require shear and torsional analysis of the total section to resist the total shear transmitted by all the double-T supported stems and the torsional moment caused by the application of the *eccentric* load from the supported stems. The designed ledge reinforcement area discussed in this example is in addition to the shear and torsional reinforcement required for the total beam.

Figure 10.8 gives the details of the ledge connection reinforcement, but does not include the shear and torsional reinforcement that has to be designed for the entire L-beam.

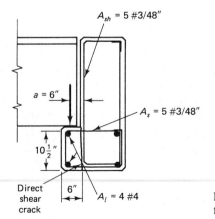

$A_{sh} = 5 \text{ #3/48"}$

$a = 6"$

$A_s = 5 \text{ #3/48"}$

$10\frac{1}{2}"$

Direct shear crack

$6"$

$A_l = 4 \text{ #4}$

Figure 10.8 Ledge connection reinforcement in Example 10.3.

10.8 SELECTED CONNECTION DETAILS

As discussed in the Section 10.1, connections are the major links in the overall structural system whose performance determines whether the structure will be safe and stable. Consequently, the design engineer has to be particularly cautious in the design and selection of the appropriate section for reasons of both safety and economy. Details of the design of numerous types of connections are given in Refs. 10.1 and 10.2, and Figures 10.9 through 10.16 show typical details of selected connections from these two references. The diagrams illustrate the application of the theories presented in this chapter.

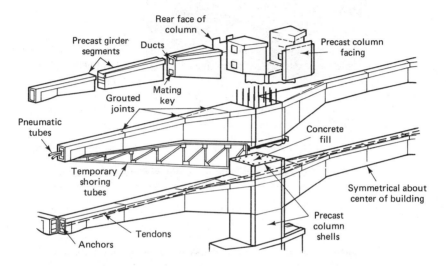

Beam assembly showing precast elements, temporary truss and prestressing

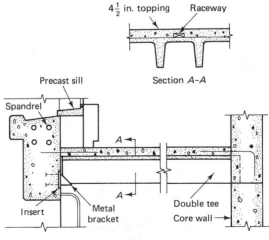

Typical floor construction

Figure 10.9 Construction method and typical joint details in Gulf Life Prestressed Building, Jacksonville, Florida.

Installation of prestressed double-T elements at the rod-suspended support, Civil Engineering Laboratory, Rutgers University.

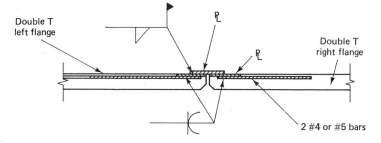

Double T
left flange

℞

℞

Double T
right flange

2 #4 or #5 bars

Figure 10.10 Connection of double-T flanges if no composite-action concrete topping is used.

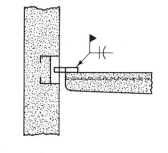

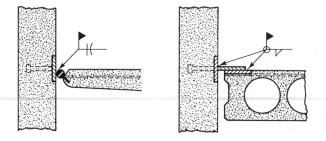

Figure 10.11 Typical precast floor-to-wall connections.

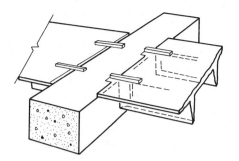
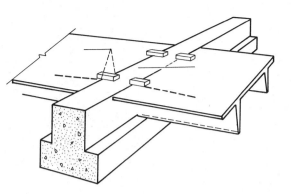
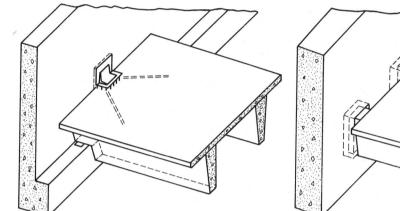
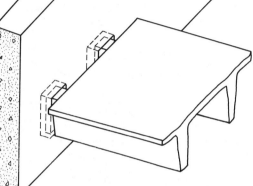
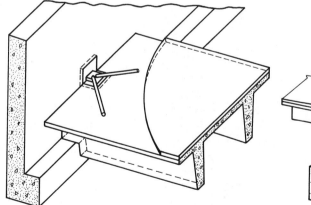
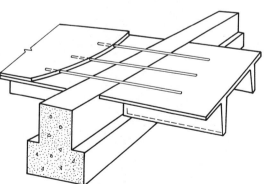

Figure 10.12 Typical double-T connections.

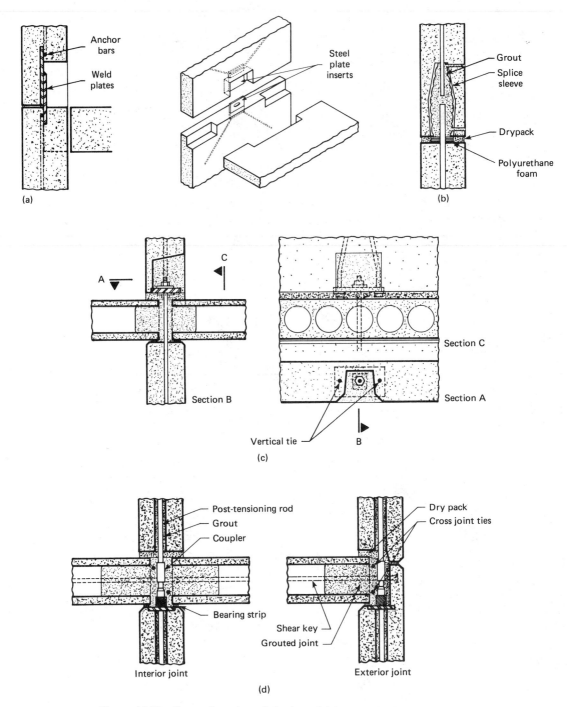

Figure 10.13 Connections through horizontal joints. (a) Weld plates. (b) Grouted splice sleeve. (c) Plate-bolt connections. (d) Post-tensioned connection.

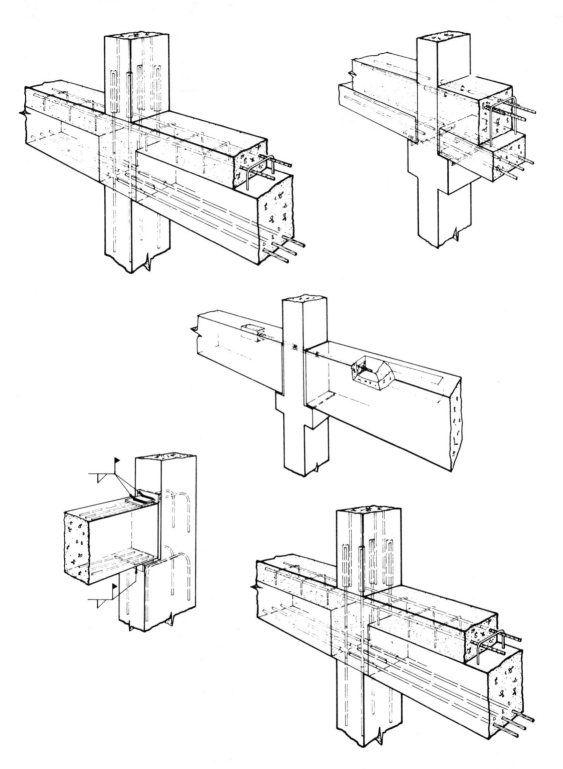

Figure 10.14 Moment connections.

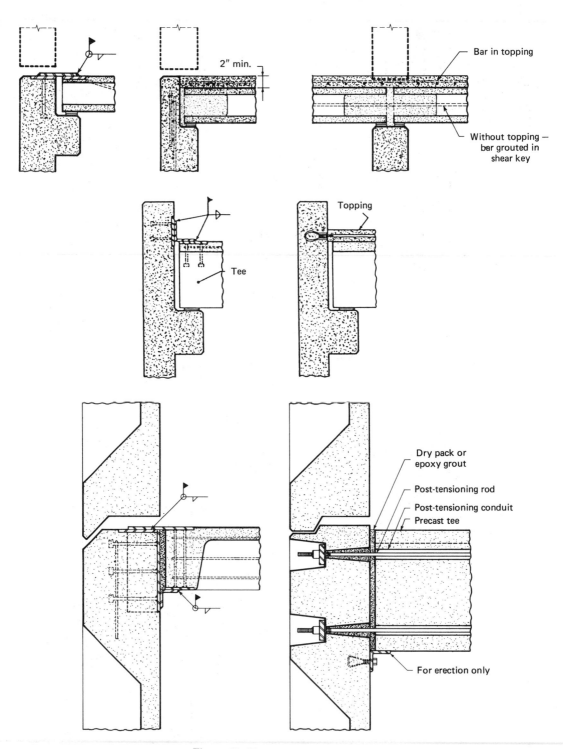

Figure 10.15 Floor-to-bearing wall connections.

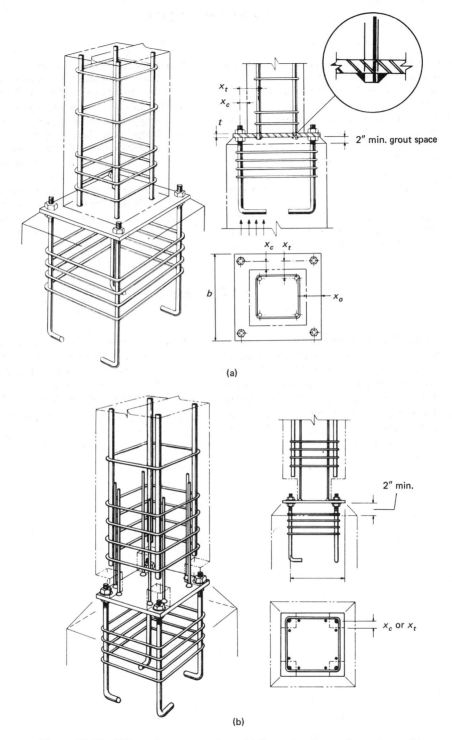

Figure 10.16 Column base connections. (a) Base plate larger than column. (b) Flush base plate.

Sec. 10.8   Selection Connection Details

REFERENCES

10.1 Prestressed Concrete Institute. *PCI Design Handbook—Precast and Prestressed Concrete.* 3d ed. Chicago: Prestressed Concrete Institute, 1985.

10.2 PCI Committee on Connection Details. *Manual on Design and Detailing of Connections for Precast and Prestressed Concrete.* Chicago: Prestressed Concrete Institute, 1988.

10.3 Post-Tensioning Institute. *Post-Tensioning Manual.* 4th ed. Phoenix: Post-Tensioning Institute, 1985.

10.4 Shaikh, A. F., and Yi, W. "In-Place Strength of Welded Headed Studs." *Journal of the Prestressed Concrete Institute* 30 (1985): 56–81.

10.5 Mattock, A. H., and Chan, T. C. "Design and Behavior of Dapped-End Beams." *Journal of the Prestressed Concrete Institute* 30 (1985): 28–45.

10.6 Mirza, S. A., and Furlong, R. W. "Serviceability Behavior of Concrete Inverted T-Beam Bridge Bent Caps." *Journal of the American Concrete Institute* 80 (1983): 294–304.

10.7 Clough, D. P. "Design of Connections for Precast Prestressed Concrete Buildings for the Effect of Earthquake." *PCI Technical Report No. 5.* Chicago: Prestressed Concrete Institute, 1985.

10.8 PCI Erectors Committee. *Recommended Practice for Erection of Precast Concrete.* Chicago: Prestressed Concrete Institute, 1985, pp. 1–198.

PROBLEMS

10.1 Design the end reinforcement required to resist the bearing stresses caused by an end vertical shear $V_u = 125,000$ lb (556 kN) and a horizontal tensile force $N_u = 24,000$ lb (107 kN) in a PCI 16RB28 rectangular beam supported at its ends on 5 in. × 5 in. pads. Take $f'_c = 5,000$ psi (34.47 MPa), normal weight, and $f_y = 60,000$ psi (413.7 MPa).

10.2 If the beam in Problem 10.1 is dapped at its ends and rests on concrete corbels, design the flexure and direct shear reinforcement for the dapped ends in order to prevent potential shear cracking and failure.

10.3 Solve Example 10.3 for factored loads $V_u = 29,000$ lb (128 kN) and factored $N_u = 2,500$ lb (11.1 kN).

CHAPTER 11

Prestressed Concrete Circular Storage Tanks and Shell Roofs

11.1 INTRODUCTION

Prestressed concrete circular tanks are usually the best combination of structural form and material for the storage of liquids and solids. Their performance over the past half-century indicates that, when designed with reasonable skill and care, they can function for 50 years or more without significant maintenance problems.

The first effort to introduce circumferential prestressing into circular structures was that of W. S. Hewett, who applied the tie rod and turnbuckle principle in the early 1920s (Ref. 11.6). But the reinforcing steel available at that time had very low yield strength, limiting the applied tension to not more than 30,000 to 35,000 psi (206.9 to 241.3 MPa). Indeed, significant long-term losses due to concrete creep, shrinkage, and steel relaxation almost neutralized the prestressing force. As higher strength steel wires became available, J. M. Crom, Sr., in the 1940s, successfully developed the principle of winding high-tensile wires around the circular walls of prestressed tanks. Since that time, over 3,000 circular storage structures have been built of various dimensions up to diameters in excess of 300 feet (92 m.)

The major advantage in performance and economy of using circular prestressing in concrete tanks over regular reinforcement is the requirement that no cracking be allowed. The circumferential "hugging" hoop stress in compression provided by external winding of the prestressing wires around the tank shell is the natural technique for eliminating cracking in the exterior walls due to the internal liquid, solid, or gaseous loads that the tank holds. Other techniques of circumferential prestressing using *individual* tendons which are anchored to buttresses have been more widely used

Two 583,000-bbl (92,500-m^3) double-wall prestressed concrete tanks for liquefied natural gas storage, Philadelphia. (*Courtesy*, N.A. Legatos, Preload Technology, Inc., New York.)

in Europe than in North America for reasons of local economy and technological status.

Containment vessels utilizing circumferential prestressing, which can be either situ-cast or precast in segments, include water storage tanks, wastewater tanks and effluent clarifiers, silos, chemical and oil storage tanks, cryogenic vessels, and nuclear reactor pressure vessels. All these structures are considered as thin shells because of the exceedingly small ratio of the container thickness to its diameter. Because no cracking at working-load levels is permitted, the shells are expected to behave elastically under working-load and overload conditions.

11.2 DESIGN PRINCIPLES AND PROCEDURES

11.2.1 Internal Loads

Consideration of the behavior of circular tanks involves an examination of both the interior pressure due to the material contained therein acting on a thin-walled cylindrical shell cross section and the exterior radial and sometimes vertical prestressing forces balancing the interior forces. The interior pressure is horizontally radial, but varies vertically depending on the type of material contained in the tank. If the material is water or a similar liquid, the vertical pressure distribution against the tank walls is *triangular,* with maximum intensity at the base of the wall. Other liquids which are accompanied by gas would give a *constant* horizontal pressure throughout the height of the wall. The vertical pressure distribution in tanks used for storage of granular material such as grain or coal would be essentially similar to the gas pressure distribution, with a constant value along most of the depth of the material contained. Figure 11.1 shows the pressure distributions for these three cases of loading.

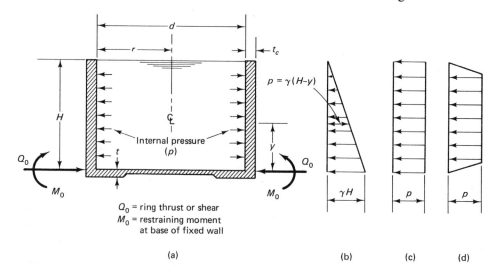

Figure 11.1 Tank internal pressure diagrams. (a) Tank cross section, showing radial shear Q_o and restraining moment M_o at base for fixed-base walls. (b) Liquid pressure, triangular load. (c) Gaseous pressure, rectangular load. (d) Granular pressure, trapezoidal load.

The basic elastic theory of cylindrical shells applies to the analysis and design of the walls of prestressed tanks. A ring force causes ring tension in the thin cylindrical walls, assumed unrestrained at the ends at each horizontal section. The magnitude of the force is proportional to the internally applied pressure, and *no* vertical moment is produced along the height of the walls. If the wall ends are restrained, the magnitude of the ring force changes and a bending moment is induced in the vertical section of the tank wall. The magnitudes of the ring forces and vertical moments are thus a function of the degree of restraint of the cylindrical shell at its boundaries and are computed from the elastic shell theory and its simplifications and idealizations to be discussed subsequently.

Liquid Load and Freely Sliding Base. From basic mechanics, the ring force is

$$F = \frac{pd}{2} = pr \qquad (11.1\ a)$$

and the ring stress is

$$f_R = \frac{pd}{2t} = \frac{pr}{t} \qquad (11.1\ b)$$

where d = diameter of cylinder
r = radius of cylinder
t = thickness of wall core
p = unit internal pressure at wall base = γH
for γ = unit weight of material contained in vessel.

The tensile ring stress *at any point below the surface* of the material contained in the vessel becomes

$$f_R = \gamma(H - y)\frac{d}{2t} = \gamma(H - y)\frac{r}{t} \qquad (11.2\ a)$$

where H is the height of the liquid contained and y is the distance *above* the base. The corresponding ring force is

$$F = \gamma(H - y)r \qquad (11.2\ b)$$

The maximum tensile ring stress at the base of the freely sliding tank wall for $y = 0$ becomes, as in Equation 11.1 b,

$$f_R(\text{max}) = \frac{\gamma H d}{2t} = \frac{\gamma h r}{t} \qquad (11.2\ c)$$

Gaseous Load on Freely Sliding Base. Again from basic principles of mechanics, the constant tensile ring stress is

$$f_R = \frac{pd}{2t} = \frac{pr}{t} \qquad (11.3)$$

Note that while theoretically the centerline diameter dimension is more accurate to use, the ratio t/d is so small that the use of the internal diameter d is appropriate.

Liquid and Gaseous Load on a Restrained Wall Base. If the base of the wall is fixed or pinned, the ring tension at the base vanishes. Because of the restraint

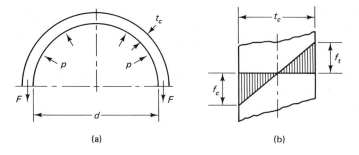

Figure 11.2 Ring tension and flexural stresses. (a) Flexural stress due to bending moment M in the wall thickness of the vertical section. (b) Ring tension internal force F in the horizontal section.

imposed on the base, the simple membrane theory of shells is then no longer applicable, due to the imposed deformations of the restraining force at the wall base. Instead, bending modifications to the membrane stresses become necessary (see Refs. 11.2 and 11.6), and the deviation of the ring tension at intermediate planes along the wall height must be approximated as in Ref. 11.2 and the discussion in Sec. 11.3.

If the vertical bending moment in the horizontal plane of the wall at any height is M_y, the flexural stress in compression or tension in the concrete becomes

$$f_t = f_c = \frac{M_y}{S} = \frac{6M_y}{t^2} \text{ per unit height} \tag{11.4}$$

The distribution of the flexural stress across the thickness of the tank wall is shown in Figure 11.2.

11.2.2 Restraining Moment M_0 and Radial Shear Force Q_0 at Freely Sliding Wall Base Due to Liquid Pressure

11.2.2.1 Membrane theory. The study of forces and stresses in a circular uncracked tank wall is an elasticity problem in cylindrical shell analysis. If the shell is free to deform under the influence of the internal liquid pressure, the basic membrane equations of equilibrium apply. The longitudinal unit force N_y, the "hugging" circumferential unit force N_θ, and the central unit shears $N_{y\theta}$ and $N_{\theta y}$ are shown in the differential element of Figure 11.3(b). Note that these *four* unknowns all act in the plane of the shell.

The basic three equations of equilibrium for these four unknown unit forces are

$$\frac{\partial N_\theta}{\partial \theta} + r\frac{\partial N_{y\theta}}{\partial y} + p_\theta r = 0 \tag{11.5 a}$$

$$r\frac{\partial N_y}{\partial y} + \frac{\partial N_{\theta y}}{\partial \theta} + p_y = 0 \tag{11.5 b}$$

$$\frac{N_\theta}{r} = p_z = 0 \tag{11.5 c}$$

where $\partial N_{y\theta} = \partial N_{\theta y}$ due to loading symmetry. The unknowns are thus reduced to three, representing a statically *determinate* structure subjected to direct forces only.

For axisymmetrical loading as in Figure 11.3(c), $p_\theta = p_y = 0$ and $p_z = p \cdot f(y)$, independent of θ. Hence,

$$p_z = -\gamma(H - y) \tag{11.6}$$

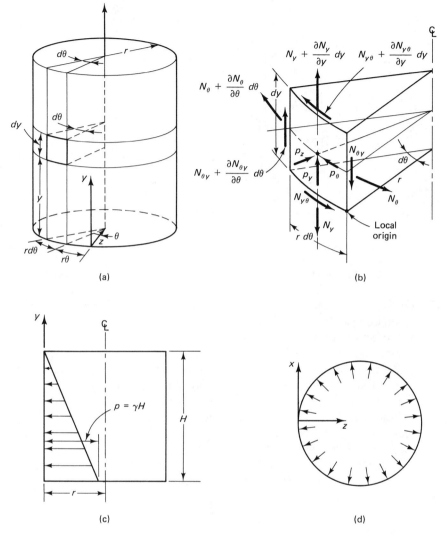

Figure 11.3 Membrane forces in cylindrical tank. (a) Tank shell geometry. (b) Shell membrane forces. (c) Liquid-filled tank elevation. (d) Axisymmetrical internal pressure at any horizontal plane.

and the solution to Equation 11.5 is

$$N_{y\theta} = N_y = 0$$

and

$$N_\theta = \gamma(H - y)r \tag{11.7}$$

11.2.2.2 Bending theory. The introduction of restraint at the boundary of the vessel induces radial ring horizontal shear and vertical moments in the shell. Consequently, the membrane force equations presented in the previous section have to be modified by superimposing these additional moments and shears. The modified ex-

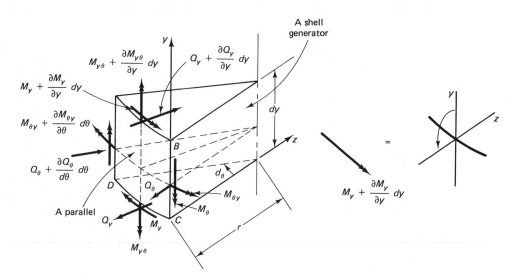

Figure 11.4 Bending moments and normal shears in a cylindrical shell wall.

pressures are denoted the *bending theory of circular shells;* the theory accounts for strain compatibility requirements in the induced deformations caused by the induced shears and moments.

The bending moments and central shears in the axisymmetrically loaded cylindrical shell are shown by force and moment vectors in Figure 11.4. The infinitesimal element *ABCD* shows the points of application and sense of the unit moments M_y about the *x*-axis and M_θ about the *y*-axis, the circumferential unit moments $M_{y\theta}$ and $M_{\theta y}$, the unit normal shear Q_y acting in the plane of the vertical shell generator and perpendicularly to the shell axis, and the unit radial shear Q_θ acting through the shell radius in the plane of the shell parallels.

Superposition of the moments and shears in Figure 11.4 on the forces in Figure 11.3(b) results in the following equilibrium equations:

$$\frac{\partial N_\theta}{\partial \theta} + \frac{\partial N_{y\theta}}{\partial y} - Q_\theta + p_\theta r = 0 \qquad (11.8\ a)$$

$$\frac{\partial N_y}{\partial y} r + \frac{\partial N_{\theta y}}{\partial \theta} + p_y r = 0 \qquad (11.8\ b)$$

$$\frac{\partial Q_\theta}{\partial \theta} + \frac{\partial Q_y}{\partial y} r + N_\theta + p_z r = 0 \qquad (11.8\ c)$$

$$-\frac{\partial M_y}{\partial y} r + \frac{\partial M_{y\theta}}{\partial y} + Q_y r = 0 \qquad (11.8\ d)$$

$$\frac{\partial M_\theta}{\partial \theta} + \frac{\partial M_{y\theta}}{\partial y} r - Q_\theta r = 0 \qquad (11.8\ e)$$

Due to symmetry of loading, $N_{y\theta} = N_{\theta y} = M_{y\theta} = M_{\theta y} = 0$, and dQ_θ can be disregarded, reducing the partial differential equations 11.8 to the set of ordinary

Two hundred fifty thousand-bbl (39,750-m³) prestressed concrete propane gas storage container, Winnipeg, Manitoba, Canada. (*Courtesy*, N.A. Legatos, Preload Technology, Inc., New York.)

differential equations

$$\frac{dN_y}{dy}r + p_y r = 0 \tag{11.9 a}$$

$$\frac{dQ_y}{dy}r + N_\theta + p_z r = 0 \tag{11.9 b}$$

$$-\frac{dM_y}{dy}r + Q_y r = 0 \tag{11.9 c}$$

With the central membrane forces N_y constant and taken to be zero (see Refs. 11.1 and 11.3), the remaining equations 11.9 b and 11.9 c can be written in the following simplified form having the three unknowns N_θ, Q_y, and M_y:

$$\frac{dQ_y}{dy} + \frac{1}{r}N_\theta = -p_z \tag{11.10 a}$$

$$\frac{dM_y}{dy} - Q_y = 0 \tag{11.10 b}$$

In order to solve these equations, displacements have to be considered and equations of geometry developed.

Force Equations. If v and w are the displacements in the y and z directions, then the unit strains in these directions are, respectively,

$$\epsilon_y = \frac{dv}{dy}$$

and

$$\epsilon_\phi = -\frac{w}{r}$$

which give

$$N_y = \frac{Et}{1 - \mu^2}(\epsilon_y + \mu\epsilon_\phi) = \frac{Et}{1 - \mu^2}\left(\frac{dv}{dy} - \mu\frac{w}{r}\right) = 0 \qquad \text{(11.11 a)}$$

and

$$N_\phi = \frac{Et}{1 - \mu^2}(\epsilon_\phi + \mu\epsilon_y) = \frac{Et}{1 - \mu^2}\left(-\frac{w}{r} + \mu\frac{dv}{dy}\right) \qquad \text{(11.11 b)}$$

where μ = Poisson's ratio
   t = thickness of the wall core.

From Equation 11.11 a,

$$\frac{dv}{dy} = \mu\frac{w}{r} \qquad \text{(11.12 a)}$$

From Equation 11.11 b,

$$N_\theta = -Et\frac{w}{r} \qquad \text{(11.12 b)}$$

Moment Equations.   Due to symmetry, there is no change in curvature in the circumferential direction; hence, the curvature in the y direction has to be equal to $-d^2v/dy^2$. Using the same moment expressions for thin elastic plates results in

$$M_\phi = \mu M_y \qquad \text{(11.13 a)}$$

$$M_y = -D\frac{d^2w}{dy^2} \qquad \text{(11.13 b)}$$

where $D = Et^3/12(1 - \mu^2)$ is the shell or plate flexural rigidity.
   Introducing Equations 11.12 and 11.13 into Equations 11.10 results in

$$\frac{d^2}{dx^2}\left(D\frac{d^2w}{dy^2}\right) + \frac{Et}{a^2}w = p_z \qquad \text{(11.14)}$$

If the wall thickness t is constant, Equation 11.14 becomes

$$D\frac{d^4w}{dy^2} + \frac{Et}{r^2}w = p_z \qquad \text{(11.15)}$$

Letting

$$\beta^4 = \frac{Et}{4r^2D} = \frac{3(1 - \mu^2)}{(rt)^2}$$

Equation 11.15 becomes

$$\frac{d^4w}{dy^4} + 4\beta^4w = \frac{p_z}{D} \qquad \text{(11.16)}$$

Equation 11.16 is the same as is obtained for a prismatic bar with flexural rigidity D supported by a continuous elastic foundation and subject to the action of a unit load intensity p_z. The general solution to this equation (Ref. 11.1) for the *radial* displacement in the z-direction is

$$w = e^{\beta y}(C_1 \cos \beta y + C_4 \sin \beta y)$$
$$+ e^{-\beta y}(C_3 \cos \beta y + C_4 \sin \beta y) + f(y) \tag{11.17}$$

where $f(y)$ is the particular solution of Equation 11.16 as a membrane solution giving displacement

$$w = \frac{p_z r^2}{Et}$$

11.2.3 General Equations of Forces and Displacements

Solving Equation 11.17 and introducing the notation

$$\Phi(\beta y) = e^{-\beta y}(\cos \beta y + \sin \beta y)$$

$$\psi(\beta y) = e^{-\beta y}(\cos \beta y - \sin \beta y)$$

$$\theta(\beta y) = e^{-\beta y} \cos \beta y$$

$$\zeta(\beta y) = e^{-\beta y} \sin \beta y$$

the expression for radial deformation in the z direction and its consecutive derivatives at any height y above the wall base can be evaluated from the following simplified expressions as a function of the wall base unit moments M_0 and unit radial shears Q_0:

$$\text{Deflection } w = -\frac{1}{2\beta^3 D}[\beta M_0 \psi(\beta y) + Q_0 \theta(\beta y)] \tag{11.18 a}$$

$$\text{Rotation } \frac{dw}{dy} = \frac{1}{2\beta^2 D}[2\beta M_0 \theta(\beta y) + Q_0 \Phi(\beta y)] \tag{11.18 b}$$

$$\frac{d^2 w}{dy^2} = -\frac{1}{2\beta D}[2\beta M_0 \Phi(\beta y) + 2Q_0 \zeta(\beta y)] \tag{11.18 c}$$

$$\frac{d^3 w}{dy^3} = \frac{1}{D}[2\beta M_0 \zeta(\beta y) - Q_0 \psi(\beta y)] \tag{11.18 d}$$

The shell functions $\Phi(\beta y)$, $\psi(\beta y)$, $\theta(\beta y)$, and $\zeta(\beta y)$ are given in the standard influence coefficients of Table 11.1 (Ref. 11.1), for a range $0 \le \beta y \le 3.9$.

The maximum radial displacement or deflection at the restrained wall base, from Equation 11.18a, is

$$(w)_{y=0} = -\frac{1}{2\beta^3 D}(\beta M_0 + Q_0) \tag{11.19 a}$$

and the maximum rotation of the wall at the base, from Equation 11.18 b, becomes

$$\left(\frac{dw}{dy}\right)_{y=0} = \frac{1}{2\beta^2 D}(2\beta M_0 + Q_0) \tag{11.19 b}$$

TABLE 11.1   TABLE OF FUNCTIONS Φ, ψ, θ, AND ζ

βy	Φ	ψ	θ	ζ
0	1.0000	1.0000	1.0000	0
0.1	0.9907	0.8100	0.9003	0.0903
0.2	0.9651	0.6398	0.8024	0.1627
0.3	0.9267	0.4888	0.7077	0.2189
0.4	0.8784	0.3564	0.6174	0.2610
0.5	0.8231	0.2415	0.5323	0.2908
0.6	0.7628	0.1431	0.4530	0.3099
0.7	0.6997	0.0599	0.3798	0.3199
0.8	0.6354	−0.0093	0.3131	0.3223
0.9	0.5712	−0.0657	0.2527	0.3185
1.0	0.5083	−0.1108	0.1988	0.3096
1.1	0.4476	−0.1457	0.1510	0.2967
1.2	0.3899	−0.1716	0.1091	0.2807
1.3	0.3355	−0.1897	0.0729	0.2626
1.4	0.2849	−0.2011	0.0419	0.2430
1.5	0.2384	−0.2068	0.0158	0.2226
1.6	0.1959	−0.2077	−0.0059	0.2018
1.7	0.1576	−0.2047	−0.0235	0.1812
1.8	0.1234	−0.1985	−0.0376	0.1610
1.9	0.0932	−0.1899	−0.0484	0.1415
2.0	0.0667	−0.1794	−0.0563	0.1230
2.1	0.0439	−0.1675	−0.0618	0.1057
2.2	0.0244	−0.1548	−0.0652	0.0895
2.3	0.0080	−0.1416	−0.0668	0.0748
2.4	−0.0056	−0.1282	−0.0669	0.0613
2.5	−0.0166	−0.1149	−0.0658	0.0492
2.6	−0.0254	−0.1019	−0.0636	0.0383
2.7	−0.0320	−0.0895	−0.0608	0.0287
2.8	−0.0369	−0.0777	−0.0573	0.0204
2.9	−0.0403	−0.0666	−0.0534	0.0132
3.0	−0.0423	−0.0563	−0.0493	0.0071
3.1	−0.0431	−0.0469	−0.0450	0.0019
3.2	−0.0431	−0.0383	−0.0407	−0.0024
3.3	−0.0422	−0.0306	−0.0364	−0.0058
3.4	−0.0408	−0.0237	−0.0323	−0.0085
3.5	−0.0389	−0.0177	−0.0283	−0.0106
3.6	−0.0366	−0.0124	−0.0245	−0.0121
3.7	−0.0341	−0.0079	−0.0210	−0.0131
3.8	−0.0314	−0.0040	−0.0177	−0.0137
3.9	−0.0286	−0.0008	−0.0147	−0.0140

where M_0 and Q_0 are respectively the restraining moment and the ring shear at the base shown in Figure 11.1.

For tanks with constant wall thickness, the unit forces along the wall height are as follows:

$$N_\theta = -\frac{Etw}{r} \qquad (11.20\ a)$$

$$Q_y = -D\frac{d^3w}{dy^3} \tag{11.20 b}$$

$$M_\theta = \mu M_y \tag{11.20 c}$$

$$M_y = -D\frac{d^2w}{dy^2} \tag{11.20 d}$$

From Equations 11.18 c, 11.18 d, 11.20 b, and 11.20 d, the expressions for vertical moments and horizontal radial shears at the base of the wall, where y is zero, become (Ref. 11.1)

$$\left(M_y\right)_{y=0} = M_0 = \left(1 - \frac{1}{\beta H}\right)\frac{yHrt}{\sqrt{12(1-\mu^2)}} \tag{11.21 a}$$

$$(Q_y)_{y=0} = Q_0 = -(2\beta H - 1)\frac{\gamma rt}{\sqrt{12(1-\mu^2)}} \tag{11.21 b}$$

The expression for the vertical moment at any level y above the wall base can be obtained from

$$M_y = -\frac{1}{\beta}[\beta M_0 \Phi(\beta y) + Q_0 \zeta(\beta y)] \tag{11.22}$$

The *offset* ring shear force ΔQ_y corresponds to a radial displacement w_y of the wall at a height y above the base when the tank is *empty* and the values of Q_0 and M_0 due to a full liquid or full gas load are *induced*, as shown in Figure 11.5. This force can be expressed as either

$$\Delta Q_y = +\frac{Et}{r}(w_y)$$

or

$$\Delta Q_y = \frac{Et}{2r\beta^3 D}[\beta M_0 \psi(\beta y) + Q_0 \theta(\beta y)]$$

or

$$\Delta Q_y = +\frac{16(1-\mu^2)}{\beta^3 rt^2}[\beta M_0 \psi(\beta y) + Q_0 \theta(\beta y)] \tag{11.23}$$

The ring shear Q_y at a plane y above the base would be equal to the difference between the ring force for a freely sliding base and ΔQ_y:

$$Q_y = F - \Delta Q_y \tag{11.24}$$

It is important to be consistent in the sign convention used throughout a solution. The easiest approach is to draw the deflected shape of the wall and use a positive $(+)$ notation for the following conditions:

1. Moment causing tension on the outside extreme fibers.
2. Ring tension radial forces.

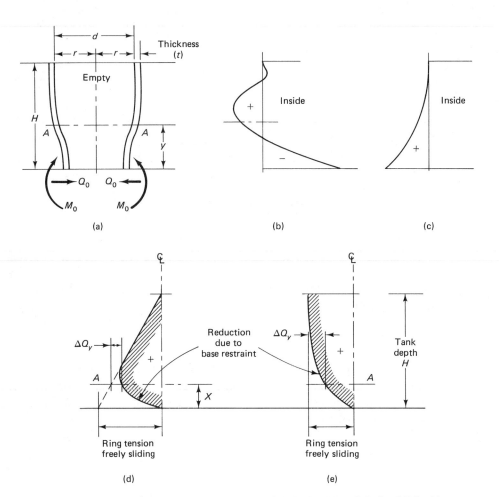

Figure 11.5 Wall base restraint in empty tank inducing M_o and Q_o for full liquid or gas pressure. (a) Deformed walls of empty tank. (b) Moment along vertical section (+ represents tension on outside). (c) Ring tension force F in horizontal section (always positive). (d) Offset ΔQ_y for liquid pressure. (e) Offset ΔQ_y for gas pressure.

3. Thrust inwards toward the vertical axis. Here, the same sense is used as for ring tension forces in order to draw the diagram for the balancing prestressing forces on the same side as the ring tension forces for comparison.

4. Lateral wall movement *inwards* toward the vertical axis.

5. Anticlockwise rotation.

Pinned Wall Base, Liquid Pressure. When the wall base is *pinned* and carrying a liquid load moment $M_0 = 0$ at the base,

$$Q_0 = +\frac{2\beta^3 \gamma H (rt)^2}{12(1 - \mu^2)}$$

or

$$Q_0 = + \frac{\gamma H}{[12(1 - \mu^2)]^{1/4}} \left(\frac{rt}{2}\right)^{1/2} \qquad (11.25)$$

The value of the shell constants β, β^2, and β^4 for use in the preceding equations can easily be calculated from the expression for β^4 as follows:

$$\beta^4 = \frac{Et}{4r^2 D} = \frac{3(1 - \mu^2)}{(rt)^2} \qquad (11.26 \text{ a})$$

$$\beta^3 = \frac{[3(1 - \mu^2)]^{3/4}}{(rt)^{3/2}} \qquad (11.26 \text{ b})$$

$$\beta^2 = \frac{[3(1 - \mu^2)]^{1/2}}{(rt)} \qquad (11.26 \text{ c})$$

$$\beta = \frac{[3(1 - \mu^2)]^{1/4}}{(rt)^{1/2}} \qquad (11.26 \text{ d})$$

11.2.4 Ring Shear Q_0 and Moment M_0, Gas Containment

If the edges of the shell are free at the wall base, the internal pressure produces only hoop stress $f_R = pr/t$ and the radius of the cylinder increases by the amount

$$w = \frac{rf_r}{E} = \frac{pr^2}{Et} \qquad (11.27)$$

Also, for full restraint at the wall base,

$$(w)_{y=0} = \frac{1}{2\beta^3 D}(\beta M_0 + Q_0) \qquad (11.28 \text{ a})$$

and

$$\left(\frac{dw}{dy}\right)_{y=0} = \frac{1}{2\beta^2 D}(2\beta M_0 + Q_0) = 0 \qquad (11.28 \text{ b})$$

Solving for M_0 and Q_0 gives

$$M_0 = -2\beta^2 Dw = -\frac{p}{2\beta^2} = \frac{prt}{\sqrt{12(1 - \mu^2)}} \qquad (11.29 \text{ a})$$

and

$$Q_0 = +4\beta^3 Dw = +\frac{p}{\beta} = +\frac{p(2rt)^{1/2}}{[12(1 - \mu^2)]^{1/4}} \qquad (11.29 \text{ b})$$

Pinned Wall Base, Gas Pressure. If the wall base is *pinned* and carrying a gas load moment $M_0 = 0$ at the base,

$$Q_0 = 2\beta^3 D \left(\frac{pr^2}{Et}\right)$$

TABLE 11.2 EQUATIONS FOR LIQUID-RETAINING TANKS

Parameter	Equation	Number
Flexural rigidity, D	$Et^3/[12(1 - \mu^2)]$	
Ring stress, f_R	$\gamma(H - Y)r/t$	11.2 a
Ring force, F	$\gamma(H - y)r$	11.2 b
Pressure, P_z	$\gamma(H - y)$	11.2 b
Radial deflection, w	$\dfrac{1}{2\beta^2 D}[\beta M_0 \psi(\beta y) + Q_0 \theta(\beta y)]$	11.18 a
Rotation $\dfrac{dw}{dy}$	$\dfrac{1}{2\beta^3 D}[2\beta M_0 \theta(\beta y) + Q_0 \Phi(\beta y)]$	11.18 b
Maximum deflection, $(w)_{y=0}$	$\dfrac{1}{2\beta^3 D}(\beta M_0 + Q_0)$	11.19 a
Maximum rotation $\left(\dfrac{dw}{dy}\right)_{y=0}$	$\dfrac{1}{2\beta^3 D}(2\beta M_0 + Q_0) = 0$	11.19 b
$M_0 = (M_y)_{y=0}$	$-\left(1 - \dfrac{1}{\beta H}\right)\dfrac{\gamma H r t}{\sqrt{12(1 - \mu^2)}}$	11.21 a
$Q_0 = (Q_y)_{y=0}$	$+ (2\beta H - 1)\dfrac{\gamma r t}{\sqrt{12(1 - \mu^2}}$	11.21 b
M_y	$+\dfrac{1}{\beta}[\beta M_0 \Phi(\beta y) + Q_0 \zeta(\beta y)]$	11.22
Empty tank offset, ΔQ_y	$+\dfrac{6(1 - \mu)^2}{\beta^3 r t^2}[\beta M_0 \psi(\beta y) + Q_0(\beta y)]$	11.23
Q_y	$+ (F - \Delta Q_y)$	11.24
Q_0 when $M_0 = 0$ (Pinned base)	$+\dfrac{\gamma H \sqrt{rt/2}}{[12(1 - \mu^2)]^{1/4}}$	11.25
Tank Constants: β^3	$[3(1 - \mu^2)]^{3/4}/(rt)^{3/2}$	11.26 b
β^2	$[3(1 - \mu^2]^{1/2}/rt$	11.26 c
β	$[3(1 - \mu^2)]^{1/4}/(rt)^{1/2}$	11.26 d

TABLE 11.3 EQUATIONS FOR GAS-RETAINING TANKS

Parameter	Equation	Number
Maximum deflection $(w)_{y=0}$	$\dfrac{1}{2\beta^3 D}(\beta M_0 + Q_0)$	11.28 a
Maximum rotation $\left(\dfrac{dw}{dy}\right)_{y=0}$	$\dfrac{1}{2\beta^3 D}(2\beta M_0 + Q_0) = 0$	11.28 b
$M_0 = (M_y)_{y=0}$	$-\dfrac{prt}{\sqrt{12(1 - \mu^2)}}$	11.29 a
$Q_0 = (Q_y)_{y=0}$	$+\dfrac{p\sqrt{2rt}}{[12(1 - \mu^2)]^{1/4}}$	11.29 b
Q_0 when $M_0 = 0$ (Pinned base)	$+\dfrac{p\sqrt{rt/2}}{[12(1 - \mu^2)]^{1/4}}$	11.30

Note: Values of β, β^2, and β^3 constants as in Table 11.2.

or

$$Q_0 = \frac{p}{[12(1 - \mu^2)]^{1/4}} \left(\frac{rt}{2}\right)^{1/2} \tag{11.30}$$

Table 11.2 presents a summary of the design equations for liquid-retaining tanks, and Table 11.3 gives a similar summary for gas-retaining tanks.

11.3 MOMENT M_0 AND RING FORCE Q_0 IN LIQUID-RETAINING TANK

Example 11.1

A prestressed concrete circular tank is fully restrained at the wall base. It has an interior diameter $d = 125$ ft (38.1 m) and retains water having height $H = 25$ ft (7.62 m). The wall thickness $t = 10$ in. (25 cm). Compute (a) the unit vertical moment M_0 and the radial ring force Q_0 at the base of the wall, and (b) the unit vertical moment M_y at $7\frac{1}{2}$ ft (2.29 m) above the base. Use Poisson's ratio $\mu = 0.2$ and unit water weight $\gamma = 62.4$ lb/ft^3 (1,000 kg/m^3).

Solution

(a) *At Wall Base*

$$r = \frac{1}{2} \times 125 = 62.5 \text{ ft } (11.9 \text{ m})$$

$$t = 10 \text{ in.} = 0.83 \text{ ft } (.25 \text{ m})$$

From Equation 11.26 d,

$$\beta = \frac{[3(1 - \mu^2)]^{1/4}}{(rt)^{1/2}} = \frac{[3(1 - 0.2 \times 0.2)]^{1/4}}{(62.5 \times 0.83)^{1/2}} = 0.181$$

From Equation 11.21 a,

$$M_0 = -\left(1 - \frac{1}{\beta H}\right) \frac{\gamma H r t}{\sqrt{12(1 - \mu^2)}}$$

$$= -\left(1 - \frac{1}{0.181 \times 25}\right) \times \frac{62.4 \times 25 \times 62.5 \times 0.83}{\sqrt{12(1 - 0.04)}}$$

$$= -18{,}574 \text{ ft-lb/ft (kN-m/m) of circumference}$$

From Equation 11.21 b,

$$Q_0 = +(2\beta H - 1)\frac{\gamma r t}{\sqrt{12(1 - \mu^2)}}$$

$$= +(2 \times 0.181 \times 25 - 1) + \frac{62.4 \times 62.5 \times 0.83}{\sqrt{12(1 - 0.04)}}$$

$$= +7{,}677 \text{ lb/ft (112 kN/m) of circumference}$$

(b) *$7\frac{1}{2}$ ft above Wall Base*

$$y = 7.5 \text{ ft}$$

$$\text{Water height} = (H - y) = 25 - 7.5 = 17.5 \text{ ft } (5.33 \text{ m})$$

$$\text{Height ratio} = \left(1 - \frac{y}{H}\right) = 1 - \frac{7.5}{25} = 0.7$$

$$\beta y = 0.181 \times 7.5 = 1.36$$

From Equation 11.22,

$$M_y = +\frac{1}{\beta}[\beta M_0 \Phi(\beta y) + Q_0 \zeta(\beta y)]$$

From Table 11.1 for $\beta y = 1.36$,

$$\Phi = 0.311$$

$$\xi = 0.252$$

$$M_y = +\frac{1}{0.181}(-0.181 \times 18,574 \times 0.311 + 7,677 \times 0.252)$$

$$= +4,912 \text{ ft-lb/ft of circumference}$$

11.4 RING FORCE Q_Y AT INTERMEDIATE HEIGHTS OF WALL

Example 11.2

Compute the radial ring force Q_y in Example 11.1 at (a) $y = 7\frac{1}{2}$ ft (2.29 m) and (b) $y = 10$ ft (3.05 m) above the wall base.

Solution The freely sliding base ring force $F = \gamma H r = 62.4 \times 25 \times 62.5 = 97,500$ lb/ft (1,423 kN/m). From Equation 11.23, the ring force offset is

$$\Delta Q_y = +\frac{6(1 - \mu^2)}{\beta^3 r t^2}[\beta M_0 \psi(\beta y) + Q_0 \theta(\beta y)]$$

From Example 11.1, $\beta = 0.181$; hence, $\beta^3 = 0.0059$.

(a) *Q_y at 7.5 ft above Wall Base*

$$\beta y = 0.181 \times 7.5 = 1.36$$

From Table 11.1 for $\beta y = 1.36$,

$$\psi = -0.1965$$

$$\theta = +0.0543$$

$$\Delta Q_y = +\frac{6(1 - 0.04)}{0.0059 \times 62.5(0.83)^2}$$

$$\times [0.181(-18,574)(-0.1965) + 7,677(+0.0543)]$$

$$= 24,431 \text{ lb/ft } (356 \text{ kN/m})$$

From Equation 11.2b, the ring force $F = \gamma(H - y)r = 62.4 \times (25 - 7.5) \times 62.5 = 68,250$ lb/ft. So $Q_{7.5} = F - \Delta Q_y = 68,250 - 24,431 = 43,819$ lb/ft (705 kN/m) of circumference, as shown in Figure 11.6(a): (a) At $7\frac{1}{2}$ ft above the base; (b) At 10 ft above the base.

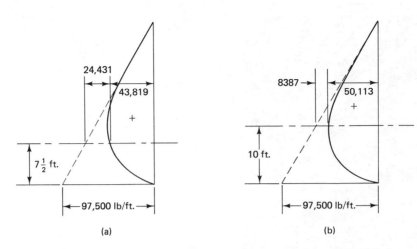

Figure 11.6 Radial ring force profile. (a) At $7\frac{1}{2}$ ft above the base. (b) At 10 ft above the base in Ex. 11.1.

(b) *Q_y at 10.0 ft above Wall Base*

$$\beta y = 0.181 \times 10 = 1.81$$

From Table 11.1 for $\beta y = 1.81$,

$$\psi = -0.1984$$

$$\theta = -0.0387$$

$$\Delta Q_y = \frac{6(1 - 0.04)}{0.0059 \times 62.50(0.83)^2}$$

$$\times\ [0.181(-18,574)(-0.1984) + 7,677(-0.0387)] = 8,387\ \text{lb/ft}$$

The ring force $F = \gamma(H - y)r = 62.4(25 - 10)62.5 = 58,500$ lb/ft. So $Q_{10} = F - \Delta Q_y = 58,500 - 8,387 = 50,113$ lb/ft (731 kN/m) of circumference, as shown in Figure 11.6(b). Compare how close this value is to $Q = 50,115$ lb/ft obtained by using membrane coefficients in Example 11.3.

11.5 CYLINDRICAL SHELL MEMBRANE COEFFICIENTS

The bending moment at any level along the height above the base of a cylindrical tank can be computed from the bending moment expression for a cantilever beam. This is accomplished by multiplying the cantilever moment values by coefficients whose magnitudes are functions of the geometrical dimensions of the tank and which are termed *membrane coefficients*. The basic moment expressions developed in Section 11.2 for the circular container can be rearranged into a factor H^2/dt denoting *geometry* and a factor γH^3 or pH^2 denoting *cantilever effect*, for liquid and gaseous loading, respectively (Ref. 11.2).

The tank constant β in Equation 11.26 d is a function of rt or dt, where d is the

Two-and-a-half-million-gallon tendon prestressed concrete tank with the horizontal and vertical tendons utilizing plastic sheathing to protect the prestressing steel from seepage through the wall. (*Courtesy*, Jorgenson, Hendrickson and Close, Denver, Colorado.)

tank diameter. Using Poisson's ratio $\mu \cong 0.2$ for concrete, we have

$$\beta = \frac{[3(1 - \mu^2)]^{1/4}}{(rt)^{1/2}} = \frac{1.30}{(rt)^{1/2}} = \frac{1.84}{(dt)^{1/2}}$$

The factor $1/\beta H$ used in the basic bending expressions of Section 11.2 can be rewritten in terms of $(dt/H^2)^{1/2}$ since $\beta = 1.84/(dt)^{1/2}$. The product βy can also be rewritten in terms of $\lambda(H^2/dt)^{1/2}$ using $y = \lambda H$, where y is the height above the base.

Consequently, the moment M_y of Equation 11.22 in a wall section a distance y above the base can be represented in terms of the form factor H^2/dt and the cantilever factor γH^3 or pH^2 as follows:

$$M_y = \text{numerical variant} \times \text{form factor} \times \text{cantilever factor}$$

or

$$M_y = \left[\text{variant} \times \frac{H^2}{dt} \right] \times [\gamma H^3 \quad \text{or} \quad pH^2] \tag{11.31}$$

Prestressing preload circular tank wall with wire winder. (*Courtesy*, N.A. Legatos, Preload Technology, Inc., New York.)

The form factor H^2/dt is constant for the particular structure being designed. Hence, the product of the variant and the form factor produces the membrane coefficient C, so that Equation 11.31 becomes

$$M_y = C\gamma H^3 \qquad (11.32\ a)$$

for a liquid load and

$$M_y = CpH^2 \qquad (11.32\ b)$$

for a gaseous load.

Tables 11.4–11.16 from Ref. 11.5 give the membrane coefficients C for various form factors H^2/dt and most expected boundary and load conditions. They significantly reduce the computational efforts that are normally required in the design and analysis of shells, without loss of accuracy in the results. Using the membrane coefficients for the solution of the circular tank forces and moments should give results reasonably close to those obtained from the bending solutions presented in Section 11.2 and the sets of equations listed in Tables 11.2 and 11.3.

TABLE 11.4 MOMENT INFLUENCE COEFFICIENTS, TRIANGULAR LOAD

Moments in Cylindrical Wall
Triangular Load
Fixed Base, Free Top
Mom. = coef. × γH^2 ft. lb. per ft.
Positive sign indicates tension in the outside

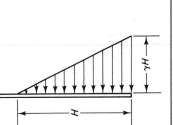

Liquid Load

$\dfrac{H^2}{dt}$	\multicolumn{10}{c}{Coefficients at point}
	0.1H	0.2H	0.3H	0.4H	0.5H	0.6H	0.7H	0.8H	0.9H	1.0H
0.4	+.0005	+.0014	+.0021	+.0007	−.0042	−.0150	−.0302	−.0529	−.0816	−.1205
0.8	+.0011	+.0037	+.0063	+.0080	+.0070	+.0023	−.0068	−.0224	−.0465	−.0795
1.2	+.0012	+.0042	+.0077	+.0103	+.0112	+.0090	+.0022	−.0108	−.0311	−.0602
1.6	+.0011	+.0041	+.0075	+.0107	+.0121	+.0111	+.0058	−.0051	−.0232	−.0505
2.0	+.0010	+.0035	+.0068	+.0099	+.0120	+.0115	+.0075	−.0021	−.0185	−.0436
3.0	+.0006	+.0024	+.0047	+.0071	+.0090	+.0097	+.0077	+.0012	−.0119	−.0333
4.0	+.0003	+.0015	+.0028	+.0047	+.0066	+.0077	+.0069	+.0023	−.0080	−.0268
5.0	+.0002	+.0008	+.0016	+.0029	+.0046	+.0059	+.0059	+.0028	−.0058	−.0222
6.0	+.0001	+.0003	+.0008	+.0019	+.0032	+.0046	+.0051	+.0029	−.0041	−.0187
8.0	.0000	+.0001	+.0002	+.0008	+.0016	+.0028	+.0038	+.0029	−.0022	−.0146
10.0	.0000	.0000	+.0001	+.0004	+.0007	+.0019	+.0029	+.0028	−.0012	−.0122
12.0	.0000	−.0001	+.0001	+.0002	+.0003	+.0013	+.0023	+.0026	−.0005	−.0104
14.0	.0000	.0000	.0000	.0000	+.0001	+.0008	+.0019	+.0023	−.0001	−.0090
16.0	.0000	.0000	−.0001	−.0002	−.0001	+.0004	+.0013	+.0019	+.0001	−.0079

Notes: 1-Tables 11.4–11.16 Adapted from Ref. 11.5.
2-0.0H is the bottom and 1.0H is the top of the wall.
3-Shear acting inwards is positive; moment applied at an edge is positive when outward rotation results at that edge

TABLE 11.5 MOMENT INFLUENCE COEFFICIENTS, RECTANGULAR LOAD

Moments in Cylindrical Wall
Rectangular Load
Fixed Base, Free Top
Mom = coef. × pH^2 ft. lb. per ft.
Positive sign indicates tension in the outside

Gas Load

$\dfrac{H^2}{dt}$	Coefficients at point									
	0.1H	0.2H	0.3H	0.4H	0.5H	0.6H	0.7H	0.8H	0.9H	1.0H
0.4	−.0023	−.0093	−.0227	−.0439	−.0710	−.1018	−.1455	−.2000	−.2593	−.3310
0.8	.0000	−.0006	−.0025	−.0083	−.0185	−.0362	−.0594	−.0917	−.1325	−.1835
1.2	+.0008	+.0026	+.0037	+.0029	−.0009	−.0089	−.0227	−.0468	−.0815	−.1178
1.6	+.0011	+.0036	+.0062	+.0077	+.0068	+.0011	−.0093	−.0670	−.0529	−.0876
2.0	+.0010	+.0036	+.0066	+.0088	+.0089	+.0059	−.0019	−.0167	−.0389	−.0719
3.0	+.0007	+.0026	+.0051	+.0074	+.0091	+.0083	+.0042	−.0053	+.0223	−.0483
4.0	+.0004	+.0015	+.0033	+.0052	+.0068	+.0075	+.0053	−.0013	−.0145	−.0365
5.0	+.0002	+.0008	+.0019	+.0035	+.0051	+.0061	+.0052	+.0007	−.0101	−.0293
6.0	+.0001	+.0004	+.0011	+.0022	+.0036	+.0049	+.0048	+.0017	−.0073	−.0242
8.0	.0000	+.0001	+.0003	+.0008	+.0018	+.0031	+.0038	+.0024	−.0040	−.0184
10.0	.0000	−.0001	.0000	+.0002	+.0009	+.0021	+.0030	+.0026	−.0022	−.0147
12.0	.0000	.0000	−.0001	.0000	+.0004	+.0014	+.0024	+.0022	−.0012	−.0123
14.0	.0000	.0000	.0000	.0000	+.0002	+.0010	+.0018	+.0021	−.0007	−.0105
16.0	.0000	.0000	.0000	−.0001	+.0001	+.0006	+.0012	+.0020	−.0005	−.0091

TABLE 11.6 MOMENT INFLUENCE COEFFICIENTS, TRAPEZOIDAL LOAD

Moments in Cylindrical Wall
Trapezoidal Load
Hinged Base, Free Top
Mom. = coef. × (γH^2 + pH^2) ft. lb. per ft.
Positive sign indicates tension in the outside

$\dfrac{H^2}{dt}$	0.1H	0.2H	0.3H	0.4H	0.5H	0.6H	0.7H	0.8H	0.9H	1.0H
					Coefficients at point					
0.4	+.0020	+.0072	+.0151	+.0230	+.0301	+.0348	+.0357	+.0312	+.0197	0
0.3	+.0019	+.0064	+.0133	+.0207	+.0271	+.0319	+.0329	+.0292	+.0187	0
1.2	+.0016	+.0058	+.0111	+.0177	+.0237	+.0280	+.0296	+.0263	+.0171	0
1.6	+.0012	+.0044	+.0091	+.0145	+.0195	+.0236	+.0255	+.0232	+.0155	0
2.0	+.0009	+.0033	+.0073	+.0114	+.0158	+.0199	+.0219	+.0205	+.0145	0
3.0	+.0004	+.0015	+.0040	+.0063	+.0092	+.0127	+.0152	+.0153	+.0111	0
4.0	+.0001	+.0007	+.0016	+.0033	+.0057	+.0083	+.0109	+.0118	+.0092	0
5.0	.0000	+.0001	+.0006	+.0016	+.0034	+.0057	+.0080	+.0094	+.0078	0
6.0	.0000	.0000	+.0002	+.0008	+.0019	+.0039	+.0062	+.0078	+.0068	0
8.0	.0000	.0000	−.0002	.0000	+.0007	+.0020	+.0038	+.0057	+00.54	0
10.0	.0000	.0000	−.0002	−.0001	+.0002	+.0011	+.0025	+.0043	+.0045	0
12.0	.0000	.0000	−.0001	−.0002	.0000	+.0005	+.0017	+.0032	+.0039	0
14.0	.0000	.0000	−.0001	−.0001	−.0001	.0000	+.0012	+.0026	+.0033	0
16.0	.0000	.0000	.0000	−.0001	−.0002	−.0004	+.0008	+.0022	+.0029	0

TABLE 11.7 MOMENT INFLUENCE COEFFICIENTS, EMPTY TANK (SHEAR APPLIED AT TOP, BASE FIXED)

Moments in Cylindrical Wall
Shear Per Ft., Q, Applied at Top
Fixed Base, Free Top
Mom = coef. $\times$ VH ft. lb. per ft.
Positive sign indicates tension in outside

Empty Tank

$\dfrac{H^2}{dt}$	Coefficients at point									
	0.1H	0.2H	0.3H	0.4H	0.5H	0.6H	0.7H	0.8H	0.9H	1.0H
0.4	+0.093	+0.172	+0.240	+0.300	+0.354	+0.402	+0.448	+0.492	+0.535	+0.578
0.8	+0.085	+0.145	+0.185	+0.208	+0.220	+0.224	+0.223	+0.219	+0.214	+0.208
1.2	+0.082	+0.132	+0.157	+0.164	+0.159	+0.145	+0.127	+0.106	+0.084	+0.062
1.6	+0.079	+0.122	+0.139	+0.138	+0.125	+0.105	+0.081	+0.056	+0.030	+0.004
2.0	+0.077	+0.115	+0.126	+0.119	+0.103	+0.080	+0.056	+0.031	+0.006	-0.019
3.0	+0.072	+0.100	+0.100	+0.086	+0.066	+0.044	+0.025	+0.006	-0.010	-0.024
4.0	+0.068	+0.088	+0.081	+0.063	+0.043	+0.025	+0.010	-0.001	-0.010	-0.019
5.0	+0.064	+0.078	+0.067	+0.047	+0.028	+0.013	+0.003	-0.003	-0.007	-0.011
6.0	+0.062	+0.070	+0.056	+0.036	+0.018	+0.006	0.000	-0.003	-0.005	-0.006
8.0	+0.057	+0.058	+0.041	+0.021	+0.007	0.000	-0.002	-0.003	-0.002	-0.001
10.0	+0.053	+0.049	+0.029	+0.012	+0.002	-0.002	-0.002	-0.002	-0.001	0.000
12.0	+0.049	+0.042	+0.022	+0.007	0.000	-0.002	-0.002	-0.001	0.000	0.000
14.0	+0.046	+0.036	+0.017	+0.004	-0.001	-0.002	-0.001	-0.001	0.000	0.000
16.0	+0.044	+0.031	+0.012	+0.001	-0.002	-0.002	-0.001	0.000	0.000	0.000

TABLE 11.8 MOMENT INFLUENCE COEFFICIENTS, EMPTY TANK (SHEAR APPLIED AT TOP, HINGED BASE)

Moments in Cylindrical Wall
Moment Per Ft., M, Applied at Base
Hinged Base, Free Top
Mom. = coef. × M ft. lb. per ft.
Positive sign indicates tension in outside

Empty Tank

$\dfrac{H^2}{dt}$	Coefficients at point									
	0.1H	0.2H	0.3H	0.4H	0.5H	0.6H	0.7H	0.8H	0.9H	1.0H
0.4	+0.013	+0.051	+0.109	+0.196	+0.296	+0.414	+0.547	+0.692	+0.843	+1.000
0.8	+0.009	+0.040	+0.090	+0.164	+0.253	+0.375	+0.503	+0.659	+0.824	+1.000
1.2	+0.006	+0.027	+0.063	+0.125	+0.206	+0.316	+0.454	+0.616	+0.802	+1.000
1.6	+0.003	+0.011	+0.035	+0.078	+0.152	+0.253	+0.393	+0.570	+0.775	+1.000
2.0	−0.002	−0.002	+0.012	+0.034	+0.096	+0.193	+0.340	+0.519	+0.748	+1.000
3.0	−0.007	−0.022	−0.030	−0.029	+0.010	+0.087	+0.227	+0.426	+0.692	+1.000
4.0	−0.008	−0.026	−0.044	−0.051	−0.034	+0.023	+0.150	+0.354	+0.645	+1.000
5.0	−0.007	−0.024	−0.045	−0.061	−0.057	−0.015	+0.095	+0.296	+0.606	+1.000
6.0	−0.005	−0.018	−0.040	−0.058	−0.065	−0.037	+0.057	+0.252	+0.572	+1.000
8.0	−0.001	−0.009	−0.022	−0.044	−0.068	−0.062	+0.002	+0.178	+0.515	+1.000
10.0	0.000	−0.002	−0.009	−0.028	−0.053	−0.067	−0.031	+0.123	+0.467	+1.000
12.0	0.000	0.000	−0.003	−0.016	−0.040	−0.064	−0.049	+0.081	+0.424	+1.000
14.0	0.000	0.000	0.000	−0.008	−0.029	−0.059	−0.060	+0.048	+0.387	+1.000
16.0	0.000	0.000	+0.002	−0.003	−0.021	−0.051	−0.066	+0.025	+0.354	+1.000

TABLE 11.9 SHEAR Q INFLUENCE COEFFICIENTS

Shear at Base of Cylindrical Wall

$Q = \text{coef.} \times \begin{cases} \gamma H^2 \text{ lb. (triangular)} \\ pH \text{ lb. (rectangular)} \\ M/H \text{ lb. (mom. at base)} \end{cases}$

Positive sign indicates shear acting inward

$\dfrac{H^2}{dt}$	Triangular load, fixed base	Rectangular load, fixed base	Triangular or rectangular load, hinged base
0.4	0.436	0.755	0.245
0.8	0.374	0.552	0.234
1.2	0.339	0.460	0.220
1.6	0.317	0.407	0.204
2.0	0.299	0.370	0.189
3.0	0.262	0.310	0.158
4.0	0.236	0.271	0.137
5.0	0.213	0.243	0.121
6.0	0.197	0.222	0.110
8.0	0.174	0.193	0.096
10.0	0.158	0.172	0.087
12.0	0.145	0.158	0.079
14.0	0.135	0.147	0.073
16.0	0.127	0.137	0.068

TABLE 11.10 RING TENSION INFLUENCE COEFFICIENTS, TRIANGULAR LOAD (FIXED BASE)

Tension in Circular Rings
Triangular Load
Fixed Base, Free Top
F = coef. × γHR lb. per ft.
Positive sign indicates tension

'Liquid Load'—Fixed

Coefficients at point

$\dfrac{H^2}{dt}$	0.0H	0.1H	0.2H	0.3H	0.4H	0.5H	0.6H	0.7H	0.8H	0.9H
0.4	+0.149	+0.134	+0.120	+0.101	+0.082	+0.066	+0.049	+0.029	+0.014	+0.004
0.8	+0.263	+0.239	+0.215	+0.190	+0.160	+0.130	+0.096	+0.063	+0.034	+0.010
1.2	+0.283	+0.271	+0.254	+0.234	+0.209	+0.180	+0.142	+0.099	+0.045	+0.016
1.6	+0.265	+0.268	+0.268	+0.266	+0.250	+0.266	+0.185	+0.134	+0.075	+0.023
2.0	+0.234	+0.251	+0.273	+0.285	+0.285	+0.274	+0.232	+0.172	+0.104	+0.031
3.0	+0.134	+0.203	+0.267	+0.322	+0.357	+0.362	+0.330	+0.262	+0.157	+0.052
4.0	+0.067	+0.164	+0.256	+0.339	+0.403	+0.429	+0.409	+0.334	+0.210	+0.073
5.0	+0.025	+0.137	+0.245	+0.346	+0.428	+0.477	+0.469	+0.398	+0.259	+0.092
6.0	+0.018	+0.119	+0.234	+0.344	+0.441	+0.504	+0.514	+0.447	+0.301	+0.112
8.0	+0.011	+0.104	+0.218	+0.335	+0.443	+0.534	+0.575	+0.530	+0.381	+0.151
10.0	−0.011	+0.098	+0.208	+0.323	+0.437	+0.542	+0.608	+0.589	+0.440	+0.179
12.0	−0.005	+0.097	+0.202	+0.312	+0.429	+0.543	+0.628	+0.633	+0.494	+0.211
14.0	−0.002	+0.098	+0.200	+0.306	+0.420	+0.539	+0.639	+0.666	+0.541	+0.241
16.0	0.000	+0.099	+0.199	+0.304	+0.412	+0.531	+0.641	+0.687	+0.582	+0.265

TABLE 11.11 RING TENSION INFLUENCE COEFFICIENTS, RECTANGULAR LOAD (FIXED BASE)

Tension in Circular Rings
Rectangular Load
Fixed Base, Free Top
F = coef. × pR lb. per ft.
Positive sign indicates tension

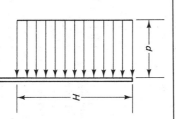

'Gas' Load—Fixed

Coefficients at point

$\dfrac{H^2}{dt}$	0.0H	0.1H	0.2H	0.3H	0.4H	0.5H	0.6H	0.7H	0.8H	0.9H
0.4	+0.582	+0.505	+0.431	+0.353	+0.277	+0.206	+0.145	+0.092	+0.046	+0.013
0.8	+1.052	+0.921	+0.796	+0.669	+0.542	+0.415	+0.289	+0.179	+0.089	+0.024
1.2	+1.218	+1.078	+0.946	+0.808	+0.665	+0.519	+0.378	+0.246	+0.127	+0.034
1.6	+1.257	+1.141	+1.009	+0.881	+0.742	+0.600	+0.449	+0.294	+0.153	+0.045
2.0	+1.253	+1.144	+1.041	+0.929	+0.806	+0.667	+0.514	+0.345	+0.186	+0.055
3.0	+1.160	+1.112	+1.061	+0.998	+0.912	+0.796	+0.646	+0.459	+0.258	+0.081
4.0	+1.085	+1.073	+1.057	+1.029	+0.997	+0.887	+0.746	+0.553	+0.322	+0.105
5.0	+1.037	+1.044	+1.047	+1.042	+1.015	+0.949	+0.825	+0.629	+0.379	+0.128
6.0	+1.010	+1.024	+1.038	+1.045	+1.034	+0.986	+0.879	+0.694	+0.430	+0.149
8.0	+0.989	+1.005	+1.022	+1.036	+1.044	+1.026	+0.953	+0.788	+0.519	+0.189
10.0	+0.989	+0.998	+1.010	+1.023	+1.039	+1.040	+0.996	+0.859	+0.591	+0.226
12.0	+0.994	+0.997	+1.003	+1.014	+1.031	+1.043	+1.022	+0.911	+0.652	+0.262
14.0	+0.997	+0.998	+1.000	+1.007	+1.022	+1.040	+1.035	+0.949	+0.705	+0.294
16.0	+1.000	+0.999	+0.999	+1.003	+1.015	+1.032	+1.040	+0.975	+0.750	+0.321

TABLE 11.12 RING TENSION INFLUENCE COEFFICIENTS, TRIANGULAR LOAD (PINNED BASE)

Tension in Circular Rings
Triangular Load
Hinged Base, Free Top
F = coef. × γHR lb. per ft.
Positive sign indicates tension

'Liquid Load'—Pinned

Coefficients at point

$\dfrac{H^2}{dt}$	0.0H	0.1H	0.2H	0.3H	0.4H	0.5H	0.6H	0.7H	0.8H	0.9H
0.4	+0.474	+0.440	+0.395	+0.352	+0.308	+0.264	+0.215	+0.165	+0.111	+0.057
0.8	+0.423	+0.402	+0.381	+0.358	+0.330	+0.297	+0.249	+0.202	+0.145	+0.076
1.2	+0.350	+0.355	+0.361	+0.362	+0.358	+0.343	+0.309	+0.256	+0.186	+0.098
1.6	+0.271	+0.303	+0.341	+0.369	+0.385	+0.385	+0.362	+0.314	+0.233	+0.124
2.0	+0.205	+0.260	+0.321	+0.373	+0.411	+0.434	+0.419	+0.369	+0.280	+0.151
3.0	+0.074	+0.179	+0.281	+0.375	+0.449	+0.506	+0.519	+0.479	+0.375	+0.210
4.0	+0.017	+0.137	+0.253	+0.367	+0.469	+0.545	+0.579	+0.553	+0.447	+0.256
5.0	−0.008	+0.114	+0.235	+0.356	+0.469	+0.562	+0.617	+0.606	+0.503	+0.294
6.0	−0.011	+0.103	+0.223	+0.343	+0.463	+0.566	+0.639	+0.643	+0.547	+0.327
8.0	−0.015	+0.096	+0.208	+0.324	+0.443	+0.564	+0.661	+0.697	+0.621	+0.386
10.0	−0.008	+0.095	+0.200	+0.311	+0.428	+0.552	+0.666	+0.730	+0.678	+0.433
12.0	−0.002	+0.097	+0.197	+0.302	+0.417	+0.541	+0.664	+0.750	+0.720	+0.477
14.0	0.000	+0.098	+0.197	+0.299	+0.408	+0.531	+0.659	+0.761	+0.752	+0.513
16.0	+0.002	+0.100	+0.198	+0.299	+0.403	+0.521	+0.650	+0.764	+0.776	+0.543

TABLE 11.13 RING TENSION INFLUENCE COEFFICIENTS, RECTANGULAR LOAD (HINGED BASE)

Tension in Circular Rings
Rectangular Load
Hinged Base, Free Top
F = coef. × pR lb. per ft.
Positive sign indicates tension

'Gas' Load—Pinned

Coefficients at point

$\dfrac{H^2}{dt}$	0.0H	0.1H	0.2H	0.3H	0.4H	0.5H	0.6H	0.7H	0.8H	0.9H
0.4	+1.474	−1.340	+1.195	+1.052	+0.903	+0.764	+0.615	+0.465	+0.311	+0.154
0.8	+1.423	+1.302	+1.181	+1.058	+0.930	+0.797	+0.649	+0.502	+0.345	+0.166
1.2	+1.350	+1.255	+1.161	+1.062	+0.958	+0.843	+0.709	+0.556	+0.386	+0.198
1.6	+1.271	+1.203	+1.141	+1.069	+0.985	+0.885	+0.756	+0.614	+0.433	+0.224
2.0	+1.205	+1.160	+1.121	+1.173	+1.011	+0.934	+0.819	+0.669	+0.480	+0.251
3.0	+1.074	+1.079	+1.081	+1.075	+1.049	+1.006	+0.919	+0.779	+0.575	+0.310
4.0	+1.017	+1.037	+1.053	+1.067	+1.069	+1.045	+0.979	+0.853	+0.647	+0.356
5.0	+0.992	+1.014	+1.035	+1.056	+1.069	+1.062	+1.017	+1.906	+0.703	+0.394
6.0	+0.989	+1.003	+1.023	+1.043	+1.063	+1.066	+1.039	+0.943	+0.747	+0.427
8.0	+0.985	+0.996	+1.008	+1.024	+1.043	+1.064	+1.061	+0.997	+0.821	+0.486
10.0	+0.992	+0.995	+1.000	+1.011	+1.028	+1.052	+1.066	+1.030	+0.878	+0.523
12.0	+0.998	+0.997	+0.997	+1.002	+1.017	+1.041	+1.064	+1.050	+0.920	+0.577
14.0	+1.000	+0.998	+0.997	+0.999	+1.008	+1.031	+1.059	+1.061	+0.952	+0.613
16.0	+1.002	+1.000	+0.998	+0.999	+1.003	+1.021	+1.050	+1.064	+0.976	+0.543

TABLE 11.14 EMPTY TANK RING TENSION INFLUENCE COEFFICIENTS, FIXED BASE

Tension in Circular Rings
Shear per Ft., Q, Applied at Top
Fixed Base, Free Top
$F = \text{coef.} \times VR/H$ lb. per ft.
Positive sign indicates tension

Empty Tank

$\dfrac{H^2}{dt}$	Coefficients at point									
	0.0H	0.1H	0.2H	0.3H	0.4H	0.5H	0.6H	0.7H	0.8H	0.9H
0.4	− 1.57	− 1.32	− 1.08	− 0.86	− 0.65	− 0.47	− 0.31	− 0.18	− 0.08	− 0.02
0.8	− 3.09	− 2.55	− 2.04	− 1.57	− 1.15	− 0.80	− 0.51	− 0.28	− 0.13	− 0.03
1.2	− 3.95	− 3.17	− 2.44	− 1.79	− 1.25	− 0.81	− 0.48	− 0.25	− 0.10	− 0.02
1.6	− 4.57	− 3.54	− 2.60	− 1.80	− 1.17	− 0.69	− 0.36	− 0.16	− 0.05	− 0.01
2.0	− 5.12	− 3.83	− 2.68	− 1.74	− 1.02	− 0.52	− 0.21	− 0.05	+ 0.01	+ 0.01
3.0	− 6.32	− 4.37	− 2.70	− 1.43	− 0.58	− 0.02	+ 0.15	+ 0.19	+ 0.13	+ 0.04
4.0	− 7.34	− 4.73	− 2.60	− 1.10	− 0.19	+ 0.26	+ 0.38	+ 0.33	+ 0.19	+ 0.06
5.0	− 8.22	− 4.99	− 2.45	− 0.79	+ 0.11	+ 0.47	+ 0.50	+ 0.37	+ 0.20	+ 0.06
6.0	− 9.02	− 5.17	− 2.27	− 0.50	+ 0.34	+ 0.59	+ 0.53	+ 0.35	+ 0.17	+ 0.01
8.0	−10.42	− 5.36	− 1.85	− 0.02	+ 0.63	+ 0.66	+ 0.46	+ 0.24	+ 0.09	+ 0.01
10.0	−11.67	− 5.43	− 1.43	+ 0.36	+ 0.78	+ 0.62	+ 0.33	+ 0.12	+ 0.02	0.00
12.0	−12.76	− 5.41	− 1.03	+ 0.63	+ 0.83	+ 0.52	+ 0.21	+ 0.04	− 0.02	0.00
14.0	−13.77	− 5.34	− 0.68	+ 0.80	+ 0.81	+ 0.42	+ 0.13	0.00	− 0.03	− 0.01
16.0	−14.74	− 5.22	− 0.33	+ 0.96	+ 0.76	+ 0.32	+ 0.05	− 0.04	− 0.05	− 0.02

Ring tension

Moment

TABLE 11.15 EMPTY TANK RING TENSION INFLUENCE COEFFICIENTS, HINGED BASE

Tension in Circular Rings
Moment per Ft, M, Applied at Base
Hinged Base, Free Top
F = coef. × MR/H² lb. per ft.
Positive sign indicates tension

Coefficients at point

$\dfrac{H^2}{dt}$	0.0H	0.1H	0.2H	0.3H	0.4H	0.5H	0.6H	0.7H	0.8H	0.9H
0.4	+ 2.70	+ 2.50	+ 2.30	+ 2.12	+ 1.91	+ 1.69	+ 1.41	+ 1.13	+ 0.80	+ 0.44
0.8	+ 2.02	+ 2.06	+ 2.10	+ 2.14	+ 2.10	+ 2.02	+ 1.95	+ 1.75	+ 1.39	+ 0.80
1.2	+ 1.06	+ 1.42	+ 1.79	+ 2.03	+ 2.46	+ 2.65	+ 2.80	+ 2.60	+ 2.22	+ 1.37
1.6	+ 0.12	+ 0.79	+ 1.43	+ 2.04	+ 2.72	+ 3.25	+ 3.56	+ 3.59	+ 3.13	+ 2.01
2.0	− 0.68	+ 0.22	+ 1.10	+ 2.02	+ 2.90	+ 3.69	+ 4.30	+ 4.54	+ 4.08	+ 2.75
3.0	− 1.78	− 0.71	+ 0.43	+ 1.60	+ 2.95	+ 4.29	+ 5.66	+ 6.58	+ 6.55	+ 4.73
4.0	− 1.87	− 1.00	− 0.08	+ 1.04	+ 2.47	+ 4.31	+ 6.34	+ 8.19	+ 8.82	+ 6.81
5.0	− 1.54	− 1.03	− 0.42	+ 0.45	+ 1.86	+ 3.93	+ 6.60	+ 9.41	+11.03	+ 9.02
6.0	− 1.04	− 0.86	− 0.59	+ 0.05	+ 1.21	+ 3.34	+ 6.54	+10.28	+13.08	+11.41
8.0	− 0.24	− 0.53	− 0.73	− 0.67	− 0.02	+ 2.05	+ 5.87	+11.32	+16.52	+16.06
10.0	+ 0.21	− 0.23	− 0.64	− 0.94	− 0.73	+ 0.82	+ 4.79	+11.63	+19.48	+20.87
12.0	+ 0.32	− 0.05	− 0.46	− 0.96	− 1.15	− 0.18	+ 3.52	+11.27	+21.80	+25.73
14.0	+ 0.26	+ 0.04	− 0.28	− 0.76	− 1.29	− 0.87	+ 2.29	+10.55	+23.50	+30.34
16.0	+ 0.22	+ 0.07	− 0.08	− 0.64	− 1.28	− 1.30	+ 1.12	+ 9.67	+24.53	+34.65

Ring tension

Moment

Empty Tank

TABLE 11.16 SUPPLEMENTARY INFLUENCE COEFFICIENTS FOR VALUES OF H^2/dt GREATER THAN 16 FOR TABLES 11.4–11.15

Table 11.4a

$\dfrac{H^2}{dt}$	Coefficients at point				
	.80H	.85H	.90H	.95H	1.00H
20	+.0015	+.0014	+.0005	−.0018	−.0063
24	+.0012	+.0012	+.0007	−.0013	−.0053
32	+.0007	+.0009	+.0007	−.0008	−.0040
40	+.0002	+.0005	+.0006	−.0005	−.0032
48	.0000	+.0001	+.0006	−.0003	−.0026
56	.0000	.0000	+.0004	−.0001	−.0023

Table 11.5a

$\dfrac{H^2}{dt}$	Coefficients at point				
	.80H	.85H	.90H	.95H	1.00H
20	+.0015	+.0013	+.0002	−.0024	−.0073
24	+.0012	+.0012	+.0004	−.0018	−.0061
32	+.0008	+.0009	+.0006	−.0010	−.0046
40	+.0005	+.0007	+.0007	−.0005	−.0037
48	+.0004	+.0006	+.0006	−.0003	−.0031
56	+.0002	+.0004	+.0005	−.0001	−.0026

Table 11.6a

$\dfrac{H^2}{dt}$	Coefficients at point				
	.75H	.80H	.85H	.90H	.95H
20	+.0008	+.0014	+.0020	+.0024	+.0020
24	+.0005	+.0010	+.0015	+.0020	+.0017
32	.0000	+.0005	+.0009	+.0014	+.0013
40	.0000	+.0003	+.0006	+.0011	+.0011
48	.0000	+.0001	+.0004	+.0008	+.0010
56	.0000	.0000	+.0003	+.0007	+.0008

Table 11.7a

$\dfrac{H^2}{dt}$	Coefficients at point				
	.05H	.10H	.15H	.20H	.25H
20	+0.032	+0.039	+0.033	+0.023	+0.014
24	+0.031	+0.035	+0.028	+0.018	+0.009
32	+0.028	+0.029	+0.020	+0.011	+0.004
40	+0.026	+0.025	+0.015	+0.006	+0.001
48	+0.024	+0.021	+0.011	+0.003	0.000
56	+0.023	+0.018	+0.008	+0.002	0.000

Table 11.8a

	Coefficients at point				
$\frac{H^2}{dt}$	.80H	.85H	.90H	.95H	1.00H
20	−0.015	+0.095	+0.296	+0.606	+1.000
24	−0.037	+0.057	+0.250	+0.572	+1.000
32	−0.062	+0.002	+0.178	+0.515	+1.000
40	−0.067	−0.031	+0.123	+0.467	+1.000
48	−0.064	−0.049	+0.081	+0.424	+1.000
56	−0.059	−0.060	+0.048	+0.387	+1.000

Table 11.9a

	Coefficients at point		
$\frac{H^2}{dt}$	Tri. Fixed	Rect. Fixed	T. or R. Hinged
20	+0.114	+0.122	+0.062
24	+0.102	+0.111	+0.055
32	+0.089	+0.096	+0.048
40	+0.080	+0.086	+0.043
48	+0.072	+0.079	+0.039
56	+0.067	+0.074	+0.036

Table 11.10a

	Coefficients at point				
$\frac{H^2}{dt}$	.75H	.80H	.85H	.90H	.95H
20	+0.716	+0.654	+0.520	+0.325	+0.115
24	+0.746	+0.702	+0.577	+0.372	+0.137
32	+0.782	+0.768	+0.663	+0.459	+0.182
40	+0.800	+0.805	+0.731	+0.530	+0.217
48	+0.791	+0.828	+0.785	+0.593	+0.254
56	+0.763	+0.838	+0.824	+0.636	+0.285

Table 11.11a

	Coefficients at point				
$\frac{H^2}{dt}$	.75H	.80H	.85H	.90H	.95H
20	+0.949	+0.825	+0.629	+0.379	+0.128
24	+0.986	+0.879	+0.694	+0.430	+0.149
32	+1.026	+0.953	+0.788	+0.519	+0.189
40	+1.040	+0.996	+0.859	+0.591	+0.226
48	+1.043	+1.022	+0.911	+0.652	+0.262
56	+1.040	+1.035	+0.949	+0.705	+0.294

TABLE 11.16 (continued)

Table 11.12a

$\frac{H^2}{dt}$	Coefficients at point				
	.75H	.80H	.85H	.90H	.95H
20	+0.812	+0.817	+0.756	+0.603	+0.344
24	+0.816	+0.839	+0.793	+0.647	+0.377
32	+0.814	+0.861	+0.847	+0.721	+0.436
40	+0.802	+0.866	+0.880	+0.778	+0.483
48	+0.791	+0.864	+0.900	+0.820	+0.527
56	+0.781	+0.859	+0.911	+0.852	+0.563

Table 11.13a

$\frac{H^2}{dt}$	Coefficients at point				
	.75H	.80H	.85H	.90H	.95H
20	+1.062	+1.017	+0.906	+0.703	+0.394
24	+1.066	+1.039	+0.943	+0.747	+0.427
32	+1.064	+1.061	+0.997	+0.821	+0.486
40	+1.052	+1.066	+1.030	+0.878	+0.533
48	+1.041	+1.064	+1.050	+0.920	+0.577
56	+1.021	+1.059	+1.061	+0.952	+0.613

Table 11.14a

$\frac{H^2}{dt}$	Coefficients at point				
	.00H	.05H	.10H	.15H	.20H
20	-.16.44	- 9.98	- 4.90	- 1.59	+ 0.22
24	-18.04	-10.34	- 4.54	- 1.00	+ 0.68
32	-20.84	-10.72	- 3.70	- 0.04	+ 1.26
40	-23.34	-10.86	- 2.86	+ 0.72	+ 1.56
48	-25.52	-10.82	- 2.06	+ 0.26	+ 1.66
56	-27.54	-10.68	- 1.36	+ 1.60	+ 1.62

Table 11.15a

$\frac{H^2}{dt}$	Coefficients at point				
	.75H	.80H	.85H	.90H	.95H
20	+15.30	+ 25.9	+ 36.9	+ 43.3	+ 35.3
24	+13.20	+ 25.9	+ 40.7	+ 51.8	+ 45.3
32	+ 8.10	+ 23.2	+ 45.9	+ 65.4	+ 63.6
40	+ 3.28	+ 19.2	+ 46.5	+ 77.9	+ 83.5
48	- 0.70	+ 14.1	+ 45.1	+ 87.2	+103.0
56	- 3.40	+ 9.2	+ 42.2	+ 94.0	+121.0

11.6 PRESTRESSING EFFECTS ON WALL STRESSES FOR FULLY HINGED, PARTIALLY SLIDING AND HINGED, FULLY FIXED, AND PARTIALLY FIXED BASES

The liquid or gas contained in a cylindrical tank exerts *outward* radial pressure γh or p on the tank walls, inducing ring tensions in each horizontal section of wall along its height. This ring tension in turn causes tensile stresses in the concrete at the *outside* extreme wall fibers, resulting in impermissible cracking. To eliminate this cracking that causes leaks and structural deterioration, external *horizontal* prestressing is applied which induces *inward* radial thrust that can balance the outward radial tension. Additionally, in order to prevent the development of cracks in the inside walls when the tank is empty, *vertical* prestressing is induced to reduce the residual tension within the range of the modulus of rupture of the concrete and with an adequate safety factor.

In order to ensure against the development of cracking at the outside face of the tank wall, it is good practice to apply somewhat larger horizontal prestressing forces than are required to neutralize or balance the outward radial forces caused by the internal liquid or gas, thereby producing *residual* compression in the tank when it is full (Ref. 11.2). Such an increase in circumferential prestressing forces through the use of additional horizontal prestressing steel, and sometimes mild vertical steel, also counteracts the effects of temperature and moisture gradients across the wall thickness in an adverse environment.

11.6.1 Freely Sliding Wall Base

When the boundary condition is such that the wall at its base can freely slide when the tank is internally loaded, there is no moment in the vertical wall due either to liquid load or to prestressing when the tank is totally filled to height H. Only a small nominal moment develops when the tank is partially filled, partially prestressed, or empty, and *no* vertical prestressing is necessary. The deflected shape of the freely sliding tank is shown in Figure 11.7.

While free sliding is an ideal condition that renders the structure statically determinate and hence most economical, it is difficult to achieve in practice. Frictional forces produced at the wall base after the tank becomes operational and the difficulty of achieving liquid tightness render this alternative essentially unimplementable.

11.6.2 Hinged Wall Base

For walls with a hinged connection to the base, the maximum radial forces due to the liquid retained and the prestressing at the critical section a distance y above the base are almost equal to those in the freely sliding case at height y. But vertical moments are introduced, and vertical prestressing becomes necessary to reduce the tensile stresses in the concrete at the outer wall face.

The deflected shape of the hinged wall is shown in Figure 11.8. Note that the critical section for ring forces is not necessarily at the same height as the moment critical section.

In order to minimize the possibility of cracking, a residual ring compression of a minimum value of 200 psi (1.38 MPa) is necessary for wire-wrapped prestressed

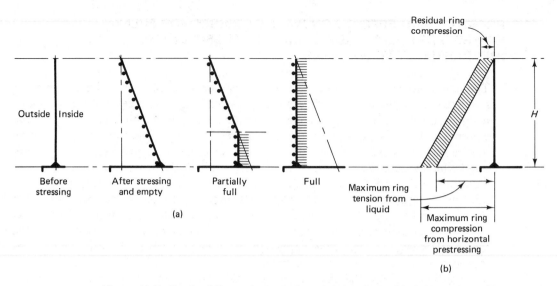

Figure 11.7 Freely sliding tank. (a) Deflected shape. (b) Residual ring compression.

tanks without diaphragms, and 100 psi (0.7 MPa) for tanks with a continuous metal diaphragm. The maximum tension at the inside face of the wall should not exceed $3\sqrt{f'_c}$ at working-load level as given in Table 11.17 in a later section. The deflected shape of the tank walls and the stress variations in the concrete across the thickness of the section when the tank is empty and when it is full are shown in Figure 11.8. For tanks prestressed with pretensioned and post-tensioned tendons, the minimum residual compressive stress should be as stipulated in Section 11.10.

11.6.3 Partially Sliding and Hinged Wall Base

A partially sliding and hinged wall-base system is accomplished by providing a slot in the wall-base supporting slab such that the wall can slide within its base during the prestressing. After prestressing and all losses due to creep, shrinkage, and relaxation have taken place, the slot is sealed and the tank wall behaves as hinged under service-load conditions. The magnitude of sliding can be controlled such that either full or partial sliding is allowed before hinging is accomplished. A partial slide of about 50 percent of the full slide with hinging at the end of the wall movement has the structural advantages of both full sliding and hinging, and the sealing of the wall-base slab-pinned joint against leakage of liquids or gases is more dependable than if full sliding prior to anchorage is allowed. The deformed shape of the wall during the prestressing procedure, together with the ring forces, vertical moments, and concrete stress variations across the wall thickness, is shown in Figure 11.9. The vertical prestress needed for the partial slide-pinned case can be considerably smaller than the fully pinned case without sliding.

11.6.4 Fully Fixed Wall Base

Full fixity of the wall at its base means full restraint against rotation at the wall base. This condition can be accomplished if the lower segment of the wall is cast mono-

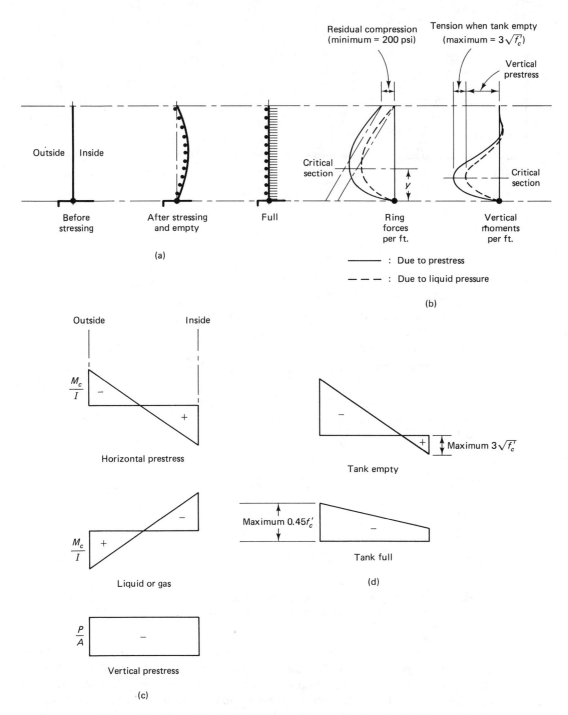

Figure 11.8 Hinged-base tank. (a) Deflected shape of tank wall. (b) Horizontal ring forces and vertical moments. (c) Concrete stresses across wall thickness. (d) Resultant wall stresses.

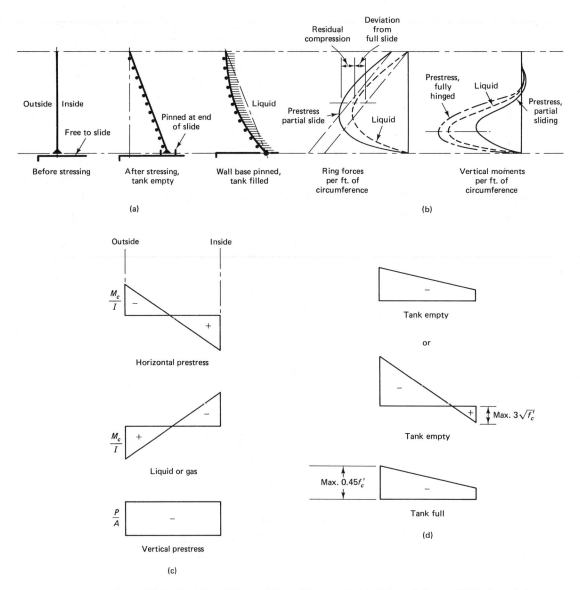

Figure 11.9 Partially sliding and hinged-base tank. (a) Deflected shape. (b) Horizontal ring forces and comparative vertical moments. (c) Concrete stresses across wall thickness. (d) Resultant wall stresses.

lithically and is well anchored into a base slab of a similar stiffness. But such an indeterminate system is difficult to fully achieve and is not economical as well, since a tank base area is very large and partial fixity becomes necessary (see shortly). The radial horizontal forces from both prestressing and the contained internal pressure are unchanged from the triangular form for liquid, rectangular for gas, and trapezoidal for granular contained material. The restraint imposed by the horizontal slab base, however, modifies the ring forces and introduces additional moment in the vertical section of the wall. Because of fixity at the base, no displacement takes place at either the

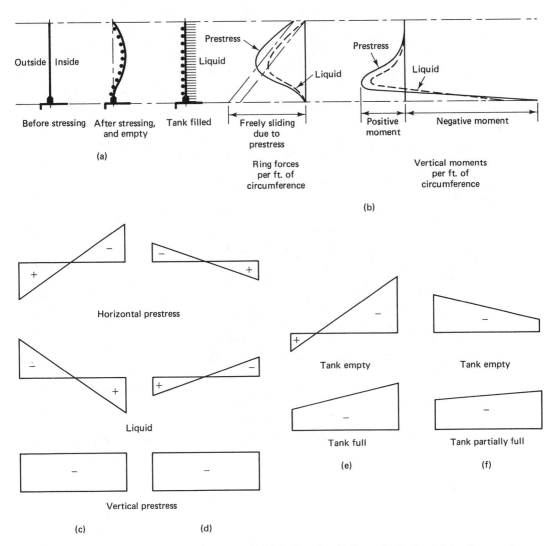

Figure 11.10 Fully fixed-base tank. (a) Deflected wall shape. (b) Horizontal ring forces and vertical moments. (c) Concrete stresses across wall for full tank. (d) Concrete stresses across wall for partially full tank. (e) Resultant stresses, full tank. (f) Resultant stresses, partially full tank.

bottom or the top of the wall, and a change in curvature along the height of the wall above the base takes place when the tank is empty, as is shown in Figure 11.10. Note that the wall should be designed to become essentially vertical, with a maximum residual compressive stress due to prestressing of 200 psi as in the previous cases. The vertical prestress needed for tanks with fully fixed wall bases is considerably greater than the vertical prestress needed for the other boundary conditions. This is necessary in order to offset the high tensile stresses in the wall base at the *outside* face caused by the large negative movement at the base (see Figure 11.10(a) and (b)) and the reverse curvature near it. It is sometimes more economical to use mild steel reinforcement at the lower portion of the wall in addition to prestressing, in order to be able to use lesser vertical prestressing and assign the excess negative movement to the nonprestressed reinforcement. The tensile stresses in the concrete can also be reduced by using *eccentric* vertical prestressing with the appropriate eccentricity achieved by trial and adjustment, as well as by using additional mild steel. Vertical prestressing in tanks is expensive, however, due to the required anchorages at the top and bottom of the tank wall. Thus, reducing the level of vertical prestress needed in the design adds to the economy of the total design of the system.

11.6.5 Partially Fixed Wall Base

11.6.5.1 Rotational restraint. As indicated previously, full restraint against rotation at the wall base is difficult to achieve. The reasons are essentially threefold: (1) one has to provide the necessary stiffness in the tank floor slab at the wall junction for total fixity; (2) subsoil movement under the wall can cause rotation of the wall base; and (3) a concentration of anchorages is required, for both the vertical prestressing of the wall and the horizontal circumferential prestressing of the wall-base segment since the wall and base ring are separately prestressed.

Because the floor slab area is large, its restraining or stiffening influence is limited to the narrow peripheral toe cantilevering from the wall bottom. The choice of the correct width of the toe or base ring determines whether or not the assumed degree of fixity of the wall base gives the correct stiffness values in the design. Figure 11.11 schematically demonstrates the effect of the base ring width on the rotation of the wall and the deformation of the ring. Part (c) of the figure gives an equilibrium state where the tip of the ring is at the same level as the bottom of the wall, whereas the conditions represented in parts (a) and (b) involve deformations below the bottom of the wall and are consequently unsatisfactory.

The theoretical formulation of the solution to the critical ring base width can be

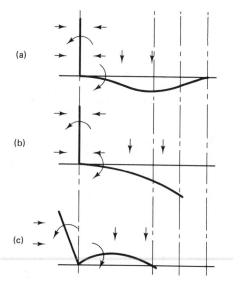

Figure 11.11 Base ring effective width. (a) Full base slab. (b) Large cantilever. (c) Equilibrium condition.

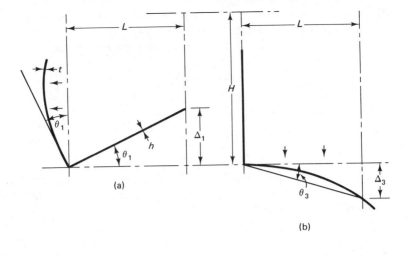

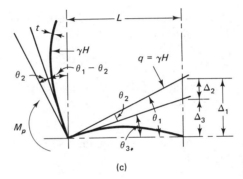

Figure 11.12 Deformation and rotation of wall base. (a) Fully free wall.
(b) Fully fixed wall. (c) Superposition of (a) and (b).

attained through the use of the principle of superposition by combining the case of a
freely rotating wall with that of a totally fixed wall as shown in Figure 11.12. Let

M_0 = theoretical fully fixed moment at the wall base

M_p = partial moment at the wall base caused by the loaded cantilever toe

θ_1 = free rotation of wall base when pinned only, corresponding to deflection
Δ_1 of a stiff unloaded toe

θ_2 = wall base rotation due to restraining moment M_p, corresponding to
deflection Δ_2 of a straight unloaded toe

θ_3 = rotation of the tip of the stiffening toe as a cantilever under vertical load,
corresponding to deflection Δ_3 of the toe tip due to the vertical load

L = width of stiffening toe

q = unit load applied to the stiffening toe = γH, where H is the height of a tank
whose diameter is d, whose wall thickness is t, and whose base slab
thickness is h.

Then the unit rotation θ of the wall at its base due to moment M_0, but without radial displacement, can be obtained from Equation 11.18 a by setting $w = 0$ to get $Q = -\beta M$. Equation 11.18 b for unit rotation then becomes

$$\theta_1 = \frac{M_o}{2\beta D}, \qquad \theta_2 = \frac{M_p}{2\beta D} \qquad (11.33)$$

Hence, we have

$$\Delta_1 = \frac{LM_o}{2\beta D}, \qquad \Delta_2 = \frac{LM_p}{2\beta D} \qquad (11.34)$$

If the stiffening wall toe is considered a cantilever subjected to a transverse load γH, the maximum cantilever moment M_p and the corresponding deflection Δ_3 are, respectively,

$$M_p = \frac{\gamma HL^2}{2}, \qquad \Delta_3 = \frac{3\gamma HL^3}{2Eh^3} \qquad (11.35)$$

The moment at the fixed wall base can be obtained using the membrane coefficient C from Table 11.4 for the applicable form factor H^2/dt and type of load. For liquid load,

$$M_o = C\gamma H^3 \qquad (11.36)$$

The deflected form due to full load, from Figure 11.12(c), is

$$\Delta_1 = \Delta_2 + \Delta_3$$

Substituting for Δ_2 and Δ_3 from Equation 11.34 into Equations 11.35 and 11.36 and rearranging terms gives

$$L^2 = \frac{2CH^2}{1 + \dfrac{(t/h)^3}{(dt)^{1/2}}} \qquad (11.37)$$

and

$$M_0 = \frac{\gamma HL^2}{2} \qquad (11.38)$$

Now let the term

$$S = \frac{(t/h)^3}{(dt)^{1/2}} \qquad (11.39)$$

in Equation 11.37 be designated a *modifying factor for partial fixity*. This factor is normally small and represents the difference between the total fixity movement M_0 and the partial restraint movement M_p. Hence,

$$M_p = M_o(1 - S) \qquad (11.40)$$

If the value of S is very small, as is the case in large-diameter tanks (diameter larger than 125 to 150 ft), the expressions for L and M_p become expressions for full fixity,

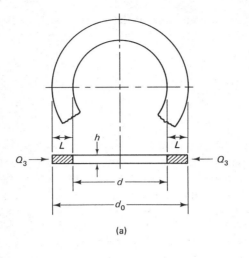

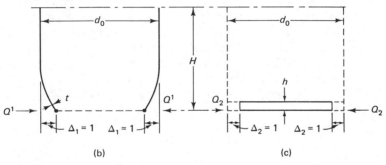

Figure 11.13 Deformation of circular wall base ring. (a) Ring plan and cross section. (b) Deflected wall bottom due to radial force Q'. (c) Deflected ring base due to radial force Q_2.

namely,

$$L^2 = 2CH^2$$

and

$$M_p = C\gamma H^3$$

11.6.5.2 Base radial deformation. The radial deformation Δ_s of the base ring subjected to radial force in its plane can be obtained from the theory of circular plates with concentric holes. The expression for the deflection of the plate shown in Figure 11.13(a) is

$$\Delta_s = \frac{d_o Q}{2hE}\left(\frac{d_o^2 + d^2}{d_o^2 - d^2} - \mu\right) \tag{11.41}$$

where μ = Poisson's ratio ~ 0.2 for concrete and E is the modulus. The horizontal radial thrust per unit of circumference required to induce unit displacement in a solid circular slab is

$$Q_2 = \frac{2.5hE}{d_o} \tag{11.42}$$

and the corresponding value of the radial thrust applied to the outer ring is

$$Q_3 = \frac{2hE}{d_o K}$$

(11.43)

where

$$K = \left(\frac{d_o^2 + d^2}{d_o^2 - d^2} - \mu\right)$$

and d = inside diameter of base ring = $(d_o - 2L)$.

The relative stiffness of the wall to the base is determined in terms of the force required to produce a *unit* deformation in the wall and the base slab from the principles of virtual work as shown in Figures 11.13(b) and (c). The distribution of the prestressing energy between the wall and base slab ring is a function of their relative radial stiffness; hence, determining the relative stiffness is necessary. In doing so, however, one must keep in mind that the stiffness response of the base ring in a prestressed tank to radial compression in its own plane is considerably larger than the response of the cylindrical wall of the tank under radial internal pressure. Thus, the loss of prestress from the difference in stiffness is insignificant in large-diameter tanks (Ref. 11.2), but should be considered in small-diameter tanks.

The unit deformation Δ due to the radial force Q' per unit of circumference *without rotation* at the foot base can be obtained from Equation 11.18 b using $2\beta M = -Q$ for rotation $dw/dy = 0$. The unit deflection Δ in Equation 11.18 a becomes

$$\Delta = \frac{3Q}{4\beta^3 D}$$

or

$$\Delta = \frac{Q'}{4\beta^3 D}$$

(11.44)

where

$$D = \frac{Et^3}{12(1 - \mu^2)}$$

Using $\mu \sim 0.2$, Equation 11.44 for unit radial displacement of the wall at the wall base without rotation becomes

$$Q' = 2.2 \, E\left(\frac{t}{d}\right)^{3/2}$$

(11.45)

where E is the modulus of concrete. From Equation 11.42, the radial force per unit of circumference required to produce unit radial displacement in the solid circular slab is

$$Q_2 = 2.5E\left(\frac{h}{d_o}\right)$$

(11.46)

By superimposing Q' on Q_2, the total force exerted at the wall-slab base junction is distributed to the wall and the slab base in proportion to the relative energy required to produce unit deformation in each.

Six-million-gallon tendon-prestressed circular tank seen from inside with situ-cast walls. (*Courtesy*, Jorgensen, Hendrickson and Close, Inc., Denver, Colorado.)

The proportion of the total force $Q' + Q_2$ to be carried by the wall is

$$R = \frac{Q'}{Q' + Q_2}$$

say

$$\frac{1}{1 + S}$$

Rearranging terms while combining Equations 11.45 and 11.46 results in

$$S = \frac{2.5(h/d)}{2.2(t/d)^{3/2}}$$

assuming that $d \sim d_o$, or

$$S = 1.1\left(\frac{h}{t}\right) \times \left(\frac{d}{t}\right)^{1/2} \qquad (11.47)$$

If S_1 is small, the proportion of the horizontal force transferred from the slab base to the wall can be taken, with sufficient accuracy, to be

$$R = \frac{100}{S} \text{ percent} \qquad (11.48)$$

When only the outer ring of the slab is compressed by radial thrust at the rim, the value of Q_2 has to be modified from that obtained by Equation 11.42, and S_1 in Equation 11.48 becomes

$$S_1 = \frac{1}{K}\left(\frac{h}{t}\right) \times \left(\frac{d}{t}\right)^{1/2} \qquad (11.49)$$

where, from before,

$$K = \left(\frac{d_0^2 + d^2}{d_0^2 - d^2} - \mu\right)$$

in which d is the inner slab ring diameter $= d_o - 2L$ and d_o is the outer diameter.

11.7 RECOMMENDED PRACTICE FOR SITU-CAST AND PRECAST PRESTRESSED CONCRETE CIRCULAR STORAGE TANKS

11.7.1 Stresses

General guidelines for situ-cast and precast prestressed concrete circular storage tanks are provided by the Prestressed Concrete Institute (Ref. 11.6), the American Concrete Institute (Refs. 11.7–11.9), and the Post-Tensioning Institute (Ref. 11.10) for choosing the applicable allowable stresses, dimensioning, minimum wall thickness, and construction and erection procedure. The allowable stresses in concrete and shotcrete are given in Table 11.17 (Ref. 11.7), with modifications to accommodate the recommended stresses in Ref. 11.6. Allowable stresses in the reinforcement are given in Table 11.18.

TABLE 11.17 ALLOWABLE CONCRETE STRESSES IN CIRCULAR TANKS

	Concrete situ-cast and precast		Shotcrete situ-cast	
Type and limit of stress	Temporary[a] stresses f_{ci}, psi	Service load stresses f_c, psi	Temporary[a] stresses f_{gi}, psi	Service load stresses f_g, psi
Axial compression, f_c	$0.55 f'_{ci}$	$0.45 f'_c$	$0.45 f'_{gi}$ but not more than $1{,}600 + 40 t_c$ psi	$0.38 f'_g$
Axial tension	0	0	0	0
Flexural compression, f_c	$0.55 f'_{ci}$	$0.4 f'_c$	$0.45 f'_{gi}$	$0.38 f'_{gi}$
Maximum flexural tension[b], f_t	$3\sqrt{f'_{ci}}$	$3\sqrt{f'_c}$		
Minimum residual compression, f_{cv}	$200\left(\dfrac{f_{ci}}{f_c}\right)$	200 psi	$200\left(\dfrac{f_{ci}}{f_c}\right)$	200 psi

[a] Before creep and shrinkage losses.
[b] Fiber stress in precomposed tension zone.

TABLE 11.18 STRESSES IN REINFORCEMENT

Type of stress	*Max allowable stress
Tendon jacking force	$0.94f_{py} \leq 0.85f_{pu}$
Immediately after prestress transfer	$0.82f_{py} \leq 0.75f_{pu}$
Post-tensioning tendons at anchorage and couplers, immediately after tendon anchorage	$0.70f_{pu}$
Service load stress, f_{pe}	$0.55f_{pu}$
Nonprestressed mild steel at initial prestressing, f_{si}	$f_y/1.6$
Final service load stress, f_s (psi), potable water storage,	
60 grade steel	24,000
corrosive storage	18,000
dry storage	$f_y/1.8$

* 1,000 psi = 6,895 Pa.

11.7.2 Required Strength Load Factors

The structure, together with its components and foundations, would have to be designed so that the design strength exceeds the effect of factored load combinations specified by ACI 318, ANSI A58.1, or as justified by the engineer based on rational analysis, with the following exceptions:

Feature	Load factor
Initial liquid pressure	1.3
Internal lateral pressure from dry material	1.7
Prestressing forces:	
Final prestress after losses	1.7
Strength reduction factor for both reinforcement and concrete, ϕ	0.9

The nominal moment strength equation M_n is similar to the one used for linear prestressing, i.e.,

$$M_n = A_{ps}f_{ps}\left(d_p - \frac{a}{2}\right) \qquad (11.50\ a)$$

or

$$M_n = A_{ps}f_{ps}\left(d_p - \frac{a}{2}\right) + A_s f_y\left(d - \frac{a}{2}\right) \qquad (11.50\ b)$$

when mild vertical steel A_s is used and

where A_{ps} = vertical prestressing steel per unit width of circumference, in^2
  f_{ps} = stress in prestressed reinforcement at nominal strength, psi
  f_y = yield strength of mild steel, psi

11.7.3 Minimum Wall Design Requirements

11.7.3.1 Circumferential forces

Liquid

$$\text{Initial } F_i = \gamma r H_y \frac{f_{pi}}{f_{pc}} \text{ per foot of wall} \tag{11.51 a}$$

Backfill

$$\text{Initial } F_{bi} = p(r + t) \tag{11.51 b}$$

where t is the total wall thickness.

11.7.3.2 Thickness and stresses

Core Wall Thickness

$$t_c = \frac{F_i}{f_{ci}} \tag{11.52}$$

but not less than the minimum wall thickness to be set out in subsection 11.7.3.6.

Final Stress Due to Backfill and Initial Prestress

$$f = \frac{F_{bi}}{t} + \frac{F_i}{t_c} \frac{f_{pe}}{f_{pi}} \tag{11.53}$$

11.7.3.3 Deflections.
The unrestrained initial elastic radial deflection of the wall due to initial prestessing is

$$\Delta_i = \frac{F_i r}{t_{co} E_c} \tag{11.54}$$

where r = tank inner radius

t_{co} = thickness of wall core at top or bottom of wall

$E_c = 57,000\sqrt{f_c'}$ psi for both normal-weight concrete and shotcrete.

The final radial deflection Δf may reach 1.5 to 3 times the initial unrestrained deflection. For normal conditions, the final permitted radial deflection can be taken as

$$\Delta f = 1.7\Delta_i \tag{11.55}$$

11.7.3.4 Restraint effects

Maximum Vertical Wall Bending Due to Radial Shear

$$M_y = 0.24Q_0\sqrt{rt_{co}} \tag{11.56 a}$$

This moment occurs at a distance

$$y = 0.68\sqrt{rt_{co}} \tag{11.56 b}$$

from the base or top edge.

Radial Shear for Monolithic Base Details Which May be Assumed to Provide Hinged Connection

$$Q_o = 0.38F_i\sqrt{\frac{t_{co}}{r}} \tag{11.57}$$

This type of detail should be used only with situ-cast tanks which incorporate a diaphragm in their wall construction.

11.7.3.5 Mild Steel for Base Anchorage.

If a diaphragm is used, extend the full area of the inside bars in a U-shape a distance

$$y_1 = 1.4\sqrt{rt_{co}} \tag{11.58 a}$$

above the base. If no diaphragm is used, extend to

$$y_2 = 1.8\sqrt{rt_{co}} \tag{11.58 b}$$

above the base. Note that anchorage length has to be added to y_1 or y_2. The minimum area of nominal vertical steel at the base region is

$$A_s = 0.005t_{co} \tag{11.59}$$

and should be extended above the base a distance of 3 ft or

$$y_3 = 0.75\sqrt{rt_{co}} \tag{11.60}$$

whichever is greater.

11.7.3.6 Minimum wall thickness

Situ-Cast Walls

Type of tank	Minimum wall thickness
Shotcrete-steel diaphragm tanks	$3\frac{1}{2}$ in.
Tanks without vertical prestressing	8 in.
Tanks with vertical prestressing	7 in.

Precast Walls

Type of tank	Minimum wall thickness
Tanks with vertical pretensioning and external circumferential prestress	5 in.
Tanks with vertical pretensioning and internal circumferential prestress	6 in.
Tanks with vertical post-tensioning and internal circumferential prestress	7 in.

It should be noted that for tanks prestressed with tendons, a thickness not less than 9 in. is advisable for practical considerations.

11.8 CRACK CONTROL IN WALLS OF CIRCULAR PRESTRESSED CONCRETE TANKS

Vessey and Preston in Ref. 11.14 recommend the following expression based on Nawy's work in Ref. 11.15 for the maximum crack width at the exterior surface of the prestressed tank wall:

$$w_{max} = 4.1 \times 10^{-6} \epsilon_{ct} E_{ps} \sqrt{I_x} \tag{11.61}$$

where ϵ_{ct} = tensile surface strain in the concrete

I_x = grid index = $\dfrac{8}{\pi}\left(\dfrac{s_2 s_1 t_b}{\phi_1}\right)$

s_2 = reinforcement spacing in direction "2"
s_1 = reinforcement spacing in perpendicular direction "1"
t_b = concrete cover to center of steel
ϕ_1 = diameter of steel in main direction "1".

The tensile strain can be calculated from

$$\epsilon_{ct} = \frac{\alpha_t f_{pi}}{E_{ps}} \tag{11.62}$$

where α_t = stress parameter $\cong f_p / f_{pi}$
f_p = actual stress in the prestressing steel
f_{pi} = initial prestress before losses.

For liquid-retaining tanks, the maximum allowable crack width is 0.004 in.

11.9 TANK ROOF DESIGN

Roofs for storage tanks are constructed in the form of a shell dome or as flat roofs supported internally on columns. The cost of the roof is generally about one-third of the overall cost of the structure. In the case of flat roofs, whether precast or situ cast, the design follows the normal design principles of floor systems for reinforced or prestressed concrete one-way- or two-way-action floors as stipulated in the ACI 318 Code. If the roof is made out of precast prestressed elements, and the tank diameter is not exceedingly large, no interior columns are necessary. Otherwise, the added cost of interior columns and the accompanying footings would increase the cost of the overall structure.

A shell roof in the form of a dome has distinct advantages for tanks not exceeding 150 ft. in diameter, namely, that the dome does not need supporting interior columns and can also be economical in underground storage tanks in withstanding backfill load. Hence, the shell form and the manner of its connection to the tank walls have a significant effect on cost. Preferably, the roof shell should be supported by tank walls with a completely *flexible* joint; otherwise the design of both the tank wall and the roof dome will have to be modified in relation to their degree of interrestraint and relative stiffness, with the concomitant added construction cost.

A spherical shell of low rise-to-diameter ratio h'/d of approximately $\frac{1}{8}$ is reasonable to use. Such a flat dome or axisymmetrical shell introduces outward horizontal thrust at the springing, which has to be resisted by a properly designed prestressed ring beam at the support level. The type of support of the ring beam determines the extent to which redundant reactions and moments due to end restraint impose additional direct and bending stresses in the shell near the springing. In other words, the membrane solution has to be adequately modified by superimposing on it the bending effects determined by the strain compatability requirements of the bending theory.

11.9.1 Membrane Theory of Spherical Domes

11.9.1.1 Shell of revolution. The basic membrane equations of equilibrium for the direct forces in a shell of revolution as shown in Figure 11.14 are used for defining the unit meridional forces N_ϕ, unit tangential forces N_θ, and unit central shears $N_{\phi\theta}$ and $N_{\theta\phi}$ in terms of the gravity loads p_ϕ, p_θ, and p_z. These equations are as follows:

$$\text{Tangential:} \quad \frac{\partial(N_\phi r_o)}{\partial \phi} - N_\theta \frac{\partial r_o}{\partial \phi} + \frac{\partial N_{\theta\phi}}{\partial} + p_\phi r_o r_1 = 0 \tag{11.63 a}$$

$$\text{Meridional:} \quad \frac{\partial N_\theta}{\partial \theta} r_o + N_{\theta\phi} \frac{\partial r_o}{\partial \phi} + \frac{\partial N_{\theta\phi}}{\partial \phi} + p_\theta r_2 r_1 = 0 \tag{11.63 b}$$

$$z\text{-direction:} \quad \frac{N_\phi}{r_2} + \frac{N_\theta}{r_2} + p_z = 0 \tag{11.63 c}$$

Because of loading symmetry, all terms involving $\partial\theta$ vanish, and those involving $\partial\phi$ can be rewritten as total differentials $d\phi$ since nothing varies with respect to θ. Also, the circumferential load component $p_\theta = 0$, as the shear resultants vanish along the meridional and parallel circles. Hence, Equations 11.63 can be rewritten as

$$\frac{d}{d\phi}(N_\phi r_o) - N_\theta r_1 \cos\phi + p_y r_1 r_o = 0 \tag{11.64 a}$$

$$\frac{N_\theta}{r_1} + \frac{N_\theta}{r_2} + p_z = 0 \tag{11.64 b}$$

11.9.1.2 Spherical dome

Membrane Analysis of the Equilibrium Forces. The spherical dome has a uniform curvature. Consequently, $r_1 = r_2 = r_o$. Assuming that the radius of the sphere $= a$, then $r_o = a \sin\phi$ in Figure 11.14(c), and, setting $p_z = w_D$ for self-weight, the general equilibrium equations 11.64 become

$$N_\theta = aw_D\left(\frac{1}{1 + \cos\phi} - \cos\phi\right) \tag{11.65 a}$$

and

$$N_\phi = -\frac{aw_D}{1 + \cos\phi} \tag{11.65 b}$$

where w_D is the intensity of self-weight per unit area. It is plain from Equation 11.65 b that the meridional force N_ϕ is always negative. Therefore, *compression* develops

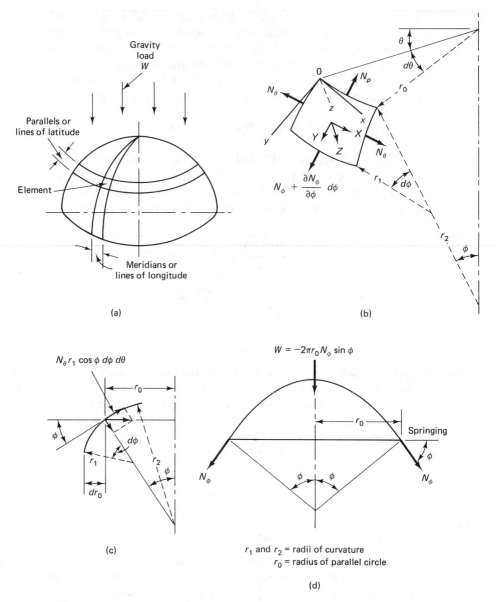

Figure 11.14 Membrane forces in a shell of revolution. (a) Meridian and parallel lines. (b) Membrane forces on infinitesimal surface element. (c) Component of force $N_\theta r_1 d\phi$ in the y direction needed to simplify the basic equation 11.63a. (d) Dome cross section with total gravity load W.

along the meridians and increases as the angle ϕ increases: when $\phi = 0$, $N_\phi = -\frac{1}{2}aw_D$; and when $\phi = \pi/2$, $N_\phi = -aw_D$.

The tangential force N_θ is negative, i.e., compressive, only for limited values of the angle ϕ. Setting $N_\theta = 0$ in Equation 11.65 a, $1/(1 + \cos \phi) - \cos \phi = 0$ gives $\phi = 51°49'$. This determination indicates that for ϕ greater than $51°49'$, tensile stresses develop in the direction perpendicular to the meridians. The distribution of the

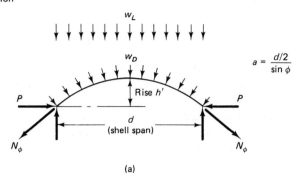

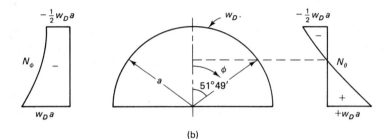

(b)

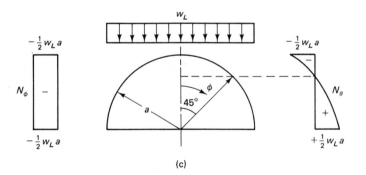

(c)

Figure 11.15 Gravity membrane force distribution in a spherical dome. (a) Flat dome segment of rise h'. (b) Membrane stresses due to self-weight w_D ($N_\theta = 0$ for $\phi = 50°, 49'$). (c) Membrane stresses due to snow load w_L ($N_\theta = 0$ for $\phi = 45°$).

meridional stresses N_ϕ and the tangential stresses N_θ for both the self-weight w_D and the external live load w_L is shown in Figure 11.15.

If the external load is uniform, such as snow giving a projection intensity w_L, the meridional force N_ϕ is obtained from free-body equilibrium by equating the external load to the internal meridional force, i.e., $-\pi(d/2)^2 w_L = 2\pi(a \sin \phi)N_\phi$. Since $d/2 = a \sin \phi$, we obtain

$$N_\phi = \frac{w_L d}{2} \qquad (11.66\text{ a})$$

Hence, N_ϕ is constant throughout the shell depth, as is plain in Figure 11.15.
N_θ due to the live load w_L is

$$N_\theta = -aw_L \cos^2 \phi + \frac{aw_L}{2} = aw_L\left(\frac{1}{2} - \cos^2 \phi\right) = \frac{aw_L}{2}\cos 2\phi \qquad (11.66\text{ b})$$

For the case of $N_\theta = 0$, the shell angle $\phi = 45°$. Consequently, shell stresses due to tangential forces N_θ for ϕ less than 45 degrees are compressive, eliminating cracking. From the distribution of the tangential forces N_θ, it can be concluded that roofs of storage tanks should be *flat*, i.e., the ratio h'/d in Figure 11.15(b) should not exceed $\frac{1}{8}$, so that the concrete will be totally in compression due to both N_ϕ and N_θ, as angle ϕ is less than 51°49′ for meridional forces and 45° for tangential forces.

As discussed at the outset, the support type at the springing level, if restrained, introduces indeterminate reactions that result in direct and bending stresses in the shell near the springing level. Accordingly, the bending theory, a rigorous procedure beyond the scope of this text, has to be applied. Refs. 11.1 and 11.3, on the subject of plates and shells, can be used for determining the resulting bending stresses. The following covers the design of the prestressed ring beam at the springing level to counter the horizontal component of the meridional compressive thrust N_ϕ which causes the edge of the dome to move inwards.

From Equations 11.65 b and 11.66 a, the meridional thrust, N_ϕ, for self-weight w_D per unit surface area and uniform live load w_L per unit projected area can be written as

$$N_\phi = -a\left(\frac{w_D}{1 + \cos \phi} + \frac{w_L}{2}\right) \qquad (11.67)$$

where $a = d/\sin \phi$ is the radius of the shell.

Note that the thrust, N_ϕ, becomes vertical at the springing ($\phi = \pi/2$) of a hemispherical dome and is equal to $W = a/2(w_D + w_L)$ per unit width. At other values of ϕ, N_ϕ is inclined and the value of its horizontal component is needed for the design of the prestressed ring beam at the springing level, namely, the shell rim. This horizontal component is $p = N_\phi \cos \phi$. If P is the prestressing force per beam height in the ring beam, then $P = pd/2$ from Equation 11.1 a, and

$$P = \frac{d}{2}(N_\phi \cos \phi) \qquad (11.68)$$

Evidently, if the force P could be applied directly to the dome rim, the stresses in the dome would be those defined by Equation 11.67. This is usually not feasible, since the large amount of prestressing steel needed due to P cannot be accommodated in the small thickness of the shell, and the stress in the concrete in the rim zone would be very high indeed. Thus, an edge beam has to be provided, transforming the shell into a statically determinate structure.

Prestressing the Statically Indeterminate Flat Dome. The simplest boundary condition is obtained when the edge beam reaction is vertical and without any support restraint, as shown in Figure 11.16, where the dome thrust N_ϕ passes through the beam centroid. If an imaginary cut along line A-A is made, the horizontal thrust $N_\phi \cos \phi$

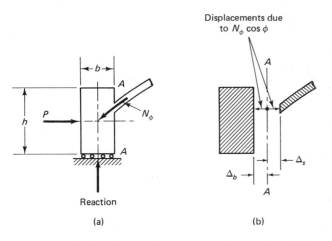

Displacements due to $N_\phi \cos \phi$

Figure 11.16 Ring beam effects. (a) Simply supported beam with thrust line passing through ring beam centroid. (b) Shell displacements at rim; rotations disregarded.

causes the dome edge to move *inwards* a distance (Ref. 11.16)

$$\Delta_s = \frac{d}{2Et}(N_\theta - \mu N_\phi) \tag{11.69}$$

where μ = Poisson's ratio ~ 0.2 for concrete
d = shell span

and the tangential unit force is obtained from Equation 11.65 a as

$$N_\theta = \frac{w_D d}{2 \sin \phi}\left(\frac{1}{1 + \cos \phi} - \cos \phi\right) - \frac{w_L d}{2 \sin \phi}(\cos 2\phi) \tag{11.70}$$

Conversely, the meridional thrust N_ϕ causes the ring beam to move *outwards* a distance

$$\Delta_b = \frac{N_\phi(\cos \phi)d^2}{4Ebh} \tag{11.71}$$

The prestressing force must therefore be sufficient to move the ring beam *inwards* a total distance

$$\Delta_T = \Delta_s + \Delta_b$$

so that the total force acting on the ring beam cross section is

$$P = \frac{bh}{t}(N_\phi - \mu N_\phi) + \frac{d(N_\phi \cos \phi)}{2} \tag{11.72}$$

where h is the total ring beam depth. A comparison of Equations 11.72 and 11.68 shows that the effective prestressing force needed in the former is greater than that required in the latter. The magnitude of this increase is about 5 to 10 percent. The same conditions also hold true for domes in which the line of thrust from the dome does not pass through the centroid of the ring beam and the beam itself is rigidly attached to the wall as in Figure 11.17(a). The required prestressing force P can be obtained approximately by increasing the value of P in Equation 11.68 by 10 percent (Ref. 11.16). In such a case, the stresses in the shell itself at the springing level zone can

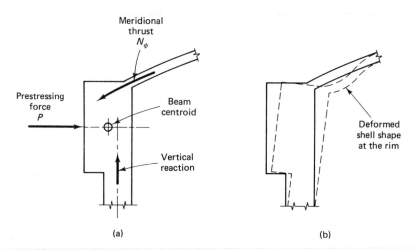

Figure 11.17 Edge ring beam monolithic with tank wall. (a) Thrust N_ϕ not passing through ring beam centroid—general case. (b) Shell deformed shape due to excessive prestressing.

significantly differ from those obtained in the membrane solution, and the bending solution modifications have to be made as in Ref. 11.1 or 11.3.

If the horizontal radial prestressing force in the ring beam is larger than required, excessive bending deformation develops in the shell rim, as is shown in Figure 11.17(b), with a significant increase in the value of the tangential force N_θ as compared to the increase in the meridional force N_ϕ. As a result, the bending stresses in the concrete in the affected zone could exceed the maximum allowable at service load. If the intitial prestress before losses is P_i, the area of the beam cross section is

$$A_c = \frac{P_i}{f_c} \qquad (11.73)$$

where P_i = initial prestressing force P/γ
f_c = allowable compressive stress in the concrete
$\overline{\gamma}$ = residual stress percentage.

It is desirable to maintain a low value of f_c, about $0.2f_c'$ and not exceeding 800–900 psi, in order to minimize any excessive strain that develops in the edge ring beam, which in turn could produce high stresses in the shell at the springing zone.

The area of the prestressing steel in the dome ring is

$$\text{Unit } A_{ps} = \frac{P_i}{f_{pi}} \qquad (11.74\ a)$$

where f_{pi} is the allowable stress, in psi, in the prestressing reinforcement before losses. If accurate analysis to determine A_{ps} is not required, the steel area can be taken as

$$A_{ps} = \frac{W \cot \phi}{2\pi f_{pe}} \qquad (11.74\ b)$$

where W = total dead and live load on the dome due to $w_D + w_L$
f_{pe} = effective steel prestress after losses, psi.

The minimum thickness of the dome required to withstand buckling (Ref. 11.7) may be taken to be

$$\text{Min } h_d = a\sqrt{\frac{1.5P_u}{\phi\beta_i\beta_c E_c}} \qquad (11.75)$$

where a = radius of dome shell
P_u = ultimate uniformly distributed design unit pressure due to dead load and live load = $(1.4D + 1.7L)/144$
ϕ = strength reduction factor for material variability = 0.7
β_i = buckling reduction factor for deviations from true spherical surface due to imperfections
β_i = $(a/r_i)^2$, where $r_i \le 1.4a$
β_c = buckling reduction factor for creep, material nonlinearity, and cracking = $0.44 + 0.003W_L$, but not to exceed 0.53
E_c = initial modulus of concrete = $57{,}000\sqrt{f_c'}$ psi.

11.10 PRESTRESSED CONCRETE TANKS WITH CIRCUMFERENTIAL TENDONS

Instead of wrapping the prestressing wires or strands, as is done in the Preload System, internal or external horizontal tendons are used. These tendons are stressed after they are placed within or on the wall. Vertical post-tensioning is incorporated in the walls as part of the vertical reinforcement. The concrete walls are either cast in place or precast, and the core wall is considered to be the portion of the concrete wall that is circumferentially prestressed. No steel diaphragms are used in this type of construction as compared with wrapped-wire prestressing, where the tank walls can be either with or without steel diaphragms.

The internal prestressed reinforcement is protected by the concrete cover as required in ACI 318, and the ducts or sheathing have to be filled with corrosion-inhibiting materials or grouted. The bonded post-tensioned tendon reinforcement has to be protected by portland cement grout as required in the ACI 318 code, and external tendons should be protected by a shotcrete cover of 1 in. (25 mm) minimum thickness.

The wall design procedures are similar to those of circular tanks prestressed by wire or strand wrapping, and the same requirements for crack control and water or liquid tightness apply. A minimum residual compressive stress of 200 psi (1.4 MPa) in the concrete wall after all prestress losses has to be provided in the design when the tank is filled to the design level. If the tank is not covered, a residual compressive stress of 400 psi (2.8 MPa) has to be provided at the wall top, reducing linearly to not less than 200 psi at $0.6\sqrt{Rh}$ from the top of the liquid level.

Typical Wall Base and Dome Roof Connections. From the foregoing discussions, it is clear that the boundary conditions at the base of the circular prestressed tank and at the ring beam support for the roof dome determine the practicality, economy, and success of the entire design. Consequently, accumulated experience in developing the connections at these boundary conditions is invaluable. A selection of connection details taken from Refs. 11.6 to 11.9 is given in Figures 11.18 through 11.22.

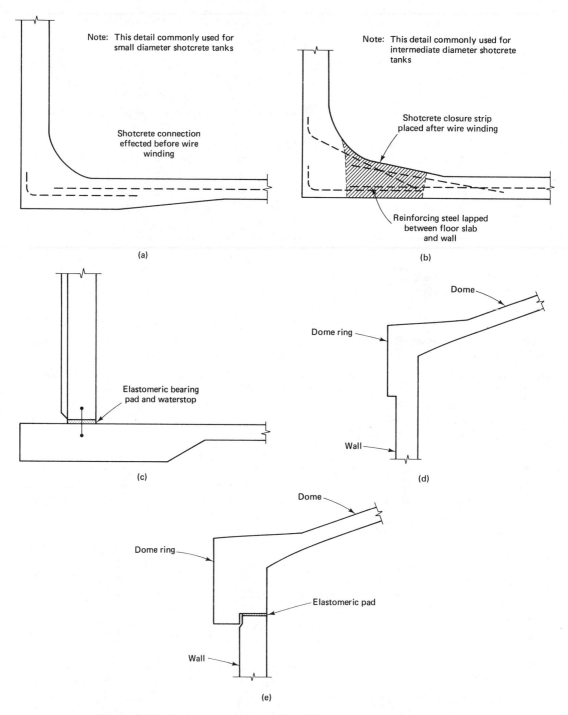

Figure 11.18 Cast-in-place tanks. (a) Monolithic base joint; monolithic and fully restrained against translation before and after wire winding. (b) Monolithic base joint; hinged with limited restraint against translation during wire winding, and monolithic and fully restrained against translation after wire winding. (c) Separated base joint, allows translation, rotation, or both (e.g., elastometric bearing pad). (d) Monolithic dome-wall connection. (e) Separated dome-wall connection.

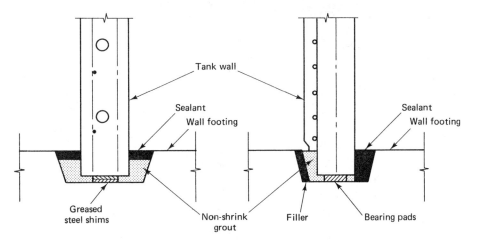

Figure 11.19 Wall base joints for precast tanks.

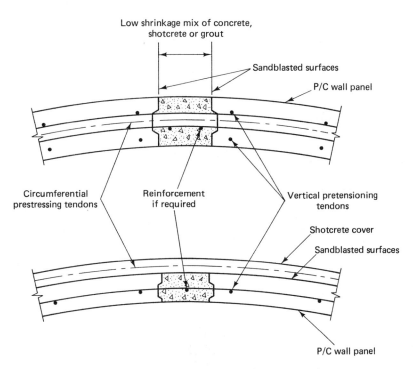

Figure 11.20 Vertical wall joints for precast tanks.

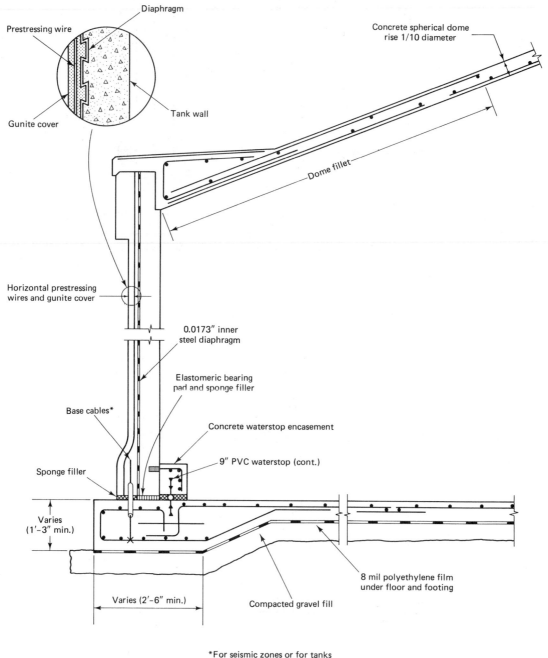

Diaphragm

Prestressing wire

Gunite cover

Tank wall

Concrete spherical dome
rise 1/10 diameter

Dome fillet

Horizontal prestressing
wires and gunite cover

0.0173″ inner
steel diaphragm

Elastomeric bearing
pad and sponge filler

Base cables*

Concrete waterstop encasement

9″ PVC waterstop (cont.)

Sponge filler

Varies
(1′–3″ min.)

8 mil polyethylene film
under floor and footing

Varies (2′–6″ min.)

Compacted gravel fill

*For seismic zones or for tanks
with unequal backfill only

Figure 11.21 Typical tank section of a domed preload prestressed concrete tank with an inner steel diaphragm (*Courtesy*, Preload Technology, Inc., New York).

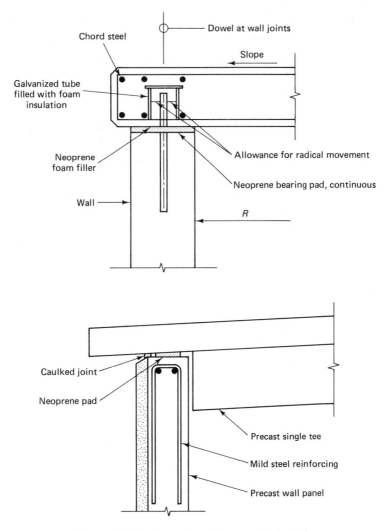

Figure 11.22   Connections for precast tank roofs.

11.11 STEP-BY-STEP PROCEDURE FOR THE DESIGN OF CIRCULAR PRESTRESSED CONCRETE TANKS AND DOME ROOFS

The following trial-and-adjustment procedure is recommended for designing a prestressed concrete circular tank and its roof shell:

1. Select the prestressing system, the type of prestressing wire, the concrete strength, and the type of restraint that can be accomplished under local conditions.

2. Determine the contained material pressure on the wall: γH for liquid and p for gas. Use the trapezoidal distribution for granular or solid containment.

Find the unit ring force $F = \gamma(H - y)r$ for a completely sliding base, where r is the radius of the tank and y is the distance above the base.

3. Choose, from Tables 11.4 through 11.16, the applicable vertical moment coefficients for the particular load type and wall base restraint condition caused by liquid pressure

$$M_y = +\frac{1}{\beta}[\beta M_o \phi(\beta y) + Q_o \zeta(\beta y)]$$

and determine the corresponding horizontal radial ring tensions

$$Q_o = +(2\beta H - 1)\frac{\gamma rt}{\sqrt{12(1 - \mu^2)}}$$

and $Q_y = (F - \Delta Q_y)$, where the offset

$$\Delta Q_y = +\frac{6(1 - \mu^2)}{\beta^3 rt^2}(\beta M_o \psi(\beta y) + Q_o \theta(\beta y)$$

and

$$\beta = \frac{[3(1 - \mu^2)]^{1/4}}{(rt)^{1/2}}$$

where $\mu \sim .20$ for concrete.

4. Find the applicable membrane coefficients C from Tables 11.4 through 11.16. Compute the applicable ring force $F = C\gamma Hr$.

5. Compute the critical vertical moments in the wall using the applicable membrane coefficient C. The equation for moment due to liquid load is

$$M_y = C(\gamma H^3 + pH^2)$$

or

$$M_y = CpH^2$$

due to gas load if applicable. Compute the moment at the base, where applicable, and at the critical y plane above the base.

6. Choose the level of vertical prestressing force.

7. Compute the concrete stresses across the thickness of the wall both for the condition when the tank is empty and for when it is totally full. Allow maximum residual axial compressive stress $f_{cv} = 200$ psi at service and a maximum tensile stress $f_t = 3\sqrt{f_c'}$ as shown in Table 11.17.

8. Design both the horizontal and the vertical prestressing steel limiting stresses to those given in Table 11.18.

9. Compute the factored moment M_U using the applicable load factors given in subsection 11.7.2. The required $M_n = M_U/\phi$, where $\phi = 0.9$. Compute the available nominal moment strength $M_n = A_{ps}f_{ps}(d_p - a/2)$, or $M_n = A_{ps}f_{ps}(d_p - a/2) + A_s f_y(d - a/2)$. The available M_n has to be greater than or equal to the required M_n.

10. Design the length L of the annular ring at the base of the wall from the equation

$$L^2 = \frac{2CH^2}{1 + \dfrac{(t/h)^3}{(dt)^2}}$$

where t is the thickness of the wall and h the thickness of the base slab.

11. Compute the percentage of prestress in the base to be transferred to the wall from the formula

$$\text{Percentage } R = \frac{1}{1 + S}$$

where $S = 1.1(h/t) \times (d/t)^{1/2}$.

When only the outer rim of the slab ring is compressed by radial thrust at the rim, the value of S is modified to

$$S_1 = \frac{1}{K}\left(\frac{h}{t}\right)\left(\frac{d}{t}\right)^{1/2}$$

where

$$K = \left(\frac{d_o^2 + d^2}{d_o^2 - d^2} - \mu\right)$$

in which d_0 = outer diameter
d = inner slab ring diameter = $d_o - 2L$.

12. Check the minimum wall thickness requirements, and evaluate the unrestrained initial elastic radial deflection

$$\Delta_i = \frac{F_i r}{t_{co} E_c}$$

where $E_c = 57{,}000\sqrt{f_c'}$
t_{co} = thickness of wall core at top or bottom of wall
$r = \frac{1}{2}d$.

The final radial deflection $\Delta_f = 1.7\Delta_i$.

13. Anchor the steel from the base to the wall such that the steel extends into the wall a distance $y_2 = 1.8\sqrt{rt_{co}}$ or 3 ft, whichever is greater. Also, ensure that the minimum nominal vertical steel at the base region is

$$A_s = 0.005 t_{co}$$

14. Verify the maximum crack width $w_{\max} = 4.1 \times 10^{-6}\epsilon_{ct} E_{ps}\sqrt{I_x}$,

where ϵ_{ct} = tensile surface strain in the concrete = $(\lambda_t f_p)/(E_{ps})$
f_p = actual stress in the steel
f_{pi} = initial prestress before losses
$\lambda_t \sim f_p/f_{pi}$

$$I_x = \text{grid index} = \frac{8}{\pi}\left(\frac{s_2 s_1 t_b}{\phi_1}\right)$$

s_1 = spacing of reinforcement in direction "1"

ϕ_1 = diameter of steel in direction "1"

s_2 = spacing of reinforcement in direction "2"

t_b = concrete cover to center of steel, in.

Note that allowable w_{max} = 0.004 in. for liquid-retaining tanks.

15. Design the roof cover dome after selecting the type of connection at the top of the tank wall. Limit the ratio of the rise h' of the dome to its base d such that h'/d does not exceed $\frac{1}{8}$.

Compute the required horizontal radial prestressing force P for the edge beam from the equation

$$P = \frac{bh}{t}(N_\theta - \mu N_\phi) + \frac{d(N_\phi \cos\,\phi)}{2}$$

where

$$N_\theta = \frac{w_D d}{2\sin\,\phi}\left[\frac{1}{1 + \cos\,\phi} - \cos\,\phi\right] - \frac{w_L d}{2\sin\,\phi}(\cos\,2\phi)$$

$$N_\phi = -a\left(\frac{w_D}{1 + \cos\,\phi} + \frac{w_L}{2}\right)$$

and

h = total depth of rim beam

b = ring beam width

w_D = intensity of self-weight of shell per unit area (dead load)

w_L = intensity of live-load projection.

16. Compute the ring edge beam cross section

$$A_c = \frac{P_i}{f_c}$$

where P_i = initial prestressing force = $P/\bar{\gamma}$

$\bar{\gamma}$ = residual stress percentage

f_c = allowable compressive stress in the concrete, not to exceed $0.2f'_c$, but not more than 800–900 psi, in the edge beam.

17. Compute the area of the edge beam prestressing tendon

$$A_{ps} = \frac{P_i}{f_{si}}$$

where f_{si} is the allowable stress in the prestressing steel before losses, or

$$A_{ps} = \frac{W \cot\,\phi}{2\pi f_{pe}}$$

Two prestressed concrete anaerobic digester tanks during construction (*Courtesy,* N.A. Legatos, Preload Technology, Inc., New York.)

if accurate analysis is not performed. In the latter, W is the total dead and live load on the dome due to $w_D + w_L$ and f_{pe} is the effective prestress after losses.

18. Check the minimum dome thickness required to withstand buckling, i.e.,

$$\text{Min. } h_d = a\sqrt{\frac{1.5p_u}{\phi\beta_i\beta_c E_c}}$$

where a = radius of dome shell

P_u = ultimate uniformly distributed design unit pressure due to dead load and live load = $(1.4D + 1.7L)/144$

ϕ = strength reduction factor for material variability = 0.7

β_i = buckling reduction factor for deviations from true spherical surface due to imperfections

$\beta_i = (a/r_i)^2$, where $r_i \leq 1.4a$

β_c = buckling reduction factor for creep, material nonlinearity, and cracking = $0.44 + 0.003W_L$, but not to exceed 0.53

E_c = initial modulus of concrete = $57,000\sqrt{f_c'}$ psi.

Figure 11.23 gives a step-by-step flowchart for a recommended sequence of operations to be performed in the design of circular prestressed concrete tanks and their shell roofs.

START

1 Input: d, H, r, h, a, $\angle\phi$, h', γ, p, W_D, W_L, f'_c, f'_{ci}, f_t, f_c, f_{cv}, f_{pu}, f_{pi}, f_{py}, f_{ps}, f_{pe}

2 Assume wall thickness t and type of wall base joint. Compute $F = \gamma(H - y)r$ for freely sliding base. Select membrane coefficient C from Tables 10.4–10.16

$$\beta = \frac{[3(1 - \mu^2)]^{1/4}}{(rt)^{1/2}}$$

Compute max. M_y at y above base

$M_y = c(\gamma H^3 + pH^2)$

M_o, Q_o, ΔQ_y and Q_y

$$Q_o = +(2\beta H - 1)\ \frac{\gamma rt}{\sqrt{12(1 - \mu^2)}}$$

$$\Delta Q_y = +\ \frac{6(1 - \mu^2)}{\beta^3 rt^2}\ [\beta M_o\,\psi(\beta y) + Q_o\,\theta(\beta y)]$$

$Q_y = F - \Delta Q_y$

3 Choose vertical prestress P_v. Compute concrete fiber stresses at critical base section when tank is empty and when full

$$f = -\frac{P_v}{A} \pm \frac{M_L c}{I} + \frac{M_p c}{I}$$

where M_L = liquid load vertical unit moment

M_p = prestress vertical unit moment

Max $f_c = 0.45 f'_c$

Min $t = 7$ in. with vertical prestress

Max. allow. residual axial $f_{cv} = 200$ psi

Max. allow. tensile stress $f_t = 3\sqrt{f'_c}$

4 Revise wall section details ← No — t adequate? — Yes →

5 Compute factored moment M_u using load factors:

Initial liquid pressure	1.3
Internal lateral pressure from dry material	1.7
Final prestress after losses	1.7
Strength reduction factor	0.9

Rqd $M_n \leq$ available $M_n = \dfrac{M_U}{0.9}$

Avail. $M_n = A_{ps} f_{ps}\left(d_p - \dfrac{a}{2}\right) + A_s f_y\left(d - \dfrac{a}{2}\right)$

Figure 11.23 Flowchart for the design of circular prestressed tanks and their flat dome roofs.

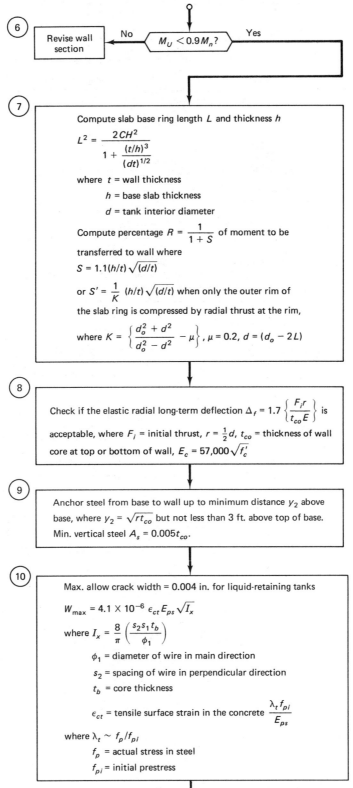

6 | Revise wall section ←— No —— $M_u < 0.9 M_n$? —— Yes —→

7

Compute slab base ring length L and thickness h

$$L^2 = \frac{2CH^2}{1 + \dfrac{(t/h)^3}{(dt)^{1/2}}}$$

where t = wall thickness

h = base slab thickness

d = tank interior diameter

Compute percentage $R = \dfrac{1}{1+S}$ of moment to be transferred to wall where

$S = 1.1(h/t)\sqrt{(d/t)}$

or $S' = \dfrac{1}{K}\,(h/t)\sqrt{(d/t)}$ when only the outer rim of the slab ring is compressed by radial thrust at the rim,

where $K = \left\{ \dfrac{d_o^2 + d^2}{d_o^2 - d^2} - \mu \right\}$, $\mu = 0.2$, $d = (d_o - 2L)$

8

Check if the elastic radial long-term deflection $\Delta_f = 1.7\left\{ \dfrac{F_i r}{t_{co} E} \right\}$ is acceptable, where F_i = initial thrust, $r = \frac{1}{2}d$, t_{co} = thickness of wall core at top or bottom of wall, $E_c = 57,000\sqrt{f_c'}$

9

Anchor steel from base to wall up to minimum distance y_2 above base, where $y_2 = \sqrt{rt_{co}}$ but not less than 3 ft. above top of base. Min. vertical steel $A_s = 0.005t_{co}$.

10

Max. allow crack width = 0.004 in. for liquid-retaining tanks

$$W_{max} = 4.1 \times 10^{-6}\, \epsilon_{ct} E_{ps} \sqrt{I_x}$$

where $I_x = \dfrac{8}{\pi}\left(\dfrac{s_2 s_1 t_b}{\phi_1} \right)$

ϕ_1 = diameter of wire in main direction

s_2 = spacing of wire in perpendicular direction

t_b = core thickness

ϵ_{ct} = tensile surface strain in the concrete $\dfrac{\lambda_t f_{pi}}{E_{ps}}$

where $\lambda_t \sim f_p/f_{pi}$

f_p = actual stress in steel

f_{pi} = initial prestress

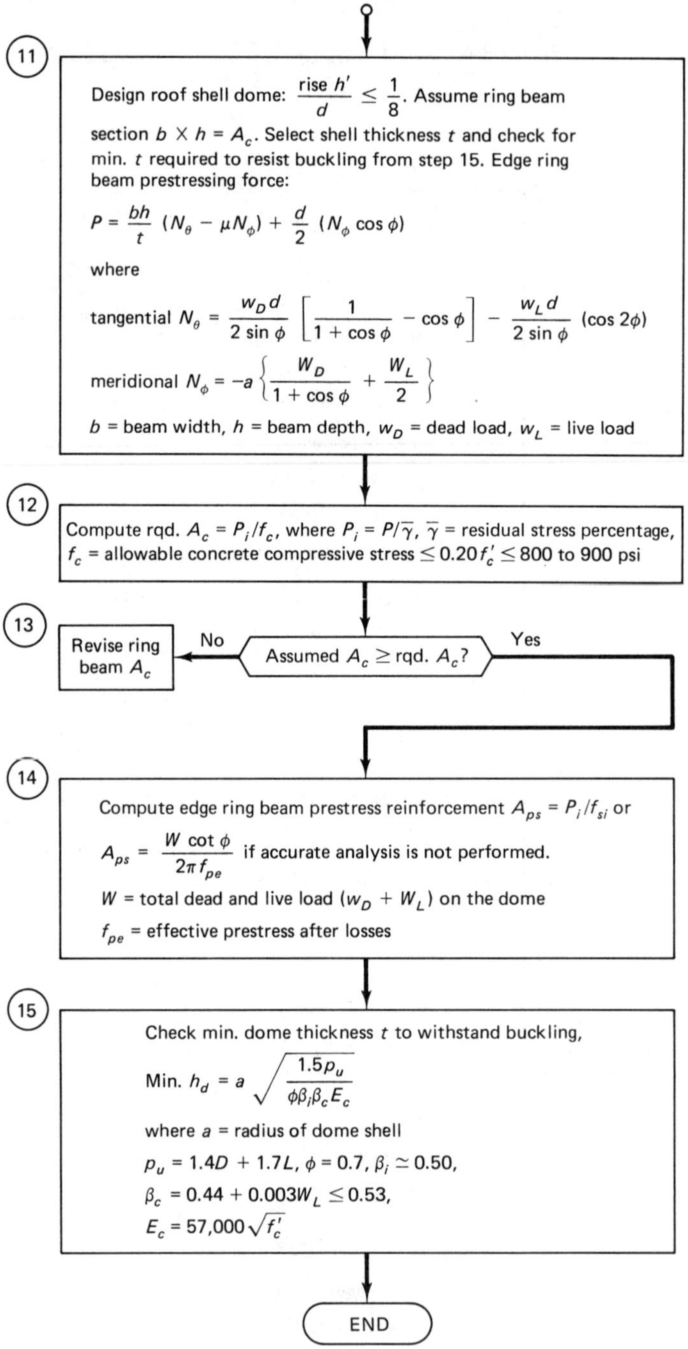

11

Design roof shell dome: $\dfrac{\text{rise } h'}{d} \leq \dfrac{1}{8}$. Assume ring beam

section $b \times h = A_c$. Select shell thickness t and check for min. t required to resist buckling from step 15. Edge ring beam prestressing force:

$$P = \frac{bh}{t}(N_\theta - \mu N_\phi) + \frac{d}{2}(N_\phi \cos \phi)$$

where

tangential $N_\theta = \dfrac{w_D d}{2 \sin \phi} \left[\dfrac{1}{1 + \cos \phi} - \cos \phi \right] - \dfrac{w_L d}{2 \sin \phi}(\cos 2\phi)$

meridional $N_\phi = -a \left\{ \dfrac{W_D}{1 + \cos \phi} + \dfrac{W_L}{2} \right\}$

b = beam width, h = beam depth, w_D = dead load, w_L = live load

12

Compute rqd. $A_c = P_i/f_c$, where $P_i = P/\overline{\gamma}$, $\overline{\gamma}$ = residual stress percentage, f_c = allowable concrete compressive stress $\leq 0.20 f_c' \leq 800$ to 900 psi

13

Revise ring beam A_c ◄── No ─── Assumed $A_c \geq$ rqd. A_c? ─── Yes ───►

14

Compute edge ring beam prestress reinforcement $A_{ps} = P_i/f_{si}$ or

$A_{ps} = \dfrac{W \cot \phi}{2\pi f_{pe}}$ if accurate analysis is not performed.

W = total dead and live load $(w_D + W_L)$ on the dome

f_{pe} = effective prestress after losses

15

Check min. dome thickness t to withstand buckling,

Min. $h_d = a \sqrt{\dfrac{1.5 p_u}{\phi \beta_i \beta_c E_c}}$

where a = radius of dome shell

$p_u = 1.4D + 1.7L$, $\phi = 0.7$, $\beta_i \simeq 0.50$,

$\beta_c = 0.44 + 0.003 W_L \leq 0.53$,

$E_c = 57,000 \sqrt{f_c'}$

END

Figure 11.23 (*continued*)

11.12 DESIGN OF CIRCULAR PRESTRESSED CONCRETE WATER-RETAINING TANK AND ITS DOMED ROOF

Example 11.3

Determine the maximum horizontal ring forces and vertical moments, and design the wall prestressing reinforcement, for a circular prestressed concrete tank whose diameter $d = 125$ ft (38.1 m) and which retains a water height $H = 25$ ft (7.62 m) for the following conditions of wall base support: (a) hinged, (b) fully fixed, (c) semisliding, and (d) partially fixed. Also, design the prestressed concrete ring edge beam for the domed roof shell assuming that the shell rise-span ratio $h'/d = \frac{1}{8}$. Use a flat shell roof having shell angle $\phi = 36°$, and find the area of prestressing reinforcement for both wire-wrapped and tendon reinforced conditions. Given data are as follows:

$f'_c = 5,000$ psi (34.47 MPa), normal-weight concrete

$f'_{ci} = 3,750$ psi (25.86 MPa)

$f_t = 212$ psi (0.86 MPa) $\leq 3\sqrt{f'_c}$

$f_c = 0.45 f'_c = 2,250$ psi (15.51 MPa)

residual $f_{cv} = 225$ psi (1.55 MPa)

f_{pu} (wire) $= 250,000$ psi (1,724 MPa)

f_{pu} (strands and tendons) $= 250,000$ psi (1,724 MPa)

$f_{pi} = 0.7 f_{pu} = 175,000$ psi (1,207 MPa)

$f_{ps} = 220,000$ psi (1,517 MPa)

$w_L = 15$ psf (718 Pa) for snow load on dome

Assume 26 percent total loss in prestress for all long-term effects.

Solution Disregard the weight of the wall and the roof dome effect as insignificant on the stresses as compared to the effect of the vertical prestress forces. Consider the water pressure distribution shown in Figure 11.24 on the tank wall giving

$$\gamma = 62.4 \text{ lb/ft}^3 \ (1,000 \text{ kg/m}^3)$$

$$r = \frac{d}{2} = \frac{125}{2} = 62.5 \text{ ft } (19.1 \text{ m})$$

Assume the wall thickness $t = 10$ in. $= 0.83$ ft (25.4 cm). Then the form factor

$$\frac{H^2}{dt} = \frac{25 \times 25}{125 \times 0.83} = 6$$

and $\gamma H r = 62.4 \times 25 \times 62.5 = 97,500$ lb/ft of circumference.

Basic Forces and Moments. Tables 11.19 through 11.21 give the basic forces and moments in the tank wall.

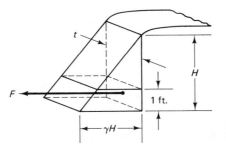

Figure 11.24 Liquid ring tension F, wall base freely sliding.

TABLE 11.19 MAXIMUM RING TENSION $F = C(\gamma Hr)$ LB/FT CIRCUMFERENCE, EXAMPLE 11.3

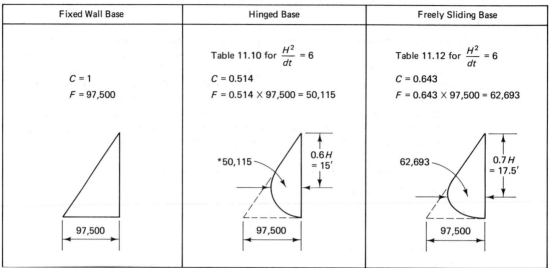

Fixed Wall Base	Hinged Base	Freely Sliding Base
$C = 1$ $F = 97,500$	Table 11.10 for $\dfrac{H^2}{dt} = 6$ $C = 0.514$ $F = 0.514 \times 97,500 = 50,115$	Table 11.12 for $\dfrac{H^2}{dt} = 6$ $C = 0.643$ $F = 0.643 \times 97,500 = 62,693$

*Compare with 50,113 lb/ft in the detailed method of Example 11.2.

TABLE 11.20 VERTICAL MOMENTS $M = C(\gamma H^3)$ FT-LB/FT, EXAMPLE 11.3. POSITIVE (+) = TENSION IN OUTSIDE FACE

Freely Sliding Base	Fixed Wall Base	Hinged Base
$M_y = M_o = 0$	Table 11.4 $C = +0.0051$ for $0.7H = 17.5$ ft $C = -0.187$ for $1.0H = 25$ ft $M_y = +0.0051 \times 62.4(25)^3$ $\quad = +4,973$ $M_o = -0.0187 \times 62.4(25)^3$ $\quad = -18,233$	Table 11.6 $C = +0.0078$ for $0.8H = 20$ ft $C = 0$ for $1.0H$, or full height $M_y = +0.0078 \times 62.4(25)^3$ $\quad = +7,605$ $M_o = 0$

*This moment value is very close to the value obtained by using the detailed method and the moment functions of Table 11.1 and Example 11.1 ($M_o = -18,574$).

TABLE 11.21 PRESTRESSING EFFECTS USING 225-PSI RESIDUAL RADIAL COMPRESSION, EXAMPLE 11.3. RING FORCES Q LB/FT, VERTICAL MOMENTS M_y FT-LB/FT

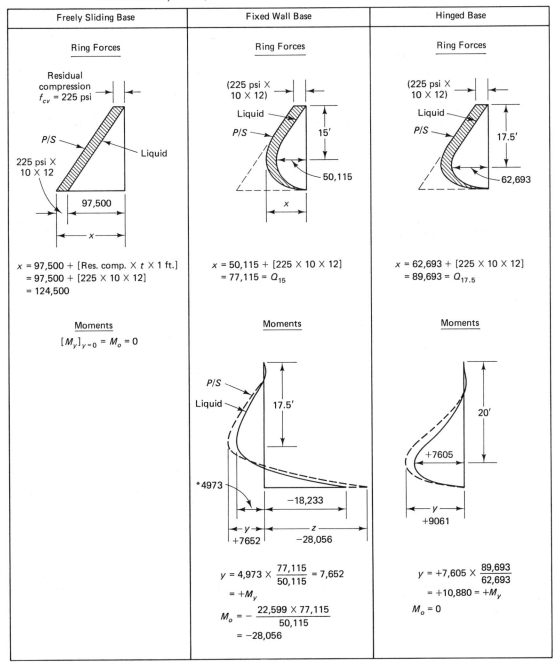

Freely Sliding Base	Fixed Wall Base	Hinged Base
Ring Forces	**Ring Forces**	**Ring Forces**
Residual compression f_{cv} = 225 psi	(225 psi × 10 × 12)	(225 psi × 10 × 12)
P/S	Liquid	Liquid
225 psi × 10 × 12	P/S 15'	P/S 17.5'
Liquid	50,115	62,693
97,500	x	
$x = 97,500 +$ [Res. comp. × t × 1 ft.]	$x = 50,115 +$ [225 × 10 × 12]	$x = 62,693 +$ [225 × 10 × 12]
$= 97,500 +$ [225 × 10 × 12]	$= 77,115 = Q_{15}$	$= 89,693 = Q_{17.5}$
$= 124,500$		
Moments	**Moments**	**Moments**
$[M_y]_{y=0} = M_o = 0$	P/S Liquid 17.5'	20'
	*4973	+7605
	−18,233	
	y z	y
	+7652 −28,056	+9061
	$y = 4,973 \times \dfrac{77,115}{50,115} = 7,652$	$y = +7,605 \times \dfrac{89,693}{62,693}$
	$= +M_y$	$= +10,880 = +M_y$
	$M_o = -\dfrac{22,599 \times 77,115}{50,115}$	$M_o = 0$
	$= -28,056$	

*Compare with the value $M = +4,912$ ft-lb/ft obtained by the detailed method of Example 11.1.

Wall Maximum Concrete Stresses at 20 ft from Top: Hinged Base. By trial and adjustment, provide vertical concentric prestress $P_v = 50{,}000$ lb/ft (729.5 N/m) of circumference. Then for a wall thickness $t = 10$ in. calculate the resulting stresses as shown in Figure 11.25.

$$f_+ = \frac{M}{S} = \frac{10{,}880 \times 12}{\dfrac{12(10)^2}{6}} = \mp 653 \text{ psi}$$

$$f_+ = \frac{M}{S} = \frac{7{,}605 \times 12}{\dfrac{12(10)^2}{6}} = \pm 456 \text{ psi}$$

$$f_v = \frac{P_v}{A_c} = \frac{50{,}000}{12 \times 10} = -417 \text{ psi}$$

$$f_4 = \boxed{1} + \boxed{3}$$

Max. $f_t = 236$ psi $\cong 3\sqrt{5{,}000}$

$\qquad \cong 212$ psi, O.K.

Max. $f_c = -1{,}070$ psi $< 0.45 f'_c$, O.K.

$$f_5 = \boxed{1} + \boxed{2} + \boxed{3}$$

Max. $f_c = -614$ psi $< 0.45 f'_c$, O.K.

Outside   Inside

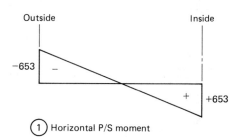

① Horizontal P/S moment

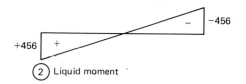

② Liquid moment

③ Vertical P/S

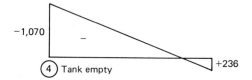

④ Tank empty

⑤ Tank full

Figure 11.25 Stress at maximum moment, 20 ft from top, psi. Negative $(-)$ = compression, positive $(+)$ = tension.

Wall Maximum Concrete Stress at 17 ft 6 in. from Top: Fully Fixed Base. The maximum positive moment M_y is at 17 ft 6 in. from the top of the wall. By trial and adjustment, use *eccentric* vertical prestressing $P_v = 100{,}000$ lb/ft closer to the outer face ($e = 1.05$ in. (26.7 m). Then calculate the resulting stresses in the wall as shown in Figure 11.26.

$$f_+ = \frac{M}{S} = \frac{7{,}652 \times 12}{\dfrac{12(10)^2}{6}} = \mp370 \text{ psi}$$

$$f_+ = \frac{M}{S} = \frac{4{,}973 \times 12}{\dfrac{12(10)^2}{6}} = \pm298 \text{ psi}$$

$$f_v = \frac{P_v}{A_c} = \frac{100{,}000}{12 \times 10} = -833 \text{ psi}$$

$$f_v = \frac{P_v e(c)}{I} = \frac{100{,}000 \times 1.05}{\dfrac{12(10)^2}{6}} = \mp525 \text{ psi}$$

$$f_5 = \textcircled{1} + \textcircled{3} + \textcircled{4}$$

Max. $f_t = +151$ psi $< 3\sqrt{f_c'} = 212$, O.K.

Max. $f_c = -1{,}817$ psi $< 0.45 f_c' = -2{,}250$ psi, O.K.

$$f_6 = \textcircled{1} + \textcircled{2} + \textcircled{3} + \textcircled{4}$$

Max. $f_c = -1{,}519$ psi $< 0.45 f_c'$, O.K.

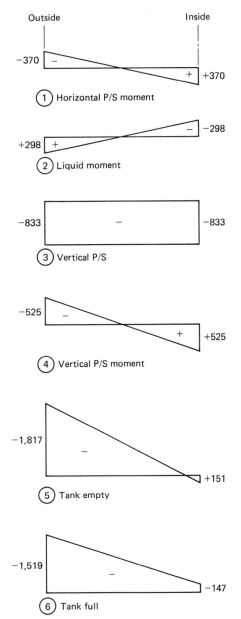

Figure 11.26 Stresses at maximum positive (+) moment, 17 ft, 6 in. from top, psi. Negative (−) = compression, positive (+) = tension.

Wall Maximum Concrete Stress at Base: Fully Fixed Base. Use eccentric vertical prestress $P_v = 100,000$ lb closer to the outer face ($e = 1.05$ in.). Then calculate the resulting stresses in the wall as shown in Figure 11.27.

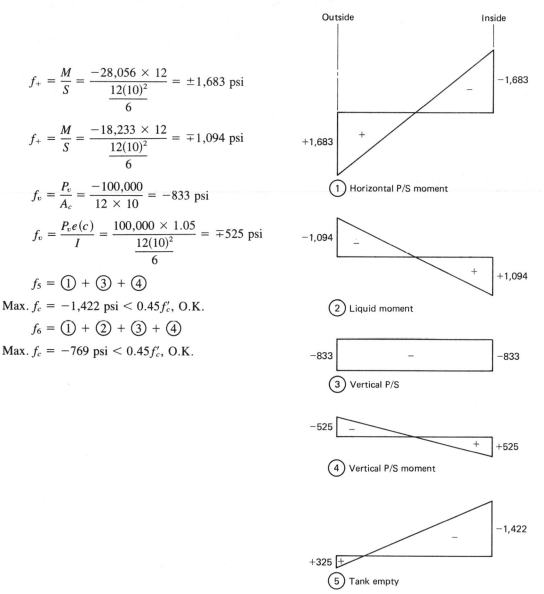

$$f_+ = \frac{M}{S} = \frac{-28,056 \times 12}{\dfrac{12(10)^2}{6}} = \pm 1,683 \text{ psi}$$

$$f_+ = \frac{M}{S} = \frac{-18,233 \times 12}{\dfrac{12(10)^2}{6}} = \mp 1,094 \text{ psi}$$

$$f_v = \frac{P_v}{A_c} = \frac{-100,000}{12 \times 10} = -833 \text{ psi}$$

$$f_v = \frac{P_v e(c)}{I} = \frac{100,000 \times 1.05}{\dfrac{12(10)^2}{6}} = \mp 525 \text{ psi}$$

$f_5 = \text{①} + \text{③} + \text{④}$

Max. $f_c = -1,422$ psi $< 0.45 f_c'$, O.K.

$f_6 = \text{①} + \text{②} + \text{③} + \text{④}$

Max. $f_c = -769$ psi $< 0.45 f_c'$, O.K.

Figure 11.27 Stresses at maximum negative (−) moment at wall base, psi. Negative (−) = compression, positive (+) = tension.

Wall Maximum Concrete Stress: Semisliding Base. By trial and adjustment, use concentric vertical prestress $P_v = 20,400$ lb/ft (297 kN/m). Then semislide $M = \frac{1}{2}(+10,880) = 5,440$ ft-lb/ft, and calculate the resulting stresses in the wall as shown in Figure 11.28.

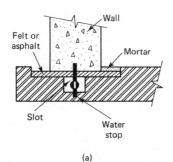

(a)

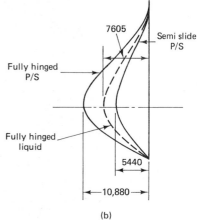

(b)

$$f_+ = \frac{M}{S} = \frac{5,440 \times 12}{\dfrac{12(10)^2}{6}} = \mp 326 \text{ psi}$$

$$f_+ = \frac{M}{S} = \frac{+7,605 \times 12}{\dfrac{12(10)^2}{6}} = \pm 456 \text{ psi}$$

$$f_v = \frac{-P_v}{A_c} = \frac{-20,400}{12 \times 10} = -170 \text{ psi}$$

$$f_4 = \text{①} + \text{③}$$

Max. $f_c = 496$ psi $< 0.45 f_c'$, O.K.

Max. $f_t = 156$ psi $< 3\sqrt{f_c'}$, O.K.

$$f_5 = \text{①} + \text{②} + \text{③}$$

Max. $f_c = 300$ psi $< 0.45 f_c'$, O.K.

Max. $f_t = -40$ psi $< 3\sqrt{f_c'}$, O.K.

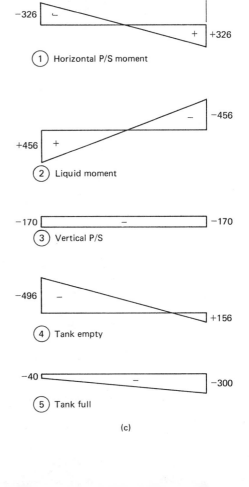

Figure 11.28 Stresses at maximum positive (+) moment, psi. (a) Wall base details. (b) Semislide moment, ft-lb/ft. (c) Concrete stresses, psi.

(c)

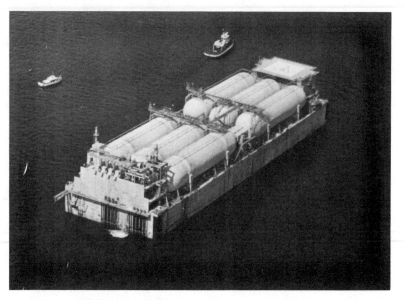

Arco Floating LPG Barge: ABAM-designed largest floating prestressed hull in the world. (*Courtesy*, ABAM Engineers, Tacoma, Washington.)

Partial Fixity at the Wall Base. The restraint moment is $M_P = M_o(1 - S)$, where the full fixity moment $M_o = 18,233$ ft-lb/ft. The modifying factor for partial fixity $S = (t/h)^3/(dt)^{1/2}$.

Figure 11.29 shows the deformed shape of the base slab. If the base slab thickness $h = 10$ in., then, from Equations 11.39 and 11.40,

$$S = \frac{(10/10)^3}{(125 \times 0.83)^{1/2}} = 0.10$$

and

$$M_p = M_o(1 - S) = 18,233(1 - 0.1) = 16,410 \text{ ft-lb/ft.}$$

The moment loss due to partial fixity $= 18,233 - 16,410 = 1,823$ ft-lb/ft. From Equation 11.37 for the base ring width L,

$$L^2 = \frac{2CH^2}{1 + s}$$

Also, from Table 11.4, the membrane coefficient at the base for form factor $\frac{H^2}{dt} = 6$ is $C = -0.0187$. Thus, we have

$$L^2 = \frac{2 \times 0.0187(25)^2}{1 + 0.1} = 21.25$$

and it follows that

$$L = 4.61 \text{ ft} = 4 \text{ ft } 7\tfrac{1}{2} \text{ in.}$$

Accordingly, use a ring slab base width $L = 4$ ft 9 in. (145 cm). Since for large-diameter tanks S has a very small value, the degree of fixity, as the solution shows, is almost the same for both fully fixed and partially fixed wall bases.

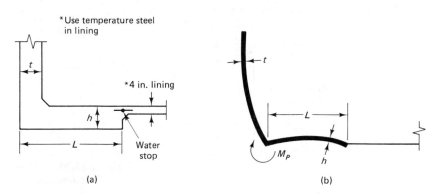

*Use temperature steel in lining

t

*4 in. lining

h

Water stop

M_P

t

L

h

(a)

(b)

Figure 11.29 Deformed shape of base slab. (a) Wall base. (b) Deformed section.

From Equations 11.47 and 11.48, the percent R of prestress in the base that is transferred to wall $= 100/S_1$, where

$$S_1 = 1.1\left(\frac{h}{t}\right)\left(\frac{d}{t}\right)^{1/2} = 1.1\left(\frac{10}{10}\right)\left(\frac{125}{0.83}\right)^{1/2} = 13.50\%$$

Consequently,

$$R = \frac{100}{13.50} = 7.4\%$$

which means that the required design prestress for the wall can be slightly reduced, as some compression is available from the base ring.

Design of Prestressing Reinforcement

Horizontal Prestressing. Use the same size wire to wrap the circular wall, varying the spacing of the wire hoops in 5-ft bands along the tank height. In the case of the freely sliding tank wall, the minimum spacing is in the lowest band at the base, as presented graphically in Figure 11.30.

In order to determine the variation of wire pitch throughout the height of the wall, additional computations of the horizontal ring thrust Q_y have to be made at the bottom of each band. Consequently, only one typical calculation of size and wire distribution will be made for purposes of illustration.

Taking the case of the fixed wall base from Table 11.21, the maximum $Q_{15} = 77,115$ lb/ft of circumference per foot height of wall. So trying 0.192 in. dia (4.88 mm) prestressing 250 K wire, we obtain $A_{ps} = 0.0289$ in² per wire and $f_{pi} = 0.7 f_{pu} = 0.7 \times 250,000 = 175,000$ psi (1,207 MPa).

Now assume 26 percent prestress loss for elastic shortening, seating, creep, shrinkage, and steel relaxation. Then

$$f_{pe} = 0.74 \times 175,000 = 129,500 \text{ psi (893 MPa)}$$

$$A_{ps} = \frac{77,115}{129,500} = 0.60 \text{ in}^2 \text{ per 1 ft of wall height}$$

$$\text{No. of wire loops in 5-ft band} = \frac{0.60 \times 5}{0.0289} = 104$$

Hence, use 104 wire loops in the 5-ft wall band whose base is 15 ft below the top of the water level. Also, use 2-in. shotcrete to cover the wrapped horizontal 0.192 in. dia wires.

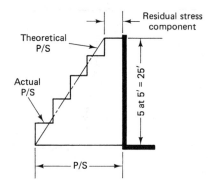

Residual stress
component

Theoretical
P/S

5 at 5' = 25'

Actual
P/S

P/S

Figure 11.30 Horizontal-prestress wire distribution bands.

If the tank were prestressed with $\frac{1}{2}$ in. dia 250 K seven-wire-strand tendons, A_{ps} would be 0.144 in^2/tendon and the required number of tendons in a 5-ft-height band would be 0.60 × 5/0.144 ≅ 20 tendons.

Vertical Prestressing. For proportioning the vertical prestressing reinforcement, P_v = 100,000 lb/ft at e = 1.05 in. (1,459 N/m at e = 26.7 mm) on the outer force side. Hence, try $\frac{1}{2}$ in. dia (17.7 mm dia) seven-wire 250 K strands. We obtain

$$A_{ps} = 0.144$$

$$f_{pu} = 250,000 \text{ psi } (1,724 \text{ MPa})$$

$$f_{pi} = 0.7f_{pu} = 0.7 \times 250,000 = 175,000 \text{ psi } (1,207 \text{ MPa})$$

Assume 26 percent total prestress loss. Then f_{pe} = 0.74 × 175,000 = 129,500 psi (889 MPa), the required A_{ps} per foot of circumference = 100,000/129,500 = 0.772 in^2 (4.98 cm^2), and the number of vertical strands per foot of circumference = 0.772/0.144 = 5.36. Thus, use $\frac{1}{2}$ in. dia seven-wire 250 K strands for vertical prestressing at $2\frac{1}{4}$ in. center-to-center spacing = 0.769 in^2 ≅ 0.772 in^2, O.K.

Nominal Moment Strength Check of Tank Wall. The maximum wall vertical moment for a fixed-base wall, from Table 11.21, is M = 28,056 ft-lb/ft or in.-lb/in. of circumference. We thus have:

$$\text{S.F.} = 1.3 \text{ (step 5 of flowchart)}$$

$$M_u = 1.3 \times 28,056 = 36,473 \text{ in.-lb/in.}$$

$$\text{Rqd. } M_n = \frac{M_u}{0.9} = \frac{36.473}{0.9} = 40,525 \text{ in.-lb/in.}$$

$$d = \frac{10}{2} + 1.05 = 6.05 \text{ in. } (15.37 \text{ cm})$$

$$A_{ps} = \frac{0.144}{2.25} = 0.064 \text{ in}^2/\text{in. width}$$

$$a = \frac{A_{ps}f_{ps}}{0.85f_c'b} = \frac{0.064 \times 220,000}{0.85 \times 5,000 \times 1} = 3.31 \text{ in.}$$

$$\text{Available } M_n = A_{ps}f_{ps}\left(d - \frac{a}{2}\right) = 0.064 \times 220,000\left(6.05 - \frac{3.31}{2}\right)$$

$$= 61,882 \text{ in.-lb/in.} \gg \text{Rqd. } M_n = 40,525 \text{ in.-lb/in., O.K.}$$

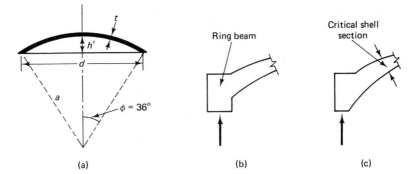

Figure 11.31 Tank dome shell roof. (a) Geometry of dome. (b) Edge ring beam. (c) Equivalent ring beam.

The wall design should include a check of the deflection as described in step 8 of the flowchart. Also, a determination should be made of the anchor steel at the base of the wall as well as the crack width w_{max} in step 9 of the flowchart. Finally, a check of temperature and creep effects has to be made to ascertain whether any additional nonprestressed mild steel has to be added to the prestressed wall reinforcement.

Design of Roof Dome Prestressed Edge Ring Beam. Use a rise-span ratio $h'/d = \frac{1}{8}$. Also, choose a freely supporting reaction at the top of the tank wall, using a neoprene pad under the edge ring beam. The shell would then have the form shown in Figures 11.31 and 11.32.

Since $d = 125$ ft., $h' = 125/8 = 15.63$ ft (4.76 m). Also, since $\phi = 36°$ is less than $51° 49'$, the entire shell would be in compression, and only temperature reinforcement is needed. The shell radius is

$$a = \frac{d/2}{\sin \phi} = \frac{62.5}{0.588} = 106 \text{ ft (32.3 m)}$$

From Equation 11.75, the minimum shell thickness to withstand buckling is

$$h_d = a\sqrt{\frac{1.5P_u}{\phi \beta_i \beta_c E_c}}$$

Hence, assuming that $t = 3.0$ in., we have

$$P_u = 1.4D + 1.7L = 1.4\left(\frac{3}{12} \times 150\right) + 1.7 \times 15 = 78 \text{ lb/ft}^2$$

$$\phi = 0.7$$

$$\beta_i = (a/r_i)^2 = \left(\frac{106}{1.4 \times 106}\right)^2 = 0.51$$

$$\beta_c = 0.44 + 0.003 \times 15 = 0.49 < 0.53, \text{ use } \beta_c = 0.53$$

$$E_c = 57,000\sqrt{5,000} = 4.03 \times 10^6 \text{ psi}$$

$$\text{Min } h = a\sqrt{\frac{1.5P_u}{\phi \beta_i \beta_c E_c}} = 106\sqrt{\frac{1.5 \times 78}{0.7 \times 0.51 \times 0.49 \times 4.03 \times 10^6}}$$

$$= 1.36 \text{ in. (3.5 cm)} < 3 \text{ in., O.K.}$$

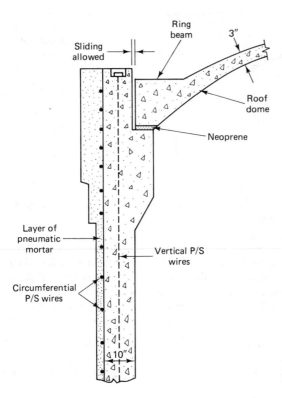

Figure 11.32 Dome prestressed ring beam support detail in Example 11.3.

So use a shell $t = 3$ in. (7.6 cm). Then $\sin \phi = \sin 36° = 0.59$, $\cos \phi = \cos 36° = 0.81$, and a = sphere radius = 106 ft.

From Equation 11.70, the tangential force per unit length of circumference is

$$N_\theta = \frac{W_D d}{2 \sin \phi} \left[\frac{1}{1 + \cos \phi} - \cos \phi \right] - \frac{W_L d}{2 \sin \phi} (\cos 2\phi)$$

$$= \frac{37.5 \times 125}{2 \times 0.59} \left[\frac{1}{1 + 0.81} - 0.81 \right] - \frac{15 \times 125}{2 \times 0.59} (0.31)$$

$$= -1,516 \text{ lb/ft}$$

From Equation 11.67, the meridional force per unit length of circumference, with $a = 106$ ft, is

$$N_\phi = -a \left(\frac{w_D}{1 + \cos \phi} + \frac{w_L}{2} \right)$$

$$= -106 \left(\frac{37.5}{1.81} + \frac{15}{2} \right) = -2,991 \text{ lb/ft (43.6 kN/m)}$$

From Equation 11.72, the radial prestressing force in the ring beam required to produce compatibility of deformation with the shell rim is

$$P = \frac{bh}{t} (N_\phi - \mu N_\phi) + \frac{d}{2} (N_\phi \cos \phi)$$

To determine the cross-sectional area bh of the ring beam, use $P = (d/2)(N_\phi \cos \phi)$ for

Olympic oval at the University of Calgary, Calgary, Canada. Structural Engineers: Simpson, Lester, Goodrich; Calgary, Alberta, Canada. (*Courtesy*, Prestressed Concrete Institute.)

the first trial, since the first term of the equation has less than 10 percent of the total value of P (see the discussion accompanying Equation 11.62). We obtain

$$P = \frac{d}{2}(N_\phi \cos \phi) = \frac{125}{2}(-2{,}991 \times 0.81) = -151{,}149 \text{ lb per ft}$$

Given that the total prestress loss is 26 percent, it follows that

$$\bar{\gamma} = 1 - 0.26 = 0.74$$

and

$$P_i = -\frac{151{,}419}{0.74} = 204{,}620 \text{ lb/ft}$$

Use a maximum concrete compressive stress $f_c = 800$ psi (5.52 MPa) in order to minimize excess strain in the edge beam, which could produce high stresses in the shell rim. The required cross-sectional area of the prestressed ring beam is

$$A_c = bh = \frac{P_i}{f_c} = \frac{204{,}620}{800} = 256 \text{ in}^2$$

Try $b = 14$ in. and $h = 20$ in. Then $A_c = 280$ in^2. Substituting into Equation 11.72, we get

$$P = \frac{280}{3.0}[-1{,}516 - 0.2(-2{,}991)] + \frac{125}{2}(-2{,}991 \times 0.81)$$

$$= -85{,}661 - 151{,}419 = -237{,}080 \text{ lb/ft}$$

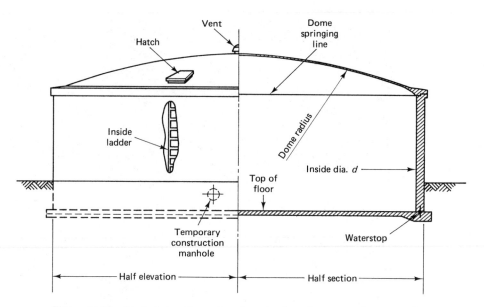

Figure 11.33 Typical elevation and section of a domed prestressed concrete circular tank.

Use

$$P_i = \frac{237,080}{0.74} = 320,378 \text{ lb (981 kN)}$$

From before,

$$f_{pi} = 0.7 f_{pu} = 175,000 \text{ psi}$$

so

$$A_{ps} = \frac{P_i}{f_{pi}} = \frac{320,378}{175,000} = 1.83 \text{ in}^2 \text{ (8.13 cm}^2\text{)}$$

Trying $\frac{1}{2}$ in. dia (12.7 mm) seven-wire 250 K strands, we obtain

$$A_{ps}/\text{strand} = 0.153 \text{ in}^2$$

and

$$\text{No. of strands} = \frac{1.83}{0.153} = 11.96$$

If the prestress loss is slightly more than 26 percent, the number of strands should be approximately 12. Hence, use twelve $\frac{1}{2}$ in. dia seven-wire-strand tendon to prestress the edge ring beam.

Check the Concrete Stress in the Critical Section t = 3 in. of the Shell Rim. The meridional compression $N_\phi = -2,991$ lb/ft of circumference, and the compressive stress $f_c = 2,991/(12 \times 3) = 83$ psi only, which is satisfactory. The support details of the edge ring beam and the roof are shown in Figure 11.32. Note that the ring beam is supported vertically on a neoprene pad, which enables sliding. A typical elevation and section of a domed prestressed circular tank is shown in Figure 11.33.

REFERENCES

11.1 Timoshenko, S., and Woinowsky-Krieger, S. *Theory of Plates and Shells*. 2d ed. New York: McGraw Hill, 1959. 580 pp.

11.2 Creasy, L. R. *Prestressed Concrete Cylindrical Tanks*. New York: Wiley, 1961. 215 pp.

11.3 Billington, D. P. *Thin Shell Concrete Structures*. 2d ed. New York: McGraw Hill, 373 pp., 1982.

11.4 Ghali, A. *Circular Storage Tanks and Silos*. London: E. & F. N. Spon Ltd., 1979, 210 pp.

11.5 PCA, "Circular Concrete Tanks without Prestressing", Concrete Information Series ST-57, Portland Cement Association, Skokie, Ill., 1957, 32 pp.

11.6 PCI Committee on Precast Prestressed Concrete Storage Tanks. "Recommended Practice for Precast Prestressed Concrete Circular Storage Tanks." Chicago: Prestressed Concrete Institute, 1987.

11.7 ACI Committee 344. *Design and Construction of Circular Prestressed Concrete Structures, ACI 344R*. Detroit: American Concrete Institute, 1970.

11.8 ACI Committee 344. *Design and Construction of Circular Wire and Strand Wrapped Prestressed Concrete Structures, ACI 344-IR*. Detroit: American Concrete Institute, 1989.

11.9 ACI Committee 344. *Design and Construction of Circular Prestressed Concrete Structures with Circumferential Tendons, ACI 344.2R*. Detroit: American Concrete Institute, 1989.

11.10 Post-Tensioning Institute. *Post-Tensioning Manual*. 4th ed. Phoenix: Post-Tensioning Institute, 1985.

11.11 Prestressed Concrete Institute. *PCI Design Handbook*. 3d ed. Chicago: Prestressed Concrete Institute, 1985.

11.12 Tadros, M. K. "Expedient Service Load Analysis of Cracked Prestressed Concrete Sections." *Journal of the Prestressed Concrete Institute*, Chicago, Vol. 27, No. 6, Nov-Dec (1983): 137–158.

11.13 Brondum-Nielsen, T. "Prestressed Tanks." *Journal of the American Concrete Institute*, Detroit, July-August 1985, pp. 500–509.

11.14 Vessey, J. V., and Preston, R. L. *A Critical Review of Code Requirements for Circular Prestressed Concrete Reservoirs*. Paris: F.I.P., 1978.

11.15 Nawy, E. G., and Blair, H., *Further Studies of Flexural Crack Control in Structural Slab Systems*. Detroit: American Concrete Institute, SP-30, 1971.

11.16 Abeles, P. W., and Bardhan-Roy, B. K. *Prestressed Concrete Designer's Handbook*. 3d ed. London: Viewpoint Publications, 1981, 556 pp.

PROBLEMS

11.1 Solve Example 11.3 if the tank diameter is 120 ft (36.6 m) and the water height is 30 ft (9.1 m). Assume that the total prestress loss is 20 percent, and use a rise-span ratio $h'/d = \frac{1}{10}$ for the roof dome, assuming that half the shell angle is $\phi = 45°$.

11.2 A circular prestressed concrete shell has an internal diameter $d = 85$ ft (26 m) and retains water to a height $H = 22$ ft (6.7 m). Determine the maximum horizontal ring forces and vertical moment, and design the prestressing reinforcement using both horizontal and vertical prestressing. Also, design a roof dome shell for the tank assuming a rise-span ratio $h'/d = \frac{1}{8}$ and half shell angle $\phi = 30°$. Solve for (a) hinged, (b) partially fixed, and (c) sliding wall base fixity, and design the prestressing reinforcement for both wire-wrapped and tendon prestressed conditions. Given data are:

$f'_c = 6,000$ psi (41.37 MPa), normal weight

$f'_{ci} = 4,250$ psi (29.30 MPa)

$f_t \leq 3\sqrt{f'_c} = 230$ psi (1.59 MPa)

$f_c = 0.45f'_c = 2,700$ psi (18.61 MPa)

$f_{cv} = 250$ psi (1.72 MPa)-residual compressive stress

f_{pu} for both wire and strand or tendon = 250,000 psi (1,724 MPa)

$f_{pi} = 0.7f_{pu} = 175,000$ psi (1,207 MPa)

Snow load intensity $w_L = 20$ lb/ft^2 (958 Pa)

Assume 20 percent total loss in prestress.

Computer Programs in BASIC

The programs presented in this appendix are furnished as a guide to the development of a user's own programs for the design of prestressed concrete members. While every effort has been made to utilize the existing state of the art and to assure accuracy of the analytical solution and design techniques, neither the author nor the publisher make any warranty, either expressed or implied, regarding the use of these programs for other than informational purposes. The user of the program is responsible for the final evaluation as to the validity, accuracy, and applicability of any results obtained. The listed programs can be procured on $5\frac{1}{4}$ in. or $3\frac{1}{2}$ in. diskette from PC SOFT-WARE, Box 161, East Brunswick, New Jersey 08816.

A-1 COMPUTER PROGRAM EGNAWY10 FOR ESTIMATION OF TIME-DEPENDENT LOSSES IN PRESTRESSED CONCRETE BEAMS

A computer program in BASIC for IBM PC XT/AT or PS/2 attempts to analyze the partial loss of prestress in pretensioned and post-tensioned beams due to time-dependent effects. It is based on the step-by-step flow chart Figure 3.9 and the discussions and examples in Chapter 3.

The program uses equations for incremental time-dependent steps that assess steel relaxation, shrinkage and creep. It also differentiates between stress-relieved and low relaxation strands. Input values are A_c, I_c, f_{pu}, E_{ps}, f'_c, f'_{ci}, span L, A_{ps}, W_D, W_{SD}, W_L and anchorage seating loss Δ_A.

Equations 3.7 and 3.8 give the loss due to relaxation as a function of time range $(t_2 - t_1)$.

A basic maximum creep coefficient $C_u = 2.35$ is used in the analysis that is reduced by a time function such that

$$C_t = \left(\frac{t^{0.60}}{1 + t^{0.60}}\right)C_u$$

as discussed in Equations 3.9a, 3.9b and 3.10. The user can input other maximum creep values into the programs which are different than $C_u = 2.35$ and the computer run would base the computations on the user's value of C_u.

Shrinkage loss is computed using time-dependent Equation 3.15a for moist-cured concrete with maximum $\epsilon_{SH} = 800 \times 10^{-6}$ in./in. and Equation 3.15b for steam-cured concrete with maximum $\epsilon_{SH} = 730 \times 10^{-6}$ in./in.

$$\text{Moist-cured:} \qquad \epsilon_{SH,t} = \left(\frac{t}{t + 35}\right)\epsilon_{SH}$$

$$\text{Steam-cured:} \qquad \epsilon_{SH,t} = \left(\frac{t}{t + 55}\right)\epsilon_{SH}$$

The user can input other shrinkage strain values directly into the program that are normally lower than $\epsilon_{SH} = 0.0008$ in./in.

For post-tensioned beams, the program computes frictional losses due to six different types of sheathing for tendons or strands using the governing ACI Code coefficients μ for curvature friction effect, and K for wobble effect. Then, it sums up all the losses in prestress and tabulates them as seen in one typical computer run of a double-T beam presented in this appendix.

```
llist
1 REM -- EGNAWY10 "TIME DEPENDANT LOSSES IN PRESTRESSED CONCRETE BEAMS"
2 REM -- **************************************************************
3 REM -- COPYRIGHT 1988 BY DR. EDWARD G. NAWY
4 REM -- ALL RIGHTS RESERVED
6 CLS: LOCATE 5,33: PRINT "PROGRAM EGNAWY10"
7 LOCATE 7,16: PRINT "TIME DEPENDENT LOSSES IN PRESTRESSED CONCRETE BEAMS"
8 LOCATE 11,33: PRINT "COPYRIGHT 1988"
9 LOCATE 13,32: PRINT "DR. EDWARD G. NAWY"
10 LOCATE 15,32: PRINT "ALL RIGHTS RESERVED"
14 J=0: FOR I=1 TO 1500: J=J+1: NEXT I
20 SCREEN 1
30 PAINT (15,15),2
35 SCREEN 2
40 PRINT
41 PRINT "**********************************************************************"
42 PRINT "* PROGRAM TO CALCULATE TIME DEPENDANT LOSSES   OF           *"
43 PRINT "*                                                           *"
44 PRINT "*    PRESTRESSED CONCRETE SIMPLY SUPPORTED BEAMS            *"
45 PRINT "*                                                           *"
46 PRINT "**********************************************************************"
47 PRINT :PRINT :PRINT
48 INPUT "PRESS (RETURN)   KEY TO CONTINUE ";PPPPPP
49 SCREEN 1
50 PAINT (15,15),2
51 PRINT:PRINT :PRINT
60 SCREEN 2
100 DIM K(6),MU(6),PP(10),TIME(50),LOSS(50),DC(50),DS(50),DR(50)
299 PRINT"------------------------------------------"
300 PRINT "READ SECTION DIMENSIONS AND SPAN"
305 PRINT"---------------------------- ":PRINT :PRINT
306 INPUT "THE BEAM SPAN IS (FEET)   = ";L:PRINT :PRINT
307 INPUT"DO YOU WANT TO INPUT SECTION PROPERTIES(Y/N)";XX$:IF XX$="N" THEN 310S
SAVE "B:EGNAWY10"
308 INPUT"AC =",AC: INPUT"IC =",IC: INPUT"CT =",CT: INPUT"CB =",CB: INPUT"WD =",
```

```
WD: GOTO 635
310 PRINT TAB(5);"*** THE SECTIONS MENU ***"
320 PRINT TAB(5);"***************************":PRINT : PRINT
330 PRINT TAB(2);"1-RECTANGULAR SECTION":PRINT TAB(2);"2-T-SECTION"
340 PRINT TAB(2);"3-I-SECTION":PRINT
350 INPUT "WHAT IS YOUR CHOICE ? ";WW : PRINT
360 PRINT :PRINT"INPUT DATA ":PRINT"--------------------":PRINT
370 ON WW GOTO 510,450,380
380 INPUT "THE TOP FLANGE WIDTH        =";B1:PRINT
390 INPUT "THE TOP FLANGE DEPTH        =";H1:PRINT
400 INPUT "THE BOTTOM FLANGE WIDTH     =";B2:PRINT
410 INPUT "THE BOTTOM FLANGE DEPTH     =";H2:PRINT
420 INPUT "THE TOTAL BEAM DEPTH        =";H:PRINT
430 INPUT "THE WEB WIDTH               =";BW:PRINT
435 INPUT"DO YOU WANT TO CORRECT DIMENSIONS (Y/N) ";A$:IF A$="Y" THEN 380
440 GOTO 570
450 INPUT "THE FLANGE WIDTH            =";B1:PRINT
460 INPUT "THE FLANGE DEPTH            =";H1:PRINT
470 INPUT "THE TOTAL DEPTH             =";H:PRINT
480 INPUT "THE WEB WIDTH               =";BW:PRINT

490 B2=BW
495 INPUT"DO YOU WANT TO CORRECT DIMENSIONS (Y/N) ";A$:IF A$="Y" THEN 450
500 GOTO 570
510 INPUT "THE WIDTH                   =";B1:PRINT
520 INPUT "THE TOTAL DEPTH             =";H:PRINT
530 BW=B1
540 B2=B1
550 H1=H
560 INPUT"DO YOU WANT TO CORRECT DIMENSIONS (Y/N) ";A$:IF A$="Y" THEN 510
569 REM ----------------------------------------------------------
570 REM CALCULATION OF SECTION GEOMETRIC PROPERTIES AND ITS WEIGHT
580 REM ----------------------------------------------------------
590 AC= (H*BW)+((B1-BW)*H1)+((B2-BW)*H2)
600 CT=((BW*H^2/2)+((B1-BW)*H1^2/2)+((B2-BW)*H2*(H-H2/2)))/AC
610 IC=(BW*H^3/12+BW*H*(H/2-CT)^2)+((B1-BW)*H1^3/12+((B1-BW)*H1*(CT-H1/2)^2))+(((
B2-BW)*H2^3/12+((B2-BW)*H2*((H-H2/2)-CT)^2))
620 CB=H-CT
630 WD=AC*150/144
635 RR = IC/AC
640 ST=IC/CT
650 SB=IC/CB
764 PRINT"---------------------"
765 PRINT" CONCRETE PROPERTIES  "
770 PRINT"---------------------"
775 PRINT : INPUT "SPECIFIED CONCRETE COMPRESSIVE STRENGTH f'c (psi) ? =";FPC
780 PRINT : INPUT "CONCRETE STRENGTH AT INITIAL PRESTRESS  f'ci(psi) ? =";FPCI
782  PRINT:INPUT"DO YOU WANT TO CORRECT THE PREVIOUS DATA (Y/N)?";B$:IF B$="Y"TH
EN 775
899 PRINT"-------------------------"
900 PRINT"DESIGN LOAD,ECCENTRICITIES  "
910 PRINT"-------------------------" :PRINT:PRINT
920 PRINT:INPUT"THE IMPOSED DEAD LOAD (LB/FT)        =";WSD
930        INPUT"THE LIVE LOAD          (LB/FT)      =";WL
940        INPUT"TENDONS ECCENTRICITY AT MIDSPAN     =";ECEN
950        INPUT"TENDONS ECCENTRICITY AT SUPPORT     =";ESUP
980 PRINT:INPUT"DO YOU WANT TO CORRECT THE PREVIOUS DATA (Y/N)?";B$:IF B$="Y" TH
EN 920
1049 REM --------------------------------------
1050 REM CONCRETE ALLOWABLE STRESS CALCULATIONS
1060 REM --------------------------------------
1110 EC=57000!*SQR(FPC)
1120 ECI=57000!*SQR(FPCI)
1190 PRINT"----------------"
1200 PRINT"STEEL PROPERTIES"
1210 PRINT"----------------"
1220 PRINT
1230 INPUT "ULTIMATE STRENGTH OF PRESTRESSING STEEL  fpu (psi)?   ="; FPU
1240 INPUT "INITIAL PRESTRESSING STRESS              fpi (psi) ?  ="; FPI
1250 INPUT "YIELD STRENGTH OF PRESTRESSING STEEL   fpy (psi)  ?  ="; FPY
1260        INPUT"YOUNG'S MODULUS OF PRESTRESSED STEEL  (psi)    =";EPS
1262 PRINT:INPUT "AREA OF PRESTRESSED STEEL          =";APS
1267 INPUT       "NUMBER OF PRESTRESSED TENDONS      =";N
1269 PRINT:INPUT"DO YOU WANT TO CORRECT THE PREVIOUS DATA (Y/N)";B$:IF B$="Y" TH
EN 1220
1270 PRINT:PRINT"----------------------"
```

```
1280 PRINT       " ADDITIONAL DATA       "
1290 PRINT       "---------------------":PRINT
1300 INPUT "PRETENSIONED TYPE (1) , POST-TENSIONED TYPE (2)      =";S
1310 PRINT "MOIST CURED FOR 7 DAYS TYPE 1"
1312 PRINT "STEAM CURED FOR 3 DAYS TYPE 2"
1315 INPUT SSS
1320 IF S=1 THEN 1340
1330 INPUT "NUMBER OF TENDONS JACKED AT THE TIME       ="; NN
1340 INPUT " ANCHORAGE SLIP                  (in)      =";SL
1350 PRINT:PRINT
1365       INPUT "DEPTH OF PRESTRESSED STEEL      =";DPS
1368 PI= APS*FPI
1369 FOR I=1 TO 6
1370 READ K(I), MU(I)
1380 NEXT I
1390 IF S= 1 THEN 1550
1400 REM -------------------------
1410 REM   FRICTION LOSS
1420 REM -------------------------
1430 PRINT TAB (10) "TENDONS TYPE MENU "
1440 PRINT "1-TENDONS IN FLEXIBLE METAL SHEATING (WIRE TENDONS )     "
1450 PRINT "2-TENDONS IN FLEXIBLE METAL SHEATING ( 7 WIRE STRANDS)   "
1460 PRINT "3-TENDONS IN FLEXIBLE METAL SHEATING (HIGH STRENGTH BAR) "
1470 PRINT "4-TENDONS IN RIGID METAL DUCT, 7 WIRE STRAND     "
1480 PRINT "5-PREGREASED TENDONS , WIRE TENDONS , 7 WIRE STRAND    "
1490 PRINT "6-MASTIC COATED TENDONS , WIRE TENDONS AND 7 WIRE STRANDS "
1500 PRINT:INPUT"TYPE YOUR CHOICE "; I :PRINT:PRINT
1510 X=8*ECEN/(L*12)
1520 DF = FPI*((MU(I)*X)+(K(I)*L))
1530 PP(1) =  (DF/FPI*100)
1540 PRINT "LOSSES DUE TO FRICTION" TAB(50) "= " DF " psi " TAB(67) PP(1) " %"
1550 REM ---------------------
1560 REM ANCHORAGE SLIP LOSS
1570 REM ---------------------
1580 DA = SL*EPS/(L*12)
1590 PP(2)= (DA/FPI*100)
1600 PRINT "LOSSES DUE TO ANCHORAGE SLIP"TAB(50)" = " DA " psi " TAB(67) PP(2)"
%"
1610 FFI = FPI-DF-DA
1620 IF S=2 THEN 1690
1630 PRINT:PRINT
1640 INPUT "NUMBER OF DAYS BETWEEN JACKING AND TRANSFER =";TTRANSFER
1650 TTRANSFER= TTRANSFER*24
1660 DR=(LOG(TTRANSFER)/LOG(10))
1670 DR=DR*FFI*((FFI/FPY)-.55)/10
1675 PRINT"LOSSES DUE TO RELAXATION BETWEEN JACKING AND TRANSFER =";DR " psi":PR
INT:PRINT:PRINT
1680 FFI = FFI-DR
1690 REM ---------------------
1700 REM ELASTIC SHORTENING LOSS
1710 REM ---------------------
1720 MD=(WD*L^2)*12/8
1730 MSD=(WSD*L^2)*12/8
1740 ML=(WL*L^2)*12/8
1770 X= MD*ECEN/IC
1780 AA= FFI*APS*(1+(ECEN^2/RR))
1790 AA= AA/AC
1800 X= X-AA
1810 DE= ABS(X*EPS/ECI)
1820 IF S=1 THEN 1920
1830 IF NN=N THEN 1910
1840 NN=N/NN
1850 SUM=0
1860 FOR I=1 TO (NN-1)
1870 SUM=SUM + I/(NN-1)
1880 NEXT I
1890 DE=SUM *DE/NN
1900 GOTO 1920
1910 DE =0
1920 PP(3)= (DE/FPI*100)
1930 PRINT"LOSSES DUE TO ELASTIC SHORTENING"TAB(50)" = "DE" psi "TAB(67) PP(3)"
%"
```

```
1931 PRINT "DO YOU WANT TO INPUT SPECIAL SHRINKAGE STRAIN VALUE(Y/N)"
1932 INPUT OOO$:IF OOO$="N" THEN 1935
1933 INPUT "SHRINKAGE STRAIN VALUE";SHRINKS
1935 INPUT"ARE STRANDS STRESS RELIEVED";QQQ$
1940 REM ---------------------
1950 REM CREEP LOSS
1960 REM ---------------------
1961 INPUT"DO YOU WANT TO INPUT CREEP FACTOR";YYY$:IF YYY$="N" THEN 1965
1962 INPUT "ULTIMATE CREEP FACTOR IS ";CU
1963 GOTO 1970
1965 CU=2.35
1970 I=0
1975 I=I+1
1980 INPUT "TIME AFTER JACKING THAT LOSSES ARE NEEDED (DAYS)":TIME(I)
2050 MOMENT = MD
2080 X=MOMENT*ECEN /IC
2090 AA=FFI*APS*(1+(ECEN^2/RR))
2100 AA=AA/AC
2110 X=X-AA
2120 CTIME=(TIME(I)^.6)/(10+TIME(I)^.6)*CU
2130 DC(I)=CTIME*ABS(X)*EPS/EC
2140 PP(4)=INT(DC(I)/FPI*100)
2150 PRINT"LOSSES DUE TO CREEP"TAB(50)" = " DC(I)" psi " TAB(67) PP(4)" %"
2160 REM -----------------
2170 REM SHRINKAGE LOSS
2180 REM -----------------
2181 IF SHRINKS=0! THEN 2190
2184 X=TIME(I)/(TIME(I)+35)*SHRINKS
2185 GOTO 2230
2190 ON SSS GOTO 2200,2220
2200 X=TIME(I)/(TIME(I)+35)*8.000001E-04
2210 GOTO 2230
2220 X=TIME(I)/(TIME(I)+55)*7.300001E-04
2230 DS(I)=X*EPS
2240 PP(5)=(DS(I)/FPI*100)
2250 PRINT"LOSSES DUE SHRINKAGE"TAB(50)" = " DS(I)" psi " TAB(67)PP(5)" %"
2260 REM -----------------
2270 REM RELAXATION STEEL LOSS
2280 REM -----------------
2290 FSTEEL=FFI-DE
2300 TIME(I)=TIME(I)*24
2310 IF S=2 THEN 2340
2320 X=(LOG(TIME(I))/LOG(10))-(LOG(TTRANSFER)/LOG(10))
2330 GOTO 2345
2340 X=(LOG(TIME(I))/LOG(10))
2345 Y=FSTEEL/FPY
2347 IF Y>.55 THEN 2350
2348 Y=.6
2350 DR(I)=X*FSTEEL*((Y)-.55)/10
2351 IF QQQ$="Y" THEN 2360
2355 DR(I)=XX*FSTEEL*((Y)-.55)/45
2360 PP(6)=(DR(I)/FPI*100)
2370 PRINT "LOSSES     TO STEEL RELAXATION"TAB(50)" = "DR(I)" psi "TAB(67)PP(6)"
%"
2380 LOSS(I) = DF+DA+DE+DC(I)+DS(I)+DR(I)
2450 INPUT "DO YOU NEED LOSSES AFTER ANOTHER TIME INTERVAL";YY$
2460 IF YY$="Y" THEN 1975
2500 PRINT TAB(10) "------------------------------"
2510 PRINT TAB(10) "      TABLE OF FINAL RESULTS    "
2520 PRINT TAB(10) "------------------------------"
2530 PRINT"TIME(DAYS)"TAB(15)"TOTAL"TAB(25)"FRICTI."TAB(35)"ANCHOR."TAB(45)"ELAS
T."TAB(55)"CREEP"TAB(65)"SHRI."TAB(75)"RELAX."
2535 PRINT          TAB(15)"LOSSES"TAB(25)"LOSS"TAB(35)"LOSS"TAB(45)"LOSS"TAB(
55)"LOSS"TAB(65)"LOSS"TAB(75)"LOSS"
2540 FOR K=1 TO I
2550 PRINT TIME(K)/24 TAB(15) LOSS(K) TAB(25) DF TAB(35) DA TAB(45) DE TAB(55) I
NT(DC(K)) TAB(65) INT(DS(K)) TAB(75) INT(DR(K))
2560 PRINT
2570 NEXT K
8000 DATA 0.0015,0.25,0.002,0.25,0.0006,0.3
8010 DATA 0.0002,0.25,0.002,0.15,0.002,0.15
9999 END
```

Example A-1   Time Dependent Partial Losses in Pretensioned Prestressed Concrete Beam

Compute the long-term partial losses in prestress for the pretensioned T beam in example 3.8 using the computer program EGNAWY10.

Input

Beam Span L	= 70	ft.
Flange width b	= 120	in.
Flange depth h_1	= 2	in.
Total depth h	= 32	in.
Web width b_w	= 12.5	in.
f'_c	= 5,000	psi
f'_{ci}	= 3,500	psi
W_{SD}	= 250	plf.
W_L	= 400	plf.
Midspan tendon eccentricity ϵ_c	= 18.73	in.
Support tendon eccentricity ϵ_e	= 12.98	in.
f_{pu}	= 270,000	psi
f_{pi}	= 189,000	psi
f_{py}	= 229,500	psi
E_{ps}	= 28,000,000	psi
A_{ps}	= 1.836	in^2
No of tendons	= 12	
Depth of prestressing steel d_p	= 28.75	in.
Number of days between jacking and transfer	= 0.75	

Output

```
******************************************************************
* PROGRAM TO CALCULATE TIME DEPENDANT LOSSES   OF               *
*                                                               *
*    PRESTRESSED CONCRETE SIMPLY SUPPORTED BEAMS                *
*                                                               *
******************************************************************


PRESS (RETURN)  KEY TO CONTINUE ?


----------------------------------------
READ SECTION DIMENSIONS AND SPAN
----------------------------------------


THE BEAM SPAN IS (FEET)   = ? 70
                ----------------------------------
                    TABLE OF FINAL RESULTS
                ----------------------------------
```

TIME(DAYS)	TOTAL LOSSES	FRICTI. LOSS	ANCHOR. LOSS	ELAST.) LOSS	CREEP LOSS	SHRI. LOSS	RELAX. LOSS
.75	9787.352	0	0	8236.696	1256	293	0
1	10553.68	0	0	8236.696	1472	388	455
7	18048.41	0	0	8236.696	3939	2333	3539
14	22179.41	0	0	8236.696	5305	4000	4637
21	24973.09	0	0	8236.696	6206	5250	5280
30	27586.71	0	0	8236.696	7043	6461	5845
45	30621.6	0	0	8236.696	8022	7875	6487
90	35587.91	0	0	8236.696	9684	10080	7586
365	43368.96	0	0	8236.696	12552	12775	9804
730 10903	46091.9	0	0	8236.696	13592	13359	
1825 12355	48911.89	0	0	8236.696	14583	13736	

Ok

A-2 COMPUTER PROGRAM EGNAWY12 FOR SERVICE LOAD ANALYSIS AND DESIGN IN FLEXURE OF PRESTRESSED CONCRETE BEAMS

This computer program in BASIC for IBM PC XT/AT or PS/2 is intended to proportion prestressed concrete T and I beams data pretensioned and post-tensioned. The flow chart Figure 4.29 and the discussions and example solutions of chapter 4 are the background of the program.

The input data comprise $f_{pu}, f'_c, f_{ci}, W_D, W_{SD}, W_L$, span L and effective prestress coefficient γ. It computes the service load level flexural moments M_D, M_{SD}, M_L and M_I and selects the cross-sectional dimensions of the prestressed beam on the basis of the maximum allowable service load stresses in the concrete and the prestressing steel as set by the ACI 318 Code.

The basis for the section selection is the section moduli values S_b and S^t obtained from the computed moment values and the allowable concrete stresses at service load. By trial and adjustment, the program iterates to the closest T or I beam section that can sustain the service level load within the maximum allowable service load stresses at effective prestress both for tension and compression.

Standard PCI sections are contained in the program when standard sections are to be chosen by the user. The computer run of one typical example in this appendix illustrates the output resulting from use of this program.

Example A-2 Service Load Proportioning of Prestressed Concrete Beams

Select the appropriate section of example 4.2 of a prestressed pretensioned beam having a span of 65 ft. (19.8 in.) using the computer program EGNAWY12.

Input

f_{pu}	=	270,000 psi
f'_c	=	6,000 psi
f'_{ci}	=	4,500 psi
Span L	=	65 ft.
W_{SD}	=	100 plf.
W_L	=	1,100 plf
Effectiveness Ratio γ	=	0.82

Trial I section chosen

Top flange width	b_1	= 17 in.
Top flange thickness	h_1	= 4.85 in.
Bottom flange width	b_2	= 18 in.
Bottom flange thickness	h_2	= 7 in.
Total depth	h	= 40 in.
Web width	b_w	= 6 in.
Initial prestressing force		= 376,110 lb.
Depth of prestressing steel		= 36.15 in.

Output

```
**********************************************
*                                            *
*        SERVICE LOAD DESIGN OF               *
*                                            *
*   SIMPLY SUPPORTED PRESTRESSED BEAMS        *
*                                            *
**********************************************


PRESS (RETURN)  KEY TO CONTINUE ?

RANGE OF INIT.PRES.FORCE      PRES.STEEL DEPTH
   474155  TO   299175            38
   492498  TO   310749            37
   512316  TO   323253            36
   533796  TO   336806            35
   557156  TO   351546            34
   582654  TO   432208            33

INPUT YOUR CHOSEN INITIAL PRESTRESSED FORCE THEN THE CORRESPONDING DEPTH OF PRES
TRESSED STEEL
? 376110, 36. 15
PERMISSIBLE LINEAR STRESSES
---------------------------

F.in.top= 402.4923
F.in.bot=-2700
F.top=-2700
F.bot= 929.5161

ACTUAL LINEAR STRESSES
----------------------
F.in.top=-67.87824
F.in.bot=-1821.373
F.top=-2450.798
F.bot= 632.9773

DO YOU WISH TO REENTER OTHER DIMENSIONS(Y/N)     ? Y
```

```
RANGE OF INIT.PRES.FORCE     PRES.STEEL DEPTH
   171381  TO -159926             38
   196223  TO -166113             37
   229488  TO -172797             36
   276334  TO -180042             35
   347210  TO -187921             34
   466987  TO -196522             33
   490540  TO -205947             32
   515251  TO -216321             31

INPUT YOUR CHOSEN INITIAL PRESTRESSED FORCE THEN THE CORRESPONDING DEPTH OF PRES
TRESSED STEEL
? 376110, 33.65
PERMISSIBLE LINEAR STRESSES
--------------------------

F.in.top= 402.4923
F.in.bot=-2700
F.top=-2700
F.bot= 929.5161

ACTUAL LINEAR STRESSES
---------------------

F.in.top= 396.8533
F.in.bot=-2233.981
F.top= 325.4197
F.bot=-1831.865

 ACTUAL STRESSES CALCULATED AT END SECTION

DO YOU WISH TO REENTER OTHER DIMENSIONS(Y/N)
? N
Ok
```

A-3 COMPUTER PROGRAM EGNAWY14 FOR THE STRENGTH ANALYSIS AND DESIGN IN FLEXURE OF PRESTRESSED CONCRETE BEAMS

This computer program in BASIC for IBM PC XT/AT or PS/2 analyzes the sections already proportioned by the service load level requirements in flexure. It follows the step-by-step flow chart Figure 4.46 and the discussions and example solutions of chapter 4.

The input data formats the section type, mainly, whether it is a T, I or rectangular section, the dimensions b, d, d_p, the maximum stresses f'_c, f_{pu}, f_{py}, f_{ps}, and reinforcement moduli E_s and E_{ps}. The strain-compatibility solution gives the moment strength M_n for flanged sections where the neutral axis falls within or outside the flange for typical T-beams sections. It also verifies that the reinforcement, both mild and prestressed, is within the maximum allowable by the ACI-318 Code for both under-reinforced and over-reinforced sections.

For under-reinforced rectangular sections, it computes the flexural moment strength from

$$M_n = A_{ps}f_{ps}\left(d_p - \frac{a}{2}\right) + A_s f_y\left(d - \frac{a}{2}\right) + A'_s f_y\left(\frac{a}{2} - d\right)$$

For under-reinforced flanged sections, it computes the flexural moment strength from

$$M_n = A_{pw}f_{ps}\left(d_p - \frac{a}{2}\right) + A_s f_y(d - d_p) + 0.85 f'_c(b - b_w)h_f\left(d_p - \frac{h_f}{2}\right)$$

where

$$A_{pw}f_{ps} = A_{ps}f_{ps} - 0.85\,f'_c(b - b_w)h_f$$

A strength reduction factor $\phi = 0.9$ is used in the program. The computer run gives a typical output for the evaluation of the nominal moment strength of prestressed sections.

A-4 COMPUTER PROGRAM EGNAWY16 FOR THE SHEAR STRENGTH DESIGN AND SHEAR REINFORCEMENT SELECTION IN PRESTRESSED CONCRETE BEAMS

This is a computer program in BASIC for IBM PC XT/AT or PS/2 intended to determine the shear strength and shear reinforcement required in prestressed concrete beams. It follows the step-by-step flow chart Figure 5.16 and the discussions of flexure shear V_{ci} and web shear W_{cw} in chapter 5 and the detailed example solutions and diagrams.

The input data compares the load data W_D, W_{SD} and W_L and the stresses f'_c, f_{pu}, f_y, f_{pe}, effective prestressing force P_e and the section dimensions of the top flange, b_1 and h_1, total depth h and the web width b_w.

It computes the ACI approximate shear strength:

$$V_c = b_w d_p \left(0.6\,\lambda\sqrt{f'_c} + 700\,\frac{V_u d}{M_u} \right)$$

and the other alternate more refined solution:

flexure shear:

$$V_{ci} = 0.6\,\lambda\sqrt{f'_c}\,b_w d + V_d + \frac{V_i M_{cr}}{M_{max}} - 1.7\,\lambda\sqrt{f'_c}\,b_w d_p$$

and the web shear:

$$V_{cw} = (3.5\,\lambda\sqrt{f'_c} + 0.3\bar{f_c})b_w d_p + V_p$$

where λ is a function of type of concrete used, namely, normalweight, sand-lightweight, or all-lightweight.

Based on choosing the controlling shear strength, the program proceeds to compute the required area of the web stirrups for shear, checking for the minimum area A_v as required by the ACI Code. Then it selects the required spacing of the stirrups based on the stirrup size being used in the design.

A-5 COMPUTER PROGRAM EGNAWY18 FOR THE DESIGN OF CONCRETE BRACKETS AND CORBELS

This computer program in BASIC for IBM PC XT/AT or PS/2 is intended to proportion brackets or corbels that are essential components in precast prestressed concrete prestressing systems. The step-by-step procedure given in the flow chart Figure 5.27 forms the basis of the program as well as the discussion and example calculations in Chapter 5.

The input data comprise the allowable stress values f'_c, f_y, the vertical shear load V_u, the horizontal frictional force N_{uc}, and the section dimensions, namely, the depth h, the effective depth d, the moment arm a, and the width b_w of the section.

By trial and adjustment the program computes areas of the main steel A_s and the horizontal steel closed steel stirrups A_h. The program uses the ACI 318 Code requirements for determining the areas of the steel reinforcement, and assumes that the minimum horizontal frictional force N_{uc} is 20 percent of the vertical shear V_u acting on the corbel.

A-6 COMPUTER PROGRAM EGNAWY20 FOR MOMENT CURVATURE ANALYSIS OF BONDED PARTIALLY PRESTRESSED BEAMS

A computer program in BASIC for IBM PC XT/AT or PS/2 is intended to evaluate the moment-curvature relationships in a bonded, prestressed concrete beam. It computes these values for three loading stages:

1. Linear Uncracked Stage
2. Linear Cracked Stage
3. Non-linear Cracked Stage

The user should use the "interactive format" response to a prompt in the program for both T and I beam sections.

Input data comprise section dimensions b_1 and h_1 for the top flange; b_2 and h_2 for the bottom flange; total depth h and web width b_w. It also includes input of the effective prestressing force P_e after losses, prestressing a steel area A_{ps}, prestressing steel depth d_p, mild steel area A_s, mild steel depth d, the yield strengths f_{py} and f_y and the ultimate prestressing strength f_{pu}.

The program internally generates the co-ordinates of the moment-curvature relationships using strain increments of $\epsilon = 0.0005$ in./in. beyond zero strain, starting at a strain $\epsilon_1 = 0.001$ in./in. for the linear post-cracking stage through the non-linear cracking stages up to a strain of $\epsilon_c = 0.003$ in./in. It is intended to evaluate the curvature of individual sections of prestressed concrete elements. The moment-curvature plot ($M - \phi$) between the stages $\epsilon_1 = 0.001$ and $\epsilon_c = 0.003$ in./in., if assumed a straight line, permits the user to interpolate intermediate $M - \phi$ values when necessary.

The program prompts input of data on the number of layers of prestressing steel reinforcement and the computer run checks the stress level in the concrete section at the extreme fibers at each loading and cracking stage.

A-7 COMPUTER PROGRAM EGNAWY22 FOR TIME DEPENDENT DEFLECTION EVALUATION OF PRESTRESSED CONCRETE SIMPLY SUPPORTED BEAMS

This is a computer program in BASIC for IBM XT/AT or PS/2 intended to evaluate the time-dependent deflection and camber in simply supported bonded prestressed concrete beams by the approximate time step method. It follows the steps shown in the flow chart Fig. 7.18 and requires the input of the value of the effective prestressing force P_e and the prestress loss ΔP obtained from Program EGNAWY10 on time-dependent losses.

Input data comprise section properties A_c, I_c, c_t, c_b, self-weight W_D, super-imposed dead load W_D and live load W_L. Also, the tendon eccentricities e_c and e_e at midspan and support sections are input into the program.

The program prompts the user to input any time intervals for which deflection is to be computed starting from the prestress transfer load application, then throughout the time-history of the prestressed member. A Table of final results is given in the output.

Example A-7 Time Dependent Deflection Computation for Prestressed Concrete Beams

Compute by the approximate incremental time-step method the long-term deflections of the beam in example 7.9 for the time intervals of 7, 30, 90, 365 and 1825 days (5 years) using the computer program EGNAWY22. Assume a maximum shrinkage coefficient $\epsilon_{SH} = 0.0005$ in./in. and a maximum creep coefficient $C_u = 2.35$.

Input

Beam Span L	= 65 ft.
Beam Section	= T
Flange width	= 120 in.
Flange depth h_1	= 2.25 in.
Total depth h	= 48 in.
Web width b_w	= 8 in.
f'_c	= 5,000 psi.
f'_{ci}	= 3,750 psi.
W_{SD}	= 100 plf.
W_L	= 1,100 plf.
Midspan tendon eccentricity ϵ_c	= 33.14 in.
Support tendon eccentricity ϵ_c	= 20.0 in.
f_{pi}	= 189,000 psi.
f_{pe}	= 154,900 psi.
E_{ps}	= 27,500,000 psi.
E_s	= 29,000,000 psi.
A_{ps}	= 2.142 in.
d_p	= 45.95 in.
A_s	= 0
d	= 0
Prestress force after transfer	= 331,967 lb.
Time from casting to curing	= 3 days
Time from casting to prestressing	= 4 days
Time from casting to load application	= 30 days

Tendon profile	= harped
Distance from support to harped point	= 32.5 ft.
Curing process	= steam

Output

```
***************************************************************
* PROGRAM TO CALCULATE TIME DEPENDANT DEFLECTION OF           *
*                                                             *
*    PRESTRESSED CONCRETE SIMPLY SUPPORTED BEAMS              *
*                                                             *
***************************************************************
PRESS ANY KEY TO CONTINUE ?
-------------------------------------
 READ OF SECTION DIMENSION AND SPAN
-------------------------------------


THE BEAM'S SPAN IS (FEET)   =? 65


     DATA OBTAINED FROM LOSSES PROGRAM
     **********************************


PRESTRESSING FORCE AFTER TRANSFER (after elastic loss)IN LBS=? 331967
     TIME SCHEDULE FOR CONSTRUCTION
     ****************************

TIME FROM SECTION CASTING TO END OF CURING (DAYS)       =? 3
TIME FROM SECTION CASTING TO PRESTRESSING(posttensioned) OR
          OR TRANSFER (pretensioned)  (DAYS)           =? 4
TIME FROM SECTION CASTING TO APPLICATION OF SUPERIMPOSED DEAD
          LOAD AND LIVE LOAD (DAYS)                       =? 30



SELECT THE PRESTRESSED CABLES LAYOUT

1-PARABOLIC CABLES
2-HARPED CABLES
3-STRAIGHT CABLES

? 2
DISTANCE FROM SUPPORT TO HARPED POINT (FT.)   = ? 32.5

ENTER 1 FOR MOIST CURED CONCRETE AND 2 FOR STEAM-CURED CONCRETE ? 2

 DO YOU WISH TO CALCULATE DEFLECTION AFTER ANOTHER TIME INTERVAL? N


     ***********************************************
     *         TABLE OF FINAL RESULTS              *
     ***********************************************


 TIME (days)              DEFLECTION

    7                     -1.160123

    30                    -.744374

    90                    -.9480114

   365                    -1.081456

   1825                   -1.079121

Ok
```

Unit Conversions, Design Information, Properties of Reinforcement

Selected from the Prestressed Concrete Institute Design Handbook

TABLE B-1 CONVERSION TO INTERNATIONAL SYSTEM OF UNITS (SI)

To convert from	to	multiply by
Length		
inch (in.)	millimeter (mm)	25.4
inch (in.)	meter (m)	0.0254
foot (ft)	meter (m)	0.3048
yard (yd)	meter (m)	0.9144
Area		
square foot (sq ft)	square meter (sq m)	0.09290
square inch (sq in.)	square millimeter (sq mm)	645.2
square inch (sq. in.)	square meter (sq m)	0.0006452
square yard (sq yd)	square meter (sq m)	0.8361
Volume		
cubic inch (cu in.)	cubic meter (cu m)	0.00001639
cubic foot (cu ft)	cubic meter (cu m)	0.02832
cubic yard (cu yd)	cubic meter (cu m)	0.7646
gallon (gal) Can. liquid*	liter	4.546
gallon (gal) Can. liquid*	cubic meter (cu m)	0.004546
gallon (gal) U.S. liquid*	liter	3.785
gallon (gal) U.S. liquid*	cubic meter (cu m)	0.003785

To convert from	to	multiply by
Force		
kip	kilogram (kgf)	453.6
kip	newton (N)	4448.0
pound (lb)	kilogram (kgf)	0.4536
pound (lb)	newton (N)	4.448
Pressure or Stress		
kips/square inch (ksi)	megapascal (MPa)**	6.895
pound/square foot (psf)	kilopascal (kPa)**	0.04788
pound/square inch (psi)	kilopascal (kPa)**	6.895
pound/square inch (psi)	megapascal (MPa)**	0.006895
pound/square foot (psf)	kilogram/square meter (kgf/sq m)	4.882
Mass		
pound (avdp)	kilogram (kg)	0.4536
ton (short, 2000 lb)	kilogram (kg)	907.2
ton (short, 2000 lb)	tonne (t)	0.9072
grain	kilogram (kg)	0.00006480
tonne (t)	kilogram (kg)	1000
Mass (weight) per Length		
kip/linear foot (klf)	kilogram/meter (kg/m)	0.001488
pound/linear foot (plf)	kilogram/meter (kg/m)	1.488
pound/linear foot (plf)	newton/meter (N/m)	14.593
Mass per volume (density)		
pound/cubic foot (pcf)	kilograpm/cubic meter (kg/cu m)	16.02
pound/cubic yard (pcy)	kilograpm/cubic meter(kg/cu m)	0.5933
Bending Moment or Torque		
inch-pound (in.-lb.)	newton-meter	0.1130
foot-pound (ft-lb)	newton-meter	1.356
foot-kip (ft-k)	newton-meter	1356
Temperature		
degree Fahrenheit (deg F)	degree Celsius (C)	$t_C = (t_F - 32)/1.8$
degree Fahrenheit (deg F)	degree Kelvin (K)	$t_K = (t_F + 459.7)/1.8$
Energy		
British thermal unit (Btu)	joule (j)	1056
kilowatt-hour (kwh)	joule (j)	3,600,000
Power		
horsepower (hp) (550 ft lb/sec)	watt (W)	745.7
Velocity		
mile/hour (mph)	kilometer/hour	1.609
mile/hour (mph)	meter/second (m/s)	0.4470
Other		
Section modulus (in.^{3})	mm^3	16.387
Moment of inertia (in.^{4})	mm^4	416.231
Coefficient of heat transfer (Btu/ft^2/h/°F)	W/m^2/°C	5.678
Modulus of elasticity (psi)	MPa	0.006895
Thermal conductivity (Btu-in./ft^2/h/°F)	Wm/m^2/°C	0.1442
Thermal expansion (in./in./°F)	mm/mm/°C	1.800
Area/length (in.^{2}/ft)	mm^2/m	2116.80

*One U.S. gallon equals 0.8321 Canadian gallon   **A pascal equals one newton/square meter

TABLE B-2 RECOMMENDED MINIMUM FLOOR LIVE LOADS*

Uniformly Distributed Loads

Occupancy or Use	Live Load (psf)
Apartments (*see* Residential)	
Armories and drill rooms	150
Assembly halls and other places of assembly:	
Fixed seats	60
Movable seats	100
Platforms (assembly)	100
Balcony (exterior)	100
On one and two family residences only and not exceeding 100 sq ft	60
Bowling alleys, poolrooms, and similar recreational areas	75
Corridors:	
First floor	100
Other floors, same as occupancy served except as indicated	
Dance halls and ballrooms	100
Dining rooms and restaurants	100
Dwellings (*see* Residential)	
Fire escapes	100
On multi- or single-family residential buildings only	40
Garages (passenger cars only)	50

Occupancy or Use	Live Load (psf)
Residential (cont.)	
Hotels and multifamily houses:	
Private rooms and corridors serving them	40
Public rooms and corridors serving them	100
Schools:	
Classrooms	40
Corridors above first floor	80
Sidewalks, vehicular driveways, and yards, subject to trucking (2)	250
Stadiums and arena bleachers (3)	100
Stairs and exitways	100
Storage warehouse:	
Light	125
Heavy	250
Stores:	
Retail:	
First floor	100
Upper floors	75
Wholesale, all floors	125
Walkways and elevated platforms (other than exitways)	60

Concentrated Loads

For trucks and buses use AASHTO lane loads (1)

	Load (lb)
Grandstands (see Stadium and arena bleachers)	
Gymnasiums, main floors and balconies	100
Hospitals:	
Operating rooms, laboratories	60
Private rooms	40
Wards	40
Corridors, above first floor	80
Hotels (see Residential)	
Libraries:	
Reading rooms	60
Stack rooms (books & shelving at 65 pcf) but not less than	150
Corridors, above first floor	80
Manufacturing:	
Light	125
Heavy	250
Marquees and canopies	75
Office buildings:	
Offices	50
Lobbies	100
File and computer rooms require heavier loads based upon anticipated occupancy	
Penal institutions:	
Cell blocks	40
Corridors	100
Residential:	
Dwellings (one-and two family)	
Uninhabitable attics without storage	10
Uninhabitable attics with storage	20
Habitable attics and sleeping areas	30
All other areas	40

Location	Load (lb)
Elevator machine room grating (on area of 4 sq in)	300
Finish light floor plate construction (on area of 1 sq in)	200
Garages	(4)
Office floors	2000
Scuttles, skylight ribs, and accessible ceilings	200
Sidewalks	8000
Stair treads (on area of 4 sq in at center of tread)	300

(1) American Association of State Highway and Transportation Officials.

(2) AASHTO lane loads should also be considered where appropriate.

(3) For detailed recommendations, see Assembly Seating, Tents and Air Supported Structures, ANSI/NFPA 102-1978 [220.3].

(4) Floors in garages or portions of buildings used for storage of motor vehicles shall be designed for the uniformly distributed live loads shown or the following concentrated loads: (1) for passenger cars accommodating not more than nine passengers, 2000 pounds acting on an area of 20 sq in; (2) mechanical parking structures without slab or deck, passenger cars only, 1500 pounds per wheel; (3) for trucks or buses, maximum axle load on an area of 20 sq in.

* Source: American National Standard ANSI A58.1-1982
Local building codes take precedence.

TABLE B-3 DEAD WEIGHTS OF FLOORS, CEILINGS, ROOFS, AND WALLS

	Weight (psf)
Floorings	
Normal weight concrete topping, per inch of thickness	12
Sand-lightweight (120 pcf) concrete topping, per inch	10
Lightweight (90-100 pcf) concrete topping, per inch	8
7/8″ hardwood floor on sleepers clipped to concrete without fill	5
1 1/2″ terrazzo floor finish directly on slab	19
1 1/2″ terrazzo floor finish on 1″ mortar bed	30
1″ terrazzo finish on 2″ concrete bed	38
3/4″ ceramic or quarry tile on 1/2″ mortar bed	16
3/4″ ceramic or quarry tile on 1″ mortar bed	22
1/4″ linoleum or asphalt tile directly on concrete	1
1/4″ linoleum or asphalt tile on 1″ mortar bed	12
3/4″ mastic floor	9
Hardwood flooring, 7/3″ thick	4
Subflooring (soft wood), 3/4″ thick	2 1/2
Asphaltic concrete, 1 1/2″ thick	18
Ceilings	
1/2″ gypsum board	2
5/8″ gypsum board	2 1/2
3/4″ plaster directly on concrete	5
3/4″ plaster on metal lath furring	8
Suspended ceilings	2
Acoustical tile	1
Acoustical tile on wood furring strips	3
Roofs	
Ballasted inverted membrane	16
Five-ply felt and gravel (or slag)	6 1/2
Three-ply felt and gravel (or slag)	5 1/2
Five-ply felt composition roof, no gravel	4
Three-ply felt composition roof, no gravel	3

724

Asphalt strip shingles — 3
Rigid insulation, per inch — 1/2
Gypsum, per inch of thickness — 4
Insulating concrete, per inch — 3

Walls	Un-Plastered	One side Plastered	Both sides Plastered
4" brick wall	40	45	50
8" brick wall	80	85	90
12" brick wall	120	125	130
4" hollow normal weight concrete block	28	33	38
6" hollow normal weight concrete block	36	41	46
8" hollow normal weight concrete block	51	56	61
12" hollow normal weight concrete block	59	64	69
4" hollow lightweight block or tile	19	24	29
6" hollow lightweight block or tile	22	27	32
8" hollow lightweight block or tile	33	38	43
12" hollow lightweight block or tile	44	49	54
4" brick 4" hollow normal weight block backing	68	73	78
4" brick 8" hollow normal weight block backing	91	96	101
4" brick 12" hollow normal weight block backing	119	124	129
4" brick 4" hollow lightweight block or tile backing	59	64	69
4" brick 8" hollow lightweight block or tile backing	73	78	83
4" brick 12" hollow lightweight block or tile backing	84	89	94
4" brick, steel or wood studs, 5/8" gypsum board	43		
Windows, glass, frame and sash	8		
4" stone	55		
Steel or wood studs, lath, 3/4" plaster	18		
Steel or wood studs, 5/8" gypsum board each side	6		
Steel or wood studs, 2 layers 1/2" gypsum board each side	9		

TABLE B-4   AREA OF BARS IN A 1-FOOT-WIDE SLAB STRIP

Cross section area of bar, A_s (or $A_s{}'$), in.^{2}

| Spacing, in. | Bar size | | | | | | | | | | | | Spacing, in. |
	#3	#4	#5	#6	#7	#8	#9	#10	#11	#14	#18	
4	0.33	0.60	0.93	1.32	1.80	2.37	3.00	3.81	4.68			4
4½	0.29	0.53	0.83	1.17	1.60	2.11	2.67	3.39	4.16	6.00		4½
5	0.26	0.48	0.74	1.06	1.44	1.90	2.40	3.05	3.74	5.40	9.60	5
5½	0.24	0.44	0.68	0.96	1.31	1.72	2.18	2.77	3.40	4.91	8.73	5½
6	0.22	0.40	0.62	0.88	1.20	1.58	2.00	2.54	3.12	4.50	8.00	6
6½	0.20	0.37	0.57	0.81	1.11	1.46	1.85	2.34	2.88	4.15	7.38	6½
7	0.19	0.34	0.53	0.75	1.03	1.35	1.71	2.18	2.67	3.86	6.86	7
7½	0.18	0.32	0.50	0.70	0.96	1.26	1.60	2.03	2.50	3.60	6.40	7½
8	0.17	0.30	0.47	0.66	0.90	1.19	1.50	1.91	2.34	3.38	6.00	8
8½	0.16	0.28	0.44	0.62	0.85	1.12	1.41	1.79	2.20	3.18	5.65	8½
9	0.15	0.27	0.41	0.59	0.80	1.05	1.33	1.69	2.08	3.00	5.33	9
9½	0.14	0.25	0.39	0.56	0.76	1.00	1.26	1.60	1.97	2.84	5.05	9½
10	0.13	0.24	0.37	0.53	0.72	0.95	1.20	1.52	1.87	2.70	4.80	10
10½	0.13	0.23	0.35	0.50	0.69	0.90	1.14	1.45	1.78	2.57	4.57	10½
11	0.12	0.22	0.34	0.48	0.65	0.86	1.09	1.39	1.70	2.45	4.36	11
11½	0.11	0.21	0.32	0.46	0.63	0.82	1.04	1.33	1.63	2.35	4.17	11½
12	0.11	0.20	0.31	0.44	0.60	0.79	1.00	1.27	1.56	2.25	4.00	12
13	0.10	0.18	0.29	0.41	0.55	0.73	0.92	1.17	1.44	2.08	3.69	13
14	0.09	0.17	0.27	0.38	0.51	0.68	0.86	1.09	1.34	1.93	3.43	14
15	0.09	0.16	0.25	0.35	0.48	0.63	0.80	1.02	1.25	1.80	3.20	15
16	0.08	0.15	0.23	0.33	0.45	0.59	0.75	0.95	1.17	1.69	3.00	16
17	0.08	0.14	0.22	0.31	0.42	0.56	0.71	0.90	1.10	1.59	2.82	17
18	0.07	0.13	0.21	0.29	0.40	0.53	0.67	0.85	1.04	1.50	2.67	18

TABLE B-5 PROPERTIES AND DESIGN STRENGTHS OF PRESTRESSING STRAND AND WIRE

Seven-Wire Strand, f_{pu} = 270 ksi

Nominal Diameter, in.	3/8	7/16	1/2	9/16	0.600
Area, sq in.	0.085	0.115	0.153	0.192	0.215
Weight, plf	0.29	0.40	0.53	0.65	0.74
0.7 f_{pu} A_{ps}, kips	16.1	21.7	28.9	36.3	40.7
0.75 f_{pu} A_{ps}, kips	17.2	23.3	31.0	38.9	43.5
0.8 f_{pu} A_{ps}, kips	18.4	24.8	33.0	41.4	46.5
f_{pu} A_{ps}, kips	23.0	31.0	41.3	51.8	58.1

Seven-Wire Strand, f_{pu} = 250 ksi

Nominal Diameter, in.	1/4	5/16	3/8	7/16	1/2	0.600
Area, sq in.	0.036	0.058	0.080	0.108	0.144	0.215
Weight, plf	0.12	0.20	0.27	0.37	0.49	0.74
0.7 f_{pu} A_{ps}, kips	6.3	10.2	14.0	18.9	25.2	37.6
0.8 f_{pu} A_{ps}, kips	7.2	11.6	16.0	21.6	28.8	43.0
f_{pu} A_{ps}, kips	9.0	14.5	20.0	27.0	36.0	53.8

Three- and Four-Wire Strand, f_{pu} = 250 ksi

Nominal Diameter, in.	1/4	5/16	3/8	7/16
No. of wires	3	3	3	4
Area, sq in.	0.036	0.058	0.075	0.106
Weight, plf	0.13	0.20	0.26	0.36
0.7 f_{pu} A_{ps}, kips	6.3	10.2	13.2	18.6
0.8 f_{pu} A_{ps}, kips	7.2	11.6	15.0	21.2
f_{pu} A_{ps}, kips	9.0	14.5	18.8	26.5

Prestressing Wire

Diameter	0.105	0.120	0.135	0.148	0.162	0.177	0.192	0.196	0.250	0.276
Area, sq in.	0.0087	0.0114	0.0143	0.0173	0.0206	0.0246	0.0289	0.0302	0.0491	0.0598
Weight, plf	0.030	0.039	0.049	0.059	0.070	0.083	0.098	0.10	0.17	0.20
Ult. strength, f_{pu}, ksi	279	273	268	263	259	255	250	250	240	235
0.7 f_{pu} A_{ps}, kips	1.70	2.18	2.68	3.18	3.73	4.39	5.05	5.28	8.25	9.84
0.8 f_{pu} A_{ps}, kips	1.94	2.49	3.06	3.64	4.26	5.02	5.78	6.04	9.42	11.24
f_{pu} A_{ps}, kips	2.43	3.11	3.83	4.55	5.33	6.27	7.22	7.55	11.78	14.05

TABLE B-6 PROPERTIES AND DESIGN STRENGTHS OF PRESTRESSING BARS

Smooth Prestressing Bars, f_{pu} = 145 ksi*

Nominal Diameter, in.	3/4	7/8	1	1 1/8	1 1/4	1 3/8
Area, sq in.	0.442	0.601	0.785	0.994	1.227	1.485
Weight, plf	1.50	2.04	2.67	3.38	4.17	5.05
$0.7\,f_{pu}\,A_{ps}$, kips	44.9	61.0	79.7	100.9	124.5	150.7
$0.8\,f_{pu}\,A_{ps}$, kips	51.3	69.7	91.0	115.3	142.3	172.2
$f_{pu}\,A_{ps}$, kips	64.1	87.1	113.8	144.1	177.9	215.3

Smooth Prestressing Bars, f_{pu} = 160 ksi*

Nominal Diameter, in.	3/4	7/8	1	1 1/8	1 1/4	1 3/8
Area, sq in.	0.442	0.601	0.785	0.994	1.227	1.485
Weight, plf	1.50	2.04	2.67	3.38	4.17	5.05
$0.7\,f_{pu}\,A_{ps}$, kips	49.5	67.3	87.9	111.3	137.4	166.3
$0.8\,f_{pu}\,A_{ps}$, kips	56.6	77.0	100.5	127.2	157.0	190.1
$f_{pu}\,A_{ps}$, kips	70.7	96.2	125.6	159.0	196.3	237.6

Deformed Prestressing Bars

Nominal Diameter, in.	5/8	1	1	1 1/4	1 1/4	1 3/8
Area, sq. in.	0.28	0.85	0.85	1.25	1.25	1.58
Weight, plf	0.98	3.01	3.01	4.39	4.39	5.56
Ult. strength, f_{pu}, ksi	157	150	160*	150	160*	150
$0.7\,f_{pu}\,A_{ps}$, kips	30.5	89.3	95.2	131.3	140.0	165.9
$0.8\,f_{pu}\,A_{ps}$, kips	34.8	102.0	108.8	150.0	160.0	189.6
$f_{pu}\,A_{ps}$, kips	43.5	127.5	136.0	187.5	200.0	237.0

Stress-strain characteristics (all prestressing bars):

For design purposes, following assumptions are satisfactory:

E_s = 29,000 ksi

f_y = 0.95 f_{pu}

*Verify availability before specifying

TABLE B-7 MOMENTS IN BEAMS WITH FIXED ENDS

Loading	Moment at A	Moment at center	Moment at B
(1)	$-\dfrac{Pl}{8}$	$+\dfrac{Pl}{8}$	$-\dfrac{Pl}{8}$
(2)	$-Pla(1-a)^2$		$-Pla^2(1-a)$
(3)	$-\dfrac{2Pl}{9}$	$+\dfrac{Pl}{9}$	$-\dfrac{2Pl}{9}$
(4)	$-\dfrac{5Pl}{16}$	$+\dfrac{3Pl}{16}$	$-\dfrac{5Pl}{16}$
(5)	$-\dfrac{Wl}{12}$	$+\dfrac{Wl}{24}$	$-\dfrac{Wl}{12}$
(6)	$-\dfrac{Wl(1+2a-2a^2)}{12}$	$+\dfrac{Wl(1+2a+4a^2)}{24}$	$-\dfrac{Wl(1+2a-2a^2)}{12}$
(7)	$-\dfrac{Wl(3a-2a^2)}{12}$	$+\dfrac{Wla^2}{6}$	$-\dfrac{Wl(3a-2a^2)}{12}$
(8)	$-\dfrac{Wla(6-8a+3a^2)}{12}$		$-\dfrac{Wla^2(4-3a)}{12}$
(9)	$-\dfrac{5Wl}{48}$	$+\dfrac{3Wl}{48}$	$-\dfrac{5Wl}{48}$
(10)	$-\dfrac{Wl}{10}$		$-\dfrac{Wl}{15}$
W = Total load on beam			

TABLE B-8 CAMBER (DEFLECTION) AND ROTATION COEFFICIENTS FOR PRESTRESS FORCE AND LOADS*

Prestress Pattern	Equivalent Moment or Load	Equivalent Loading	Camber $+$	End Rotation $+$	End Rotation $+$
(1)	$M = Pe$		$+\dfrac{Ml^2}{16\,EI}$	$+\dfrac{Ml}{3\,EI}$	$-\dfrac{Ml}{6\,EI}$
(2)	$M = Pe$		$+\dfrac{Ml^2}{16\,EI}$	$+\dfrac{Ml}{6\,EI}$	$-\dfrac{Ml}{3\,EI}$
(3)	$M = Pe$		$+\dfrac{Ml^2}{8\,EI}$	$+\dfrac{Ml}{2\,EI}$	$-\dfrac{Ml}{2\,EI}$
(4)	$N = \dfrac{4Pe'}{l}$		$+\dfrac{Nl^3}{48\,EI}$	$+\dfrac{Nl^2}{16\,EI}$	$-\dfrac{Nl^2}{16\,EI}$
(5)	$N = \dfrac{Pe'}{bl}$		$+\dfrac{b(3-4b^2)\,Nl^3}{24\,EI}$	$+\dfrac{b(1-b)\,Nl^2}{2\,EI}$	$-\dfrac{b(1-b)\,Nl^2}{2\,EI}$

	w formula	Loading	Col A	Col B (+)	Col C (−)
(6)	$w = \dfrac{8\,Pe'}{l^2}$		$+\dfrac{5wl^4}{384\ EI}$	$+\dfrac{wl^3}{24\ EI}$	$-\dfrac{wl^3}{24\ EI}$
(7)	$w = \dfrac{8\,Pe'}{l^2}$		$+\dfrac{5\,wl^4}{768\ EI}$	$+\dfrac{9\,wl^3}{384\ EI}$	$-\dfrac{7\,wl^3}{384\ EI}$
(8)	$w = \dfrac{8\,Pe'}{l^2}$		$+\dfrac{5\,wl^4}{768\ EI}$	$+\dfrac{7\,wl^3}{384\ EI}$	$-\dfrac{9\,wl^3}{384\ EI}$
(9)	$w = \dfrac{4\,Pe'}{(0.5 - b)\,l^2}$ $\quad w_1 = \dfrac{w}{b}\,(0.5 - b)$		$\left[\dfrac{5}{8} - \dfrac{b}{2}\,(3 - 2b^2)\right]\dfrac{wl^4}{48\ EI}$	$+\dfrac{(1 - b)\,(1 - 2b)\ wl^3}{24\ EI}$	$-\dfrac{(1 - b)\,(1 - 2b)\ wl^3}{24\ EI}$
(10)	$w = \dfrac{4\,Pe'}{(0.5 - b)\,l^2}$ $\quad w_1 = \dfrac{w}{b}\,(0.5 - b)$		$\left[\dfrac{5}{16} - \dfrac{b}{4}\,(3 - 2b^2)\right]\dfrac{wl^4}{48\ EI}$	$\left[\dfrac{9}{8} - b\,(2 - b^2)\right]\dfrac{wl^3}{48\ EI}$	$\left[\dfrac{7}{8} + b\,(2 - b^2)\right]\dfrac{wl^3}{48\ EI}$
(11)	$w = \dfrac{4\,Pe'}{(0.5 - b)\,l^2}$ $\quad w_1 = \dfrac{w}{b}\,(0.5 - b)$		$\left[\dfrac{5}{16} - \dfrac{b}{4}\,(3 - 2b^2)\right]\dfrac{wl^4}{48\ EI}$	$\left[\dfrac{7}{8} - b\,(2 - b^2)\right]\dfrac{wl^3}{48\ EI}$	$\left[\dfrac{9}{8} + b\,(2 - b^2)\right]\dfrac{wl^3}{48\ EI}$

* The tabulated values apply to the effects of prestressing. By adjusting the directional notation, they may also be used for the effects of loads.

For patterns 4–11, superimpose on 1, 2, or 3 for other C.G. locations.

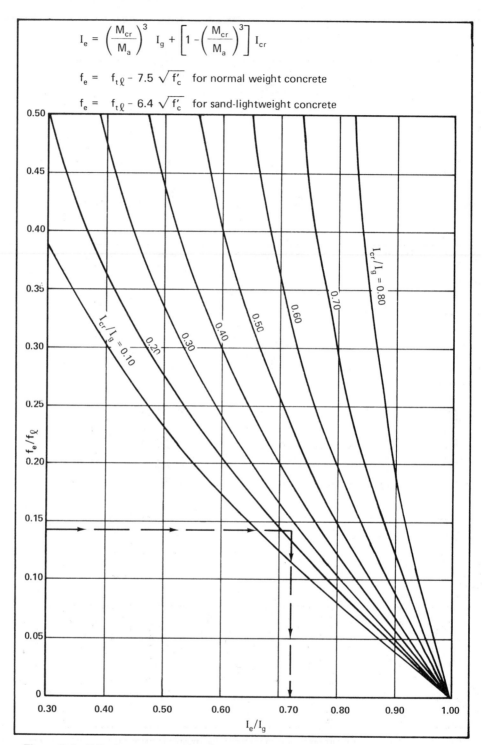

Figure B-1 Effective moment of inertia

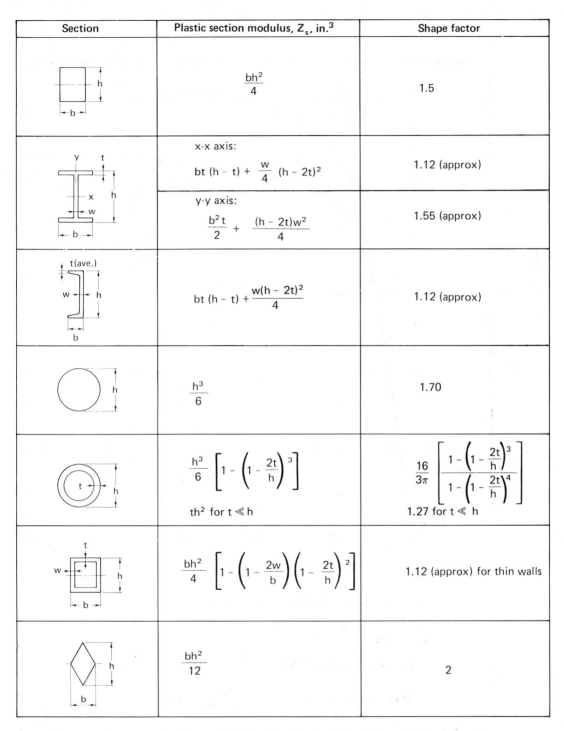

Section	Plastic section modulus, Z_s, in.^{3}	Shape factor
(rectangle, b wide, h tall)	$\dfrac{bh^2}{4}$	1.5
(I-beam; y, t, x, h, w, b)	x-x axis: $bt(h-t) + \dfrac{w}{4}(h-2t)^2$	1.12 (approx)
	y-y axis: $\dfrac{b^2 t}{2} + \dfrac{(h-2t)w^2}{4}$	1.55 (approx)
(channel; t(ave.), w, h, b)	$bt(h-t) + \dfrac{w(h-2t)^2}{4}$	1.12 (approx)
(solid circle, h)	$\dfrac{h^3}{6}$	1.70
(hollow circle; t, h)	$\dfrac{h^3}{6}\left[1-\left(1-\dfrac{2t}{h}\right)^3\right]$ th^2 for $t \ll h$	$\dfrac{16}{3\pi}\left[\dfrac{1-\left(1-\dfrac{2t}{h}\right)^3}{1-\left(1-\dfrac{2t}{h}\right)^4}\right]$ 1.27 for $t \ll h$
(hollow rectangle; t, w, h, b)	$\dfrac{bh^2}{4}\left[1-\left(1-\dfrac{2w}{b}\right)\left(1-\dfrac{2t}{h}\right)^2\right]$	1.12 (approx) for thin walls
(diamond; h, b)	$\dfrac{bh^2}{12}$	2

Figure B-2 Plastic section moduli and shape factors

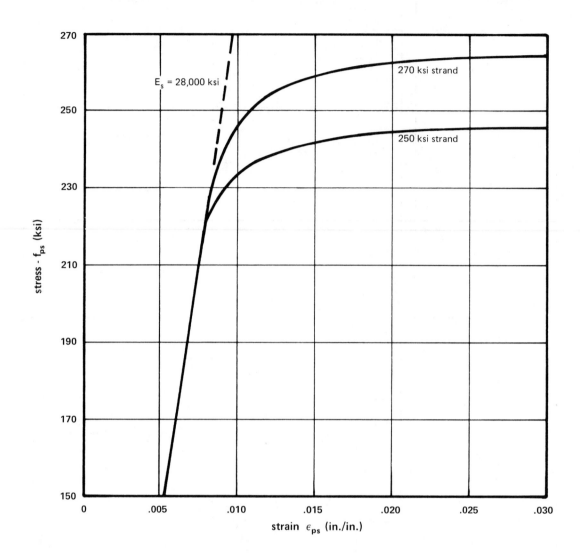

These curves can be approximated by the following equations:

$\epsilon_{ps} \leqslant 0.008$: $f_{ps} = 28{,}000\,\epsilon_{ps}$ (ksi)

$\epsilon_{ps} > 0.008$:

250 ksi strand: $f_{ps} = 248 - \dfrac{0.058}{\epsilon_{ps} - 0.006} < 0.98\,f_{pu}$ (ksi)

270 ksi strand: $f_{ps} = 268 - \dfrac{0.075}{\epsilon_{ps} - 0.0065} < 0.98\,f_{pu}$ (ksi)

Figure B-3 Typical stress-strain curve, 7-wire stress-relieved and low-relaxation prestressing strand

$$I_g = K_{i4}\left(\frac{1}{12}\,b_w h^3\right)$$

$$K_{i4} = 1 + (\alpha_b - 1)\beta_h^3 + \frac{3(1 - \beta_h)^2(\beta_h)(\alpha_b - 1)}{1 + \beta_h(\alpha_b - 1)}$$

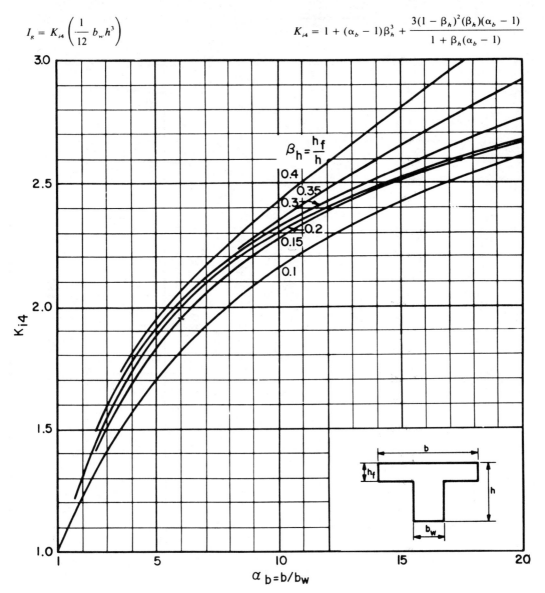

Example: For the T-beam shown, find the moment of inertia I_g:

$$\alpha_b = b/b_w = 143/15 = 9.53$$

$$\beta_h = h_f/h = 8/36 = 0.22$$

Interpolating between the curves for $\beta_h = 0.2$ and 0.3, read $K_{i4} = 2.28$

$$I_g = K_{i4}\frac{b_w h^3}{12} = 2.28\,\frac{15(36)^3}{12} = 133{,}000 \text{ in.}^4$$

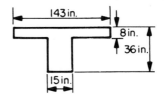

Figure B-4   Gross moment of inertia of T sections

Index